The Global Cryosphere, Second Edition

Recent studies indicate that – due to climate change – the Earth is undergoing rapid changes in all cryospheric components, including polar sea ice shrinkage, mountain glacier recession, thawing permafrost, and diminishing snow cover. This textbook provides a comprehensive summary of all components of the Earth's cryosphere, reviewing their history, physical and chemical characteristics, geographical distributions, and projected futures states. This new edition has been completely updated throughout, and provides state-of-the-art data from GlobSnow-2 CRYOSAT, ICESAT, and GRACE. It includes a comprehensive summary of cryospheric changes in land ice, permafrost, freshwater ice, sea ice, and ice sheets. It discusses the models developed to understand cryosphere processes and predict future changes, including those based on remote sensing, field campaigns, and long-term ground observations. Boasting an extensive bibliography, over 120 figures, and end-of-chapter review questions, it is an ideal resource for students and researchers of the cryosphere.

Roger G. Barry was Director of the National Snow and Ice Data Center (NSIDC) at the University of Colorado, Boulder from 1977 to 2008, and Professor of Geography from 1968 to 2010. He was appointed a University of Colorado Distinguished Professor in 2004. From 2012 to 2014 he was Director of the International CLIVAR Project Office, National Oceanography Centre, Southampton, UK. He was a Fellow of the American Geophysical Union, and his awards included the Founder's Medal of the Royal Geographical Society, the Nobel Peace Prize (as part of the Intergovernmental Panel on Climate Change team), a Guggenheim Fellow, a Fulbright Fellow, a Humboldt Prize Fellow, and a Foreign Member of the Russian Academy of Environmental Sciences (RAEN). Roger passed away peacefully in Boulder, Colorado in March, 2018.

Thian Yew Gan is a Professor at the University of Alberta, Canada, research ambassador of the German Academic Exchange Service, a fellow of American Society of Civil Engineers, and a lead author for the Sixth Assessment Report (AR6-WGI) of the Intergovernmental Panel of Climate Change (IPCC). Dr. Gan is internationally renowned for his innovative contributions to hydroclimatology, cryospheric science, climate change, remote sensing, and water resources management. He has received eleven international fellowships, an award from the Association of Science and Engineering Technology Professionals of Alberta (ASET), and has been a visiting professor to Germany, France, Finland, Switzerland, Sweden, Singapore, Malaysia, Japan, Hong Kong, Philippines, Thailand, Australia, and New Zealand.

The Global Cryosphere

Past, Present, and Future

SECOND EDITION

Roger G. Barry
University of Colorado at Boulder

Thian Yew Gan
University of Alberta

CAMBRIDGE
UNIVERSITY PRESS

CAMBRIDGE
UNIVERSITY PRESS

University Printing House, Cambridge CB2 8BS, United Kingdom

One Liberty Plaza, 20th Floor, New York, NY 10006, USA

477 Williamstown Road, Port Melbourne, VIC 3207, Australia

314–321, 3rd Floor, Plot 3, Splendor Forum, Jasola District Centre,
New Delhi – 110025, India

103 Penang Road, #05–06/07, Visioncrest Commercial, Singapore 238467

Cambridge University Press is part of the University of Cambridge.

It furthers the University's mission by disseminating knowledge in the pursuit of
education, learning, and research at the highest international levels of excellence.

www.cambridge.org
Information on this title: www.cambridge.org/9781108487559
DOI: 10.1017/9781108767262

First published 2011

Second Edition 2022

Printed in the United Kingdom by TJ Books Limited, Padstow Cornwall

A catalogue record for this publication is available from the British Library.

Library of Congress Cataloging-in-Publication Data
Names: Barry, Roger G. (Roger Graham), 1935– author. | Gan, Thian-Yew author.
Title: The global cryosphere : past, present and future / Roger G Barry, Thian Yew Gan.
Description: Cambridge, United Kingdom ; New York, N.Y. : Cambridge University Press, 2021. | Includes
bibliographical references and index.
Identifiers: LCCN 2020058381 (print) | LCCN 2020058382 (ebook) | ISBN 9781108487559 (hardback) | ISBN
9781108720588 (paperback) | ISBN 9781108767262 (ebook)
Subjects: LCSH: Cryosphere – History. | Cold regions – History. | Glaciers – History. | Ice sheets – History.
Classification: LCC QC880.4.C79 B37 2021 (print) | LCC QC880.4.C79 (ebook) | DDC 551.31–dc23
LC record available at https://lccn.loc.gov/2020058381
LC ebook record available at https://lccn.loc.gov/2020058382

ISBN 978-1-108-48755-9 Hardback
ISBN 978-1-108-72058-8 Paperback

Additional resources for this publication at www.cambridge.org/globalcryosphere2

Contents

Color plates between pages 372 and 373

Preface

This text aims to fill a long-standing gap in the scientific literature. While there are many texts on individual components of the cryosphere – snow cover, glaciers, ice sheets, lake and river ice, permafrost, sea ice, and icebergs – there is no comprehensive account. The text is aimed at upper division undergraduates and beginning graduate students in environmental sciences, geography, geology, glaciology, hydrology, water resources engineering, and ocean sciences, as well as providing a reference source for scientists in all environmental science and engineering disciplines.

The text builds on an introductory graduate-level course "Topics in snow and ice" taught by the late Roger G. Barry (RGB) at the Geography Department, University of Colorado, Boulder, over the last 30 years, and on part of a graduate-level course "Advanced surface hydrology" taught by Thian Yew Gan (TYG) as a professor of hydrology and water resources engineering at the Department of Civil/Environmental Engineering, University of Alberta, Edmonton, for almost 30 years. The former course in turn built on RGB's widening exposure to snow and ice data and literature through the work of the National Snow and Ice Data Center (NSIDC) from 1981 on. Barry's earlier field experience at the McGill Subarctic Research Laboratory, Schefferville, PQ, Canada in 1957–1958, Tanquary Fiord, Ellesmere Island, Arctic Canada in summer 1963 and spring 1964, Baffin Island, Arctic Canada in 1967 and 1970, and participation in a summer school on the Russian icebreaker *Kapitan Dranitsyn* in autumn 2005 provided additional insights, as did leaves at the Alfred Wegener Institute for Polar and Marine Research in 1994, the Geographical Institute, ETH, Zurich in 1997, and the Laboratoire de Glaciologie et Géophysique in Grenoble in 2004. RGB stepped down from the Directorship of NSIDC in May 2008 and worked half-time from January 2009–December 2010. This phase of the writing was greatly assisted by RGB being a recipient of a Humboldt Foundation Prize Award in 2009–2011. He spent May–October 2009 and August–October 2010 as a visitor at the Kommission für Glaziologie of the Bavarian Academy of Sciences in Munich (BASM), courtesy of its Director, Dr. Ludwig Braun. TYG began his collaboration with RGB during his visit to NSIDC as a CIRES (Cooperative Institute of Research in Environmental Science) visiting fellow in 2007, and worked with RGB on the first edition of this book at Boulder in 2008 and at BASM in 2009. RGB and TYG designed the second edition at Boulder in 2017, and TYG wrote the second edition in 2018–2020, especially at the Hong Kong Baptist University in 2019 as a recipient of a university fellowship of the Sir Run-Run Shaw Foundation. Between 1992 and 2008, TYG has had field experience conducting snow measurement in the

Canadian high Arctic and in the Canadian Prairies (CP), and also monitoring river ice breakup in the Northwest Territories of Canada, remote sensing of snow, and modeling of snowmelt in CP and Swiss Alps. Readers can find a solutions manual for the review questions and color versions of all the figures in the book at the following web address: cambridge.org/globalcryosphere2.

Acknowledgments

First and foremost, thanks are due to Hong Kong Baptist University for their award of a University Fellowship by their Sir Run Run Shaw Endowment Fund to Thian Yew Gan (TYG) in 2019, to the Chinese University of Hong Kong for its International Mobility Program of 2017 and 2019, and to the Southern University of Science and Technology, China for its Nashan professorship of 2019 to TYG. These supports enabled TYG to work on the book without much distraction. Furthermore, thanks are also due to the Humboldt Foundation of Germany for their award of a Humboldt Prize Fellowship in 2009–2011 that enabled Roger G. Barry (RGB) to work on the first edition of the book at the Kommission fuer Glaziologie (Commission for Geodesy and Glaciology) of the Bavarian Academy of Sciences, Munich, and thanks go to its Director Dr. Ludwig Braun for his hospitality and help.

Thanks to a CIRES (Cooperative Institute of Research in Environmental Science) fellowship that supported TYG's 2007 visit, and to the National Science and Engineering Research Council (NSERC) of Canada that supported his 2008 visit, to the National Snow and Ice Data Centre (NSIDC), University of Colorado, Boulder, and to NSIDC for providing the necessary facilities for working on the first edition of the book.

TYG is indebted to his wife, Esther Lynn, and children, Hanah Kai Lynn, Jonathan Kai Ernn, and Johanna Kai Yee, for their selfless support, encouragement, inspiration, patience, and motivation throughout this book project over the past two years.

We are also indebted to the following chapter reviewers for the first edition of the book for their suggestions. Any remaining errors are our own.

Chris Hiemstra, U.S. Army Corps of Engineers, CRREL, Ft. Wainright, AK (Ch. 2A)
Karl Birkekand, U.S.D.A. Forest Service National Avalanche Center, Bozeman, MT
  (Ch. 2B)
Jack D. Ives, Carleton University, Ottawa (Ch. 3)
Mark F. Meier, INSTAAR, University of Colorado, Boulder (Ch. 3)
Ted Scambos, NSIDC, University of Colorado, Boulder (Ch. 4 and Ch. 8)
Fritz Nelson, University of Delaware (Ch. 5)
Glen Liston, Colorado State University (Ch. 6A)
Spyros Beltaos, National Water Research Institute, Burlington, Ontario (Ch. 6B)
Norbert Untersteiner, University of Washington, Seattle, WA (Ch. 7)
Klaus Heine, Department of Geography, University of Regensburg (Ch. 9)

We thank Drs. Richard Armstrong, Mary Brodzik, Faye Hicks, Jack Ives, Adina Racoviteanu, Vladimir Romanovsky, Nikolai Shiklomanor, Koni Steffen, and Heidi Sevestre for photographs, NSIDC student helpers Sam Massom, Yana Duday, and Mike Laxer for illustration assistance; and we also thank Matt Lloyd and Sarah Lambert of Cambridge University Press for their enthusiastic support of the project.

Our thanks go to individuals, societies, and organizations for their permission to reproduce figures from books and journals.

1 Introduction

1.1 Definition and extent

The *cryosphere* is the term which collectively describes the portions of the Earth's surface where water is in its frozen state – snow cover, glaciers, ice sheets and shelves, freshwater ice, sea ice, icebergs, permafrost, and ground ice. The word *kryos* is Greek meaning icy cold. Dobrowolski (1923, p. 2, Barry *et al.*, 2011) introduced the term cryosphere and this usage was elaborated by Shumskii (1964, pp. 445–55) and by Reinwarth and Stäblein (1972). Shumskii included atmospheric ice, but this has generally been excluded. The cryosphere is an integral part of the global climate system. It has important linkages and feedbacks with the atmosphere and hydrosphere that are generated through its effects on surface energy and on moisture fluxes, by releasing large amounts of fresh water when snow or ice melts (which affects thermohaline oceanic circulations), and by locking up fresh water when they freeze. In other words, the cryosphere affects atmospheric processes such as clouds and precipitation, and surface hydrology through changes in the amount of fresh water on lands and oceans. Slaymaker and Kelly (2006) published a study of the cryosphere in the context of global change, while Bamber and Payne (2004) detail the mass balance of glaciers, ice sheets, and sea ice. The discipline of *glaciology* encompasses the scientific study of snow, floating ice, and glaciers, while the study of permafrost (*cryopedology*) has largely developed independently.

In a report on the International Polar Year, March 2007–March 2009, the World Meteorological Organization (2009) identified the following important foci of cryospheric research: rapid climate change in the Arctic and in parts of the Antarctic; diminishing snow and ice worldwide (sea ice, glaciers, ice sheets, snow cover, permafrost); the contribution of the great ice sheets to sea-level rise and the role of subglacial environments in controlling ice-sheet dynamics; and methane release to the atmosphere from melting permafrost. These topics will be discussed, but in each case we first survey the basic characteristics and processes at work for each cryospheric element. We also consider the past cryosphere throughout geological time and model simulations of future cryospheric states and their significance. In the concluding Chapter 11, practical applications of snow and ice research are presented. We begin by considering the dimensions of the cryosphere.

Dimensions of the cryosphere

Table 1.1 shows the major characteristics of the components of the cryosphere.

Figure 1.1 illustrates the global distribution of these components.

The cryosphere has seasonally varying components and more permanent features. Snow cover has the second largest extent of any component of the cryosphere, with a mean annual area of approximately 26 million km^2 (Table 1.1). Almost all of the Earth's snow-covered land area is located in the Northern Hemisphere, and temporal variability is dominated by the seasonal cycle. The Northern Hemisphere mean snow cover extent ranges from ~46 million km^2 in January to 3.8 million km^2 in August. Sea-ice extent in the Southern Hemisphere varies seasonally by a factor of five, from a minimum of 3–4 million km^2 in February to a maximum of 17–20 million km^2 in September (Gloersen *et al.*, 1993; Stroeve *et al.*, 2012). The seasonal variation is much less in the Northern Hemisphere where the confined nature and high latitudes of the Arctic Ocean result in a much larger perennial ice cover, and the surrounding land limits the equator-ward extent of wintertime ice. Northern Hemisphere ice extent varies by only a factor of two, from a minimum of 7–9 million km^2 in

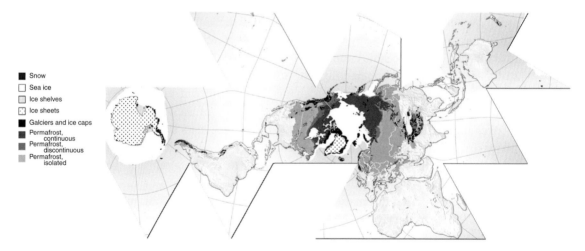

■ Snow
□ Sea ice
▨ Ice shelves
▧ Ice sheets
■ Galciers and ice caps
■ Permafrost,
    continuous
■ Permafrost,
    discontinuous
▨ Permafrost,
    isolated

Figure 1.1 The global distribution of the components of the cryosphere (from Hugo Ahlenius, courtesy UNEP/GRID-Arendal, Norway) (http://upload.wikimedia.org/wikipedia/commons/b/ba/Cryosphere_Fuller_Projection.png) (A black and white version of this figure will appear in some formats. For the color version, please refer to the plate section.)

Table 1.1 Areal and volumetric extent of major components of the cryosphere (updated after Goodison *et al.*, 1999)

Component	Area (10^6 km^2)	Ice volume (10^6 km^3)	Sea-level equivalent (m) [a]
LAND SNOW COVER [b]			
Northern Hemisphere			
Late January	46.5	0.002	
Late August	3.9		
Southern Hemisphere			
Late July	0.85		
Early May	0.07		
SEA ICE			
Northern Hemisphere			
Late March	14.0 [c]	0.05	
Early September	6.0 [c]	0.02	
Southern Hemisphere			
Late September	15.0 [d]	0.02	
Late February	2.0 [d]	0.002	
PERMAFROST (underlying the exposed land surface, excluding Antarctica and S. Hemisphere high mountains)			
Continuous [e]	10.69	0.0097–0.0250	0.024–0.063
Discontinuous and sporadic	12.10	0.0017–0.0115	0.004–0.028
CONTINENTAL ICE AND ICE SHELVES			
East Antarctica[f]	10.1	21.7	52
West Antarctica[f] and Antarctic	2.3	3.0	5
Peninsula Greenland[g]	1.7	2.85	7.3
Small ice caps and[h] mountain glaciers	0.74	0.24	0.6
Ice shelves[f]	1.5	0.66	

[a] Sea-level equivalent does not equate directly with potential sea-level rise, as a correction is required for the volume of the Antarctic and Greenland ice sheets that are presently below sea level. 400,000 km^3 of ice is equivalent to 1 m of global sea level.

[b] Snow cover includes that on land ice, but excludes snow-covered sea ice (Robinson *et al.*, 1995).

[c] Actual ice areas, excluding open water. Ice extent ranges between approximately 7.0 and 15.4 × 10^6 km^2 for 1979–2004 (Parkinson *et al.*, 1999a).

[d] Actual ice area excluding open water (Gloersen *et al.*, 1993). Ice extent ranges between approximately 3.8 and 18.8 × 10^6 km^2. Southern Hemisphere sea ice is mostly seasonal and generally much thinner than Arctic sea ice.

[e] Data calculated using the Digital Circum-Arctic Map of Permafrost and Ground-Ice Conditions (Brown *et al.*, 1997) and the GLOBE-1 km Elevation Data Set (Zhang *et al.*, 1999).

[f] Ice-sheet data include only grounded ice. Floating ice shelves, which do not affect sea level, are considered separately (Drewry *et al.*, 1982; Huybrechts *et al.*, 2000; Lythe *et al.*, 2001)

[g] Dahl-Jensen *et al.* (2009)

[h] Radić and Hock (2010).

September to a maximum of 14–16 million km^2 in March during 1979–2004. Subsequent years have seen much smaller areas in late summer (Kay *et al.*, 2011).

Ice sheets are the greatest potential source of fresh water, holding approximately 77% of the global total. Fresh water in ice bodies corresponds to 71 m of world sea-level equivalent, with Antarctica accounting for 90% of this and Greenland almost 10%. Other ice caps and glaciers account for about 0.5% (Table 1.1).

The *World Atlas of Snow and Ice Resources* (Kotlyakov, 1997) provides maps of climatic factors (air temperature, solid precipitation), snow water equivalent (SWE), runoff, glacier morphology, mass balance and glacier fluctuations, river freeze-up/breakup avalanche occurrence, and many other variables. The maps range from global, at a scale 1:60 million, to regional maps at 1:5 million to 1:10 million and local maps of individual glaciers at 1: 25,000 to 1:100,000.

Permafrost (perennially frozen ground) may occur where the mean annual air temperature (MAAT) is less than −1 °C and is generally continuous where MAAT is less than −7 °C. It is estimated that permafrost underlies about 22 million km^2 of exposed Northern Hemisphere land areas (Table 1.1), with maximum areal extent between about 60° and 68° N. Its thickness exceeds 600 m along the Arctic coast of northeastern Siberia and Alaska, but permafrost thins and becomes horizontally discontinuous toward the margins. Only about 2 million km^2 consists of actual ground ice ("ice-rich"). The remainder (dry permafrost) is simply soil or rock at subfreezing temperatures. A map of Northern Hemisphere permafrost and ground ice (1:10 million) was published by Brown *et al.* (2001) and is available electronically at http://nsidc .org/data/ggd318.html.

Seasonally frozen ground, not included in Table 1.1, covers a larger expanse of the globe than snow cover. Its depth and distribution varies as a function of air temperature, snow depth and vegetation cover, ground moisture, and aspect. Hence it can exhibit high temporal and spatial variability. The area of seasonally frozen ground in the Northern Hemisphere is approximately 55 million km^2 or 58% of the land area in the hemisphere (Zhang *et al.*, 2003b).

Ice (see Note 1.1) also forms on rivers and lakes in response to seasonal cooling. The freeze-up/breakup processes respond to large-scale and local weather factors, producing considerable interannual variability in the dates of appearance and disappearance of the ice. Long series of lake-ice observations can serve as a climatic indicator; and freeze-up and breakup trends may provide a convenient integrated and seasonally specific index of climatic perturbations. The total area of ice-covered lakes and rivers is not accurately known and hence this element has not been included in Table 1.1.

1.2   The role of the cryosphere in the climate system

The elements of the cryosphere play several critical roles in the climate system (Barry, 1987, 2002b; Barry and Gan, 2011). The primary one operates through the ice–albedo feedback mechanism. This concerns the expansion of snow and ice cover increasing the albedo, thereby increasing the reflected solar radiation and lowering the temperature, thus enabling the ice and snow cover to expand further. At the present day this effect is working in the opposite direction with the shrinkage of snow and ice cover lowering the albedo and increasing the absorption of solar radiation, thereby raising the temperature and further reducing the snow and ice cover. On a global scale, the ice–albedo effect amplifies climate sensitivity by about 25–40% (depending on cloudiness changes).

A second major influence is the insulation of the land surface by snow cover and of the ocean (as well as lakes and rivers) by floating ice. This insulation greatly modifies the temperature regime in the underlying land or water. The difference in the temperature of air overlying bare ground versus snow-covered ground is of the order of 10 °C based on winter measurements in the Great Plains of North America. The absence of snow cover could mean higher mean-annual surface air temperature, but severe wintertime cooling, and a substantial increase in permafrost areas over high-latitude regions of the Northern Hemisphere such as Siberia (Vavrus, 2007).

A third effect is on the hydrological cycle due to the storage of water in snow cover, glaciers, ice caps, and ice sheets and associated delays in freshwater runoff. The time scales involved range from weeks to months in the case of snow cover, decades to centuries for glaciers and ice caps, and to 10^5–10^6 years in the case of ice sheets and permafrost. The more permanent features of the cryosphere accordingly have a great influence on eustatic changes in global sea level (see Table 1.1). A 1-mm rise in eustatic sea level requires the melting of 360 Gt of ice.

A fourth effect is related to the latent heat involved in phase changes of ice/water. This applies to all elements of the cryosphere. It is estimated, for example, that a 10-cm snow cover over England has a latent heat of fusion of 10^{15} kJ; melting the Greenland Ice Sheet would require ~10^{21} kJ. Ohmura (1987) calculated that the melting of ice since the Last Glacial Maximum about 20 ka accounted for 26–39×10^3 MJ m^{-2}, of similar magnitude to the total energy stored in the climate system (30–60×10^3 MJ m^{-2}).

A fifth effect is caused by seasonally frozen ground and permafrost modulating water and energy fluxes, and the exchange of carbon (especially methane), between the land and the atmosphere.

1.3   The organization of snow and ice observations and research

The organization of cryospheric data began during the International Geophysical Year (IGY), 1957–1958, with the establishment of the World Data Center (WDC) system.

WDCs for Glaciology were designated in the United States, the Soviet Union, and the United Kingdom. In 1976, World Data Center-A for Glaciology was transferred from the US Geological Survey in Tacoma, WA, to the National Oceanic and Atmospheric Administration (NOAA) in Boulder, CO, where it has subsequently been operated by the

University of Colorado (Barry, 2002a). The scope of its operations expanded to address data on all forms of snow and ice and in 1981 the National Environmental Satellite Data and Information Service (NESDIS) of NOAA designated a National Snow and Ice Data Center (NSIDC). Its financial support was greatly augmented by contracts and grants from the National Aeronautics and Space Agency (NASA) and the National Science Foundation. Roger G. Barry served as Director from 1976 until 2008 and was succeeded by Mark Serreze. Details on its data holdings and research activities may be found at http://nsidc.org. World Data Center-C for Glaciology addresses bibliographic data and is operated by the Scott Polar Research Institute at Cambridge, UK, World Data Center-D for Glaciology was established at the Laboratory for Glaciology and Geocryology, Lanzhou, China in 1986. The letter designations were dropped in 1999 and in 2009 the International Council of Science (ICSU) decided to convert the WDC system into a World Data System. This is not yet operational but in the interim the WDCs continue to function as before. The International Science Council's (ISC) General Assembly created the ISC-WDS (World Data System) in Tokyo, Japan in October 2008 (https://www.worlddatasystem.org/).

Over the last few years, major advances have occurred in the organization of snow and ice observations and research. Initially, the organization took place within the various cryospheric subfields (snow, avalanches, glaciers and ice sheets, freshwater ice, sea ice, and permafrost). Then, beginning in the 1990s, the Global Climate Observing System (GCOS), and its partners the Global Ocean Observing System (GOOS) and Global Terrestrial Observing System (GTOS), defined Essential Climate Variables (ECVs) (Barry, 1995; GCOS, 2004). For the cryosphere, these include snow cover, glaciers, permafrost, and sea ice. Global Terrestrial Networks (GTN) were specified for glaciers (GTN-G) and permafrost (GTN-P) (http://gosic.org/ios/GTOS_observing_system.asp).

At a higher level, the Integrated Global Observing System (IGOS) initiated the preparation of a report on a cryosphere theme (Key *et al.*, 2007) that documented the available and needed cryospheric data sets. In May 2007, the 15th Congress of the World Meteorological Organization (WMO) received a proposal from Canada to create a Global Cryosphere Watch (GCW), analogous to the Global Atmosphere Watch (GAW). In 2011, the 16th World Meteorological Congress (WMC) approved the GCW Implementation Strategy. In 2015, the 17th WMC implemented GCW in WMO Programmes (https://globalcryospherewatch.org/).

In July 2007, at the XXIVth General Assembly of the International Union of Geophysics and Geodetics (IUGG) in Perugia, Italy, the IUGG Council launched the International Association of Cryospheric Sciences (IACS) as the eighth IUGG Association. IAGS was developed from the International Commission of Snow and Ice of the International Association of Hydrological Sciences (IAHS) via the transitional Union Commission for the Cryospheric Sciences (UCCS). This superseded the International Commission for Snow and Ice (ICSI) (Radok, 1997). IACS has the following five divisions: snow and avalanches, glaciers and ice-sheets, marine and freshwater ice, cryosphere, atmosphere and climate, and planetary and other ices of the solar system. www.iugg.org/associations/iacs.php

On the research side, the World Climate Research Programme (WCRP) established a Climate and Cryosphere (CliC) Project in 2000 (Allison *et al.*, 2001; Barry, 2003) that has four thematic areas: interactions between the atmosphere, snow, and land; interactions between land ice and sea level; interactions between sea ice, oceans, and the atmosphere; and cryosphere-ocean/cryosphere-atmosphere interactions on a global scale (http://clic .npolar.no). The CliC project is directed by a Science Steering Group and regularly organizes workshops and conferences.

In 2019, Intergovernmental Panel of Climate Change published the Special Report on the Ocean and Cryosphere in a Changing Climate (SROCC) that summarizes the characteristics and interconnection of ocean and cryosphere and their importance in the earth system, particularly in light of climate change impact.

1.4   Remote sensing of the cryosphere

Cryospheric science has benefitted enormously from the ready availability of satellite data since the mid-1960s. In recent years, there has been significant progress in the observation of the cryosphere by satellites, such as temporal and regional changes in ice sheets, trends and anomalies in Arctic sea ice cover, substantial decline of Arctic summer sea ice extent in 2012, pan-Arctic measurements of changes in ice thickness using satellite altimetry, and changes of its volume and mass. As a result, recent fluctuations in the cryosphere have been mapped with increasing certainty, demonstrating the potential for rapid loss in sea ice and snow cover, and the associated sea-level rise. Remote-sensing measurements of regional glacier volume change are also now available widely and modeling of glacier mass change has improved considerably. We will briefly summarize the main instruments that have operated and some of their applications. Further details are provided in the relevant chapters.

The hemispheric analysis of snow cover extent began in October 1966 from NOAA's polar orbiting Very High Resolution Radiometer (VHRR) and continued with the use of the Advanced VHRR (AVHRR) and other visible-band satellite data. Global snow cover maps are now available from the Moderate Resolution Imaging Spectroradiometer (MODIS) on Terra (February 2000–present) and Aqua (July 2002–present). In December 1972, the NASA launched the Electrically Scanning Microwave Radiometer (ESMR) on Nimbus 5 enabling all-weather mapping of sea ice extent.

Passive microwave (PM) remote sensing has been in the core attention for obtaining representative retrieval of SWE data, which plays an important role in hydrological process. Despite of coarse resolutions and large uncertainties from systematic and random error in SWE data retrieved from spaceborne microwave sensors, such multifrequency satellite data are widely used in retrieving snow information of large coverage because the sensors are of wide swath, which produces many repeated observations and are insensitive to illumination and minimal influence of cloud cover. In October 1978, the Scanning Multichannel Microwave Radiometer (SMMR) launched on Nimbus 7 allowed sea ice concentrations and SWE to be delimited. SMMR operated until August 1987 and the records continued to the present with the Special Sensor Microwave Imager (SSM/I) on Defense Meteorological Satellite Program (DMSP) satellites. The Advanced Microwave Scanning Radiometer – Earth (AMSR-E) observing system on board the Aqua satellite of NASA launched in 2002 provides a higher spatial resolution than SMMR and SSM/I (http://weather .msfc.nasa.gov/AMSR/). The frequency of the SMMR sensor is 6.6–37 GHz, the SSM/I sensor is 19–85.5 GHz, and the AMSR-E sensor is 6.9–89 GHz. Most retrieval algorithms for SWE exploit the negative spectral gradient between ~37 GHz frequency sensitive to snow grain volume scattering and a ~19 GHz frequency largely insensitive to snow (Takala *et al.*, 2011). The larger the difference between brightness temperature (TB) measurements at these two frequencies, the higher the SWE estimated.

AMSR-E has mapped SWE data sets over northern and southern hemisphere producing AE_5DSno SWE products since 2002. However, due to spatial variability of climate and terrain properties, the application of AE_5DSno SWE products (and similar PM data) and that of SMMR and SSM/I in mountainous areas has been argued. These products have 25 × 25 km fixed-grid cell resolution, which is very coarse for rugged surface topography of mountainous areas. However, there are classic downscaling methods based on uniform and nonuniform redistributions of SWE within a coarse pixel, and elevation and aspect direction maps developed to generate SWE products in 500 m spatial resolution, utilizing daily Moderate Resolution Imaging Spectroradiometer (MODIS) snow cover images with cloud covers removed by, say, a time neighborhood method.

The Landsat series began in 1972 and in April 1999 Landsat 7 was launched. The Multispectral Scanner (MSS) at 80 m resolution operated through the mid-1990s, but with Landsat 4 (1982) and Landsat 5 (1984) the Thematic Mapper (TM) with 30 m resolution came into use. With the Landsat 7 launched in April 1999, the Enhanced TM (ETM) could provide data at 15–30 m resolution. The latest Landsat 8 satellite launched in February of 2013 carries two sensors, the Operational Land Imager (OLI) and Thermal Infrared Sensor (TIRS) which acquire images in nine spectral bands with a spatial resolution of 30 meters. Landsat data have been widely used for mapping mountain glaciers. Together with 15 m resolution data from the Advanced Spaceborne Thermal Emission and Reflection Radiometer (ASTER) instrument (http://asterweb.jpl.nasa.gov/asterhome/), aboard the Terra satellite, outlines for over 83,000 glaciers have been compiled into the database of the Global Land Ice Measurement from Space (GLIMS) project at the NSIDC.

Extensive synthetic aperture radar (SAR) data have been obtained since the 1990s. The European Space Agency's (ESA) Earth Remote Sensing (ERS)-1 active microwave instrument operated between 1992 and 1996 and ERS-2 has been operating since 1996. The available time series has been used to determine ice-sheet mass balances. The Canadian RADARSAT-I sensor has been providing SAR coverage of Arctic sea ice since 1995. In 1997, RADARSAT was rotated so that the first high-resolution mapping of the entire

Antarctic continent could be performed. The RADARSAT-II mission launched in late 2007, which carries a C-band SAR offering multiple modes of operation including quad polarization, ensures the continuity and improvement of SAR coverage of Arctic sea ice. The NASA scatterometer on QuikSCAT has operated since 1999 providing another view of sea ice extent.

ERS radar altimetry has been used to estimate ice thickness in both polar regions. In 1997 interferometry with SAR was used to obtain ice velocity vectors over the East Antarctic ice streams. NASA's Geoscience Laser Altimeter System (GLAS) on the Ice, Cloud, and Land Elevation Satellite (ICESat) was used to measure ice-sheet elevations and changes in elevation, as well as sea ice freeboard from February 2003 through November 2009. ICESat-2, part of NASA's Earth Observing System launched in September 2018, measures ice-sheet elevation and sea ice thickness, land topography, vegetation characteristics, and clouds. In April 2010 ESA's Earth Explorer CryoSat mission, which carries a SAR Interferometric Radar Altimeter (SIRAL), was launched. It is dedicated to precise monitoring of changes in the thickness of sea ice in the polar oceans and variations in the thickness of the Greenland and Antarctic ice sheets.

The Gravity Recovery and Climate Experiment (GRACE) was a joint mission of NASA and the German Aerospace Center. Since March 2002 to October 2017, by sensing gravity, the attraction between two objects from space, the twin satellites of GRACE have been mapping Earth's gravitational field anomalies determined by mass every 30 days (Tapley *et al.*, 2004), which is very useful for the scientific understanding of ice-sheet dynamics, land-ice response, and sea-level change to climate warming (Velicogna and Wahr, 2006). The movements of water and ice would cause a change in the Earth's mass, which causes the distance between the twin satellites to change. By tracking changes in the distance between both satellites, we can calculate how gravity on Earth is changing regionally, from which GRACE can identify the shrinkage of ice sheets and land ice (ice sheets, icefields, ice caps, and glaciers) worldwide. GRACE data show that ice sheets of Greenland and Antarctica have lost 280 ± 58 and 67 ± 44 Gt of ice per year between 2003 and 2013, respectively, which equate to a total of 0.9 mm yr^{-1} of sea-level rise.

GRACE can also determine the cause of sea-level rise, whether it is the mass of melting glaciers being added to the ocean, or from thermal expansion of warming water. Sea level has been rising globally given loses of land ice from meltwater runoff and calving of solid ice into the ocean are greater than gaining land ice through precipitation. In recent years, scientists estimated losses from land ice have been responsible for two-thirds of the observed sea-level rise (Cazenave *et al.*, 2018; Chen *et al.*, 2013). For example, using RL05 GRACE data for 2003–2012, Velicogna and Wahr (2013) found a mass loss of 258 ± 41 Gt yr^{-1} for Greenland, with an acceleration of 31 ± 6 Gt yr^{-2}, and a loss that progressively affect the entire periphery. For Antarctica, they found a loss of 83 ± 49 and 147 ± 80 Gt yr^{-1}, with an acceleration of -12 ± 9 Gt yr^{-2} and a dominance from the southeast pacific sector of West Antarctica and the Antarctic Peninsula. The successor of GRACE, GRACE-FO, or GRACE Follow-On was launched on May 2018 to continue monitoring Earth's gravity and tracking changes in global sea levels, glaciers, and ice sheets, as well as large lake and river water levels, and soil moisture.

The recently launched ICESat-2 (Ice, Cloud, and Land Elevation Satellite-2) satellite carries a photon-counting laser altimeter to measure the height of a changing Earth, for example, the elevation of ice sheets, glaciers, sea ice, and more in unprecedented details by 10,000 laser pulses a second. Our planet's frozen and icy polar regions, called the cryosphere, are particularly sensitive to a warming climate. For instance, Arctic amplification is climate change impact amplified in the Arctic, which has been warming at about twice the global rate since 1980s, caused by several feedback mechanisms, namely, ice–albedo, temperature (Planck and Lapse rate), water vapor, CO_2, and cloud feedbacks. Therefore, how much and how fast is our cryosphere changing under a warmer climate is of significant interest to us. ICESat-2 has four science objectives: (1) estimate how melting ice sheets contribute to sea-level rise; (2) measure signatures of ice-sheet changes to assess the mechanisms driving those changes; (3) estimate sea ice thickness to examine ice/ocean/atmosphere exchanges of energy, mass, and moisture; and (4) measure vegetation canopy height to estimate large-scale biomass changes in ecosystems worldwide (IceSat-2, 2018).

CryoSat-2 is a European Space Agency environmental research satellite and part of the CryoSat programme to study the Earth's polar ice caps (Pessina *et al.*, 2011). It was launched on April 8, 2010 to understand how climate change is affecting marine ice floating in the Arctic Ocean, how the thickness of the ice, both on land and floating in the sea is changing for in recent years, the summer Arctic sea ice extent has been diminishing drastically, since early 2000. CryoSat-2 is used to determine if there is a trend toward diminishing ice cover by providing data about the polar ice caps and changes in the ice thickness with a resolution of about 1.3 cm. On October 22, 2010, CryoSat-2 was declared operational following six months of on-orbit testing. The main instrument of CryoSat-2 is an interferometric radar range-finder with twin antennas, which measures the height difference between the upper surface of floating ice and surrounding water often known as "free-board."

Note 1.1

Ice: Ice is the solid phase, usually crystalline, of water. The word derives from Old English *is*, which has Germanic roots. There are other ices – carbon dioxide ice (dry ice), ammonia ice, and methane ice – but these will not concern us here. Ice is transparent or an opaque bluish-white color depending on the presence of impurities or air inclusions. Light reflecting from ice often appears blue, because ice absorbs more of the red frequencies than the blue ones. Ice at atmospheric pressure is approximately 9% less dense than liquid water. It is the only known nonmetallic substance to expand when it freezes.

REVIEW QUESTIONS

(1) List five key components of the cryosphere.

(2) Seasonally which component of the cryosphere undergoes the largest change, and the difference of the component (in ratio) between winter and summer?

(3) Explain how the cryosphere is an integral part of the global climate system in terms of the five major influences it exerts on the climate system.

(4) Globally what are the greatest potential sources of freshwater, in terms of the estimated volume in km^3, and in terms of the global sea-level equivalent in m?

(5) How has the cryosphere been undergoing rapid changes during the twentieth and the early twenty-first centuries or what are key indicators of cryospheric change?

(6) In your opinion, what is one of the most striking physical manifestations of climate warming in recent decades?

(7) Explain possible implications to the future global cryosphere under the potential impact of climate change, and how cryospheric changes will impact the global climate system.

Part I

The terrestrial cryosphere

The terrestrial cryosphere forms the largest element of the overall cryosphere of the Earth (Table 1.1). It embraces seasonal snow cover (including avalanches), glaciers and ice caps, and the two large ice sheets of Greenland and Antarctica. It also includes perennially and seasonally frozen ground and freshwater ice in lakes and rivers. Each of these major components is treated in separate chapters.

2A   Snowfall and snow cover

Terrestrial snow cover occupies higher latitude areas of the Northern Hemisphere (NH) from several up to 9 months each year in the Arctic land surface, with significant influence on the surface energy budget, subsoil thermal regime, and the freshwater storage. Snow cover also interacts with vegetation and affects terrestrial habitats and species.

2.1   History

The hexagonal form of snowflakes was first noted by Johannes Kepler in 1611. Robert Hooke revealed the variety of crystalline structures as seen through a microscope in 1665. Similar studies were performed in the mid-eighteenth century in France and England. Bentley and Humphries (1931) published a book with over 2,500 illustrations of snowflake photographs of variety of snow crystals.

The earliest snow surveys were made at Mt. Rose, Nevada, in 1906 by James Church, and by 1909–1910 a network of stations was being surveyed by him. Snow surveys provide an inventory of the total amount of snow covering a drainage basin or a given region. Church also invented the Mt. Rose sampler – a hollow steel tube designed so that each inch of water in the sample weighs 1 ounce (28.35 g). Snow surveying began at locations in several western states between 1919 and 1929 and in the latter year California organized cooperative snow surveys (Stafford, 1959).

In 1931, a permanent Committee on the Hydrology of Snow was organized in the Hydrology section of the American Geophysical Union, chaired until 1944 by Dr. Church. By 1951, there were about one thousand snow courses in the western states and British Columbia. A snow course comprises an area demarcated for measuring the snow periodically during each snow season. Usually three to eight samples are taken and averaged to determine the snow depth and snow water equivalent (SWE) for that location. Stream flow forecasting to assess water supply is the primary objective. In remote locations, aerial markers were installed; these are vertical markers with equally spaced crossbars. The depth of snow is determined by visual observation from low-flying aircraft. The number of snow courses has declined considerably in recent years in part due to the extension of the Snow Telemetry (SNOTEL) network. These are automated weather stations designed to operate in severe, remote mountainous environments. Most sites collect daily, or even hourly, SWE and precipitation, and relay it by meteor burst technology to collection stations in Boise, Idaho, or Portland, Oregon.

Remote sensing of snow cover by the Very High Resolution Radiometer (VHRR) of National Oceanic and Atmospheric Administration (NOAA) that began in 1966 and its continuation – the Advanced VHRR (AVHRR) – provides the longest time series of hemispheric snow cover data. Spaceborne passive microwave measurements were applied to estimate snow depth and SWE in the late 1970s, as discussed later in this chapter. The Cold Land Processes Experiment (CLPX) of NASA took place in the winter of 2002 and spring of 2003, in the central Rocky Mountains of the western United States where there is a rich array of different terrain, snow, soil, and ecological characteristics to test and improve algorithms for mapping snow. Through the field campaigns of CLPX, algorithms for SWE retrieval and soil freeze–thaw status from spaceborne passive microwave sensors, and radar retrieval algorithms for snow depth, density, and wetness were evaluated and improved. The data were also used to improve spatially distributed, uncoupled snow/soil models and coupled cold land surface schemes.

The National Operational Hydrologic Remote Sensing Center (NOHRSC) of NOAA in the United States developed the airborne mapping of SWE using surface-emitted gamma radiation from potassium, uranium, and thorium radioisotopes in the soil. Gamma radiation is attenuated by snow cover and absorbed by water in the snowpack (NWS, 1992), and so to estimate SWE, both gamma counts and soil moisture over snow and bare ground are needed. Such SWE data had been used to develop passive microwave retrieval algorithms (e.g., Singh and Gan, 2000). Snow depth can also be estimated by microwave radiation transfer models, such as that of Chang *et al.* (1987), even though such models may underestimate the snow depth, as Butt (2009) found in a study in the United Kingdom.

2.2   Snow formation

Snow

The creation of saturation conditions necessary for the formation of water droplets or ice particles occur mainly through convection or updraft, cyclonic cooling induced by circulation, frontal or non-frontal lifting of warm air, or orographic cooling by mountain barriers. Snow forms primarily through *heterogeneous nucleation*. This process involves air that is saturated with a temperature below 0 °C. Water vapor condenses and solidifies, or vapor is deposited on nuclei, which grow into ice and snow crystals. These freezing nuclei may be clay mineral dust (kaolinite, e.g., becomes active at −9 °C), aerosols, pollutants, ice crystal splinters from clouds above, or artificial seeding agents (solid CO_2 or "dry ice," silver iodide, or urea). The crystals may continue growing through interactions between crystals (crystal aggregation) or with supercooled water droplets, a process called *riming* (the capture of supercooled cloud droplets by snow crystals) to form snow pellets and/or snowflakes (Mosimann *et al.*, 1993). The minimum size of ice crystals involved in riming is ~60 μm diameter for hexagonal plates and 30 μm width and 60 μm length for columnar ice crystals (Ávila *et al.*, 2009). Under extremely low temperatures (below −40 °C), ice particles can also be formed by the spontaneous freezing of water molecules, which is called "homogeneous nucleation." Homogeneous nucleation of water droplets occurs at −40 °C; at −10 °C approximately $1/10^6$ drops freeze and at −30 °C about $1/10^3$ drops freeze.

Ice crystal shapes are hexagonal in form from 0 °C to −80 °C and cubic form from −80 °C to −130 °C. The reason is that an oxygen molecule is tetrahedral; two together form a hexagon or tetrahedra offset by 60° form a cubic crystal. A cubic crystal will transform to a hexagon if warmed but not vice versa. Crystal types have a dependence on temperature and saturation vapor pressure over ice. Under various combinations of temperature and super-saturation conditions with respect to ice, a wide range of snowflakes/pellets results (Figure 2.1). In general, as the temperature decreases, plates → needles → prisms. They can be broadly classified as dendritic and sector plates that involve crystal growth on the a-axis (horizontal), or columns (prisms and needles) that involve growth on the c-axis (vertical) (Figure 2.1). Mason (1994) suggests that transitions between crystal types in clouds can lead to more effective release of precipitation through the formation of precipitation elements that have a better chance of surviving below-cloud-base evaporation.

Snowfall

Whenever snow crystals grow to a size when gravitational pull exceeds the buoyancy effect of air, snowfall occurs. Snowfall typically reaches the ground when the freezing level is not higher than about 250 m above the surface, and the surface air temperature averages ≤1.2 °C. Snow may fall as snowflakes, snow grains (the solid equivalent of drizzle; white, opaque ice particles ≤1 mm in diameter), or graupel (snow pellets of opaque conical or rounded ice particles 2–5 mm in diameter formed by aggregation).

Snowflakes

Snowflakes can be classified into many types (Grey and Prowse, 1993; Sturm *et al.*, 1995). Snowflakes form through the growth of ice crystals by the accretion of water vapor and by their aggregation in branched clusters. The saturation vapor pressure is lower over an ice surface than a water surface, reaching a maximum difference of 0.12 mb at −12 °C. As a result, in a mixed phase cloud, supercooled water droplets tend to evaporate and vapor is deposited onto ice crystals. This is known as the Bergeron–Findeisen process after its

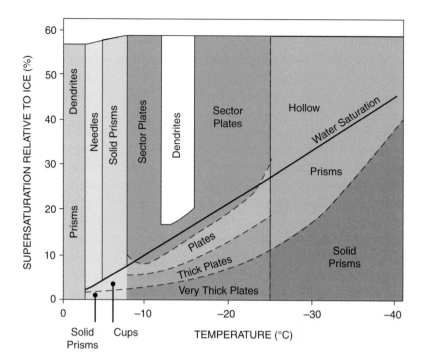

Figure 2.1 Types of snow crystals resulting from various combination of temperature and supersaturation (D. Kline, after Kobayashi, 1961).

discoverers. Snowflakes grow in small cap clouds over elevated terrain when ice crystals falling from an upper cloud layer seed them. This is known as the seeder–feeder mechanism (Barry, 2008, p. 273). Ice crystals may float in the atmosphere as "diamond dust" when the air temperature is ≤−40 °C. The design and variations of snowflakes are way beyond human imaginations, as some examples in Figure 2.2 that show needle, sheath, and varieties of stellar crystals with plates, dendritic, and sector-like branches. Bentley, who was born in 1865, even believed that no two snowflakes are exactly alike (Teel, 1994).

Depth hoar

Other than in permafrost areas (high latitudes or high elevations in middle latitudes), the ground is mostly warm or near freezing when the ground is snow covered. This is true even when the air is very cold, because snow is a good insulator. Therefore, there will usually be liquid water in the snowpack, and it is common for the snow near the ground to remain damp for most of the winter. *Depth hoar* forms at the base of a snowpack, as a result of large temperature gradients between the warm ground and the cold snow surface, when rising water vapor freezes onto existing snow crystals. It usually requires a thin snowpack combined with a clear sky or low air temperature, and it grows best at snow temperatures from −2 °C to −15 °C.

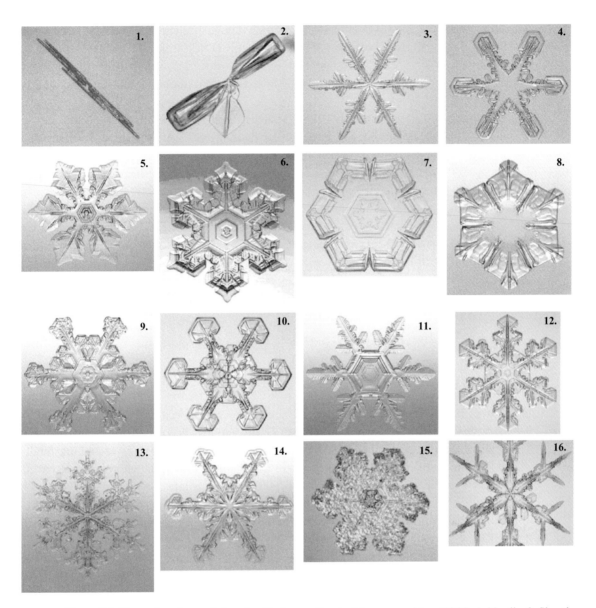

Figure 2.2 Examples of snowflakes classified according to Magono and Lee (1966): 1. Needle, 2. Sheath, 3. Stellar crystal, 4. Stellar crystal with sector-like ends, 5. Stellar crystal with plates at ends, 6. Crystal with broad branches, 7. Plate, 8. Plate with simple extension, 9. Plate with sector-like ends, 10. Rimed plate with sector-like ends, 11. Hexagonal plate with dendritic extensions, 12. Plate with dendritic extensions, 13. Dendritic crystal, 14. Dendritic crystal with sector-like ends, 15. Rimed stellar crystal with plates at ends, and 16. Stellar crystal with dendrites (A black and white version of this figure will appear in some formats. For the color version, please refer to the plate section.)

Therefore, the occurrence of depth hoar is common in high Arctic regions such as Alaska, the Northwest Territory, Nunavut, and northern Siberia (Derksen *et al.*, 2009). Depth hoar constitutes about 20% of all snow layers in the high Arctic (46% in sub-Arctic), an average grain size of 6.5 mm (long-axis) and 2 mm (short-axis), and about 0.23 gm cm^{-3} in density (Derksen *et al.*, 2014).

Depth hoar consists of sparkly, large-grained, faceted, cup-shaped ice crystals up to 10 mm in diameter. Beginning and intermediate facets are 1–3 mm square; advanced facets can be cup-shaped 4–10 mm in size. Larger-grained depth hoar is more persistent and can last for weeks. Depth hoar is strong in compression but not so in shear, and hence often behaves like a stack of champagne glasses; it can fail in the form of collapsing layers, or in shear, with fractures often propagating long distances and around corners. Even though the stability of the snowpack depends on the cohesion between layers of snow and meteorological factors, almost all catastrophic avalanches, which involve the entire season's snow cover, fail on depth hoar layers (Tremper, 2008). Seismic sensors have been used for the remote detection of snow avalanches for they can also estimate avalanche velocity, size, and type (Lacroix *et al.*, 2012).

2.3   Snow cover

Introduction

Snow is an integral component of the global climate system because of its linkages and its feedback between surface energy, moisture fluxes, clouds, precipitation, hydrology, and atmospheric circulation (King *et al.*, 2008). It is the most spatially extensive and seasonally variable component of the global cryosphere (see Table 1.1). On an average, snow covers almost 50% of the Northern Hemisphere's land surface in late January, with an August minimum of about 1%. In addition, there is perennial snow cover over the Antarctic ice sheet (12 million km^2) and at the higher elevations of the Greenland Ice Sheet (about 0.6 million km^2) (Figure 2.3).

Since snow produces substantial changes in the surface characteristics, and the atmosphere is sensitive to physical changes of the earth surface, its presence over large areas of the Earth for at least part of the year exerts an important influence on the climate, both locally and globally. The best-known effect involves the albedo–temperature positive feedback, whereby an expanded (reduced) snow cover increases (decreases) the reflection of incoming solar radiation, reducing (increasing) the temperature, and thereby encouraging an expansion (reduction) of the snow cover. Fresh snow has a spectrally integrated albedo of 0.8–0.9, making it the most reflective natural surface. This value decreases with age to 0.4–0.7 as the snow density increases through settling and snow metamorphism and is reduced still further by impurities in or on the snow (e.g., mineral dust, soot, aerosols, biogenic matter) (see Figure 2.4). The cooling effect of snow cover is illustrated by the example that, in the Upper

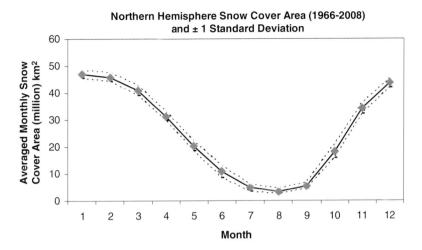

Figure 2.3 Averaged monthly snow cover area of Northern Hemisphere in ($\times 10^6$) km^2 calculated from weekly snow cover extent maps produced primarily from daily visible satellite imagery of NOAA-AVHRR by the Rutgers Global Snow Lab

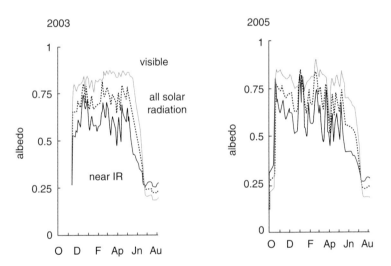

Figure 2.4 Field measurements of broadband albedo at Mammoth Mountain in the Sierra Nevada for (a) 2003 and (b) 2005, showing albedo in the visible, near-infrared and all solar radiation (adapted from Dozier *et al.*, 2009)

Midwest of the United States, winter months with snow cover are about 5–7 °C colder than the same months without snow cover. Snow, a poor conductor of heat, also insulates the soil surface and sea ice. Therefore, a better knowledge of the snow cover and its properties over large regions will lead to a better understanding of our climate.

Snow is held in cold storage until there is sufficient energy to melt it to water or to sublimate it to water vapor. The storage of water in the seasonal snow cover introduces into the hydrological cycle an important delay of weeks to months, causing a peak in the annual runoff in spring and early summer when the river water is agriculturally more valuable. It is highly beneficial to be able to estimate the amount and timing of release of this stored precipitation to spring runoff, which allows a better management of water resources for irrigation and hydroelectric production planning. The dynamics of water storage in seasonal snowpack is also critical to the effective management of water resources globally. Snow water accumulated in winter in the Arctic river basins is critical for the springtime snowmelt, and the freshwater from its river systems accounts for about 50% of the net flux of freshwater into the Arctic Ocean (Barry and Serreze, 2000), which is a large percentage when compared to the freshwater inputs to the tropical oceans, where freshwater input is dominated by direct precipitation. Frozen soil affects the snowmelt runoff and soil hydrology by reducing the soil permeability. Runoff affects ocean salinity and sea ice conditions (Peterson *et al.*, 2002), and the degree of surface freshening can affect the global thermohaline circulation (Broecker, 1997). In future, we expect even larger freshwater input from the Arctic as the global hydrologic cycle accelerates, higher precipitation is expected in high latitudes and higher runoff from Arctic river basins (Nummerlin *et al.*, 2016).

Snowpacks affect energy and water exchanges, and so both snow cover and SWE are important climatic and hydrologic variables. In particular, snow controls the climate and hydrology of the cryosphere and higher latitude regions significantly, and the amount and distribution of snow is affected by the climate and vegetation types. In the Canadian Prairies, mixed precipitation can occur within a certain range of temperature (Kienzle, 2008), but on a whole approximately one-third of its annual precipitation occurs as snowfall, and the shallow snow cover generates as much as 80% of the annual surface runoff. In the Colorado Rockies, the Sierra Nevada of California, and the Cascade Mountains of Washington, snowmelt can account for up to 65–80% of the annual water supply (Serreze *et al.*, 1999), but shifts in snowmelt runoff due to climate warming is expected to affect water supply in snow-dominated river basins (Fritz et al., 2011).

The snow covers of North America (NA) and Eurasia change seasonally, in accordance with the position of the Sun that shines directly at the Tropic of Cancer in the Northern Hemisphere (NH) on June 21 (summer solstice) and then moves southward, reaching the Tropic of Capricorn in the Southern Hemisphere on December 22 (winter solstice), before moving northward for the next 6 months; the cycle repeats itself on an annual time scale. The extent of snow cover in the NH lands reaches an average maximum of about $46.8 \times 10^6 \text{ km}^2$ in January and February, and an average minimum of about $3.4 \times 10^6 \text{ km}^2$ in August (see Figure 2.5) (Ropelweski, 1989; Brown and Armstrong, 2008; Robinson, 2008), which constitute 8% and about 0.5% of the Earth's surface, respectively. From 1966 to 2008, the maximum January snow cover of NH ranged from as low as $42 \times 10^6 \text{ km}^2$ (in 1982) to as high as $50.1 \times 10^6 \text{ km}^2$ (in 2008) (GSL, 2008). For 1966–2008, the mean annual NH snow extent was $25.5 \times 10^6 \text{ km}^2$ (Robinson, 2008). The 1967–2016 mean annual NH snow extent was lower after mid-1980s relative to what it was before the mid-1980s. Even though the linear trend estimated was − 25,000 km^2 per year, this is due to a step-like drop in the mid-1980s, for the decline of NH snow cover is nonlinear, and above-average snow cover was even observed in 2015–2016 (Connolly *et al.*, 2019).

In the NH, most mid-summer snow cover is found over the Greenland and some parts of the Canadian High Arctic (Figure 2.5a), while about 60% of winter snow cover is found over Eurasia and 40% over Canada and the upper portion of the United States, sometimes down to latitude 30° N (Figure 2.5c). Figure 2.6 is a composite monthly NOAA-AVHRR image of NA that shows large seasonal variations in snow cover between the four seasons. In contrast, in South America, there is only a small area covered with snow in July.

Snow cover is observed *in situ* at hydrometeorological stations, from daily depth measurements, (monthly) snow courses, and in special automated networks such as about 730 SNOwpack TELemetry (SNOTEL) automated systems of snow pressure pillows, sonic snow depth sensors, precipitation gauges, and temperature sensors distributed across the United States. The extent of snow cover is also observed and mapped daily (since June 1999) over the NH from the operational satellites of the NOAA, United States.

Canada has extensive *in situ* snow depth and snow course networks that are a valuable database for monitoring cryospheric changes and for validating satellite data such as those shown in Figures 2.5 and 2.6. However, most of the field observations are concentrated in southern latitudes and lower elevations, where the majority of the population lives. At many northern sites, manned stations have been replaced by automatic weather station (AWS) that use acoustic sounders to measure the height of the snow surface.

Besides seasonal variability, snow cover is subject to inter-annual fluctuations but only about 40% of these have been found to be associated with continental to hemispheric scale forcing (Robinson *et al.*, 1995), and the rest could be partly attributed to regional forcings or "coherent" regions. Regarding the boreal winter variability, Saito and Cohen (2003) noted that snow cover at continental scale varies similarly as the inter-annual to inter-decadal oscillations of an internal mode of the atmosphere, but it leads the atmosphere by several months through their mutual oscillations. By principal component analysis (PCA) and composite analysis, Frei and Robinson (1999) found that over western NA, snow cover extent (SCE) is associated with the longitudinal North American ridge, the Pacific North America (PNA) index, while over eastern NA, it is associated with the meridional oscillation of the 500-mb geopotential height, the North Atlantic Oscillation (NAO), and the teleconnection patterns are coupled to tropospheric variability during autumn and winter. Gobena and Gan (2006) found during El Niño winters, the southeasterly flow of warm dry Pacific air and the northwesterly flow of cool dry Arctic air will be the dominant flow over western Canada and Pacific Northwest (PNW) of the United States, giving rise to drier climate (less snowfall) over these regions. On the other hand, La Niña winters are associated with an erosion of the western Canadian ridge and strengthening of the Pacific Westerly, giving rise to greater moisture supply and so more winter snowpack in western Canada and the PNW of the United States. For Europe, Henderson and Leathers (2010) found that large (small) snow

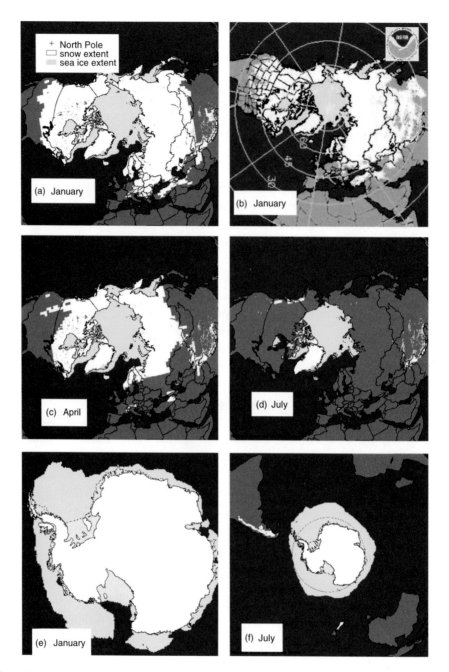

Figure 2.5 Seasonal variation in the mean monthly snow and sea ice cover extent for January, April, and July over Northern Hemisphere using data of NSIDC over 1967–2005 for snow and 1979–2005 for ice; for January and July over Antarctic/Southern Hemisphere over 1987–2002 for snow and 1979–2003 for ice (Maurer, 2007) by Lambert Azimuthal Equal-Area (http://nsidc.org/data/atlas) projection; January 31, 2008 snow and ice chart of NH adapted from NOAA-AVHRR image of NOAA (http://wattsupwiththat .com/2008/02/09/jan08-northern-hemisphere-snow-cover-largest-since-1966/) (A black and white version of this figure will appear in some formats. For the color version, please refer to the plate section.)

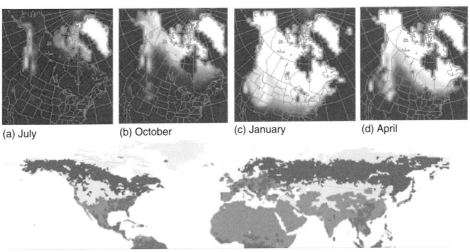

(a) July   (b) October   (c) January   (d) April

(e) Snow-free forests (dark green), unforested areas with snow cover (gray), and forests
with snow cover (red) are shown

Figure 2.6 Seasonal variation in the mean monthly snow cover extent for (a) July, (b) October, (c) January, and (d) April over North America computed from snow charts derived from weekly visible satellite images of NOAA-AVHRR over 1972–1993 (www.tor.ec.gc.ca/CRYSYS/cry-edu.htm); (e) Northern Hemisphere snow and forest covers for January, 2005 computed from the NSIDC Equal-Area Scalable Earth Grid (EASE-Grid) snow cover product (Armstrong and Brodzik, 2005) and the University of Maryland global land cover classification (Hansen *et al.*, 2000) (taken from Rutter *et al.*, 2009) (A black and white version of this figure will appear in some formats. For the color version, please refer to the plate section.)

extent is associated with negative (positive) 850 hPa zonal wind anomalies, negative (positive) phase of the North Atlantic Oscillation, negative (positive) 1,000–500 hPa thickness anomalies, and generally positive (negative) Northern European precipitation anomalies. Besides solar radiation, snowpacks are related to surface air temperature, precipitation, storm tracks, and mid-tropospheric geopotential heights at 500 mb. Frei and Robinson (1999) postulate that snow extent, by exerting an influence on lower tropospheric dynamics (e.g., air temperature), could even modulate atmospheric circulations.

Brown (2000) observed some decline in NH snow cover in recent decades, but the declines are not statistically significant. From 1972 to 2000, using weekly NH snow cover data of high latitude and high elevation areas derived from visible bands of NOAA satellite observations, Dye (2002) found that the week of the last-observed snow cover in spring shifted earlier by 3–5 days per decade estimated from a linear regression analysis, and the duration of the snow-free period increased by 5–6 days per decade, primarily as a result of earlier snow cover disappearance in spring. Similarly, based on the 1966–2007 snow cover data of NOAA satellites and simulations from the Coupled Model Intercomparison Project Phase 3 Model (CMIP3), on the response of NH land area with seasonal snow cover to warming and increasing precipitation, Brown and Mote (2009) found the largest decrease in snow cover duration (SCD) was concentrated in zones where seasonal mean air temperatures were in the range of −5 to +5 °C, which extended around the mid-latitudinal coastal margins of the continents. Regional studies in the western United States (e.g., Adam *et al.*, 2009) show that losses of snowpack associated with warming trends have been ongoing since the mid-twentieth century, especially near boundaries of areas that currently experience substantial snowfall. These findings very likely reflect clear signals of human-induced impact on the climate shown by the changing snowpacks of NH and by the river flows of western United States (Barnett *et al.*, 2008), and Brown and Robinson (2011) detected significant contractions of snow cover in NH over 1922–2010.

According to the CMIP5 Report, observed SCE has decreased in the NH, especially in spring (Vaughan *et al.*, 2013). Satellite records indicate that over the period 1967–2012, the annual mean SCE negative trend is statistically significant, with the largest change of −53% (−40% to −66%) occurred in June. Over 1922–2012, the available snow cover data for March and April show a 7% decline and a strong negative (−0.76) correlation with the March–April 40° N to 60° N land temperature. The spring SCE in Arctic land areas north of 60° N has significantly declined since 1960s, estimated respectively at −3.5% in May and −13.4% in June per decade between 1981 and 2018, relative to the 1981–2010 mean, from multiple data sets (SROCC, 2019). From surface observations, satellite data, and model-based analyses, the snow cover duration has also become shorter, between −0.7 and −3.9 days per decade depending on region and time period, but all spring snow cover duration trends from all data sets are negative (Bulygina *et al.*, 2011; Liston and Hiemstra, 2011; Estilow *et al.*, 2015). These same multisource data sets also identify reductions in autumn snow extent and duration at −0.6 to −1.4 days per decade (Brown *et al.*, 2017). The CMIP6 report will provide further updates on observed snow cover changes (Eyring *et al.*, 2016).

The mountain snow cover is characterized by a very strong interannual and decadal variability (Mankin and Diffenbaugh, 2015). Long-term *in situ* snow cover data are limited in some regions, particularly in High Mountains of Asia and Northern Asia. For key mountainous regions, Stewart (2009) found that higher temperatures have decreased snowpack and resulted in earlier melt in spite of precipitation increases at mid-elevation regions but not at high-elevation regions, which remain well below freezing during winter. Mountain snow cover at lower elevations has generally declined in duration by about several days per decade mainly due to more precipitation falling as rain and to higher melt rate at most elevations, mostly due to increased air temperature (Marty *et al.*, 2017). The mean snow depth and SWE have declined since the mid-twentieth century, with regional variations but at higher elevation, snow cover trends are generally insignificant (SROCC of IPCC, 2019).

At lower elevations in the European Alps, Western North America, Himalaya, and subtropical Andes, the snow depth or mass is projected to decline by 25% between the recent past period (1986–2005) and the near future (2031–2050). By 2081–2100, reductions of up to 80% are expected under RCP8.5, 50% under RCP4.5, and 30% under RCP2.6 climate scenarios of CMIP5 (SROCC of IPCC, 2019, ch. 2). At higher elevations, projected reductions are smaller as temperature increases at higher elevations mainly affect the ablation component of snow mass evolution. The projected increase in wintertime snow accumulation may even result in a net increase in winter snow mass. All elevation levels and mountain regions are projected to exhibit sustained interannual variability of snow conditions throughout the twenty-first century.

Snow cover, depth distribution, and blowing snow

At continental scale or larger, snow cover distribution primarily depends on latitude and seasons (Figures 2.5 and 2.6). At the macro or regional scale, for areas up to 10^6 km^2, and distances from 10 to 1,000 km, snow cover distribution depends on latitude, elevation, orography, and meteorological factors. For example, snowfall caused by orographic cooling tends to increase with a rise in elevation, and frontal activities involving cold fronts generally produce more intense snowfall over relatively smaller areas as against warm fronts that produce moderate or light snowfall over larger areas, because the former has relatively steep leading edge while the latter has mild leading edge. On the mesoscale, with distances of 100 m to 10 km, snow distribution depends on the blowing effect of wind, relief, and vegetation patterns, while on the microscale, 10–100 m, the influencing factors are more local. Over highly exposed terrain, the effects of meso- and microscale differences in vegetation and terrain features may produce wide variations in accumulation patterns and snow depths.

Blowing snow occurs when the force of wind exceeds the shear strength of the snowpack surface that resists snow particles to move. Blowing snow increases with wind speeds and the amount of snowfall but decreases with increasing surface roughness. The effects of wind on the accumulation and distribution of a snowpack are most pronounced in open environments, for example, the Canadian Prairies or Siberian steppes, with three modes of snow particle movement: snow particles begin in motion by creeping or rolling on snowpack surface, then by saltation or bouncing when wind speed increases, and finally in turbulent diffusion or snow particles suspended in the air under high wind speed. These three modes of transport typically occur less than 1 cm above ground under a low wind speed $U < 5$ m s^{-1}, between 1 and 10 cm for $U = 5$–10 m s^{-1}, and between 1 and 100 m for $U > 10$ m s^{-1}, respectively.

Based on wind tunnel studies with surface wind speeds of up to 40 m s^{-1}, Dyunin et al. (1977) argued that saltation accounts for most drifting snow at all conceivable wind speeds. However, Budd et al. (1964) found that turbulent suspension was the primary mechanism from snowdrift studies at Byrd Station, Antarctica. Suspension increases at about U^4, whereas saltation increases linearly with U at high wind speeds (at which most transport occurs), so suspension dominates the overall effect of wind (Pomeroy and Gray, 1990; Pomeroy, p.c. December 2009). At low wind speeds, saltation is the dominant process.

Blowing snow is important in open environments, especially for high elevation alpine areas above treeline, in the Prairies, and in the tundra of the North American Arctic and

Siberia. In these regions snow depth variation depends mainly on terrain features because without the hindering effect of vegetation cover, wind causes snow drift and redistribution to smooth topography, so that mountain tops and plateau tend to have thin snowpack as snow tends to be blown to valleys and low-lying areas which as a result tend to have relatively thick snowpack. In the coastal tundra and open subarctic forest near Churchill, Manitoba of Canada, Kershaw and McCulloch (2007) found that snowpack characteristics measured from 2002 to 2004 also depend on vegetation characteristics, ecosystems, and associated micro-climates. Ecosystems that dominate the circumpolar north are such as wetland, black and white spruce forest, burned forest, forest-tundra transition, and tundra. In lower latitudes as the forest canopy density generally increases, higher snow accumulation has been found in forests of medium density (25–40%) than large open areas because of reduced wind effects by densely forested areas (e.g., Veacth *et al.*, 2009). Although forest structure and canopy interception, and local terrain characteristics will influence snow retention at local scales, Lundquist *et al.* (2013) show that where the mean DJF temperatures exceed −1 ° C, forests with lower total canopy cover are likely to enhance snow retention by minimizing mid-winter and early spring melt.

Gordon *et al.* (2009) developed a camera system to measure the relative blowing snow density profile near the snow surface in Churchill, Manitoba, and Franklin Bay, Northwest Territory. Within the saltation layer, they found that the observed vertical profile of mass density is proportional to $\exp(-0.61z/H)$, where H, the average height of the saltating particles, varies from 1.0 to 10.4 mm, while z, the extent of the saltation layer, varies from 17 to over 85 mm. At greater heights, $z > 0.2$ m, the blowing snow density varies according to a power law ($\rho_s \propto z^{-\gamma}$), with a negative exponent $0.5 < \gamma < 3$. Between these saltation and suspension regions, results suggest that the blowing snow density decreases following a power law with an exponent possibly as high as $\gamma \approx 8$.

2.4   Snow cover modeling in land surface schemes of GCMs

Snow cover is treated in land surface models (LSMs), but snow and ice albedo parameter-izations differ widely in their complexity (Barry, 1996), and more physically based schemes should generally result in better snow and ice albedo parameterization, which ideally should be validated against field measurements (Pirazzini, 2009). The Snow Model Intercomparison Project (SnowMIP) was conducted using 24 snow cover models developed in ten different countries (Essery and Yang, 2001). The models differ from single versus multi-layers, with and without a soil model, variable versus constant heat conductivity and snow density, and the treatment of liquid storage. Only 4 of the 24 models met all the five criteria. However, more recent models have been developed to incorporate snow dynamics affected by soil–snow–vegetation interactions in forests (Essery, 2013).

Twenty seven atmospheric general circulation models (GCMs) were run under the aus-pices of the Atmospheric Model Intercomparison Project (AMIP)-I. The GCMs of AMIP-I reproduced a seasonal cycle of snow extent similar to the observed cycle, but they tend to underestimate the autumn and winter snow extent (especially over North America) and overestimated spring snow extent (especially over Eurasia). The majority of models display less than half of the observed interannual variability. No temporal correlation is found between simulated and observed snow extent, even when only months with extremely high or low values are considered (Frei and Robinson, 1995). The second-generation AMIP-II simulations gave better results (Frei *et al.*, 2003).

Slater *et al.* (2001) found that various snow models in land surface schemes could model the broad features of snow cover and snowmelt processes for open grasslands on both intra- and interannual basis. On the other hand, modeling the spatial variability of snow cover is more problematic because this requires careful consideration of blowing snow transport and sublimation, canopy interception, and patchy snow conditions which are difficult to parameterize accurately. Woo *et al.* (2000) made some progress in understanding some such processes at a local scale, but it is still a long way to incorporating field observations into land surface schemes and climate models where such processes have to be extended to spatial scales of the order of 10–100 km. Until now, most land surface schemes and climate models do not account for the subgrid variability of snow cover in each grid cell. There are new snow models developed in land surface schemes such as that of the European Centre for Medium-Range Weather Forecasts (ECMWF) that includes a new parameterization of snow density, incorporating a liquid water reservoir, and revised formulations for the subgrid snow cover fraction and snow albedo. The new scheme reduces the end of season ablation biases from 10 to 2 days in open areas and from 21 to 13 days in forest areas, and the albedo bias, and so reducing the average surface net shortwave radiation bias by 5.2 W m^{-2} in 14% of the NH land (Dutra et al., 2010).

To realistically simulate grid-averaged surface fluxes, Liston (2004) developed a Subgrid SNOW Distribution (SSNOWD) submodel that explicitly considers the changes of snow-free and snow cover areas (SCAs) in each surface grid cell as the snow melts, by assuming SWE distributes according to a lognormal distribution and the snow-depth coefficient of variation (CV). Using a dichotomous key based on air temperature, topographic variability, and wind speed, Liston proposed a nine-category, global distribution of subgrid snow-depth-variability, each category being assigned a CV value based on published data. The SSNOWD then separately computed surface-energy fluxes over the snow-covered and snow-free portions of each model grid cell, weighing the resulting fluxes according to these fractional areas. Using a climate version of the Regional Atmospheric Modeling System (ClimRAMS) over a North American domain, SSNOWD was compared with a snow-cover formulation that ignores sub-grid snow-distribution. The results indicated that accounting for snow-distribution variability has a significant impact on snow-cover evolution and associated energy and moisture fluxes.

Nitta *et al.* (2014) incorporated SSNOWD into the Minimal Advanced Treatments of Surface Interaction and Runoff (MATSIRO) land surface model. Two 29-year global offline simulations, with and without SSNOWD, were performed while forced with the Japanese 25-

year Reanalysis (JRA-25) data set combined with an observed precipitation data set. The snow cover fraction was improved by including SSNOWD, particularly for the accumulation season and/or regions *with* relatively small amounts of snowfall. In the NH, the daily snow-covered area simulated largely agree with the Interactive Multisensor Snow and Ice Mapping System (IMS) snow analysis data sets, and the seasonal cycle in the NH was improved because SSNOWD formulates the snow cover fraction differently for the accumulation and ablation seasons, and represents the hysteresis of the snow cover fraction between different seasons.

Modeling blowing snow

Pomeroy *et al.* (1993) developed the first comprehensive blowing snow model for the prairies environment. It estimates saltation, suspension and sublimation using readily available meteorological data. They show that within the first 300 m of fetch, transport removes 38–85% of the annual snowfall. However, beyond 1 km of fetch, sublimation losses from blowing snow dominate over transport losses. In Saskatchewan, sublimation losses are 44–74% of annual snowfall over a 4-km fetch. Subsequently, Pomeroy (2000) showed that the ratio of snow removed and sublimated by blowing snow to that transported at prairies (arctic) sites was 2:1 (1:1), respectively.

Essery *et al.* (1999) developed a distributed model of blowing snow transport and sublimation to consider physically based treatments of blowing snow and wind over complex terrain for an Arctic tundra basin. By considering sublimation, which typically removes 15–45% of the seasonal snow cover, the model is able to reproduce the distributions of snow mass, classified by vegetation type and landform, which they approximated with lognormal distributions. The representation used for the downwind development of blowing snow with changes in wind speed and surface characteristics is shown to have a moderating influence on snow redistribution. Spatial fields of snow depth have power spectra in one and two dimensions that occur in two frequency intervals separated by a scale break between 7 and 45 m (Trujillo *et al.*, 2007). The break in scaling is controlled by the spatial distribution of vegetation height when wind redistribution is minimal and by the interaction of the wind with surface concavities and vegetation when wind redistribution is dominant.

Liston and Sturm (1998) developed a SnowTran-3D that simulates wind-driven snow-depth evolution over topographically variable terrain and tested it in an arctic–tundra landscape, for it is generally applicable to treeless areas characterized by strong winds, below-freezing temperatures, and solid precipitation. Liston et al. (2007) extended the SnowTran-3D to version 2.0 that simulates wind-related snow distributions over the range of topographic and climatic environments globally. This version includes three primary enhancements to the original model: (1) an improved wind sub-model, (2) a two-layer sub-model describing the spatial and temporal evolution of friction velocity that must be exceeded to transport snow, and (3) a 3-D, equilibrium-drift profile sub-model that forces snow accumulations to duplicate observed drift profiles.

In mountainous regions, wind plays a prominent role in determining snow accumulation patterns and turbulent heat exchanges, strongly affecting the timing and magnitude of snowmelt runoff. Winstral and Marks (2002) use digital terrain analysis to quantify aspects of the upwind topography related to wind shelter and exposure. They develop a distributed time-series of snow accumulation rates and wind speeds to force a distributed snow model. Terrain parameters were used to distribute rates of snow accumulation and wind speeds at an hourly time step for input to ISNOBAL, an energy and mass balance snow model that accurately modeled the observed snow distribution (including the formation of drifts and scoured wind-exposed ridges) and snowmelt runoff. In contrast, ISNOBAL forced with spatially constant accumulation rates and wind speeds taken from the sheltered meteorological site at Reynolds Mountain in southwest Idaho, a typical snow-monitoring site, over-estimated peak snowmelt runoff, and underestimated snowmelt inputs prior to the peak runoff.

2.5   Snow interception by canopy

Snowfall can be intercepted by an over-story canopy and below the treeline, snow depth variation depend more on land use or vegetation types such as coniferous or broadleaf forests with different canopy structure (Gan, 1996). Snow falling into a canopy is influenced by two possible phenomena: (1) turbulent air flow above and within the canopy may lead to variable snow input rates and microscale variation in snow loading on the ground, (2) direct interception of snow by the canopy elements may either sublimate or fall to the ground. Interception processes are related to vegetation type (deciduous or evergreen), vegetation density, needle characteristics, canopy form and area, branch orientation, leaf area index (LAI), and the presence of nearby open areas. Increasing air temperature tends to increase the cohesiveness of snow and so increase the amount of intercepted snow retained in canopy. For forested environments, most studies show greater snow accumulation in open areas than in forest even though redistribution of intercepted snow by wind to clearings is not typically a significant factor. Instead, interception by canopy and subsequent sublimation, which constitutes the interception loss, are major factors contributing to the difference. Intercepted snow can also melt and flow down the stems of plants as stemflow.

Snow intercepted by the canopy also constitutes part of the overall accumulation of snowfall. Snow is intercepted and stored at different levels of vegetation until the maximum interception storage capacities are reached. Maximum interception storage capacities associated with different vegetation are determined from projected LAI from canopy top to ground per unit of ground area, or LAI (Dickinson *et al.*, 1991). An example algorithm to estimate snow intercepted by canopy is

$$I = c_{su}(I^* - I_o) \left(1 - e^{\frac{C_c P_s}{I^*}} \right), \tag{2.1}$$

where I (kg m^{-2}), the snow interception, is related to a snow unloading coefficient, c_{su}, the maximum snow load, I^*, initial snow load, I_o (kg m^{-2}), an exponential function of snowfall, P_s (kg m^{-2} per unit time), snow density ρ_s^f and the canopy density, C_c, which depends on vegetation species, and $I^* = S_p \text{LAI} \left(0.27 + \frac{46}{\rho_s^f} \right)$. Cumulative snow interception on isolated coniferous trees has been shown to follow a number of probability distributions, ranging from linear to a logistic distribution of the form (Satterlund and Haupt, 1967),

$$I = \frac{I^*}{1 + e^{-K(Ps - Ps,ip)}} \tag{2.2}$$

Here, K = rate of interception storage (mm^{-1}), Ps = SWE of a snowfall event (mm), and Ps,ip = SWE of snowfall at inflection point on a sigmoid growth curve (mm).

The canopy of certain forest types can intercept substantial amount of snowfall (Figure 2.7), which alters both the accumulation of snow on the ground as well as snowmelt rates (Hardy and Bistow, 1990). Therefore the distribution of snow on the forest floor is affected differently depending on the tree species and the prevailing forest structure (Golding and Swanson, 1986). While coniferous forests typically form tree wells around the stems during winter, leafless deciduous forests give rise to snow cones at tree trunks (Sturm, 1992). The overall effect of most forest canopies is a snowpack with spatially heterogeneous depth and SWE. Pomeroy and Schmidt (1993) observed that SWE beneath the tree canopy is equal to 65% of the undisturbed snow in the boreal forest. In contrast, Hardy et al. (1997) measured 60% less snow in boreal jack pine tree wells than in forest openings at maximum accumulation.

Hedstrom and Pomeroy (1998) developed a physically-based snowfall interception model that scales snowfall interception processes from branch to canopy, and takes account of the persistent presence and subsequent unloading of intercepted snow in cold climates. To investigate how snow is intercepted at the forest stand scale, they collected measurements of wind speed, air temperature, above- and below-canopy snowfall, accumulation of snow on the ground and the load of snow intercepted by a suspended, weighed, full-size conifer from spruce and pine stands in the southern boreal forest. Interception efficiency is found to be particularly sensitive to snowfall amount, canopy density, and time since snowfall. Further work resulted in process-based algorithms describing the accumulation, unloading and sublimation of intercepted snow in forest canopies (Pomeroy et al., 1998). These algorithms scale up the physics of interception and sublimation from small scales, where they are well understood, to forest stand-scale calculations of intercepted snow sublimation. However, under windy and dense vegetation environments, blowing snow and canopy

Figure 2.7 Snow intercepted by canopy

interception of snow are two key factors contribute to the redistribution of snowfall that are still challenging in snow hydrologic applications. Using aerial LiDAR data, Moeser *et al.* (2015) developed canopy parameters (LAI, canopy closure, distance to canopy, gap fraction, and tree size parameters) and integrated these canopy metrics and the underlying efficiency distribution to a conceptual model based on snow interception measurements at Davos, Switzerland. Their model performed better at both point and larger grid scales when compared to previous models based on canopy closure and LAI to partition interception from snowfall and the interception efficiency as an exponential decrease of interception efficiency with increasing precipitation.

2.6   Sublimation of snow

Beside redistribution, another major influence of the wind transport of snow is sublimation, a special form of evaporation, whereby solid ice is transformed directly to atmospheric water vapor. Sublimation involves the latent heat of fusion (l_{fs} = 333 kJ kg^{-1}) for ice to water plus the latent heat of vaporization for water to vapor ($l_v \approx$ 2,501 kJ kg^{-1}). Hence it requires ~7.5 times the amount of energy required for snowmelt. Sublimation, that depends on ground surface conditions, wind speed, humidity, net solar radiation, and atmospheric stability, may account for less than 10% of the annual snowfall, but could increase substantially under dry, warm, and windy winter conditions, with snowpack losses reaching 80% under extreme situations (Beaty, 1975). For a given weather condition, forest cover (types and densities) could reduce sublimation on ground by controlling the amount of net solar radiation reaching the ground and by reducing the wind speed. On the other hand, sublimation of canopy-intercepted snow tends to increase with denser stands, high LAI, and tall trees.

Furthermore, strong positive net radiation alone tends to increase melting than sublimation, and the effect of forest cover diminishes during atmospheric inversions.

Snow sublimation occurs from the ground and the forest canopy, but most efficiently from wind-induced, turbulent snow transport. Sublimation from blowing snow can consume about 20% of the snow in the Sierra Nevada (Kattelmann and Elder, 1991), 30–50% in Colorado (Berg, 1986), and 10–90% in Alpine mountains when snow was under turbulent suspension on wind-exposed mountain ridges (Strasser *et al.*, 2008). In western Canada, snow sublimation during winter can amount to 40% of the seasonal snowfall, or 30% of the annual snowfall (Woo *et al.*, 2000). In the Canadian Prairies, sublimation may amount to over 50 mm of SWE per year. Zhang *et al.* (2004) noted that in the taiga of eastern Siberia, the Tianshan, eastern Tibetan Plateau, and Mongolia, sublimation could be large, in particular under neutral atmospheric conditions. Hood *et al.* (1999) calculated sublimation from the seasonal snowpack for 9 months during 1994–1995 at Niwot Ridge in the Colorado Front Range using the aerodynamic profile method. They calculated latent heat fluxes at ten-minute intervals and converted them directly into sublimation or condensation at three heights above the snowpack. The total net sublimation for the snow season was estimated at 195 mm of water equivalent (w.e.) or 15% of the maximum snow accumulation; monthly sublimation during fall and winter ranged from 27 to 54 mm w.e., and daily sublimation often showed a diurnal periodicity with higher rates of sublimation during the day. Sexstone *et al.* (2018) simulated the snow sublimation across the north-central Colorado Rocky Mountains to about 28% of the winter precipitation, and the highest relative snow sublimation fluxes occurred during the lowest snow years. Snow sublimation from forested areas accounted for the majority of sublimation fluxes, highlighting the importance of canopy and sub-canopy surface sublimation in this region.

Sublimation of blowing snow within the near-surface atmospheric boundary layer can deplete the snow mass flux, especially under relatively arid, warm, and windy winter conditions. It is also sensitive to air temperature, wind speed, particle size, relative humidity, and terrain features. Often, for extensively flat areas fully covered with snow, the atmospheric boundary layer near the surface is usually sufficiently developed to assume a steady mass flux of blowing snow.

A popular algorithm for estimating snow sublimation is in the form of Dalton's law. In this, the depth of snow sublimation, D_s (cm) is a function of average wind speed ($\bar{u}_b$) at height z_b above snowpack, the vapor pressures (e_s and e_a) at snowpack level and at height z_a above the snowpack, ρ_w is the density of water,

$$D_s = \frac{E_e}{(l_v + l_{fs})\rho_w} \tag{2.3}$$

where E_e is the energy used for snow sublimation, given as

$$E_e = \frac{k_1}{6}\left[\frac{0.622}{P_a}\right]\bar{u}_b(e_s - e_a)(z_a z_b)^{-1/6}(\Delta t) \tag{2.4}$$

The constant, $k_1 = 0.00651$ cm m$^{-1/3}$ hr day^{-1} mb^{-1} km^{-1}, Δt the time step, and P_a the atmospheric pressure. The snowpack depth change due to sublimation (ΔD_s) is given as

$$\Delta D_s = \frac{\rho_w}{\rho_s}D_s, \tag{2.5}$$

where ρ_s is the density of the snowpack. A simpler way to estimate E_e is

$$E_e = B_e U(e_s - e_a), \tag{2.6}$$

where B_e is the bulk transfer coefficient for turbulent exchange above the melting snow. Equations (2.3) to (2.6) are designed to estimate snow sublimation in windy environments. Snow models that simulate snow sublimation include the Alpine MUltiscale Numerical Distributed Simulation Engine (AMUNDSEN) of Strasser et al. (2008), and the SnowTran-3D of Liston et al. (2007).

2.7   Snow metamorphism

Over time, a snowpack will undergo compaction as ice crystals metamorphose, and settle, which is partly due to increasing overburden load as snowfall occurs. Partly due to compaction, snow depth will decrease while snow density will increase as snow metamorphoses from low density, fine grains to high density, coarse grains, isothermal snowpack with higher liquid permeability and thermal conductivity. Changes to snowpack properties via metamorphism vary between wet and dry snow, but the amount of SWE should theoretically remain unchanged, unless it is reduced by sublimation. As vapor pressure is higher in warmer than in cooler snowpack, and over convex than concave ice surfaces because of difference in the radius of curvature, there will be vapor diffusion from warmer to cooler locations, over crystal surfaces and between snow grains, resulting in irregular ice crystals transforming into well-rounded, coarser grains, even depth hoar. Mass and energy transfer by vapor pressure and temperature gradient can also give rise to faceted snow crystals of various shapes and patterns.

The freeze–thaw cycles of snowpack dictated by the diurnal temperature cycle (warm day and cold night) causes melting of small grains and then refreezing to rounded, large-grained snowpacks, and possibly the formation of firn and glacial ice. In wet snow, small ice crystals tend to melt first, and when the meltwater refreezes, it is absorbed by the larger snow grains which tend to grow more rapidly under more liquid water since water is a better conductor of heat than air. Under increasing pressure, snow is compressed and slowly deforms to firn and then to ice.

By definition, the density of snow ρ_p is: $\rho_p = \rho_i\left[1 - \dfrac{W_{liq}}{100}\right](1 - \varnothing) + \rho_w\left[\dfrac{W_{liq}}{100}\right]\varnothing$, where ρ_i is the density of ice, ϕ the porosity of snowpack, ρ_w the density of water and W_{liq} the liquid water content in the snowpack. Newly fallen snow normally has a density ρ_p of about 100 kg m^{-3} or less, an albedo of 90% ($\alpha = 0.9$) or higher, and grain size of 50 μm to about 1 mm, but the grain size and density will increase as snow ages. Snow grains are considered very fine if it is less than 0.2 mm, 0.2–0.5 mm as fine, 0.5–1 mm as medium, greater than 1 mm as coarse, and greater than 2 mm as very coarse (Fierz et al., 2009). Snow hardness, which can be measured by the force in Newton (N) needed to penetrate with an object such as the SWISS rammsonde, or by a hand hardness index (De Quervain, 1950), is expected to increase as snow settles. Snow hardness ranges from very soft with the hardness index ranges from 1 (penetration force <50 N), to 5 or very hard (up to 1,200 N), respectively. Table 2.1 gives a breakdown of snow types and typical densities, and snow grain shapes encountered during the process of metamorphosis shown in Figure 2.8. According to Sturm et al. (1997), the thermal conductivity of snow is primarily dependent on snow density even though ice grain structure and temperature are also controlling factors. Sublimation will cause a thinner snow cover, or reduced SWE, but not necessarily reduce the SCA. Hence, it is difficult to detect the effect of sublimation from snow cover data.

There is a strong connection between snow properties and land surface water and energy fluxes that influence weather and climate all over the cryosphere. The variability of the snowpack significantly influences the water cycle globally, and especially high latitudes. SCA exhibits a fairly wide range of spatial and temporal fluctuations seasonally, which in turn affects the variability in the surface albedo and radiation balance, vapor fluxes to the atmosphere through sublimation and evaporation, and meltwater infiltrating into the soil and river systems. This seasonal and interannual variability of snowpacks affects the general circulation of the atmosphere (Walland and Simmonds, 1997).

Table 2.1 Density of typical snow covers

Snow type	Density ρ_p (kg m^{-3})
Wild snow	10–30
Ordinary new snow immediately after falling in still air	50–65
Settling snow	70–100
Average wind-toughened snow	250–300
Hard wind slab	320–400
New firn snow	400–550
Advanced firn snow	550–650
Thawing firn snow	>600

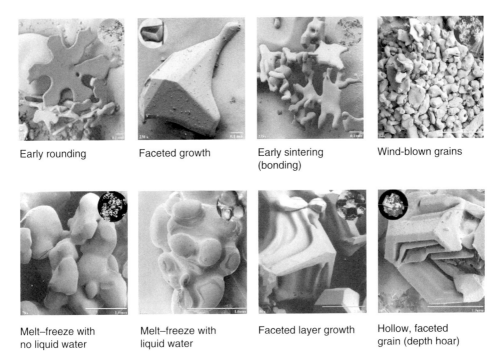

Early rounding	Faceted growth	Early sintering (bonding)	Wind-blown grains
Melt–freeze with no liquid water	Melt–freeze with liquid water	Faceted layer growth	Hollow, faceted grain (depth hoar)

Figure 2.8 Snow grain shapes under different stages of metamorphosis (Don Cline, NOHRSC, National Weather Service, USA)

SCE has been shown to exhibit a close negative relationship with hemispheric air temperature over the post-1971 period (Robinson and Dewey, 1990). The snow-temperature relationship is strongest in March, when the largest warming and most significant reduction in SCE have been observed in both Eurasia and North America since 1950 (Brown, 2000). The Arctic summer warming mainly results from the increase of snow-free days and the transition from tundra to forest (Chapin *et al.*, 2005). However, under climate warming, the snow cover over the Arctic is projected to increase by 0–15% for the maximum SWE due to increased atmospheric moisture, but decreases in SCD by about 10–20% over much of the Arctic (Callaghan *et al.*, 2011a).

Snowfall can be intercepted by an over-story canopy and then released to the ground snowpack through meltwater drip, mass release, or throughfall. The ground snowpack can exist in a number of layers, with the surface layer subjected to high frequency energy and water exchanges with the lower atmosphere, while the lower layers undergo heat exchanges through conduction and infiltration of meltwater flow downwards. Snow grains become coarser and its density increases as the snowpack ages and compressed by further snowfall.

In terms of wetness, snow is classified as dry if its liquid water content (W_{liq}) or the percent of liquid water by weight in the snowpack is near 0% and there is little tendency for snow grains to stick together, which usually happens when the snowpack temperature $T_p \leq 0$ °C. When W_{liq} reaches about 3%, snow is considered moist and it has a distinct tendency to stick together, and $T_p \approx 0$ °C. Beyond 3–8% of W_{liq}, snow is considered wet, 8–15% of W_{liq} as very wet when water can be squeezed out by hand, and slushy or soaked when W_{liq} exceeds 15% and $T_p > 0$ °C (Fierz *et al.*, 2009). When $T_p > 0$ °C, the pores can hold water mostly by capillarity and tension. Because of liquid water, it can be shown that

$$\frac{L_{ms}}{L_{fs}} = 1 - \frac{W_{liq}}{100},$$

(2.7)

where L_{fs} = Latent heat of fusion of pure ice, and L_{ms} = Latent heat of fusion of snow. Because of the presence of liquid water in most snowpacks, L_{ms} is usually less than L_{fs} which is about 333 kJ kg^{-1}.

The ground snowpack can exist in a number of layers, with the surface layer subjected to high frequency energy and water exchanges with the atmosphere, while the lower layers undergo heat exchanges through conduction and the infiltration of meltwater flowing downwards. Snow grains become coarser and, as the snowpack ages, its density increases and it becomes compressed by further snowfall. However, density could decrease over time if there were a substantial amount of depth hoar in the snowpack (Hiemstra, personal communication).

2.8   *In situ* measurements of snow

Ground snowfall data are collected using a ruler, a snow board or a snow pillow, non-recording snow gauges such as the MSC snow gauge with a Nipher shield of the shape of an inverted bell to reduce wind effects on precipitation collectors, the Swedish SMHI precipitation gauge, and the USSR Tretyakov Gauge. Non-recording gauges can be read daily or over a period time, such as monthly or by seasons, but that will require anti-freeze such as propylene glycol mixed with ethanol and evaporation suppressants such as mineral oil, and such gauges are elevated to prevent them from being inundated by possible heavy accumulation of snow. Weighing-type, self-recording snow gauges such as the Fisher Porter and the universal gauges that measure temporal snowfall data by a spring and transmitting the data via satellite to a data collection center, or lately by tipping buckets connected to data-loggers from which recorded data can be downloaded. With ground measurements of snowfall, the catch of solid and mixed precipitation in precipitation gauges is melted and total precipitation is usually reported. Even though such gauges can operate unattended up to a year, they should be serviced periodically to ensure collecting reliable precipitation data.

Owing to the huge cost in collecting ground measurements of snow, and the harsh environment in remote areas such as mountains dominated by snowpack where more than

Figure 2.9 (a) Western Snow Conference (WSC) snow sampler. (b) Meteorological Service of Canada (MSC) snow sampler. (c) Snow gauges with and without Nipher shield (foreground) and Tretyakov shield (background).

70% of snow could accumulate above the mean elevation of snow gauging stations (Gillian *et al.*, 2010), we cannot rely on snow gauges or ground-based, snow course measurements (Figure 2.9a) to estimate the SCA or the amount of SWE at the regional scale, yet seasonal snow mass variations at mid- to high-latitudes are the largest signals in the changes of terrestrial water storage (Niu *et al.*, 2007). Information on snow cover has been collected routinely at hydrometeorological stations, with records beginning in the late-nineteenth century at a few stations, and more widely since the 1930s–1950s. The ground is considered to be snow-covered when at least half of the area visible from an observing station has snow cover. However, it is also possible to install snow stakes or aerial markers in relatively inaccessible sites by which snow depth can be visually observed from a low-flying aircraft.

Other than being point measurements, it is well know that snow gauges, even mounted with shields such as the Nipher shield (Figure 2.9c), suffer from under-catch problems especially under windy conditions, where gauge totals may underestimate snowfall by 20–50% or more. For example, the catch ratios of Wyoming fence to WMO-DFIR (World Meteorological Organization-Double Fence Inter-Comparison Reference) were 89% and 87% at Regina and Valdai, respectively (Figure 2.10a). Yang *et al.* (2000) found that the mean catch of snowfall for the US 8″ gauge at Valdai was 44%. For the Tretyakov and Hellmann gauges, the mean catch of snowfall was 63–65% and 43–50%, respectively at the northern test sites of the WMO experiment. For the WMO site set up at the Reynolds Creek Experimental Watershed in southwest Idaho, Hanson *et al.* (1999) found that an unshielded universal recording gauge measured 24% less snow than was measured by the Wyoming shielded gauge. In a mountainous watershed in NW Montana, Gillian *et al.* (2010) found

(a)

(b)

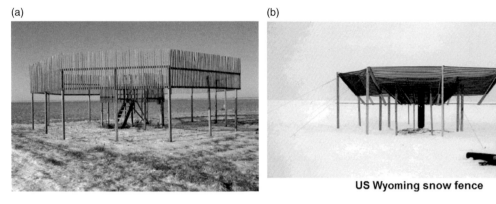

US Wyoming snow fence

Figure 2.10 (a) On the basis of the WMO Double Fence Intercomparison Reference (DFIR) (DFIR at Saskatchewan taken from figure 3 of Rasmussen *et al.*, 2013), the mean catch for (b) Wyoming snow fence was 89% of snowfall at Regina (Canada) and 87% at Valdai (Russia) (figure 1 of Yang *et al.*, 2000)

greater than 25% of the basin's SWE accumulates above the highest measurement station. At the Marshall Field Site located south of Boulder, Colorado, Rasmussen *et al.* (2013) found that the single Alter-shielded gauge accumulates ~50% less precipitation than the same GEONOR gauge in the DFIR, showing the strong wind undercatch. The double Alter-shielded gauge is slightly better with ~55% undercatch. Without wind shield, snow under-catch problems can be partly corrected by applying adjustment coefficients to snow gauge data as a function of wind speed.

The Pan-Arctic Snowfall Reconstruction (PASR) used a land surface model of NASA to reconstruct solid precipitation from observed snow depth and surface air temperatures for the pan-arctic region during 1940–1999, with the objective of correcting cold season precipitation gauge biases (Cherry *et al.*, 2007). Reconstructed snowfall at test stations in the United States and Canada is either higher or lower than gauge observations, and is consistently higher than snowfall from the 40-yr ECMWF Re-Analysis data (ERA-40), which has been replaced by the ERA-Interim reanalysis data of higher resolution (Dee *et al.*, 2011). PASR snowfall does not have a consistent relationship with snowfall derived from the WMO Solid Precipitation Intercomparison Project correction algorithms.

In Canada, snow depth and the corresponding *snow-water equivalent* (SWE) are measured at ground stations. Depth is routinely measured at fixed stakes, or by a ruler inserted into the snowpack, and this depth is reported in daily weather observations at 0900 hours. Average maximum snow depths vary from 30 to 40 cm on Arctic Sea ice to several meters in maritime climates such as the mountains of western North America. The SWE along snow courses is measured from depth and density determinations made at weekly to monthly time intervals. Such snow course networks are decreasing because of their cost and the data may not be truly representative.

From analyzing 848 stations across Canada that were reporting daily snowfall and daily precipitation from October 2004 to February 2005, Cox (2005) found that the histogram of the frequency of snowfall events by snow depth/SWE ratio is dominated by a spike at the 10:1 ratio, a bias caused by the 10:1 approximation being used in place of actual measurements (Figure 2.11a). Recognizing the inadequacy of this 10:1 ratio, for climate stations only equipped with a snow ruler, Mekis and Hopkinson (2004) proposed an alternative for more accurately estimating the SWE at a station based on a factor called the Snow Water Equivalent Adjustment Factor (SWEAF) which can range from 0.6 to 1.8, with SWEAF generally increases with latitude; the province of British Columbia tends to have SWEAF less than 1 (Figure 2.11b).

The Canadian Meteorological Centre (CMC) Daily Snow Depth Analysis Data set consists of NH snow depth data obtained from surface synoptic observations, meteorological aviation reports, and special aviation reports acquired from the WMO information system (http://nsidc.org/data/nsidc-0447.html). In the USSR and Russian Federation, snow depth has been measured daily as the average of three fixed stakes at hydrometeorological stations. The Historical Soviet Daily Snow Depth (HSDSD) data begin in 1881 through 1995 at 284 WMO stations throughout Russia and the former Soviet Union; other parameters include snow cover percent, snow characteristics, and site characterization (Armstrong, 2001). The HSDSD data have been updated in 2015 (https://catalog.data.gov/dataset/histor ical-soviet-daily-snow-depth-hsdsd614b3). They are available at http://nsidc.org/data/g010 92.html. Snow measurements were also performed at fixed intervals over a 1–2 km transect, by taking an average snow depth for 100–200 points, and an average SWE determined for 20 points. At some locations transects are made in fields and in forests, separately. The snow measurements were carried out at 10-day intervals and are available at 1,345 sites for 1966–1990 at http://nsidc.org/data/g01170.html.

2.9   Remote sensing of snowpack properties, snow cover area, and snow water equivalent

Given the high albedo of snow compared to other natural surfaces, remotely sensed data can provide useful information on the distribution of snow cover, optical properties of snow cover, and in some instances, the SWE, even in a forest environment (Veacth et al., 2009). The visible band has the largest application in the SCE mapping because of snow's high albedo to reflected (visible) sunlight that makes snow cover easily identifiable from space, while the infrared red band has minimal application in snow cover mapping because the snow's surface temperature is similar to other surfaces.

Since 1966, the SCA of the NH has been monitored from space platforms by the US National Oceanic and Atmospheric Administration's (NOAA) National Environmental Satellite Data and Information Service (NESDIS) using Very High Resolution

(a)
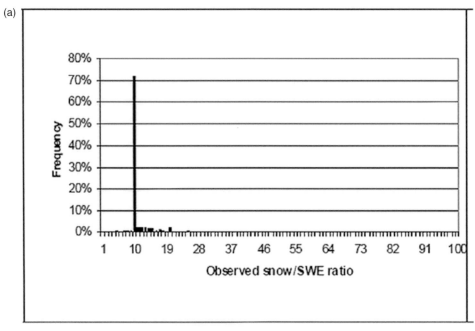

(b)
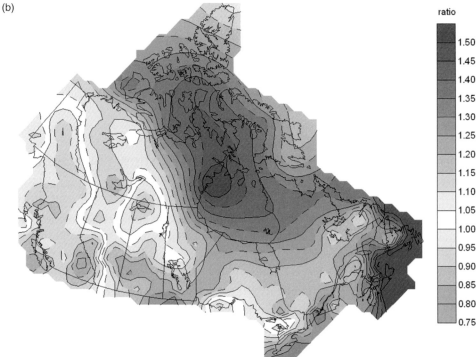

Figure 2.11 (a) The frequency of snowfall events by snow/SWE ratio collected across Canada for October 2004 to February 2005 is dominated by a spike at the 10:1 ratio, a bias caused by the 10:1 approximation being used in place of actual measurements (Cox, 2005). (b) Snow water equivalent adjustment factor map used for adjusting the snow ruler measurements to more accurately estimate the SWE of Canada (Mekis and Vincent, 2011)

Radiometer (VHRR) sensors in the visible bands (0.58–0.68 μm, red band). These data are limited by illumination and cloud cover, and are of 1-km resolution. Reliable hemispheric snow-cover data have been available since 1972 from the NOAA-AVHRR satellites. The visible images are interpreted manually, and snow extent is mapped over the NH on a daily basis since 1999 (formerly weekly). The charts have been digitized for grid boxes varying in size from 16,000 to 42,000 km^2, and these data have also been remapped to a 25×25 km Equal Area Scalable Earth (EASE) grid for 1978–1995 and combined with the extent of Arctic sea ice mapped from passive microwave data to display the seasonal cryosphere in the NH (https://nsidc.org/data/nsidc-0046). There is a more limited record from AVHRR data for 1974–1986 in the Southern Hemisphere, where the SCE in South America varies between about 1.2 million and 0.7 million km^2 in July. There is negligible snow cover in January in the Southern Hemisphere apart from Antarctica. Since the early 2000s, the multi-frequency, dual-polarized MODIS instruments onboard NASA's EOS Terra and Aqua satellites, and the Medium-Resolution Imaging Spectrometer (MERIS) onboard of ESA's ENVISAT also provide snow cover maps (Seidel and Martinec, 2004).

In the last three decades, through models and advances in remote sensing, especially new satellite sensors and imaging spectrometers, we have made progress in the interpretation of snow optical properties such as spectral and broadband albedo, fractional snow-covered area, grain size, liquid water content in the near-surface layer, concentration of snow algae, and radiative forcing caused by impurities such as dust (Dozier et al., 2009). All of these results from imaging spectrometry have been verified with surface field measurements or, in the case of fractional SCA, with high-resolution aerial photography.

The presence of tree canopy, cloud cover, and a high incident angle in alpine areas could obscure the view and can lead to the under-estimation of snow cover. The AVHRR sensors have produced global observations of SCA of 1-km resolution, while MODIS produces SCA of 500 m resolution, and such data encompass a variety of temporal and spatial compositions (Hall and Riggs, 2007). Figure 2.12 shows the NH monthly snow cover frequency derived from NOAA-AVHRR data of 1966–2003 (Armstrong et al., 2005, 2006). These products can be processed with cloud discrimination algorithms (Ackerman et al., 1995, 1998) to maximize snow cover information and minimize the interference from cloud cover. Hans et al. (2019) developed a cloud detection algorithm for 1-km resolution Sentinel-2 snow/ice images.

Using the daily MODIS/Terra snow cover product, Parajka et al. (2010) developed a method for mapping snow cover with cloudiness by reclassifying pixels assigned as clouds to snow or land according to their positions relative to the regional snow-line elevation. Essentially, the elevation of each pixel classified as clouds is compared with the mean elevation of all snow (μ_S) and land (μ_L) pixels, respectively. In the case where the elevation of the cloud-covered pixel is above the μ_S of the regional snow-line, the pixel is assigned as snow covered. If the elevation is below the μ_L of the regional land-line, the pixel is assigned as land. Where the elevation is in between μ_S and μ_L, the pixel is assigned as partially snow

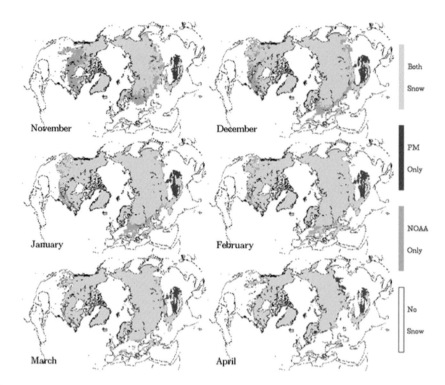

Figure 2.12 Monthly (November–April) snow cover extent climatology for Northern Hemisphere derived from long-term snow cover data of NOAA and passive microwave over 1978–2005 (taken from Armstrong *et al.*, 2006) (A black and white version of this figure will appear in some formats. For the color version, please refer to the plate section.)

covered. They found this method to produce robust snow cover maps for a study site at Austria, up to a cloud cover as large as 85%.

In contrast to low resolution but high observation frequency satellites (e.g., two passes every 24-hr for AVHRR sensors), there are high-resolution satellites (20–80 m) such as the American Landsat-TM, the French Spot, and the ASTER sensor of Terra/Aqua satellites, which could provide a strong contrast between snow and snow-free areas, leading to more accurate snow cover maps to be produced, such as the mapping of montane snow cover at subpixel resolution from the Landsat Thematic Mapper using decision tree classification models by Rosenthal and Dozier (1996). However, the drawback is their low observation frequency of every 16–18 days (Rango, 1993), which may not be sufficient to monitor the distribution of snow cover particularly in cloudy areas, or mountain basins, where optical sensors may not be able to obtain usable observations for several passes.

Mapping snow cover can also use microwave data that can penetrate cloud cover, produce data in all weather conditions and at night, and have good observation frequency passing every 1–2 days. Unfortunately they are of coarse resolution of about 10–25 km and so only large areas can be mapped to any accuracy. Furthermore, because microwaves penetrate thin layers of snow cover with little absorption, microwaves generally under-predict the extent of snow partly because they cannot discriminate light snow cover and other surface features, particularly over high, rugged terrain and stratified snowpacks (Chang *et al.*, 1991).

For the past several decades, numerous large-scale field data collections through radar and microwave sensors and experiments have been conducted, including SIRC/X SAR, QuikSCAT and CLPX (Ulaby *et al.*, 1984; Kendra *et al.*, 1998; Nghiem and Tsai, 2001; Cline *et al.*, 2007). The optimal frequency range with the necessary sensitivity to volumetric snowpack properties is passive microwave at 8–37 GHz (X-, Ku-, and Ka-bands; 2–5.6 cm wavelengths). Long-term record of remotely sensed SWE information has been derived from low-resolution (about 25 km) passive microwave measurements in the 18–40 GHz range (K- and Ka-bands) as explained below. On the other hand, Lieven *et al.* (2019) have recently used C-band backscatter images of much higher resolution (1 km), of the Sentinel-1 satellites 1A and 1B, the SAR mission of ESA and Copernicus, to map snow depth of about 4,000 sites across the NH mountains.

Remote sensing of SWE

Because snowpack can attenuate gamma radiation, over a thousand flight lines have been conducted mainly in USA and Canada to collect airborne gamma data for SWE survey during winters and distributed by the NOHRSC of NOAA-USA (www.nohrsc.noaa.gov/ snowsurvey). However, as airborne data, gamma radiation data are not as readily available as space borne data and also because gamma radiation is attenuated by water in all phases, the effect of soil moisture has to be accounted for to avoid inaccurate estimation of SWE.

Since 1978, spatially distributed SWE data have been retrieved from the brightness temperature (TB) in Kelvin from passive microwave remote sensing sensors such as the Television Infrared Observation Satellite (TIROS-N) launched in 1978 (Dozier *et al.*, 1981), and Scanning Multichannel Microwave Radiometer (SMMR) flew on NASA's Nimbus 7 from October 25, 1978 to August 20, 1987, and the Special Sensor Microwave Imager (SSM/ I) mounted on the Defense Meteorological Satellite Program satellites since September 7, 1987. Since May 2002, TBs retrieved from the Advanced Microwave Scanning Radiometer-EOS (AMSR-E) sensor aboard the Aqua satellite have been used to estimate SWE. The AMSR-E sensor uses one of the largest ever microwave antennas to detect faint microwave emissions from the Earth's surface. AMSR-E produces global and continuous daily SWE and snow depth data sets at 25-km spatial resolution. The frequencies (resolution) of SSM/I are 6.6 (150 km), 19, 22, 37 (25 km) to 85 (12.5 km) GHz, while that of AMSR-E are 6.9 (50 km) to 89 GHz (5 km) resolution. These TBs are provided in the form of horizontal (H) and vertical (V) polarizations.

The basis of this approach is that microwaves are not sensitive to water vapor or liquid water in the troposphere, especially at long wavelength, and microwave radiation emitted by

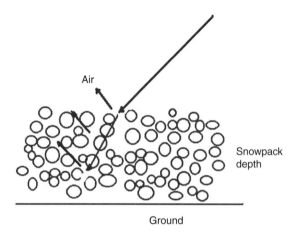

Figure 2.13 In the volumetric scattering of snow, some fractions of the radiation can enter beyond the boundary into the lower, layered snowpacks. The reflected radiation will consist of that reflected at the top planar boundary, and that which is first partially transmitted into the second medium and reflected at the second planar boundary and partially transmitted back to the first medium, and so on

the land surface is attenuated by snow cover. However, the effects of any liquid water due to snowmelt, the obscuring of the snow surface by the vegetation canopy, changes in the grain size of the snow, and terrain irregularities greatly complicate such determinations. Moreover, the typical satellite footprint is of the order of 12–25 km, meaning that the signal is a complex spatial average and hard to relate to point measurements. Nevertheless, such methods are being used for operational SWE mapping over the high plains and prairies of North America (Goodison *et al.*, 1990; Singh and Gan, 2000, 2009), and other parts of NH (Takala *et al.*, 2011), unless the snow packs are wet, reasonably accurate SWE can mostly be estimated from the passive microwave data of SMMR, SSM/I, and AMSR-E. Passive microwave data represent an important resource for monitoring trends and spatial variability of terrestrial snowpacks (Wulder *et al.*, 2007).

Because of volumetric scattering (Figure 2.13), the dominant loss mechanism for microwave radiation greater than 15 GHz incident on a snowpack, it is possible to relate empirically the brightness temperature (TB) of certain frequencies and polarizations (H or V) to the SWE of the snow packs (e.g., Singh and Gan, 2000; Armstrong *et al.*, 2001; Derksen *et al.*, 2009). The volumetric scattering of snow on the incident microwave radiation depends on the snow grain size, snow density, depth, SWE, temperature, degree of metamorphism, nocturnal crust development, ice lenses, and other factors (Mätzler, 1994). In theory, the greater the depth of snow, the lower should be the TB if climatic and snowpack conditions remain the same, but some studies have indicated that TB can increase with depth beyond a certain snow depth (Mätzler *et al.*, 1982; Schanda, 1983). The snow scattering property is affected by factors such as snow metamorphism, which dictates the internal structure of snow as it ages, the multiple melt-freeze cycles, which together contribute to the complicated physical processes in the formation of snow structure; and snow redistribution that depend mainly on wind, terrain features, and land use (Rott and Nagler, 1995; Rosenfeld and Grody, 2000).

Shi *et al.* (2009) used SNTHERM to simulate the snow profiles of the 1200-km transect of snow stratigraphy measured in the Alaska region, called the Snow Science Traverse-Alaska Region (SnowSTAR2002), and the Microwave emission Model for Layered Snowpacks (MEMLS) to simulate the 19- and 37-GHz TB for both SNTHERM-MEMLS and SnowSTAR2002-MEMLS cases, which matched well with the passive microwave data. They concluded that the simulation of snow microphysical profiles is a viable strategy for retrieving SWE from passive microwave data.

Over the past several decades, various snow retrieval algorithms for passive microwave data have been developed and applied (e.g., Chang *et al.*, 1982; Hallikainen and Jolma, 1992; Walker and Goodison, 1993; Ferraro *et al.*, 1994; Gan, 1996; Wilson *et al.*, 1999; Singh and Gan, 2000; Gan *et al.*, 2009). In general, microwave-derived maps tend to underestimate snow extent during fall and early winter, due to a weak signal from shallow and intermittent snow cover, and the underestimation can be as much as 20%, decreasing to a few percent during mid-winter and spring. Conversely, a thinner atmosphere between the snow-covered surface and satellite could lead to over-estimating SCE, as is the case for the Tibetan Plateau 3,000–5,000 m above sea level (Savoie *et al.*, 2009). SWE products for NH from the National Snow and Ice Data Center (NSIDC) and GlobSnow from the European Space Agency (ESA) have been developed using different algorithms. By comparing the data with historical snow depth measurements obtained from ~7,400 meteorological stations across the NH, Liu *et al.* (2014) show that for SWEs between 30 and 200 mm, GlobSnow products agree better with ground measurements than NSIDC products which tend to suffer from "microwave saturation," underestimating SWEs over 120 mm. However, GlobSnow products tend to overestimate SWE less than 30 mm marginally more often than that of NSIDC products. Using the new GlobSnow 3.0 dataset, Pulliainen et al. (2020) show that the 1980–2018 annual maximum snow mass in the NH was about 3,062 ± 35 gigatons with different continental trends over this study period, e.g., snow mass decreased by 46 gigatons per decade across NA but had a negligible trend across Eurasia.

Linear and nonlinear regression algorithms to estimate SWE

The 1979–1987 SMMR SWE data provided by the NSIDC (National Snow and Ice Data Center) were retrieved from Equation 2.8,

$$SWE(mm) = 4.77(T_{BH19} - T_{BH37}) \qquad (2.8)$$

where TB is in K. The SSM/I SWE data of NSIDC were retrieved using Equation 2.9, and false SWE signals from deserts are filtered (Armstrong and Brodzik, 2002). To ensure that snow packs are detectable by passive microwave data, SWE less than 7.5 mm is considered unreliable and set to zero.

$$SWE(mm) = 4.77(T_{BH19} - T_{BH37} - 5) \qquad (2.9)$$

The daily SWE is adjusted for surface forest cover (A_F in percent) using the BU-MODIS (NSIDC, 2005) land cover data so that,

$$SWE(mm) = \frac{SWE}{(1 - A_F)} \qquad (2.10)$$

Any pixels with forest cover higher than 50% are set to the 50% threshold, so that the forest correction by Equation (2.10) is limited to a maximum factor of 2, given that microwave data can only detect snowpack properties at canopy densities less than 60–70% (Pulliainen et al., 2001). Using measurements from snow course transects in the former Soviet Union, Armstrong and Brodzik (2002) reported a general tendency for nearly all of the algorithms tested to underestimate SWE, especially when the forest-cover density exceeded 30–40%. Ideally, specific SWE retrieval algorithms should be derived for different vegetation types such as boreal forest or tundra (Derksen et al., 2005) but that is only feasible with adequate field campaigns which are expensive and time consuming.

Hallikainen (1989) developed an algorithm similar to Equation 2.9 to account for the surface effect of land cover when there is no snow. Goodison et al. (1990) used an algorithm similar to Equation (2.9) but vertically polarized data TB $_{V19}$ and TB_{V37}, to model the snow of Canadian prairies. Derksen et al. (2009) found TB $_{V37}$ of AMSR-E to be the appropriate polarization for remote sensing of the SWE of tundra snowpack in the Arctic.

The retrieval of SWE from passive microwave data based on Equations 2.8–2.10 applied in a continental framework is expected to have varying accuracy, depending on the density and types of vegetation, and snowpack characteristics, frozen and unfrozen water, north- or south-facing slopes, topographic variability, mountains versus flat plains, and so on (Chang et al., 1997; Tait, 1998; Goita et al., 2003; Takala et al., 2011), and the possible effect of wet snow albeit even though NSIDC selectively used "cold" pass daily TB to prepare these monthly SWE data. Lake ice could also cause low SWE values and the presence of snowmelt, ice lenses, surface hoar, depth hoar and very deep snow packs could confound the interpretation of TB for SWE. In view of the aforementioned limitations, SWE data derived from passive microwave data should be more dependable in the Canadian Prairies dominated by grassland than in the Canadian Arctic with countless frozen water bodies or in forested, mountainous areas such as the Rocky Mountains.

Besides the effect of forest cover, various screening criteria have been proposed to eliminate data affected by depth hoar, wet snow, and water bodies. Chang et al. (1982) indicated the possibility of discriminating the effect of depth hoar and the underlying ground condition (frozen or unfrozen) using the polarization factor, p-factor = (V37 − H37)/(V37 + H37). Singh and Gan (2000) used a p-factor >0.026 to eliminate the SSM/I data that were affected by depth hoar and the presence of a water body of significant size in the vicinity of the footprint that has the effect of causing an underestimation of predicted SWE because of the high dielectric constant of the water body (or the presence of water in the snowpack due to above-freezing temperatures). Other recommended screening criteria are such as TB $_{V37}$ < 250 K, (TB $_{V19}$ − TB $_{V37}$) ≥ 9 K, TB $_{V37}$ − TB $_{H37}$ ≥ 10 K, TB $_{V37}$ > 225 K (e.g., Goodison and Walker, 1994).

The traditional TB difference between 19 and 37 GHz has been shown to be inappropriate for the lake-rich environment in the Arctic, and retrieving tundra SWE can be challenging because the SWE versus 37-GHz TB relationship could have a reversed slope that occurs beyond a theoretical limit of approximately 120-mm SWE (Derksen et al., 2009). In the northern environment, where a footprint could include both frozen water bodies (A_W) and tundra (A_{TUNDRA}), Gan (1996) assumed microwave emission of the former to be related to air temperature (T_a),

$$SWE = K_1(A_{TUNDRA})(T_{BH19} − T_{BH37}) + K_2(A_W)(T_a) + K_3 \qquad (2.11)$$

Unlike Equations (2.8) and (2.9), coefficients K_1, K_2, and K_3 for Equation (2.11) will be region or basin dependent.

The above algorithms assume relatively simple TB-SWE relationships, even though TB is influenced by many snow parameters, and so more complicated relationships have been developed, such as Equation (2.12) that Singh and Gan (2000) applied to the Red River Basin of North Dakota and Minnesota,

$$SWE(mm) = K_4(T_{BV19} − T_{BH37}) + K_5(AMSL) + K_6(1 − A_F) + K_7(1 − A_W) + K_8 TPW$$
$$(2.12)$$

where AMSL is the average elevation above mean sea level and TPW is the total precipitable water. Again, the coefficients are site specific. For the Red River Basin, by first screening SSM/I data to remove data representing wet snow, and calibrate the screened SSM/I data (dry snow) with corresponding airborne SWE data, they found K_4 = 0.2357, K_5 = 0.0064, K_6 = 4.0399, K_7 = 20.0287, and K_8 = 1.0825. Singh and Gan (2000) also developed another algorithm similar to Equation (2.11) that was based on the surface/ground TB converted from the at-satellite TB by applying the atmospheric attenuation model of Choudary (1993) which accounted for the attenuation of atmosphere water vapor (based on TPW in cm) on the satellite TB data.

Above are parametric regressions relating TB to SWE. The non-parametric approach has also being used, such as the Projection Pursuit Regression (PPR) of Friedman and Stuetzle (1981) by Gan et al. (2009). PPR models the response variable as a sum of functions of linear combinations of predictor variables. Suppose y and x's denote response and predictor vectors respectively, PPR finds the number of terms M_o, direction vectors ($a_1, a_2, . . ., a_{Mo}$), and nonlinear transformations ($\Phi_1, \Phi_2, . . ., \Phi_{Mo}$) as shown in Equation (2.13),

$$\hat{y} \approx \bar{y} + \sum_{m=1}^{Mo} \beta_m \phi_m (\alpha_m^T x). \tag{2.13}$$

Through minimizing the expected distance or mean square error between y, which is the observed SWE, and $\hat{y}$, the estimated SWE, using Equation (2.14), the model parameters β_m (the response linear combinations), a_m (the direction vectors), Φ_m (the predictor functions), for $m = 1, 2, \ldots, M_o$ are obtained.

$$L_2(\beta, \alpha, \phi, x, y) = E[y - \hat{y}]^2 \tag{2.14}$$

A successful application of PPR lies in selecting an optimum number of terms, M_o, determined by trial and error, often by starting the algorithm with a large M_o and then decreasing M_o such that the increase in accuracy due to an additional term is not worth the increased complexity (Friedman, 1985). The optimum M_o is determined in terms of the fraction of variance it cannot explain (Friedman and Stuetzle, 1981). From Equation (2.14), this unexplained variance, U is given as

$$U = \frac{L_2(\beta, \alpha, \phi, x, y)}{\mathrm{Var}(y)}. \tag{2.15}$$

Artificial neural network (ANN) algorithms to estimate SWE

ANN has been widely used in many fields of research because of its ability to model nonlinear and poorly understood systems with their inherent non-linearity and complex internal structure. Other than some drawbacks such as being classified as black box models, the problem of over-fitting and tedious training, ANN can approximate almost any function. Gan *et al.* (2009) used the Modified Counter Propagation Network (MCPN) to model the SWE for the Red River basin of North Dakota and Minnesota from SSM/I data. The MCPN that makes use of the self-organizing feature map (SOFM) learning algorithm (Kohonen *et al.*, 1996), consists of an interconnected network of three layers, namely, the input, hidden, and output layers.

The unsupervised clustering procedure of SOFM performs the input-hidden layer transformation ($SD_i \rightarrow IP_j$), the non-linear part of the Input/Output mapping. The training of SOFM is carried out by computing the distance d_j between the normalized input vector (the input snow data SD_i, given in Table 2.2) and the weighting vector w_{ji} as

$$d_j = \sqrt{\left[\sum_{i=1}^{no} (SD_i - w_{ji})^2 \right]} \qquad j = 1, \ldots, n_1 \tag{2.16}$$

where n_o is the number of input variables, and n_1 is the number of hidden nodes, among which the winning node (c) has the smallest d_j ($d_c = \min(d_j), j = I_c$). To complete the SOFM training, the updating of weights w_{ji} is performed only for the hidden nodes in the neighborhood Λ_c surrounding the winner node as

$$
\begin{aligned}
w_{ji}(t) &= w_{ji}(t-1) + \eta(t)(SD_i - w_{ji}(t-1)), \qquad \text{for } j \in \Lambda_c(t) \\
&\qquad\qquad\qquad\qquad\qquad\qquad i = 1, 2, \ldots, n_0, \quad 0 < \eta(t) < 1 \quad (2.17) \\
w_{ji}(t) &= w_{ji}(t-1) \qquad\qquad\qquad\qquad\qquad\qquad\qquad \text{otherwise,}
\end{aligned}
$$

with t being the iteration counter for the training process, and $\eta(t)$ is the learning rate which together with $\Lambda_c(t)$ are decreased iteratively from their initial settings of $\eta_0 = 0.2 \sim 0.5$ and $\Lambda_0 = n_1/2$. Before the hidden-output layer transformation, the intermediate output parameters, IP_j, corresponding to the input vector, SD_i, are computed as

$$
\begin{aligned}
IP_j &= 1 - d_j \quad \text{for } j \in \Omega \\
IP_j &= 0 \qquad\quad \text{otherwise,}
\end{aligned} \qquad (2.18)
$$

where Ω the size of hidden nodes centered on the neighborhood of I_c, should be equal to or greater than that of SD_i. The training of the weights (M_{kj}) required for the hidden-output transformation ($IP_j \rightarrow SWE_k$) is performed by a simple recursive gradient search. In the neighborhood of Ω surrounding the active node, these weights are recursively updated as

$$
v_{kj}(t) = v_{kj}(t-1) + \beta(TSWE_k - SWE_k)y_j, \quad \text{for } j \in \Omega, \ k = c, \qquad (2.19)
$$

where β is the learning step size ($0 \le \beta \le 1$), and $TSWE_k$ is the observed (target) SWE and SWE_k is its estimated value after each iteration. Adjustments are made to M_{kj} to obtain their final values, which together with IP_j are used to compute the output, SWE_k

$$
\begin{aligned}
SWE_k &= \sum_j M_{kj} IP_i \quad \text{for } j \in \Omega, \ k = c \\
SWE_k &= \varnothing \qquad\qquad \text{for } k \ne c.
\end{aligned} \qquad (2.20)
$$

At 25×25 km resolution, we expect the micro-scale spatial variability of snowpack to be mostly averaged out, and so we can mostly expect to detect meso- to macro-scale variability of snowpack from the above data. Figure 2.14 shows the monthly distributions of SWE in NH for November 2003–February 2004 derived from SSM/I (left) and AMSR-E (right), which are very similar to each other with limited differences. As expected, such SWE data are subjected to errors, particularly SWE values from mountainous areas with large topographic variability due to a possible mixture of deep snow on north-facing slopes but almost snow free on south-facing slopes, or forested areas because of mixed microwave emissions from trees, snow canopy and ground surface. Figure 2.15 shows some obvious differences between the Northern Hemisphere SWE map with enlarged area in Northern Europe and Western Russia for October 24–31, 2003, derived from (a) the NSIDC Global Monthly EASE-Grid

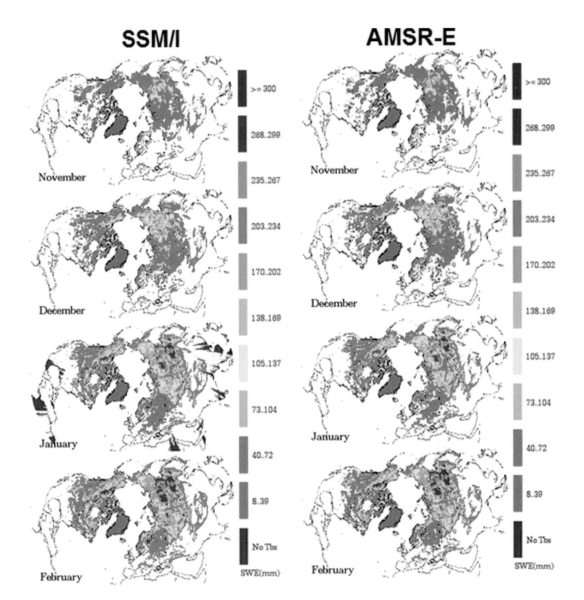

Figure 2.14 Northern Hemisphere monthly average snow water equivalent (SWE) derived from SSM/I (left) and AMSR-E (right), November 2003–February 2004 (Courtesy of Brodzik, M. J., NSIDC/CIRES) (A black and white version of this figure will appear in some formats. For the color version, please refer to the plate section.)

SWE Climatology, (b) with additional snow-covered area (red) determined from at least 25% of component MODIS CMG pixels indicating snow cover, and (c) by combining 89 GHz data with MODIS snow-covered area (selected day with maximum difference in $[37 - 85]$ GHz) (Armstrong *et al.*, 2006). Figure 2.15c shows an improvement in shallow snow SWE

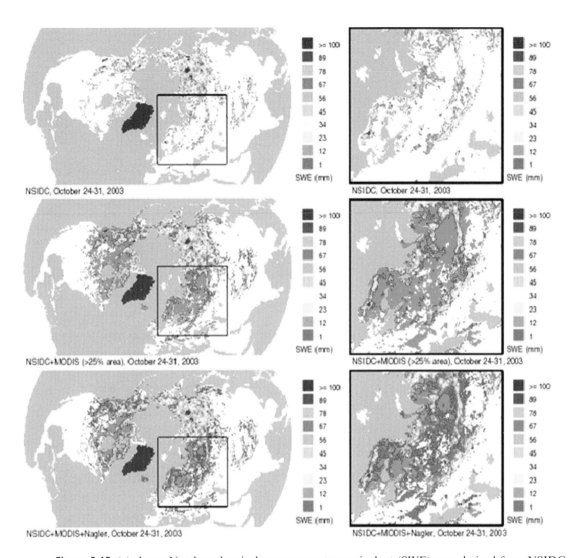

Figure 2.15 (a) shows Northern hemisphere snow water equivalent (SWE) map derived from NSIDC Global Monthly EASE-Grid SWE Climatology, with enlarged area in Northern Europe and Western Russia for October 24–31, 2003; (b) shows additional snow-covered area (red), determined from at least 25% of component MODIS CMG pixels indicating snow cover; (c) shows improvement in shallow snow SWE estimates by combining 89 GHz data with MODIS snow-covered area (select day w/ max diff 37 – 85 GHz) (Courtesy of Brodzik, M. J., NSIDC/CIRES) (A black and white version of this figure will appear in some formats. For the color version, please refer to the plate section.)

estimates for in passive microwave imagery, wet areas containing melting snow or wet snow packs could return low or zero SWE values. Singh and Gan (2000) employed screening procedures to eliminate potentially erroneous SWE data in their analysis.

Gan *et al.* (2009) retrieved the SWE for the Red River basin of North Dakota and Minnesota using SSM/I, physiographic and atmospheric data by MCPN, a Projection Pursuit Regression (PPR) and a nonlinear regression. They used the airborne gamma-ray measurements of SWE as the observed SWE. They screened the SSM/I data for the presence of wet snow, large water bodies like lakes and rivers, and depth hoar. They found MCPN to produce encouraging results in both calibration and validation stages (R^2 was about 0.9 for both calibration (C) and validation (V)), better than PPR (R^2 was 0.86 for C and 0.62 for V), which in turn was better than the multivariate nonlinear regression at the calibration stage (R^2 was 0.78 for C and 0.71 for V). MCPN is probably better than the linear and nonlinear regression counterparts because of its parallel computing structure and its ability to learn and generalize information from the SWE-SSM/I relationships.

Takala *et al.* (2011) developed an algorithm assimilating synoptic snow depth data of weather stations with passive microwave radiometer data to produce a 30-year time-series of seasonal SWE for the NH, which was validated against independent SWE data from Russia, the former Soviet Union, Finland and Canada. The results show an overall strong performance in the SWE retrieved with root mean square errors below 40 mm when SWE <150 mm, but the retrieval uncertainty increases when SWE is above 150 mm. The SWE estimates obtained by this assimilation approach are better than SWE obtained from typical stand-alone retrieval algorithms shown above.

Active microwave data

In contrast to passive microwave data, active microwave data acquired by synthetic-aperture radar (SAR) sensors can attain much higher spatial resolutions. SWE can also be estimated using active microwave data such as the SIR-C/X-SAR data by Shi and Dozier (2000), or possible future data from the dual-frequency (X-/Ku-band) synthetic aperture radar (SAR) satellite proposed by the Cold Land Processes Working Group (CLPWG) of NASA. Active and passive microwave sensors have different sensitivities to the same snowpack properties, and so they can provide complementary information. With improved resolution (about 10 km), passive microwave measurements can help supporting the high-resolution radar measurements, make linkages across process scales, and help relating future, dual-frequency SAR missions such as the Snow and Cold Land Process (SCLP) mission to the long-term microwave record of snow.

2.10   Snowmelt modeling

Spring snowmelt forms a major hydrological event of the year and it is the major source of fresh water for municipal and industrial water supply, irrigation, and hydropower generation over regions of mid- and upper latitudes. Globally more than one billion people depend on melting snow or glaciers for their primary water resource. The amount of snowmelt depends upon the energy available at the snow surface and the SWE present in the basin. An accurate estimate of the highly variable, basin to regional scale melt process is still a great challenge. This problem is more severe in open environments where blowing snow is a dominant winter phenomenon that prevents the computation of an accurate annual water balance. On the other hand, in a forest environment, it is also found that as canopy density increases, penetration of radiation and snowmelt rate decreases. However, under some conditions, snowmelt had been found to increase under dense canopy due to decreased terrestrial radiative losses (Yamazaki and Kondo, 1992). Under leafless deciduous canopies, the net radiation alone is a good predictor of snow ablation (Price, 1988) but it is not adequate to estimate snowmelt under the influence of canopy in the boreal forest (Metcalfe and Buttle, 1995).

The principles behind snowmelt modeling are first described below, then the intercomparisons of snowmelt models conducted in recent years are discussed. Snowmelt models developed for open and/or forest sites vary over a wide range of complexity, ranging from simple models such as the SNOW-17 of Anderson (1976) which uses the degree-day method and a simple approach to consider canopy's hindrance to snowfall, to land surface schemes in GCMs of intermediate complexity, for example, SSiB3 of Xue *et al.* (2003), to complex canopy-atmosphere-soil models, for example, ACASA of Pyles *et al.* (2000). Modeling snow processes has been identified as an area of continuing weakness in global land surface models (Dirmeyer *et al.*, 2006), because large discrepancies remain in albedo for forests under snowy conditions, due to difficulties in estimating the extent of masking of snow by vegetation (IPCC, 2007; Essery *et al.*, 2009).

Empirical snowmelt-runoff models

Empirical or statistical models (Equation 2.8) such as linear (all $\gamma_i = 1$, $i = 1, 2, \ldots, n$) or nonlinear (at least one or more of $\gamma_i \neq 1$, $i = 1, 2, \ldots, n$) regressions based on SWE measured using snow pillows or snow course samplers at selected sites and/or baseflow measured for winter months such as November to March, as independent variables, for example, B_1, B_2, $\ldots$, B_n, have been popularly used to forecast spring snowmelt of river basins. These models assume that under average snowpack conditions, empirical relationships derived between spring snowmelt runoff and measured snowpack data in the past are applicable in future years.

$$Q = a_o + a_1 B_1^{\gamma_1} + a_2 B_2^{\gamma_2} + \ldots\ldots + a_n B_n^{\gamma_n}. \tag{2.21}$$

Another statistical method for predicting spring snowmelt runoff is to relate SCA with spring snowmelt.

Degree-day or temperature index (TI) method

Many operational snowmelt runoff models use the degree day (temperature index) approach that estimates the snowmelt rate M (mm d^{-1}) as the difference between the mean daily air temperature (T_a) and a melt-threshold or base temperature (T_{thm}) adjusted by some optimized melt factors m_f in mm d^{-1} °C^{-1}, and a depletion curve that relates how of the original snow cover remains versus mean areal extent of snow cover (e.g., WMO, 1986). m_f depends on vegetation types, the slope and aspect of the land surface, percent of snow cover, time of the year and the climatic regime (Frank and Lee, 1966). So in applying Equation (2.22) to each basin zone, a different m_f is usually used to reflect the vegetation characteristics and climate of each zone. The TI method has been widely used in operational snowmelt forecasting because by adjusting the degree-day, ($T_a - T_{thm}$), with an appropriate m_f, it can approximately represent the daily energy supply for melting snowpack on a regional basis.

$$M = m_f(T_a - T_{thm}). \tag{2.22}$$

In addition, m_f has been found to vary seasonally, and tends to increase as the season progresses because of a decrease in snow albedo, and an increase in incoming solar radiation; in some cases the melt factor is allowed to vary through the melt season, such as a form of sinusoidal function in Equation 2.23,

$$m_f = \frac{m_{f\,max} + m_{f min}}{2} + \sin\left(\frac{2\pi n}{366}\right)\left[\frac{m_{f\,max} - m_{f min}}{2}\right], \tag{2.23}$$

where $m_{f max}$ and $m_{f min}$ are the maximum and minimum melt rate, and n is the Julian date.

Even though the degree-day approach may work well in mid-latitude or temperate environments when there is a strong correlation between T_a and the dominant energy responsible for snowmelt, it does not adequately account for many climatic factors related to snowmelt. For example, Male and Granger (1981) showed that in open, non-forested areas the short wave radiation exchange is a dominant melt-producing energy flux, but it generally does not correlate well with air temperature. Moreover, the consideration of a uniform snow accumulation in the whole elevation range followed by a snowmelt process based on an assumed areal depletion curve is hardly imaginable in a mountain basin (Martinec, 1980). Some newer models use satellite images to update its areal snow cover distribution and do not solely rely on an areal depletion curve. Examples of more widely used degree-day models are such as the NWSRFS (National Weather Service River Forecast System) (Anderson, 1976), SRM (Snowmelt Runoff Model) (Martinec et al., 1998), HBV (Swedish Meteorological and Hydrological Model) (Bergström, 1995; Seibert and Vis, 2012), and others.

Modified degree-day or modified temperature index (MTI) method

Singh *et al.* (2005) introduced a modified degree-day method that includes the near-surface soil temperature (T_g) as an additional predictor,

$$m = m_f(m_{rf})(T_r - T_{thm}) \tag{2.24}$$

where m_{rf} is an adjustment factor for m_f so as to better capture the onset of initial snowmelt and is estimated from

$$m_{rf} = [\beta_1 + \beta_2(\tan^{-1}T_g + \beta_3)]^\psi, \tag{2.25}$$

where β_1, β_2, β_3, and ψ are model parameters derived through calibration (see Singh *et al.*, 2005). As a tangent function, m_{rf} is relatively small when $T_g < 0$ °C, and reaches an upper limit of one when $T_g \geq 0$ °C. The effect of m_{rf} is "felt" mostly during the onset of snowmelt because its value approaches 1.0 when T_g approaches 0 °C. The desired rate of change of m_{rf} can be achieved by adjusting the parameter ψ. T_r is a reference temperature computed as a weighted average of T_g and T_a,

$$T_r = \chi T_a + (1 - \chi)T_g, \tag{2.26}$$

where χ is also a model parameter. The rationale for the proposed modification comes from past studies that pointed out the importance of T_g as an indicator of spring snowmelt (Woo and Valverde, 1982), as well as the analysis of data observed at the Paddle River Basin (PRB) in central Alberta by Singh *et al.* (2005). Analysis of hourly data observed for 6 years during the spring season (March 1–April 30) at PRB shows that there is a significant correlation between net radiation (R_n) and T_g at daily time step. Moreover, the data revealed that Pearson's correlation coefficient between T_g and R_n (ranging from 0.62 to 0.89) was mostly higher than that between T_a and R_n (ranging from 0.47 to 0.87). Since R_n generally dominates the energy balance during spring snowmelt in the Canadian Prairies (Shook, 1995), adding T_g as another predictor should improve the performance of Equation 2.22.

Modeling snowmelt by energy balance method (EBM)

A more physically based snowmelt modeling approach is an one-dimensional (1-D) mass and energy balance model, such as the US Army Cold Regions Research and Engineering Laboratory (CRREL) model, SNTHERM developed by Jordan (1991) for predicting snow-pack properties and temperature profiles (Figure 2.16). SNTHERM calculates energy exchange at the surface and bottom of the snowpack, grain growth, densification and settlement, melting and liquid water flow, heat conduction and vapor diffusion, and solar absorption. The model's surface boundary conditions require incoming solar and longwave radiation; wind speed, air temperature and humidity at some reference height; and

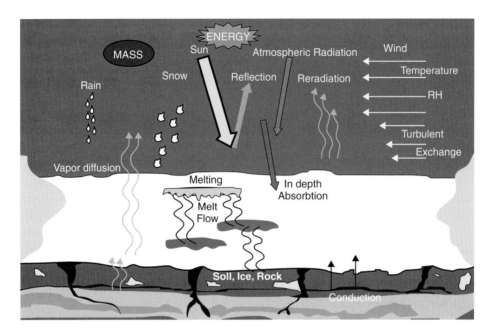

Figure 2.16 Energy fluxes considered in SNTHERM, a physically based 1-D energy balance model for snow and soil by Jordon (1991) (figure taken from SNTHERM fact sheet)

precipitation. Lower boundary conditions include soil textural properties (currently clay or sand used as defaults), wetness and temperature profile.

SNTHERM accounts for changes in albedo due to grain growth, sun angle and cloud cover but it does not account for the decrease in effective albedo when the snow depth is shallow and when radiation penetrates through the snowpack to the underlying soil. It is expected that the snow albedo decreases exponentially to the soil albedo when the radiation penetrates through the snowpack to the underlying soil, and in a forest environment, the accumulation of forest litter could reduce snow albedo. In the energy balance approach, which is mostly considered in the vertical direction only, both the energy content of the snowpack plus a soil layer underneath that interacts thermally with the snowpack should be considered. This procedure provides a simple approximation of the effects of frozen ground, or snow falling on warm ground. The model output provides snow depth, profiles of snow temperature, water content, density, grain size, and surface fluxes of sensible heat and evaporation. Shi *et al.* (2009) used SNTHERM to simulate the snow profiles of a 1,200-km transect of snow stratigraphy measured in Alaska. SNTHERM has been extensively tested for the prediction of temperature, snow depth, SWE, and snow-covered sea ice in sites such as the California Sierra Nevada (Mark, 1988), Greenland (Rowe *et al.*, 1995), the Canadian Boreal forest (Hardy *et al.*, 1998), the Arctic (Jordan *et al.*, 1999), Antarctic (Andreas *et al.*, 2005), southern Finland (Koivusalo and Burges, 1996), and other Nordic environments (Langlois et al., 2009).

A comparative study of three snow models with different complexities was carried out by Jin *et al.* (1999) to assess how a physically detailed snow model can improve snow modeling within general circulation models. The three models were (a) SNTHERM; (b) a simplified three-layer model, Snow–Atmosphere–Soil Transfer (SAST), which includes only the ice and liquid-water phases; and (c) the snow submodel of the Biosphere-Atmosphere-Transfer-Scheme (BATS), which calculates snowmelt from the energy budget and snow temperature by the force–restore method. SNTHERM gave the best match to observations with the SAST simulation being close. BATS captured the major processes in the upper layers of a snowpack where solar radiation is the main energy source and gave satisfactory seasonal results.

CROCUS is a model of the Centre d'Etudes de la Neige, Grenoble (Brun *et al.*, 1992). It is a 1-D physical model that determines mass and energy balance for a snow cover and is used for operational avalanche forecasting. The snow cover is represented as a pile of layers parallel to the ground. Energy exchanges are projected orthogonally to the slope. The model describes the evolution of the internal state of the snow cover as a function of meteorological conditions. The variables describing the snow cover are temperature, density, liquid water content, and snow type of each layer. To match the natural layers, the thickness and number of layers are adjusted by the model. The model simulates the heat conduction, melting/refreezing of snow layers, settlement, metamorphism, and percolation. It simulates dry and wet snow metamorphism with experimental laws derived from laboratory data. Snow grains are characterized by their size and type. This allows an accurate albedo of the snow cover to be calculated.

Bartelt and Lehning (2002) presented a 1-D physical model of the snowpack (SNOWPACK) with equations for heat transfer, water transport, vapor diffusion, and mechanical deformation. In their model, new snow, snow drift, and ablation are considered, and the snow layers are treated in terms of height, density, and microstructure (grain size, shale, and bonding). The model is used for avalanche warnings in Switzerland. Langlois *et al.* (2009) simulated the SWE in southern Quebec using SNOWPACK, CROCUS, and SNTHERM models driven by local and reanalysis meteorological data, which are in agreement with ground measurements of SWE in winter seasons of 2004/2005, 2005/2006, and 2007/2008, with correlation coefficients ranging between 0.72 and 0.99.

One-dimensional vertical EBM

A 1-D energy balance snowmelt model developed by Singh *et al.* (2009), which they refer to as SDSM-EBM, is briefly described. The model incorporates processes for snow interception by forest canopy, separate snowpack energy and mass balance for open and forested areas, separate water balance for liquid and ice phases, snow sublimation, compaction, refreezing, and so on. The snow interception capacity at different levels of canopy is estimated as a function of LAI, forest types (Hardy and Hansen-Bistow, 1990), tree species, and prevailing forest structure (Golding and Swanson, 1986). Hedstrom and Pomeroy (1998) tested the snow interception model for the Canadian southern boreal forest, which Singh *et al.* (2009) implemented in SDSM-EBM.

The transfer of energy at the snow surface and the snow/soil interface governs the snowmelt. The amount of energy available for melting snow is determined from the 1-D, energy equation (Equation 2.27) applied to a control volume of snow having upper and lower interfaces with air and ground, respectively. During melting, the snowpack is isothermal at 0 °C ($T_{sp}^{t+1} = 0$ °C) and the heat for snowmelt, Q_o, can be calculated as

$$Q_o = Q_n \pm Q_h \pm Q_e \pm Q_g + Q_v + Q_f - Q_{cc}, \qquad (2.27)$$

where

Q_n = Net radiation (short-wave and long-wave) absorbed by the snow,
Q_h = Convective or turbulent sensible heat flux exchanged at the surface due to a difference in temperature at the snow-air interface,
Q_e = Convective or turbulent latent heat exchanged at the surface due to vapor movement as a result of a difference in vapor pressure (evaporation, sublimation, condensation) at the snow-air interface,
Q_g = Ground heat flux at snow-interface by conduction,
Q_v = Advective heat of precipitation,
Q_f = Energy released by freezing of liquid water content, and
Q_{cc} = Cold content of snowpack in the previous time step.

Part of Q_o may be used to overcome the cold content of the snowpack, Q_{cc} (see Equation 2.37). All fluxes are computed in terms of W m^{-2} or older units such as langley per min. The net shortwave radiation ($Q_{sn} = Q_{si} - Q_{so}$) can either be measured by a net radiometer or by two pyrometers measuring the incoming short wave radiation (Q_{si}) and the reflected short wave radiation (Q_{so}) which depends on the albedo of the snow-covered area, and the net longwave radiation ($Q_{ln} = Q_{li} - Q_{lo}$) can be similarly measured using two pyrgeometers with one inverted. By knowing the air temperature and the surface temperature of snowpack, and the emissivity of the atmospheric medium, it is also possible to estimate Q_{li} and Q_{lo} using the Stefan-Boltzmann's constant and the blackbody theory.

$$Q_n = Q_{sn} + Q_{ln} \qquad (2.28)$$

Q_s can also be converted to Q_{sn} [$= Q_s(1 - \alpha)$] as a function of the areal albedo of a partially ablated snow cover (α), which can be taken as the larger of that retrieved from satellite images such as that of NOAA-AVHRR or MODIS, or computed from

$$\alpha = \alpha_{sn} A_{sn} + \alpha_g (1 - A_{sn}) \qquad (2.29)$$

where α_{sn} is the snow albedo that can be estimated from an albedo decay function (Riley *et al.*, 1972), α_g is the albedo of the ground surface and A_{sn} is the fraction of snow-covered area, which can be tracked using either a linear, $A_{sn} = \dfrac{SD}{h}$, or a non-linear, parameterizing the SWE probability distribution based on the basin's topography and vegetation in a dimensionless form, $A_{sn}(D) = \oint \left(\dfrac{D}{D_{max}}\right)$, depletion curve. SD ($D$) is the snow depth (SWE), and h (D_{max}) is the depth (SWE) at which snow cover is complete (or the depth (SWE) below which bare patches start to emerge). In computing A_{sn}, $\dfrac{D}{D_{max}}$ is replaced by $\dfrac{SD}{h}$. From snow data collected at the Paddle River Basin (PRB), Singh *et al.* (2009) found that h ranges between 0.07 and 0.1 m as the cutoff for a partial snow cover. In forest-covered areas, α_{sn} can be further modified to account for the effect of litter fall fraction (Hardy *et al.*, 1998).

Q_s can also be estimated empirically (e.g., Bras, 1990) but it is subjected to errors because of possible cloud cover effects and air pollution problems. Instead of measuring snow albedo, it can also be estimated from the age of the last snowfall since α_{sn} declines with age and α_{sn} ranges from about 0.9 or higher for freshly fallen snow to less than 0.4 for shallow, dirty, and wet snow. Snowpack albedo has also being related to the cumulative maximum air temperature (US Army Corps of Engineers, 1956), snow grain size, or snow surface density which tends to increase with time.

The other important components of energy balance are Q_h, Q_e and Q_v. Q_g, usually being the smallest, is often ignored. The sensible heat, Q_h, is due to the turbulent flux of energy exchanged at the snow surface, as the result of a difference in temperature between air (T_a) and snow surface (T_s), in contrast to Q_e that is caused by the vapor pressure difference between air and the snow surface. As a turbulent heat transport through convection (like Q_e), Q_h is affected by wind, such that

$$Q_h = l_v k_2 \left[\frac{P_a}{P_o}\right] \bar{u}_b (T_a - T_s)(z_a z_b)^{-1/6}, \tag{2.30}$$

where $k_2 = 0.00357$ cm m$^{1/3}$ hr day^{-1} °C^{-1} km^{-1}, P_o (P_a) is the standard (actual) atmospheric pressure, $\bar{u}_b$ is the wind speed (m s^{-1}), z_a and z_b = heights where air temperature and wind speed are measured respectively, and $l_v \approx 250$ kJ kg^{-1}. A simpler way to estimate Q_h transported through convection is

$$Q_h = C_h \rho_a C_p \bar{u}_b (T_a - T_s), \tag{2.31}$$

where ρ_a is the air density, C_p is the specific heat of air (1,004 J kg^{-1} °C^{-1}), and C_h is a bulk transfer coefficient that depends on displacement, roughness height, and atmospheric stability. An even simpler expression to estimate Q_h is

$$Q_h = B_h \bar{u}_b (T_a - T_s), \tag{2.32}$$

where B_h is the bulk transfer coefficient that replaces $C_h \rho_a C_p$ given in Equation (2.31). In SDSM-EBM, T_s is computed using one of three simplified heat flow models, namely the force-restore method, the surface conductance method, and the Kondo and Yamazaki method (Singh and Gan, 2005).

The latent heat flux (Q_e) is the sum of surface sublimation/condensation ($Q_{e,surf}$) and blowing snow sublimation ($Q_{e,bss}$). When water vapor is transported to the snow surface, it changes phase to either liquid or solid, releasing $Q_{e,surf}$ at the snow surface,

$$Q_{e,surf} = l_e E = C_e \rho_a \left(\frac{0.622 l_e}{P_o} \right) \bar{u}_b (e_a - e_s), \tag{2.33}$$

where E is the rate of vapor transfer, C_e the bulk transfer coefficient for water vapor, l_e is the latent heat of sublimation (2,836 kJ kg^{-1}) or the latent heat of vaporization (2,501 kJ kg^{-1}), e_a and e_s are the actual and snow surface vapor pressure (assumed saturated at T_s) in Pascal, respectively, and P_o is the standard atmospheric pressure at sea level ($\approx$101.33 kPa). e_s is estimated as

$$e_s = 611 \exp \left(\frac{17.27 T_s}{237.3 + T_s} \right). \tag{2.34}$$

The bulk transfer coefficient under neutral condition is computed from Brutsaert (1982),

$$C_e = \frac{\kappa^2}{[\ln((z_r - d_0)/z_0)]^2}, \tag{2.35}$$

where κ is the von Karman constant, z_r is the reference height, d_0 is the zero displacement height and z_0 is the roughness height. The model can be run assuming neutral conditions or one of three options, namely that of Price and Dunne (1976), Louis (1979), and Morris (1989) can be used to adjust C_e.

If the precipitation is rainfall, there is positive advective heat Q_v that contributes toward melting the snowpack, Q_o, given as

$$Q_v = c_s \rho_w RT, \tag{2.36}$$

where R = intensity of rainfall in cm hr^{-1}, T = temperature of rainfall in °C, and c_s = specific heat (2.093 kJ kg^{-1} °C^{-1} for snow, 4.186 kJ kg^{-1} °C^{-1} for water).

The cold content of a snowpack is the heat required to raise the temperature of the snowpack to 0 °C, if the temperature of the snowpack, T_p, which can be different from the snowpack surface temperature T_s, is less than 0 °C.

$$Q_{cc} \approx -\rho_p c_s D T_p, \tag{2.37}$$

where ρ_p is the snowpack density, c_s is the snowpack specific heat, D is the snowpack depth in cm, and T_p is the snowpack temperature in °C. If D_{cc} is the cold content expressed in terms of the equivalent depth of ice at 0 °C that can be melted to water at 0 °C,

$$D_{cc} = \frac{Q_{cc}}{-\rho_w l_{fs}} = \frac{\rho_p C_s D T_p}{\rho_w l_{fs}}. \tag{2.38}$$

In computing Q_{cc}, the heat capacity of the entrapped air is neglected, and the snowpack temperature T_p may be assumed as T_s if no information is available. In other words, T_p is assumed to be independent of depth below snow surface (z).

After overcoming the cold content of the snowpack, Q_{cc}, the depth of snowmelt, M, due to Q_o acting for a time interval Δt, is then

$$M = \frac{(Q_o \Delta t - Q_{cc})}{\rho_w l_{fs} \theta}, \tag{2.39}$$

where ρ_w is the density of water and θ is the thermal quality of snow, which is the fraction of ice in a unit mass of wet snow, or the ratio of the heat necessary to produce a given amount of water from snow to the amount of heat needed to produce the same quantity of melt from pure ice at 0 °C. θ usually ranges between 0.95 and 0.97.

The total energy needed to melt a snowpack is $Q_o \Delta t = \rho_p D l_{ms} + Q_{cc}$, where l_{ms} is the actual latent heat to melt the snowpack of depth D. If Q is the amount of energy needed to produce the same amount of melt from pure ice at 0 °C, such that $Q = \rho_p D l_{fs}$ then by definition, the thermal quality of snow, θ is given by

$$\theta = \frac{l_{ms}}{l_{fs}} - \frac{c_s T_p}{l_{fs}} \tag{2.40}$$

In theory θ can also be estimated if we know the liquid water content (W_{liq}) of a snowpack since by knowing that we can calculate l_{ms} (Equation 2.7), which will be easier than trying to measure l_{ms}. Among various methods available to determine W_{liq} of a snowpack, a popular approach is to measure W_{liq} by a time-domain reflectometry because of the large difference in dielectric constant between water and ice (Schneebeli et al., 1998).

Given the insulating effect of vegetation cover, Q_g, the exchange of energy between the snowpack and the underlying ground by conduction is often ignored except where the ground is frozen, or in tundra. If Q_g needs to be accounted for, it can be computed as

$$Q_g = \lambda_g \left(\frac{\partial T_g}{\partial z} \right), \tag{2.41}$$

where λ_g, the thermal conductivity of soil, is about 0.4–2.1 W m^{-1} °C^{-1} for unfrozen silt and clay, and T_g is the ground temperature that changes with elevation z.

Snowpack water balance

The water balance equations can be expressed in terms of water and ice at both canopy and ground levels as

$$\rho_w c_s W^{t+\Delta t} T_{sp}^{t+\Delta t} = \rho_w c_s W^t T_{sp}^t + (Q_n + Q_h + Q_e + Q_g + Q_p) + Q_o. \tag{2.42}$$

Here, $W^{t+\Delta t}$ accounts for both the addition of precipitation (P_r or P_s) during the time step and the change in water and ice mass due to Q_e (sublimation or freezing) depending on whether T_{sp}^t is isothermal at zero or less than zero,

$$W_{liq}^{t+\Delta t} = \begin{cases} W_{liq}^t + P_r + \dfrac{Q_e}{\rho_w l_v} & if \quad T_{sp}^t = 0 \\ W_{liq}^t + P_s & if \quad T_{sp}^t < 0 \end{cases} \tag{2.43}$$

$$W_{ice}^{t+\Delta t} = \begin{cases} W_{ice}^t + P_s & if \quad T_{sp}^t = 0 \\ W_{ice}^t + P_s + \dfrac{Q_e}{\rho_w l_{fs}} & if \quad T_{sp}^t < 0 \end{cases}. \tag{2.44}$$

The net energy exchange in the snowpack (Q^*) is then equal to

$$Q^* = (Q_n + Q_h + Q_e + Q_g + Q_p). \tag{2.45}$$

If $Q^* < 0$, the snowpack is losing energy to the atmosphere (cooling), and some liquid water (if available) may be re-frozen. The amount of energy released to the snowpack (positive value) by re-freezing liquid water is given by

$$Q_o = \min(-Q^*, \rho_w l_{fs} W_{liq}^{t+\Delta t}). \tag{2.46}$$

The resulting changes in the liquid and ice phases are given by

$$W_{liq}^{t+\Delta t} = W_{liq}^{t+\Delta t} - \dfrac{Q_o}{-_w l_{fs}} \tag{2.47}$$

$$W_{ice}^{t+\Delta t} = W_{ice}^{t+\Delta t} + \dfrac{Q_o}{-_w l_{fs}} \tag{2.48}$$

$$W^{t+\Delta t} = W_{liq}^{t+\Delta t} + W_{ice}^{t+\Delta t}. \tag{2.49}$$

The negative snowpack temperature, $T_{sp}^{t+\Delta t}$ (associated with its cold content), is then updated from Equation (2.42). If $Q^* > 0$, the snowpack is gaining energy from the atmosphere (heating), and in the process the negative $T_{sp}^{t+\Delta t}$ will increase until it just reaches the isothermal condition ($T_{sp}^{t+\Delta t} \to 0$). When $T_{sp}^{t+\Delta t}$ becomes positive, it is set equal to zero and Q_o is computed by Equation (2.46) and applied to Equation (2.47) and Equation (2.48) to compute the new liquid and ice components of SWE.

At each time step, the compaction of snowpack, S_{comp} is based on the present snowpack density ρ_{sp} ($= W^{t+\Delta t} / SD^{t+\Delta t}$), maximum $\rho_{s,max}$, and a settlement constant, c_s (Riley *et al.*, 1972), as

$$S_{comp} = SD^{t+\Delta t} \, c_s \left(1 - \frac{\rho_{sp}}{\rho_{s,\max}}\right). \tag{2.50}$$

The depth of snowpack after compaction is the difference between SD and S_{comp}. $\rho_{s,max}$ and c_s are manually calibrated such that the simulated SD matches the corresponding snow course data for a given land-use. The effect of rain compaction on snow is also based on Equation (2.50), where SD^t replaces $SD^{t+\Delta t}$ when precipitation is in the form of rain.

During melt (positive Q_o) and $T_{sp}^{t+\Delta t}$ is isothermal at 0 °C, water is removed as meltwater (M_{ij}) when the liquid phase increases beyond the current liquid water holding capacity ($LWHC$) of the snowpack at the expense of the ice phase, or M_{ij} is held within the pack when snowmelt first appears at the bottom of the snowpack.

$$M_{ij} = W_{liq}^{t+\Delta t} - (\mathrm{LWHC}) W^{t+\Delta t} \tag{2.51}$$

where i is the sub-basin number and j is the land use type, and recommended values for LWHC, a function of snowpack properties and the presence of depth hoar, are $0.02W$ to $0.05W$ (US Army Corps of Engineers, 1956), and $0.05W$ for $\rho_{sp} < 400$ kg m^{-3} (Riley $et\ al.$, 1972). As meltwater contributes runoff at the bottom of snowpack, the new $W_{liq}^{t+\Delta t}$,

$$W_{liq}^{t+\Delta t} = W_{liq}^{t+\Delta t} - M_{ij}. \tag{2.52}$$

The final SWE is computed from Equation (2.49). Routing the meltwater through the snowpack is usually neglected because the routing time for moderately deep snow covers is usually less than the hourly time step of the 1-D energy balance snowmelt model.

Singh $et\ al.$ (2009) tested the standard and modified temperature index methods, and the 1-D EBM on a small watershed called the Paddle River Basin (PRB) (53° 52′ N, 115° 32′ W), a sub-basin of the Mackenzie River basin, in a semi-distributed approach whereby PRB was divided into units that have similar drainage patterns identified from digital elevation model data. This semi-distributed approach provides a trade-off between modeling resolution, complexity and data availability (Biftu and Gan, 2001) so that the snowmelt (M_i) for sub-basin i and at each time step is the sum of melt from each land cover, weighted by its corresponding areal fraction ϕ_j as

$$M_i = \sum_{j=1}^{n} \phi_j m_{ij}, \tag{2.53}$$

where n is the total number of land cover classes considered. They found that both EBM and MTI models show good agreements between their simulated and observed values of PRB of 2006/2007 based on $\rho_{s,max}$ of 250 kg m^{-3} for the deciduous forest (DF) areas (Figure 2.17a) and the coniferous forest (CF) (Figure 2.17b), respectively. Apparently better simulations of the snow depth could be achieved by varying $\rho_{s,max}$ with time, which is in line with the anticipated increase in snow density with time. $\rho_{s,max}$ continues to change due to interaction of the snowpack with freshly fallen snow and settlement, and usually reaches the highest value at the end of snow accumulation period. When compared to the snow pillow data collected, EBM reproduced the snow accumulation between January 1, 2007 and March 13, 2007 almost perfectly. After March 13, the model overestimated snow accumulation and similarly lagged the ablation by about 2 days. On the other hand, MTI slightly overestimated snow accumulation up to the beginning of March and then reproduced the late season accumulation and ablation almost perfectly. Similarly, the time series of SWE estimated from the daily snow depth averaged from hourly automated snow depth measurements and the mean measured ρ_s from bi-weekly snow surveys conducted for the BERMS (Boreal Ecosystem Research and Monitoring Sites) sites located within the southern boreal forest of Saskatchewan for 2006/2007 shown in Figure 2.17c, and for 2007/2008 in Figure 2.17d, respectively, are compared with SWE retrieved by the GlobSnow project (Luojus *et al.* 2010). Despite of snow interactions with mixed vegetation and relatively dense forest, the BERMS data generally compare well with the GlobSnow data.

As an energy balance model, EBM accounts for the Q_{cc} explicitly but as an empirical model, MTI does not account for the Q_{cc} of the snowpack and so occasional warm air masses occurring in the winter may cause MTI to over-simulate snowmelt runoff. This is the limitation of the temperature index (TI) approach, where without considering the effect of T_g, the model performance could be rather poor in both the calibration and the validation stages, as shown in Figure 2.17c, when the effect of T_g in the MTI model (Equation 2.23) was completely ignored by setting χ to 1 and ψ to 0, making it a standard TI model (Equation 2.21). In order to appreciate better the improvement achieved by introducing T_g and also to perform a fair comparison between TI and MTI, Singh *et al.* (2009) re-calibrated the m_f of TI (Equation 2.21) based on T_a only. It was necessary to use an artificially low m_f for TI to perform well at the calibration stage (figure not shown), for example, m_f of 0.04 mm hr^{-1} °C^{-1} for DF, which is a very low melt factor. However, the model performance decreased considerably when these m_f values were used for all validation stages (Figure 2.17c). Apparently in a Prairie environment where the seasonal snow cover is shallow to moderately deep, where T_g is found to have a fairly strong correlation with net radiation (Q_n), and the onset of major snowmelt usually happens when T_g approaches 0 °C, using both T_g and T_a in a temperature index approach (e.g., MTI) should generally lead to more accurate results than using T_a alone, which is much less data demanding compared to the EBM model.

Singh *et al.* (2009) further assessed the three snowmelt models in terms of hourly simulated runoff at the outlet of the PRB within the framework of the semi-distributed hydrologic model DPHM-RS (Biftu and Gan, 2001). At the calibration stage (figure not shown), EBM ($R^2 = 0.85$, and RMSE = 1.01) and MTI ($R^2 = 0.79$, RMSE = 1.24) performed reasonably well even though in terms of the timing of peak flows, both EBM and MTI show a tendency to somewhat lag the peak flows observed. At the validation stage, they found the results to be less satisfactory ($R^2 = 0.5$ for both EBM and MTI) partly because of errors in the lapse rate and gradient used to distribute point temperature and precipitation measurements to sub-

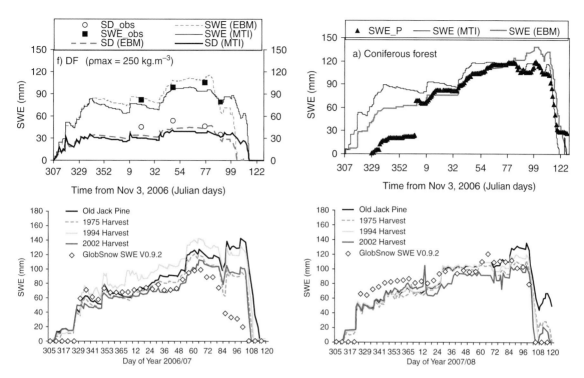

Figure 2.17 Comparison of observed SWE and snow depth (SD) to those obtained by EBM (dashed line) and MTI (solid line) of a sub-basin of PRB for 2006/2007 for (a) the deciduous forest (DF) and (b) the coniferous forest (CF) with daily snow pillow data, respectively; and comparison of observed SWE for the BERMS sites located within the southern boreal forest of Saskatchewan for (c) 2006/2007 and (d) for 2007/2008, respectively with the GlobSnow data (taken from Takala *et al.*, 2011)

basins, errors in the rating curve-discharge relationships due to icing, and the "regulatory" effects of beaver dams in PRB's streamflow.

Even though the degree-day or temperature index (TI) approach has been a popular tool in modeling the spring snowmelt in temperate climates, it seems that either the data intensive, energy balance or the modified TI approach of Singh *et al.* (2009) should be applied to watersheds of northern climates. More extensive research on modeling the snowmelt processes in the cryosphere using the MTI approach should be done since the energy balance approach is generally not practical because its data requirements are usually not met except in watersheds chosen for intensive research studies.

Most energy based snowmelt models consider energy balance in the vertical direction only which may not be sufficient in a mountainous watershed where the effects of terrain features can be important. A snowpack may receive radiation reflected from a surrounding slope or shielded from the incoming radiation by adjacent terrains. For example, if the diffused radiation from the atmosphere to a horizontal plane is Q_D, the diffused radiation received by a slope of inclination angle $\acute{Z}$ from the atmosphere will be $0.5Q_D\cos^2\acute{Z}$ where $0.5\cos^2\acute{Z}$ is the sky-view factor (DeWalle and Rango, 2008). If there is an adjacent terrain of albedo α, the incoming shortwave radiation reflected by the adjacent terrain and received by this slope will be $Q_{si}\alpha(1 - 0.5\cos^2\acute{Z})$.

The effects of forest cover on snowmelt processes can be approximately accounted for by estimating the amount of shortwave radiation penetrated through the canopy, which, similar to the interception of snow by canopy, is a function of the LAI. If the incoming short wave radiation is Q_{si}, then the amount of radiation reaching the snowpack becomes $Q_{si}\exp$ ($-\xi$LAI) where ξ is the extinction coefficient for shortwave radiation in a forest. Values of ξ depend on the forest types, for example, Baldocchi *et al.* (1984); Chen *et al.*, (1997). The incoming longwave radiation, Q_{li}, for a snowpack under forest cover will be the weighted sum of incoming longwave radiation that penetrated through the overhead canopy and the longwave radiation emitted by the forest.

Two-dimensional energy balance approach

The energy balance model presented above and most if not all energy balance models do not directly account for the effects of the distribution of two-dimensional (2-D) patchy snowcover on the local advective energy of melt processes. This limits their application in a shallow snow cover (<60 cm) environment particularly in late melting periods when beside radiative fluxes, turbulent fluxes should also be accounted for. It has been found that the maximum snowmelt rate occurs when the land is only partially snow-covered, and often when it is slightly less than 60% (Shook and Grey, 1997). Because of lower albedo, the bare ground absorbs a larger amount of solar radiation than the adjacent snow patches. The energy imbalance induces an advective, turbulent transfer of latent and sensible heat from the bare ground to snow patches, enhancing the melt rate. Since advective melting is the greatest along the leading edge of a snowfield, under constant climatic conditions, the melt rate of a patchy snow cover should be related to the perimeter of the patches (Shook, 1993), or linearly relate the effective patch size for a regular pattern to the average patch size of a complex snow-cover pattern (Essery *et al.*, 2006).

Besides radiative, sensible and latent heat fluxes, there can be local advection (turbulent energy) due to large area of bare patches within a snow field, which may significantly alter the energy balance, and becomes increasingly important to melt as the snow cover dwindles. The proportion between radiation and turbulent energy sources to melt depend on the size of the snowfield. The smaller snowpatches will be dominated by turbulent melt throughout the season, or until they disappear. For large snowfields, melting is dominated by radiative melt early in the season and turbulent melt late in the season as they decrease in area (Shook and Gray, 1997). Near the leading edge of an alpine snowfield, Olyphant and Isard (1988) found that advection may contribute more than 30 MJ m^{-2} d^{-1} of melt energy on a very windy day and more than 12 MJ m^{-2} d^{-1} on a relatively windless day. They also found that the corresponding advective energy at 1 km from the leading edge decreases to 5 and 2 MJ m^{-2} d^{-1} on windy and windless days, respectively. Further, the effects of wind on a snowpack are more pronounced on the windward than on the leeward slopes.

The sensible heat and latent heat fluxes are primarily related to wind speed, atmospheric stability, temperature and vapor pressure of air and snow surfaces, when they are considered as 1-D flux elements. However, near the edge of patchy snow cover, such simplification may not give good estimation and thus requires the use of a 2-D model that considers the development of the boundary layer beginning at the edge of the snowpack. A 2-D turbulent diffusion model, proposed by Weisman (1977), considers this aspect of energy exchange when air flows over a snow cover. The model assumes a steady turbulent flow of warm moist air mass moving from a homogeneous surface onto a ripe snowpack (isothermal at 0 °C), no change in albedo over the snowpack and so no change of the radiation terms from point to point. Also the model assumes that albedo does not vary over the snowpack, and so Q_n remains constant. This assumption implies that only Q_e and Q_h vary over the snowpack. The model, therefore, considers the 2-D aspect of Q_e and Q_h, and quantify the snowmelt M', due to condensation or sublimation and sensible heat, as,

$$M' = Q_e + Q_h. \tag{2.54}$$

The steady-state equations that describe the mean airflow over the snow are continuity, conservation of momentum, sensible heat and water vapor.

$$\text{Continuity}: \qquad \frac{\partial u}{\partial x} + \frac{\partial w}{\partial z} = 0 \tag{2.55}$$

$$\text{X momentum}: \qquad u\frac{\partial u}{\partial x} + w\frac{\partial w}{\partial z} = \frac{1}{\rho}\frac{\partial \tau}{\partial z} \tag{2.56}$$

$$\text{Sensible heat}: \qquad u\frac{\partial T}{\partial x} + w\frac{\partial T}{\partial z} = \frac{1}{\rho c_p}\frac{\partial H}{\partial z} \tag{2.57}$$

$$\text{Vapor}: \qquad u\frac{\partial q}{\partial x} + w\frac{\partial q}{\partial z} = \frac{1}{\rho}\frac{\partial V}{\partial z}, \tag{2.58}$$

where u, w are mean wind components in the x and z directions, T is temperature, q is specific humidity, ρ is air density, c_p is specific heat of air at constant pressure, and τ, H and V are the turbulent fluxes of momentum, sensible heat, and water vapor, respectively. The molecular diffusion, lateral and forward turbulent diffusion, and the pressure term in x-momentum equation have all been neglected. The problem of airflow over a sudden change in surface temperature and humidity has been solved using the mixing length theory. A vapor diffusion equation is included in the set of conservation equations and a vapor buoyancy term is included in the stability length.

Weisman (1977) found that a stable atmospheric condition dampens the turbulent diffusion of fluxes, and the melt rate is at a maximum near the leading edge and decreases by one-third approximately 15–20 m from the leading edge. He provides an approximation for the advection of energy that relates the dimensionless melt energy at a point ($\hat{M}'$) and total average melt over the snowpack $\hat{M}'$ as a function of dimensionless horizontal downwind distance ($\hat{x}$) and dimensionless snowpack fetch $\hat{x}$, respectively.

$$\hat{M}' = a\hat{x}^{-b} \qquad (2.59)$$

$$\hat{M}' = c\hat{x}_o^{-d}. \qquad (2.60)$$

The constants a, b, c and d depend on the stability parameters, A_*, which is associated with temperature change, and B_*, which is associated with specific humidity change. The snow surface temperature T_s can be retrieved from NOAA-AVHRR, MODIS or Landsat-TM data. Once the dimensionless melt is known, it can be converted to dimensional melt energy M' by multiplying $\hat{M}'$ by the energy gradient at the leading edge of the snowpack. Tables 2.2 and 2.3 list some typical values. The above equation underestimates the melt flux for values of $\hat{x}$ less than 10^4, which comes out to be about 25 m from the leading edge for an average roughness of snow (0.002 m).

The stability parameters, A^* is associated with temperature change and B^* is associated with specific humidity change, given as

$$A^* = \frac{kgz_o(T_o - T_{osoil})}{u_{*a}^2 T_{osoil}} \qquad (2.61)$$

$$B^* = -0.61\frac{kgz_o(q_o - q_{osoil})}{u_{*a}^2}, \qquad (2.62)$$

Table 2.2 Values of constants a and b (Weisman, 1977)

A^*	$B^*=0$ A	$B^*=0$ b	$B^* = 0.001$ A	$B^* = 0.001$ B
0	0.516	0.125	0.422	0.125
0.0001	0.394	0.125	0.380	0.125

Table 2.3 Values of constants c and d (Weisman, 1977)

A^*	B*=0 C	$B^*=0$ d	$B^* = 0.001$ c	$B^* = 0.001$ d
0	0.516	0.110	0.317	0.106
0.0001	0.346	0.105	0.316	0.100

where k = von Karman constant, 0.4, z_o roughness height (m), u_{*a}= frictional velocity upwind of the leading edge of sow (m s^{-1}), T_o = temperature of snow surface (0 °C), T_{osoil}= temperature of soil surface upwind of leading edge (0 °C), q_o= surface specific humidity at snow surface temperature (assume saturation) and q_{osoil} = surface specific humidity at the soil surface temperature, upwind of leading edge. The knowledge of soil surface temperature upwind of the leading edge of the snow patch can also be obtained from the empirical equations for vapor pressure deficit and psychrometric equations.

Once the dimensionless melt is known, it can be converted to dimensional melt energy M' by multiplying $\hat{M}'$ by the energy gradient at the leading edge of the snowpack.

$$M' = \hat{M}'[\rho u_{*a}c_p(T_o - T_{osoil}) + \rho u_{*a}l_v(q_o - q_{osoil})], \qquad (2.63)$$

where ρ is air density (kg m^{-3}), and c_p is specific heat of air (J kg^{-1} °C^{-1}).

Olyphant and Isard (1988) found that Equation (2.27) does not provide a good estimate near the leading edges of discontinuous melting snow surfaces where the advective heat contributes greatly to the energy balance of late-lying snow, and they introduced a modified version of Weisman (1977)'s 2-D boundary layer model. The model pertains to a unit width of snow surface aligned with the prevailing wind. The horizontal pressure gradient is assumed to be negligible, as is the forward component of turbulent diffusion. Time derivatives have been added to the momentum and heat flux equations in an effort to simulate time-dependent responses to the diurnal variation of ambient forcing. The relevant boundary layer equations include one continuity and three conservation equations (momentum, sensible heat and latent heat). The estimation of fetch length, which fluctuates with the wind direction, and the irregular shape of snow patches can be attempted empirically. So far, there have been limited attempts at 2-D snowmelt modeling, which is predominantly 1-D.

Intercomparison of snowmelt models

Since the intercomparison project by the World Meteorological Organization (WMO) on snowmelt runoff models conducted in 1986, some other model intercomparison studies have been conducted, ranging from small scale, individual studies, to international effort, of which the latest and most comprehensive is the SnowMIP2 (Rutter *et al.*, 2009). Singh *et al.* (2009), for example, compared three semi-distributed snowmelt models, namely, a temperature-Index or degree-day model based on air temperature (T_a), a modified temperature index model based on T_a and near ground-surface temperature (T_g), and an energy-balance model that considers liquid and ice phases separately. For the Canadian Prairies where snowpack is shallow to moderately deep, and winter is relatively severe, apparently the modified temperature index model based on T_a and T_g can perform comparably well with the more complex energy balance model, because the advantage of using both T_a and T_g is partly attributed to T_g showing a stronger correlation with solar radiation than T_a during the spring

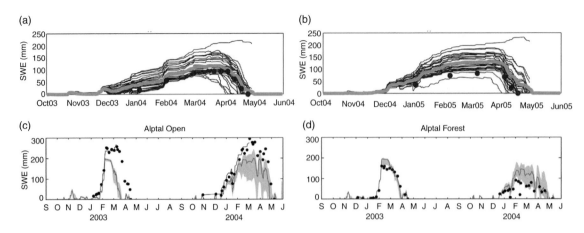

Figure 2.18 Observed SWE (black dots) and the average of thirty-three modeled estimates for (a) (gray line) of forest site at Hyytiälä (61° 51′ N, 24° 17′ E) of Finland (taken from Rutter *et al.*, 2009), and (b) (green line) of open and forest site at Alptal (47° 03′ N, 8° 48′ E) of Switzerland (taken from Essery *et al.*, 2009)

snowmelt season, and partly to the onset of major snowmelt which usually happens when T_g approaches 0 °C.

In the SnowMIP2 project, thirty-three snowpack models of varying complexity were compared across a wide range of hydrometeorological and forest canopy conditions, for up to two winter snow seasons applied to five northern sites in the NH: Alptal of Switzerland; Boreal Ecosystem Research and Monitoring Sites (BERMS) of Canada; Fraser of USA; Hitsujigaoka of Japan and Hyytiälä of Finland (see Figure 2.11). As expected, Rutter *et al.* (2009) concluded that it is easier to model SWE at open than at forested sites, where precipitation phase and duration of air temperatures >0 °C are key factors to the divergence and convergence of modeled estimates of the sub-canopy snowpack, and more consistent results are found between open than forested sites (Figure 2.18). Furthermore, a model that estimates snowpack accurately at a forest site may not do well at an open site and vice versa. Calibrated models at forest sites perform better than uncalibrated models, though the benefits of calibration may not translate to subsequent years, nor to open conditions, which is expected since forests have large influences on snow dynamics and many recent models have included vegetation canopy (e.g., Niu and Yang, 2004; Bartlett *et al.*, 2006).

Chen *et al.* (2014) evaluated the snowpack simulated by six LSMs against one year of SWE data at 112 SNOw TELemetry (SNOTEL) sites in the Colorado River Headwaters region and 4-year flux tower data at two AmeriFlux sites. While all models captured the key characteristics of the seasonal SWE evolution well at all SNOTEL sites, no single model was the best capturing the combined features of the peak SWE, the timing of peak SWE, and the length of snow season. The models responded differently to different forest coverage, and deficient in the treatment of snow albedo and its cascading effects on surface energy deficit, surface temperature, stability correction, and turbulent fluxes, for all models substantially overestimated (underestimated) radiative flux (heat flux). There are also significant inter-model differences in modeling snowmelt and sublimation efficiency, for models with high snow accumulation and melt rates were able to reproduce the observed seasonal evolution of SWE. It seems the parameterization of cascading effects of snow albedo and below-canopy turbulence and radiation transfer is critical to simulate the SWE and the winter land-atmosphere interactions correctly.

An integrated approach to modeling snow accumulation and ablation processes

With the availability of geographical information systems (GIS), digital terrain elevation (DTED) data, snow products such as that of NOHRSC, re-analysis data, spatially distributed remotely sensed (RS) data to augment our limited ground-based, point observations, and an exponential growth in computing power, it will be desirable to integrate distributed, physically based snow accumulation and ablation processes with RS data, and ground measurements of snowpack to better model snowmelt processes. The idea is to progress from empiricism (e.g., degree-day method) to a discipline of applied science, and to model hydrological processes from measurable causative factors. The building and applications of distributed snow models depend largely on our ability to retrieve useful snow accumulation and snowmelt information of reasonable resolution from RS data in order to augment limited ground measurements.

Ideally, such distributed models should be developed in the direction of sub-grid parameterizations rather than the traditional quest for refining the resolution of small-scale parameterizations. This means finding a trade-off between the resolution of processes modeled, the types of data available and the information contained in the data, and the accuracy required. Otherwise, such models may be difficult to apply because of excessive data demand and the difficulty in obtaining the parameters required at all grid elements. On the other hand, some distributed snow models are based on rectangular or square grid elements of constant size without considering sub-grid parameterization, irrespective of the terrain features. This means that process descriptions may become "artificial," since nature does not behave as a system of symmetrical grids placed side by side.

There have also been developments in land surface schemes using existing Surface Vegetation Atmosphere Transfer Schemes (SVATS), such as the Land Data Assimilation Systems (LDAS) (http://ldas.gsfc.nasa.gov/) which are forced with gauged precipitation, RS data, radar precipitation, and output from numerical weather prediction models. Then *in situ* or remotely sensed measurements of LDAS storages (such as snow), water and energy fluxes will be used to validate and constrain the LDAS predictions using certain data assimilation techniques. However, one of the possible drawbacks of LDAS is the discretization of grids symmetrically without including sub-grid parameterization. Even with subgrid parameterizations in distributed models, ideally it may be more desirable to develop semi-distributed models that discretize river basins according to terrain characteristics, and that are designed to model snow, water, and energy dynamics with practical details under the forcing of fluxes and the influence of terrain and vegetation characteristics, in a framework that mimics nature as much as possible.

2.11   Recent observed changes in snowpack and snow cover

In spite of uncertainties associated with snow maps derived from passive microwave images, the detection of large scale changes to snowpack in higher latitude regions in relation to possible global warming is only possible through satellite images. Due to their effects on energy and moisture budgets, and surface temperature being highly dependent on snow cover, snow cover trends serve as key indicators of climate change (Armstrong and Brun, 2008). Observational records from satellites indicate that the annual SCA in the NH has been decreasing since 1960s due to increasing air temperature. The decline in seasonal snow cover is marked in spring and summer, but fall and winter snow cover has increased. The observed changes in the hemispheric snow cover correspond to warming and the feedback effects on the energy balance (Barry, 2009). Global-warming tends to affect winter temperatures, and in high latitudes, warmer winters may be snowier as a result of increased atmospheric moisture content.

Snowmelt in Greenland and Antarctic

About 20 years of SSM/I passive microwave data has been used to monitor the regional snowmelt of Greenland since 1988 (Tedesco, 2007; Tedesco *et al.*, 2008). Defining a melting index (MI) as melting area multiplied by melting days, 2005 was the year with the highest MI, followed by 2002, 1998, 2004, and 2007. In 2007, areas higher than 2,000 m in southern Greenland experienced about 30 more days of melting than the study period of 1988–2007, and 53% higher MI than the average, even though overall Greenland had a MI about 20% higher than the average. In contrast, in 2008 northern Greenland experienced record melting, and they attributed record melting of snow in different parts of Greenland to warmer surface temperature.

Snow cover extent

Based on the 12-month running means, the monthly SCE over NH lands (including Greenland), Eurasia and NA for about 40 years (1966–2006) mainly show negative snow cover anomalies since the late 1980s, with the exception of 1996 and 2002 for NH and Eurasia, and 1997 for Eurasia and NA (Robinson, 2008). The SCE over NA has been shown to increase during autumn and early winter (November–January), but decrease over early spring (Frei and Robinson, 1999; Dyer and Mote, 2006), which indicates an earlier onset of the spring snowmelt. Callaghan et al. (2011b) found Arctic snow cover showing different regional responses to climate warming and increasing winter precipitation that has characterized the Arctic climate in recent decades, with the largest and most rapid decreases in SWE and snow cover duration (SCD) observed over maritime regions of the Arctic with the highest precipitation amounts. Eurasian Arctic region has experienced larger declines in SCD (12.6 days), compared to the North American Arctic region (6.2 days) between 1982 and 2011 (Barichivich *et al.*, 2013). Further, the North American sector of the Arctic exhibiting decreases in SCE and snow depth when *in situ* observations became available since 1950s, while widespread decreases in SCE are only apparent over Eurasia in the Arctic after 1980s, but snow depths are increasing in many regions of Eurasia. Warming and more frequent winter thaws are contributing to changes in snowpack structure.

Even though the SCE in January 2008 of 50.1×10^6 km^2 was the largest recorded SCE for January in the last 40 years, the 2008 annual SCE over NH lands averaged 24.4×10^6 km^2, which is 1.1 million km^2 less than the 39-year average, putting 2008 as the fourth least extensive snow cover on record, and the 12-month running means ran below the long-term average throughout 2008, following the generally lower than average extents in 2007. The SCE in January has been declining consistently since the notable minimum in the late 1980s and early 1990s over Eurasia, even though in North America, the SCE anomalies have somewhat rebounded over the course of 2008 from a 2005–2007 minimum. From a snow model that generated monthly SCE of NH from 1905 to 2002, McCabe and Wolock (2010) found a substantial decrease in the March SCE of NH since 1970s attributed to an increase in the mean winter temperature and a contraction of the circumpolar vortex and a poleward movement of storm tracks, which resulted in decreased precipitation (and snow) in the low- to mid-latitudes and an increase in precipitation (and snow) in high latitudes.

Are these generally negative snow cover anomalies since the late 1980s an indication of the warming trends detected for North America, Europe, and Asia in NH? From the Mann–Kendall's trend analysis of SCE of NH over 1972–2006, Dery and Brown (2007) found significant declines in SCE during spring over North America and Eurasia, with lesser declines during winter and evidence of a poleward amplification of decreasing SCE trends during spring over Eurasia and North America, which is consistent with an enhanced snow-albedo feedback over northern latitudes that act to reinforce an initial anomaly in the cryospheric system.

In recent years, the fraction of precipitation in mid- and higher-latitude regions falling as snow has decreased, the fall accumulation has been occurring later while the spring ablation earlier, and the northern hemisphere's (NH) SCE and snow depth have decreased, which would have significant socioeconomic consequences as more than one-sixth of the world's population depend on meltwater for their water supply. Climate warming has been the dominant factor in the observed decrease in winter SCE and earlier spring ablation over North America (NA) (Dyer and Mote, 2006), given air temperature anomalies over NH's mid-latitude land areas has been shown to explain about 50% of the observed variability in SCE (Brown and Robinson, 2011).

Numerous observations have shown that the SCE of NH has been decreasing rapidly in recent decades, especially in spring, and in warmer locations where SCE are closely linked to temperature variability, for example, March–April SCE of NH is decreasing at 3.4%±1.1% per decade (1979–2005) (Brown and Robinson, 2011; Hernández-Henríquez et al., 2015). Over 1967–2012, the annual mean SCE has significantly decreased, with the largest change, −53% [−40 to −66%], occurred in June. Over 1922–2012, a much smaller 7% [4.5–9.5%] decline for March–April was found. Majority of climate stations show decreasing trends, and stations at lower elevation or higher average temperature were the most liable to show a decrease.

The satellite-based visible SCE of National Oceanic and Atmospheric Administration (NOAA) climate data record (CDR) developed from weekly Northern Hemisphere SCE data span from October 4, 1966 to the present. Satellite data incorporated into the NH SCE CDR at various starting dates include platforms of the Environmental Science Services Administration (ESSA) series, NOAA Polar-orbiting Operational Environmental Satellites (POES), NOAA Geostationary Operational Environmental Satellites (GOES), the Defense Meteorological Satellite Program (DMSP) series, the Meteosat series, Geostationary Meteorological Satellites (GMS), Aqua, Terra, and the Multi-functional Transport Satellites (MTSAT) (Estilow et al., 2015). Starting in June 1999 the CDR is completely derived using the Interactive Multisensor Snow and Ice Mapping System (IMS) involving a diverse set of products: satellite imagery, snow and ice analysis maps, National Centers for Environmental Prediction (NCEP) model data, and surface observations, and improved discrimination between snow and cloud covers. With an 89 × 89 Cartesian grid laid over an NH polar stereographic projection and a cell size of 190.6 km × 190.6 km, cell areas range from ~10,700 km^2 near the Equator to ~41,800 km^2 near the pole.

Despite of many up and down cycles in snowfall that had occurred over the NH in last five decades, Figure 2.19 of the Rutgers University Snow Lab based on this CDR data set, the NH SCE exhibits an overall increasing trend since the late 1980s, in contrast to the overall decreasing trend observed between late 1960s and 1980s. The increasing trend in the SCE of the NH observed in the last three decades are attributed to increasing SCE in the Fall and winter seasons, especially the Fall season, which together had offset the overall decreasing trend in the spring SCE of the NH shown in Figure 2.19d. Trend estimates for annual precipitation of the 1951–2008 time series of CRU TS 3.10.01 (updated from Mitchell and Jones, 2005) are 5.82 ± 2.72 and 1.13 ± 2.01, for GHCN V2 (updated through 2011; Vose et al., 1992) are 4.52 ± 2.64 and 1.39 ± 1.98, and for GPCC V6 (Becker et al., 2013) are 2.69 ± 2.54 and 1.50 ± 1.93 mm yr^{-1}, for both latitudinal bands 30° N–60° N and 60° N–90° N, respectively (Hartmann et al., 2013). Even though confidence in precipitation change is medium for the years after 1950 because of insufficient data, the non-significant positive annual precipitation trends over both latitudinal bands of the NH should have contributed to increasing trends in the Fall and Winter SCE of NH of the last five decades. However, likely because of more rainfall than snowfall in spring and earlier onset of spring snowmelt,

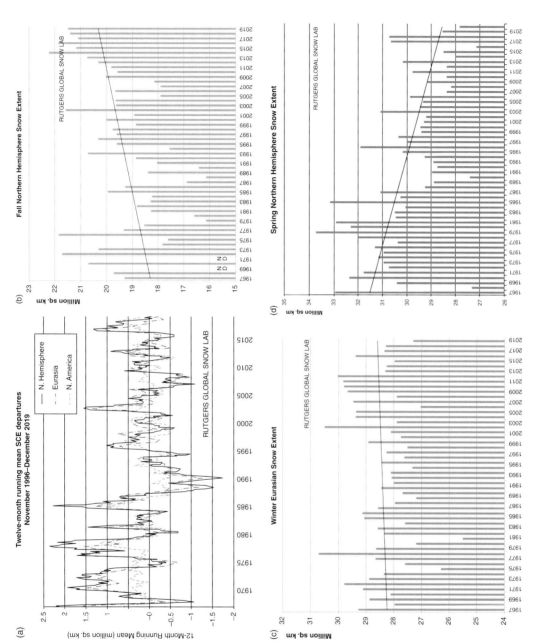

Figure 2.19 (a) Twelve-month running anomalies of monthly snow extent plotted on the seventh month using values from November 1966 to December 2018 for NH (https://climate.rutgers.edu/snowcover/chart_anom.php?ui_set=0&ui_region=nhland&ui), and seasonal snow cover extent of NH for (b) Fall, (c) Winter, and (d) Spring. The mean 1981–2010 SCE for the NH in Fall, Winter, and Spring are 18,912, 45,480, and 29,788 km², respectively

the Spring SCE of NH generally show decreasing trends despite of positive annual precipitation trends at latitudinal bands north of 30° N.

Snow water equivalent

SWE of snowpacks, the depth of liquid water produced by a snowpack after complete melting, can be estimated from *in situ* depth (d) and density (ρ_s) in kg m^{-3} measurements, SWE = d(ρ_s/ρ_w), where the density of water ρ_w is about 1,000 kg m^{-3}. So far passive microwave radiometry is the only spaceborne technique proven useful in extracting SWE data (Walker and Silis, 2002). SWE has been estimated from the brightness temperature (TB) of microwave frequency and polarization (H or V) of the TIROS-N sensor, the Scanning Multichannel Microwave Radiometer (SMMR) from the Nimbus 7 satellite with data between October 25, 1978 to August 20, 1987; the Special Sensor Microwave/Imager (SSM/I) data from the Defense Meteorological Satellite Program (DMSP) satellite with data available since September 7, 1987 (Armstrong *et al.*, 2006), and succeeded by the Special Sensor Microwave Imager/Sounder (SSMIS) in 2005, and the Advanced Microwave Scanning Radiometer-Earth Observing System (AMSR-E) data from the Aqua satellite with data available between 2002 and 2011 or the Advanced Microwave Scanning Radiometer 2 (AMSR2) on the JAXA GCOM W1 spacecraft, launched May 18, 2012 and is still operating. Limitations to such SWE data are a coarse resolution (typically 25 km), underestimated in forests which mask the snow cover and vegetation suppresses the scattering signal, as does liquid water (wet snow). From retrievals of SMMR-SWE over Canada, Dong *et al.* (2005) concluded that for practical applications, the uncertainty of such SWE data is acceptable for regions more than 200 km away from large open water bodies, daytime air temperatures lower than 2 °C, and SWE values are above 100 mm.

Mote et al. (2005) detected widespread declines in modeled April 1 SWE over the Western United States and Canada between 1950 and 1997. Gan *et al.* (2013) detected significant decreasing trends in SWE in NA over 1979–2007, which are more extensive in Canada than in the United States where such decreasing trends are mainly found along the Rocky Mountains. However, because of uncertainties associated with SWE data retrieved from passive microwave data, results obtained for mountainous or forested areas of NA, the tundra in Arctic Canada with many frozen lakes or snow packs that consist of depth hoar and wind slab, should be treated with caution. To assess the possible impact of climatic change to the snowpack of North America, Gan *et al.* (2013) analyzed the trends of temperature and precipitation data and then the SWE-air temperature and SWE-precipitation relationships. Using the gridded, 2-m air temperature data of the North American Regional Reanalysis (NARR) (Mesinger *et al.*, 2006) and that of the University of Delaware (Willmott and Robeson, 1995), they detected statistically significant temperature trends for 1979–2007 (January–April) mainly in the southern states of USA. They also detected significant decreasing precipitation trends from the University of Delaware and the NARR data but there are limited agreements between the locations of significant decreasing precipitation trends and significant decreasing SWE trends. However, extensive areas of negative correlations between SWE and temperature exist both across the United States and Canada, and the distribution of these areas of negative correlation closely follows the areas of the decreasing trends detected from the SWE data, which for Canada is mainly east of the Canadian Rocky Mountains, while for the United States is mainly on the American Rocky. Apparently, extensive decreasing trends in SWE data of passive microwave detected in Canada and parts of USA are caused more by increasing temperatures than by decreasing precipitation. Higher air temperature means more rainfall and less snowfall especially in areas and seasons where the average air temperature is close to 0 °C, earlier onset of spring snowmelt, and generally less snowpack.

From SWE estimated from satellite measurements for October to April over 1979–2007 in NA, Gan et al. (2013) found that about 25–30% of the pixels covered with snow showed statistically significant negative trends, compared with 5–10% that showed increasing trends. The overall mean trend magnitudes estimated are about -0.4 to -0.5 mm a^{-1} which means the overall SWE of NA had decreased by about 10–13 mm or possibly more, or in terms of snow depth could range from about 4 to 13 cm (depending on the snow density) over 1979–2007. The detected changes should have significant impacts on the spring snowmelt of, say, the Canadian Prairies and the Washington Cascades. Using a combination of satellite data, in situ data and land surface model simulations, Park et al. (2012) found negative trends in snow depth more pronounced in North America than in Eurasia between 1948 and 2006. Dyer and Mote (2006) found significant (>1 cm yr^{-1}) negative regional snow depth trends in central Canada along a southeast line from the Yukon Territory to the Great Lakes using in situ snow depth data for 1960–2000 for North America.

Global Cryosphere Watch (2015) has compiled a comprehensive snow data set from passive-microwave, in situ, and analysis/reanalysis products for non-alpine regions of the NH. The data set at 25 km resolution is available at daily, weekly and monthly time steps since 1979. By combining different data sources, GlobSnow is probably better than SWE data derived from stand-alone remote sensing data sets (Frei et al., 2012).

Based on the non-parametric Mann–Kendall test at 0.05 significant levels, more snow-covered pixels of the monthly GlobSnow SWE data set of NH for November–April (1988–2017) show decreasing than increasing trends (Figure 2.20). From the total number of snow-covered pixels analyzed, up to 15.5% (7.7%) of the pixels show statistically significant decreasing (increasing) trends. December has the largest SCE and the greatest percentage of statistically significant decreasing trends, of which the majority are located north of 55° latitude. April exhibits the greatest percentage of statistically significant positive trends and most of these pixels are located in Asia.

Figure 2.21 shows spatial distributions of grids of the Northern Hemisphere with significant SWE (5-day averages) trends at $p < 0.05$ significant level over 1988–2017 using the Mann–Kendall test. There are more statistically significant negative than positive SWE trends especially across Canada, the high Arctic, and Europe, with scattered positive trends in Russia.

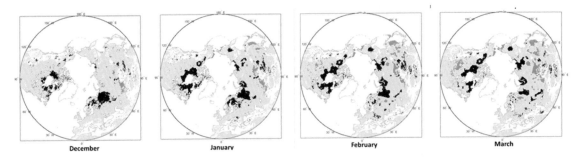

Figure 2.20 Northern Hemisphere monthly snow cover for December–March of GlobSnow SWE data set of 1988–2017 that exhibit increasing (red) and decreasing (black) trends statistically significant at the 0.05 level using Mann–Kendall test

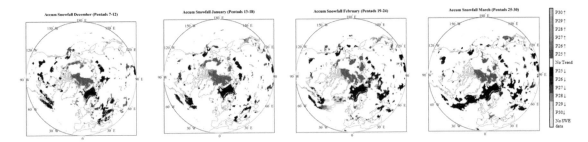

Figure 2.21 Spatial distributions of grids of the Northern Hemisphere GlobSnow data set with significant SWE (5-day averages) trends at $p < 0.05$ significant level over December–March of 1988–2017 using the Mann–Kendall test, with more statistically significant negative than positive SWE trends especially across Canada, the high Arctic, and Europe, while scatted positive trends in Russia (A black and white version of this figure will appear in some formats. For the color version, please refer to the plate section.)

As expected, based on the Mann–Kendall's test applied to the ERA-Interim reanalysis data, Figure 2.22 shows predominantly warming trends over the Northern Hemisphere in the last 30 years (1988–2017). On the other hand, some scattered cooling trends are still detected in the northern Pacific Ocean in January and March, over some eastern parts of Siberia in February, and a small part of the north Atlantic Ocean in March.

The Spearman's rank correlation between SWE and air temperature across the Northern Hemisphere is predominantly negative, which is expected because under the effect of climate warming, there should be an overall less snowpack partly attributed to more rainfall than snowfall, if the total precipitation will remain unchanged or even higher (Figure 2.23).

The Self Organizing Map (SOM) (Kohonen, 2001) is an artificial neural network developed to uncover an underlying structure in a data set through an unsupervised learning to produce a 2-D array of nodes, called a map, that are organized in such a way that similar items are placed close to each other. By applying the SOM and K-means clustering analysis,

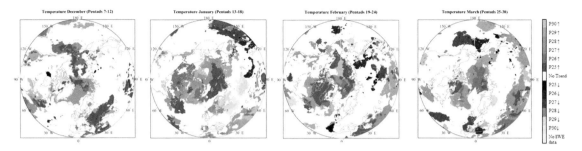

Figure 2.22 The spatial distribution of grids with statistically significant ($p < 0.05$) temperature Pentad (5-day averages) trends by the Mann–Kendall test in the Northern Hemisphere based on the 1988–2017 ERA-Interim reanalysis temperature data (A black and white version of this figure will appear in some formats. For the color version, please refer to the plate section.)

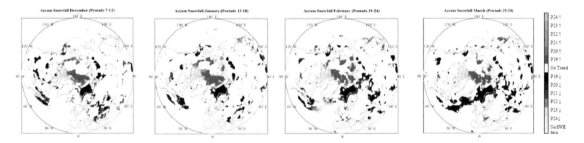

Figure 2.23 Spatial distributions of grids of the Northern Hemisphere with significant Spearman's rank correlation between SWE and air temperature over 1988–2017 at $p < 0.05$ significant level (A black and white version of this figure will appear in some formats. For the color version, please refer to the plate section.)

Northern Hemisphere grids were grouped into 20 different clusters based on the SWE niveograph of November–March for 1988–2017, for example, one grid can be assigned to a different cluster in different years. Figure 2.24 shows that grids in snow-dominated areas of Canada, Siberia, and northern Europe are generally assigned to more clusters (up to 20) than other regions of NH that also experience snowfall.

Figure 2.25 shows some limited statistically significant Spearman's rank correlation between SWE of the Northern Hemisphere and large-scale climate anomalies such as AO, PNA, NAO, and PDO, respectively. There are more negative (black) than positive (red) correlation especially between PNA and SWE, which means less snowpack during the warm than the cold phase of PNA. For example, in western Canada, Gan *et al.* (2007) show that El Niño (La Niña) leads to a 14% decrease (20% increase) in the mean winter precipitation, strong positive (negative) PNA leads to a 12% decrease (9% increase) in mean winter precipitation, and PDO is associated with an 8% decrease (9% increase) in mean winter precipitation. The detected teleconnections could occur at interannual or interdecadal levels depending on the climate anomaly, and their strength changes in time and space.

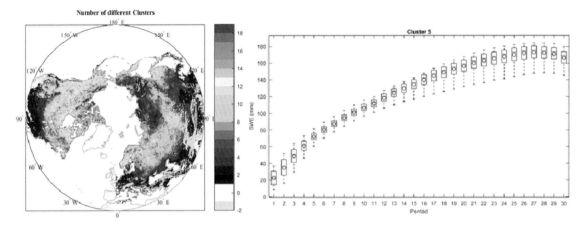

Figure 2.24 (a) Number of different clusters derived from the SOM and K-means clustering analysis that each grid in Northern Hemisphere is assigned over 1988–2017, and an example of (b) the SWE in boxplots for all grids of NH assigned to Cluster #5 (A black and white version of this figure will appear in some formats. For the color version, please refer to the plate section.)

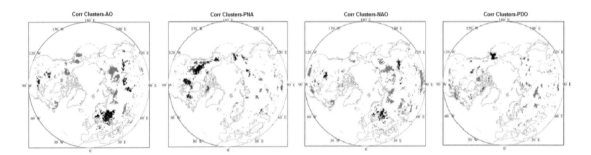

Figure 2.25 The spatial distribution of grids with significant ($p < 0.05$) positive (red) and negative (black) Spearman's rank correlation between cluster areas, AO, PNA, NAO, and PDO, respectively (A black and white version of this figure will appear in some formats. For the color version, please refer to the plate section.)

Climate indices significantly correlated with any of the first four PCs of the SWE of NH at 0.05 significance level are shown in Table 2.4. Apparently AO, NAO and PNA exert a stronger influence on the snowpack of the NH between January and April. PC1 is more strongly correlated with AO and NAO, while PDO and PNA are more correlated with PC2 to PC4 in some months. Therefore, some of the observed changes in SWE are partly attributed to the influence of large-scale climate anomalies but climate warming has undoubtedly played a bigger role, especially in the detected negative trends of SWE of NH.

Wu *et al.* (2018) found slower snowmelt rates over the entire NH over 1980–2017, but with higher ablation rates in locations with large SWE, because of the reduction of SWE in deep snowpack regions. However, moderate and high snow ablation rates show a decreasing trend. Kapnick and Hall (2012) detected significant losses of spring mountain snowpack at

Table 2.4   Climate indices statistically significantly correlated to PC1 to PC4 of November–April 1979–2014 SWE of Northern Hemisphere

	Nov	Dec	Jan	Feb	Mar	Apr
PC1			AO, NAO, PNA	AO, NAO	AO, NAO	AO
PC2			PDO	PDO	PNA	PNA
PC3	PDO		AO	PNA	PDO	
PC4			PNA		SOI	PDO

Note: AO = Arctic Oscillation, NAO = North Atlantic Oscillation, PDO = Pacific Decadal Oscillation, SOI = Southern Oscillation Index, PNA = Pacific North American Index

western United States over past several decades. For Canada, there has been extensive decreasing snow depths and snow cover duration and extent since the mid-1970s, with the largest declines in western Canada and proportionally greater changes later in winter and spring (DeBeer *et al.*, 2016). Berghuijs *et al.* (2014) show that in conterminous USA, catchments with a higher fraction of precipitation falling as snow tends to have higher mean streamflow, which is likely to decrease in catchments that experience significant reductions in the fraction of precipitation falling as snow because of a warmer climate.

Most studies also show negative trends in snow depth and snow duration over past decades in the mountain cryosphere of Europe (Beniston *et al.*, 2018), with less pronounced changes at high elevations (Terzago *et al.*, 2013). Spring SWE shows a decreasing trend in Alps (Marty *et al.*, 2017). However, there were positive trends of maximum snow depth and SWE in higher and colder parts of the Fennoscandian Mountains although it turns out to be negative trends in recent years (Kivinen and Rasmus, 2015). Matti *et al.* (2017) show that the flood seasonality for snowmelt-dominated Scandinavian catchments have changed over the twentieth century with statistically significant decreasing (increasing) trends in summer (winter and spring) maximum and mean daily flows in some catchments. Changes in annual flood occurrences generally point toward a shift from snowmelt-dominated to rainfall-dominated flow regimes in some regions, with flood peaks due to spring snowmelt consistently occurring earlier, often at the expense of the much needed summer streamflow for growing crops (Kerkhoven and Gan, 2011).

There are other cryospheric observations also support the effect of a warmer climate in the recent past. For example, the number of snow days in Switzerland decreased abruptly at the end of the 1980s. Marty (2008) found that records at 34 long-term stations between 200 m and 1,800 m asl for 1948–2007 show an unprecedented series of low winter snow in the last 20 years. The abrupt change in 1988 resulted in a loss of 20–60% of the total number of snow days with no clear trend since then. The decrease is shown to be correlated with an increase in winter temperatures.

SCE over North America has been shown to increase during autumn and early winter (November-January), although it also decreased in early spring (Frei and Robinson, 1999;

Dyer and Mote, 2006), which indicates an earlier onset of the spring snowmelt. Some parts of the Canadian Arctic stretching from western Hudson's Bay to the North Slope of Alaska shows a persistent pattern of high SWE (Armstrong and Brodzik, 2002; Andreadi and Lettenmaier, 2006). However, as noted earlier, SWE data derived from passive microwave for the tundra in Arctic Canada should be treated with caution (Derksen *et al.*, 2009). Less snowfall has been observed in the lower Missouri River Basin (Berger et al., 2002) and in New England (Huntington *et al.*, 2004).

Mote *et al.* (2005) found evidence of declining snowpack in the western United States, except in the southern Sierra Nevada. Bedford and Douglass (2008) analyzed daily SWE from 28 SNOTEL stations in the Great Salt Lake Basin for 1982–2007. They found an advance of about 15 days in the date of peak SWE as well as (less robust) evidence of a decrease in peak and April 1 SWE amounts.

Through three spring indicators – lilacs, honeysuckles, and streamflow – Cayan *et al.* (2001) found earlier onset of the spring season by up to three weeks in the western North America since the 1970s. By simulating the snow energy balance to climatic changes projected by nine regional climate models to the end of the twenty-first century in the Pyrenees, Moreno et al. (2008) concluded that the most significant changes to future snowpack processes are related to temperature. Tedesco (2007) concluded that record melting of snow in southern Greenland in 2007 was caused by higher surface temperature. The above documented observations about the cryosphere agree with the enhanced increasing trends of surface air temperature anomaly changes over other continents, especially enhanced temperature trends occurred in the last several decades since 1960s reported by IPCC (2013) (see Figure 2.26 adapted from Jones *et al.*, 2013). In Figure 2.26, multi-model means are shown as thick lines, and 5–95% ranges shown as thin light lines, and as black lines for HadCRUT4. Mean temperatures are shown for Antarctica and six continental regions. Temperatures are shown with respect to 1880–1919 for all regions apart from Antarctica where temperatures are shown with respect to 1950–2010.

REVIEW QUESTIONS

(1) Name three atmospheric conditions necessary for the process of snow crystals forming in the atmosphere to the occurrence of snowfall on land.

(2) Explain the basic differences in the formation of snow crystals through heterogeneous and homogeneous nucleation. Which of the two is the primary snow formation process?

(3) List out five types of snowflakes under different combinations of temperature and supersaturation conditions with respect of ice. Do you believe that no two snowflakes are exactly alike?

(4) What are the primary factors that affect the snow cover distribution at continental scale, macro or regional scale, mesoscale scale, and micro scale?

(5) As one of two major hydrologic influences of wind transport of snow, blowing snow and sublimation, show that the amount of energy (latent heat) required for snow to sublimate is about 7.5 times that for snow to melt.

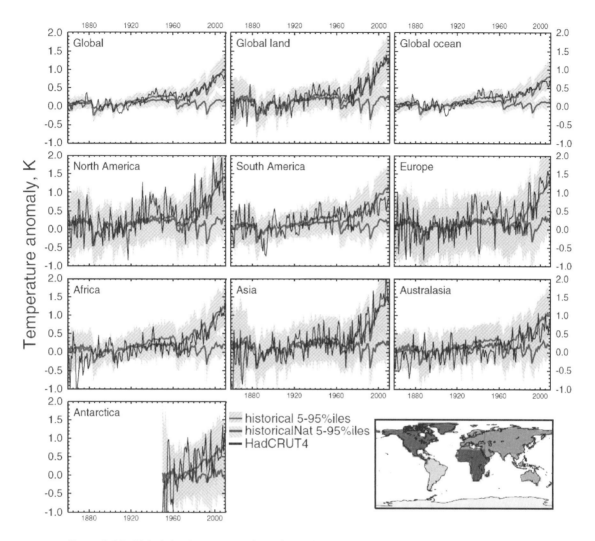

Figure 2.26 Global, land, ocean, and continental annual mean temperatures anomaly with respect to 1880–1919 for CMIP3 and CMIP5 historical (red) and historicalNat (blue) simulations, and Hadley Centre/Climatic Research Unit gridded surface temperature data set 4 (HadCRUT4, black). Weighted model means shown as thick dark lines and 5–95% ranges shown as shaded areas (taken from Jones *et al.*, 2013)

(6) Describe how the property of a snowpack will change as the winter progresses, for example, how a snowpack metamorphoses over time, in terms of, say, snow depth, snow density, snow grain shape and size, and albedo?

(7) Estimate the maximum snow load of the canopy of pine trees in $kg\,m^{-2}$, if the tree species coefficient S_p for pine is 6.6, the LAI of pine is 2, and the density of snow is $100\,kg\,m^{-3}$. What will likely happen to snow intercepted by canopy?

(8) What are the problems of *in situ* measurements of snow, such as using a snow sample, or a snow gauge? What devices have been used to correct under-catch problems of snow gauges under windy conditions, and generally how effective are such devices?

(9) What are the advantages of remotely sensed snowpack data over *in situ* ground measurements of snowpacks? Briefly describe the properties of optical sensors of NOAA-AVHRR and MODIS satellites to estimate the spatial distribution of snow cover extent (SCE), and passive microwave sensors of SMMR, SSM/I and AMSR-E to estimate snow water equivalent (SWE) data at regional to continental scales? What are the limitations or possible errors of snowpack information retrieved from satellite data?

(10) Compare the advantages and disadvantages of snowmelt modeling between the simple, empirical, temperature index or degree-day methods and the more physically-based, energy balance methods. Which of the two approaches will you prefer and why?

(11) Estimate the daily snowmelt in mm per day of a snowpack located in a boreal forest environment of North America using the degree-day method, if the mean daily temperature T_a is 7 °C, the threshold temperature T_{thm} is 3.5 °C, and the recommended melt factor m_f is 1.83 mm per °C per day (Equation 2.22)?

(12) A homogeneous, 60-cm-deep (*D*) snowpack with a density $\rho_p = 250$ kg cm^{-3} is in thermal equilibrium at a uniform temperature T_p of −5 °C, and the snowpack surface temperature $T_s = T_p$. What is the cold content (Q_{cc}) of the snowpack in J m^{-2} (Equation 2.37)? Note that C_s, the specific heat of snowpack, is 2,093 J kg^{-1}°C^{-1}.

(13) Suppose the incoming solar radiation is 218 W m^{-2} while the surface albedo of the 60-cm-deep snowpack α_{sn} is 0.75, the fractional snow cover area A_{sn} is 0.8 (Equation 2.28), and the ground albedo α_g is 0.20, air temperature T_a is 8 °C, relative humidity of air is 50%, and wind velocity (*U*) is 9 km hr^{-1}. Rain, assuming with a temperature the same as T_a, falls on the snowpack at a constant rate of 1 cm hr^{-1}. Ignore the cloud cover effect and assume an atmospheric emissivity of 0.76, compute the advective heat (Q_v) of the rainfall event (Equation 2.36) occurring over the snowpack, and the time to overcome the Q_{cc} before melting begins? Note that melting begins as soon as Q_{cc} is overcome.

(14) Next, find the thermal quality θ of the snowpack (Equation 2.40) before and after Q_{cc} is overcome, given that the latent heat of fusion of pure ice is 333,000 J kg^{-1}, while the latent heat of fusion of this snowpack, l_{ms}, is 316,350 J kg^{-1}. What is the liquid water content (%) of this snowpack (Equation 2.7)?

(15) Estimate the time to melt the snowpack completely **after** overcoming Q_{cc}, assuming the energy available for melting, Q_o (Equation 2.27) remains unchanged throughout the period of melting. Only consider Q_n, Q_h, and Q_e in computing Q_o (ignore Q_g and Q_v, rain has stopped) (Equations 2.30–2.35 and Equation 2.39). Note that after overcoming Q_{cc}, the snowpack temperature T_s will be 0 °C, not −2 °C. P_a is the atmospheric pressure (≈1,013 mb). The bulk transfer coefficients for sensible heat transfer, $D_h = 2$ J m^{-3}°C^{-1}, and for latent heat transfer, $D_e = 0.1$ J m^{-3} Pa^{-1}.

2 B   Avalanches

2.12 History

The word avalanche is derived from the French "avaler" (to swallow). An avalanche involves the rapid flow of a mass of sow down a slope, triggered by either natural processes or human activity. Avalanches have long been feared in Alpine countries. On March 1, 1910, on the Great Northern Railway line thorough the Cascade Range at Stevens Pass, WA, northeast of Seattle, 96 passengers and crew were killed by a massive avalanche that struck a stationary train. Three days later in Rogers Pass, British Columbia, an avalanche running from the opposite slope killed 57 workmen, who were clearing a previous slide from the rail lines. During World War I some 50,000 troops were killed by avalanches in the Italian Alps that were triggered by artillery fire.

Avalanche research began with the establishment of the Eidgenössische Instituts für Schnee- und Lawinenforschungs (EISLF) in Davos, Switzerland, in 1936. This led to the construction of the Weissflujoch Research Station at 2,680 m in 1943.

Concurrently, Seligman (1936) in England published the classic introduction to the scientific study of snow and avalanches, followed closely by Paulcke (1938) who summarized a decade of work in which he recognized the occurrence of depth hoar and the role of snow types in avalanche formation. Bader *et al*. (1939) laid the foundations of snow mechanics and understanding of avalanche formation.

McClung (1981) developed the first model of dry slab avalanche release based on fracture mechanics. The first field measurements of snow stability in a spatial context were made by Conway and Abrahamson (1984) who analyzed shear strength along the fracture lines of slab avalanches shortly after triggering, and on slopes that had not failed. This work led to questioning the validity of point tests of stability and studies of the spatial variability of various snowpack properties at the slope scale.

Avalanche and weather condition information for states in the USA as well as for Austria, Canada, France, Finland, Germany, Italy, New Zealand, Norway, Scotland, Slovenia, and Spain can be accessed via www.avalanche-center.org.

For France (the Alps, Pyrenees, and Savoie), there are detailed records of avalanche events (magnitude and extent) at individual sites (with site maps) for all regions of avalanche occurrence, and weather conditions for the preceding 3 days and 4 hours prior to the event (Jamard et al., 2002); these data from the Enquête Permanente sur les Avalanches (EPA) are available on line (www.avalanches.fr). Records began in 1891 in Savoie, 1905 in the Alps, and 1965 in the Pyrenees.

In the former Soviet Union during the 1970–1980s there was widespread mapping of avalanches along highways and railways, and at mountain resorts and centers of mining, together with studies of the physical mechanical properties of snow, based at the Central Asian Hydrometeorological Research Institute (SANIGMI) in Tashkent, Uzbekistan. Avalanche observations were collected throughout the USSR (Avalanche Cadasters, 1984–1991) and forecasting methods for different avalanche hazard regions were developed (Kanaev et al., 1987; Moskalev, 1997). After the breakup of the USSR, avalanche observations in Uzbekistan were assembled in a data bank (Batirov et al., 2003) and analyzed with GIS tools for avalanche hazard mapping and risk assessment (Semakova et al., 2009).

In China, avalanche research began in the 1960s at the Lanzhou Institute of Glaciology and Cryopedology with work in the Tian Shan. Areas of seasonal and perennial avalanches are mapped by Zeng et al. (2008). They are mostly around the margins of the Tibet Plateau, in the western mountains, and the northeast of the country, while high hazard regions are identified in the western Tian Shan, the central Himalayas, northern Yunnan, and western Sichuan centered on Gongga Mountain (29.5° N, 101.9° E) in the Hengduan Mountains. The lowest altitude of seasonal avalanche release ranges from 1,700 m in the western Tian Shan to 3,700 m on Gongga Mountain, and 1,500 m on Chola Mountain (31.8° N, 99.1° E).

2.13 Avalanche characteristics

Avalanches share certain common elements: a trigger that initiates the avalanche, a starting zone from where the slide originates, a slide path along which the avalanche flows, a run-out zone where the avalanche comes to rest, and a debris deposit which is the accumulated mass of avalanched snow and associated debris once it has come to rest. Avalanches tend to run in the same paths year after year.

The morphological characteristics used to classify avalanches include the type of snow, the nature of the failure, the sliding surface, the propagation mechanism of the failure, the slope angle, direction, and elevation. Avalanche size, mass, and destructive potential are rated on logarithmic scales, typically with four to seven categories; for example, Canada and the United States recently agreed a new danger scale from one (low) to five (extreme) (see Table 11.1).

Avalanches range in size from sluffs with a volume of <10 m^3 to extreme releases of 10^7–10^8 m^3; corresponding impact pressures range from <10^3–10^6 Pa. There are two main categories of avalanche – loose snow avalanches and slab avalanches. Commonly, they begin with the failure of snow layers with densities less than 300 kg m^{-3} and they continue until the kinetic angle of repose is reached at an angle of about 17° (Perla, 1980; McClung and Schaerer, 2006).

An avalanche path comprises a starting zone, the track, and a runout–deposition zone. Loose snow avalanches occur in freshly fallen snow that has low density with little internal cohesion among individual snow crystals, and they are most common on steep terrain. Loose snow avalanches are initiated when the angle of repose, ~45°, is exceeded. The angle increases as the temperature rises due to increased cohesion. In fresh, loose snow the release is usually at a point and the avalanche then gradually widens down slope as more snow is entrained. Observations show that snow entrainment is determined by the along-track availability of snow mass as well as by snowcover structure, while topographic features and flow variables are of lesser importance (Bianchi Janetti et al., 2008). Downslope propagation typically continues onto level terrain; the kinetic angle of repose is about 17°.

Slab avalanches occur when a cohesive slab is released over an extensive plane of weakness on slopes of 35–40°. Slab thicknesses are 0.1–4 m and have a mean density of ~200 kg m^{-3}. A slab avalanche originates in snow with sufficient internal cohesion to enable a snow layer, or layers, to behave mechanically as a single entity. A slab ~1–2 m thick breaks free via brittle failure along a characteristic fracture line that may span an entire slope. Such avalanches occur when a snowpack has a weak layer below a slab of cohesive snow. When snow has fallen the crystals undergo a sintering process (particles adhere to one another) forming bonds within the snowpack that cause the snow particles to become rounder. Vapor is transferred from grains to bonds and necks that connect them, from smaller to larger grains, and from convexities to concavities due to differential vapor pressure gradients (Colbeck, 1983). Consolidated snow is less likely to sluff than either loose powdery layers or wet isothermal snow. Low air temperatures at the snow surface produce a temperature gradient in the snow, because the ground temperature at the base of the snowpack is typically close to freezing. When a temperature gradient of >10 °C per m is sustained for more than a day, *depth hoar* (large faceted grains up to 10 mm diameter) will form in the snowpack, through the transport of moisture upward, away from the depth hoar, along the temperature gradient. This process is known as temperature gradient (TG) metamorphism. Depth hoar makes a very weak layer in the snowpack that is highly susceptible to failure when loaded. Strong temperature gradients can also form in near-surface layers through a variety of mechanisms, resulting in the rapid formation of weak faceted crystals that are significant weak layers once buried (Birkeland, 1998).

The variables of interest for forecasting are velocity, run-out distance and impact pressure. According to Perla (1980), maximum velocities range from 20 to 30 m s^{-1} for path lengths up to 500 m and slope angles of 25–35°. The mean run-out length on 67 Colorado avalanche paths was 380 m (Bovis and Mears, 1976). Occasionally, the run-out may cross a valley floor and continue up the facing slope. Impact pressures are a maximum at 1–2 m above the surface and range in value from about $1–10 \times 10^5$ N m^{-2} (Perla, 1980).

Mears (1976) notes that in *hard slab avalanches* the snow is bonded strongly together. Just after release, the majority of the avalanche mass is composed of relatively large snow blocks (~10–100 cm in length) that slide, roll, and collide with one another. Because of their large size and high free fall velocity they never become suspended due to turbulence much above ground level. Instead, the mass moves as a cascade of discrete blocks of snow. In the case of *soft slab* release, disintegration of the slab into increasingly smaller fragments and air entrainment of fine particles occur rapidly and much of the mass becomes airborne well above ground level. As a result of the increased mean distance between snow particles in the snow/air suspension, flow height and velocity increase, and the avalanche behaves like a fluid; sometimes the airborne part separates from the bulk of the avalanche and travels further as a powder snow avalanche with velocities >60 ms^{-1} and depths of 20–100 m.

A third type is a wet snow or isothermal avalanche, which occurs when the snowpack becomes saturated by water. These tend to start and spread out from a point. When the percentage of water is very high they are known as slush flows and they can move on very shallow slopes (<10°). Wet slab avalanches may also occur where the slab is moist and water moves along a weak interface that eventually fractures.

The length/thickness ratio of a slab avalanche is ~100 according to McClung (2009) and the width and length are closely comparable. Approximate estimates of avalanche mass – which is related to destructive potential – for average characteristic dimensions based on slab depth D (the only length possibly known prior to avalanching) are given by McClung (2009). These are shown in Table 2.5 for five different size classes. The depth guidelines are found to be close to, but somewhat below, the medians of depths estimated by guides in the field. Typically, the median values from the guides' estimates are within 20 cm of the upper limits of the guidelines.

Slab fracture surfaces are designated as follows: the upper fracture surface is called the crown surface, the sides are the flank surfaces, and the forward part of the slab is termed the stauch wall. Snow slabs typically fracture in slope-parallel shear at a weak layer in the snowpack according to Johnson *et al.* (2004) (see Figure 2.28). However, Heierli *et al.* (2008) note that the critical crack length for shear crack propagation along such layers should increase as the slope decreases, whereas experiments show that the critical length of artificially introduced cracks remains constant with decreasing slope. This results from volumetric collapse of the weak layer, leading to the formation and propagation of mixed-mode anticracks (see Glossary), which are driven

Table 2.5 Avalanche size classification system from Canada and the United States with depth guidelines for different sizes of avalanche based on mass (from McClung, 2009)

Size description	Typical mass (t)	Typical path length (m)	Typical impact pressure (kPa)	Typical slab depth D (m)
1	<10	10	1	0.15–0.20
2	100	100	10	0.30–0.40
3	1,000	1,000	100	0.60–0.80
4	10,000	2,000	500	1.3–1.7 [*]
5	100,000	3,000	1,000	2.6–3.5[*]

1 Relatively harmless to people
2 Could bury, injure, or kill a person
3 Could bury a car or destroy a small building or a few trees
4 Could destroy a railway car, large truck, several buildings, or a forest with area up to 4 ha
5 Largest snow avalanches known could destroy a village or forest of 40 ha
* Sizes with mass expected to be affected by entrainment.

Figure 2.27 Conceptual model of dry snow slab avalanche release (from Schweizer *et al.*, 2003)

simultaneously by slope-parallel and slope-normal components of gravity. The critical crack lengths are a few centimeters for slab thicknesses of 1–2 m. The model indicates that there is no threshold in slope angle for the tendency of a snow slope to nucleate and propagate cracks.

Weak layers originate most commonly as near-surface faceting of the snow, or through the growth of surface hoar, that becomes buried. Faceting processes above crusts and wet layers are the most efficient way to develop weak layers. Rapid loading produces instability while gradual loading promotes increased strength through a pressure-sintering process.

The ductile to brittle fracture transition occurs at a strain rate of about 10^{-4}–10^{-3} s^{-1} (Schweizer, 1998). In ductile fracture, extensive plastic deformation takes place before fracture, while in brittle fracture this does not occur. In either case, at the slab thickness scale, shear is essential for fracture propagation in a layered sloping snowpack according to McClung (1981). McClung (1987) suggested that fracture follows the initiation of a shear band at a stress concentration in the weak layer. When a critical downslope length is reached, the shear band propagates rapidly. This length is estimated to be between 0.1 and 10 m, with 0.1–1 m for fast loading (Schweizer et al., 2003). The model of Heierli et al. (2008) discussed earlier, however, does not require the presence of shear. The ductile to brittle fracture transition occurs at a strain rate of about 10^{-4}–10^{-3} s^{-1} (Schweizer, 1998). (In ductile fracture, extensive plastic deformation takes place before fracture, while in brittle fracture this does not occur.)

According to Schweizer et al. (2003) the failure of a snow slope needs to be considered from a fracture mechanics view focusing on three critical variables: stress, flaw size, and toughness. Snow fracture toughness in tension ranges from 0.1 to 1.5 kPa m$^{0.5}$ depending on snow density. Bazant et al. (2003) related the avalanche release process to fracture toughness. They found that fracture toughness in shear is approximately proportional to snow thickness to the power of 1.8. External stress acting on the snowpack is required to trigger avalanches. External stresses include: additional snowfall or rainfall, wind loading, radiative and convective heating, rock fall, corniche collapse, icefall, and other sudden impacts. The effect of a rise in temperature on snow stability is complex. There are immediate decreases in stability due to decreased hardness and increased toughness of the slab, on the one hand, and a delayed increase in stability due to an increase in bond formation and a temperature gradient decrease in the snow that leads to increased strength, on the other (Schweizer et al., 2003) Wind transport of snow generates differential loading on slopes. Top loading occurs when wind deposits snow perpendicular to the fall-line on a slope while cross loading occurs when wind deposits snow parallel to the fall-line. Lee mountain slopes commonly experience top-loading. Human triggers include skiers, snowboarders, snowmobiles, and controlled explosives. McClung (2002) notes that the proximate cause of most dry slab avalanches is overloading while that of most wet snow avalanches is internal changes in snow properties. Sturm and Benson (2004) point out that external agents causing snowpack variability during deposition are precipitation, sublimation and wind, and after deposition are primarily radiation, temperature and wind. Following deposition, the major internal driver is snow metamorphism. The interaction of these drivers with topography and vegetation cover is the primary source of spatial variability in the snow cover. In southwest Montana, Birkeland (2001) showed that stability is only weakly linked to terrain, snowpack, and snow-strength variables on a day preceded by consistently stormy weather conditions, whereas a day preceded by more variable weather had a stronger relationship between stability and the other variables. On both days stability decreased on high-elevation, north-facing slopes. Schweizer *et al.* (2008) address the spatial variation of layer properties such as thickness, density, grain size and strength in both the slope-perpendicular and the lateral directions. These properties are used to define layer boundaries for study of slope instability. At the slope scale, critical weak layers are often spatially continuous. Slope scale studies have measured shear strength, a derived stability index, penetration resistance (using the snow micro-penetrometer (SMP) with mm resolution), and the Rutschblock score (see Glossary), and reported a wide range of coefficients of variation, that relate to the ratio of grain size to density (a texture index). Sub-slope variability may occur due to the presence of rocks, where there is often a weaker snowpack than in the surroundings. Snow stability on slopes is controlled by the mean slope stability, by the distribution of stability over the slope, and by the scale of patterns of strong and weak areas on the slope (Schweizer *et al.*, 2003). Sturm and Benson (2004) show that the heterogeneity of snow stratigraphy increases up to a scale length of about 100 m, after which it remains relatively constant through scales two orders of magnitude greater. At the regional scale, a study in the Columbia Mountains of British Columbia shows that weak layers were consistently present in certain aspects and elevations over hundreds of kilometers across an entire mountain range (Hägeli and McClung, 2003).

Once an avalanche is in motion, the weight of snow falling downslope is counteracted by several interacting components. These are: friction between the avalanche and the ground surface; fluid-dynamic drag at the leading edge of the avalanche; shear resistance between the avalanche and the air through which it is passing; and shear resistance between the fragments within the avalanche itself.

Armstrong and Armstrong (1987) analyze avalanche occurrence in three climatic zones in the western United States (see Figure 2.28). They find that the weather and snow conditions of the continental zone have a greater potential to develop and maintain a snow cover with low bulk strength. This is because depth hoar forms more readily where there are stronger temperature gradients due to low air temperatures and thin snowpacks. It is hypothesized that this weaker snow structure will

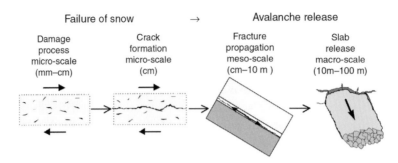

Figure 2.28 Conceptual model of dry snow slab avalanche release (from Schweizer et al., 2003) [Courtesy of American Geophysical Union].

contribute to a lower overall snow stability for longer periods of time following storm periods compared to the maritime zone. The intermountain zone is intermediate in its characteristics between the continental and maritime zones. Mock and Birkeland (2000) expanded on this work, showing the variability within the three zones, as well as how the zones change from year to year, depending on the winter temperature and snowfall distribution.

In Switzerland, based on 50 years of data on snow profiles below 2,500 m on Weissflujoch, 65% of avalanches occurred on lee north- and east-facing slopes as a result of wind loading by snow drifting (DeQuervain and Meister, 1987). Twenty percent of avalanches were of the wet snow type and twenty-six percent were unrelated to meteorological conditions. Avalanche frequency is found to be high on treeless slopes, with highly concave cross-slope curvature in combination with a mean slope angle >36° (Schweizer *et al.*, 2003). In the Davos region of Switzerland, avalanche risk is high with 3-day new snow totals of ~40–60 cm and very high with ≥100 cm.

The yield of avalanches was measured at 45 avalanche paths in Rogers Pass, British Columbia over a period of 19 years. Schaerer (1988) shows that the percentage of snow removed annually by avalanches (the yield ratio) had a mean value of 11% and a 30-year maximum of 31%. The figures agree well with Alix (1924) who estimated that the snow volume carried downhill by avalanches was between 10% and 25% of the total snow accumulation on slopes in the French Alps. De Scally (1992) estimates that avalanched snow in the Himalaya may represent 6% of annual runoff when there are high yield coefficients. There tends to be a delay in the melting of avalanched snow (De Scally, 1992), mainly as a result of the relatively small area of the deposited snow (Keylock, 1997).

2.14 Avalanche models

Among the earliest papers on avalanche forecasting is one by LaChapelle (1966). He introduced the distinction between meteorological and structural forecasts and their relation to climate. He also distinguishes between hard and soft slabs, and between direct and climax avalanches.

Direct action avalanches, which are the most common type, occur as a result of snow conditions developed over a few hours or at the most over a few days of prolonged snowfalls. However, forecasting of delayed-action climax avalanches is the most challenging issue facing avalanche forecasters. In practical terms, a climax avalanche often refers to one where the bed surface is the ground. It may be triggered by a new snowfall or a skier, but it involves snow layers at the release point that have been deposited by more than one storm. The basic weaknesses may be established weeks or even months in advance of actual release. McClung and Schaerer (2006) state that the most common trigger of climax avalanches is the failure of a buried layer of snow crystals that were produced by fast growth rates such as surface hoar, depth hoar, or recrystallization by radiation conditions. Climax avalanches tend to be a result of a major snowstorm. However, Schweizer et al. (2009) show that forecasting the avalanche occurrence on a major avalanche path based on new snow amounts involves high uncertainty. For example, the return period of an avalanche reaching a road was about 5 years, while the return period for the corresponding new snow depth was substantially smaller, slightly less than 2 years. The return period of the critical new snow depth was about two to five times smaller than the return period of the avalanche, implying a large number of false alarms.

There are two main types – loose snow avalanches and slab avalanches. Commonly, they begin with the failure of snow layers with densities less than 300 kg m^{-3}. An avalanche path comprises a starting zone, the track, and a runout–deposition zone. Loose snow avalanches are initiated when the angle of repose is exceeded – about 45°. The angle increases as temperatures rise due to increased cohesion. Slush avalanches can occur on slopes <10°. Slab avalanches occur when a cohesive slab is released over an extensive plane of weakness on slopes of 35–40°. Slab thicknesses are 0.1–4 m and have a mean density of ~200 kg m^{-3}. Bed surface temperatures are near 0 °C.

The variables of interest for forecasting an avalanche are velocity, run-out distance and impact pressure. The acceleration of an avalanche is resisted by surface friction, air drag at the front and upper boundary, and ploughing at the advancing front and underneath surface. According to Perla (1980), maximum velocities range from 20 to 30 m s^{-1} for path lengths up to 500 m and slope angles of 25–35°. The mean run-out length on 67 Colorado avalanche paths was 380 m (Bovis and Mears, 1976). On occasion, the run-out may cross a valley floor and continue up the facing slope.

Impact pressures are a maximum at 1–2 m above the surface and range in value from about 1–10 × 10^5 N m^{-2} (Perla, 1980).

There have been a number of models developed to describe avalanche flow. Voellmy (1955) used a simple empirical formula, treating an avalanche as a sliding block of snow moving with a drag force (F) that is proportional to the square of the speed of its flow (V):

$$F = 0.5 \, \rho \, V^2.$$

Voellmy's (1955) equation for the maximum velocity an avalanche will reach on a uniform track inclined at an angle α is

$$V^2 = \xi \, h'(\sin \alpha - \mu \cos \alpha), \qquad (2.64)$$

where h' is the flow height, ξ is the coefficient of turbulent friction, and μ is a coefficient of sliding friction. For avalanches confined to a channel the flow height, h', is replaced by the hydraulic radius, R (the ratio of the cross-sectional area of the channel in which the fluid is flowing to the wetted inner perimeter). ξ ranges from 300 to 500 for a rough boulder- or timber-covered slope, and from 500 to 800 for a typical unconfined slope. The dynamic friction coefficient μ varies between about 0.1 and 0.3 depending on avalanche velocity.

Voellmy also provides a formula for the distance, S, an avalanche will travel in its decelerating phase in the runout zone.

$$S = V^2/2g[\mu \cos \beta - \tan \beta + V^2/2 \, \xi \, h'], \qquad (2.65)$$

where β is the slope of the runout zone.

The model requires the specification of a reference point at which deceleration starts, but the need to specify this was eliminated by Perla et al. (1980).

The original depth-averaged two-parameter models of the Voellmy type are currently implemented within a hydraulic-continuum framework and incorporate terms to deal with active/passive pressure conditions (Salm et al., 1990). These models can predict run-out distances, flow and deposit depths, velocities and pressures along the path. They have been variously developed in 1-D and 2-D versions, including a pseudo 2-D version by Bartelt et al. (1999).

Bakkehøi et al. (1983) developed an alpha–beta statistical model of runout distance where the alpha point is the maximum extent of avalanche debris over a period of ~50–300 years and beta is the point on the path profile where the slope drops to 10°. The alpha and beta angles are measured from these two points to the starting zone. A linear regression is fitted to measured alpha and beta angles and alpha, the maximum runout distance, is determined from beta. A runout ratio model was proposed by McClung and Leid (1987), where Δx is the horizontal distance between alpha and beta, X_β is the horizontal distance between beta and the starting zone, and δ is the angle from the alpha point to the beta point:

$$\Delta x / X_\beta = [\tan \beta - \tan \alpha] / [\tan \alpha - \tan \delta]. \tag{2.66}$$

Values of this ratio for a mountain range follow a Gumbel-type extreme value distribution. Martinelli (1986) and McClung and Mears (1991) illustrate these two approaches to estimate maximum runout.

Borstad and McClung (2009) suggest that the precision in the friction coefficient necessary for using a dynamics model to predict runout distances is higher than the current state of modeling technique, which starts numerical simulations at the midpoint of the length of the avalanche path at maximum speed. The Coulomb friction coefficient is chosen to produce knowledge about avalanche resistance mechanisms. They develop a new, unique speed profile from this new starting point at maximum speed to a state of rest at an empirically pre-determined runout position. The technique reproduces the observed sharp deceleration of avalanche flow in the runout zone.

A simulation model in current use is SNOWPACK (Lehning *et al.*, 2002). It is a 1-D snow cover model based on finite-element methods and is used operationally by the Swiss Federal Institute for Snow and Avalanche Research. It uses the input data from some 50 automatic weather and snow stations in the Swiss Alps. The model calculates the snow cover evolution during the winter: its stratification, density, crystal structure, snow water equivalent and runoff. The model is physically based: energy balance, mass balance, phase changes, water and water vapor movement, and wind transportation are included. Most of the calculations are based on elements of the snow microstructure (crystal size and form, bond size, and number of bonds per crystal). An important characteristic is that the amount of new snow is determined from the measured total snow depth and the model-calculated settling rate together with an estimation of the new snow density. Using an improved formulation for snow metamorphism and linking the rate of snow metamorphism to the viscosity and thermal conductivity, the mass and energy balance in the model compare well with independent measurements. It is shown that the model can be used to determine high Alpine snowfall rates and the spring ablation period is also modeled correctly. As input, the model needs air temperature, relative humidity, wind speed and direction, shortwave and longwave radiation, snow depth or precipitation, and if possible, surface and ground temperatures. The time resolution of the data is between 30 minutes and 6 hours.

Lundy *et al.* (2001) validate SNOWPACK statistically for Montana. Snowpack temperatures are predicted reasonably accurately, the modeled and observed densities correlate well, but the model typically underestimates snowpack settlement. Its application is also illustrated by Hirashima *et al.* (2008) for western Japan during the snowy winter of 2005/2006. The equations for the stability index were found to be unsuitable for the study area considered. High avalanche danger continued for more than 2 months in the model, due to inaccurate parameterization of the shear strength for the snow conditions. As a result, Hirashima *et al.* (2008) developed more appropriate parameterizations for western Japan.

Models have also been developed for predicting avalanches using statistical methods, as illustrated by Bovis (1977) who used linear discriminant function analysis with data on snow cover and weather conditions to assess avalanche occurrence in the San Juan Mountains, Colorado. Föhn *et al.* (1977) compare conventional (empirical) and statistical forecasting methods for regional scales (entire mountain ranges) where a spectrum of avalanche events is of interest, including the number, size, type, altitude zone, aspect and slope angle. The data for Weissflujoch, Switzerland span 20 years but the predictions are for three seasons. They combine weather variables and snow condition variables using principal component analyses and then discriminant analysis to identify wet and dry avalanche types and non-avalanche days. Statistical models I and II employ, respectively, 7 and 14 variables, while model II uses only gridded meteorological data at 700 mb and at the surface over Western Europe. In February all models have difficulty predicting short wet/dry avalanche cycles mixed with "safe" periods. In March models I and II overestimate the avalanche probability whereas the conventional method and model I capture the variability well. A new approach to statistical forecasting is offered by Schirmer *et al.* (2009). They use the SNOWPACK model to simulate snow stratigraphy for a site near Davos and incorporate meteorological data from an AWS. The best results (73% accuracy) were obtained with a nearest-neighbor method using the avalanche danger of the previous day as an additional input.

Another approach is statistical avalanche forecasting as illustrated by Bovis (1977) who used linear discriminant function analysis with data on snow cover and weather conditions to assess avalanche occurrence in the San Juan Mountains, Colorado. Fõhn *et al.* (1977) compare conventional (empirical) and statistical forecasting methods for regional scales (entire mountain ranges) where a spectrum of avalanche events is of interest, including the number, size, type, altitude zone, aspect and slope angle. The data for Weissflujoch, Switzerland span 20 years but the predictions are for three seasons. They combine weather variables and snow condition variables using principal component analyses and then use discriminant analysis to identify wet and dry avalanche types and non-avalanche days. Statistical models I and II employ, respectively, 7 and 14 variables, while model II uses only gridded meteorological data at 700 mb and the surface over Western Europe. In February all models have difficulty predicting short wet/dry avalanche cycles mixed in with "safe" periods. In March models II and II overestimate the avalanche probability whereas the conventional method and Model I capture the variability well.

Another approach examines the state of the snow and the weather conditions at a representative snowfield on a given day. The variables considered are the ones found to give the best results in statistical methods. Then the records are checked to find the

10 days (nearest neighbors) that best match these conditions and check whether or not an avalanche subsequently occurred. Buser (1983) illustrates a flow diagram on this basis:

Variables from measurements and observation
    ↓ a weighting vector
Parameters related to the event (avalanche)
    ↓
Calculate distances between the actual day and each of the past ones
    ↓
Take nearest ten cases (from Euclidean distance)
    ↓
Check avalanche records of these days
    ↓
Compare with the actual situation
    ↓
If satisfied, exit; if not, what is wrong?

This approach has been adopted in a number of studies; Purves *et al.* (2003) demonstrate it for Scottish conditions.

McClung (2002) lays out methods of applied avalanche forecasting taking account of human and physical factors. A practical guide to avalanche forecasting is available at www.meted.ucar.edu/afwa/avalanche/. Jamieson *et al.* (2008) undertake a verification of avalanche bulletins issued in western Canada (the Coast, Columbia, and Rocky Mountain ranges) during winters 2004–2005 and 2005–2006. The regional bulletins are issued three to seven times per week and cover areas ranging from 1,000 to 29,000 km^2. For 192 cases, there was 59–64% agreement between the regional danger ratings and the local ratings of current avalanche danger. The level is closely similar to that found by Elder and Armstrong (1987) for three forecast regions in Colorado.

2.15   Trends in avalanche conditions

Schneebeli *et al.* (1997) found no change in the frequency of extreme snowfall events, nor snow depth, (reflecting potential avalanche activity) in the Davos region of the Swiss Alps from 1896 to 1993, although temperatures showed an increase. The frequency of destructive avalanches in the Swiss Alps for 1947–1993 showed no evidence of trends, although at Davos their frequency declined over this period. Eckert *et al.* (2010) analyzed 60 years of avalanche data for the French Alps using a database presented by Jamard *et al.* (2002). They found that while the runout distance had not changed over 60 years, the runout altitude had decreased from ~1,400 m in 1948 to 1,350 m in 1977, and then recovered by 2006. The changes appear to reflect the quantity of available snow since there were high accumulations of cold snow around 1980, with subsequent decreases.

At Rogers Pass of British Columbia, major avalanche winters need not be big snow years. Avalanches occur either during winters with zonal flow and numerous storms, or with cold meridional airflow and strong changes in temperature during incursions of Pacific air (Fitzharris and Schaerer, 1980). Sawyer and Butler (2006) examined avalanche occurrences near the southern boundary of Glacier National Park, Montana, USA from newspaper reports for 1982–2005, updating a chronology available from the same source since 1946. Trends for the entire 1946–2005 data set showed a marked decrease in reported avalanches from the 1960s onward, compared with 1949–1957.

From analyzing 21 snow and weather variables of 189 naturally released forest avalanches by a hierarchical clustering method, Teich et al. (2012) tested for long-term trends in meteorological conditions favorable for two types of forest avalanches over winters of 1970/1971–2010/2011) by a logistic regression model, which are forest avalanches triggered (1) under heavy snowfall, stormy and permanently cold conditions, and (2) after periods of high insolation and an increase in air temperature. The number of potential forest avalanche days decreased at 11 of 14 snow and weather stations in the Swiss Alps for Type 1 avalanches and at 12 of 14 stations for Type 2 avalanches, independent from elevation and climatic region. The negative trends show a decrease of snow and weather conditions associated with avalanche releases in forests under climate change impact. With the currently observed increase in forest cover density in the Swiss Alps, it is likely that avalanche releases in forested terrain will become less frequent.

REVIEW QUESTIONS

(1) Define an avalanche. What natural factors and/or human activities that usually triggers an avalanche, and the range in size of an avalanche in terms of mass, path length, impact pressure, and slap depth (see Table 2.5)?

(2) Explain differences in the conditions and process of occurrence of three types of avalanches: loose snow avalanches, slab avalanches, and wet snow or isothermal avalanche.

(3) Name three common elements of avalanches, such as a trigger, morphological characteristics used to classify avalanches, and type of snow. Does the density of snow play a part in the failure of a snow layer on slopes of mountains?

(4) What is the typical sequence of dry snow slab fracture in slope-parallel shear at a weak layer in a snowpack leading to the avalanche release shown in Figure 2.27? In your opinion, should the critical crack length for shear crack propagation along weak layers increase or remain constant as the slope decreases?

(5) How does wind load, snow drift, formation of depth hoar under strong temperature gradients in snowpack, amount of snowfall in a few days, lee north- versus east-facing slopes, and wet versus dry snow affect the bulk strength and overall stability of snow slabs, and the occurrence of avalanches?

(6) Avalanche models range from empirical models, such as treating an avalanche as a sliding block of snow moving down a slope, to statistical and physically based models such as the SNOWPACK, 1-D snow cover model based on finite-element methods. Explain the advantages/disadvantages between simpler to physically based models in terms of input data requirements and the reliability of forecast.

(7) Under climate change impact, would we expect increasing or decreasing trends in avalanche conditions across northern Europe and North America and the reasons for the expected trends?

3   Glaciers and ice caps

3.1   History

The word "Gletscher" (glacier) first appeared on a map of the Alps in 1538, but the term "Ferner" for old snow was used in the Tyrol in 1300 and "Kees" (ice) in 1533 and on a map from 1604 (Klebelsberg, 1948, pp. 1–2).

A sketch map of the Vernagtferner Glacier in the Ötztal of Austria dates from 1601 (Nicolussi, 1990). In 1705, Johann Jakob Scheuchzer, after visiting the Rhône Glacier, confirmed that glaciers are formed by the accumulation, compaction, and recrystallization of snow, and they move and flow under the influence of gravity. In 1792–1794, Sveinn Pálsson made the earliest known scientific study of glaciers, including a glacier sketch map of the Vatnajökull glaciers in Iceland. Excerpts of his 1795 report were published in Danish between 1881 and 1884, but it was only translated in full in 1945 and then in Icelandic. So the work remained totally unknown by most glaciologists until recently translated into English by Williams and Sigurðsson (Pálsson, 2004).

Accounts of Alpine glaciers began with H. B. de Saussure's *Voyages dans les Alpes* (1779–1796), but the first measurements of glacier velocity were made by Franz Hugi between 1827 and 1831 on the Unteraar Glacier (Steiner *et al.*, 2008). Attention intensified in the 1830s and 1840s with the writings of Louis Agassiz (1840, 1967) on the last Ice Age. The first scientific maps of glaciers in the Alps date from the 1840s (Mayer, 2010). Forbes (1859) published a collection of papers addressing glacier theory. From visits to the Alps, Tyndall (1860) observed that glaciers were flowing rivers of ice receiving precipitation in their upper parts and discharging it below. In the 1890s, S. Finsterwalder (1897) made photogrammetric studies of Vernagt- and Hintereisferner in the Ötztal, Tyrol. The first books on glacier science appeared in German around the end of the nineteenth century by Heim (1885) and Hess (1904), and the journal *Zeitschrift für Gletscherkunde* began publication in 1906–1907. Reid (1896a) undertook pioneering surveys of Muir Glacier in the St. Elias Range (58.8° N) in the early 1890s. In coastal Alaska, Tarr and Martin (1914) undertook several expeditions to Glacier Bay and Yakutat Bay and around the enormous Malaspina Glacier piedmont lobe. Hamberg (1910) worked on glaciers in the Sarek, northern Sweden, and during the 1920s there was research in Iceland and Svalbard. The 1920–1930s period saw work in the upper Indus and Karakoram by Visser (1928) and on the 70-km-long Fedtschenko Glacier in the Pamir by R. Finsterwalder (1932). Stern (1926) made some of the first glacier thickness measurements on the tongue of the Hintereisferner, Tyrol, (ice 10–38 m thick) using differences in electrical conductivity and dielectric constant. Mothes (1926) made the first seismic (echo sounding) measurements on the same glacier and recorded 293 m thickness; then he determined a thickness of 729 m on the Aletsch Glacier at Konkordiaplatz (Mothes, 1929). He was followed over the next decade by many others (Brockamp and Mothes, 1930; Klebelsbrg, 1948, p. 211). Hess (1935) made calculations of internal ice motion also on the Hintereisferner using measurements on a 214 m borehole sunk in 1899 and one of 153 m sunk in 1902. He also showed that the height of the ice surface decreased from 2,610 m in July 1904 to 2,455 m in July 1933. Ahlmann (1935, 1948) and Sverdrup (1935) began the study of glaciers around the North Atlantic and made the first measurements of glacier ablation and energy balance parameters in Svalbard. Müller (1959) carried out the first measurements on the Khumbu Glacier, Nepal. The first velocity profile through a glacier was obtained by Gerrard et al. (1952). Glen (1953) made fundamental experiments on ice, and Nye (1953) and Perutz (1953) made calculations for glaciers that laid the basis for ice dynamics and the physical properties of ice as a material. Their results indicated that at the bottom of a glacier the shear stress is about 100 kPa.

Internationally coordinated glacier monitoring activities began in 1894 when the Council of the sixth International Geological Congress decided to create an International Glacier Commission (Radok, 1997). Data on glacier fluctuations, beginning in 1881, were first published by Forel (1895) for the Commission and the records have been maintained in the series *Fluctuations of glaciers*; the latest issue covers 2000–2005. In 1948, at the first post-war Assembly of the International Association of Scientific Hydrology (IASH), the IASH Commission on Glaciers and Snow, established in 1939, was renamed the International Commission on Snow and Ice (ICSI). ICSI officers led several programs during UNESCO's International Hydrological Decade, 1965–1974. In the 1960s, the Permanent Service on Fluctuations of Glaciers (PSFG) was established under Peter Kasser, together with the World Glacier Inventory (WGI) under Fritz Müller, both in Zurich, Switzerland, and a network of glacier stations for the measurement of heat, ice, and water balances in representative glacier basins was started under Mark Meier. In 1986, the PSFG and WGI merged in the World Glacier Monitoring Service (WGMS), directed by Wilfried Haeberli until May 2010, when he was succeeded by Michael Zemp (www.geo.unizh.ch/wgms/).

Glacier photographs and data began to be organized through the World Data Centers (WDCs) for Glaciology, established as part of the World Data Center system that was created during the 1957–1958 International Geophysical Year (IGY). The history of the WDCs is summarized in Chapter 1, Section 1.3. The National Snow and Ice Data Center has a glacier photograph collection (https://nsidc.org/data/glacier_photo/index.html) with dates ranging from mid-1800s to the present. As of June 2010, more than 15,000 glacier photographs are in the database, mostly of glaciers in the Rocky Mountains, the Pacific Northwest, Alaska, and Greenland. However, the collection does include a smaller number of photos of glaciers in Europe, South America, the Himalayas, and Antarctica. There is also a special collection of Repeat Photography of Glaciers taken from the same vantage point and at the same time of year, but taken many years apart. These photographs can show evidence of glacier and climate changes over time.

The first measurements of mass balance were made on the Rhône Glacier, Switzerland, in 1874. Chen and Funk (1990) rescued the measurements of annual mass balance for 1882–1883 to 1908–1909 from earlier literature. Continuous measurements began in 1914 at two sites on Claridenfirn, Switzerland. Mass balance data were collected on the Kårsa Glacier in Sweden in 1925–1926 by Ahlmann and Tryselius (1929) and followed by five years of measurements between 1941–1942 and 1947–1948 by Wallén (1948). Continuous annual measurements of glacier-wide mass balance began on the Storglaciären, Sweden, in 1945–1946. These were expanded to Storbreen, Norway, and glaciers in the Alps, western North America and the former USSR in the 1950s. Other field studies in the 1940s involved diurnal temperature measurements in the Bergschrunds (a crevasse that forms where the moving glacier ice separates from the stagnant ice above) of glacier cirques to assess the role of freeze–thaw processes in headwall erosion (Battle and Lewis, 1951) and the rotational-slip hypothesis of glacial movement (Lewis, 1949) based on observations of thrust planes in glaciers in Iceland and the Jotunheimen.

Modern glacier research largely originated in the 1940s in Great Britain and the Alps. Notable names are Perutz and Seligman (1939) on glacier flow, Bader et al. (1939) and Haefeli (1940) on snowpack stress and strain relationships. Thorarinsson (1943) worked independently on glaciers in Iceland. Early texts include Dryglaski and Matchatschek's (1942) on *Gletscherkunde* and Klebelsberg's (1948/1949) two-volume German text on glaciology and glacial geology. A landmark meeting of metallurgists and glaciologists in Britain in 1948 set the stage for the development of theories of ice flow, initially by E. Orowan (British Glaciological Society, 1949). The first modern textbook on glaciers was published by Lliboutry (1965) in French; Paterson (1994, first published in 1969) published one in English, followed by Post and LaChapelle (2000). Other general works are by Knight (1999), Nesje and Dahl (2000), and Hambrey and Alean (2004) and on glacier fluctuations by Oerlemans (1989). Texts for specific regions include Hope et al. (1976) for New Guinea (Irian Jaya) and Hastenrath (1981) for Ecuador. Kaser and Osmaston (2002) published a book that focused on the tropical glaciers of East Africa, Irian Jaya, and the northern tropical Andes. For benchmark glaciers in the Former Soviet Union, there are individual monographs in Russian detailing their characteristics and associated meteorological data. These include the Abramov in the Pamir-Alai (Glazyrin et al., 1993) and the Fedchenko in Tajikistan (Shul'tz, 1962). Björnsson (2009) published on Icelandic glaciers.

Various regional and local glacier maps for Eurasia are contained in Kotlyakov (1997) and a concise glacier inventory with small-scale maps (Shi, 2008b) and a 1:4 million glacier map have been published for China (Tao, 2006), as well as an atlas for some Indian glaciers (Raina and Srivastava, 2008) and a comprehensive inventory and atlas for the Indian Himalaya by Sangewar and Shukla (2009), although the maps lack geographical coordinates.

3.2 Definitions

Glaciers are large masses of ice that are formed where the accumulation of snowfall constantly exceeds the snowmelt and sublimation; glaciers move slowly away from the center of accumulation, or down a mountain valley, due to the stresses caused by their weight. The word "glacier" is derived from the Latin *glacies* meaning ice. Compacted dry snow first undergoes grain settling leading to a density of ~550 kg m^{-3} and a porosity of about 40%. Basically, densification occurs due to the reduction in pore space by the weight of the overlying snow. Wind, solar radiation, and vertical temperature gradient may also accelerate the densification and grain growth in the near-surface **firn** layer. Firn (or névé) is granular, partially consolidated snow that has passed through one summer's melt season but is not yet glacier ice. It has a density between 400 and 650 kg m^{-3} while thawed and refrozen firn ranges between 600 and 830 kg m^{-3}. **Sintering** – the bonding of snow particles that is produced by the diffusion of water molecules to particle contacts – is a process that produces rounded grains that allow closer packing and intergranular bonding. Colbeck (1997) gives a review of sintering in seasonal dry and wet snow covers and in laboratory experiments. Freeze–thaw cycles are another major factor in grain rounding. Gradually, re-crystallization by molecular diffusion enables the closing-off of air bubbles with a density of ~830 kg m^{-3}. The firn eventually becomes ice with a density approaching 917 kg m^{-3} over a time interval of between ~150 and 300 years. Typical depths for firn to become ice are 25 m on the Tibetan Plateau, 70 m at Summit in Greenland, and 100–150 m in Antarctica.

A NASA-funded project on Global Land Ice Measurements from Space (GLIMS) has mapped worldwide glacier outlines from ASTER and Landsat data through a network of regional centers around the world (Raup *et al.*, 2007). Currently, there are over 546,300 outlines in the GLIMS database (http://glims.colorado.edu/glacierdata/). The GLIMS project, with participation of more than 60 institutions in 28 nations, has assembled a baseline study to quantify the areal extent of existing glaciers (Kargel *et al.*, 2014). For satellite mapping, the GLIMS project (www.glims.org/) defines a glacier as body of ice that is observed at the end of the melt season or, in the case of tropical glaciers, after the melt of transient snow cover (Racoviteanu *et al.*, 2009). At a minimum, this includes all tributaries and connected feeders that contribute ice to the main glacier, plus all debris-covered parts of it. Stagnant ice and ice above the bergschrund that is still in contact with the glacier are considered to be part of the glacier.

3.3   Glacier characteristics

Glaciers assume a variety of forms: they may be conical on volcanic peaks, valley glaciers with a tongue of variable length, cirque glaciers without a tongue, or piedmont lobes. Ice caps may occur on plateaus or extend over several mountain peaks. Mountain glaciers and ice caps outside the two major ice sheets and the Antarctic Peninsula cover some 785,000 km^2 (4.2% of the global ice area), according to Dyurgerov and Meier (2005) (see Table 3.1). Antarctica has 21.5% of the global total area, the Canadian Arctic Archipelago has 19%, and Asia has 15%. The unmeasured area is considerable and hence this estimate is still subject to revision. Also, the time interval to which the data refer is not well defined – around 1980 is the approximate date cited by Williams and Ferrigno (1998) for tabulations in the "Satellite Image Atlas of Glaciers of the World." Estimates of the number of glaciers are equally uncertain – around 170,000 is a suggested number (with 85% of these in the Northern Hemisphere) (Dyurgerov, 2001), but the estimate of glaciers around the Antarctic margins is highly uncertain (Dyurgerov and Meier, 2005). There is a problem with the cut-off used for small-sized ice bodies. Kääb (2002) find that glaciers between 0.01 and 1 km^2 in area account for 25% of the glacierized area in the Berne-Valais region of Switzerland, for example.

The WGI is based largely on data from the WGMS in Zurich, Switzerland. It now contains information on more than 100,000 glaciers throughout the world: http://nsidc.org/data/G01130.html.

Parameters within the WGI include geographic location, area, length, orientation, elevation, classification of morphological type, and date of observation, and hence the inventory is not homogeneous. An example for selected fields given for the Aletsch Glacier, Switzerland, is as follows:

CH4N01336026 ALETSCHGL. 46.5012° N, 8.0390° E, area 86.76 km^2, mean width 1.2 km, mean length 22.6 km, mean elevation 3,140 m, mapped in 1969.

Ohmura (2009) points out that glaciers still needing to be inventoried include: ~54,000 in the Canadian Cordillera, 9,600 in the Canadian Arctic, 11,600 in South America, 6,100 in India and Pakistan, and 5,300 in Alaska, but there is an even greater need for an inventory of glaciers around Antarctica. Recommendations for improving the inventory procedure from digital sources are given by Paul et al. (2009). Also, Cogley (2009b) reports on an extended WGI that contains records for 131,000 glaciers and nearly half of the global glacier area. It is available at www.trentu.ca/geography/glaciology/glaciology.htm.

A new global data set of glacier outlines (Randolph Glacier Inventory (RGI)) was compiled using various data sources from 1950 to 2010 with varying levels of detail and quality by Arendt et al. (2015. There is also an inventory of peripheral glaciers separated from the Greenland ice sheet published by Rastner et al. (2012), but peripheral glaciers documented for the Antarctica are only partially complete, such as glaciers on the islands in the Antarctic and subantarctic (Bliss et al., 2013). Globally, uncertainties in the total number of glaciers estimated are not known because of regional uncertainties, regional variation in the minimum size of glaciers included in the inventory, and the subdivision of contiguous ice masses. The present best estimate is about 170,000 occupying a total area of about 730,000 km^2, with almost 80% of the glacier area found in the Canadian Arctic (regions 3 and 4), Alaska (region 5), high mountains of Asia (regions 13, 14, and 15), Greenland (region 17), Antarctic, and subantarctic (region 19) of Table 3.1.

Table 3.1 The global glacier outline inventory taken from the RGI 2.0 (Arendt *et al.*, 2015) divided into 19 RGI regions with their respective numbers of glaciers and area (km^2). The minimum and maximum values of glacier mass are the minimum and maximum of the estimates given in Grinsted (2013), Huss and Farinotti (2012), Marzeion *et al.* (2012) and Radić *et al.* (2014) . The mean sea-level equivalent (SLE) of the mean glacier mass is the mean of estimates from the same four studies, converted using an ocean area of 362.5 × 10^6 km^2 (adapted from Vaughan *et al.*, 2013). Regional mass change rates (ΔMass) in Gt yr^{-1} for 2003–2009 are taken from Gardner *et al.* (2013), with Central Asia (region 13), South Asia West (region 14), and South Asia East (region 15) merged into a single region.

Region	Region name	Number of glaciers	Area (km^2)	Mass (Gt) (minimum)	Mass (Gt) (maximum)	Mean SLE (mm)	ΔMass (Gt yr^{-1})
1	Alaska	23,112	89,267	16,168	28,021	54.7	−50 ± 17
2	Western Canada & United States	15,073	14,503.5	906	1,148	2.8	−14 ± 3
3	Arctic Canada North	3,318	103,990.2	22,366	37,555	84.2	−33 ± 4
4	Arctic Canada South	7,342	40,600.7	5,510	8,845	19.4	−27 ± 4
5	Greenland	13,880	87,125.9	10,005	17,146	38.9	−38 ± 7
6	Iceland	290	10,988.6	2,390	4,640	9.8	−10 ± 2
7	Svalbard	1,615	33,672.9	4,821	8,700	19.1	−5 ± 2
8	Scandinavia	1,799	2,833.7	182	290	0.6	−2 ± 0
9	Russian Arctic	331	51,160.5	11,016	21,315	41.2	−11 ± 4
10	North Asia	4,403	3,425.6	109	247	0.5	−2 ± 1
11	Central Europe	3,920	2,058.1	109	125	0.3	−2 ± 0
12	Caucasus	1,339	1,125.6	61	72	0.2	−1 ± 0
13	Central Asia	30,200	64,497	4,531	8,591	16.7	−26 ± 12
14	South Asia (West)	22,822	33,862	2,900	3,444	9.1	
15	South Asia (East)	14,006	21,803.2	1,196	1,623	3.9	
16	Low Latitudes[a]	2,601	2,554.7	109	218	0.5	−4 ± 1
17	Southern Andes[a]	15,994	29,361.2	4,241	6,018	13.5	−29 ± 10
18	New Zealand	3,012	1,160.5	71	109	0.2	0 ± 1
19	Antarctic and subantarctic	3,274	13,2267.4	27,224	43,772	96.3	−6 ± 10
	Total	168,331	726,258.3	113,915	191,879	412.0	−259 ± 28

[a] Shapefiles were created from late-summer, cloud-free Landsat 7 ETM+ imagery acquired prior to the 2003 scan line corrector (SLC) failure.

Glaciers occur on high mountains in all latitudes. Glaciers on Mt. Kenya (at 4,800 m) lie on the Equator and on the Ruwenzori peaks (4,600–4,800 m) at 1° S; Kibo summit on Kilimanjaro (5,895 m) is at 3° N latitude. In Irian Jaya, the remnant glaciers on Puncak Jaya are at 4° S, 4,700–4,800 m. In mainland Asia, glaciers in northern Burma (Myanmar) are located in the eastern Himalaya at 28.3° N, 97.5° E (above 4,700 m) where the Languela Glacier forms the headwaters of the Irrawady River. On Hkakabo Razi (white snow peak) (5,880 m) there is an ice cap and outlet glaciers above 5,300 m (Kingdon-Ward, 1949). The southernmost glacier in Asia is the Baihuhe Glacier 1 on Mt. Yulong (27.4° N, 100.0° E, 5,596 m) (Liu *et al.*, 2008). In the northern Andes, small glaciers are present on the Pico Bolivar and Pico Humboldt in the Sierra Nevada de Merida, Venezuela, at 4,900 m, 8.5° N (Schubert, 1992). In Colombia, there are mountain glaciers on the Ruiz-Tolima Massif in the Cordillera Central (4.8° N, 75.3° W), the Sierra Nevada del Cocuy region of the Cordillera Oriental (6.5° N), and the Sierra Nevada de Santa Marta of the Cordillera Central (10.5° N) (Morris, 2006). At 19° N in Mexico, the Pico de Orizaba (5,600 m) and Popocatepetl (5,400 m) both have ice cover. Palacios and Vázquez-Selem (1996) report on the Jamapa Glacier on Orizaba situated at about 4,700–5,000 m.

Glaciers – masses of long-lasting land ice – gain and lose ice in the accumulation and ablation zones, respectively. In the accumulation zone, snow accumulates, whereas in the ablation zone it melts and bare ice is exposed. As glaciers occur in climatic conditions where snow accumulates for over one or more years and slowly transforms into firn and eventually into ice, they are sensitive climate indicators because they adjust their size in response to changes in climate. On an annual basis there is a summer **snow line** demarcating the boundary between the two zones. The **equilibrium line altitude** (**ELA**) is the altitude on a glacier where the annual accumulation of mass is exactly compensated by the annual ablation of mass (i.e. there is zero specific balance). The distinction between the two zones is complicated by the role of glacial meltwater. Müller (1962) identifies additional zones: below the dry snow zone is the **percolation zone**, where some meltwater penetrates into the glacier where it refreezes. In the **wet snow (or saturation) zone**, all the seasonal snow melts. The meltwater either percolates into the depths of the glacier or flows down-glacier where it might refreeze as **superimposed ice**. On Arctic ice caps and the two large ice sheets, all five zones are present. A glacier's equilibrium line is located at the lower limit of the wet snow zone. On equatorial glaciers the altitudinal variation of the snow line seasonally may only be ~500 m, whereas in the Himalaya it is over 2,000 m. The seasonal variation is least on hyper-maritime glaciers and largest on continental interior glaciers.

In some situations, wind-blown snow and avalanching contribute substantially to the mass balance of small cirque glaciers. Hughes (2009) indicates that both sources help explain the persistence of four small glaciers (2–5 ha area) in the Prokletije Mountains of northern Albania. These are situated between 1,980 and 2,420 m altitude, among the lowest elevations for glaciers at 42.5° N. The same processes help account for the presence of small cirque glaciers in the Rocky Mountains of Colorado well below the snow line altitude (Hoffman *et al.*, 2007). In Rocky Mountain National Park, the accumulation on Andrews Glacier was

Table 3.2 Areal extent (km^2) and distribution of glaciers and ice caps (Dyurgerov and Meier, 2005; Berthier *et al.*, 2010)

Canadian Arctic Archipelago	151,433
Asia	120,680
Arctic islands	92,386
Greenland outside the GIS	76,200
Alaska	87,862
Continental North America outside Alaska	49,660
Europe	17,290
Antarctica outside the Antarctic IS*	169,000
South America	15,000
Subantarctic islands	7,000
New Zealand	1,160
East Africa	6
Irian Jaya	2

Radić and Hock (2010) recalculated the glacier area and volume for 19 regions. Greenland outside the GIS has 54,400 km^2, Alaska 79,260 km^2, western North America outside Alaska 21,480 km^2, South America 34,700 km^2, and the Subantarctic islands 3,740 km^2; other values are comparable.
* Shumskiy, 1969

eight times the regional snow accumulation, and on Tyndall Glacier four times, based on measurements from 1962 (Outcalt and MacPhail, 1965). Dadic *et al.* (2010) show that the local wind velocity over glacierized basins is a function of the small-scale topography and needs to be modeled accurately to reliably determine the mass balance distribution on small alpine glaciers.

Glacial ice often appears blue when it has become very dense. Years of compression gradually increase the ice density, forcing out the tiny air pockets between ice crystals. When glacier ice becomes extremely dense, the ice absorbs all other colors in the spectrum and reflects primarily blue light.

Common surface forms in snow and glacier ice are the nieve penitente, or pinnacle, and the sun cup. These are described in Box 3.1.

Defining glaciers accurately

The GLIMS project has developed tools and methods that can be used to create accurate glacier outlines and resultant measures of glacier extent. The importance of this is illustrated by a study of the Bering Glacier system (BGS), AK, by Beedle *et al.* (2008). Previously published measurements of BGS surface area vary from 1,740 to 6,200 km^2, depending on how the boundaries of this system are defined (Post and Meier, 1980). Their preferred value of

Box 3.1 Nieves penitentes and sun cups

These forms were first reported in the dry tropical Andes in 1835 by C. Darwin (1839), and Troll (1942), and then in the Himalaya (Workman, 1914). They comprise snow or ice pinnacles 1–5 m high, with spacing between the pinnacles comparable to their height (Figure 3.1). Workman noted that the orientation of the long axes of the pinnacles coincides with the direction of the slope inclination, and the apices and their steepest, most sharply cut sides usually point up slope. Matthes (1934) observed them tilted towards the elevation of the midday sun. The apices commonly lean over giving the "penitent" (hooded monk) form. They are mainly confined to lower latitudes and high altitudes. In the Andes they form above about 3,600 m, while on Mt. Kilmanjaro and in the Himalaya they are located above 4,600 m (Matthes, 1934). Lliboutry (1954) points out that the key climatic condition for the differential ablation that leads to the formation of penitentes is that the dew point temperature is always below freezing. Thus, snow/ice will sublimate, which requires 7.8 times more energy than does melting. The basic mechanism is that surface depressions absorb more radiation than high points. After differential ablation starts, the surface geometry of the evolving penitente produces a positive feedback, and radiation is trapped by multiple reflections between the walls. The steep walls intercept a minimum of solar radiation while in the troughs higher temperatures and humidity enhance ablation, leading to a downward growth of the penitentes. Penitentes tend to lower the net ablation rate and thus act to preserve high-altitude glaciers.

A 1-D model of their formation has been formulated by Betterton (2001) who analyzed the wavelength (~2 cm) of the fastest growing disturbance. Snow ablation is related to absorbed radiation, with a small-scale cut-off length for the surface height of the penitente that approximates the optical extinction depth (~1 cm). Laboratory experiments by Bergeron et al. (2006) confirm that penitente initiation (1–5 cm high) and coarsening requires low temperatures.

Penitentes differ from 5- to 10-cm-wide sun cups that form on melting snowfields and temperate glaciers (Rhodes et al., 1987) where the uniform temperatures mean that no surface temperature gradients are present. Dirt layers exceeding a few centimeters on the surface decreases the amount of reflected radiation and prevents the concentration of solar radiation in hollows. Maximum ablation is recorded for thin dirt layers 0.5–5-cm thick (depending on the thermal properties of the dirt). When dirt adheres to the snow surface it tends to become concentrated on the highest local points (Betterton, 2001).

ice draining to the piedmont lobe is 5,000 km^2. Using Landsat images from 2000 and 2001, Beedle et al. constructed a new outline with an area of 3,632 km^2. Three different models of BGS net balance led to an estimate of -1.2 m yr^{-1} w.e. and a total volume change of -4.2 km^3 yr^{-1} for 1950–2004. These values represent a contribution to sea level of 0.0236 mm a^{-1}.

Even though Himalayan Mountains have the third largest deposit of ice and snow in the world, after Antarctica and the Arctic, the largest area of poorly known glaciers are in the Himalayas (Shrestha, 2005). Based on visible satellite imagery, Qin (1999) estimates that

Figure 3.1 Penitentes 0.5–1-m tall on the slope of the volcanic peak Nevado Coropuna, Peru (15.5° S) at about 5,300 m altitude in August 2003 [Courtesy Adina Racoviteanu, INSTAAR and Department of Geography, University of Colorado]

18,065 glaciers occupy an area of 34,660 km^2 with an ice volume of 3,735 km^3; 58% of the glacier area is in the central Himalaya 28–32° N (the Ganges drainage), 30% in the western Himalaya, 32–36° N (the Indus drainage), and 12% in the eastern Himalaya *ca.* 28 ° N (the Brahmaputra drainage). However, in recent years, glaciers of Himalayas are melting at alarming rates (Maurer *et al.*, 2019).

Glacier types

Glaciers are classified according to their morphology and their temperature characteristics. The most common forms of glacier are cirque glaciers, hanging glaciers, simple valley glaciers, multiple branch valley glaciers, transection glaciers, and piedmont lobes (see Figure 3.2). Glaciers that calve in to the sea are termed tidewater glaciers. Thermodynamically, glaciers are classified as temperate, polythermal, or polar. Temperate glaciers are at the pressure melting point, except at the surface in winter, and have meltwater present. Cold, polar glaciers are usually frozen to their bed and have no surface melt. Polythermal glaciers are more complex; they may be subfreezing in the ablation zone and temperate to some depth in the accumulation zone; or they may be mostly subfreezing, but temperate at their sole. They may have supraglacial water channels during the melt season. However, in practice many glaciers have different thermal characteristics throughout their length if they originate in high mountains and descend to the lowlands. The morphological classification is more widely used. Cirque glaciers occupy armchair-shaped basins in alpine areas and have a width exceeding their length. Their size is 0.1 to ~3 km^2. Hanging glaciers

Figure 3.2 Types of glaciers. (a) A cirque glacier in Itirbilung Fiord (69.3° N, 68.7° W), northern Labrador in 1966. The summit ice cap is at about 1,600 m. [Courtesy: Dr. Jack D. Ives, Carleton University, Ottawa] (b) Small, unnamed hanging glaciers descending towards the Harvard Glacier, Alaska. February 14, 1994. [Courtesy Dr. Austin Post, NSIDC Glacier Photograph Collection, 94 V_037] (c) The Variegated Glacier, Alaska. (photographer unknown). August 22, 1965. (d) Part of the piedmont lobe of the Malaspina Glacier, Alaska (photographer unknown), 17 September 1966. (c) and (d) Courtesy of National Snow and Ice Data Center/World Data Center for Glaciology, Boulder, CO, USA.

exist on steep mountain slopes and do not connect with the valley glacier below. Instead they contribute mass by icefalls and avalanches. Valley glaciers issue from cirques and descend into the lower part of the valley, or they may radiate from an ice cap or ice field that covers a mountain range or high plateau. They have variable lengths from a few kilometers to more than 100 km. One of the longest glaciers is the Hubbard in Alaska which stretches ~120 km. In many cases tributary glaciers may join the main valley glacier making a multiple branch glacier. Transection glacier refers to a form that occupies a dissected mountain range and flows in several directions. Piedmont lobes form when a large valley glacier spreads out on the adjacent lowlands (Figure 3.2d): the Malaspina Glacier in Alaska is the best-known example.

Taylor Glacier exemplifies a little-studied type of outlet glacier that flows slowly through a region of rugged topography and dry climate. It connects the East Antarctic Ice Sheet with the McMurdo Dry Valleys. Kavanaugh *et al.* (2009a, 2009b) show that it is in a state of near zero mass balance in the lower half. Sublimation accounts for most of the ablation from all sectors of the glacier along the 80-km-long ablation zone. The mean accumulation rate in the catchment for Taylor Glacier on the north side of Taylor Dome is only 3–5 cm a^{-1}. Taylor Glacier flows at only 5–15 m a^{-1}. The flow of the glacier over major bed undulations can be regarded as a "cascade"; it speeds up over bedrock highs and where the valley narrows and slows down over deep basins and in wide spots. This pattern is an expected consequence of mass conservation for a glacier near steady state.

Tidewater glaciers

Glaciers that terminate in the ocean are termed tidewater glaciers (Figure 3.3). Apart from tidal and wave action, glaciers terminating in lakes experience similar processes. In contrast to glaciers that terminate on land, where flow velocities decline as the terminus is approached, tidewater glaciers accelerate at the terminus. The floating or grounded tongue is subject to additional forces that lead to ice loss through calving. The calving rate is largely controlled by the water depth and the ice velocity at the calving front. This process is discussed in Chapter 3, Section 3.8. Post and Motyka (1995) illustrate the effect of water depth >300 m in the case of the Le Conte Glacier, Alaska. Calving balances the high flow rate and this will continue until a terminal shoal can form that inhibits calving.

Tidewater glaciers exhibit recurring periods of advance alternating with rapid retreat, and punctuated by periods of stability. The best examples are located in southern Alaska and Patagonia. Molnia (2007) identifies 51 active and 9 former tidewater glaciers in Alaska. From studies on Columbia Glacier, Alaska, Post (1975) characterized the tidewater calving glacier advance/retreat cycle as: (1) advancing, (2) stable-extended, (3) drastically retreating, or (4) stable-retracted. During the advance phase the glacier builds a terminus shoal of sediment. The glacier is not very sensitive to climate during the advance as its Accumulation Area Ratio (AAR) is high (~0.7). The glacier again becomes sensitive to changing climate when it is at the maximum extended position. As the glacier retreats from the shoal into deeper water, the calving rate

Figure 3.3 The calving terminus of Harvard Glacier, located at the head of College Fiord, Prince William Sound, AK (61.7° N, 147.7° W), 3 September 2000. Photographer: Dr. Bruce Molnia, USGS. [Source: NSIDC Glacier Photograph Collection]

increases. Muir Glacier on Glacier Bay, Alaska, retreated 33 km from 1886 to 1968 with extensive calving, and a further 7 km by 2001. Between 1941 and 2004 the Muir Glacier retreated more than 12 km and thinned by more than 800 m. Now, the glacier is near the head of its fiord, with minimal calving. Pfeffer (2003) points out that Columbia Glacier (61° N, 147° W) has retreated more than 13 km from its pre-1980 position, and continues to retreat at ~0.5 km a^{-1}, while Hubbard Glacier (61° N, 140° W) has been advancing since 1980. Meier and Post (1987) show that the fast flow of temperate calving glaciers is almost entirely due to basal sliding. Accelerated flow at the glacier terminus causes thinning of adjacent upstream ice, and thinning in turn increases flow to the terminus by reducing effective pressure at the bed. Pfeffer (2007) points out that, once initiated, retreat appears to be irreversible in nearly all cases and it continues until the terminus reaches shallow water. While Meier and Post (1987) considered that tidewater glacier retreat is not directly driven by climatic change, Pfeffer argues that climatically induced long-term thinning triggers retreat through alterations of glacier geometry that reduce the resistive stresses.

Accumulation Area Ratio

The **Accumulation Area Ratio (AAR)** is the fraction of a glacier surface that has net accumulation. It is closely related to the vertical profile of mass balance. A landmark study of AAR values for a single year on 475 glaciers in western North America by Meier and Post (1962) showed that values ranged from >0.6 in the Pacific Northwest where mass budgets were positive; to 0.25–0.5 in the Rocky Mountains of Canada, northwest Montana, and the Cascade Range of Washington where budgets were negative; to <0.2 in the western

Alaska Range and the Wyoming Rocky Mountains where glaciers were stagnant or retreating. For the glacier inventory of the entire former Soviet Union (24,000 glaciers), which is based on surveys spanning mainly the 1960s–1970s, Bahr *et al.* (1997) obtain a mean AAR of 0.578 and for 5,400 glaciers in the European Alps a value of 0.58. They show that this relationship can be derived theoretically (see below). They express the AAR as

$$\text{AAR} = \left[\frac{(1)}{(m+1)} \right]^{1/m} \tag{3.1}$$

where m is the balance rate exponent. The mass balance rate has the approximate form:

$$b = -c_m x^m + c_o \tag{3.2}$$

where c_m and c_o are balance profile parameters and x is the length along "horizontal" axis. For $m = 1$, AAR = 0.5, for $m = 2$, AAR ~0.58 (corresponding to the observational data); and for $m = 3$, AAR = 0.707.

The conventional AAR value (wrongly) assumed for steady state glaciers is 0.65. Kamniansky and Pertziger (1996) argue that the area accumulation ratio (AAR) is approximately linear with net balance and this is near zero for AAR = 0.65. However, Dyurgerov and Bahr (1999) dispute this finding. Xie *et al.* (1996) determine that the glacier median altitude approximates the "steady-state" equilibrium line altitude (ELA$_0$) and show that the net balance at this altitude closely corresponds to the mean specific balance of the whole glacier. Dyurgerov and Bahr instead suggest that the terminus balance is well correlated with the height difference between the mean glacier altitude and the height of the terminus. Such altitude data are readily available in existing glacier inventories and can now be determined from ground and airborne surveys.

Furbish and Andrews (1984) and Osmaston (2005) describe methods of ELA estimation via Area-Altitude Balance Ratios (AABRs). Benn and Lehmkuhl (2000) provide an overview of several different methods of ELA estimation: the Balance ratio (BR) method of Furbish and Andrews (1984), the Accumulation-area ratio (AAR) method, the maximum elevation of lateral moraines (MELM), the toe-to-headwall altitude ratios (THAR), and the toe-to-summit altitude method (TSAM). They conclude that the BR, AAR, and MELM methods yield results which bear some relationship to the concept of the steady-state ELA as applied to modern glaciers, whereas THAR, TSAM, and cirque-floor methods simply summarize some aspects of the glaciated catchment. The results of THAR and TSAM methods should be termed "Glacier Elevation Indices" (GEIs). Benn and Lehmkuhl (2000) recommend that multiple methods be employed in estimating ELAs.

Rea (2009) examines Area-Altitude Balance Ratios (AABRs) providing an empirically derived set of ratios that can be used for ELA estimation in paleo-glacier reconstructions and for quantifying paleo-climatic conditions.

$$\text{AABR} = b_{nab}/b_{nac} = (\bar{z}_{ac}\,A_{ac})/(\bar{z}_{ab}\,A_{ab}), \qquad (3.3)$$

where b_{nab} and b_{nac} are the net mass balance gradients in the ablation and accumulation zones respectively, $\bar{Z}_{ac}$ and $\bar{Z}_{ab}$ are the area-weighted mean altitudes of the accumulation and ablation areas respectively and A_{ac} and A_{ab} the areas of accumulation and ablation, respectively. Representative values are given as follows: a global AABR = 1.75 ± 0.71; mid-latitude maritime = 1.9 ± 0.81; high latitude = 2.24 ± 0.85; North America–West Coast = 2.09 ± 0.93; North America–Eastern Rockies = 1.11 ± 0.1; Canadian Arctic = 2.91 ± 0.35; Svalbard = 2.13 ± 0.52; Western Norway = 1.5 ± 0.4; European Alps = 1.59 ± 0.6; Central Asia = 1.75 ± 0.56; Kamchatka = 3.18 ± 0.16.

Ohmura (2001) reviews the use of summer temperature as a melt index, tracing the origin of this approach to Ahlmann (1924). He shows it to be a satisfactory method because downward longwave radiation is the predominant heat source (from 50 to 80%) compared with absorbed solar radiation (15 to 34%) . He notes that most of the atmospheric radiation received at the surface comes from the near-surface layer of the atmosphere (90% is from the lowest kilometer for cloud-free conditions and 70% under overcast sky).

Glacier limits

Hastenrath (2009) determined the altitude of the mean annual freezing level based on a 1958–1997 global data set and the mean equilibrium line altitude (ELA) for the first half of the twentieth century for tropical glaciers. The freezing level is around 4,000–5,000 m, with lower levels in the outer tropics. The mean ELA is reached in the Australasian sector on four mountains, and in Africa on three mountains, near the Equator. In the American cordilleras many peaks are still glaciated above 0 °C, but in the arid southern tropical Andes even summits above 6,000 m do not reach the mean ELA. There have been many regional studies of the altitude of the ELA and the glaciation level (GL), which is determined from the arithmetic mean elevation of the lowest peak with ice and the highest summit without ice. This method was first applied to the coastal areas of the North Atlantic by Ahlmann (1948) and then maps were constructed for Fennoscandinavia (Østrem, 1964), British Columbia and Alaska (Østrem, 1966, 1972) and Baffin Island (Andrews and Miller, 1972). Miller *et al.* (1975) mapped the glaciation level (GL) and ELA for the Queen Elizabeth Islands in the high Canadian Arctic, for example, where the ELA is about 100–200 m below the GL. The GL is at 300 m asl along the northwestern margin and has a very steep gradient of 15 m km^{-1}.

In western China, the ELA rises from 2,800 m in the Altai Mountains (49° N) to 5,400 m on the northern slopes of the Kun Lun (37° N), and to 6,000 m on the north slope of Mt. Everest (28° N) (Lin *et al.*, 2008). The ELA rises about 152 m per degree of latitude. On the south slopes of the central Himalaya the ELA drops to ~5,400 m as a result of the monsoon precipitation. The aridity of the Tibetan Plateau and the summer heat source (Barry, 2008, pp. 65–7) cause the ELA to be highest in the west of the plateau where it is around 5,800–6,000 m.

3.4   Mass balance

Cogley (2005) reviews the mass and energy budgets of glaciers and ice sheets. The principal mass balance components are positive contributions from snowfall, snow drifting, and condensation, and negative contributions from snow and ice melt, sublimation, wind scour, and iceberg calving.

Methods of measuring mass balance components in the field are detailed by Østrem and Brugman (1991) and Kaser *et al.* (2003). The **direct glaciological method** relies on repeated measurements at stakes and snow pits on the glacier surface to determine annual mass balance. The **annual balance** is calculated for fixed dates (e.g. 1 October in the Northern Hemisphere) while the **net balance** is the minimum mass at the end of each summer. This definition may be inappropriate for tropical glaciers. The specific mass balance (kg m^{-2} or mm w.e.) is determined by dividing the mass balance by the glacier area, which allows the comparison of different glaciers. Following procedures developed in the 1940s–1950s, and documented first by Meier (1962), the stakes are usually arrayed in longitudinal and cross-glacier profiles (Østrem and Brugman, 1991). This may now be combined with high-precision geodetic and photogrammetric techniques for determination of mass and volume changes with high spatiotemporal resolution. Tests carried out on the Abramov Glacier, Pamir-Alatau, in 1979–1980 indicate that a regular network of stakes gives more accurate mass balance data than the usual longitudinal and cross-sectional stake arrays (Kamniansky and Pertziger, 1996). Fountain and Vecchia (1999) show that 5–10 stakes are sufficient for determining mass balance on small (<10 km^2) glaciers and this number seems to be scale invariant up to some unknown limit.

There is a need to determine both winter and summer balances in order to understand glacier changes and their causes. However, long series of such records are few and mainly from Europe (Vincent *et al.*, 2004). Dyurgerov and Meier (1999) note that the summer balance controls the recent negative trend in mass balance with generally little change in the winter balance. However, Dyurgerov (2001, 2003) draws attention to the increase in both winter (positive) and summer (negative) balances indicating intensification of the water cycle with significant mass loss since the end of the 1980s.

The **indirect (or geodetic) method** of mass balance determination usually involves geodetic determinations that are based on the bedrock as a fixed reference surface, measured from boreholes in the glacier, whereas the direct survey method is referenced to the previous balance year's summer surface. Hubbard *et al.* (2000) combine Digital Elevation Model (DEM) and photogrammetric data with ice flow modeling for analysis of the Haut Glacier d'Arolla.

Several authors note differences between geodetic and glaciological estimates of mass balance (Elsberg *et al.*, 2001; Braithwaite *et al.*, 2002). A number of workshops have been held over the years to examine methods of mass balance determination and the errors involved. The findings of recent deliberations are reported by Fountain *et al.* (1999).

Often, little attention is paid to error assessments, but for the Storglaciären, Sweden, Jansson (1999) finds that uncertainties in measurements likely translate into uncertainties in the mass balance of about 0.1 m w.e. a^{-1}. Krimmel (1999) compares the direct and geodetic method for the South Cascade Glacier, WA and finds that the latter gave a systematically larger estimate of 0.25 m w.e. a^{-1}. Errors through neglected basal and internal melt due to infiltration (Bazhev, 1997), density assumptions, and so on, are estimated to give a total error of only 0.09 m w.e. a^{-1} and the discrepancy is attributed rather to sinking of stakes and the area integration procedure. For Hintereisferner, Austria, Kuhn *et al.* (1999) found that the two approaches agreed closely, although Fischer (2010) reports a 24% larger value from the geodetic method compared with the direct mass balance measurements there over 50 years. Cogley (2009a) provides extensive global data sets of direct and geodetic mass balance measurements and finds a negligible mean difference between them.

A different approach is to determine the hydrological mass balance. This is obtained from data on precipitation, glacier runoff, and condensation/sublimation. The hydrological mass balance (bn_H) is given by Sicart *et al.* (2007) as

$$Bn_H = P - 1/Sg \, [D - (S - Sg)cP] - E \tag{3.4}$$

where Sg is the glacier surface area (m^2), S is the total surface area of the basin (m^2), c is the runoff coefficient of the surface not covered by the glacier (0.5–1.0), P is the annual precipitation (m a^{-1}), D is the runoff at the outlet of the basin (m^3 a^{-1}), and E is the glacier sublimation (m a^{-1}). The second term in the equation is the fraction of total discharge that comes from the glacier melting. The runoff coefficient mainly depends on evaporation and infiltration in rocky areas.

The sensitivity of mass balance to temperature and precipitation has been investigated by numerous authors (Oerlemans *et al.*, 1998; Braithwaite *et al.*, 2002). de Woul and Hock (2005) use a degree-day temperature model for 42 Arctic glaciers and show low sensitivity to a temperature increase in continental climates (~−0.2 m a^{-1} K^{-1}) and high sensitivity (up to −2 m a^{-1} K^{-1}) in maritime climates. The sensitivity to a 10% precipitation increase ranged from +0.03 to +0.36 m a^{-1} offsetting about 20% of a 1 K temperature increase (de Woul, 2008). For 88 glaciers worldwide, based on European Centre for Medium-Range Weather Forecasts (ECMWF) re-analysis (ERA)-40 data, de Woul (2008, table 1) shows a range in temperature sensitivity of −0.2 to −2.93 m a^{-1} K^{-1} and a sensitivity to a 10% precipitation increase of +0.01 to +0.43 m a^{-1}. Dyurgerov and Meier (2005) show that the sensitivity of mass balance to air temperature was highly variable and was increasing from the 1960s to 1979 and then it stabilized.

3.5   Remote sensing

Partly because of the cost of fieldworks, manpower, and difficulty to gain access to many glaciers for direct measurements, satellite data provides a practical alternative to study the characteristics, dynamics, future changes, and responses of glaciers to climate warming impact. In addition to complementing field observations in space and time, repeated images of glaciers enable us to assess the impact of their changes on society at global (sea-level rise), regional (streamflow, hydro-power), and local (hazards such as flooding) scales. Remote sensing of glaciers and ice caps began with both vertical and oblique aerial photography. A campaign to photograph all land areas of the Canadian Arctic was undertaken in the late 1940s–1950s (Dunbar and Greenway, 1956). The US Geological Survey (USGS) photographed all of Alaska after World War II and continued in western North America and Alaska from the 1950s (Post, 2005): http://earthweb.ess.washington.edu/EPIC/Collections/Post/index.htm. The photographs are archived at the University of Alaska, Fairbanks, and some are held at NSIDC.

Overviews of remote sensing of glaciers are provided by Hall and Martinec (1985) and Pellikka and Rees (2010). The parameters that can be determined using remote sensing (optical, thermal infrared, passive microwave, radar, altimetry and gravity) include glacier extent, area, topography, thickness, elevation, volume, surface velocity and snow lines; as well as surface reflectance, temperature and melt extent; glacier zonation (facies) mass balance, temporal and volumetric changes. In recent years, space platforms and techniques to study glaciers have been growing: optical, microwave and altimetry sensors such as Landsat, ASTER, Sentinel 1/2/3, SRTM, ALOS PALSAR, TerraSAR-X, ICESat, CryoSat-2, and so on, are now available. Even though new satellite sensors provide opportunities for extracting more accurate information, modification of existing algorithms to the more advanced capabilities of newer sensors are often necessary.

Glacier mapping from Landsat Thematic Mapper (TM) data has followed several approaches. Paul (2000) notes that these include: (1) manually delineating the glacier outline by cursor tracking (2) segmentation of ratio images, and (3) unsupervised or supervised classification. The first approach has been used to determine glacier length changes. Various combinations of ratios have been used. Bayr *et al.* (1994) derive a glacier mask by using thresholds with ratio images of raw digital numbers from TM channels 4 and 5. Paul states that neither unsupervised nor supervised classification methods proved to be suitable for glacier mapping. Williams and Ferrigno (2012) utilized maps, aerial photographs, Landsat 1, 2, and 3 MSS images and Landsat 2 and 3 RBV images to compile a "Satellite image atlas of glaciers of the world." in the US Geological Survey Professional Paper 1386 which contains eleven volumes: an overview volume (1386-A) and ten regional volumes (1386-B to -K) on glaciers of different regions of the Earth. Some later contributors also used Landsat 4 and 5 MSS and Thematic Mapper (TM), Landsat 7 Enhanced Thematic Mapper-Plus (ETM+), and other satellite images. The ten regional volumes are Antarctica (1386-B), Greenland (1386-C), the glaciers of Iceland (1386-D), the glaciers of Europe (1386-E), the glaciers of Asia (1386-F), the glaciers of the Middle East and Africa (1386-G), the glaciers of Irian Jaya, Indonesia, and New Zealand (1386-H), the glaciers of South America (1386-I), the glaciers of North America (1386-J), and the glaciers of Alaska (1386-K). The results compiled in the ten volumes provide data for practically all glaciated areas on Earth. 1386-A (483 pp.) has five main chapters: Components of the Earth system and their relationship, the state of the cryosphere at the beginning of the twenty-first century for glaciers, snow, floating ice including sea ice and lake and river ice, and permafrost and periglacial environments. The largest chapter on glaciers (243 pages) also treats approaches to classifying glaciers, global glaciations through geologic time, ice cores and climate, changes in glaciers and sea level, mass balance measurements, glacier discharge impact on ocean salinity, glaciological hazards, etc. The section on ice cores primarily deals with tropical high mountain glaciers. The section on glaciers and sea level hardly uses the Gravity Recovery and Climate Experiment (GRACE) tandem satellite observations and does not discuss the effects of geoid changes on the regional sea level when a large ice sheet like Greenland undergoes significant mass loss.

Racoviteanu *et al.* (2008) summarize the major instruments with medium and high spatial resolution applied in optical remote sensing studies of glaciers (glacier delineation, DEM construction, volume–area scaling and accumulation area ratio/ELA methods for mass balance determination). Problems of automated mapping of glaciers from space are reviewed and summarized by Racoviteanu *et al.* (2009) who note that the delineation of ice bodies is generally time consuming and often error prone. However, for debris-free ice bodies classification algorithms can be largely successful as illustrated in Figure 3.4. A GLIMS workshop addressed the mapping of: clean ice and lakes; ice divides, and debris-covered ice; problems in assessing changes in glacier area and elevation through comparison with older data; the generation of digital elevation models (DEMs) from satellite stereo pairs; and accuracy and error analysis.

Bamber and Kwok (2004) review remote sensing approaches to estimating glacier mass balance. Methods are available to determine surface elevation changes (laser altimetry or stereo-photogrammetry); for estimation of ice flux (feature tracking with visible imagery or interferometric SAR) and ice thickness (with radio echo sounding, see Box 3.2); and repeat measurements of changes in spatial extent, snowline elevation, and accumulation–ablation area ratio (with visible imagery). Examples are provided for Andean glaciers by Bamber and Rivera (2007). Glacier mass balance has been estimated for three glaciers in the French Alps using a combination of SAR data and surface stakes by Dedieu *et al.* (2005). The essential key is to obtain a high-resolution digital elevation model (DEM). This is possible from ASTER data using stereoscopy, provided there are ground control points (GCPs) on and around the glaciers. The value of ASTER DEMs for glacier volume change determination has been demonstrated by Miller *et al.* (2009) for a glacier in Spitsbergen. Airborne lidar data validated the annual elevation changes to within 0.6% in terms of volume.

Glacier velocity can be determined using sequential Landsat imagery with tracking of features such as crevasses (König *et al.* (2001). Scambos and Bindschadler (1993) illustrate this approach for Antarctica. The images are split into long and short wavelength using high- and low-pass filters. The long wavelength image contains topographic effects only, and is used to coregister images. The short wavelength image is used to track small, sharp features.

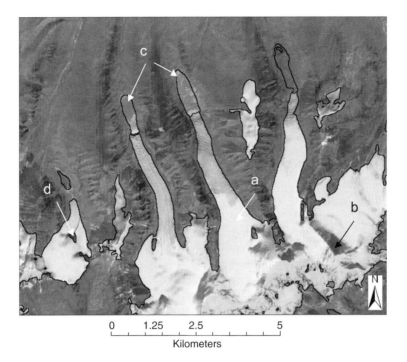

Figure 3.4 A classification algorithm for clean ice in northern Sikkim/China for 2001 ASTER imagery (from Racoviteanu *et al.*, 2008b) Arrows show (a) clean snow and ice and (b) shadowed glaciers, both correctly classified, (c) proglacial lakes misclassified as glaciers, and (d) internal rock areas correctly delineated. [Source: Sensors 8 (2008) figure 5, p. 3373. http://www.mdpi.org/sensors/ by Molecular Diversity Preservation International (MDPI), Basel, Switzerland).

Glacier velocity can also be determined using interferometrice SAR (InSAR). This is described in Section 4.3. The application of SAR and InSAR to mapping glaciers and glacier facies, DEM generation and glacier velocity studies is detailed by Høgda *et al.* (2010).

3.6 Glacier flow and flow lines

As the snow and ice on a slope thickens, a point is reached where they begin to move, due to a combination of the surface slope and the weight of the overlying snow and ice. On steep slopes this can occur with as little as 15–20 m of ice. Ice flows downhill due to the internal deformation of ice and gravity. When its thickness exceeds about 50 m, the pressure on the ice below that depth causes plastic flow. The upper 50 m or so of the glacier is under less pressure and this gives rise to the *fracture zone*. The upper layer often forms deep cracks known as **crevasses** that are typically up to ~50 m deep (Figure 3.5). Some crevasses are transverse to the

Box 3.2 Radio echo sounding of glaciers and ice sheets

Radio echo sounding (RES) is used to measure ice thickness; to detect the conditions at the base of an ice sheet including the presence of subglacial lakes, subglacial debris, bed roughness and basal crevasses; and to identify the internal structure of ice masses, including internal reflecting horizons, and internal reflections from hydrological features (Plewes and Hubbard, 2001). RES characteristics differ greatly for "cold" ice versus wet, and often crevassed glaciers and this determines the equipment that is used.

The propagation of radar signals is determined by the relative permittivity and electrical conductivity of the medium. Relative permittivity (the dielectric constant) relates to a material's ability to transmit ("permit") an electrical field. Pure ice has a permittivity of 3.15 but this can increase to ~80 in the presence of free water and impurities. The ability of a material to conduct an electric current is termed electrical conductivity, which for ice is ~0.01 mS m^{-1} (S = Siemens). Radar measurements of ice thickness depend on accurate knowledge of the radar wave velocity in ice since this determines the two-way travel time of the signal in the ice. Measured values of this velocity in ice range from $1.65-1.72 \times 10^8$ m s^{-1}. Wave refraction may occur where snow and firn overlay the ice because the wave velocity is greater in those materials. A signal (backscatter) may be modified by attenuation due to reflection, refraction and diffraction/A radio echo sounding system comprises a transmitter that generates a pulse of radio waves, a receiver that collects the return signal that is reflected back to the surface and a dipole antenna connecting the two pieces of equipment. Initial UHF pulsed systems (300–1,000 MHz) used from the 1950s could penetrate about 1,000 m of ice. While higher frequencies gave improved resolution, greater penetration was obtained with VHF (30–100 MHz) systems in the 1960s–1970s. In the mid-1970s impulse radars were developed to study temperate ice; in the case of wet, active tidewater glaciers, impulse radars with frequencies as low as 1 MHz have been used. These instruments were developed in research institutes and government laboratories. Since the early 1970s ground-penetrating radar has become widely available commercially. Their high frequency offers significantly improved vertical resolution.

There have been extensive airborne surveys of the thickness of the Antarctic and Greenland ice sheets used to construct digital elevation models (DEM) of the ice sheets, as well as airborne and surface based observations on temperate glaciers. Laterally extensive internal reflection horizons in ice sheets may arise through variations in ice density or crystal anisotropy as well as through variations in chemical impurities such as volcanic sulfates and ash layers in the ice. These horizons tend to be parallel or subparallel to the bedrock topography. Englacial channels and subglacial conduits have also been identified in temperate glaciers.

motion where the ice accelerates over a bump in the bedrock. Longitudinal ones occur where the glacier widens, and marginal (mainly transverse) ones form near the edge of the glacier, due to the reduction in ice velocity caused by friction along the valley walls.

Figure 3.5 Crevassed outlet glaciers draining from the Greenland inland ice towards the low-lying coastal region close to Kangerlussuaq on the southwest coast of Greenland [Courtesy Dr. K. Steffen, CIRES]

Below ice falls there may be **ogives** – alternating dark and light bands of ice that occur as narrow wave crests and troughs on the glacier surface (King and Ives, 1956; King and Lewis, 1961) (see Figure 3.6). The bands form due to the icefall creating broken-up fast moving ice, which greatly increases the ablation surface area in the summer. Nye (1958) showed that the waves are due to a combination of plastic deformation, as the ice is stretched in the icefall, and increased ablation, producing troughs in summer. In winter there is no ablation so a ridge forms. The combination of a dark and light band together represents the annual movement of the glacier.

Ice velocity depends on slope angle, longitudinal confinement, ice thickness, snow accumulation, basal temperature, meltwater production, and bed hardness. It ranges up to 2–3 m per day on Byrd Glacier in Antarctica and reaches 20–30 m per day on the Jakobshavn Isbræ in West Greenland. Longitudinally, glacier surface velocity is a maximum near the firn line (for a glacier of constant width) and decreases progressively towards the terminus, unless the ice calves in to the ocean or a lake. In the accumulation zone the motion is obliquely downward and in the ablation zone is obliquely upward, in valley glaciers and cirque glaciers (Reid, 1896; McCall, 1952).

The concept of basal sliding of glaciers, due to meltwater under the ice acting as a lubricant, was noted by de Saussure in the late eighteenth century and again pointed out by Wallace (1871), but many aspects of glacier motion remained unresolved until the mid-twentieth century. Early work on glacier flow was undertaken by Perutz and Seligman (1939) who measured crystal textures of snow and ice from different parts of the Aletsch glacier and made strain measurements within the glacier. They showed that whereas firn is deformed by

Figure 3.6 Ogives on the surface of the Morsarjøkull Glacier, southeast Iceland, 1953. There are two sets, side by side, one set below a connected ice fall, the other below avalanche cones where the original icefall severed (about 1937) [Courtesy Dr. J. D. Ives, Carleton University, Ottawa]

the relative motion of individual ice crystals, glacier ice yields by crystal deformation and growth, but also by slip over large thrust planes. Glaciers flow faster in the center than at the sides, owing to lateral drag, and likewise faster at the surface than in their interior. Crystals are oriented so that their basal planes are parallel to the ice motion and most are also parallel to the surface slope of the ice, illustrating plastic deformation. Glaciers yield to stresses by the deformation and growth of individual large crystals, and by the development of thrust planes where the motion is intermittent, depending on the stress exceeding some critical value.

Mechanisms of glacier flow include: intercrystalline gliding favored by the strong fabrics in glacier ice (here recrystallization is vital element), transfer of material associated with changes of state, slippage along shear planes in the glacier, and basal slip on the subglacial floor. The last of these accounts for up to 90% of the movement of thin ice on steep slopes and 20–50% of the movement in valley glaciers (Sharp, 1954). However, water-saturated sediment also can account for 90% or more of basal ice movement. Annual and diurnal variations in flow speed in the ablation zone largely reflect meltwater amounts.

The theory of ice flow was developed by Nye (1953) and Glen (1958) building on laboratory experiments of ice deformation by Glen (1952) and observations of glacier boreholes and tunnels in the Alps. Glacier flow is determined from a relation between the shear strain rate ($\partial \varepsilon_{xy}/\partial t$) and shear stress ($\tau_{xy}$) known as Glen's flow law, or the constitutive law (from materials science):

$$(\partial \varepsilon_{xy}/\partial t) = A\tau_{xy}^{n}, \qquad (3.5)$$

where t = time, $n \sim 3$, A depends on ice temperature, impurities and crystal orientation. Recommended values of A decrease from $6.8 \times 10^{15}\,\mathrm{s}^{-1}\,\mathrm{kPa}^{-3}$ at $0\,^{\circ}\mathrm{C}$ to 3.6×10^{-18} at $-50\,^{\circ}\mathrm{C}$ (Paterson, 1999, table 5.2). Stress causes ice to deform by extension/compression, and to shear leading to rotation.

Studies by Petrovic (2003) indicate that the tensile strength of ice varies from 70–310 kPa and the compressive strength varies from 5 to 25 MPa over the temperature range $-10\,^{\circ}\mathrm{C}$ to $-20\,^{\circ}\mathrm{C}$. The ice compressive strength increases with decreasing temperature and increasing strain rate, but ice tensile strength is relatively insensitive to these variables. The tensile strength of ice decreases with increasing ice grain size.

Paterson (1999, ch. 7) showed that the sliding velocity:

$$U = \mathrm{constant}\,(\tau^{0.5}/R)^{4}, \qquad (3.6)$$

where R = bed roughness and τ = basal shear stress. Hence, the sliding velocity varies as the square of the shear stress and inversely as the fourth power of the roughness. The basal shear stress under four Antarctic ice streams ranged from ~35 to 100 kPa (Vaughan et al., 2003). However, Lliboutry (1979) showed that basal friction depends not only on bed roughness, but on effective pressure – the difference between the overburden pressure and basal water pressure. Also, at sufficiently high rates of sliding cavities open in the lee of bedrock humps and serve to enhance the sliding rate (Lliboutry, 1968).

The pressure melting point (PMP) of ice plays a critical role in the basal processes of a glacier or ice sheet. The melting point of ice decreases at ~$0.7\,^{\circ}\mathrm{C}\,\mathrm{km}^{-1}$ of overlying ice due to the increasing pressure. Weertman's (1957) theory of sliding involves pressure melting and plastic deformation of ice. He argues that where basal ice is at the pressure melting point, heat flows from the low-pressure (downstream) side of a protuberance to the high-pressure (upstream) side where it melts the ice. The water so formed flows to the low-pressure (downstream) side where it re-freezes giving up its latent heat. This process is termed **regelation**. The energy to maintain the cycle is supplied by the basal shear stress. For pressure melting, the speed of sliding decreases as the obstacle size increases. Ice also deforms plastically, but this is less important than Weertman envisaged. Near a bump, the longitudinal stress in the ice and, therefore, the strain are above average. The greater the distance over which the stress is enhanced, the greater is the ice velocity. The sliding velocity resulting from differential stress concentrations around obstacles increases as the obstacle size increases according to Weertman. However, Kamb and LaChapelle (1964) showed that a "controlling obstacle spacing" is more appropriate than a controlling obstacle size. This spacing corresponds to the transition from regelation slip to plastic slip, and has a value of about 0.5–1 m. In a tunnel at the base of the Blue Glacier, WA, they observed the regelation of ice around bedrock obstacles and the formation of a regelation layer that incorporated debris particles from the bed. They also observed plastic deformation of the basal ice in warping of foliation planes and of the regelation layer. They propose that basal sliding is determined by (1) regelation slip, involving melting of the basal ice at points of increased pressure and refreezing at points of decreased pressure and (2) plastic flow, involving deformation of the ice due to stress concentrations, that is an order of magnitude smaller then regelation slip. They state that regelation slip involves (1) heat transport from points of local freezing to points of local melting; (2) mass transport of a thin basal layer of liquid water from points of melting to points of refreezing; and (3) bulk transport of the overlying ice mass, resulting from the operation of (1) and (2).

Sliding generates heat that causes basal ice to melt. For a glacier motion of 40 m a^{-1}, enough heat is released annually to produce 1 cm^3 of water per cm^2 of surface area if the shear stress is 1 bar. This is comparable to that melted due to geothermal heat. The geothermal heat flux ($\sim$0.1 W m^{-2}) melts about 6 mm of ice annually according to Paterson (1999, p.112).

When water accumulates at the base of a glacier, water pressure builds up and partly offsets the weight of the glacier allowing it to slide forward. If the glacier bed consists of a layer of deformable till, high pore water pressure may squeeze the till up into basal cavities and channels in the ice, blocking them off (Hooke, 1989).

Glacier response time

Numerous attempts to model the adjustment timescale of glaciers have appeared since the work of Nye (1961). Jóhannesson *et al.* (1989) give the response time for the glacier volume, τ, as

$$\tau = H / - b_t, \tag{3.7}$$

where H is a thickness scale of the glacier and $-b_t$ is the scale of the ablation at its terminus. On this basis, Paterson (1994, p. 320) finds that temperate maritime glaciers respond over 15–60 years, ice caps in Arctic Canada over 250–1,000 years, and the Greenland Ice Sheet over 3,000 years. Harrison *et al.* (2001) show that when a glacier changes slowly, a single timescale can be used. Their timescale includes the effects of surface elevation on net balance rate, which can increase the timescale or give rise to an unstable response. It is worth noting that their time constant determines both the rate and the magnitude of the response to a climate change.

Glacier hypsometry plays a major role in modifying glacier response to similar climate forcing (Furbush and Andrews, 1984) and Haeberli (1990) emphasizes the influence of the surface slope of the glacier. This is illustrated by the terminus behavior of 38 glaciers in the North Cascades, Washington, since 1890 (Pelto and Hedlund, 2001). They identified three different response patterns: Type 1: continuous retreat from the Little Ice Age (LIA) positions until 1950, followed by and advance until 1976 and subsequent retreat; Type 2: rapid retreat from 1890 until 1950 then slow retreat or stable until 1976 and rapid subsequent retreat; and Type 3: continuous retreat from the 1890s to the present. Despite differences in radiation due to aspect and slope, microclimates are much less important than hypsometry. Type 1 glaciers have steep slopes and extensive crevassing, with high velocities near the terminus. Their response time is 20–30 years. Type 3 glaciers have low slopes, moderate crevassing and low, terminal velocities; their response time is 60–100 years. Type 2 glaciers have intermediate characteristics and response times of 40–60 years.

Hoelzle *et al.* (2003) examine size classes of 90 glaciers worldwide and show that the mass balance change is proportional to length, which mainly reflects glacier slope. For 68 Swiss glaciers, they identify five slope and size classes and determine changes since about 1900. Long, flat glaciers have undergone constant retreat since the late nineteenth century; valley and mountain glaciers of intermediate size and slope show strong fluctuations with up to three periods of advance and retreat since 1880; steep mountain glaciers show moderate fluctuations and strong individual reactions; flat mountain glaciers have weak fluctuations but a clear overall trend; and very small steep glaciers have high-frequency variability and moderately large amplitudes. The sample of worldwide glaciers shows similar behavior. Overall, since 1900, large glaciers with lengths >8 km show greater losses of mass (-0.25 m a^{-1}) than glaciers <2.5 km (-0.14 m a^{-1}).

Raper and Braithwaite (2009) develop a new formulation for glacier volume response time (VRT) that depends directly upon the mean glacier thickness, and indirectly on glacier altitude range and vertical mass balance gradient. They treat climatic and topographic parameter separately. The former are expressed by mass balance gradients derived from degree-day modeling and the latter are quantified with data from the World Glacier Inventory. They establish a new scaling relationship between glacier altitude range and area, which accounts for the mass balance–elevation feedback, and evaluate it for seven regions. As a result of variations in this scaling parameter, the VRT can increase with glacier area (Axel Heiberg Island and Svalbard), hardly change (northern Scandinavia, southern Norway and the Alps), or even get smaller (Caucasus and New Zealand). The VRT can range from decades for glaciers in maritime (wet-warm) climates to thousands of years in continental (dry-cold) climates. In other words, wet-warm glaciers with a high mass balance sensitivity tend to have a short response time whereas dry-cold glaciers with a low mass balance sensitivity tend to have a long response time. The response times determined by Raper and Braithwaite are shown to be 2.9 times those using Jóhannesson *et al.* (1989) formula, due to the mass balance–elevation feedback.

Surging glaciers

Some temperate and cold-based glaciers exhibit surge behavior, flowing normally for many years and then suddenly accelerating to 10–100 times the speed in the quiescent phase. Ice flow rates during the active phase may range from about 150 m a^{-1} to >6 km a^{-1}, and horizontal displacements may range from <1 to >11 km (Meier and Post, 1969). Brúarjökull in Iceland has surged about every 80 years since 1625, the latest being in 1963 (Björnsson

Figure 3.7 Wahlenbergbreen glacier in Svalbard, Norway, was seen in September 2013, as its surge began speeding up. A glacier's shape, the nature of its bed, and climate change can interact to trigger normally quiescent glaciers to surge (Sevestre, 2017)

et al., 2003; Björnsson, 2009). The Variegated Glacier in Alaska (Figure 3.2c) surged in 1946/1947, 1964/1965, and 1982/1983 (Eisen *et al.*, 2001), and the Wahlenbergbreen glacier in Svalbard, Norway (Figure 3.7), that began surging in September 2013, peaked in early 2015 at 9 m per day, bulldozing everything along its path (Sevestre, 2017). Large volumes of ice are transferred and the glacier terminus may advance several kilometers in a few months. This behavior has been attributed to increased basal meltwater. Post (1969) identified 204 glaciers with unusual flow in western North America. Their restricted distribution – in the Alaska Range, eastern Wrangell Mountains, eastern Chugach Mountains, Icefield Ranges, and the St Elias Mountains near Yakutat and Glacier Bay – is not related to topography, bedrock type, altitude, orientation, or size of glacier. Post suggested that possible causes are unusual bedrock roughness or permeability, anomalously high groundwater temperatures, and/or abnormal geothermal heat flow. Proposed trigger mechanisms include fluctuations in thermal or hydrological conditions, or in deformable subglacial sediment, acting alone or in combination (Raymond, 1987). Lingle and Fatland (2004) propose that englacial water storage drives surges in temperate glaciers. The downward movement of this water overwhelms the basal drainage system and forces failure of the subglacial till.

About 4% of all glaciers are known to surge. Surging glaciers are to be found in Iceland, the Tien Shan, the Karakorum, Alaska, the Canadian Arctic Archipelago, Svalbard, and Novaya Zemlya. In the Canadian High Arctic, Copland *et al.* (2003) have identified 51 surging polythermal glaciers using aerial photographs from 1959/1960 and Landsat 7 imagery from 1999/2000. Björnsson *et al.* (2003) list 24 surging glaciers in Iceland. All but six of them have low slopes (1.3–4.3°) and they suggest that in consequence they move too slowly to remain in balance, given their high accumulation rates. In regions of lower accumulation, such as Svalbard and the Canadian High Arctic, there seems to be long-lived and less intense surge behavior. Grant *et al.* (2010) document 32 potential surging glaciers in Novaya Zemlya representing 4.6% of the archipelago's glaciers but 18% of the glacier area. They are typically long, large outlet glaciers, with relatively low overall surface slopes (median slope 1.7°), predominantly located on the more maritime western side of the island, and they tend to terminate in the sea or a lake. Outstanding problems relate to the reasons why in some areas most glaciers surge, but in others none do, and the underling differences between rapid and slow surges.

3.7   Scaling

Generalized relationships between glacier properties – length (L), area (A) and volume (V) – have been developed by Bahr (1997a, 1997b). The general dimensions of these relationships are

$$A \propto L^2 \text{ (length} \times \text{width)} \tag{3.8}$$

and

$$V \propto L^3 \text{ (area} \times \text{thickness)}. \tag{3.9}$$

To first order, $V \propto A^{1.5}$, but Chen and Ohmua (1990) and Bahr *et al.* (1997) find from observations that the exponent is 1.36 for glaciers and 1.25 for ice sheets. For reasonable closure assumptions in the model, the exponent was determined to be 1.375 for glaciers. Bahr *et al.* (1997) showed that these relationships are consistent with theory for the known properties of glacier ice (the conservation equations for mass and momentum). Van de Wal and Wild (2001) indicate an error range associated with the exponent of ± 0.125.

These scaling relationships are extended in the context of global glacier monitoring by Bahr and Dyurgerov (1999). For data from 68 valley and cirque glaciers, they find that balance at the glacier terminus B_T is a function of L^m, where $m \sim 1.7$, if L depends on the mass balance, while B_T is a function of area$^{1.09}$. For 303 Eurasian glaciers, Bahr (1997b) found that area $\propto L^{1.6}$ and the later study confirmed this relationship. Bahr and Dyurgerov propose that in the above power law relating terminal balance and length, m has bounds of 0.5–2.0. They go on to demonstrate for 80 glaciers with data for 1961–1990 that, whereas terminus balance is well correlated with mean glacier elevation minus terminus elevation, the correlation between glacier mass balance and ELA on a global (or large region) basis is poor. This latter finding is in contrast to the close relationship that holds for individual glaciers. One-hundred year simulations of six glaciers show that scaling underestimates the volume loss by up to 47% for V–A scaling but only 18% for V–L scaling (Radic *et al.*, 2008). The choice of scaling constants for V–L scaling has relatively little effect on the volume evolution; in the relationship width $\propto L^q$, the scaling constant q ~ 2.2 varied from 1.5 to 3.2. This is important since the scaling constant is not generally known. Möller and Schneider (2010) point out that the V–A relationship assumes a steady-state glacier; otherwise a weighting factor is required. They propose a new approach using data on past ice extent at three or four time points and a DEM of the glacier to calibrate the V–A relationship.

Farinotti *et al.* (2009) use Bahr *et al.* (1997) scaling relationship for glaciers smaller than 3 km^2, together with a mass conservation and ice flow dynamics approach to estimate ice volume for 62 glaciers in the Swiss Alps. They estimate a volume of 74 ± 9 km^3 in 1999.

3.8   Glacier modeling

There are two general categories of glacier model. One considers the glacier mass balance, and the rate of change of total mass, while the other treats the glacier dynamics and interactions between the ice and the bed. The input data for glacier modeling include the glacier and bed geometry, the climatic boundary conditions at the surface, and the basal conditions. Additional information on englacial temperature distribution and ice velocity fields is required to validate the model output (Greve and Blatter, 2009).

Most glacier models derive from the classic work of Nye (1960). He treats a 1-D glacier with accumulation and ablation. The ice volume passing any point is a function of the ice thickness and surface slope. It is shown that a region of uniform longitudinal strain rate is temporarily unstable. The response of the glacier to a sudden change in accumulation is examined via kinematic wave theory. The lower part of the glacier thickens unstably as a kinematic wave arrives; these travel at 2–5 times the surface speed of the ice. In a later paper Nye (1987) considered the effects of kinematic wave diffusion on the solutions.

One of the earliest analytical studies of glacier change was carried out by Allison and Kruss (1977). For the glaciers of Irian Jaya they use a 3-D model of the central flow line developed by Budd and Jenssen (1975). The net balance is increased in order to reproduce the maximum LIA glacier as a steady-state length. They propose a new approach to calibrate the relationship. The glacier is then caused to shrink by reducing the mass balance. The changes involve either a shift in the ELA or a change in the accumulation. The retreat rate is well matched by a rise in the ELA of ~80 m per century, corresponding to a temperature rise of 0.6 °C. Oerlemans (1997) developed a flow line model for Nigardsbreen, Norway, and showed that it closely simulated the glacier length changes since AD 1748.

A numerical ice flow model has recently been used to study the advance of tidewater glaciers into a deep fiord (Nick *et al.*, 2007). The results suggest that irrespective of the calving criterion and the accumulation rate in the catchment, the glacier cannot advance in to deep water (>300 m) unless sedimentation at the glacier front is included.

Using a first-order theory of glacier dynamics, Oerlemans (2005) related changes in glacier length to changes in air temperature. He constructed a temperature history for different parts of the world from 169 records of glacier length. The reconstructed warming for the first half of the twentieth century was 0.5 °C. The warming signals from glaciers at low and high elevations appear to be very similar.

Changes in glacier thickness can be modeled using distributed glacier mass balance models. These are based either on a degree-day temperature index approach (Hock, 2003) or the more data intensive energy balance method (Oerlemans, 1991). For a region in the eastern Valais, Switzerland, Paul *et al.* (2007) model an area with about 50 glaciers. They input a DEM, gridded shortwave radiation, albedo and precipitation, and parameterized air temperature, pressure, relative humidity, and clouds, to calculate the mass balance of each grid cell. Results show a reasonable mass balance distribution and ELA for the Findelen Glacier. A glacier mass balance model – the Precipitation Temperature Area-Altitude (PTAA) model – was developed by Tangborn (1999) that requires only daily temperature and precipitation data from a weather station and the area-altitude distribution of the glacier. The meteorological data are converted via algorithms into snow accumulation and snow/ice ablation. The daily values of variables such as the elevation of the snowline and zero balance, glacier balance, balance flux, and AAR are correlated during the ablation season using polynomial regressions to obtain the minimum fitting error. The model is illustrated for the South Cascade Glacier, WA, for 1959–1996 (Tangborn, 1999) and has been applied by Beedle *et al.* (2007) to the Bering Glacier, AK.

It is particularly important to resolve the elevation-dependence of the primary forcing fields, temperature and precipitation, including the elevation of the freezing temperature line and the rain/snow boundary. One approach that has been used to achieve the required resolution of forcing data is downscaling of global model output by use of a regional atmospheric model. For the glacierized region of southeast Alaska, Bhatt *et al.* (2007) use high-resolution model-derived forcing to drive a mass balance model for various glaciers, most of which have been retreating (e.g., Bering glacier) but a few of which have been growing (e.g., Hubbard glacier). The global model output was obtained from the Community Climate System Model (CCSM). The results of simulations of past and future mass balances suggest that the Bering glacier will lose significant mass and that Hubbard glacier (also a tidewater glacier) will grow more slowly in the near future than in the recent past.

The most well-studied glaciers have typical dimensions of 0.1–10 km, far below the resolution capabilities of global climate models. Recently, however, a glacier parameterization scheme has been developed and implemented in a regional climate model for the Alps (Kotlarski *et al.*, 2008). The scheme interactively simulates the glacier mass balance as well as changes of the areal extent of glaciers on a sub-grid scale. The temporal evolution and the magnitude of the simulated glacier mass balance match glacier observations for the period 1958–1980, but the subsequent strong mass loss (to 2003) is systematically underestimated.

3.9   Ice caps

An ice cap is an ice mass with radial outflow that covers less than 50,000 km² of land (usually a highland area). Some are dome-shaped, with lobes and outlet glaciers in which the ice drains away. The dome of the ice cap is usually centered over the highest point of a mountain range. Ice flows away from this **ice divide** towards the ice cap's periphery where there is melting or calving into lakes or the ocean. Ice caps are common in the Arctic islands of Canada, Svalbard, Novaya Zemlya, Severnaya Zemlya, Franz Josef Land, around the margins of the Greenland Ice Sheet, the Antarctic Ice Sheet, Antarctic Peninsula, and Patagonia. Others are found in Tibet, the Tien Shan, Pamir, and Iceland. Box 3.3 describes a selection of ice caps around the world.

3.10 Glacier hydrology

The hydrology of glaciers has received considerable attention because it is a key to understanding glacier behavior. Glacier hydrology controls many of the major dynamical processes acting in glaciers (Knight, 1999). Moreover, the timescale of glacial hydrologic processes is of the order of 10^{-6}–10^0 yr compared with 10^1–10^4 yr for glaciers themselves (Clarke, 2005). The behavior of water in glaciers yields information about the structure of the ice, and of the glacier, on a variety of space and timescales. The role of water at the base of the glacier in glacier sliding plays a critical role in glacier surges and in the mechanics of ice streams. However, a large part of our understanding of englacial and subglacial drainage is based on theoretical modeling in the absence of direct observations. The determination of englacial and subglacial hydrology depends on indirect methods such as dye tracing, water-pressure monitoring, and chemical analysis of meltwater. The theoretical basis of these approaches is fully reviewed by Clarke (2005).

Liquid water is delivered to a glacier as rainfall, which may refreeze. Most liquid water in a glacier system, however, is acquired by the melting of snow and ice at the glacier surface. Water may flow on the surface in supraglacial channels, that are prominent features of subpolar glaciers and ice caps during the melt season and these may terminate in ponds and lakes on the ice surface. The water in some channels may enter a **moulin** (a French word for mill) and descend into the ice body. A moulin is a narrow tube or shaft up to ~10-m wide that forms in zones of transverse crevasses (Figure 3.9). It may penetrate to 10–40 m (the typical depth of crevasses) or descend hundreds of meters to the base of the ice. Hence, the water may become englacial or subglacial.

Studies in Greenland and Svalbard show that surface-to-bed drainage systems re-form annually by hydrologically driven fracture propagation (Benn *et al.*, 2009). On Hansbreen, Svalbard, fracturing occurred due to a combination of extensional ice flow and abundant surface meltwater at a glacier confluence. They show that englacial drainage systems in Khumbu Glacier, Nepal, and Matanuska Glacier, Alaska, consist of vertical slots that plunge down-glacier at angles of 55° or less. Surface-to-bed drainage appears to occur wherever high meltwater supply coincides with ice that is subjected to sufficiently large tensile stresses.

Hooke (1989) considers that englacial drainage systems are made up of an arborescent network of passages. Millimeter-sized tributaries coalesce downward into larger conduits. These tend to close off in the winter and reopen in the spring-summer when channels are

Box 3.3 Ice caps

Ice caps range from simple domes to complex forms where they overlie a mountain range or ranges. Here we give a few illustrations of this variety.

Devon Island's ice cap (74.5–75.8° N, 80–86° W) rests on an upland plateau dissected by steep-sided valleys that control the locations of its major outlet glaciers (Dowdeswell *et al.*, 2004). The ice cap proper, in the eastern part of the island, occupies 12,050 km^2 and has a volume of 3,980 km^3 (about 10 mm sea-level equivalent). Boon *et al.* (2010) cite values of 14,400 km^2 and, excluding the southwest arm, 12,794 km^2. The crest of the ice cap reaches 1,920 m and the maximum thickness determined by airborne ice-penetrating radar is 880 m. For 1999–2005 the mean areas occupied by the different facies were: 35% for glacier ice, 21% each for the superimposed ice zone and the saturation zone, and 22% for the percolation zone, based on QuikSCAT (Wolken *et al.*, 2009). About 50% of the ice cap is frozen to its bed (Burgess *et al.*, 2005). Typical of many Arctic ice caps, the velocity structure shows fast-flowing units within slower-moving ice. Outlet glacier velocities are 7–10 times those in the undifferentiated flow. There were 42 melt days on average for 2000–2004 (Sharp and Wang, 2009). JJA air temperatures at 700 mb are significantly positively correlated with the glacier ice area. The climate is arctic continental with very cold winters, short cool summers, and a mean annual precipitation of ~300–500 mm. During 1961–1998, 12 years had a positive mass balance and 25 were negative (Dowdeswell and Hagen, 2004). The mean balance was -0.06 ± 0.24 m a^{-1}. Two basins in the northwest sector of the ice cap are still gaining mass. Combining mass balance and calving, Mair *et al.* (2005) obtained an average value of -0.17 ± 0.06 m w.e. a^{-1}. Iceberg calving represents about 30% of the mass loss since the 1960s (Boon *et al.*, 2010). Burgess and Sharp (2008) report that volume changes derived from the basin-wide values for all drainage basins indicate a net loss of -76.8 ± 7 km^3 from the main portion of the ice cap from 1960 to 1999, contributing 0.21 ± 0.02 mm to global sea level over this time. Shepherd *et al.* (2007) using ERS interferometric SAR data indicated that the net mass balance was about half of these estimates, but assumptions made in the calculations and the neglect of the stagnant southwest arm appear to account for their lower values (Boon *et al.*, 2010).

Severnaya Zemlya is the most easterly glacierized archipelago in the Russian High Arctic located between 73–82° N and 90–110° E. Bassford *et al.* (2006) analyze the climate and mass balance of the Vavilov Ice Cap on October Revolution Island. Vavilov Station has a mean annual temperature of -16.5 °C and an annual precipitation of 423 mm. The Vavilov Ice Cap has a relatively simple form with an area of 1,771 km^2 and a summit elevation of 708 m asl; its total ice volume is 567 km^3 (Dowdeswell *et al.*, 2010). The modeled mean net balance of the entire ice cap is -2.2 cm w.e. a^{-1}, which compares closely with a measured average value of -2.8 cm w.e. a^{-1}, indicating that the ice cap was close to balance during 1974–1988. On average, 81% of the meltwater is lost from the ice cap as runoff, with the remainder refreezing as superimposed ice. Above the ELA superimposed ice makes up 40% of the total net accumulation, with the remainder coming from firn that has been densified by refreezing.

The Southern Patagonia Ice Field with an area of 11,259 km^2 is the world's third largest continental ice mass, located between 48.3° and 51.5° S at 73.5° W in the southern Andes (Figure 3.8). Aniya *et al.* (1996) point out that there is an additional area of small valley glaciers

Box 3.3 (cont.)

Figure 3.8 View of the Southern Patagonian ice field (50° S, 73.6° W) March 30, 2003 [NSIDC Glacier Photography Collection: International Space Station imagery glacier id: ISS006_E_41110. http://eol .jsc.nasa.gov/scripts/sseop/photo.pl?mission=ISS006&roll=E&frame=41110]

of 1,513 km^2. Casassa *et al.* (2002) inventory the 48 outlet glaciers, the largest of which are the Upsala (902 km², 60 km long) and Viedma (945 km², 71 km long) in the east, and the Pío XI (or Brüggen) glacier (1,265 km², 64 km long) in the west. Those flowing west terminate in fiords and those flowing east end in proglacial lakes. Annual precipitation on the western side of the ice field increases from ~3,700 mm at sea level to an estimated maximum of 7,000 mm at 700 m elevation on the ice field summits, and decreases rapidly to the east.

Gran Campo Nevado (GCN) forms an isolated ice cap on the Península Muñoz Gamero (53° S, 73° W). It has radial outlet glaciers. There are 75 glaciers organized into 16 glacier groups on the southern part of the peninsula. The largest glacier group consists of 27 drainage basins on the ice cap, which cover an overall surface area of 199.5 km^2; there are other small cirque and valley glaciers in the southern part of the peninsula totaling 53 km^2 (Schneider *et al.*, 2007). At the summit the GCN reaches approximately 1,740 m asl; the elevated plateau-like part of the ice cap is located at about 1,200 m asl. The climate is cool, very humid and extremely windy. The mean annual air temperature is 5.7 °C and there are 6,500 mm of annual precipitation at sea level, with more than 10,000 mm w.e. of solid precipitation falling at higher elevations on the ice cap. Overall glacier retreat on the ice cap amounts to an area loss of 2.4% per decade from 1942 to 2002.

The Juneau Icefield (59° N, 134.5° W) is located in British Columbia–Alaska and covers an area of 3,900 km^2 in the Coast Ranges extending 140 km north–south and 75 km east–west. An icefield differs from an ice cap in that consists of interconnected valley glaciers from which rise protruding high peaks known as nunataks – a Greenlandic term. The Juneau Icefield is the source of over 40 large valley glaciers, including the Mendenhall and Taku glaciers, and 100 smaller ones. It has been extensively studied under the Icefield Ranges Research Program but much of the data remain unpublished. For the Juneau Icefield, there is a 60-year record of mass balance for the Taku Glacier and a 53-year one for Lemon Creek Glacier, both on the southern side of the icefield (Beedle, 2005). The primary climatic driver is increased ablation season temperature beginning in 1989. The negative mass balance of Lemon Creek Glacier (11.7 km^2) increased from -0.22 m w.e. a^{-1} during 1953–1976 to -0.78 m w.e. a^{-1} for 1989–2005. Taku Glacier (671 km^2) – a former tidewater glacier that now terminates on land (Post and Motyka, 1995) – gained mass during 1946–1988 at 0.42 m w.e. a^{-1}, then its mass balance became negative -0.18 m w.e. a^{-1} for 1989–2005 (Pelto et al., 2008). Despite this difference in sign, the annual correlation between the two records is strong ($r = 0.84$), implying a regional signal. Pelto et al. (2008) show that the Taku Glacier advanced 7.5 km from 1890 to 2003, but at a slowing rate after 1988. It is 1,000–1,400-m thick around 22 km above the terminus and has a velocity of 0.5 m d^{-1}.

Vatnajökull ice cap (64° N, 17° W) is located in the southeast of Iceland, covering 8,100 km^2; it has a volume of 3,100 km^3. The average ice thickness is 400 m, with a maximum value of 1,000 m. 43 outlet glaciers drain the ice cap. Seven volcanoes are situated underneath Vatnajokull and form volcanic lakes due to melting of the basal ice. In 1996 an eruption occurred along a 6–km-long fissure on the northern rim of the Grimsvotn caldera. Meltwater (initially sub-glacial) flowed into Grimsvotn which is about 100 km^2 in area and raised the water surface until drainage began about 2 weeks later underneath Skeidarajökull and out onto the sandur in a large glacial lake outburst flood (or *jökulhlaup*). The ice cap climate is subpolar maritime with mild and windy winters and cool, wet summers with high humidity and cloudiness. Mean annual precipitation is about 750 mm.

Quelccaya ice cap (13.9° S, 70.8° W) in the Cordillera Oriental of the Peruvian Andes has an average elevation of 5,470 m and covers an area of 44 km^2. It is the largest glacierized area in the tropics. Its rim mostly forms steep ice cliffs and the ice cap feeds only a few outlet glaciers; the largest is on the western side. The ice is 160-m thick and the net accumulation is about 1,400 mm w.e. The mean annual air temperature is about -3 °C.

The Guliya ice cap (35.2° N, 81.5° E) in the far western Kunlun Shan on the Qinghai–Tibetan Plateau has a summit elevation of 6,710 m and occupies an area of 376 km^2. It resembles a "polar" ice cap having a mean annual temperature of ~-18 °C. It is 308 m thick and is surrounded by 30–40-m-high ice walls. The lowest 20 m of the ice core extracted by Thompson et al. (1997) may be more than 500,000 years old.

enlarged and new ones form. At the base of a valley glacier is a tortuous system of interlinked cavities transected by a few large, relatively straight conduits. Subglacial conduits are probably broad and low, rather than the theoretical semi-circular shape. They are usually cut into the ice (Röthlisberger channels), but occasionally they may be incised into the bedrock or subglacial sediments(Nye channels). Conduit size is determined by two opposing effects (Paterson, 1999, p. 111): (1) water flow in the conduit melts the ice in the walls by viscous dissipation in the water and heat from the friction of the water on the walls and (2) when the pressure of the

Figure 3.9 Supraglacial stream disappearing down a moulin, near the ice margin in the Ilulissat region, West Greenland. [Source: Dr. Konrad Steffen, CIRES, University of Colorado, Boulder, CO. In "The Greenland ice sheet in a changing climate" 2009 AMAP. http://amap.no/swipa/press2009/Press_Photo_3 .jpg]

overlying ice exceeds the water pressure, the conduit closes; the rate of decrease of the conduit diameter is proportional to (ice pressure − water pressure)[3] according to Nye (1953).

Following Paterson (1999) water flow is driven by the sum of the water pressure gradient and the gradient of gravitational potential energy. This gradient can be treated as the gradient of a water pressure potential, ϕ:

$$\phi = \phi_o + p + \rho_w gz \tag{3.10}$$

where ϕ_o is a constant and z is the elevation of a point in the conduit above a reference level. Assuming that the water pressure is equal to that of the overlying ice, it can be shown that

$$\phi = \phi_o + \rho_i gH + g(\rho_w - \rho_i)z \tag{3.11}$$

where H is the height of the glacier surface. Water flows in a direction perpendicular to equipotential surfaces whose slope is 11 times the surface slope and in the opposite direction. This arises because on an equipotential surface $\phi = 0$ and in two dimensions

$$\frac{dz}{dx} = \frac{\{-\rho_i\}}{\{\rho_w - \rho_i\}} \frac{dH}{dz}.$$ (3.12)

The term in brackets has a numerical value ~ -11.

Studies on the Matanuska Glacier, AK, show that early in the melt season the subglacial drainage system is not fully developed (Ensminger *et al.*, 1999). Then, as meltwater increases, the channels enlarge and increase in number and water storage increases. Late in the melt season, as melt inputs decrease, stored water drains as the channel volume increases. At the end of the melt season, ice motion leads to channel closure and drainage may stop before freeze up.

Glacial Lake Outburst Floods (GLOFs) deserve mention. There are several types. In one type, lakes from glacial meltwater form behind a moraine, which eventually fails as water pressure builds up. It is also possible that the moraine is ice cored and that the ice melts. A glacial lake can be formed by a glacier advance damming-up a river. In another type the lakes are supraglacial due to rapid ice melt at the glacier surface. The discharge may be an order of magnitude greater than normal snowmelt floods. In the Chinese part of the central Himalaya, Zeng *et al.* (2008) identify seven moraine-dammed lakes in the Pumqu and Poiqu river drainages. They are 1–2 km in length and have water volumes of 0.007–0.05 km^3. At least 20 catastrophic events have been recorded in the Himalaya over the past 50 years (Shrestha, 2005). Several events have been studied in the Nepal Himalaya. (ICIMOD rep. UNEP 2001; Ives, 1986). The evolution of Imja Lake in the Khumbu is reviewed by Watanabe *et al.* (2009) who show that the lake's water level has fallen from 5,041 to 5,004 m between 1964 and 2006, so it is no immediate hazard. The outburst from Dig Tsho glacial lake in eastern Nepal in August 1985 destroyed 14 bridges and caused severe damage to a hydropower station under construction. The peak discharge may have reached 2,000 m^3 s^{-1} (Vuichard and Zimmermann, 1986). In northwest Bhutan, GLOFs in 1957, 1969, and 1994 caused extensive damage in the Lunana district. Glacier-dammed lakes are a major hazard in the Karakorum Mountains where there are 11 surging glaciers in the upper reaches of the Indus drainage (Zeng *et al.*, 2008). On the Yarkant River on the north slope of the Karakorums, there were four major GLOF events between 1969 and 1985 with peak discharges of 4,000–6,000 m^3 s^{-1}. Ives *et al.* (2010) survey the glacial lakes in the Hindu Kush-Himalaya and discuss early warning systems for flood events and risk assessment. Details are provided on 34 GLOF events and almost 8,800 glacial lakes are identified in the mountain regions of Bhutan, China, India, Nepal and Pakistan, of which 203 are considered to be potentially dangerous.

A **jökulhlaup** (Icelandic for **glacier leap**) is a subglacial outburst flood that may be triggered by subglacial volcanic activity. The glacier is abruptly lifted ~ 1 m at the onset. Jökulhlaups occur from Vatnajökul in Iceland, which periodically cause massive flooding on the Skeidarajökul Sandur in the vicinity of Skaftafell (Björnsson, 2002; Ives, 2007). Clarke (2003) re-examines the hydraulics of jökulhlaups and argues that critical elements are the lake water temperature prior to discharge and the constriction that holds up the drainage. He re-examines the hydraulic roughness of the conduits involved and finds it to be lower than predicted. Flow constrictions are shown to be quite mobile during a flood event rather than acting as a simple bottleneck.

Floods may also occur from water stored within and beneath glaciers that is suddenly released due to some triggering event. Outburst floods thought to originate from subglacial storage have occurred frequently from South Tahoma Glacier on Mount Rainier, WA (Dridger and Fountain, 1989).

Surface melt processes

Modeling of glacier melt has typically followed two approaches: energy balance calculations or a temperature index (Hock, 2005). Escher-Vetter (1985) examined five years of energy balance measurements at Vernagtferner in the Tyrol. Melt (M) is calculated from

$$M = Q_M/\rho_w L_f \tag{3.13}$$

where Q_M is the energy used in melt, ρ_w is the density of water, and L_f is the latent heat of fusion. Q_M is determined from

$$-Q_M = R_n + H + LE + G + P \tag{3.14}$$

where R_n = net radiation, H = sensible heat, LE = latent heat of evaporation, G = heat conduction in the ice, and P = advective heat supplied by rainfall. Positive (negative) values represent heat gains (losses) by the surface.

$$H = \rho_a c_p K_H \partial \bar{\theta}/\partial z \tag{3.15}$$

$$LE = \rho_a L_v K_E \partial \bar{q}/\partial z, \tag{3.16}$$

where ρ_a = air density, c_p = specific heat at constant pressure, K_H = eddy diffusivity for heat, K_E = eddy diffusivity for vapor, θ = potential temperature, q = specific humidity, z = height, and the overbar signifies a mean value. P is determined from

$$P = \rho_w c_w R \ (T_R - T_s), \tag{3.17}$$

where ρ_w and c_w are the density and specific heat of water, respectively, R is the rain rate and T_R and T_s are the temperatures of the rain and the surface, respectively.

Table 3.3 illustrates representative values of energy budget components. On most mid-latitude glaciers R_n accounts for 50–70% of the melt. The shortwave albedo (α) term is the major determinant of the absorbed solar radiation ($S \ (1 - \alpha)$) and this depends upon the depth and age of the snow cover on the glacier surface. Fresh snow has an albedo of 0.8–0.9, whereas for melting snow it is about 0.7. Mineral dust and black carbon soot on the snow

Table 3.3 Representative values of glacier energy budget components

Location	Period	R_n	H	LE	Q_M	Source
		Units: W m^{-2} (percent of total energy)				
Aletsch Glacier	Feb–Aug 28, 1965	129 (71)	38 (21)	14(8)	−181	Röthlisberger and Lang, 1987
2,220 m ice Storglaciaren	Jul 19–Aug 27, 1994	73 (66)	33 (30)	5 (5)	−122	Hock and Holmgren,1996
1,370 m ice Ivory Glac., NZ	53 d, Jan–Feb	76(52)	44 (30)	23(16)	−147	Hay and Fitzharris, 1988
1,500 m Zongo Glac.	1972–73 Sep 1996–Aug 1997	17 (65)	6 (23)	−17(−65)	−9 (−35)	Waggon et al., 1999
5,150 m Yanlonghe Qilian Shan 1977	1977–78 Summer	139 (92)	13 (8)	−20 (−13)	−125 (−82)	Liu et al., 2008

Note: 1 W m^2 = 0.0864 MJ m^{-2} d^{-1}

lower the albedo further. Ming *et al.* (2009) estimate that carbon soot on glaciers in western China and Tibet lowers the albedo of snow-covered glaciers by 4–6%. Experiments on Alpine snow suggest a lowering of the visible albedo by ~10% by soot with a doubling of absorbed solar radiation (Sergent *et al.*, 1993). The Bowen ratio (H/LE) is conventionally used to assess the relative role of the turbulent heat transfers, but on glaciers their signs are often opposite, Accordingly, Liu *et al.* (2008) propose using the ratio of the heat used in melting to the latent heat term. They show that this ratio decreases exponentially from about 18 at 4,000 m to ~1 at 6,000 m.

Temperature index methods use degree-day calculations where heating (cooling) degree-days are summed values of the departure of the daily mean temperature above (below) a given base value such as 0 °C. These sums are used with a constant melt factor per degree-day to estimate runoff; this degree-day factor (DDF) is expressed in units of mm d^{-1} K^{-1}.

$$\Sigma M = \text{DDF}\Sigma\Delta T\Delta t, \qquad (3.18)$$

ΣM is the total melt over time interval Δt (days), and $\Sigma\Delta T$ is the accumulated positive degree-days. Values of DDF for ice range from 5.5 to 20 mm d^{-1} K^{-1} (Hock, 2003); values for selected glaciers are illustrated in Table 3.4.

Martinec and Rango (1986) developed the Snow Runoff Model (SRM), for example (see Chapter 2A), on this basis. Drawbacks to the use of temperature index models include the

Table 3.4 Selected values of degree-day factors for glaciers (after Hock, 2003)

Location	Altitude (m)	Period	DDF	Source
Alfotbreen, Norway	850–1,400	1961–1990	6.0	Laumann and Reeh, 1994
Hellstugubreen	1,400–2,000	1961–1990	5.5	Laumenn and Reeh, 1994
SW Brit. Col. 9 glacier avge		5 to 30 yrs	4.9	Shea *et al.* 2009
Satujökull, Iceland	800–1,800	1987–1992	7.7	Johannesson *et al.*, 1995
Yala Glacier, Nepal	5,120	1/6–31/7/96	9.3	Kayastha, 2001

fact that their accuracy decreases with increasing temporal resolution and spatial variability cannot be modeled accurately as melt rates may vary substantially due to topographic effects (shading, slope, and aspect angles) (Hock, 2005). The DDF also varies over the course of a melt season. Subsequently, the models have been modified to account for decreasing melt rates at higher elevations and on north-facing compared with south-facing slopes. The commonly observed success of such models is attributed to the high correlations between air temperature and infrared radiation from the atmosphere, and sensible heat flux. Ohmura (2001) shows that these two terms account for about 75% of the energy required for melt. When the sensible heat component is large there tend to be small values of DDF (Hock, 2003). This is the case in maritime environments. Sicart *et al.* (2008) examine temperature index models for glaciers in the tropical Andes (Zongo), the French Alps (Sorlin), and northern Sweden (Storglaciären). They show that the degree-day model is inappropriate for tropical glaciers due to the weak correlation of solar radiation and temperature. It works best in northern Sweden where there is a high correlation of temperature and melt energy due to variations in sensible and latent heat. Carenzo *et al.* (2009) show that a high Solar Radiation Factor (SRF) is obtained on clear-sky days on the Haut Glacier d'Arolla, whereas a higher Temperature Factor (TF) is typical of locations where glacier winds prevail and turbulent fluxes are high. Precipitation amounts have usually been treated as annual totals but Fujita (2008) shows that sensitivities are higher for the glaciers located in a summer-precipitation climate (such as Central Asia) than for those located within a winter-precipitation climate.

Glacier runoff

Glaciers play a major role in regulating water supply in many parts of the world (Meier, 1969). Water is stored over the long term ($\sim$$10^3$–$10^6$ yr) in glaciers and ice sheets and on monthly–annual timescales in seasonal snow and ice (Jansson *et al.*, 2003). Long term storage is important for water resources and global sea level while seasonal timescales affect the runoff characteristics in glacierized catchments and downstream. Typically, water is stored in glaciers, especially in the firn layer, in May–June in the Northern Hemisphere and released in June–September. Tangborn (1984) proposed a runoff model that included seasonal cyclic storage and release of water.

Kasser (1973) examined the effects of a decrease of glacier area on runoff changes for the Rhône drainage basin. The glacierized area decreased from 17.9% in 1876 to 13.6% in 1968. The influence of the change on April–September runoff for 1916–1955 was shown by a partial correlation coefficient of 0.384. Collins (2006) analyzed the influence of glacierized areas representing from 0 to 70% of basins in the upper Aare and Rhône. While unglacierized basin runoff reflected the generally increasing trend of precipitation, the more highly glacierized basins showed runoff mimicking mean May–September air temperature during two periods of warming. Runoff increased gradually from the 1900s, and then rapidly in the 1940s before decreasing to the late 1970s. Runoff increases during the subsequent warming period did not exceed those attained during the earlier warming.

A key question for hydrologists is what fraction of runoff is due to the melting and retreat of glaciers. While there was early work on this problem, it is only recently that detailed studies have been carried out to address this issue. Results using the DANUBIA hydrological model for the Upper Danube drainage system show that in the upper glacierized watersheds (Vent in the Ötztal, with 35% glacier area, for example), runoff originates almost equally from rainfall, snowmelt and ice melt (Weber *et al.*, 2011), but this decreases to 8% at Innsbruck and only 2% at Passau.

In recent decades, glaciers in many mountainous regions have shrunk, leading to changes in river streamflow attributed to glacier runoff. Predominantly in small glaciers such as in Canada, runoff from glaciers has decreased because of glacier mass loss, while runoff from larger glaciers such as Alaska has typically increased. Glacier shrinkage has resulted in significant shifts in downstream water quantity and quality (SROCC, 2019). Huss *et al.* (2008) determined the changes in runoff for scenarios SRES A2 and B2 applied to three highly glacierized basins in the Zinal valley, Valais, Swiss Alps, for 2007 to 2100. Annual runoff from the drainage basins shows an initial increase due to the release of water from storage as the glaciers shrink dramatically, but after some decades, depending on catchment characteristics and the applied climate change scenario, runoff stabilizes and then drops below the current level. Runoff increases during spring and early summer, due to earlier melt onset, whereas that in July and August decreases significantly.

Lambrecht and Mayer (2009) calculate the excess discharge resulting from glacier recession in western Austria between 1969 and 1998. The glaciers in western Austria lost about 22% of their volume during this interval (Lambrecht and Kuhn, 2007). Values of excess discharge for the catchments of the Oetztal and Zillertal ranged between 1.5 and 9% for the period depending on the degree of glacierization (4–40%). The fraction increases to 3–12% for summer months. For individual months the fraction can reach 25% for a catchment with 40% glacier coverage, but even for 8–15% glacierization it can be up to 20% of discharge.

For Nepal, where glacier ice is mainly located between 4,000 and 6,000 m asl, Armstrong *et al.* (2009) and Alford and Armstrong (2010) estimate the glacier contribution (excluding snowmelt) to runoff using a mass balance gradient of 1.4 m $(100 \text{ m})^{-1}$ below the freezing-level elevation of 5,400 m down to the glacier termini. From this they estimate for nine glacierized and gauged basins that the glacier contribution to stream flow varies from ~20% in the Budhi Gandaki Basin of the central Narayani to ~2% in the Likhu Khola Basin of the eastern Sapta Kosi, averaging approximately 10% overall. This represents ~4% of the total mean annual estimated volume for the rivers of Nepal. They also concluded that the entire volume of Nepal Himalayan ice (~480 km^3) represented only about 4% of the total annual flow of the Ganges! In a similar study for the Din Gad catchment in the Garhwal Himalaya (30.8° N, 78.7° E), which is 9.6% glacierized, Thayyen and Gergan (2010) found that glacier melt during 1994–2000 comprised 7.7–12.7% of the bulk glacier runoff.

Immerzeel *et al.* (2010) show that the glacierized area (based on GLIMS data) ranges from 2.2% of the Indus catchment to 3.1% for the Brahmaputra and 1.0% for the Ganges. The snowmelt and glacier discharge from above 2,000 m is shown to amount to 151% of the naturally generated discharge downstream on the Indus; corresponding values for the Brahmaputra and Ganges are only 27% and 10%, respectively. GCM runs for the SRES A1B scenario for AD 2050 show decreases of upstream meltwater supply of −8.4% for the Indus, −19.6% for the Brahmaputra, and −17.6% for the Ganges, but these are offset by increased precipitation (+25% for the Indus and Brahmaputra and +8% for the Ganges).

In western China, the glacial meltwater contribution ranges from 25.4% of the total discharge in Xinxiang, to 8.6% in Tibet, and 3.6% in Gansu (Kang *et al.*, 2008). In the Hexi Corridor, the meltwater from the Qilian Shan provides 14% of the discharge in the three rivers of the Corridor. The decrease in glacier volume in western China during 1980–2000 was 1.2–1.8% and the corresponding increase in the meltwater contribution was 0.8–1.3%.

3.11   Changes in glaciers and ice caps

Because they are so sensitive to temperature fluctuations, glaciers provide clues about the effects of global warming, for the recent widespread global retreat of glaciers demonstrates the effects of global warming (Roe *et al.*, 2016). With a few exceptions, glaciers worldwide have retreated at unprecedented rates over the twentieth century. Some ice caps, glaciers, and ice shelves have completely disappeared and many more are retreating so rapidly that in a few decades they may cease to exist. A study by the WGMS concluded that "rates of early twenty-first-century mass loss are without precedent on a global scale, at least for the time period observed and probably also for recorded history" (Zemp *et al.*, 2015). Pelto (2016) reported that the cumulative mass balance loss between 1980 and 2015 is "the equivalent of cutting a 20.5 m thick slice off the average glacier".

Various measurements of the mass balance of glaciers outside of Antarctica or Greenland, often known as subpolar and mountain glaciers, have been compiled and analyzed. However, current glacier retreat is a challenge for glacier monitoring. The rapid glacier retreat with large glaciers disintegrating into smaller glaciers and the appearance of periglacial lakes makes measuring changes in glacier length, area, and mass balance challenging, as locating the glacier terminus is complicated by increased debris cover and lakes (Fisher *et al.*, 2016). The ELA of some glaciers could be above summits, when the mean mass balance at all altitudinal zones is negative in the presence of accumulation, for example, when a steep east-facing slope with extremely high ablation rates borders on a shady trough with extreme accumulation rates, such as some glaciers in Austria in the last decade were mostly above summits. Therefore the ELA is above summit since there is no elevation zone with positive mean mass balances. The increase of debris cover could also change glaciers' response to climate change, and trigger rock falls from formerly ice-covered lateral rock walls.

Changes in glaciers may be with respect to the number, length, area, thickness, mass balance, or volume of the ice bodies. There are widely differing amounts of information available on these different indicators. The major sources are the WGMS in Zurich, Switzerland, and the GLIMS archive at the National Snow and Ice Data Center, Boulder, Colorado (Kargel *et al.*, 2014). The GLIMS project released Version 5.0 of the RGI in 2015 (Arendt *et al.*, 2015). The data sets assembled by Cogley (2009) and Dyurgerov (2010) are widely used as they have been subjected to some quality control. Monitoring of glacier length variations began to be coordinated in 1894 through the establishment of the International Glacier Commission in Zurich. Nevertheless, a detailed inventory of glacier location, size, and altitude extent, is available for only about 100,000 glaciers covering about 180,000 km^2. It was estimated by Meier and Bahr (1996) that there were about 160,000 glaciers in the world covering an area of 780,000 km^2, so these figures correspond to about 63% of the total number and 23% of the overall glacier area. However, Bahr *et al.* (2009) state that, taking into account the glaciers on the peripheries of Greenland and Antarctica, the total number is 300,000–400,000. Accordingly, our available database is even smaller than we believed. Glacier outlines in GLIMS are available for about 83,500 glaciers based mainly on remote sensing data from the early twenty-first century.

In all, there have been 3,385 annual mass balance measurements reported from 228 glaciers around the globe for 1946–2005 (Zemp *et al.*, 2009). However, the dataset is strongly biased toward the Northern Hemisphere and Europe; there are only 30 "reference" glaciers (in nine mountain ranges) that have uninterrupted series going back to 1976, and 12 back to 1960. Of these reference glaciers, 45% are between 0.1 and 1 km^2 and 32% between 1 and 10 km^2 in area, so the non-representative size distribution adds additional bias. The available data from the six decades indicate a strong ice loss in the 1940s and 1950s, followed by moderate mass loss until the end of the 1970s and a subsequent, continuing acceleration. The 30 "reference" glaciers with (almost) continuous measurements since 1976 show an average annual mass loss of 0.58 m w.e. for the decade 1996–2005, which is more than four times the rate for 1976–1985. Among various studies on glacier mass balance, glaciers measured are still sparsely distributed

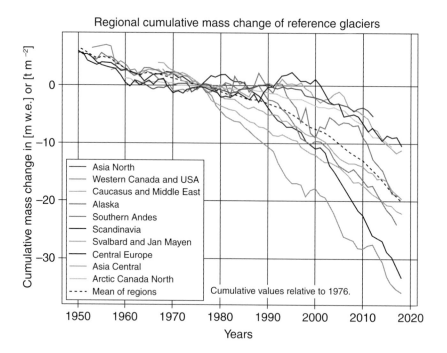

Figure 3.10 Cumulative mass change relative to 1976 given in m w.e. (water equivalent) for regional and global means based on data from a set of global reference glaciers with more than 30 continued observation years for 1950–2018. The data are compiled by the World Glacier Monitoring Service (WGMS) in annual calls for data from a scientific collaboration network in more than 40 countries worldwide. Regional values are arithmetic averages while global values are one single value averaged for each region with glaciers to avoid a bias to well-observed regions. Values before 1960 and in 2018 need to be taken with caution due to the limited sample size (https://wgms.ch/global-glacier-state/) (A black and white version of this figure will appear in some formats. For the color version, please refer to the plate section.)

over all mountain and subpolar regions, with about 70% of the observations coming from the mountains of Europe, North America, and the former Soviet Union (Gardner *et al.*, 2013; Zemp *et al.*, 2015). Figure 3.10 shows the trends in cumulative mass balance for regional means of 10 mountain regions, and the global means with respect to a set of global reference glaciers. Regional and global decreasing trends are about linear until 1990, after which losses accelerated especially in Western Canada, the United States, and Southern Andes (WGMS, 2017). In 2016/2017, observed glaciers experienced an ice loss of 0.85 m w.e.

Due to the impact of global warming since the mid-twentieth century, globally glaciers are retreating or net melting, or both (Marzeion *et al.*, 2018). The vast majority of glaciers in all high mountains have been losing mass in last several decades. The high mountain areas considered in SROCC (2019) of IPCC include ~173,000 glaciers covering an area of 252,000 km^2, spanning an altitude ranging from sea level to >8,000 m (RGI Consortium, 2017), roughly 30% of the total global glacier area, excluding those of the two ice sheets. Their mass budget is determined largely by the balance between snow accumulation and melt at the glacier surface. As many glaciers are losing mass, worldwide glaciers have significant contributed to sea-level rise (SLR) over the twentieth century (about 40%) and will continue to contribute to the SLR during the twenty-first century – around 30% (Church *et al.*, 2013; Gardner *et al.*, 2013; Marzeion *et al.*, 2014). According to the AR5 Report (Vaughan *et al.*, 2013), between 2003 and 2009, most of the ice lost was from glaciers in Alaska, the Canadian Arctic, the periphery of the Greenland Ice Sheet (GIS), the Southern Andes, and the Asian Mountains, which together account for more than 80% of the total ice loss. As glaciers are time-integrated dynamic systems, a response lag of at least 10 years to hundreds of years is observed between changes in climate forcing and glacier shape, depending on the glacier length and slope.

Figure 3.11 shows global cumulative glacier mass change for 1801–2010 set to zero mean over 1986–2005. Estimates are based on glacier length variations (Leclercq *et al.*, 2011), from area-weighted extrapolations of individual directly and geodetically measured glacier mass budgets (Cogley, 2009), and from modelling with atmospheric variables as input (Marzeion *et al.*, 2012; Hirabayashi *et al.*, 2013). Since 1801 until 2010, the global cumulative glacier mass losses is estimated to be about 3×10^4 Gt of ice. However, the earlier estimates were based on limited number of glacier mass balance measurements.

Globally total glacier areas have also decreased in all regions, with the rates of change covering a similar range of values in all regions but considerable variability in the rates of change within each region, and the rates of loss tend to be higher in recent times. As current glacier extents are out of balance with the current climate, glaciers across the world are expected to continue shrinking. Studies show that over 600 glaciers have disappeared, glaciers in the Canadian Arctic Canada (Thomson *et al.*, 2011), the Rocky Mountains (Tennant *et al.*, 2012), Patagonia (Davies and Glasser, 2012), tropical mountains (Cullen *et al.*, 2013), European Alps (Diolaiuti *et al.*, 2012), Tien Shan (Hagg *et al.*, 2012) in Asia, and in Antarctica (Carrivick *et al.*, 2012). However, due to limitations of field observations, the actual number should be much higher. Figure 3.11 shows the estimated global cumulative glacier mass change for 1801–2010 set to zero mean over 1986–2005, based on updates from Cogley (2009) and Leclercq *et al.* (2011), and from modeling (Marzeion *et al.*, 2012; Hirabayashi *et al.*, 2013). Estimates for 1801 to about 1950 involve large uncertainties because number of measured mass balance glaciers was very limited. In Figure 3.11, cumulative mass changes and corresponding rates are shown for global glaciers excluding regions 5 and 19 (bold lines) of Table 3.1, and also for global glaciers excluding only region 19 (thin lines). After 1950, mass loss rates including Greenland are all within the uncertainty bounds of those that exclude Greenland. Marzeion *et al.* simulated a rapid loss from Greenland glaciers. Even though also supported by other studies (Bjørk *et al.*, 2012), the Canadian Arctic and Iceland have mass loss anomalies an order of magnitude lower than predicted for Greenland in the same simulation. The Marzeion *et al.* rates are also considerably greater in the 1950s and 1960s than in the other studies.

We now briefly review the regional changes in glacier characteristics in the major mountain ranges of the world where adequate literature is available. For some regions only length and area change data are available; for others more comprehensive mass balance data exist.

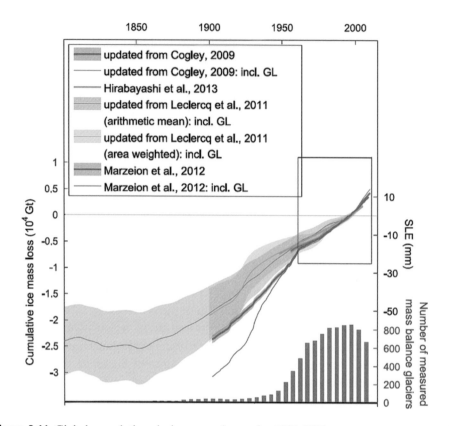

Figure 3.11 Global cumulative glacier mass change for 1801–2010 set to zero mean over 1986–2005. Estimates are based on glacier length variations (updated from Leclercq *et al.*, 2011), from area-weighted extrapolations of individual directly and geodetically measured glacier mass budgets (updated from Cogley, 2009), and from modeling with atmospheric variables as input (Marzeion *et al.*, 2012; Hirabayashi *et al.*, 2013). Uncertainties are based on comprehensive error analyses in Cogley (2009) and Marzeion *et al.* (2012) and on assumptions about the representativeness of the sampled glaciers in Leclercq *et al.* (2011), including Greenland (GL). Uncertainties are shown only for the Cogley and Marzeion curves excluding GL. The blue bars show the number of measured single-glacier mass balances per pentad in the updated Cogley (2009) time series (adapted from figure 4.12 of Vaughan *et al.*, 2013) (A black and white version of this figure will appear in some formats. For the color version, please refer to the plate section.)

Alps

Braun *et al.* (2000) consider the impacts of climate change due to CO_2 doubling on runoff in the Ötztal in the Austrian Alps. They find that in highly glacierized basins (40–80% ice), climate change scenarios suggest strongly enhanced water yields in an initial phase. This has already been observed in years with strongly negative mass balances. Higher flood peaks will result when high melt rates and heavy summer rains coincide. If glacier mass losses continue in the twenty-first century, the glacierized area will diminish and summer discharge will be gradually reduced, resulting in drastic water shortages in hot, dry summers once the glaciers have disappeared.

Farinotti *et al.* (2009) estimate the total ice volume present in the Swiss Alps in the year 1999 to be 74 ± 9 km^3; about 88% of this is stored in the 59 largest glaciers (glaciers with a surface area $A \geq 3$ km^2). For 6 of the 10 largest glaciers, which together contribute more than half of the total estimated ice volume, direct ice thickness measurements were available. Approximately 12% of this ice volume with direct ice thickness measurements made was lost between 1999 and 2008. Huss *et al.* (2010) analyzed 100-year mass balance records for 30 Swiss glaciers. They show that mass losses were particularly rapid in the 1940s and since the 1980s and that there were two short periods of mass gain in the 1910s and late 1970s. The variability was found to be anticorrelated (with a lag of one to several decades) to the Atlantic Multidecadal Oscillation (AMO) in sea surface temperatures. Positive AMO is associated with positive surface air temperature anomalies in Europe. The mean glacier mass balance in the European Alps was -0.31 ± 0.04 m w.e. a^{-1} in 1900–2011, -1 m w.e. a^{-1} over the last decade, and total ice volume change since 1900 was -96 ± 13 km^3 (Huss, 2012).

Berro *et al.* (2007) report on the changes to glaciers in the Italian Alps during 2000–2007. Retreat continued and rates have nearly doubled since 2003 compared to 1992–2002. There were huge losses of up to 3 m w.e. around 3 km altitude in the hot summer of 2003. Koboltschnig *et al.* (2009) examine the regional hydrological impact of the summer 2003 heat wave in Europe for the small, glacierized Goldbergkees basin in the Austrian Alps. It is situated directly beneath the Sonnblick observatory (3,106 m asl). The extreme anomaly of the mean summer (JJA) air temperature amount to 4.4 times the standard deviation of the long-term mean (1886–2000). The mean summer air temperature was 4.7 °C. In 2003, the solid fraction of precipitation was only 35% – the lowest value observed from 1927 to 2005. The winter balance of the Goldbergkees did not show any anomaly, but the specific net balance was -1.8 m w.e. for the 2002/2003 period – the most negative observed. During August 2003, glacier melt contributed 81% of the total runoff.

Arctic

There have been several satellite-based studies of ice extent and melt extent on Arctic ice caps. Wolken *et al.* (2008) used trimlines to map the former extent of perennial snow/ice. A trimline is a distinct line on the side of a valley formed by glacier erosion. The line marks the most recent maximum limit of the glacier. The line may be visible due to changes in color of the rock or to changes in vegetation on either side of the line. Figure 3.12 illustrated the trimline, probably from the LIA maximum, above the shrunken McCall Glacier, Alaska. Wolken *et al.* show that between the end of the LIA and 1960 the ice extent on the Queen Elizabeth Islands shrank by 37% overall. The largest areal decrease was in the eastern part,

Figure 3.12 The McCall Glacier in the Brooks Range, northern Alaska (69.3° N, 143.8° W) in (a) July 1958 (Austin Post) and (b) August 2003 (Matt Nolan). The glacier trimline is clearly visible on the later image (http://nsidc.org/cgi-bin/gpd_deliver_jpg.pl?mccall1958070001. http://nsidc.org/cgi-bin/gpd_deliver_jpg.pl?mccall2003081401)

but in central and western islands the low relief led to almost complete removal of ice and snow by 1960 with only small increases in the ELA. Meighen Island in the west was an exception with a 40% reduction. On southern Baffin Island, a new glacier inventory by Paul and Svoboda (2009) (see Figure 3.13) shows that the glacier ice area (volume) decreased by about 22 (25)% between the LIA maximum extent and the year 2000. Mair *et al.* (2009) found a significant negative mass balance in the Prince of Wales Ice field on Ellesmere Island over 1963–2003 of -2 ± 0.45 km^3 w.e. a^{-1} by iceberg discharge.

Wang *et al.* (2005) analyzed the Canadian Arctic ice caps also using QuikSCAT. For the Queen Elizabeth Islands the average melt season in 2000–2004 was only 38 days. However, ice cap margins facing either Baffin Bay or the Arctic Ocean have significantly longer melt seasons than margins facing the interior of the islands. The annual mean melt duration over the larger ice caps is positively correlated with the local 500 mb height.

Mair *et al.* (2009) analyze the mass balance of the Prince of Wales Icefield on Ellesmere Island during 1963–2003. The ice field has an area of 19,325 km^2. Nunataks and snow-covered mountains reach elevations of over 2,000 m above sea level above a broad, gently sloping central plateau ranging in altitude from 1,350 to 1,730 m asl. The ice field descends to sea level on the east coast of Ellesmere Island and has a strong east-west gradient of accumulation caused by storm tracks from Baffin Bay sweeping across the North Open Water Polynya. They find that the surface mass balance is approximately in balance, but the iceberg discharge is a highly significant component of mass loss making the overall mass balance of the ice field strongly negative (-2 ± 0.45 km^3 w.e. a^{-1}), equivalent to a mean-specific mass balance across the ice field of -0.1 m w.e. a^{-1}).

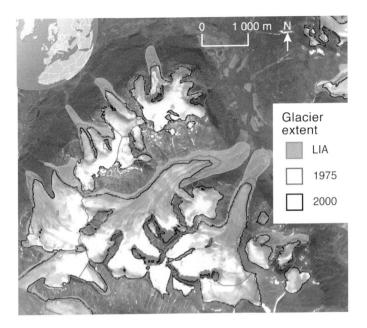

Figure 3.13 Glacier shrinkage since the Little Ice Age in the Cumberland Peninsula, Baffin Island [Source: F. Svoboda, University of Zurich] http://maps.grida.no/go/graphic/glacier-shrinking-on-cumberland-peninsula-baffin-island-canadian-arctic. Cartographer/designer Hugo Ahlenius, UNEP/GRID-Arendal. (A black and white version of this figure will appear in some formats. For the color version, please refer to the plate section.)

The Prince of Wales Icefield contributes ~ 0.005 mm a^{-1} to global eustatic SLR. Williamson *et al.* (2008) use optical satellite imagery to estimate the iceberg calving rates from Agassiz and western Grant Ice Caps on Ellesmere Island. The estimated mean annual calving rate from Agassiz Ice Cap during 1999–2002 was 0.67 ± 0.15 km^3 a^{-1}, of which $\sim 54\%$ emanated from Eugenie Glacier alone. Summer calving rates were $\sim$ 2–8 times larger than annual average rates. The average ratio of the calving flux due to terminus-volume change to that due to ice flow through the glacier terminus was ~ 0.8 for the annual rates and ~ 1.7 for summer rates.

Grant (2010) analyzed the ice cover of Novaya Zemlya using Landsat and ASTER imagery. For an area representing ~70% of the glacierized area, she found that the ice had receded from the LIA maximum by 7% in 2001. Glaciers on the northwest Barents Sea coast had retreated more than those on the Kara Sea side and steep valley glaciers terminating at low elevations had the greatest retreat rates. Between 1989 and 2001, the number of proglacial and ice contact lakes had increased by 29%.

Sharp and Wang (2009) used QuikSCAT for 2000–2004 to examine the Eurasian Arctic ice caps. They found a mean melt season ranging from 77 days on Svalbard and 75 on Novaya Zemlya to only 59 on Severnaya Zemlya. Melt duration was found to be highly correlated with

JJA 850 mb air temperature from NCEP-NCAR data. A reconstruction based on this correlation shows that the five-year period 2000–2004 was the second or third longest melt season for the ice caps during 1948–2005, with 1950–1954 the longest. Kääb (2008) compares a digital elevation model derived from 2002 ASTER satellite optical stereo, elevation data derived from 2006 GLAS laser altimetry, and contour lines from a 1: 100,000 topographic map from the 1970–1971 for two ice caps in eastern Svalbard. The resolution of ASTER imagery is about 15 m and GLAS footprints have a diameter of approximately 70 m and an along-track ground spacing of about 170 m. He obtains area changes of 17–25% and thinning rates of 0.55–0.61 m a^{-1}. Nutt *et al.* (2010) use ICESat data for 2003–2007 compared to topographic maps and digital elevation models for 1965–1990 to calculate long-term elevation changes of glaciers on the Svalbard Archipelago. The average rate of volume change over the past 40 years for Svalbard (excluding Austfonna and Kvitøya) is estimated to be −9.71 ± 0.55 km^3 a^{-1} for a contribution to global sea level of 0.026 mm a^{-1}.

In Greenland, the range of retreat varies from about 1 km to about 3 km between 1850 and 2010, over 160 years (Figure 3.14). Between the seven selected glaciers examined, Sermikavsak, Sorqaup, Tunorssuaq, Sigssarigsut, Serminguaq, Motzfeld, and Sermitslaq, the retreat of the Sorqaup glacier had been the fastest since the early 1900s (WGMS, 2008).

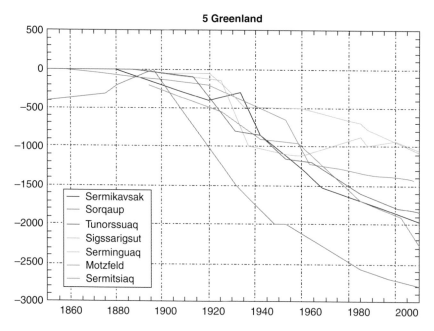

Figure 3.14 Long-term cumulative glacier length changes of the GIS compiled from *in situ* measurements (WGMS, 2008), and reconstructed data points added to measured time series of GIS from Vaughan *et al.* (2013)

Alaska

The glaciers of western North America have dominantly retreated since the end of the LIA in the nineteenth century. The area loss between 1985 and 2005 amounts to 11.5% in British Columbia and Alberta (Bolch *et al.*, 2010). In Alaska, the total area of glaciers is about 79,000 km^2. Arendt *et al.* (2002) used airborne laser altimetry to estimate volume changes of 67 glaciers from the mid-1950s to the mid-1990s. Extrapolation to all glaciers in Alaska yielded an estimated total volume change of -52 ± 15 km^3 a^{-1}, equivalent to a sea level equivalent (SLE) of 0.14 ± 0.04 mm a^{-1}. Repeat measurements on 28 glaciers from the mid-1990s to 2000–2001 indicated an increased average rate of thinning leading to an extrapolated annual volume loss from Alaskan glaciers of -96 ± 35 km^3 a^{-1} or 0.27 ± 0.10 mm a^{-1} SLE over that time interval. From 1972 to 2002, the Malaspina Glacier system in Alaska lost 156 ± 19 km^3 of ice, based on aerial photography, InSAR, and ICESat data (Muskett *et al.*, 2008). This was over an area of 3,661 km^2 representing 73% of area of the total glacier system. The Muir Glacier on Glacier Bay, Alaska, retreated more than 12 km and thinned by over 800 m between 1941 and 2004 (see Figure 3.15); until the mid-1980s it was a tidewater glacier. Berthier *et al.* (2010) revised downward the estimates of Arendt *et al.* by 34%. Combining a comprehensive glacier inventory with elevation changes derived from sequential digital elevation models, they found that between 1962 and 2006 Alaskan glaciers lost 41.9 ± 8.6 km^3 a^{-1}, and contributed 0.12 ± 0.02 mm a^{-1} SLE. The lower values are attributed to the higher spatial resolution of the glacier inventory and to the reduction of ice thinning beneath debris cover. From the Gravity Recovery and Climate Experiment (GRACE) satellite data, Luthcke *et al.* (2008) computed mass changes of the Gulf of Alaska glaciers to be -84 ± 5 Gt a^{-1} contributing 0.23 ± 0.01 mm a^{-1} to global SLR from April 2003 through March 2007.

British Columbia

The area loss between 1985 and 2005 amounts to 11.5% in British Columbia and Alberta (Bolch *et al.*, 2010). This has been accompanied by statistically detectable declines in August stream flow from glacier-fed catchments according to Moore *et al.* (2009). In contrast, glaciers in northwest British Columbia and southwest Yukon have lost mass over recent decades dominantly by thinning, with relatively low rates of terminal retreat, and glacier-fed streams there have experienced increasing flows. This represents the short-term response to glacier shrinkage.

   Glaciers in Garibaldi Provincial Park, in the southern Coast Mountains of British Columbia, were reconstructed from historical documents, aerial photographs, and fieldwork by Koch *et al.* (2009). Over 505 km^2 was covered by glacier ice at the beginning of the eighteenth century. Ice cover decreased to 297 km^2 by 1987–1988 and to 245 km^2 (49% of the early eighteenth century value) by 2005. Glacier recession was greatest between the 1920s and 1950s, with typical frontal retreat rates of 30 m a^{-1}. Many glaciers advanced between the 1960s and 1970s, but all glaciers retreated over the last 20 years. The record of twentieth century glacier fluctuations in Garibaldi

Figure 3.15 The Muir Glacier, Alaska photographed in August 1941 (left) by William O, Field and in August 2004 from the same vantage point (right) by Dr. Bruce F. Molnia, US Geological Survey. [Source: NSIDC repeat photography. http://nsidc.org/cgi-bin/gpd_deliver_jpg.pl?muir1941081301, http://nsidc .org/cgi-bin/gpd_deliver_jpg.pl?muir2004083101]

Park is similar to that in southern Europe, South America, and New Zealand. Glaciers in British Columbia and Alberta, respectively, lost $-10.8 \pm 3.8\%$ and $\sim 25.4\% \pm 4.1\%$ of their area over the period 1985–2005 according to Bolch *et al.* (2010). The least glacierized mountain ranges with smaller glaciers lost the largest percentage of their ice cover. Schiefer *et al.* (2007) use Shuttle Radar Topography Mission (SRTM) data and DEMs from aerial photography to quantify the change of glacier volume in British Columbia for 1985–1999. They find an annual volume loss of 22.48 ± 5.53 km^3 a^{-1}. The recent rate of glacier loss in the Coast Mountains is approximately double that observed for the previous two decades.

Antarctica

In the Antarctic Peninsula, 87% of 244 glaciers considered by Cook *et al.* (2005) retreated, on average, from the mid-twentieth century (*ca.* 1953) through 2004. Advances were more common through 1964 and retreats more common after 1964. In 2000–2004, 75% of the glaciers were in retreat. A major factor on the east side of the peninsula has been the breakup of small ice shelves (see Chapter 8) that led to dramatic increases in calving, acceleration, and dynamic drawdown. In the Antarctic, Kunz *et al.* (2012) found that 12 glaciers around the Antarctic Peninsula showed near-frontal surface lowering since the 1960s, with higher rates of thinning for glaciers on the north-western Antarctic Peninsula. Using a constellation of satellites, Milillo *et al.* (2019) detected a complex pattern of retreat and ice thinning of the Thwaites Glacier of West Antarctica from 1992 to 2017, with sectors retreating at 0.8 km yr^{-1} and floating ice melting at 200 m yr^{-1}, while others retreat at 0.3 km yr^{-1} with ice melting 10 times slower. It accounts for about one-third of mass loss of the Amundsen Sea Embayment of West Antarctica, which has been implicated as a dominant contributor to SLR at present and for decades to come (Deconto and Pollard, 2016).

New Zealand

For New Zealand, Hoelzle *et al.* (2007) calculated a change in glacier area from about 1850 to the mid-1970s of −49% (to 979 km^2) and a corresponding volume loss of −61% (to 67 km^3). The average mass balance for the period was determined to be about −1.25 m w.e. a^{-1} for the "wet" glaciers (with a mass balance gradient of 15 mm m^{-1}) and −0.54 m w.e. a^{-1} for the "dry" glaciers (with a mass balance gradient of 5 mm m^{-1}) of the Southern Alps. Chinn (1999) showed from aerial photography that 78 glaciers during the 1980s–1990s recorded a reversal of the previous glacier recession during the twentieth century. This reversal was associated with a snowline depression of 67 m. However, the Tasman Glacier terminus retreated 3.5 km during 1990–2007 and a large ice-contact proglacial lake developed (Quincey and Glasser, 2009). By 2007 the lake area was ~6 km^2 and it had replaced the majority of the lowermost 4 km of the glacier tongue. Surface lowering of about 1.9 ± 1.4 m a^{-1} is common in the upper areas of the glacier and around the terminal area of the glacier tongue, in areas adjacent to the lateral moraine ridges, downwasting rates reach 4.2 ± 1.4 m a^{-1}.

Andes

About 70% of tropical glaciers are in Peru (~1,370 km^2) and 20% (393 km^2) in Bolivia, with smaller areas in Ecuador and Colombia (General Secretariat of the Andean Community, 2007). For Colombia, Morris *et al.* (2006) compare historical data from 1957 to 1959, Landsat images from 1984, and ASTER images from 2001 to 2004 of glaciers in three regions. In the late 1950s, the total glacier area was 89.3 km^2 and by 2003 this had decreased by 49%. The largest decreases were in the Ruiz-Tolima Massif from 34 km^2 in 1957 to

15.8 km^2 in 2001 and in the Sierra Nevada del Cocuy from 39 to 15.3 km^2 in 2003. The Cordillera Blanca of Peru has the largest glacierized area in the tropics, stretching 180 km north–south between 8.5° S and 10° S. Glacier terminus elevations range from 4,200 to 5,370 m, with a mean of 4,880 m. On average, glacier termini are 102 m higher on the western slope of the Cordillera Blanca (4,914 m) than on the eastern slope (4,812 m) (Racoviteanu et al., 2008). Accumulation occurs during the wet summer season with airflow from the Amazon, and there is relatively little in the dry season. Ablation occurs year round, with higher rates in the wet season. Racoviteanu et al. (2008) find an overall loss in glacierized area of 22.4% from aerial photos in 1970 to Spot imagery in 2003, and an average rise in glacier terminus elevations by 113 m. Likewise, the volume changes of 21 glaciers in the Cordillera Real, Bolivia, determined photogrammetrically, has been decreasing since 1975 without any significant acceleration (Soruca et al., 2009). A notable fact is that the famous Chacaltaya Glacier in Bolivia disappeared in 2009. From the mass balance as a function of exposure and altitude, the ice volume loss of 376 glaciers in the region has been assessed. Soruca et al. show that these glaciers lost 43% of their volume between 1963 and 2006 (mainly 1975–2006), and 48% of their surface area between 1975 and 2006.

The Institute of Snow and Ice Studies (IANIGLA) in Argentina has developed inventories for areas in Southern Patagonia (49° S) and the Andes around 40° S. In the latter, the ice cover decreased about 12% since the early 1980s. Between 1945 and 1986, the Southern Patagonia Icefield lost 500 km^2, nearly 4% of its area, and a similar amount subsequently, with the largest losses from Glaciar Jorge Montt (Chile) and Glaciar Upsala (Argentina). In contrast, the surging Glaciar Pio XI (Brüggen) gained 60 km^2 (5%) over that interval and a further 8 km^2 between 1986 and 2009. Casassa et al. (2002) show that O'Higgins Glacier thinned by 3.2 m a^{-1} from 1914 to 1933 and by 6.7 m a^{-1} between 1933 and 1960. From 1975 to 1995 the rate varied from −2.5 to −11 m a^{-1}. Tyndall Glacier thinned by 2.0 m a^{-1} between 1945 and 1993 and Ameghino by 2.3 m a^{-1} between 1949 and 1993. Upsala Glacier showed a rapid increase in ice loss from −3.6 m a^{-1} during 1968–1990 to −9.5 to −14 m a^{-1} during 1991–1993. Rivera et al., (2002) estimated a volume loss for the Southern Patagonia Icefield of 401 ±174 km^3 during 1945–1996, corresponding to a SLE of 0.022 m a^{-1}, or 6% of the contribution of glaciers and ice caps. This volume loss estimated is in line with calculations of Rignot et al. (2003) based on Shuttle radar data. They showed that volume loss by glacier thinning is 4–10 times larger than that by frontal loss. During the period 1968/1975–2000, glaciers of the Southern Patagonia Icefield (SPI) lost 7.2 ± 0.5 km^3 a^{-1} over an area of 8,167 km^2 and an additional 1.3 km^3 a^{-1} of frontal loss. Scaled over the entire icefield of 13,000 km^2, this implies an ice loss of 13.5 ± 0.8 km^3 a^{-1}. For the North Patagonia Icefield, the thinning of 24 glaciers is 2.63 km^3 a^{-1} over an area of 3,481 km^2. Scaled over the entire icefield of 4,200 km^2, this implies a volume loss of 3.2 ± 0.4 km^3 a^{-1}. The combined volume loss, from both icefields is 16.7 ± 0.9 km^3 a^{-1}, equivalent to a SLR of 0.042 mm a^{-1} (9% of the global total). About half of the ice loss is attributed by Rignot et al. to temperatures rising by 0.5 ° C at 850 mb, near the ELA, over the past 40 years and a decrease in precipitation over the accumulation area of the SPI. However, a substantial part of the thinning appears to be due to ice dynamics and the fact that many of the glaciers calve into freshwater lakes, or the sea.

Among glaciers of 11 mountain regions examined, Zemp *et al.* (2019) found the negative mass budgets of 2006–2015 in the Southern Andes at $-1,200$ kg m^{-2} yr^{-1} to be the highest. Glacier retreats in central Andes and Peruvian Andes have led to reduced streamflow which impacted agriculture dependent on irrigation or municipal water supplies (Drenkhan *et al.*, 2015), hazardous landslides (Anacona *et al.*, 2015), or even population migration in Bolivian Andes (Kaenzig, 2015).

Himalaya

Glaciers in the Himalaya have received very little detailed study despite the large ice area involved (35,000 km^2 with a further 16,500 km^2 in the Karakoram). Glacier termini are retreating at rates of 10–60 m yr^{-1} and many small glaciers (<0.2 km^2) have already disappeared (Bhampri and Bolch, 2009). Shrestha (2005) gives average recession rates (no time interval is identified) for a small number of Himalayan glaciers: for 11 West Himalaya glaciers the rate averages 15.9 m a^{-1}, for 13 north-slope glaciers of the Himalayan Arc it is 9.4 m a^{-1}, for 12 south slope Trans-Himalayan glaciers it is 11.6 m a^{-1}, and for 8 East Himalaya glaciers it is 23 m a^{-1}. For three basins in Himachal Pradesh (~32° N, 77–78° E), Kulkarni *et al.* (2007) give area loss of 21% for 46 glaciers during 1962 to 2001–2004. Xu *et al.* (2007) state that the Gangotri glacier (30.5° N, 79.2° E) in India has retreated three times faster over the last 30 years than during the preceding 200 years. Dyurgerov and Meier (2005) show that many Himalayan glaciers are retreating faster than the world average and are thinning at 0.3–1 m a^{-1}. Nevertheless, in the Karakorum there is evidence from over 30 glaciers of expansion since the late 1990s (Hewitt, 2009). A first attempt to assess future change has been made by Armstrong *et al.* (2009). They show that under current climatic conditions, the glaciers of Nepal experience no melt over 50% of their surface area at any time of year. In the Dudh Khosi basin, for example, the mean altitude of the freezing level is ~5,400 m and the glacier area between this elevation and 7,150 m is substantially greater than that between 4,500 and 5,400 m. In Sagamartha National Park, Nepal, Salerno *et al.* (2008) reported a 4.9% reduction in glacier area between the between the end of the 1950s and the early 1990s. Area losses were mainly experienced by smaller, low-altitude glaciers, while high-elevation glaciers expanded due to increased monsoon precipitation. Raina and Srivastava (2008) summarize the few mass balance observations available for Indian glaciers. Data on Dokriani glacier (5.5 km^2, 30.9° N, 78.8° E between 6,000 and 3,800 m altitude) for 1992–2000 show increasingly negative mass balance from -1.54×10^6 m^3 in 1992–1993 to -2.65×10^6 m^3 in 1999–2000 and an average retreat rate of 18 m a^{-1}. Between 1962 and 1995, the glacier volume is estimated to have been reduced by about 20% (Dobhai *et al.*, 2004). Using satellite estimates of glacier thinning between 2000 and 2004, Berthier *et al.* (2007) showed that in the Spiti/Lahaul area (32.2° N, 77.6° E) of Himachal Pradesh, Western Himalaya, there was a thinning of 4–7 m between 4,400 and 5,000 m. For a glacier area of 915 km^2, they calculated a specific mass balance of -0.7 to -0.85 m w.e a^{-1}, twice the rate for 1977–1999 determined by Dyurgerov and Meier (2005).

Overall, in recent decades, Himalaya glaciers have retreated at rates ranging from a few meters to over 60 m yr^{-1}, depending on the terrain and climatic factors. The amount of glacial ice loss observed and estimated was 443±136 Gt out of 3,600–4,400 Gt of glacial ice stored in the Himalaya of India, and the mean glacier mass loss in the Indian Himalaya is accelerated from –9 ± 4 to –20 ± 4 Gt yr^{-1} between 1975–1985 and 2000–2010 (Kulkarni and Karyakarte, 2014). Using digital elevation models derived from cold war–era spy satellite film and modern stereo satellite imagery, Maurer *et al.* (2019) estimated changes in ice thickness during 1975–2000 and 2000–2016 across the Himalayas. They detected consistent ice loss along the entire 2,000-km transect for both intervals, but the average loss rate during 2000–2016 at −0.43 ± 0.14 m w.e. yr^{-1} was about double that of 1975–2000 at −0.22 ± 0.13 m w.e. yr^{-1}. The corresponding region-wide ice mass changes for full glacierized areas in the Himalayas amount to −7.5 ± 2.3 Gt yr^{-1} over 2000–2016, compared to −3.9 ± 2.2 Gt yr^{-1} over 1975–2000. Using a simple energy balance method, they estimated the atmospheric temperature change that would supply the energy by longwave radiation needed to melt the observed ice losses to be 0.4–1.4 °C warmer between the two intervals. This estimate approximately agrees with the average 1 °C warmer observed during 2000–2016 relative to 1975–2000 throughout Himalayas.

Equatorial glaciers

The typical altitude of snowlines in the Equatorial regions is above 4,000 m, which are also the altitudes where Equatorial glaciers are located. The height of the freezing level in the tropical atmosphere has increased across most of the region, particularly in the outer Tropics (Bradley *et al.*, 2009). In the southern tropical Andes, high elevation surface temperatures and upper air data show a similar trend in temperature, of +0.1 °C per decade over the last 50 years.

In Irian Jaya, Klein and Kincaid (2006) showed that by AD 2002, ice extent on Puncak Jaya (4.1° S, 137.2° E) had decreased to only 2.15 km^2, or 12% of its maximum neoglacial extent, based on IKONOS high-resolution images. Between 1992 and 2000, Meren Glacier – which had an area of ~2.2 km^2 in 1972 (Allison and Peterson, 1976) – had disappeared entirely. All remaining ice masses on Puncak Jaya continue to retreat from their neoglacial maxima. Klein and Kincaid argue that the relationship between the observed rise of ~100 m in the ELA between 1972 and 2000 and increasing air temperature may be indirect. Rather, the cause of the glacier recession appears likely to involve an altitudinally dependent change in the phase of precipitation.

Comparing a Landsat 5-TM image acquired in November 3, 1988 and a Landsat 8-OLI image acquired in December 5, 2017, out of five masses of ice rested on the mountain slopes of Puncak Jaya in 1988, the Meren, Southwall glaciers, and West Northwall Firn had disappeared in the 2017 image. The remaining Carstensz and East Northwall Firn glaciers, which are among the last glaciers in eastern Tropics, could disappear in a decade (https://earthobservatory.nasa.gov/images/91716/glaciers-in-the-tropics-but-not-for-long).

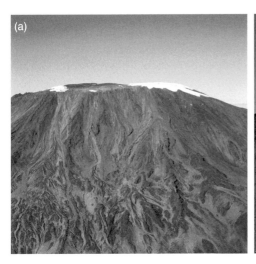

Figure 3.16 Snow and ice cover on Kibo, Mt. Kilimanjaro. (a) General view of the mountain and ice; (b) ice wall on the summit [Courtesy of Dr. Konrad Steffen, CIRES]

In Tanzania, Mt. Kilimanjaro (5,895 m) has lost 85% of its ice cover since 1912. The ice area decreased ~1% per year from 1912 to 1953 and ~2.5% per year from 1989 to 2007 (Thompson *et al.*, 2009). Of the ice cover that was present in 2000, 26% is now gone (see Figure 3.16). Ice volume changes during 2000–2007 calculated for two ice fields reveal that losses due to thinning and lateral shrinking are nearly the same. On Mt. Kenya, the Lewis Glacier had an annual precipitation of 730–1,010 mm a^{-1} during 1978–1996 and a net balance of −810 to −1,010 mm a^{-1}. For Lewis Glacier, Hastenrath (2010) show that if the air were some 0.7 °C cooler the mass budget could reach equilibrium. Observations on the secular evolution of air temperature in the areas of Mt. Kenya and Ruwenzori show comparable warming and some humidity increase. Taylor *et al.* (2006) showed from Landsat data that glaciers on Ruwenzori (Mount Stanley, Speke, and Baker) shrank from 2.01 km^2 in 1987 to 0.96 km^2 in 2003. When first surveyed by the Duke of Abruzzi in 1906, the glacier area was about 6.5 km^2. Hastenrath (2010) examines the climatic controls of glacier recession. For the summit of Kilimanjaro, above the mean freezing level where ablation is by sublimation, turbulent heat transfer processes associated with temperature differences cannot account for the imbalance of the mass budget; instead solar radiation forcing is important for both the ice thinning and the lateral retreat of ice cliffs. These reach up to 40 m in height and are mainly orientated east-west, implicating solar radiation control because in the afternoons when the Sun is to the west, the sky is cloudy. The sky is generally clear during the solstices when the Sun is to north or south and melting accounts for about 80% of the ablation on the ice walls (Kaser *et al.*, 2004; Mote and Kaser, 2007).

Central Asia

In the Aktru River basin in the central Altai Mountains of Siberia, Surazakov *et al.* (2007) used remotely sensed images with 0.6–3.0 m spatial resolution (aerial photographs, Corona and PRISM satellite images) and differential GPS data for 1952, 1966, 1975, and 2006. From 1952 to 2006 the total glacier area in the basin shrank by 7.2%. The rate of glacier area loss increased by a factor of 1.8 over the last three decades, caused mainly by an increase of summer air temperature of 1.0 °C from 1951 to 2000 at elevations below 2,500 m elevation and 0.8 °C above this level.

Kutuzov and Shagedanova (2009) evaluate changes in glacier extent and retreat rates in the eastern Terskey–Alatoo range and the Tien Shan using remote sensing data. Changes in the extent of 335 glaciers between the end of the LIA (mid-nineteenth century), 1990 and 2003 have been estimated through the delineation of glacier outlines and the LIA moraine positions on Landsat TM and ASTER images for 1990 and 2003, respectively. By 2003, the glacier surface area had decreased by 19% of the LIA value. Mapping of 10 glaciers using historical maps and aerial photographs from 1943 to 1977 shows that glacier retreat was slow in the early twentieth century but increased considerably between 1943 and 1956 and then again after 1977. The post-1990 period has been marked by the most rapid glacier retreat since the end of the LIA. Regional weather stations revealed strong warming (0.02–0.03 °C a^{-1}) during the ablation season since the 1950s. At higher elevations, represented by the Tien Shan meteorological station (41.5° N, 78° E, 3,614 m), the summer warming was accompanied by negative anomalies in annual precipitation in the 1990s that enhanced glacier retreat (Khromova *et al.*, 2003).

Table 3.5 summarizes glacier recession in central Asia. The reduction in surface area varies from ~8% to 40% over the last 40–50 years. Glaciers in the outer western Tien Shan receive more precipitation and a winter-spring maximum. Whereas those further east receive smaller amounts in spring-summer. Xu *et al.* (2010) report on the changes in the two branches of Glacier No. 1 at the head of the Urumqi River, eastern Tianshan. The glacier split into two in 1993 and in 2005 the western branch had an area of 0.6 km^2 and the eastern 1.1 km^2. The mean annual air temperature at the ELA (~4,100 m) is about −8° C and the annual precipitation is about 400–500 mm. The mean annual mass balance declined from −0.22 m a^{-1} in 1989–1998 to −0.54 m a^{-1} in 1999–2005. From 1993 to 2004, the eastern branch retreated by 38.7 m and the western branch by 64.1 m. From 1989 to 2005, the area of the eastern branch shrank by 5.3%, and the western branch by 10.3%. It is concluded that a larger slope and smaller area make the western branch more sensitive to the recent climate warming.

China

Since the LIA, the glacier area in China has decreased by about 21% (Shi *et al.*, 2008b, 2008c). The decline reached 30% for monsoon temperate type glaciers, but in glaciers of

Table 3.5 Glacier recession in central Asia (from Kutuzov and Shagedanova, 2009; Narama *et al.*, 2010). Source: *Global and Planetary Change* 69 (2009) p. 60 table 1

Region	Period	Number/area of investigated glaciers	Surface area reduction (%)	Reference
Northern Tien Shan				
Ala Archa	1963–2003	48/36.31 km^2 in 2003	15.2	Aizen *et al.* (2006)
Ili river basin	1955–2004	−/170.04 km^2 in 2004	38	Vilesov *et al.* (2006)
Malaja Almatinka	1955–1999	12/9.1 km^2 in 1955	37.6	Bolch (2007)
Bolshaja Almatinka	1955–1999	29/25.2 km^2 in 1955	34.5	
Levyj Talgar	1955–1999	42/72.3 km^2 in 1955	33.1	
Turgen	1955–1999	30/35.6 km^2 in 1955	36.5	
Upper Chon-Kemin	1955–1999	31/38.5 km^2 in 1955	16.4	
Chon-Aksu	1955–1999	48/62.8 km^2 in 1955	38.2	
Northern slopes of Zailiysky Alatau	1955–1990	307/287.3 km^2 in 1955	29.2	Vilesov and Uvarov (2001)
Tuyksu glaciers	1958–1998	7/7.74 km^2 in 1998	20.2	Hagg *et al.* (2005)
Sokoluk basin	1963–2000	77/31.7 km^2 in 1963	28	Niederer *et al.* (2007)
Central and Inner Tien Shan				
Akshiirak	1943–1977	178/317.6 km^2 in 2003	4.2	Kuzmichenok (1989) and Aizen *et al.* (2006)
	1977–2003	178/317.6 km^2 in 2003	8.7	
Western Terskey Ala-Too	1971–2002	269/226 km^2 in 2002	8	Narama *et al.* (2006)
Eastern Terskey Ala-Too	LIA–2003	335/328 km^2 in 2003	19	Kutuzov and Shahgedanova, 2009)
	1965–2003	109/120 km^2 in 1965	12.6	
Pskem	1970–2000	525/177 km^2 in 2000	19	Narama *et al.* (2010)
	2000–2007	525/169 km^2 in 2007	5	
lli-Kungöy	1970–2000	735/590 km^2 in 2000	12	Narama *et al.* (2010)
	2000–2007	735/564 km^2 in 2007	4	
SE Fergana	1970–2000	306/173 km^2 in 2000	9	Narama *et al.* (2010)
	2000–2007	306/172 km^2 in 2007	0	
Eastern Tian Shan				
No. 1 Glacier (China)	1962–2003	1/1.72 km^2 in 2003	11.8	Jing *et al.* (2006)
Middle Chinese Tian Shan	1963–2000	70/48 km^2 in 2000	13	Li *et al.* (2006)
Pamir				
Gissaro-Alay	1957–1980	4,287/2,183 km^2 in 1957	15.6	Shchetinnikov (1998)
Pamir	1957–1980	7,071/7,361 km^2 in 1957	10.5	
Pamiro-Alay	1957–1980	11,358/9,545 km^2 1957	12.5	
The Saukdara and Zulumart Ranges (eastern part of the Pamir)	1978–2001	5/33.7 km^2 in 2001	19.2	Khromova *et al.* (2006)
Muztag Ata and Konggur mountains of the eastern Pamir plateau	1962/1966–1999	302/835 km^2 in 1962/1966	7.9	Shangguan *et al.* (2006)

Table 3.5 (cont.)

Region	Period	Number/area of investigated glaciers	Surface area reduction (%)	Reference
Muksu river basin	1980–2000	–/468.4 km^2 in 1980	7.4	Desinov and Konovalov (2007)
Russian Altai				
North & South Chuya	1952-2004	21/109.8 km^2 in 1952	12.2–19.7	Shagedanova *et al.*
Djungarsky Alatau				
South Dzhungaria	1956–1990	440/218.8 km^2 in 1956	40	Vilesov and Morozova (2005)

extreme continental type it was only 9%. For nine river systems in the Qilian Shan, Tien Shan, Altai, west Kunlun, and Karakorum ranges it averaged 18%. In western China the fraction of glaciers in retreat increased from 53% during 1950–1970 to 90% in 1980–1990. For Tibet, Ding and Liu (2006) examined 5,000 glaciers and showed a decrease of 4.5% between the 1970s and 1999–2002 based on comparing aerial topographic maps and Landsat and ASTER images. Between 1966 and 2000 the A'nyemaqen Mountains in southeast Qinghai lost 17.3% of their glacier area due to warming and drying trends. In contrast between 1962 and 2000, glaciers in the eastern Tianshan and western Qilian Shan lost only 4.7% of their area perhaps due to wetter conditions, while in the eastern Pamir the decrease from 1960 to 1999 was 10.0% (Shi *et al.*, 2008b).

Since the LIA, the glacier area in China has decreased by about 21% (Shi *et al.*, 2008a, 2008b). The decline reached 30% for monsoon temperate type glaciers, but in glaciers of extreme continental type it was only 9%. For nine river systems in the Qilian Shan, Tien Shan, Altai, west Kunlun, and Karakorum ranges it averaged 18%. In western China, the fraction of glaciers in retreat increased from 53% during 1950–1970 to 90% in 1980–1990. For Tibet, Ding and Liu (2006) examined 5,000 glaciers and showed a decrease of 4.5% between the 1970s and 1999–2002 based on comparing aerial topographic maps and Landsat and ASTER images. Between 1966 and 2000 the A'nyemaqen Mountains in southeast Qinghai lost 17.3% of their glacier area due to warming and drying trends. In contrast between 1962 and 2000, glaciers in the eastern Tien Shan and western Qilian Shan lost only 4.7% of their area perhaps due to wetter conditions., while in the eastern Pamir the decrease from 1960 to 1999 was 10.0% (Shi *et al.*, 2008b).

Along Tien Shan, Central Asia's largest mountain range, glaciers have lost 27% of their mass and 18% of their area in the last 50 years, or almost 3,000 km^2 of glacier areas and an average of 5.4 Gt of ice per year have been lost since the 1960s (Farinotti *et al.*, 2015). As about half of its glacier volume could be depleted by the 2050s, the reduced snow and glacier melt from Tien Shan could significantly affect the water supply of Kazakhstan, Kyrgyzstan, Uzbekistan, and parts of China with dry winter climate. In the Tibetan Plateau (TP), the glacier mass balance has decreased by -16.3 ± 3.5 Gt yr^{-1} between 2000 and 2016, with the most significant mass loss in the Nyainqentanglha (-4.0 ± 1.5 Gt yr^{-1}), and a slight glacier mass gains in Kunlun ($+1.4 \pm 0.8$ Gt yr^{-1}) (Brun *et al.*, 2017). Glacier area in the TP has decreased from $44,366 \pm 2,827$ km^2 in 1970s to $42,210 \pm 1,621$ km^2 in 2001 and $41,137 \pm 1,616$ km^2 in 2013 from the Landsat images (Ye *et al.*, 2017).

A new glacier change index

A new index of glacier change is proposed by Dyurgerov *et al.* (2009) based on the difference between a time-averaged AAR and the equilibrium AAR_0 determined from B_n (net balance) $= 0$ in a linear regression of B_n against yearly AAR values. The regression of B_n against AAR for 86 glaciers for 1961–2004 is $R^2 = 0.55$. The glacier's displacement from equilibrium

$$\alpha_d = \frac{(AAR - AARo)}{AARo} \tag{3.19}$$

represents an undelayed response to the change in mass and energy balance components, whereas the change in glacier area is a delayed response to those changes. For 65 tropical glaciers the α_d is -65% implying rapid shrinkage, whereas for mid-latitude and polar glaciers the values are smaller. The average value for all 86 glaciers is $-15 \pm 2.2\%$. Only 11 glaciers have a positive value at any time during 1961–2004. The majority of equilibrium AAR values are 50–60%, with an average of 57%, and this has remained stable over time.

   In summary, it is worth pointing out that, in most mountain regions of the world, glacier mass loss has accelerated over the last 20–30 years compared with the preceding 30–40 years. Hence, the notion that the shrinkage is merely a "bounce" out of the LIA interval of glacier advances is no longer tenable; the recent changes are indisputably attributable to global warming.

Sea-level rise

One Gt of ice is approximately equal to 1 km^3 of freshwater (1.1 km^3 of ice), and 362.5 Gt of ice removed from the land and immersed in the oceans will cause roughly 1 mm of global SLR. However, global glacier mass balances do not correspond to the total glacier contribution to SLR partly because glacier ice below sea level does not contribute to SLR, and meltwater from glaciers could be partly stored in lakes or wetlands, intercepted by natural processes and human activities (e.g., drainage to lakes, impoundment in reservoirs, agriculture use) instead of discharging to the ocean (Loriaux and Casassa, 2013).

   The mean contribution of glaciers to SLR for 1993–2016 is estimated at 0.65 ± 0.051 mm yr^{-1} and 0.74 ± 0.18 mm yr^{-1} for 2005–2016. Without complete observed dataset for glacier mass changes, most methods in deriving glacier sea-level contribution must extend local observations to a larger region. The main source of uncertainty is that the vast majority of glaciers are unmeasured, methodological differences, such as the downscaling of atmospheric forcing required for glacier modeling, the separation of glacier mass change to other mass change in the spatial gravimetric signal, and estimating mass change from different raw measurements (e.g., length and volume changes, mass balance measurements, and geodetic methods).

Dyurgerov and Meier (2005) calculated that the contribution of glaciers and ice caps to SLR from 1960 to 2004. Meier *et al.* (2007) expand on those results and show that the ice mass change for 1995–2005 was -402 ± 95 Gt a^{-1}, of which the glaciers around the Gulf of Alaska contributed a quarter. The change of ice mass around the Gulf of Alaska increased dramatically from -40 Gt a^{-1} during 1961–1990 to -86 Gt a^{-1} for 1990–2004 (Meier *et al.*, 2007). In 2006, glaciers and ice caps were accounting for 1.8 mm a^{-1} of the 3.1 ± 0.7 mm a^{-1} of SLR.

An analysis of SLR from mountain glaciers and ice caps has been performed by Hock *et al.* (2009) for 1961–2004. They use a temperature- and precipitation-driven mass balance model on 88 glaciers for which seasonal mass balances for ≥ 5 years were available. The mass balance model is perturbed by a hypothetical uniform 1 K temperature and 10% precipitation increase in order to obtain, for each glacier, mass balance sensitivities. The contribution of each month to total mass balance sensitivity is obtained by running the model with each month perturbed individually by 1 K for all 88 glaciers. Global surface mass balance sensitivity to a uniform 1 K temperature rise, derived as an unweighted mean over all grid cells, is -0.68 m a^{-1} K^{-1}. The model is forced by daily data from the 0.5° resolution re-analysis data (ERA-40) by the European Centre for Medium-Range Forecasts (ECMWF). Temperature and precipitation changes are obtained from linear trend analyses of suitable meteorological data derived principally from the ERA-40, and expressed for each year as anomalies from 1961 to 1990 means. Hock *et al.* estimate a global surface mass loss of all mountain glaciers and ice caps as 0.79 ± 0.34 mm a^{-1} sea-level equivalent compared to only 0.50 ± 0.18 mm a^{-1} in Lemke *et al.* (2007). Glaciers and ice caps around Antarctic contributed 28% of the global total due to exceptional warming around the Antarctic Peninsula and high sensitivities of the ice masses to temperature changes.

Cogley (2009a) combines geodetic and direct measurements of glacier mass balance, mostly obtained since the 1950s. There are 344 glaciers with direct measurements for 4,146 balance years and 327 glaciers with geodetic measurements for 16,383 balance years. Moreover, the former data include 32% that are calving glaciers compared with only 7% having direct measurements. The average difference between the two sets of mass balance measurements is small although the geodetic estimates are slightly more negative. Combining the two data sets, the glacier contribution to SLR for 2001–2005 is estimated at 1.12 mm a^{-1} sea-level equivalent, compared with 0.77 mm a^{-1} of Kaser *et al.* (2006). Hirabayashi *et al.* (2010) developed a global glacier model that can be coupled to global land surface and hydrological models. They use the glacier model HYOGA to compute glacier mass balance by a simple degree-day approach for 50 m sub-grid elevation bands, modeling all glaciers within a grid cell as one glacier, with a spatial resolution of 0.5° by 0.5°. They developed the global glacier mass balance and glacier area for 1948–2006 by driving HYOGA with daily near-surface atmospheric data. Global glacier area was estimated as 534,893 km^2 in 1948, in reasonable agreement with other estimates but the rate of shrinkage to 2006 was only 20% of Cogley's (2008) value. The calculated global mass balance for 1960/1961 to 1989/1990 was 122 Gt a^{-1} and increased to 277 Gt a^{-1} for 1990/1991–2003/2004; The former closely matches the IPCC estimate of Lemke *et al.* (2007), but the latter is 20% larger. Corresponding SLRs are 0.34 and 0.76 mm a^{-1}, respectively.

Alaskan glaciers are thought to have a major contribution to SLR over the last 50 years, perhaps accounting for as much as one third of the total. New studies have an important bearing on this question. In the St. Elias Mountains of Alaska, Arendt *et al.* (2009) found a mass change over a glacierized area of 32,900 km^2 between 2001 and 2007 of −0.6 m w.e. a^{-1} using airborne laser altimetry and high-resolution estimates from GRACE. A comprehensive analysis of all Alaskan glaciers using ASTER and SPOT 5 data for 2006 versus a map-based DEM in 1962 was carried out by Berthier *et al.* (2010). The glaciers lost ~42 km^3 a^{-1} of water and d 0.12 mm a^{-1} to SLR. In the Yukon, glaciers lost 22% of their surface area since 1957–1958, which is scaled to a total mass loss of 406 ± 177 Gt, representing 1.12 ± 0.49 mm of global SLR (Barrand and Sharp, 2010). Glacier recession in British Columbia could account for ~0.67 ± 0.12 mm of SLR over 1985–1999 (Schiefer *et al.*, 2007).

Bahr *et al.* (2009) show that the current accumulation area of glaciers is linked to future changes in glacier volume and consequently to changes in sea level. Long-term mass balance data from 86 mountain glaciers and ice caps from around the world show that the equilibrium AAR averages 0.57 ±0.01, whereas the data for 53 ice caps and glaciers for 1997–2006 indicate that the current average AAR is only 0.44 ± 0.02. Because accumulation areas are too small, glaciers must lose about 27% of their volume to attain equilibrium with current climate. As a result, at least 184 ± 33 mm of SLR are necessitated by mass wastage of the world's mountain glaciers and ice caps, even if the climate does not continue to warm. If the climate continues to warm at current rates, a minimum of 373 ± 21 mm of SLR over the next 100 years is expected. These numbers are substantially higher than previous estimates. In addition, calving from tidewater glaciers can contribute disproportionately large amounts of meltwater, estimated at ~94 mm through the year 2100 (Pfeffer *et al.*, 2008).

From analyzing about 20,000 geodetic glacier observations, Zemp *et al.* (2019) found an increase in glacier mass loss over all mountain regions from 470 ± 80 kg m^{-2} yr^{-1} in 1986–2005 to 610 ± 90 kg m^{-2} yr^{-1} in 2006–2015, which is considerably more than their estimated global average. During 2006–2015 mass budgets were the most negative in the Southern Andes (−1,200 kg m^{-2} yr^{-1}) but the least negative in High Mountain Asia (−190 kg m^{-2} yr^{-1}, Figure 3.17). However, due to large ice extent, the total mass loss and corresponding SLR is largest in Alaska, followed by the Southern Andes and High Mountain Asia. All mountain regions combined contributed to a SLR at 0.41 mm yr^{-1} in 2006–2015. Based on observed volume changes from more than 19,000 glaciers worldwide, Zemp *et al.* (2019) found that mountain glaciers outside of the polar regions contributed 27 ± 22 mm to the global mean SLR from 1961 to 2016.

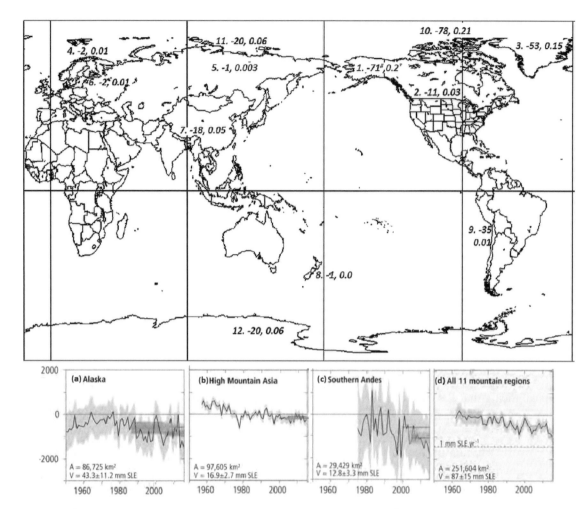

Figure 3.17 Regional budgets (mass loss) for 2006–2015 in Gt yr^{-1} and mm sea-level equivalent (SLE) per year for 1. Alaska, 2. Western Canada and USA, 3. Greenland, 4. Scandinavia, 5. North Asia, 6. Central Europe, 7. High Mountains Asia, 8. New Zealand, 9. South Andes, 10. Arctic Canada, 11. Arctic Russia, and 12. Antarctic periphery, respectively. Annual glacier mass budgets in units of kg m^{-2} yr^{-1} of (a) Alaska, (b) High Mountain Asia, (c) Southern Andes, and (d) all 11 mountain regions (taken from figure 2.4 of SROCC, 2019). Shading refers to the random error of the regional mass change. Estimates by Zemp *et al.* (2019) are based on extrapolation of glaciological and geodetic balances. Estimates by Ciraci *et al.* (2020) and Wouters *et al.* (2019) are from the Gravity Recovery and Climate Experiment (GRACE) and only shown for the regions with glacier area >3,000 km^2. Estimates by Gardner *et al.* (2013) were used in AR5. Glacier areas (*A*) and volumes (*V*) are based on RGI Consortium (2017) and Huss and Farinotti (2012) updated to the glacier outlines of RGI 6.0, respectively.

The total glacier mass loss in the world, excluding those on the periphery of the ice sheets, was very likely 275 ± 135 Gt yr^{-1} (0.76 ± 0.37 mm yr^{-1}) in 1993–2009, and 301 ± 135 Gt yr^{-1} (0.83 ± 0.37 mm yr^{-1}) between 2005 and 2009 (Vaughan *et al.*, 2013). Jacob *et al.* (2012) show that glaciers and ice caps (GIC), excluding the Greenland and Antarctic peripheral GICs, lost mass at 148 ± 30 Gt yr^{-1} (160 ± 32.6 km^3 yr^{-1}) from January 2003 to December 2010 (Figure 3.18).

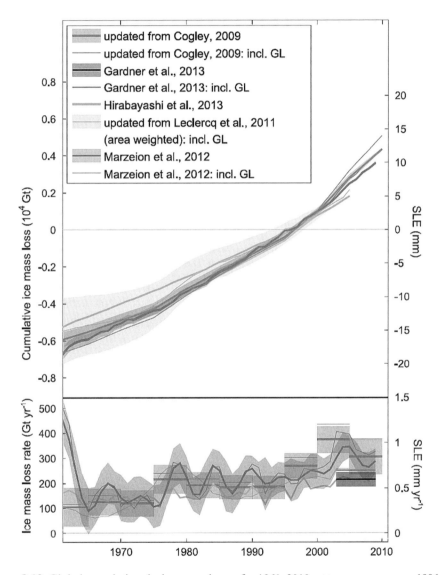

Figure 3.18 Global cumulative glacier mass change for 1961–2010 set to zero mean over 1986–2005. As in Figure 3.11, estimates are based on glacier length variations, area-weighted extrapolations of individual directly and geodetically measured glacier mass budgets, and from modelling with atmospheric variables as input (adapted from Vaughan *et al.*, 2013) (A black and white version of this figure will appear in some formats. For the color version, please refer to the plate section.)

The 5th Assessment Report (AR5) of IPCC predicted the future mass loss of glaciers based on several hundred glaciers. The Glacier Model Intercomparison Project (GlacierMIP) projected future sea level rise, for the year 2100 relative to the year 2015, driven by glacier mass loss outside the Greenland and Antarctic Ice Sheets (but also including glaciers along the peripheries of those ice sheets). In the GlacierMIP project, six global glacier models were driven with four emission scenarios: RCP2.6, RCP4.5, RCP6.0, and RCP8.5. Averaged results from the models ranged from 94 ± 25 mm for RCP2.6 to 200 ± 44 mm for RCP8.5 (Hock *et al.* 2019). In summary, there is very high confidence that globally, the mass loss from glaciers has increased since the 1960s, even though for 2003–2009, Gardner *et al.* (2013) indicate that mass loss could be overestimated in some regional and global time series, given glaciers with measured mass balances in some studies are concentrated in sub-regions with higher mass losses. The acceleration of mass loss of glaciers worldwide is consistent with climate warming observed since the twentieth century.

REVIEW QUESTIONS

1. What are two key characteristics commonly used to classify glaciers?
2. How many types of glaciers can be classified thermodynamically, and how are they different from each other in terms of surface melt, and in the ablation zone, e.g., temperate versus polar glaciers?
3. In high mountains, what are the typical elevations where tropical glaciers are located, such as in Mount Kenya? Conversely, approximately how does the mean equilibrium line altitude (ELA) of glaciers rise with a decrease in latitude, or increase in m per degree in latitude? For example, ELA rises from 2,800 m in the Altai Mountain (49° N) to 6,000 m on the north slope of Mt. Everest (28° N).
4. Glaciers are large masses of ice that form where the accumulation of snowfall constantly exceeds the snowmelt and sublimation. Likely due to climate warming impact, globally what is the rate of glacier mass loss in Gt per year in recent decades?
5. As glaciers worldwide are losing mass under the impact of climate warming, should the present global average accumulation area ratio (AAR) be larger or smaller than 0.5?
6. If sea level rises at about 3 mm yr^{-1}, and about 60% of the sea-level rise is due to glacier melting, what is the total melt rate of glaciers across the world in km^3 per year, given the total area of all the oceans is about 361,300,000 km^2?
7. What are popularly used satellite images to map glaciers, and problems associated with automated mapping of glaciers from space such as shown in Figure 3.4?
8. Using representative values of Glacier energy budget components for the Aletsch Glacier of Valais, Switzerland from Table 3.2, compute its average melt loss in m^3 per year, assuming the average daily energy fluxes is about 40% of the values given in Table 3.2, and glacier melting occurs mainly between spring, summer and early autumn (2/3 of a year) only. Also, assume that on the average only 25% of the glacier (120 km^2 of ice area, 23.6 km long, 1.6 km width, and 1 km thick), undergoes ablation; and the accumulation is

mainly due to an average winter precipitation of about 600 mm over the whole glacier. Compare the mass loss you computed with the observed glacier retreat in the last 30 years, estimated at about 27 m yr^{-1}. What could be reasons for the discrepancy between your estimate of glacier mass loss and the glacier retreat observed?

9. Suppose the mean annual air temperature (T_A) of the Aletsch Glacier of Valais, Switzerland is about 3 °C, and its mean winter (NDJF) temperature (T_W) is −2 °C. First compute its the accumulative positive degree-days, $\sum \Delta T$, using Equation 3.18, assuming $(\sum \Delta T)\Delta t = 365T_A + 120T_W$. Then using the degree-day method, compute its average melt loss in m^3 per year, if its average degree-day factor (DDF) is 4.5 mm d^{-1} K^{-1}, and glacier melting occurs mainly between spring, summer and early autumn (March to October). Also, assume that on the average only 25% of the glacier (120 km^2 of ice area, 23.6 km long, 1.6 km width, and 1 km thick), undergoes ablation; and the accumulation is mainly due to an average winter precipitation of about 600 mm over the whole glacier. Compare the mass loss you computed with the observed glacier retreat in the last 30 years, estimated at about 27 m yr^{-1}. Based on the discrepancy between your estimate of net glacier mass loss (total ablation volume − accumulation) and the glacier retreat observed, estimate the amount of glacier thinning in m per year.

10. Using the degree-day factor (DDF) of 6 mm d^{-1} °C^{-1}, compute the net glacier mass loss of the Alfotbreen Glacier of Norway based on the monthly temperature and precipitation provided in the table below.

Mon	Jan	Feb	Mar	Apr	May	Jun	Jul	Aug	Sep	Oct	Nov	Dec
Day	31	28	31	30	31	30	31	31	30	31	30	31
Temp. (°C)	−3	−4	−5	−3	1	4.5	6	6.5	3	1	−2	−3
Precip. (mm)	700	560	530	320	280	290	310	400	670	700	690	760
$\Sigma M = DDF\Sigma\Delta T\Delta t$	—	—	—	—							—	—

11. What are typical ranges of ice flow rates of moving glaciers in km per year? Explain the characteristics of surging glaciers (glaciers on the run), and what could have contributed to the sudden acceleration observed in some surging glaciers? For example, the Wahlenbergbreen glacier in Svalbard of Norway that began speeding up in September 2013. At the peak of the surge, beginning in 2015, the ice advanced at 9 m per day, bulldozing everything in its path.

12. Many river basins with contribution from glacier runoff have shown increasing trends in streamflow in recent years because many glaciers are losing mass due to climate warming impact. According to projections of CMIP5 of IPCC, roughly when will this short-term increase in streamflow response to glacier shrinkage slow down and eventually disappear?

4   Ice sheets

While ice sheets were extensive in the Northern Hemisphere during the Pleistocene glaciations, covering much of North America and Scandinavia, the two remaining continental ice sheets are in Greenland and in Antarctica. Greenland is essentially a single dome reaching above 3 km, while the Antarctic Ice Sheet (AIS) has a more complex form that rises above 4 km and is bordered by two major ice shelves and numerous smaller ones. These ice sheets have existed for millions (tens of millions in the case of Antarctica) of years. Arbitrarily, an ice sheet is defined as glacier ice extending over 50,000 km^2 in area.

Ice sheet mass balance is determined by the surface mass balance (SMB) and ice discharge. The SMB is the net outcome of snow accumulation and summer ablation below the equilibrium line altitude (ELA). Ice discharge is primarily via a small number of ice streams and major outlet glaciers that calve into the surrounding oceans, directly in Greenland or mainly from ice shelves in Antarctica.

4.1   History of exploration

A. E. Nordenskjold first explored Greenland scientifically in 1870 and 1883. Robert Peary explored the northern part of the ice sheet in 1888 and traveled inland from Thule for about 150 km; he later explored northern Greenland. Also in 1888, Fridtjof Nansen made the first crossing of the ice sheet at about 64° N from Umivik to Godthåb (Nuuk). Subsequent explorations include C. H. Ryder in East Greenland in 1892, E. von Drygalski in West Greenland in 1892–1893, J. Charcot in the Danish expedition of 1906–1908, A. de Qervain who in 1912 made the first west–east crossing of the ice, and J. P. Koch and Alfred Wegener who also crossed the ice sheet in 1912. During 1930–1931 the British Arctic Air Route Expedition of Lauge Koch and Gino Watkins and Alfred Wegener's Greenland Expedition operated their stations – Watkins Ice Cap (67.1° N, 41.8° W, 2,440 m) (Mirrless, 1932) and Eismitte (70.9° N, 40.7° W, 3,000 m) (Loewe, 1935, 1936). An extreme low of −68 °C was recorded. Post–World War II there were many expeditions to Greenland.

P.-E. Victor led the Expedition Glaciologique Internationale au Groenland (EGIG) in 1946–1948, which established a survey line across the ice sheet. The Expeditions Polaires Francaises (EPF) operated *Station Centrale* at 70.9° N, 40.6° W, 2,993 m, during 1949–1951. The British North Greenland Expedition in 1952–1954 occupied a station *North Ice* (78.1° N, 38.5° W, 2,345 m) (Hamilton *et al.*, 1956). Summit station at 72.6° N, 38.5° W, 3,278 m, has operated since 1988. The history of ice core drilling is summarized in Box 4.1.

Box 4.1 Ice cores

The first deep ice cores (300–400 m) were collected during the International Geophysical Year (IGY), 1957–1959, in Greenland and Antarctica through the foresight of Henri Bader at the Snow, Ice, and Permafrost Research Establishment (SIPRE) of the US Army Corps of Engineers (Langway, 2008). Chester Langway undertook the first core processing and scientific analysis. The first ice cores drilled to bedrock were at Camp Century, near Thule, Greenland in 1966 and Byrd Station, Antarctica (80° S, 119.5° W) in 1968. In 1992 and 1993 ice cores to bedrock (~3,000 m) were extracted at the Greenland Ice Sheet Project (GISP)-II and Greenland Ice Core Project (GRIP) drilling sites in central Greenland. Ten years later in 2003 the North-GRIP site was drilled to 3,085 m giving a record that extended into the Eemian last interglacial period (see Chapter 9). In Antarctica, major ice cores were drilled at Vostok Station, Dome Concordia, Dome Fuji, and Köhnen station. A new core is being drilled at the ice divide of the West Antarctic Ice Sheet (WAIS Divide).

Intensive ice core research began in the 1960s when Dansgaard *et al.* (1969) analyzed 1,600 ice samples from the 1,390-m ice core from Camp Century. Since then, ice cores have been extracted and analyzed from numerous other locations on the Greenland and Antarctic Ice Sheets and ice caps in the Andes, Tibet, and the Arctic islands, as well as cold glaciers in the Rocky Mountains and Alps. Cores are extracted in 2–6-m-long sections with a typical diameter of 10 cm and are stored in the United States at the National Ice Core Laboratory (NICL) of the US Geological Survey in Denver, CO. Many sections of the cores are sliced off and shipped to research institutions around the country for different analyses.

A whole suite of paleoclimatic information can be extracted from ice cores (Alley, 2000). Snowfall carries with it the compounds that are in the air at the time of deposition, ranging from sulfate, nitrate, and other ions, to mineral dust, radioactive fallout, trace metals, and pollen. Also trapped in the ice are small air bubbles that provide a sample of the air itself giving information about the composition of the atmosphere at the time the ice formed. A cornerstone of ice core research is the $\delta^{18}O$ (delta-O-18) isotopic record that reflects air temperature. Almost all oxygen has an atomic weight of 16 (^{16}O), but a small fraction is ^{18}O. Water contains 99.73% $^1H_2{}^{16}O$ and 0.20% $^1H_2{}^{18}O$, where ^1H is hydrogen (^2H is deuterium and ^3H is tritium). $^1H_2{}^{16}O$, being 12% lighter than ^{18}O, is preferentially evaporated from the ocean surface. In turn, precipitation contains slightly more of the heavier $^1H_2{}^{18}O$. Hence, as an air mass precipitates over an ice sheet interior there is an increasing deficit of $^{18}O/^{16}O$. If the signal can be attributed to temperature change alone, ignoring the effects of salinity and ice volume change, a $\delta^{18}O$ increase of 0.22 ppm is equivalent to a 1 °C cooling. In central Antarctica, the $\delta^{18}O$ is of the

Box 4.1 (cont.)

order of 50 ppm. Hence, the ratio of ^{18}O to ^{16}O varies and as air travels inland over Greenland and Antarctica there is greater depletion of ^{18}O, corresponding to lower air temperatures. In a similar way the ratio of $^1H/^2H$ (hydrogen to deuterium, D) provides data on ocean temperature and moisture content in the air mass source region. The deuterium excess, $d = \delta D - 8\delta^{18}O$, provides information on ocean surface conditions and hemispheric ocean/atmosphere circulations (Vimeux *et al.*, 1999). Interglacials are characterized by high *d* values and glacials by low ones. Annual snow layers can be counted to give a time history of mass accumulation spanning 30,000–40,000 years. Electrical conductivity measurements record sulfate deposition from volcanic eruptions. Mineral dust records wind-blown sands and aridity.

   One of the longest climate records was recovered from the European Project for Ice Coring in Antarctica (EPICA) drill site at Dome Concordia (75.1° S, 123.4° E, 3,233 m asl). The 3,139 m of core so far analyzed spans 740 ka and covers eight glacial–interglacial cycles (EPICA Community Members, 2004). The basal ice is estimated to date back to ~960 ka. The Vostok core (3,600-m deep), the previous longest record, spans 422 ka and shows a strong correlation between CO_2 and temperature (Figure 4.1). At present, the oldest ice core records of CO_2 is located at the Antarctic Dome C site (75°06′S, 123°24′E, 3,233 m asl), which extends back 800,000 years, over which atmospheric CO_2 levels estimated from the ice core data have fluctuated between 170 and 300 ppmv, corresponding with conditions of glacial and interglacial periods (Lüthi *et al.*, 2008).

The Southern Ocean was regularly visited by sealers and whalers in the late nineteenth century but Antarctic scientific exploration began only at the turn of the century. A British expedition led by C. Borchgrevink was the first to winter over on the continent in 1899 at Cape Adair. E. von Drygalski led the first German Antarctic Expedition in the ship *Gauss* during 1901–1903. There followed a series of attempts to reach the South Pole by Captain Robert F. Scott in 1902, Ernest Shackleton in 1907–1909, Roald Amundsen who was successful in December 1911, and Scott who reached the pole in January 1912 but perished on the return journey. In 1911–1912, an unsuccessful goal of Filchner's expedition was to establish whether land or sea linked West and East Antarctica (Lüdecke, 1995, p. 68). Flights were made over the continent between 1928 and 1935 by Richard Byrd. In 1947, the US Navy's "Operation Highjump" photographed and mapped most of the coastline. The Norwegian–British–Swedish expedition of 1946–1947 conducted some of the first glaciological research in the Antarctic, in Queen Maud Land (Schytt, 1954). During the July 1957–December 1958 IGY, 12 nations established over 60 stations in Antarctica including South Pole and Vostok. For a full list of Antarctic expeditions, see Headland (2009) and also http://en.wikipedia.org/wiki/List_of_Antarctic_expeditions.

   Texts on ice sheet behavior include Oerlemans and van der Veen (1984) on ice sheets and climate, a comprehensive overview by Hughes (1998), Bamber and Payne (2004) who focus

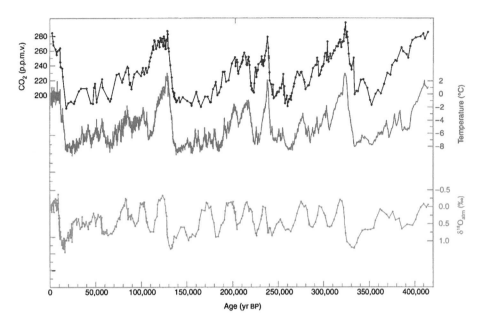

Figure 4.1 The record of temperature, δO^{18} and CO_2 from an ice core at Vostok, east Antarctica over the last 420 ka [Source: Redrawn from Wikipedia Commons. http://en.wikipedia.org/wiki/File:Vostok_420 ky_4curves_insolation.jpg or http://en.wikipedia.org/wiki/File:Vostok-ice-core-petit.png]

on mass balance, and an advanced text on modeling the dynamics of ice sheets, ice shelves, and glaciers by Greve and Blatter (2009). Research on the impact of the ice sheet includes its interactions and feedback with the Earth system (Fyke *et al.*, 2018), and the impact of its meltwater on future climate (Bronselaer *et al.*, 2018).

4.2   Mass balance

SMB varies on a number of scales, from small-scale features (sastrugi and snow dunes) to ice-sheet-scale patterns determined mainly by temperature, elevation, distance from the coast, and wind-driven processes (Eisen *et al.*, 2008). Errors in stake measurements are primarily caused by noise due to the small-scale relief-related spatial variability of snow accumulation and density. The measurement uncertainty is inversely related to the number of stakes and the period of observation. Decadal values of SMB decorrelate on the 1- to 10-km scale but co-vary over length scales of hundreds of kilometers.

The determination of ice sheet mass balance has traditionally followed the component approach where the mass input from net snowfall is compared to the losses by melting and ice flow across the grounding line. The snow input is measured by counting annual layers in snow pits and ice cores or by dating the layers by identifying known volcanic events such as the year without a summer (1816) following the great AD 1815 Tambora eruption on the Indonesian island of Sumbawa (Dai *et al.*, 2009) or radioactive horizons from thermonuclear bomb tests in the 1950s–1960s. The first major assessment of accumulation in Greenland was made from snow surveys undertaken by Carl Benson (1962). *In situ* measurements are performed at single points by stakes, snow pits, and firn and ice cores. More recently acoustic rangers for snow height have been installed at some automatic weather station (AWS) sites. Stake lines and networks have been widely installed in Antarctica (Eisen *et al.*, 2008). The conversion of height to mass changes through snow densities remains a major problem for both stakes and acoustic measurements. The mass loss from meltwater runoff is usually calculated from hydrologic models and weather data. Iceberg discharge is estimated from measurements of flow velocity at outlet glaciers and ice thickness measured by radar soundings. Formerly, repeated surveys were made of markers on the ice, but this gave way to tracking of features (such as crevasses) in the ice by high-resolution visible and radar imagery, which enables large areas to be covered.

The second method for assessing mass balance is known as the integrated approach where changes in ice surface elevation are measured by satellite altimetry or changes in ice mass are directly assessed using satellite gravity measurements. Airborne laser surveys were made over Greenland in the 1990s and satellite altimetry data from the Geoscience Laser Altimeter System (GLAS) span 2003–2009. The Gravity Recovery and Climate Experiment (GRACE) has operated since March 2002 and is providing unique ice-sheet-wide assessments of mass changes. The results from these programs are discussed later.

4.3   Remote sensing

The first remote sensing of ice sheets was airborne radio-echo sounding (RES) begun in the 1960s (Waite and Schmidt, 1961; Evans, 1967). The data display layers within the Antarctic Ice Sheet as well as ice thickness and basal characteristics such as roughness, melting, and subglacial lakes. Layering that is less than 1,000 m from the ice surface can be due to variation in ice density associated with changes in the size or shape of air bubbles within the ice and variations in ice crystal orientation and density. At ice depths below ~1,000 m, however, these layers are primarily due to ice horizons with small permittivity changes, caused by relatively high acidity, that were originally deposited on the ice surface after large volcanic eruptions. Theoretically, therefore, the internal layers identified by RES below 1,000 m should represent time horizons (isochrones) (Robin, 1975; Siegert, 1999). Impulse radars operating in the 1–20 MHz range are mainly used for ice thickness, while those in the 50–900 MHz range are used to study things such as firn and crevasse formation. Allen *et al.* (1997) used a 150-MHz system to study the Petermann Glacier and its 60-m-thick floating tongue in northwest Greenland, for example. Locational accuracy is an important attribute and now differential Global Positioning System (GPS) data are routinely collected with RES measurements providing positional uncertainties to ±2 to 5 m (Nolan *et al.*, 1995).

Remote sensing, particularly from satellites, has been the basis of recent large-scale study of ice sheets (Quincy and Luckman, 2009; Liang *et al.*, 2019). The variables that can be determined include: snow cover extent, snow volume, snow grain size, albedo and temperature, snow/ice facies, ice extent, velocities, and topography (König *et al.*, 2001). The remote sensing of snow is treated in Chapter 2 and of glaciers in Chapter 3 . Ice sheet topography is mapped by photoclinometry with visible band images, or by radar and laser altimetry. Photoclinometry (shape-from-shading) relies on the fact that surfaces facing toward the sun are brighter than ones facing away from the sun. The image brightness is approximately $A \cos \theta + B$, where θ is the angle between the surface normal and the solar direction; A and B are constants that can be determined from the heights of a few known points (Rees, 2006, p. 201). Scambos and Fahnestock (1998) demonstrate its use with Advanced Very High-Resolution Radiometer (AVHRR) data. In altimetry, a radar or lidar pulse emitted by an airborne or satellite sensor strikes the surface and a portion of the signal is returned from a near-circular area. The distance between the satellite and the ground (or range) is determined from the time delay of the return signal. The elevation is calculated from the range and the known orbital position; it is the average elevation within the pulse-limited footprint (König *et al.*, 2001). Retracking algorithms are used to detect the leading edge in the return signal. A correction needs to be applied for sloping surfaces since the first return comes from the highest terrain that is the closest point to the satellite (Bamber, 1994). Errors may be caused by inhomogeneous terrain, uncertainties in the satellite orbit, and in the case of radar, by the penetration of the pulse into the surface layer. The European Remote Sensing (ERS)-1 and -2 satellites have provided 5.7-cm wavelength radar altimetry coverage since 1991 of all of Greenland and Antarctica to 81.5° S. Envisat with Advanced Synthetic Aperture Radar (ASAR) was launched by ESA in March 2002 and provides global radar coverage also at 5.7-cm wavelength. The repeat cycle of these satellites is 35 days. Canada's RADARSAT-1 (5.7 cm) launched in 1995 provided the first high-resolution map of Antarctica when the satellite was turned to face south in September–October 1997 (Jezek, 1999). RADARSAT-2 satellite of Canada, launched in 2007, operates with a SAR sensor that transmits and receives C-band microwaves at HH-, HV-, VH-, and VV-polarizations (e.g., HV, horizontal transmit–vertical receive) are extensively used for tracking changes in winter ice cover characteristics along rivers (e.g., Lindenschmidt *et al.*, 2011). As part of NASA's Earth Observing System, the Ice, Cloud, and Land Elevation Satellite (ICESat), with a mission for measuring ice sheet mass balance and a repeat cycle of 91 days, has given lidar data (at 532 and 1,074 nm) from 2003 to 2009 covering latitudes ±86°. The radar precision of retrieved elevations over the ice sheets varies from 59 cm to 3.7 m for ERS-2 and from 28 cm to 2.06 m for Envisat according to Brenner *et al.* (2007). ICESat laser 2a data have a precision varying from 14 to 50 cm as a function of surface slope, and a lateral resolution of ~5 km (see Box 8.1). Since it was launched in September 15, 2018, the ICESat-2, that carries a photon-counting laser altimeter called "ATLAS" (Advanced Topographic Laser Altimeter System), has been measuring the height of the changing cryosphere – the elevation of ice sheets, glaciers, sea ice, and others – in unprecedented details by 10,000 laser pulses a second. Besides investigating cryospheric changes in a warming climate, ICESat-2 also measures cloud and atmospheric properties and global vegetation across temperate and tropical regions.

Interferometric SAR (INSAR) offers the opportunity to map surface topography and displacements or deformation of surface features over time (Pepe and Calo, 2017). It combines the signals received by an along-track and a cross-track radar antenna. The difference in the path (distance between satellite and ground) between two images means that there is a difference in phase (a phase shift) between the two signals. By coherently combining the signals from the two antennas, the interferometric phase difference between the received signals can be formed for each imaged point. The phase difference can be converted into surface altitude since it provides a third measurement, in addition to the along- and cross-track location of the image point. In determining surface motion, if the flight path and imaging geometry of SAR observations are identical, any interferometric phase difference is due primarily to surface motion in the direction of the radar sight line. Extraction of the full flow vector is possible if glacier flow direction can be determined from flowline features. In the first glaciological application of INSAR, Rignot et al. (1997) used ice thickness estimates at the grounding line and interferometric estimates of velocity to determine discharge for 14 outlet glaciers in northern Greenland. Joughin et al. (1999) measured discharge on the Humboldt, Petermann, and Ryder glaciers in Greenland by combining interferometrically measured velocity data with ice thickness data.

The VELMAP project at NSIDC (http://nsidc.org/data/velmap) aims to compile all ice flow data for the Antarctic continent. The project includes Landsat 7 and ASTER ice velocity maps while the second and third RADARSAT Antarctic Mapping Missions provide data north of 80° S. There are currently more than 130,000 ice vectors in the database. Image pair mapping to derive velocities for Antarctic glaciers was illustrated by Lucchita and Ferguson (1986) and Bindschadler et al. (1996), among others. Scambos et al. (1992) developed algorithms for automatic feature tracking.

Laine (2008) used data from the AVHRR Polar Pathfinder to analyze albedo changes in the Antarctic. All sectors show slight increasing spring–summer albedo trends. The steepest ice sheet albedo trend of 0.0019 ± 0.0009 yr^{-1} is found in the Ross Sea sector. Moderate Resolution Imaging Spectroradiometer (MODIS) and Quick Scatterometer (QuikSCAT) data are used by Hall et al. (2009) to map daily land surface temperature and snow albedo change, and surface melt, respectively. For 2007, the QuikSCAT product detects ~11% greater melt extent than the MODIS land surface temperature product, probably because the former is more sensitive to surface melt and can also detect subsurface melt. QuikSCAT uses a 13.4 GHz (2.2 cm) Ku band radar which is sensitive to snow wetness. The diurnal difference method was developed by Nghiem et al. (2001) to monitor the snowmelt process. The algorithm is based on diurnal backscatter difference, a relative quantity between morning and afternoon measurements.

Multi-angle Imaging SpectroRadiometer (MISR) data have been used to determine the ice albedo over Greenland (Nolin *et al.*, 2001). The MISR sensor's ability to map glacier facies and roughness was explored and documented by Nolin *et al.* (2002). MISR uses nine cameras pointed at fixed angles, ranging from nadir to ±70.5°, with four spectral bands (466–866 nm).

The MODIS Mosaic of Antarctica (MOA) image maps are derived from composites of 260 MODIS orbit swaths. The MOA provides a cloud-free view of the ice sheet, ice shelves, and land surfaces, and a quantitative measure of optical snow grain size for snow- or ice-covered areas (Scambos *et al.*, 2007) (http://nsidc.org/data/nsidc-0280.html). The USGS with the British Antarctic Survey, NASA, and the National Science Foundation have produced nine versions of the Landsat Image MOA (LIMA) from over 1,000 scenes of Landsat 7 ETM+ to latitude 82.5° S at 15-m resolution, with each version highlights different features of the Antarctic landscape (http://lima.usgs.gov/view_lima.php).

Snowmelt over the Greenland Ice Sheet has been mapped using Special Sensor Microwave Imager (SSM/I) data by Abdalati and Steffen (1997) and using Scanning Multichannel Microwave Radiometer (SMMR) and SSM/I data by Mote and Anderson (1995) and Abdalati and Steffen (2001). Mote and Anderson use a threshold value of the 37 GHz brightness temperature while Abdalati and Steffen use a cross-polarized gradient ratio (XPGR), which is a normalized difference between the 19 GHz horizontally polarized and 37 GHz vertically polarized brightness temperatures. The threshold value of XPGR is used to classify dry versus wet snow (see Abdalati, 2007).

CryoSat of ESA

The first CryoSat satellite launched in October 2005 was lost due to a launch failure. The second satellite, CryoSat-2, with a Synthetic Aperture Interferometric Radar Altimeter (SIRAL), was launched on April 8, 2010. By flying at an altitude of marginally over 700 km, CryoSat-2 reaches latitudes of 88° north and south, thus maximizing its coverage of both North and South poles. SIRAL is the first sensor designed for measuring changes at the margins of vast ice sheets and floating ice in polar oceans, for the radar altimeter can detect small changes in the height of sea ice and it can also measure sea level with an unprecedented accuracy.

GRACE of NASA

Since the GRACE satellites were launched in 2002, GRACE data have been used to assess global hydrology such as continental total water storage anomalies or changes to Greenland, and Antarctic Ice Sheets related to climate warming, variability, and human impacts (Scanlon *et al.*, 2018). GRACE satellites can be regarded as giant space sensors that monitor monthly changes in terrestrial water storages or ice masses by monitoring the time-variable gravity field. However, given the coarse spatial resolution of GRACE data ($\sim 100,000$ km^2), these data are only useful to monitor hydrologic changes at continental to global scales.

4.4   Mechanisms of ice sheet changes

Ice motion is the result of three mechanical processes: internal deformation of the ice (creep), sliding of the ice over its bed (rock or unconsolidated sediment), and shear within any underlying deformable sediment. In ice sheet interiors, where the ice is frozen to the bed, motion occurs exclusively by creep due to internal deformation. Toward the ice sheet margins, sliding and shear take over. When present, basal motion can make up 90% or more of the total ice velocity. Fast-flow features (typically with velocities >100 m a^{-1}) drain most of an ice sheet. These features are usually divided into two categories: ice streams (bounded by slower moving ice) and outlet glaciers (bounded by rock walls) that flow through deep glacially eroded troughs. Ice streams can be ephemeral, dynamic features whose location is a result of ice sheet properties, or fixed streams determined by bed properties (see Note 4.1).

Until recently, ice sheets were believed to take thousands of years to respond to external forcing. New observations, however, suggest that major changes in the dynamics of parts of the ice sheets are taking place over time scales of years (Bamber et al., 2007). The mechanisms involve: meltwater drainage lubricating the glacier bed, glacier surges, ice sheet–ocean interactions, and ice-shelf buttressing. Summer meltwater on the surface of the Greenland Ice Sheet (GIS) accumulates in supraglacial channels that drain into holes (or moulins) in the ice, and eventually reaches the bed. It has been observed at Swiss Camp in West Greenland that the rate of ice motion increases by 10–25% when surface melting begins in spring and decreases in autumn. Glacier surges involve the slow buildup of mass followed by a rapid advance of the ice. The cycle varies in duration from tens of years to a few centuries. Surges are internally driven, cyclic instabilities in ice dynamics and so may occur in different parts of an ice sheet at different times. Ice sheet–ocean interactions involve iceberg calving and the delivery of ocean heat to the underside of floating ice shelves or tongues. Recent work in Greenland discussed later, attributes rapid retreat of several ice tongues to this mechanism. Ice shelves may be buttressed by lateral stresses and bedrock rises that provide pinning points. These effects create back pressure that retards ice flow for hundreds of kilometers inland.

Analysis of the paleoclimatic history of Greenland shows that large-scale changes in the extent of the GIS are closely correlated with paleotemperature records (Alley et al., 2010; see Chapter 9). Alley et al. state that no documented major ice sheet changes have occurred independently of temperature changes. The loss of most of the ice sheet could result from a long-lasting warming of between 2°and 7° C above twentieth century means (but this threshold is poorly defined). Moreover, snowfall increases when the climate warms, but increased accumulation in the ice sheet's center is insufficient to counteract increased melting and glacier outflow near the margin. Volume changes in the Greenland and Antarctic ice sheets have played a major role in long-term variations in global sea level (see Chapter 9). The Greenland and the Antarctic Ice Sheets are important indicators of climate change and driver of sea-level rise (SLR). Measurements of temporal variations in the Earth's gravity field by the NASA/German Aerospace Center's twin GRACE satellites applied to both ice sheets show regions of ice loss in coastal Greenland and West Antarctica over the last two decades, particularly the GIS with increasing rate of loss in recent years. In the converse sense, changes in sea level have been highly significant in the history of the marine-based West Antarctic Ice Sheet, but have played little part in the long-term changes of the GIS.

4.5   The Greenland Ice Sheet

The GIS, which extends from 60° to almost 84° N, is basically a simple dome that rises to nearly 3,280 m and covers about 85% of the island. Actually, there is the higher northern dome and a southern dome that reaches 2,850 m, linked by a 2,500-m saddle. The GIS contains 2.9 million km^3 of ice representing about 7.3 m of global sea-level equivalent. Major glaciers drain to the sea (see Figure 4.2) – Hellheim in the southeast, Jakobshavn Isbrae in the southwest, and Petermann's Glacier in the northwest. The ice sheet is also surrounded by numerous small ice caps.

Figure 4.2 Aerial photograph of two outlet glaciers draining from the northwestern Greenland Ice Sheet to the sea in Baffin Bay in May 1994. The area is east of Dundas (Thule Air Base). The large outlet glacier from the Inland Ice in this part of Greenland is Kud Rasmussen Glacier whose ice front has retreated as much as 21 km since early last century. The sea ice in Wolstenholme Fjord is still intact in early May with little snow cover due to the strong katabatic winds common in this region [Courtesy Dr. K. Steffen, CIRES] (http://cires.colorado.edu/science/pro/parca/gallery/)

The surface characteristics of the ice sheet change markedly with elevation. In 1888, in traveling inland for about 150 km, Robert Peary identified five altitudinal zones of the ice sheet surface with different characteristics. Subsequently, Carl Benson (1962) defined a sequence of four snow facies from high to low altitude: (1) the dry snow zone, (2) the percolation zone, (3) the wet snow (or saturation) zone, and (4) the bare ice (or ablation) zone. He located the dry snow zone on the highest parts of Greenland with a mean annual air temperature of −28 °C, where melting is supposed never to occur. In (1) dry snow metamorphism depends on gradual compaction, due to its own weight, by gravity and wind action, recrystallization, and depth hoar development due to internal temperature and moisture gradients. In the percolation zone as surface-air temperatures increase in summer, diagenesis takes place more rapidly and dramatically by melting. Surface meltwater percolates downward and occasionally spreads laterally when it encounters a relatively impermeable layer. When the percolated water reaches a depth where temperatures are below 0 °C or when cooling affects the upper parts of the snowpack, the free water refreezes forming horizontal ice lenses. In the wet snow zone, the snow reaches melting point, and free water is present throughout the upper snowpack. Ice lenses and pipes are also present in the wet snow zone. The higher temperatures enhance compaction and cause the firn to be denser than in the percolation zone. Surface melting is normally intensive, and the wet snow zone is moist throughout the summer. The bare ice zone is where all of the winter snow accumulation melts and runs off exposing the underlying ice. Due to intensive melting, the ice layers in this zone merge into a continuous ice mass by freezing meltwater. The meltwater can potentially be discharged into the ocean resulting in a net loss of mass. Additionally, snowmelt that percolates down to the base of the ice sheet provides lubrication that enhances basal sliding and accelerates the flow of ice toward the ice sheet margins. Bader (1961) estimated that the dry snow zone occupies 30% of the ice sheet, the percolation and soaked facies 55%, and the ablation zone 15%.

Fahnestock et al. (1993) and Partington (1998) used SAR data to map snow facies on Greenland. The zones mapped were the dry snow zone, the combined percolation–wet snow zone, and the ablation zone, as well as transient melt areas. Subsequently, satellite microwave data have been used to map the snow facies distribution in Greenland and Antarctica using a clustering procedure. Tran et al. (2008) use dual channel radar (13.6 GHz Ku-band and 3.2 GHz S-band) and dual channel passive microwave (23.8 and 36.5 GHz) data from the Environmental Satellite (Envisat) for 2004. Four parameters are used for the clustering of pixel values into six classes: the Ku-band backscatter; the average microwave brightness temperature (TB); the ratio of TB (TB 23.8−TB 36.5)/(TB 23.8 + TB 36.5); and the backscatter difference (Ku− S). The results for Greenland are generally in line with earlier work. Both the brightness temperature measured by the passive sensors and the radar backscatter measured by the active sensors are affected by many factors. These include snow grain size, stratification, surface roughness, snow temperature, wetness, and density. In particular, radar is more sensitive to surface roughness while the passive sensor has a higher sensitivity to temperature.

Using MODIS imagery for 2003 and 2005–2007, Sundal *et al.* (2009) find that there is widespread supraglacial lake formation and drainage across the ice sheet. There is a 2–3-week delay in the evolution of total supraglacial lake area in the northern areas (August 3–15) compared with around July 21 in the southwest. The onset of lake growth varies by up to 1 month interannually, and lakes form and drain at progressively higher altitudes during the melt season. The maximum lake coverage for the south-western, north-western, and north-eastern study areas is 0.68%, 0.81%, and 0.58%, respectively. The similar values suggest that lake coverage is likely to be governed by factors other than climate, such as the surface topography of the ice sheet. Das *et al.* (2008) describe a supraglacial lake drainage event lasting about 1.4 hr where the water descended 980 m to the base of the ice sheet causing local vertical and horizontal movement of the ice. They propose that the integrated effect of multiple lake drainages could account for the observed summer speedup of ice motion. Using a 35-km line of GPS receivers in West Greenland, Bartholomew *et al.* (2010) measured increases of 17–40% above winter background ice motion with a 6–14% increase in annual ice flux. The highest horizontal velocities coincide with uplift of the ice sheet surface of up to 12 cm.

Christoffersen *et al.* (2018) used a well-constrained 3-D ice sheet model to demonstrate that supraglacial lakes in Greenland, which have become larger and more numerous since 2000, drain in rapid succession when tensile-stress perturbations propagate fractures that form expansive networks between far apart lakes (~80 km). Such melt-induced fractures establish surface-to-bed hydraulic pathways in areas where open crevasses are found, up to 135-km inland from the ice margin. They also noted that even though supraglacial lakes on the GIS are expanding inland, the potential impact on ice flow is ambiguous because interior surface conditions may preclude the transfer of surface water to the bed.

Greenland has an average annual precipitation of about 750 mm, which is relatively wet considering its northern latitudes of 60–84° N. The winter season is characterized by deep low-pressure systems that approach Greenland along an active North Atlantic storm track, whereas patterns in the summer months are generally weaker and systems tend to approach the ice sheet through Baffin Bay. Koyama and Stroeve (2019) found that the Arctic System Reanalysis version 1 (ASRv1) monthly precipitation over coastal or near-coastal stations in Greenland generally agrees well with the corrected gauge precipitation measured by the Danish Meteorological Institute (DMI) and precipitation retrieved from the Precipitation Occurrence Sensor System (POSS) at Summit. However, the corresponding data at Ikerasassuaq and Nuuk are overestimated compared with the POSS observations. The North Atlantic Oscillation (NAO) index and ASRv1 precipitation are moderately correlated over northern Greenland.

The accumulation rate on the ice sheet ranges from 100 kg m^{-2} a^{-1} in the northwest interior to 1,400 kg m^{-2} a^{-1} in the southeast. The overall value is approximately 300 $\pm$ 25 kg m^{-2} a^{-1}. Bales *et al.* (2009) calculate updated accumulation values based on 39 ice cores and coastal meteorological stations for 1950–2000. They find average accumulation is almost the same, but there are five coastal areas in the southwest, northwest, and eastern regions, where the accumulation values are 20–50% lower than previously estimated, and southeast and northeast regions, where the accumulation values are 20–50% higher than previously estimated. For 1958–2007, Burgess $et\ al.$ (2010) find a mean value of 337 ± 48 mm a^{-1} w.e. with rates in southeast Greenland exceeding 2,000 mm a^{-1}. Rainfall accounts for about 4% of the precipitation, but it raises the heat content of the snowpack and so contributes to melting (Dahl-Jensen $et\ al.$, 2009), Precipitation minus evaporation/sublimation for 1961–1990 was ~570 Gt a^{-1} according to Hanna $et\ al.$ (2005). The meltwater runoff is about 150 kg m^{-2} a^{-1} (Cogley, 2005). The estimated calving rate is between -73 and -132 kg m^{-2} a^{-1} with an uncertainty of $\pm70\%$ (Bigg, 1999).

In 2007, the melt extent of the GIS was 20% greater than the 1995–2006 average and it occurred up to 2,980 m asl (Mernild $et\ al.$, 2009). Record snowmelt occurred in northern Greenland during summer 2008 (Tedesco $et\ al.$, 2008). Melting lasted up to 18 days longer than the 1979–2007 average and the melting index (the area affected × the number of melt days) was three times the average. Climate stations recorded surface air maximum temperatures that were 3 °C above the 1979–2007 average for JJA. Based on QuikSCAT data, over the period 2000–2004, the mean melt duration for the ice sheet ranged from 14 to 20.5 days according to Wang $et\ al.$ (2007). The proportion of the ice sheet that experienced melting in a given year ranged from 44 to 79%; there was extensive melt in 2002 associated with a single melt event of a few days duration. The ELA fluctuates from 1,640 to 600 m. The melt area covered 29% of the GIS in 1996 and 50% in 2005. On average, the simulated non-melt area decreased $\sim 6\%$ from 1995 to 2005, similar to observed values (Abdalati and Steffen, 1997). Based on QuikSCAT data, Bhattacharya $et\ al.$ (2009) show that the melt-area time-series is characterized by a step-like increase in 1995 in association with a shift in the NAO mode from negative to positive.

Overall, in terms of melt-day area since 1978, melting on the GIS was the highest in 2012, followed by 2010, 2016, 2002, 2007, 2011, and 2019. Melting in the GIS has increased since 2000, with the melt-day area for 2019 totaling 28.3×10^6 km^2. In 2019, melting was observed over nearly 90% of GIS in one or more days, was particularly intensive along the northern edge, where melting occurred 35 days more than the 1981–2010 average (nsidc.org/green land-today/, accessed on December 17, 2019). Figure 4.3a shows that number of summer melt days are high (sometimes exceeding 100 days) along the coastal areas of Greenland, which also tend to have high anomaly of melt days with respect to the 1981–2010 average number of melt days. The melt areas (km^2) were the highest in 2012, but 2019 are also clearly higher than the median of 1981–2010, especially in June and August (Figure 4.3c). Results of the University of Liege's MAR 3.10 model show that in 2019, at slightly more than 300 Gt, the 2018–2019 SMB (total precipitation minus runoff and evaporation) departure from the average is very close to the intense melt year of 2011–2012 record of SMB (Figure 4.3d). This does not include the imbalance in discharge, or the extent to which glacier outflow exceeded remaining snowfall input. The main area of mass loss was the western side of the ice sheet. Key climatic factors for surface mass melting for GIS in 2019 are persistent anticyclonic summer conditions (high geopotential height), resulting in dry and clear sky conditions

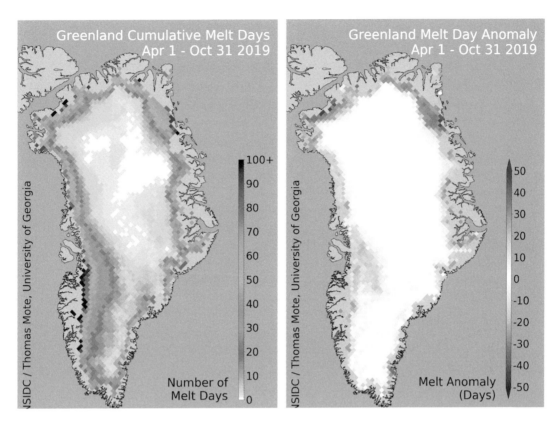

Figure 4.3 (a) Number of summer surface melt days for 2019 in the GIS, (b) Difference between 2019 summer surface melt days and the 1981–2010 average number of melt days, (c) the daily area of surface melting for 2019 and several recent years, and the median for 1981–2010, and (d) the net surface mass balance (SMB) of GIS for 2018–2019 (National Snow and Ice Data Center, https://nsidc.org/greenland-today/2019/)

conducive to surface melt enhanced by the ice–albedo feedback, and low snowfall in the 2018–2019 fall/winter seasons, particularly in the western Greenland.

Information on the mass balance of Greenland has improved over time with new techniques, at the same time as the components of the mass balance have themselves been changing. Radar altimetry from ERS-1 and -2 indicated that during the 1990s the GIS was thinning at the margins but thickening above the ELA resulting in a small overall positive mass balance (11 Gt a^{-1}) (Zwally *et al.*, 2005). Mernild *et al.* (2008) estimated an average SMB storage of 138 (±81) km^3 yr^{-1}, a loss of 257 (±81) km^3 yr^{-1}, and a runoff contribution to the ocean of 392 (±58) km^3 yr^{-1} for the period 1995–2005. Approximately 58% and 42% of the runoff came, respectively, from the western and eastern drainage areas of the GIS. In 2007, the runoff increased to 523 km^3 yr^{-1}, 35% above the 1995–2006 average (Mernild *et al.*, 2009). Dahl-Jensen *et al.*, (2009) report that the mean SMB, based on three 50-year

(c)

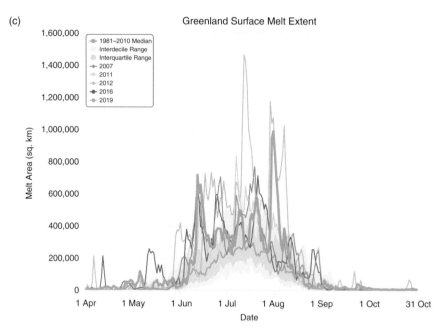

(d)

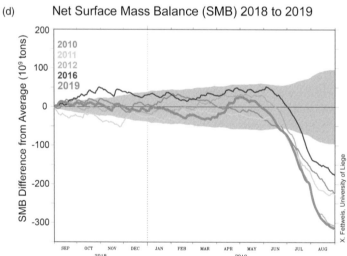

Figure 4.3 (cont.)

calculations is 284 Gt a^{-1}, with a range between estimates that is 22% of the mean. The most uncertain component is the refreezing amount. Ice discharge estimates vary widely. Reeh (1994) determined 316 Gt a^{-1}, while Bigg (1999) using calving rates obtained 170–270 Gt a^{-1}. More recent estimates by Rignot and Kanganaratnam (2006) show rates increasing from 321 in 1995, to 354 in 2000, to 421 Gt a^{-1} in 2005. They determine total ice sheet loss, combining dynamic losses and deviations from a zero-anomaly SMB, as 91 ± 31 km^3 a^{-1} of ice in 1996, 138 ± 31 km^3 a^{-1} in 2000, and 224 ± 41 km^3 a^{-1} in 2005.

Rignot *et al.* (2008) calculate that the ice sheet was losing 110 ± 70 Gt a^{-1} in the 1960s, 30 ± 50 Gt a^{-1} or near balance in the 1970s–1980s, and 97 ± 47 Gt a^{-1} in 1996, increasing rapidly to 267 ± 38 Gt a^{-1} in 2007 (100 Gt ~110 km^3). A synthesis of mass loss estimates shows a range from ~-100 to -230 Gt a^{-1} for 2003–2006 based on three different methods of assessment (Dahl-Jensen *et al.*, 2009, figure 2.13); these were mass budget, laser altimetry, and gravity determinations. Multiyear variations in ice discharge cause $60 \pm 20\%$ more variation in total mass balance than SMB, and therefore dominate the ice sheet mass budget. Averaged across ensembles of several methods, the loss of ice mass in the GIS over 1992–2011 is estimated at -142 ± 49 Gt yr^{-1} (0.39 ± 0.14 mm yr^{-1} of SLR) (Shepherd *et al.*, 2012), but at -229 [-290 to -169] Gt yr^{-1} (0.63 [0.80 to 0.47] mm yr^{-1} SLR) over 2005–2010 (Vaughan *et al.*, 2013). More recent published datasets shown in Figure 4.4 provide an average ice mass loss from Greenland of 171 Gt yr^{-1} for 1993–2016, increasing to 272 Gt yr^{-1} for 2005–2016 because of accelerating mass loss up to 2012, in which a record mass loss of over 400 Gt was estimated (van Den Broeke *et al.*, 2016), but a reduced mass loss of not more than 270 Gt yr^{-1} after that. The average trend estimate of Greenland to the sea-level budget over 2005–present is about 0.76 ± 0.1 mm yr^{-1}. However, interannual variability in mass balance of the ice sheet driven primarily by the atmospheric weather can be large.

By the mass budget method, ice loss from the GIS comes from similar amounts of surface melt and runoff and ice discharge across the grounding line but there are significant differences in the relative contributions of ice discharge and surface melt and runoff in different regions of GIS (Sasgen *et al.*, 2012). Dynamic losses dominate in southeast, central west, and in northwest Greenland, whereas in the central north, southwest and northeast sectors, surface melt and runoff appears to dominate. In the last two decades, surface melt and runoff from the GIS has increased, and that ice discharge across the grounding line has also increased due to the increased speed of some outlet glaciers. Figure 4.5 shows that ice loss has been the largest in southern Greenland where ice discharge from outlet glaciers dominate the losses. A detailed partitioning of the mass loss of the ice sheet is provided by van den Broeke *et al.* (2009). For 2000–2008, the total mass loss of ~1500 Gt is split almost equally between surface processes and ice dynamics. SMB is modeled with a regional climate model. Discharge is determined from ice flux data for 38 glacier basins covering 90% of the ice sheet, and is compared with data from the GRACE satellites, which show that the mass of the GIS has declined significantly in recent years because of iceberg calving and surface melting (Figure 4.4). However, the estimates of ice loss based on GRACE data vary between studies due to the time-variable nature of the signal, and factors such as (1) data-center-specific processing, (2) methods used to calculate the mass change, and (3) contamination by other signals within the ice sheet (e.g., glacial isostatic adjustment – GIA).

The GIS mass balance has been negative since AD 2000 owing to a decrease in mass balance and an increase in discharge. Prior to 1996, variations in precipitation accounted for SMB anomalies. Between 1996 and 2004 there were simultaneous large positive anomalies in

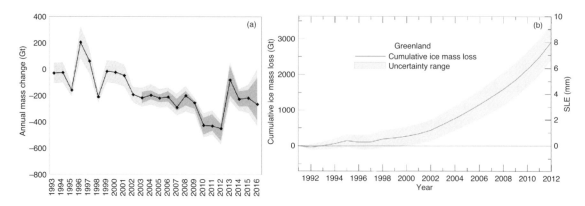

Figure 4.4 (a) Greenland annual mass change in Gt from 1993 to 2016. The medium grey region shows the range of estimates from the datasets of Barletta *et al.* (2013); Groh and Horwath (2016); Luthcke *et al.* (2013); Sasgen *et al.* (2012); Schrama *et al.* (2014); van den Broeke *et al.* (2016); Wiese *et al.* (2016a); and Wouters *et al.* (2008). The lighter grey region shows the range of estimates when stated errors are included, to provide upper and lower bounds. The dark line shows the mean mass trend (taken from Cazenave et al., 2018), (b) The cumulative annual average loss of ice mass (sea-level equivalent – SLE) of the GIS from 18 studies (taken from Vaughan *et al.*, 2013)

precipitation and runoff. After 2004, the cumulative anomaly of precipitation was constant, but runoff remained high leading to accelerated mass loss. Thirty percent of the excess liquid water refroze in the firn layer, releasing latent heat and producing warming. For 2003–2008, about half of the total ice sheet loss occurred in the wet southeast, with 70 Gt a^{-1} due to discharge.

Based on the GRACE satellite data, between 2002 and 2016, Greenland shed approximately 280 Gt of ice per year, causing global sea level to rise by about 0.8 mm yr^{-1} (Figure 4.5). In the image created from the GRACE data, orange and red shades indicate areas that lost ice mass, light blue shades indicate areas that gained ice mass, while white indicates areas where minimal change in ice mass since 2002. Higher-elevation areas near the center of Greenland experienced little to no change, but lower-elevation and coastal areas experienced up to 4 m of ice mass loss (expressed in equivalent-water-height; dark red) over the 2002–2016 period. The largest mass decreases of up to 30 cm of equivalent-water-height per year occurred along the West Greenland coast. The average flow lines of Greenland's ice converge into the locations of prominent outlet glaciers, and coincide with areas of high mass loss. Khan *et al.* (2010) show from GRACE data, compared with GPS measurements, that the well-documented changes in southern Greenland have now extended along the West Coast and into northwestern Greenland.

On the other hand, in 2017, the GIS was characterized by above-average surface albedo, below average (1981–2010) summer (JJA) melt extent, below average (2008–2017) net ablation along the margins of the ice sheet, but above the 1961–1990 average (Tedesco *et al.*, 2017). The cumulative ice sheet mass balance up until April 2017 (end of GRACE

(a)

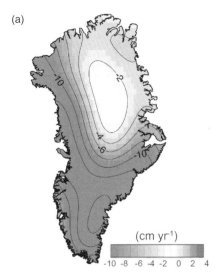

(b)

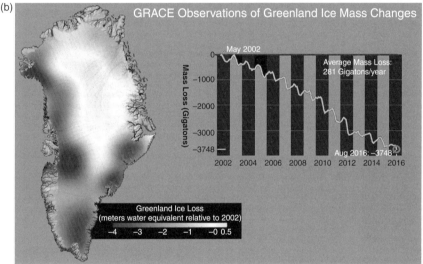

Figure 4.5 (a) Temporal evolution of mass ice loss of GIS estimated from GRACE time-variable gravity, shown in centimeters of water per year for 2003–2012, with color coded red (loss) (taken from Vaughan *et al.*, 2013), (b) GRACE observations of Greenland ice mass changes in Gt from May 2002 to August 2016. Orange and red shades showing areas with lost ice mass mainly concentrate in lower-elevation and coastal areas which experienced up to 4 m of ice mass loss (expressed in meter w.e.; dark red). Higher-elevation areas represented in white near the center of Greenland experienced minimal change in ice mass since 2002. (c) The Northern and Southern Hemisphere cryosphere in polar projection where the former shows the sea ice cover during the minimum summer extent (September 13, 2012). The yellow line is the average location of the ice edge (15% ice concentration) for the yearly minima from 1979 to 2012. Areas of continuous permafrost are shown in dark pink, discontinuous permafrost in light pink. The green line along the southern border of the map shows the maximum snow extent while the black line across North America, Europe, and Asia shows the 50% contour for frequency of snow occurrence. The Greenland Ice Sheet (blue/gray) and locations of glaciers (small gold circles) are also shown. The yellow line shows the average ice edge (15% ice concentration) during maximum extent of the sea ice cover from 1979 to 2012. Glacier locations were derived from the Randolph Glacier Inventory (Arendt *et al.*, 2012) (figure taken from Vaughan *et al.*, 2013) (A black and white version of this figure will appear in some formats. For the color version, please refer to the plate section.)

(c)

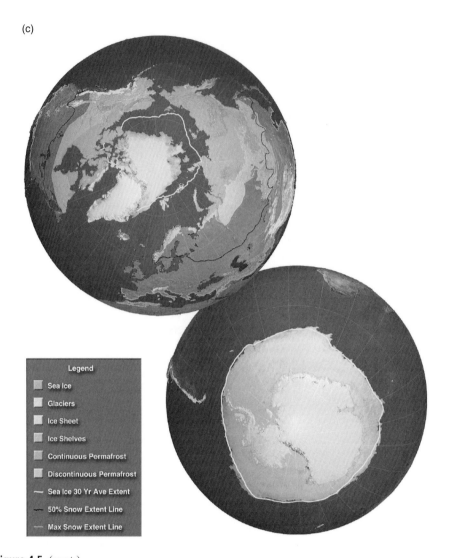

Figure 4.5 (cont.)

observations) was close to the average of 2003–2016. The glacier area in 2017 continued a period of relative stability that started in 2012–2013.

There has been extensive discussion of the causes of ice retreat in several of the major Greenland outlet glaciers. These include, in the case of the Jakobshavn Isbrae in West Greenland, the role of meltwater draining to the base of the ice and lubricating it (Zwally *et al.*, 2002); the weakening and breakup of the ice tongue that buttressed the glacier (Thomas, 2004); the thinning and the melting of the underside of the floating ice tongue by warmer ocean water (Holland *et al.*, 2008); and, in the Hellheim Glacier, ice acceleration, thinning, and retreat that started at the calving front and propagated upstream through dynamic coupling along the glacier (Nick *et al.*, 2009). Jakobshavn Isbrae has received particular attention because it drains about 7% of the entire ice sheet. Between 1850 and 1964, the calving front retreated an average of about 0.3 km a^{-1}, and was then stationary until 2000, when it rapidly accelerated to a retreat rate of 3 km a^{-1} (Korona *et al.*, 2009). The flow velocity increased from about 5.7 km a^{-1} during 1992–1997 to 9.4 km a^{-1} during 1997–2000, and reached 12.6 km a^{-1} in 2002–2003 (Joughin *et al.*, 2004). Maximum velocities in summer 2007 were 15.5 km a^{-1} (Korona *et al.*, 2009). The cause of the acceleration seems to be related to glacier thinning near the terminus, which promoted greater buoyancy and therefore increased velocity. Motyka *et al.* (2010) show that the ice loss from 1997 to 2007 totaled 160 km^3 over an area of 4,000 km^2, with two-thirds of this in grounded regions and one-third from disintegration of the floating tongue.

A study by Sundal *et al.* (2011) provides information about the relationship of meltwater and glacier motion in Greenland. They show that the development of efficient subglacial drainage associated with high meltwater input may reduce the ice velocity. Data from six major glaciers on the GIS indicate that, although the initial ice speedup was similar in 6 years of study, the glaciers experienced a dramatic late summer slowdown during the warm years when more meltwater was produced.

Much of Greenland's mass loss is by icebergs calving off the more than 200 fast-moving glaciers (0.5–13 km a^{-1}). The Jakobshavn Isbrae in West Greenland is one of the three largest outlet glaciers and drains the largest basin amounting to 6.5% of the Greenland Ice Sheet. Since 2000 AD, it has thinned dramatically and its speed nearly doubled. Comparison of summer-2007 SPIRIT (SPOT 5 stereoscopic survey of Polar Ice: Reference Images and Topographies) DTMs with October-2003 ICESat profiles shows that the thinning of Jakobshavn Isbrae (by 30–40 m in 4 years) is restricted to the fast glacier trunk (Korona *et al.*, 2009). Furthermore, >100-m thinning of the coastal section of the ice stream and the retreat of its calving front (by up to 10 km) are clearly depicted by comparing the SPIRIT DTM to an ASTER April-2003 DTM. The acceleration of Jakobshavn Isbrae has been attributed to the loss of buttressing. Joughin *et al.* (2008) show that between 1998 and 2003 the large floating ice tongue disintegrated and there was a decline in winter sea ice concentration in Disko Bay that had served to buttress the terminus by inhibiting iceberg calving. Previous studies in Greenland show that retreat of tidewater glaciers may be linked to recent increases in ice loss, thereby increasing Greenland's contribution to SLR. Moon and Joughin (2008) examined ice front changes of 203 tidewater glaciers, land-terminating glaciers, and glaciers terminating with ice shelves to understand Greenland glacier behavior over three periods: 1992–2000, 2000–2006, and 2006–2007. They observed synchronous, ice-sheet-wide increases in tidewater retreat during 2000–2006 relative to 1992–2000, coinciding with a 1.1 °C increase in mean summer temperature at coastal weather stations. Rates of retreat for the southeast and east slowed during 2006–2007 when temperatures were slightly below the 2000–2006 average. The results suggest that regional Greenland tidewater retreat responds both strongly and rapidly to climate change, with higher temperatures corresponding to increasing retreat.

Rignot *et al.* (2010) provide measurements of ocean currents, temperature, and salinity near the calving fronts of four glaciers in central West Greenland in summer 2008. They calculate submarine melt rates ranging from 0.7 to 3. 9 m d^{-1} that are two orders of magnitude larger than surface melt rates, but comparable to rates of iceberg discharge. They conclude that between 20 and 80% of the ice-front fluxes in summer 2008 were directly melted by the ocean. Simultaneously, work in East Greenland by Stranno *et al.* (2010) finds the presence of subtropical waters throughout Sermilik Fiord in summer 2008. These waters are continuously replenished through a wind-driven exchange with the shelf. The renewal of this warm water indicates that currently they cause enhanced submarine melting at the terminus of Hellheim Glacier. Nick *et al.* (2009) use numerical ice-stream model to simulate the surface evolution, flow, and stress field along a flow line on the Hellheim Glacier. They examine a reduction in back stress, showing that it increases velocity and this extends 20-km upstream from the terminus. This accelerates thinning and increases the driving stress. The terminus retreats in the model, matching the observed retreat of 7 km in the front from 2001 to 2006. Sole *et al.* (2008) also test the applicability of the various hypotheses concerning the peripheral thinning of the GIS. They report a four-fold increase in the mean thinning of marine-terminating outlet glaciers between 1993–1998 and 1998–2000, while the thinning rates for outlet glaciers that terminate on land remained statistically unchanged.

4.6   Antarctica

The Antarctic Ice Sheet (12.4 million km^2) that covers almost all of Antarctica represents about 90% of the world's fresh water. It is divided into two parts by the Transantarctic Mountains: the East Antarctic Ice Sheet (EAIS) and the much smaller West Antarctic Ice Sheet (WAIS). There are also ice caps and glaciers in the mountainous Antarctic Peninsula, which extends from 75 to 63° S. The volume of the Antarctic Ice Sheet and ice shelves is about 25.4 million km^3, and the WAIS contains 14% of this, or 3.6 million km^3. The sea-level equivalent of the Antarctic ice mass is about 57 m based on BEDMAP (Lythe *et al.*, 2001). The WAIS is bounded by the Ross Ice Shelf (RIS) covering 490,000 km^2, the Ronne–Filchner Ice Shelf (450,000 km^2) (see Chapter 8), and outlet glaciers that drain into the Amundsen Sea. Only 0.35% of Antarctica is ice-free. The average thickness of Antarctica is about 2,400 m and the maximum known thickness in Terre Adélie (69.9° S, 135.2° E) reaches 4,776 m. Most of the EAIS rests on a major landmass, but much of the WAIS is grounded below sea level, in places in excess of 2 km. The WAIS is referred to as a marine-based ice sheet.

A C-band (5.7 cm) SAR instrument on RADARSAT-1 collected the first high-resolution view of all of Antarctica in September–October 1997. The RADARSAT-1 Antarctic Mapping Project (RAMP) was followed by the Modified Antarctic Mapping Mission (MAMM) during September–November 2000 and further interferometric measurements over four fast-flowing glaciers were obtained in September–December 2004 (Jezek, 2008).

Radar "speckle" reduces the effective spatial resolution of the RADARSAT mosaic to close to 100 m, but there is a clear delineation of sharp edges, like surface crevasses or rugged topography. The initial science goals were to make a high-resolution SAR map of the continent, to examine the stability of WAIS, and to determine the mass balance and surface melt regimes. The main goals of the MAMM were to determine changes in the interior and at the coasts, to measure velocities in the interior, on ice streams and ice shelves, to calculate discharge rates for the major drainage basins, to locate grounding lines, and to determine the morphological and dynamic properties of the ice streams.

In the RAMP image, most coastal areas and much of the Antarctic Peninsula appear bright because of summer melt (Jezek, 2003). Very long curvilinear features snake across East Antarctica. These appear to follow ice divides separating the large catchment basins. It is also shown that an ice stream that extends at least 800 km into the EAIS feeds Recovery Glacier in Queen Maud Land, which enters the Filchner Ice Shelf. Terrain analysis relating ice surface plan curvature to basal topography suggests that a large subglacial basin >1,500 m below sea level underlies Recovery Glacier and its ice catchment (LeBrocq et al., 2008). Ice streams (discussed in Chapter 8) are made visible in the RAMP image by intense crevassing along shear margins. The Slessor and Bailey glaciers also extend hundreds of kilometers into East Antarctica (see Jezek, 2008, fig. 6).

Two other optical mosaics of Antarctica are available – one from MODIS (MOA) (Haran et al., 2006) and one from Landsat (LIMA) (Bindschadler et al., 2008). MODIS data have lower spatial resolution, but a wider field of view and more radiometric resolution. This often enables clearer views of extensive surface features, while LIMA's spatial detail allows smaller features to be examined.

Figure 4.6 shows a 2005 moderate resolution optical image of the MODIS MOA for the RIS with a surface textured by flow stripes, crevasses and other features related to ice flow and deformation (LeDoux et al., 2017). LeDoux et al. classified RIS into four types of structural provinces or textures arisen from ice deformation, namely, regions influenced by large outlet glaciers, shear zones, extension downstream of obstacles, and suture zones between provinces with different upstream sources.

Year-round stations have been operating at South Pole (2,835 m) and Vostok (78.4° S, 106.9° E, 3,488 m) since the IGY (1957–1958). Other permanent long-term stations are on the coast. Mayewski et al. (2010) comprehensively review the Antarctic climate system. The winter temperature at the South Pole averages $-58\,°C$ for April–September and rises sharply to $-28\,°C$ in December–January (Warren and Town, 2009). In winter a strong temperature inversion develops at the surface and the temperature typically increases by $25\,°C$ in the lowest 300 m of the atmosphere. The accumulation on Antarctica is mostly very low (Figure 4.7). Shallow firn cores collected in East Antarctica by the Norwegian–US International Polar Year traverse give accumulation rates averaged over 1815–2007 in the range of 16–32 kg m^{-2} a^{-1} (Anschutz et al., 2009). The South Pole Queen Maud Land traverses in the 1960s found very low accumulation (6 to 10 kg m^{-2} a^{-1}) concentrated around the region 81°–82° S, 20° E. There are also a few small regions with a constantly negative

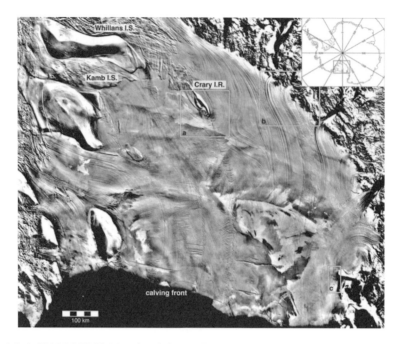

Figure 4.6 A 2005 MODIS Mosaic of Antarctica (MOA) image of the Ross Ice Shelf in which more details are given in Crary Ice Rise, coast of the Transantarctic Mountains between Beardmore and Nimrod Glaciers, and Minna Bluff (taken from LeDoux *et al.*, 2017)

SMB, due to sublimation and wind erosion, which form blue ice surfaces. Accumulation rates are larger (>500 kg m^{-2} a^{-1}) at the coasts and in the Antarctic Peninsula, but Eisen *et al.* (2008) show discrepancies between point data interpolated by passive microwave remote sensing and numerical climate models. At the South Pole, the net accumulation of ~85 kg m^{-2} a^{-1} during 1965–1994 is the highest 30-year average of the last 1,000 years at the South Pole (Mosley-Thompson *et al.*, 1999). Giovinetto (1964) estimated that the ablation zone covers about 65,000 km^2, representing <0.5% of the AIS. Ablation regions mainly form strips 10–20-km-wide adjacent to the coast, particularly from 40 to 135° E. The percolation and soaked facies cover about 10% of Antarctica, excluding the Antarctic Peninsula.

As described earlier, ice velocities can be determined from satellite imagery. Interferometric calculations indicate that the Drygalski Ice Tongue (75.4° S, 163.5° E) at a point 50 km from the coast is moving at 710 m yr^{-1} in good agreement with earlier studies (Jezek, 2003). The Lambert Glacier in East Antarctica, has velocities of 400–800 m yr^{-1} but as it extends across the Amery Ice Shelf, velocities increase up to 1,000–1,200 m yr^{-1} as the ice spreads out and thins. Pine Island Glacier has received much attention as it drains a large part of the WAIS and has been known for some time to be thinning and accelerating

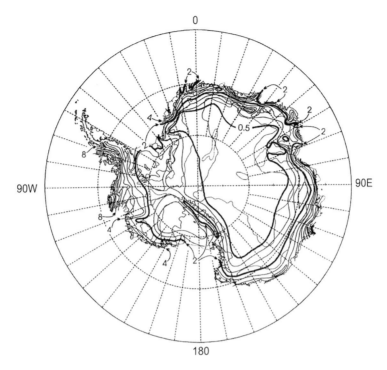

0

90W

90E

180

Figure 4.7 Mean annual accumulation (cm) over Antarctica (from Bromwich and Parrish, 1998). [Source: *Meteorology of the Southern Hemisphere*, 1998, p. 187 Fig. 4.12. Courtesy American Meteorological Society]

(Lucchita *et al.*, 1995) with the potential to make a significant contribution to SLR. Wingham *et al.* (2009) used satellite radar imagery to determine its rate of thinning between 1995 and 2008. The average rate of volume loss of the central trunk (5,400 km^2) quadrupled between 1995 and 2006 to a value in 2006 of 10.1 km^3 a^{-1}. They show that the region of lightly grounded ice is extending upstream. The changes inland are consistent with ocean-driven melting of the tongue and both observations and modeling indicate that intrusions of warm water at depth are the cause. Jenkins *et al.* (2010) used an autonomous underwater vehicle (AUV) to map the seabed beneath the Pins Island Glacier. They found that the downstream limit of grounded ice has retreated inland by 30 km, into water that is 300-m deeper than over the submarine ridge on which it was grounded until the 1970s. In the mid-1990s, the basal melt rate was estimated to exceed 50 m a^{-1} (Rignot, 1998) and the floating tongue has thinned accordingly. Water column thicknesses beneath the ice range from 400 to 600 m in the outer cavity to less than 250 m along the ridge crest. Retreat of the glacier is likely to continue for another 200-km inland until it encounters the next rise in the seabed.

It was formerly thought that the AIS was drained through a small number of fast-moving ice streams and outlet glaciers fed by relatively stable and inactive catchment areas. Active ice streams typically have velocities in the range of 100–2,000 m a^{-1}. However, Bamber *et al.* (2000) demonstrate that balance velocity estimates suggest that each major drainage basin is fed by a complex system of tributaries that penetrate up to 1,000 km from the grounding line into the interior of the ice sheet. They conclude that the distinction between the slow-moving interior of the ice sheet and fast-moving outlet glaciers and ice streams is not as clear as had been previously believed and that there is a gradation between the two extremes. Kamb (2001) found that the pressure in the water layer beneath the West Antarctic ice streams is very close to the ice flotation pressure, hence accounting for their fast flow.

A major discovery of recent years has been the existence of numerous, extensive subglacial lakes beneath the 3–4.5-km-thick AIS. Even though overlain by several kilometers of ice, changes in these subglacial lakes are observable at the ice surface using airborne radio-echo sounding. These lakes are interconnected and water flows from lake to lake. Some of these subglacial lakes may have been isolated from the outside world for millions of years, may cause changes in ice flow by draining rapidly and lubricating the ice-bed interface, allowing ice streams to flow even more rapidly.

Investigations using ICESat data for 2003–2008 have identified some 124 subglacial lakes that underlie the ice sheet, ranging widely in size (Smith *et al.*, 2009). They exist because geothermal heat flow (~0.06 W m^{-2}) warms the basal ice sufficiently to melt it locally. About 80% of the lakes are within a few hundred meters of sea level while most of the remainder are "perched" on the flanks of subglacial mountain ranges in the interior (Priscu *et al.*, 2008). The majority of lakes lie in shallow subglacial basins or in topographic depressions. Two-thirds of them are located within 50 km of a local ice divide. Wright and Siegert (2012) identified 379 subglacial lakes beneath the Antarctic continent, of which many are located in ice-stream onset zones as well as underneath slow-moving ice domes.

The largest and first subglacial lake to be identified in 1973 was Lake Vostok (Zotikov, 2006). It is about 250-km long by up to 50-km wide and hundreds of meters deep, similar in size to Lake Ontario, and is overlain by ice that is 3,750–4,150-m thick. Below ~3,550 m, the glacial ice is replaced by accretion ice from the lake water. The lake comprises two deep basins (~400 and 800-m deep) divided by a ridge where the water is about 200-m deep. The water temperature is ~ −3 °C as a result of the pressure of the overlying ice. There is debate as to whether Lake Vostok existed before continental glacierization about 15 Ma and persisted, or whether it formed by subglacial water flow into an existing and/or glacially eroded trough after the ice sheet reached its present configuration. Siegert (2005) supports the latter option and suggests that, during glacial onset, ice flow across Lake Vostok would have resembled flow across an ice marginal trough. The fresh water in the lake is replaced about every 55,000 years as the ice flows over the lake. According to Wingham *et al.* (2006), many of the subglacial lakes are also at least temporarily interconnected by subterranean river systems. Bell et al. (2007) demonstrate that subglacial lakes in the upstream region of the Recovery Glacier initiate and maintain rapid ice flow through either active modification of the basal thermal regime of the ice sheet or by the scouring of bedrock channels in periodic drainage events.

The snow facies of Antarctica as mapped by Tran *et al.* (2008) differ from those found in Greenland. The wet zone is spatially insignificant and largely confined to the Antarctic Peninsula. Most of the coastal regions of Antarctica only have narrow percolation zones. The classification partitions the ice sheet in terms of snow accumulation and/or snow drift redistribution and snow layering setup by the topographically influenced drainage winds over the ice sheet. There are seasonal changes in the distributions associated with temperature changes. Class 1 occurs along the crest line of the domes and ridges in the central East Antarctic Plateau. Dry snow, low accumulation, and very low winter temperatures characterize the region. This region changes to class 6 in summer, which is interpreted as a densification effect. Class 2 corresponds to lower altitude coastal regions of East Antarctica that are affected by strong katabatic winds that scour the surface making significant microrelief. The accumulation rate is higher and contributes to increasing the surface roughness. Class 2 is stable in its location year round. Class 3 areas are also stable in location over the year and are associated with marginal zones of the ice sheet with steep slope and mountainous terrain, such as the Transantarctic Mountains. The two main ice shelves, the Ross Ice Shelf and the Filchner–Ronne Ice Shelf, are identified as class 4 in the summer period, for the former and year round for the latter. Class 5 areas are restricted to the summer and have low accumulation and moderate winds. Class 6 is not well defined as it changes locations from winter to summer. In the central east part of the ice sheet, regions forming class 7 are located on the western side of the domes and ridges lines, characterized by low snow accumulation and moderate winds. The winds give rise to low surface roughness due to erosion and redistribution of surface snow. Class 7 becomes class 5 in summer.

A different perspective has been assembled by Frezzotti *et al.* (2002) and Scambos *et al.* (2006), both using field traverses and remote sensing. The surface microrelief comprises three categories (Goodwin, 1990): (1) depositional features formed from friable wind-transported snow (dune fields); (2) redistribution features and formed as a result of the erosion of depositional features (sastrugi, pits); and (3) erosional features formed from the long-term exposure to katabatic winds (glazed surfaces). In East Antarctica, glazed areas cover up to 30% of the surface at any given instant according to Frezzotti *et al.*; permanent areas occupy ~6% of the surface according to Scambos *et al.* These represent surfaces that have negligible accumulation and are strongly windswept. The firn is abraded by the wind. The glazed surface consists of a single snow-grain thickness layer cemented by thin (0.1–2 mm) films of regelated ice. Sastrugi glazed surfaces comprise alternating zones of wide, smooth glazed surfaces and wide, rough sastrugi zones with a distinct boundary and a nearly uniform width for considerable distances. These surfaces are typically several kilometers long and 100–200-m wide, covering several hundred km^2. In low accumulation areas, megadunes up to 2–4 m in height and spaced 1–3-km apart cover more than 500,000 km^2 of the East Antarctic Plateau (Fahnesttock *et al.*, 2000). They are perpendicular to the prevailing katabatic wind direction but present an angle in the direction of the regional surface slope turning to the left under the influence of the Coriolis force in the Southern Hemisphere (Frezzotti *et al.*, 2002). The dunes migrate very slowly up wind due to the snow accumulation on that face. The inter-dune troughs are swept by enhanced katabatic flows and the snow is blown or sublimated away. In the coastal zones of wind convergence, blowing snow removes large amounts of snow. In Terra Nova Bay during 2006–2007, Scarchilli *et al.* (2010) report that cumulative snow transport is about four orders of magnitude higher than snow precipitation. The combined processes of blowing snow sublimation and snow transport remove up to 50% of the precipitation in the coastal and slope convergence area.

Surface snow melt in Antarctica has been mapped for 1978–2004 by Liu *et al.* (2006), for 1980–2009 by Tedesco and Monaghan (2009) using passive microwave brightness temperature data from SMMR and SSM/I (figure not shown), and for 1978–2014 by Liang *et al.* (2019) using SMMR, SSM/I, and SSMIS (Special Sensor Microwave Imager Sounder) data (Figure 4.8). About 9–12% of the Antarctic surface experiences melt annually. The average snowmelt extent for the period 1980–2009 was ~1,294,000 km^2 and a snowmelt index averaged about 35 million km^2 × days. By contrast, the 2009 snowmelt extent was only ~690,000 km^2 and a snowmelt index of ~17.8 million km^2 × days. Melt is most frequent and extensive in the Antarctic Peninsula with limited occurrence on coastal areas of the Bellingshausen–Amundsen Sea and east of the Filchner–Ronne Ice Shelf. Tedesco and Monaghan show that major summer melting anomalies are related to amplified large-scale atmospheric forcing when both the Southern Annular Mode (SAM) of atmospheric circulation and the El Niño-Southern Oscillation (ENSO) are in their positive phases (i.e., strong zonal circulation in the SAM and El Niño phase of ENSO).

Liang *et al.* (2019) show that in 1978–2014, the average melting area of AIS was 1,098,000 km^2 which is less than 10% of the total area of the ice sheet, and an average melt duration of about 32 days per year, with more than 80% of the melt area experienced melting for at least 10 days per year (Figure 4.8a). Figure 4.8b shows that in 1978–2014, approximately 34% of the melt areas in AIS has lengthened (areas with green to deep blue colors), and 66% of the melt areas has shortened (areas with yellow to red colors). Therefore, the melt duration in Antarctica showed a stronger shortening than lengthening tendency for most areas over this study period.

The snowmelt variability of the GIS and AIS is related to the bipolar seesaw pattern of the Arctic and Antarctic surface air temperatures, which are correlated with the Atlantic Multidecadal Oscillation (AMO) signal based on the sea surface temperature (SST) variability in the North Atlantic. Liang *et al.* (2019) found that the residual de-trended melt extent (ME) series of AIS (GIS) are significantly anticorrelated, $\rho = -0.56$ (correlated, $\rho = 0.79$) with the AMO index at 99% level of confidence ($p < 0.01$). As AMO is related to the Atlantic Meridional Overturning Circulation (AMOC) that transports heat northward in both hemispheres, AMOC could be a possible link between the snowmelt variability of both ice sheets. When AMOC becomes stronger (weaker), northward oceanic heat flux increases (decreases), which could warm (cool) the Arctic, but cool (or warm) the Antarctica, and then cause the snowmelt to increase (decrease) in GIS and decrease (or increase) in AIS. Therefore, increasing (decreasing) AMO directly influences the snowmelt of GIS to increase (decrease), and that at AIS to decrease (increase), but the latter at a 2-year lag. Wind stress along the Antarctic Circumpolar Current dominates the upwelling around Antarctica,

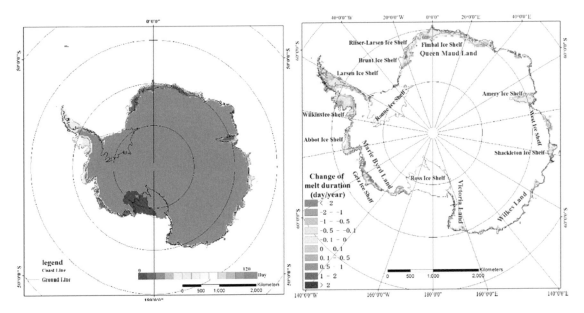

Figure 4.8 (a) Average annual melt duration (days), and (b) distribution of the rate of change of the melt duration (days) in Antarctica from 1978 to 2014 (taken from Liang *et al.*, 2019) (A black and white version of this figure will appear in some formats. For the color version, please refer to the plate section.)

drawing the deep and intermediate waters to the surface, which is warmed up by solar energy, and transported northward to the equatorial area by the Atlantic surface current. With further warming in the equator, the warm waters are transported to the Arctic area. Because this transport process takes a long time, the SST variability in the North Atlantic has a lag of about 2 years to the snowmelt of the AIS.

Compared to the Greenland, Antarctica is also losing mass but at a slower rate relatively (compare Figure 4.3b with 4.9b). The IMBIE team (2018) combined satellite data of Antarctica's changing volume, flow, and gravitational attraction, and modeling of its SMB to show that Antarctica had lost 2,720 ± 1,390 billion tons of ice between 1992 and 2017, which corresponds to a mean SLR of 7.6 ± 3.9 mm. Over 1992–2017, ocean-driven melting has caused rates of ice loss from the West Antarctica to increase from 53 ± 29 to 159 ± 26 Gt yr^{-1}; ice-shelf collapse has increased the rate of ice loss from the Antarctic Peninsula from 7 ± 13 to 33 ± 16 Gt yr^{-1}. The IMBIE team also estimated the average mass gain for the East Antarctica over 1992–2017 to be about 5 ± 46 Gt yr^{-1}, with its average rate of being the least certain because of large variations in and among model estimates of SMB and GIA for East Antarctica.

Figure 4.10 shows a broad agreement between various GRACE datasets, as most of the differences between them are caused by differences in the GIA correction, and a reasonable agreement between GRACE and the Input–Output method (IOM) estimates which indicate higher losses, which together yield an average of 0.42 mm yr^{-1} sea-level budget contribution from Antarctica over 2005–2015. All the datasets illustrate the accelerating mass loss of

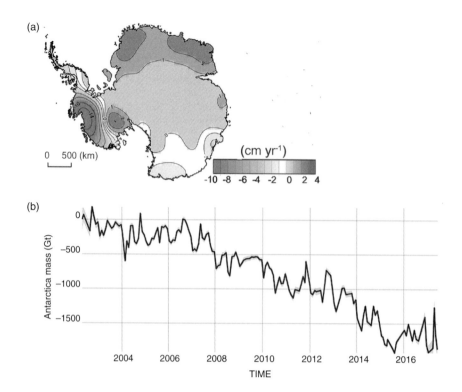

Figure 4.9 (a) Temporal evolution of mass ice loss of Antarctic Ice Sheet (AIS) estimated from GRACE time-variable gravity, shown in centimeters of water per year for 2003–2012, with color-coded red (loss) to blue (gain) (taken from Vaughan *et al.*, 2013). (b) The cumulative average loss of ice mass in Gt of the AIS from the GRACE gravity observations of 2002–2017 (A black and white version of this figure will appear in some formats. For the color version, please refer to the plate section.)

Antarctica. In 2005–2010, the ice sheet experienced ice mass loss driven by an increase in mass loss in the Amundsen Sea sector of West Antarctica (Mouginot *et al.*, 2014). The following years showed a less increase in mass loss, as colder ocean conditions prevailed in the Amundsen Sea embayment sector of West Antarctica in 2012–2013, which reduced the melting of the ice shelves in front of the glaciers (Dutrieux *et al.*, 2014). The annual turnover of mass of Antarctica is about 2,200 Gt yr^{-1} (Wessem *et al.*, 2017), with its annual mass balance essentially controlled by the balance between snowfall accumulation in the drainage basins and ice discharge along the periphery. The average trend estimate of Antarctica to the sea-level budget over 2005–present is about 0.42±0.1 mm yr^{-1}.

4.7   Overall ice sheet changes

Pritchard *et al.* (2009) show that the most profound current changes in both ice sheets are a result of glacier dynamics at ocean margins. They use ICESat laser altimetry from February

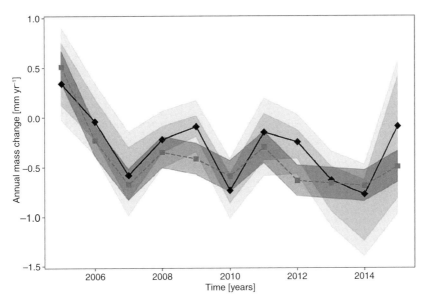

Figure 4.10 Antarctic annual sea-level contribution from 2005 to 2015. The black squares are the mean annual sea level calculated using the GRACE data sets listed in Martín-Español *et al.* (2016); Forsberg *et al.* (2017); Groh and Horwath (2016); Luthcke *et al.* (2013); Sasgen *et al.* (2013); Velicogna *et al.* (2014); Wiese *et al.* (2016b); Wouters *et al.* (2013); Rignot *et al.* (2011); and Schrama *et al.* (2014). The darker grey band shows the range of estimates from the datasets. The light grey band accounts for the error in the different GRACE estimates. The grey squares are the annual sea-level contribution and the light brown band is the associated error (taken from Cazenave *et al.*, 2018)

2003 to November 2007 to map elevation changes along the entire grounded margins of the Greenland and Antarctic Ice Sheets. They compare rates of elevation change from both fast-flowing and slow-flowing ice with those expected from surface mass-balance fluctuations. In Greenland, glaciers flowing faster than 100 m yr^{-1} thinned at an average rate of 0.84 m a^{-1}; losses were greatest in the southeast and northwestern regions, whereas there was thickening in the southwest. In the Amundsen Sea embayment of Antarctica, thinning exceeded 9.0 m a^{-1} for some glaciers. Along the Bellingshausen Sea coast slow-flowing ice caps are thickening at up to 1 m a^{-1} in strong contrast with profound dynamic thinning of up to 10 m a^{-1} on collapsed ice-shelf tributary glaciers flowing from the plateau to both the east and west coasts. Pritchard et al. conclude that dynamic thinning has intensified on key Antarctic grounding lines, is penetrating far into the interior of each ice sheet (~100–120 km in Greenland), and is spreading as ice shelves thin by ocean-driven melt.

Both ice sheets have been losing mass, and the rate of mass loss has been accelerating in recent years. Since 1992, ice loss of the GIS has accelerated, large rates of mass loss have spread to larger areas than the past, and the average loss rate has increased from 34 [−6 to 74]

Gt yr^{-1} over 1992–2001 to 215 [157 to 274] Gt yr^{-1} over 2002–2011. Ice loss from the GIS is partitioned in similar amounts between surface melt and outlet glacier discharge, and both components have increased.

Similarly, the AIS has been losing ice, mainly from the northern Antarctic Peninsula which has retreated and collapsed partially since decades ago, and from the Amundsen Sea sector of West Antarctica where ice shelves are thinning likely due to high ocean heat flux, and from the acceleration of outlet glaciers. Floating ice shelves are undergoing substantial changes in Antarctica, where the average rate of ice loss has likely increased from 30 [−37 to 97] Gt yr^{-1} over 1992–2001, to 147 [72 to 221] Gt yr^{-1} over 2002–2011. The mass loss of Antarctica, about 200 Gt yr^{-1} in recent years, is about 10% of its annual mass turnover of 2,200 Gt yr^{-1}, in contrast with Greenland where the mass loss has been growing rapidly to nearly 100% of the annual turnover of mass.

4.8   Ice sheet models

Treatment of ice sheet dynamics is beyond the scope of this book. Accounts may be found in Hughes (1998) and van der Veen and Payne (2004). In summary, we can say that ice sheet thickness and its surface slope linearly determine the stress that drives ice sheet deformation and the rate of deformation increases with the cube of the stress. The deformation rate integrated through the ice thickness determines the ice velocity, and ice flux is the depth-averaged velocity multiplied by the thickness (Alley *et al.*, 2010).

Ice sheet models make use of the shallow-ice approximation where the gravitational driving stress is locally balanced by drag at the base of the ice (Dahl-Jensen *et al.*, 2009). The ice deformation rate is related to the stress field in the ice through Glen's flow law (see Section 3.6). Mahaffy (1976) made the first numerical calculations for the Barnes ice cap, Baffin Island. Ice deformation rates are highly sensitive to ice temperature. To represent the temperature distribution in the ice sheet, ice sheet models simulate the 3-D ice thermo-dynamics – advection and diffusion of heat and strain heating due to shear deformation of the ice (the product of the shear strain rate and the driving stress on the ice sheet; see Note 4.2). In addition to ice deformation, where the bed is at the pressure-melting point, the ice also moves by basal sliding whose rate may exceed the ice motion associated with deformation by several orders of magnitude. The vertically averaged velocity is the sum of these two components. It should be noted that the physical processes that determine basal water pressure are not yet modeled or parameterized in ice sheet models. The equation for conservation of mass describes the rate of change of ice thickness (H) at each point on the ice sheet:

$$\frac{\partial H}{\partial t} = -\nabla \cdot (\bar{u} H) + b, \qquad (4.1)$$

where ∇ is the horizontal gradient operator and $\bar{u}$ is the horizontal velocity. The first term on the right describes the divergence of the ice flux and b is the local mass balance rate. Given measurements or climate model predictions of b, this equation can be integrated forward in time to simulate the evolution of ice thickness at all locations on the ice sheet.

Early ice sheet models were confined to tracing a flow-line transect. Using frictional lubrication, Budd and McInnes (1979) suggested that surging of the EAIS was possible. The surface profile of an ice sheet on a horizontal bed of half width L, thickness h, and thickness at the center H is

$$h^2 = \frac{2\tau_0}{\rho g} \,(L-x), \tag{4.2}$$

where ρ = density, g = acceleration due to gravity, τ_0 = the basal shear stress (~0–100 kPa), and $(L-x)$ is the distance from the edge measured along a flow line. The equation describes a parabola. The ice thickness at the center is $H = (2\tau_0 L/\rho g)^{0.5}$.

If the flow term $\partial q/\partial x$ is small, then the surface elevation will vary in response to the local accumulation/ablation, which will determine the profile. "Balance velocities" are steady-state velocities that are calculated from accumulation rate and surface slope. Paterson (1994) shows that for a 1,000-km radius circular ice sheet, of perfectly plastic ice with a yield stress of 100 kPa, an accumulation rate of 150 mm of ice /year, and ablation by iceberg calving at the margin, the ice in the center would be 4,700-m thick. Balance velocities would increase from 1.5 m a^{-1} at 100 km from the center to 45 m a^{-1} at 900 km and the travel time for ice to move from the center to the edge would be 150,000 years.

A hierarchy of land ice models is presented by van der Veen and Payne (2004). The simple lamellar flow model, involves a balance between driving stress and basal drag. The surface and bed topography must be nearly level for lamellar flow, which is a good approximation to conditions in the interior of an ice sheet. In cases where an ice stream is bounded by a rock wall or stagnant ice on one or both sides, lateral drag needs to be incorporated. Two general types of ice sheet model have been developed. One is prognostic, based on the original work by Budd *et al.* (1967) for Antarctica; the other category is diagnostic, addressing specific aspects of ice sheet processes. Prognostic models involve four sets of equations (van der Veen and Payne, 2004). These are: (1) diagnostic equations for the horizontal velocity components as functions of local ice geometry and ice rheology (Glen's law); (2) prognostic equations for the evolution of internal ice temperature, given appropriate boundary conditions at the upper and lower ice surfaces; (3) a diagnostic equation for ice vertical velocity via the divergence of the horizontal velocity; and (4) a prognostic equation for ice thickness based on the snow accumulation, snow/ice melt, and the divergence/convergence of horizontal ice flow. The effects of bedrock depression under the changing weight of the ice load must also be taken into account. Such models have been used to reconstruct ice sheet history over glacial cycles, as well as to assess the responses to future climate change. Diagnostic models do not address time evolution of the ice sheet and treat the internal stress regime in much greater detail, particularly the contributions of longitudinal and lateral stresses. Recently, models have been developed that do not assume negligible vertical shear.

Jenssen (1977) developed the first 3-D model that treated the coupling of ice flow and temperature for Greenland. A model that coupled the dynamical elements (ice motion) with the thermodynamics (ice heat flow and water content) was developed by Huybrechts (1992) for Antarctica. It included treatment of a coupled ice shelf, grounding-line dynamics, basal sliding, and isostatic bed adjustment. The ice sheet geometry is generated in response to changes in sea level, surface temperature, and mass balance. He shows that the coupling between ice flow and its thermodynamics causes ice sheet evolution to have a long time scale ($\sim 10^4$ a). Also, the low accumulation rates add further to this long time scale for ice sheet response. Fluctuations in the ice sheet are primarily driven by changes in eustatic sea level. Allowing the ice to be polythermal – with frozen and unfrozen parts – was addressed by Greve and Hutter (1995). Typically, the unfrozen part is adjacent to the bed of the glacier and is associated with pressure melting and meltwater formation.

A new 3-D numerical ice sheet model "Glimmer" (GENIE Land Ice Model with Multiple Enabled Regions), originally developed by A. C. Payne at the University of Bristol, has been modified to form the land-ice component of the GENIE (Grid Enabled Integrated Earth System Model) project (Rutt *et al.*, 2009). The model has the possibility of specifying several regions of the globe for simultaneous simulation. Important features of the model are that it is a regional ice model, with 3-D dynamics; the domain of each region is a projected grid, with nominally rectangular grid-boxes; and any number of active regions may be specified at run-time. Glimmer is supplied with a sophisticated module (GLINT) that allows coupling to a global climate model or reanalysis data. The core thermomechanical ice model (GLIDE) takes boundary conditions from three sources: a climate driver provides the upper surface temperature and mass balance fields, an isostasy model provides the lower surface elevation, and a geothermal model provides the geothermal heat flux through the lower ice surface. The climate driver employs an annual and a daily positive degree-day formulation for mass balance, and snow densification is modeled as a function of depth.

The evolution of the ice thickness is derived from the continuity equation for an incompressible material:

$$\frac{\partial H}{\partial t} = -\nabla \cdot (\bar{u} H) + b - S \qquad 4.3$$

where b is the SMB rate and S is the basal melt rate. Strain rates of polycrystalline ice are related to the stress tensor by the Glen flow law. Basal sliding is incorporated. Basal water is generated according to the local melt rate, and may be routed under the ice sheet. The horizontal velocity profile and the vertical ice velocity are calculated. Since the flow law depends on the ice temperature, it is necessary to determine the thermal evolution of the ice sheet. The temperature change for each column of ice is calculated from the sum of the vertical diffusion, vertical and horizontal advection, and the heat generation term. All ice in

Glimmer is considered to be grounded and four parameterization options are available to remove it at the margin. The geothermal heat flux may be specified as a global constant (of the order of 50 mW m^{-2}) or as a spatially varying field (higher in young rocks, lower in old ones). In the isostasy model, the lithosphere can be described as (1) a local lithosphere, where the flexural rigidity of the lithosphere is ignored, or (2) an elastic lithosphere, where the flexural rigidity is taken into account. The asthenosphere can be treated as (1) fluid, where isostatic equilibrium is reached instantaneously, or (2) relaxing, where the flow within the mantle is represented by an exponentially decaying hydrostatic response function.

Rutt *et al.* (2009) test Glimmer results against the European Ice Sheet Modelling Initiative (EISMINT) benchmarks of Huybrechts *et al.* (1996) for ice divide thickness and basal temperatures with uncoupled models, and for six EISMINT experiments with thermomechanically coupled models of Payne *et al.* (2000) for ice volume, ice area, fraction of base at or below melting point, ice thickness at the divide, and basal temperature at the divide. The results all show generally satisfactory agreements.

Another ice sheet model, developed by the Geophysical Institute, University of Alaska, Fairbanks, is the Potsdam Parallel Ice Sheet Model (PISM-PIK) (Winkelmann *et al.*, 2011). It is an open source, fully parallel, high-resolution ice sheet model with a hierarchy of available stress balances, including shallow ice and shelf approximations, a polythermal, enthalpy-based conservation of energy scheme, and extensible coupling to atmospheric and ocean models. PISM-PIK has been used to simulate the dynamic equilibrium of the AIS (Martin *et al.*, 2011) and in many other studies related to the AIS. Versions of PISM-PIK have been developed for Greenland and Antarctica, see www.pism-docs.org/wiki/doku.php?id=home.

Pollard (2010) provides an overview of the history of coupled ice sheet – climate modeling and includes an extensive bibliography. He notes that models have addressed many component processes including: ice stream–ice shelf flow and interactions, basal hydrology, ice-shelf calving, supra and subglacial lakes, effects of deformable sediment on basal stress, bedrock deformation, and eolian dust effects on ice sheets

4.9   Ice sheet and ice-shelf interaction

The characteristics of ice shelves are discussed in Section 8.2. Here we examine their interactions with ice sheets – notably in the Antarctic. The marine ice sheet instability hypothesis (Thomas, 1979) asserts that stable grounding lines cannot be located on an upward sloping seafloor. This is especially relevant to West Antarctica. The bed of the central WAIS is deeper than at the grounding line, suggesting that the current ice sheet may not be stable and is unlikely to be in a steady state. Observations show that the grounding line in West Antarctica has retreated up to 1,000 km since the Last Glacial Maximum from a location near the edge of the continental shelf (Conway *et al.*, 1999).

Grounded ice sheet flow is dominated by vertical shear, while ice-shelf flow is a buoyancy-driven flow dominated by longitudinal stretching and lateral shear (Schoof, 2007b). The two types of flow are coupled together near the grounding line in a complex mechanical transition zone. Ice discharge through the grounding line should increase with ice thickness there. Schoof shows that the shape of the bed is the primary control of the outflow of ice, together with ice viscosity and the slipperiness of the bed. Ice flux out of the grounded ice sheet increases when the bed at the grounding line is further below sea level, when ice viscosity is lower, or when the bed is more slippery. Schoof's numerical and analytical results confirm the marine ice sheet instability hypothesis.

The central parts of West Antarctica are significantly overdeepened and the Thwaites and Pine Island glaciers are locations in which irreversible grounding line retreat could be triggered. However, the ice shelf of Pine Island Glacier is confined in a narrow embayment. Buttressing at the sides, creating backpressure, will therefore affect longitudinal stresses and hence ice fluxes at the grounding line. This effect is not incorporated in Schoof's (2007) analysis.

4.10 Ice sheet contributions to sea level changes

John Mercer (1978) first suggested the possibility of a rapid "collapse" of much of the WAIS due to climatic warming. WAIS is a marine ice sheet and, since much of its base is well below sea level, it is considered to be inherently unstable. However, modeling and theoretical studies during the 1980s–1990s suggested that the presence of ice streams and ice sheet/ice-shelf dynamics, could stabilize the grounding line of a marine ice sheet (Vaughan, 2008) and Anandakrishnan *et al.* (2007) report the existence of a basal sedimentary wedge that stabilizes the grounding line of the Whillans Ice Stream. More recent work using ERS-1 data shows that in the Amundsen Sea sector (Pine Island, Thwaites, and Smith glacier basins) there was an indication of surface lowering of 10 cm a^{-1} and, during 1992 to1998, part of the grounding line of Pine Island Glacier was retreating inland at a rate of almost 1 km a^{-1}. Vaughan notes that this sector alone contains the potential to raise global sea level around 1.5 m, implying a reemergence of the Mercer paradigm in modified form. Bamber et al. (2009) have re-assessed the ice volume in the WAIS and find that previous determinations substantially overestimated its likely contribution to sea level. They obtain a value for the global eustatic SLR of ~3.3 m, with important regional variations.

Interferometric studies of Greenland indicate that the mass loss increased from 91 ± 31 km^3 a^{-1} in 1996 to 224 ± 41 km^3 a^{-1} in 2005, representing sea level contributions of 0.23 and 0.57 mm a^{-1}, respectively (Rignot and Kanagaratnam, 2006).

Rignot *et al.* (2008b) use interferometric SAR data for 1992–2006 to show that in West Antarctica, widespread losses along the Bellingshausen and Amundsen Sea coasts increased the ice sheet loss by 59% in 10 years to 132 ± 60 Gt yr^{-1} in 2006. In the Peninsula, losses increased by 140% to reach 60 ± 46 Gt yr^{-1} in 2006. In East Antarctica, there was a near-zero loss of 4 ± 61 Gt yr^{-1}. Measurements from the GRACE satellite program show that between April 2002 and January 2009, Antarctica lost 190 ± 77 Gt a^{-1}, with 69% of that coming from the WAIS (Chen *et al.*, 2009). Shepherd and Wingham (2007) determine that the combined imbalance of the two ice sheets is about 125 Gt a^{-1} of ice. This is only sufficient to raise sea level by 0.35 mm a^{-1} or one-tenth of the present rate of SLR of ~3.3 mm a^{-1}.

In a study of ice sheet mass balance comparing 1992–2010 modeled SMB and 2002–2010 GRACE data, Rignot *et al.* (2011) show that the acceleration in ice sheet loss over the last 18 years was 21.9 ± 1 Gt a^{-2} for Greenland and 14.5 ± 2 Gt a^{-2} for Antarctica, for a total of 36.3 ± 2 Gt a^{-2}. In 2006, the two ice sheets experienced a combined mass loss of 475 ± 158 Gt a^{-1}, equivalent to 1.3 ± 0.4 mm a^{-1} SLR.

By combining observations from multiple missions and approaches including sea-level budget analyses, Bamber *et al.* (2018) estimated lower mass losses from both ice sheets compared to estimates of GRACE-derived ocean mass change. They estimated the rate of mass losses from East Antarctica, West Antarctica, and Greenland Ice Sheets had been 80 ± 17, 197 ± 11, 320 ± 10 Gt a^{-1} over 2007–2011, and 19 ± 20, 172 ± 27, 247 ± 15 Gt a^{-1} over 2012–2016, respectively.

Shum *et al.* (2008) show that the choice of GIA model significantly affects GRACE-estimated Antarctic mass loss; the selected best model adds 0.25–0.45 mm a^{-1} to the estimate of SLR. The current estimate of Antarctica's contribution to SLR has a wide range: from +0.12 to −0.52 mm a^{-1}. The discrepancy between the observed sea-level trend of 1.8 mm a^{-1} and those estimated from various geophysical sources (2.10 ± 0.99 mm a^{-1}) is 0.30 mm a^{-1}. The role of Antarctica in SLR might be constrained better by extending satellite observations and by using long-term GPS data to discriminate subglacial vertical motion from ice mass balance.

Pfeffer *et al.* (2008) conclude from modeling that twenty-first-century increases in MSL exceeding 2 m are physically unrealistic. Even a rise of ~2 m is only possible if all glaciological variables are rapidly accelerated to very high values. The global mean sea level (GMSL) rise for 2081–2100 relative to 1986–2005 will likely be 0.26 to 0.55 m under low RCP (Representative Concentration Pathway) emission scenarios (RCP2.6), and 0.45 to 0.82 m under high emission scenarios (RCP8.5). For RCP8.5, the rise by 2100 is projected to between 0.52 and 0.98 m relative to 1986–2005 (SROCC, 2019).

Since early 1990s globally distributed sea-level measurements and SLR has been measured by high-precision altimeter satellites, the TOPEX/Poseidon (T/P), Jason-1, Jason-2, and Jason-3 altimeter satellites, and so on. Six groups (AVISO/CNES, SL_cci/ESA, University of Colorado, CSIRO, NASA/GSFC, NOAA) provide altimetry-based GMSL time series (Figure 4.11). All of them use 1-Hz altimetry measurements derived from the aforementioned reference missions which provide the most accurate long-term stability at global and regional scales (Ablain *et al.*, 2017). Complementary missions (ERS-1, ERS-2, Envisat, Geosat Follow-on, CryoSat-2, SARAL/AltiKa, and Sentinel-3A) provide increased spatial resolution and coverage of high-latitude ocean areas, poleward of 66° N–S latitude (e.g.,

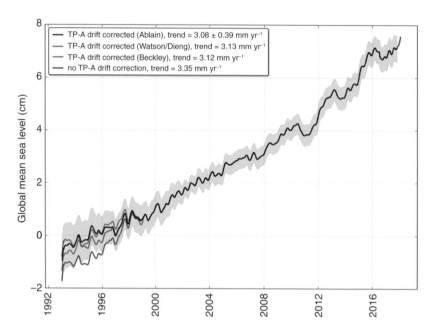

Figure 4.11 Evolution of ensemble mean global mean sea level (GMSL) time series (average of the six GMSL products from AVISO/CNES, SL_cci/ESA, University of Colorado, CSIRO, NASA/GSFC, and NOAA). On the black, red, and green curves, the TOPEX-A drift correction is applied, respectively, based on Ablain *et al.* (2017), Watson *et al.* (2015) and Dieng *et al.* (2017), and Beckley *et al.* (2017). Annual signal removed and 6-month smoothing applied; GIA correction also applied. Uncertainties (90% confidence interval) of correlated errors over a 1-year period are superimposed for each individual measurement (shaded area) (taken from Cazenave *et al.*, 2018) (A black and white version of this figure will appear in some formats. For the color version, please refer to the plate section.)

Legeais *et al.*, 2018). However, substantial differences (1–3 mm) exist between the six detrended GMSL time series.

The GMSL rise is estimated at 3.1 ± 0.3 mm yr^{-1} and acceleration of 0.1 mm yr^{-2} over 1993–present (Legeais *et al.*, 2018; Nerem *et al.*, 2018), which is over two times the rate for 1901–1990 of 1.4 mm yr^{-1} (total GMSL rise for 1902–2015 is 0.16 m) (SROCC, 2019). Ocean thermal expansion, glaciers, Greenland, and Antarctica, respectively, contribute 42%, 21%, 15%, and 8% to the global mean sea level over the 1993–present period (Cazenave *et al.*, 2018). However, Bamber *et al.* (2018) estimated a more modest global mean SLR of 1.76 ± 0.11 mm yr^{-1} over 2007–2011 and 1.85 ± 0.13 mm yr^{-1} over 2012–2016, respectively.

Note 4.1: In a remarkable finding, Bell *et al.* (2011) report that ice grows by bottom freezing in the glacial troughs of the subglacial Gamburtsev Mountains (80° S, 77° E). The mechanism involves conductive cooling of water ponded near the subglacial Gamburtsev Mountains and supercooling of water forced up steep valley walls. Persistent freeze-on thickens the ice column and, beneath Dome A, 24% of the base by area is frozen on ice. In some places, up to half the ice thickness has been added from below.

Note 4.2: Vaughan (2003) finds that threshold strain rates in Antarctic ice streams indicate yield stresses varying between 90 and 320 kPa.

Studies by Petrovic (2003) indicate that the tensile strength of ice varies from 70 to 310 kPa and the compressive strength varies from 5 to 25 MPa over the temperature range $-10\,°C$ to $-20\,°C$. The ice compressive strength increases with decreasing temperature and increasing strain rate, but ice tensile strength is relatively insensitive to these variables. The tensile strength of ice decreases with increasing ice grain size.

REVIEW QUESTIONS

(1) What is the definition of an ice sheet, and how many ice sheets are there on earth after the Pleistocene glaciations?

(2) Given the surface mass balance (SMB) of an ice sheet is the net outcome of snow accumulation and summer ablation below the equilibrium line altitude, what is the SMB of the Greenland Ice Sheet in Gt of ice per year in recent years (see Figure 4.3c)?

(3) What information about the paleoclimate record can we obtain from ice core data recovered from the Greenland and Antarctic ice sheets, and what is the longest climate record recovered from the European Project for Ice Coring in Antarctica (EPCIA)?

(4) Suppose the mass loss of the Greenland ice sheet was about 3,750 Gt over 2002–2016, how much is its annual contribution (percentage) to the global sea-level rise estimated at about 3 mm yr^{-1} in recent years given the melting of about 361.3 Gt of ice will result in about 1 mm of GMSL rise?

(5) Explain the possible key factors contributed to the surface mass loss and melting for Greenland in 2019 such as clear sunny weather related to high pressure conditions, ice–albedo feedback, low snowfall, or warm summer conditions. What is the average surface mass loss of the Greenland ice sheet in Gt of ice per year in recent years?

(6) What is the mechanism that causes expansive drainage networks to form between far apart supra-lakes across the Greenland Ice Sheet, leading to rapid draining of such lakes?

(7) If the Antarctic Ice Sheet (AIS) of about 24.7 million km^3 in volume were to undergo major melting in the next several decades, how could the sea-level rise affect coastal cities worldwide? If AIS were to melt completely, how high will the global sea-level rise?

(8) What has been the average melting area of AIS in km^2 and the average melt duration in days per year in 1978–2014? Has the melt duration in Antarctica shown a stronger shortening or a lengthening tendency for most areas over this study period?

(9) Explain how the Atlantic Meridional Overturning Circulation (AMOC) that transports heat northward in both hemispheres could be a possible link between the snowmelt variability of both ice sheets?

(10) How have the giant space sensors, NASA/German Aerospace Center's twin Gravity Recovery and Climate Experiment (GRACE) satellites monitored changes in ice masses of both ice sheets? Are temporal measurements of ice loss to both ice sheets

by various GRACE data sets show broad agreement between them and with the Input-Output method (IOM) (see Figure 4.10)?

(11) Since early 1990s, what high-precision altimeter satellites have been used to measure global mean sea-level rise?

(12) At an annual mass loss of about 200 Gt yr^{-1} in recent years, what is the annual mass loss in percentage of the annual mass turnover of AIS? In contrast, at an annual mass loss of about 280 Gt yr^{-1} in recent years with Greenland, what is the corresponding annual mass loss in percentage of its annual mass turnover?

5   Frozen ground and permafrost

5.1   History

Martin Frobisher first reported the existence of frozen ground in Baffin Island in 1577 according to Muller (2008). Tsytovich (1966) noted that Russian military reports published in 1642 contain the first mention of frozen ground in Siberia. James Isham reported the presence of frozen ground near Hudson's Bay in 1749 (Legget, 1966). Karl von Baer stated that the earliest scientific report of the existence of frozen ground was made in the mid-eighteenth century in a work on the flora of Siberia published by J.G. Gmelin in 1751 in Göttingen, but skepticism existed in scientific circles into the nineteenth century. The first widespread recognition of the occurrence of frozen ground appears in papers presented to the Royal Geographical Society of London by von Baer (1838a, 1838b). He reported on the presence of frozen ground in Siberia and noted that a merchant named Fedor Shergin sank a well at Yakutsk between 1828 and 1837 and collected temperature measurements from it for Friederich Wrangel. He found that the temperature rose steadily to near 0 °C at 116-m depth from −7.5 °C at a few meters below the surface. von Baer (1838b) noted other locations in Siberia where the ground was permanently frozen. A book prepared by him in 1842, but never published, has recently been made available in German by Tammiksaar (2001). Baer's text shows that he had already prepared a map of the distribution of permafrost in Siberia whose boundary closely resembles that of later authors. Baer also directed the expeditions of A. T. Middendorff to eastern Siberia, including temperature measurements to 116-m depth in the Shergin well at Yakutsk (Middendorff, 1844) and in other boreholes at locations east of the Yenisei River to the Pacific Ocean. Middendorff's data show that the permafrost thickness at Yakutsk was 190 m according to Shiklomanov (2005).

Systematic scientific work on permafrost in Russia dates from the late nineteenth century. The mining engineer Lopatin (1876) studied ground ice forms in the Yenisei River basin and was perhaps the first person to consider climatic change effects on permafrost. He noted that ground ice melts with warming and also pointed to the risk of coastal erosion (Shiklomanov, 2005). Wild (1882) produced a map of permafrost distribution in Russia from a conceptual model using the −2.0 °C isotherm of mean annual air temperature (MAAT) near sea level. Voeikov (1889) pointed out the importance of snow cover in influencing the occurrence of frozen ground and the effect of cold drainage on air temperatures in the mountain valleys of east Siberia. Yachevskyi (1889) analyzed the effects of climate, surface, and subsurface conditions on permafrost temperature and distribution. He also presented a more accurate map of permafrost extent across Russia (Shiklomanov, 2005, fig. 3).

There was other work in connection with railroad building and the opening up of Siberia. In 1894, the Russian Geographical Society organized a committee to develop detailed instructions for permafrost investigations along railroad lines that resulted in a manual titled *Instructions for Studying Frost in Soils*, published in Russian in 1885, with a second edition in 1912 (Yachevskyi and Vannari, 1912). Its recommendations were not implemented until the 1930s, however. Extensive work was carried out in the USSR in the 1920s–1930s and Sumgin (1927) published the first book on perennially frozen soils. He also defined permafrost as "any Earth material that remains below 0 °C for at least two years." Permafrost conferences were organized by the Soviet Academy of Sciences during the 1930s, the Permafrost Research Institute was established at Yakutsk in 1939, and by 1940 eight long-term permafrost research stations had been set up throughout Siberia.

Western knowledge about permafrost lagged considerably. Siemon Muller coined the term permafrost in a 1943 report for US wartime intelligence purposes in Alaska and published a modified version as the first widely available book in English (Muller, 1947). A later work by Muller, focused on permafrost-related engineering problems during the 1940s and 1950s, has recently been republished (Muller *et al.*, 2008). In 1979, Washburn published *Geocryology: A survey of periglacial processes and environments,* an expansion of his earlier textbook (Washburn, 1973). The first International Conference on Permafrost was held at the Purdue University, Indiana, in 1963, the second in Yakutsk, Siberia in 1973, and thereafter at 5-year intervals through 2008 in Fairbanks, Alaska. The complete proceedings of the first nine conferences are available on a DVD (University of Alaska, 2008). Yershov (1989) published a monumental five-volume text in Russian on *The geocryology of the USSR*, part of which was subsequently translated (Yershov, 1998).

Permafrost research in China started in the 1950s (Zhang, 2005b). Permafrost underlies approximately 23% of the country and about 80% of it is mountainous terrain. The Lanzhou Institute of Glaciology and Geocryology (LIGG) was founded in 1958 and was incorporated into the Cold and Arid Regions Environmental and Engineering Research Institute (CAREERI) in 1999. A permafrost observatory was established in Tumen County, south of the Tanggula Mountains in 1964 at an elevation of 4,950 m. A ground-based, long-term monitoring network has been established on the Tibetan Plateau and in northeastern China. The State Key Laboratory of Frozen Soil Engineering was established in 1991 and a permanent field research station was set up at Wudaoliang, about 150 km from Golmud. Regional maps of permafrost were published in the 1980s for northeast China and the

Qinghai-Tibetan Highway (Zhang, 2005b) and a 1:10 million-scale map of geocryological regionalization and classification in China was published by Qiu *et al.* (2000).The Qinghai-Tibet railroad, which was opened in 2007, crosses about 550 km of permafrost terrain and was specially engineered to accommodate a frozen substrate (Zhang *et al.*, 2008).

Permafrost research in Canada began in the late 1940s (Jenness, 1949). In 1952, the Division of Building Research (DBR) of the National Research Council established a permafrost research station at Norman Wells, NWT (Leggett, 1954). The DBR provided advice on the construction of northern town sites, roads, airstrips, and pipelines.

5.2   Frozen ground definitions and extent

Permafrost, or perennially frozen ground, is rock or sediment in which the temperature remains below 0 °C for 2 or more years. Typically, frozen sediment or soil contains ground ice ranging from a few tenths of a percent to 80–90% of the total permafrost volume (Romanovsky *et al.,* 2007). Seasonally frozen ground is arbitrarily defined as ground that is frozen for 2 weeks or more.

Permafrost and seasonally frozen ground regions occupy approximately 24% (22.79 $\times 10^6$ km^2, excluding glaciers and ice sheet) and 50.5% (48.12 $\times 10^6$ km^2), respectively, of the exposed land surface in the Northern Hemisphere (Zhang *et al.*, 1999, 2000; Zhang *et al.*, 2003b; Obu *et al.*, 2019) (see Figure 5.1). Permafrost is a zonal phenomenon – it is distributed geographically as a series of bands that are roughly conformable with latitude. Along a line from the polar region to the equator, these zones are usually described as continuous (>90% of the area is underlain by permafrost), discontinuous (50–90% underlain), sporadic (10–50% underlain), or isolated (<10% underlain) permafrost bodies. In continental regions of the Northern Hemisphere, the southern limit of continuous (discontinuous) permafrost is approximated by the position of the MAAT isotherm of −8 °C (−1 °C). In Canada, the transition from continuous to discontinuous permafrost occurs with a MAAT between −6 °C and −8 °C (Smith and Riseborough, 2002). The areas of the three permafrost zones are approximately 9.4, 6.7, and 9.4 million km^2, respectively. The greatest extent is in Asia – eastern Russia, Central Asia, and the Tibetan Plateau. The actual area underlain by permafrost is approximately 12–18% of the exposed land area because in the permafrost regions the frozen ground becomes discontinuous equatorward (Zhang *et al.*, 2000). Permafrost is also present under the ocean, known as subsea permafrost and is only present along shores of the Arctic Ocean (Harris, *et al.*, 2018).

There is subsea permafrost in the continental shelves of the Arctic seas because the shelves were exposed during the glacial intervals when mean sea level was ~130 m below present, enabling permafrost to develop. It is widespread in the Laptev and East Siberian seas, where the shelf is 400–700 km wide and <80 m deep; a narrower zone is present in the Beaufort Sea. Subsea permafrost can be ice-bonded (cemented by ice), ice-bearing (containing some ice) or ice-free sediment, and may contain unfrozen saline pore fluid (Rachold *et al.*, 2007). In the

Laptev Sea, where sea-bottom temperatures range from $-0.5\,°C$ to $-1.8\,°C$, permafrost is reported to be continuous to the 60 m isobath, and discontinuous to the shelf edge. Depths to the permafrost range from 2 to 10 m (Osterkamp, 2001). Mostly it is ice-bearing permafrost. Its thickness averages 300–350 over much of the Laptev and East Siberian Sea shelves and in some narrow zones over 500 m. There are several open taliks beneath the major paleo-river channels (Romanovskii et al., 2004).

In Iceland and southern Norway, the MAAT isotherm of -3 to $-4\,°C$ approximates the altitudinal lower limit of mountain permafrost (Harris et al., 2009). A 1:50,000 map of permafrost in Switzerland has been generated by the Swiss Federal Office for the Environment. Three types of ground surface were distinguished: coarse debris, bedrock, and glaciers/water. A permafrost index was based on topographic parameters. For bedrock an energy balance model was applied and for debris surfaces rules of thumb developed by Haeberli (1975) were applied.

In the Southern Hemisphere, permafrost is present in the southern Andes, the Sub Antarctic islands, and in Antarctica. More than 80% of the ice-free area of Antarctica contains continuous permafrost although permafrost soils account for only 0.35% of the Antarctic continent (Campbell and Claridge, 2009). Permafrost is generally continuous throughout East Antarctica and along the Antarctic Peninsula and the surrounding islands at elevations above 40 m asl. Along the Antarctic Peninsula, discontinuous permafrost exists at elevations between 40 and 20 m asl, and permafrost is either sporadic or lacking below 20 m asl (Bockheim et al., 2008). Ice-cemented permafrost within 70 cm of the surface, accounts for 43% of the glacier-free area in Antarctica and dry-frozen permafrost accounts for 41%.

Permafrost is widespread above about 4,500 m at 20° N, with a linear decrease in altitude to 1,000 m at 60° N (Cheng and Dramis, 1992). Much of the northern part of the Tibetan Plateau, also known as the third pole of Earth, is underlain by continuous permafrost. The lower altitudinal limit is about 4,200 m increasing to 4,800 m in the south (French, 2007). Along the Qinghai–Tibetan Plateau highway and railway, 190 boreholes show permafrost temperatures at 15 m depth exceed $-4.0\,°C$ and about half of the permafrost has a temperature above $-1.0\,°C$ (Wu et al., 2010). Temperatures depend strongly on elevation and latitude. Tibetan Plateau contains the largest area of alpine permafrost in the world (Wang et al., 2020).

The occurrence of frozen ground in the upper surface layer can be mapped using a variety of remote sensing techniques (Zhang et al., 2004). Synthetic aperture radar (SAR) provides high-resolution information on the timing and duration of the near-surface soil freeze–thaw status in cold regions, but the repeat cycle is too long when changes are rapid in spring and autumn. Passive microwave has low spatial resolution but is a promising technique for detecting near-surface soil freeze–thaw cycles over snow-free land (Zhang et al., 2004). Many procedures are indirect, relying on vegetational or geomorphic indicators. For example, Nguyen et al. (2009) show that in the Mackenzie Delta near-surface permafrost

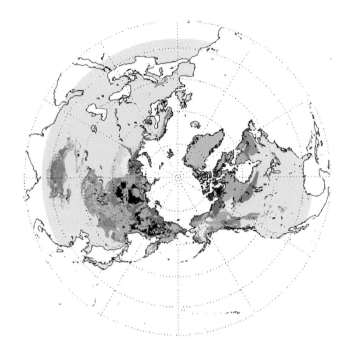

Legend for EASE-Grid Permafrost and Ground Ice Map

Permafrost Extent (percent of area)	Ground Ice Content (visible ice in the upper 10–20 m of the ground; percent by volume)				
	Lowlands, highlands, and intra- and intermontane depressions characterized by thick overburden cover (>5–10 m)			Mountains, highlands, ridges, and plateaus characterized by thin overburden cover (<5–10 m) and exposed bedrock	
	High (> 20%)	Medium (10–20%)	Low (0–10%)	High to medium (>10%)	Low (0–10%)
Continuous (90–100%)	Ch	Cm	CL	Ch	CL
Discontinuous (50–90%)	Dh	Dm	DL	Dh	DL
Sporadic (10–50%)	Sh	Sm	SL	Sh	SL
Isolated Patches (0–10%)	Ih	Im	IL	Ih	IL

Variations in the extent of permafrost are shown by the different colors; variations in the amount of ground ice are shown by the different intensities of color. Letter codes assist in determining to which basic permafrost and ground ice class any particular unit belongs. Letter codes are defined in the documentation that accompanies the data files.

Ice caps and glaciers

Figure 5.1 Map of the distribution of frozen ground in the Northern Hemisphere based on Brown *et al.* (1997) [courtesy of NSIDC, Boulder, Colorado] http://nsidc.org/data/docs/fgdc/ggd318_map_circumarc tic/index.html#format (spatial coverage map and legend) (A black and white version of this figure will appear in some formats. For the color version, please refer to the plate section.)

was found beneath all land surface types and vegetation communities except *Salix* and *Equisetum* (willow and horsetail) that could be mapped from SPOT 5 imagery. Near-surface permafrost was estimated to occur beneath 93% and 96% of the land surface within the southern and northern areas of the delta, respectively.

5.3   Thermal relationships

The relationship between air temperatures as measured in a weather shelter (Stevenson screen) at ~1.5 m height above the surface and the ground surface temperature has been described in terms of a transfer function known as an n-factor by Lunardini (1978; Klene *et al.*, 2001). The n-factor is the ratio of the surface freezing (or thawing) index to the air freezing (or thawing) index. Values range from 0.5 for moist soils to 1.0 for dry soils. In winter, due to the insulation of the ground by snow cover, the ground surface temperature exceeds the air temperature (Zhang, 2005), while in summer the situation is reversed due to the effect of vegetation cover, except in polar desert environments. Where there is snow cover and/or vegetation, temperature gradients in Arctic soils are steep whereas in the High Arctic with thin snow cover and polar desert conditions they are weak (Tarnocai, 2009). The mean annual ground surface temperature (MAGST) generally exceeds the temperature at the top of the permafrost (TTOP), or permafrost table. This positive difference is known as the **thermal offset** (see Figure 5.2). Near-surface permafrost is considered to thaw when (MAGST) at or near (the closest to) the depth of zero annual amplitude changed from ≤0 °C to >0 °C (Hjort *et al.*, 2018), where seasonal changes in ground temperature are negligible (≤0.1 °C) (Biskaborn *et al.*, 2019). Romanovsky *et al.* (2008b) show that while the MAAT decreases from −10 °C at Happy Valley in the Alaskan Coastal Plain to −16 °C at Mould Bay (Prince Patrick Island) and Isachsen (Ellef Ringnes Island), the MAGST decreases from −2 °C to −15.5 °C. In part, this is a result of decreasing snow depth northward. The mean annual temperatures at the permafrost table decrease from −3.5 °C at Happy Valley to −15.3 °C at Mould Bay. The thermal offset ranges from 0.1 °C to 1.2 °C. Its magnitude depends on the thickness of the organic layer that affects the ratio of thermal conductivity in the frozen and unfrozen states – greater thickness gives a higher ratio. The thermal state of permafrost is monitored at about 350 boreholes in North America, 180 in the former Soviet Union, and 45 in the Nordic countries (Romanovsky *et al.*, 2010). This is one component of the Global Terrestrial Network for Permafrost (GTN-P) that was established in 1999 under the Global Terrestrial Observation System (GTOS) of the World Climate Research Programme (WCRP) (see http://gosic.org/gtos/GTNet-P-data-access.htm). Records began in the late 1970s–1980s.

Not all soil water freezes at 0 °C because of dissolved mineral salts that depress the freezing point in the pore water. A layer of unfrozen ground that forms part of the permafrost where freezing is prevented by freezing-point depression in this manner is termed a *cryopeg*. This

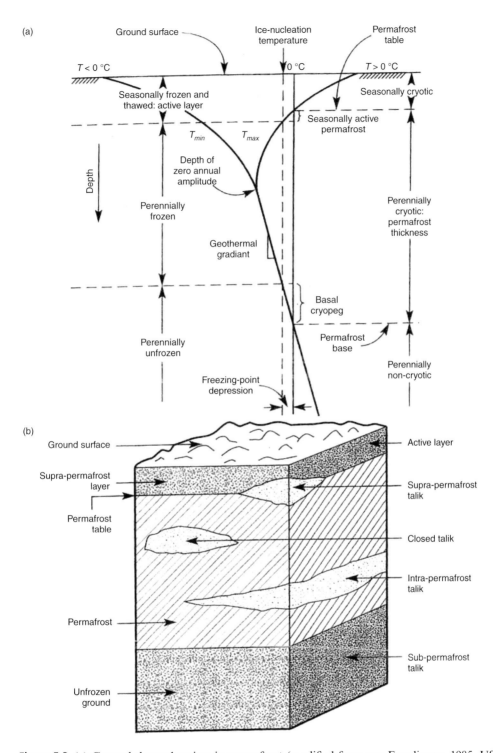

Figure 5.2 (a) Ground thermal regime in permafrost (modified from van Everdingen, 1985, US Army CRREL 85–5; (b) schematic profile of the layers in permafrost (modified from Ferrians, O. J., Kachadoorian, R., and Green, G. W. 1969. Permafrost and related engineering problems in Alaska. USGS Prof Paper 678, 37 pp.)

depression is particularly important in the case of saline groundwater. When water freezes it undergoes a volume expansion of about 9%. This expansion gives rise to frost heave of soil particles (see below) and rock splitting when water freezes in crevices. The latent heat of fusion (L_f) – the heat lost or gained by the air when liquid water changes to ice or vice versa – is 333 kJ kg^{-1}. This latent heat release causes ground temperatures initially to remain close to 0 °C during the freezing process – an effect referred to as the "zero curtain." Soil water may freeze *in situ* as pore ice or move toward the freezing plane where it forms lenses of segregated ice.

Putkonen (2008) discusses the "zero curtain" effect that refers to the persistence of soil temperatures near freezing during the autumn freeze-up in areas of permafrost. This is generally attributed to the release of latent heat, but Putkonen points out that the effect should be symmetrical in spring and autumn, which is not the case. By modeling he shows that the development of an isothermal layer in the soil near 0 °C occurs only in special cases. In the autumn, there is a soil domain between two phase-change fronts that quickly approaches 0 °C and the zero curtain effect develops as the water in the soil slowly freezes. In spring, there is an almost constant thermal gradient in the soil, which ensures that all soil layers are warming in concert and no zero curtain effect develops as there is only one phase-change front. The two conditions for the zero curtain effect are (1) the active layer contains ice or water and (2) the soil is isothermal near 0 °C. In the case of seasonally frozen ground, the zero curtain effect appears in spring when soil ice is melting, but not in autumn due to the thermal gradient effect.

The Frost Number method was developed by Nelson and Outcalt (1987) to predict the presence/absence of permafrost over large regions through calculations involving freezing and thawing degree-days (FDD and TDD). A freezing (thawing) degree-day is determined from the negative (positive) difference between the daily mean temperature and 0 °C. Values are summed on a daily basis over the season.

The *Surface Frost Number* F_+ is determined from:

$$F_+ = \frac{\mathrm{FDD_+}^{0.5}}{\mathrm{FDD_+}^{0.5} + \mathrm{TDD}^{0.5}},$$ (5.1)

where FDD$_+$ and TDD are expressed in °C days. F_+ ranges between 0 and 1; Nelson and Outcalt (1987) derived a method for establishing the geographical positions of zonal "boundaries" between regions underlain by continuous ($F_+ \geq 0.67$), discontinuous ($0.67 > F_+ \geq 0.6$), and sporadic ($0.6 > F_+ \geq 0.5$) permafrost, and no permafrost ($F_+ < 0.5$).

The occurrence of permafrost in alpine areas of Europe has been predicted using the temperature at the bottom of the winter snow cover (BTS). In the Alps and Scandinavia a BTS value below −2 °C is found in February–March beneath a snowpack of 80–100 cm, for example (Haeberli, 1973). Hoelzle (1992) states that permafrost is probable where the BTS value is <−3 °C. However, this approach is probably not applicable in dry continental mountains. The distribution of mountain permafrost depends strongly on the topoclimate and microclimate (see Barry, 2008, pp. 96–7). There are strong contrasts between sunny and shaded slopes. Ground cover of coarse blocks is important because they exert a cooling effect. This is determined by a reduced warming effect of winter snow cover, temperature-driven convection of air, and latent heat transferred by snow that penetrates voids between the rocks (Gruber and Haeberli, 2009).

5.4   Vertical characteristics of permafrost

The thickness of permafrost reaches ~700 m in the continuous zone of northern Alaska and the uplands surrounding the Mackenzie Delta (Burn and Kolkelj, 2009), while in north central Siberia – the Anabar shield – depths reach 1,000 m (Alexeev *et al.*, 2008). East of the Kolyma River in northeast Siberia, a profile along ~164–163° E (Kalinin and Yakupov, 1994) shows that average thicknesses decrease southward from about 200 m at 70° N to 75 m at 62° N. However, in the central section of the profile values decrease from ~400 m on the Jukagir Plateau in the west to 230 m in the Peculny ridge and Belsky depression in the east. Permafrost thickness along the Qinghai–Tibet highway and railway ranges from less than 10 m to over 300 m (Wu *et al.*, 2010). The majority of the permafrost on the Qinghai–Tibet Plateau is <100 m thick with substantial areas <50 m. Near the southernmost extent of continuous permafrost in North America, the thickness averages about 15–30 m according to French (2007).

The development of permafrost can involve very long time intervals. Lunardini (1995) calculates that deep permafrost (1,500 m) would require at least 100,000 years to form and possibly the entire Quaternary period. Permafrost thicknesses of around 600 m require around 50,000 years. Temperatures during the glacial intervals were ~10–12 °C less than now and the intervening interglacials had relatively brief time spans (see Chapter 9). In the Tien Shan mountains, central Asia, permafrost first formed 1.6 Ma according to Aubekerov and Gorbunov (1999) and was most extensive in the Late Pleistocene. The thickness to which permafrost develops represents a balance between surface heat loss and geothermal heat flux. The heat flow from within the Earth leads to a geothermal temperature gradient of 1 °C per 30–60 m (0.0333–0.0167 °C m^{-1}).

In the upper layers of the ground, diurnal and annual temperature fluctuations are damped with increasing depth. The depth of zero annual amplitude is typically 10–15 m below the surface. The **active layer** is the layer above the permafrost that is frozen in winter and thaws in summer. It is a type of seasonally frozen ground and lies above the permafrost table – the upper surface of the permafrost. During the first half of the thawing period, the active layer reaches 70% of its seasonal maximum at locations throughout the Alaska coastal plain and foothills according to Streletsky *et al.* (2008). Typically, the active layer reaches its maximum thickness and begins freezing upward from the bottom 1 to 2 weeks earlier than the start of freezing downward from the surface. The Circumpolar Active Layer Monitoring (CALM) observational network, established in 1991, observes the long-term response of the

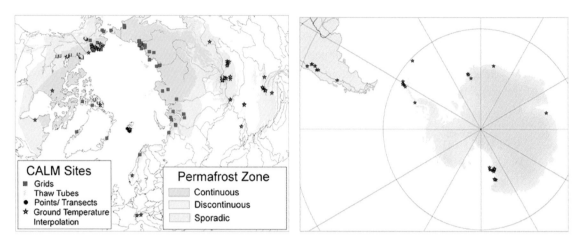

Figure 5.3 The network of CALM sites in the Northern Hemisphere (Courtesy: University of Delaware)

active layer and near-surface permafrost to changes and variations in climate at more than 125 sites in both hemispheres (Brown *et al.*, 2000; Nelson *et al.*, 2008) (see Figure 5.3). More information about CALM can be found at www.udel.edu/Geography/calm/,

Methods to survey the active layer

Most of the historical records on thaw depth in the North American Arctic have been obtained using small-diameter metal rods to probe for resistance at the bottom of the active layer. Other methods include thaw or frost tubes (Mackay, 1973) and measuring and recording ground temperature. Thaw tubes are sunk into the ground to twice the expected thaw depth. Construction details and observational procedures are provided by F. M. Nixon: www.fao.org/GTOS/doc/.../ECV-17-permafrost-ref-22.

At about 60 CALM sites active-layer thickness (ALT) is measured on grids ranging from 1 ha to 1 km², and 100 sites observe soil temperatures, including permafrost temperatures from boreholes. Grid-mean thaw depths can be used to determine ALT; these provide robust measures of the average site thaw depth at eight monitoring sites in the Mackenzie River valley. They are less than annual maximum ALTs as determined by thaw tubes (Smith *et al.*, 2009). Active-layer response to thermal forcing is found to be well-represented by grid-mean thaw depths and the square root of late-season thawing-degree days.

Active-layer thickness

ALT is dependent on a number of factors. It decreases with increasing latitude due to the shorter thawing season, lower air temperature and it decreases where there is a thick vegetation cover or snowpack. The soil moisture content is another important variable that determines the heat capacity of the soil and therefore its rate of temperature change to heating and cooling. The ALT ranges from 2 cm at 2,000 m asl on Mt. Fleming, Antarctica to 80–100 cm at warmer coastal sites farther north in Antarctica (Campbell and Claridge, 2009). The active layer at Lake Hazen on Ellesmere Island is 45-cm deep according to Tarnocai (2009), but on the sandy delta at nearby Tanquary Fiord the author measured 12 cm in 1963 and 1964. In northern Alaska, the maximum thicknesses of the active layer increases inland from about 36 cm at the coast to 62 cm at Franklin Bluffs (69.8° N, 148.7° W) (Romanovsky and Osterkamp, 1997). At Barrow, Alaska, the active layer depth is strongly related to accumulated thawing degree-days. Rather than exhibiting a simple relation with air temperature, differences in ALT between 1962–1968 and 1991–1997 may be due to changes in the stratigraphic position of segregated ice, insulation provided by surface organic matter, soil moisture, or some combination of these (Nelson *et al.*, 1997). Streletsky *et al.* (2008) show that in the Alaska foothills ALT ranges from 41 cm in moist acidic tundra to 56 cm in moist nonacidic tundra, and on the coastal plain from 40 cm on moist nonacidic tundra to 63 cm on wet graminoid tundra. In Svalbard, the active layer as measured at CALM sites ranges from ~60 to 100 cm according to O. Humlum (www.unis.no/35_staff/staff_webpages/geology/ole_humlum/CALM.htm).

However, at Jannsonhaugen, Svalbard (275 m asl) it was 3.1–3.5 m deep from 1987 to 2005 (Harris *et al.*, 2009). They also show that at the Schilthorn, Switzerland, in bedrock at 2,909 m, the depth was ~5 m in 2000 and 2004–2006 but increased to 8.6 m during the anomalously warm summer of 2003.

ALT in the Russian Arctic drainage is estimated by Zhang *et al.* (2005) from soil temperature measurements in the Lena basin at 17 stations (1956–1990), an annual thawing index based on both surface air temperature data (1901–2002), and numerical modeling (1980–2002). Based on the thawing index, the 1961–1990 average ALT is about 1.87 m in the Ob, 1.67 m in the Yenisey, and 1.69 m in the Lena basin. They show that ALT exhibits complex and inconsistent relations to variations in snow cover. On the Qinghai–Tibet Plateau, the ALT reaches 1.4 m in cold permafrost (temperature <-1 °C) and 3.5 m in warm permafrost (temperature ≥ 1 °C) (Wu and Zhang, 2010). In the former region, the maximum ALT occurs in September but is delayed until February in the warm permafrost. The duration of the active layer in the thawing state is only about four months in cold permafrost but up to 9 months in warm permafrost.

As a first approximation, the thickness of the active layer (Z) can be estimated using Stefan's equation:

$$Z = [2TK_t t/L_f]^{0.5}, \tag{5.2}$$

where T is the ground surface temperature during the thaw season, t is the thawing season duration, K_t is the thermal conductivity of unfrozen soil, and L_f is the latent heat of fusion of water (Harlan and Nixon, 1978; Nelson *et al.*, 1997). Woo *et al.* (2004) state that the rate of descent of a thawing front (dz_t/dt) is given by

$$dz_t \ /dt = \ K_t(T_f \ / \ z_f)/L_f\theta z \tag{5.3}$$

where dz_t/dt is in m s^{-1}, K_t is in J m^{-1} s^{-1} K^{-1}, T_f/z_f is the freezing temperature gradient (K m^{-1}), L_f is the volumetric latent heat of fusion of water (J m^{-1}), and θ_z is the volumetric fraction of soil moisture content at depth z. Applying this approach, Woo *et al.* (2008) modeled the maximum ALT in tundra and boreal forest environments of the Mackenzie River valley. They compare the effects of a 0.2-m and a 1.0-m-thick layer of peat overlying a mineral soil. The respective simulated maximum annual thaw depths are 0.6 and 0.36 m in the tundra (Aklavik-Inuvik) and 1.36 m and 0.65 m for the boreal forest (Fort Simpson)

Romanovsky and Osterkamp (1997) prefer the formulation of Kudryatsev *et al.* (1974) in which the major parameters are MAAT and its seasonal amplitude. Mean annual snow thickness, heat capacity, and thermal conductivity of the soil are also taken into account. The equations assume a periodic steady state with phase change (Riseborough *et al.* (2008). The details of the formulation of Kudryatsev *et al.* may be found in those sources, or in Anisimov *et al.* (1997). Romanovsky and Osterkamp point out that the assumption that the temperature in frozen ground equals 0 °C can lead to significant errors of up to 70% in using the Stefan equation in regions of cold permafrost.

The variability of seasonal thaw depth has been modeled stochastically by Anisimov *et al.* (2002). Equations for the mean, variance, and higher moments of the ALT were derived by applying stochastic averaging to the Stefan solution for the Kuparuk region of northern Alaska. The stochastic model was used with gridded climatological data and a digital land cover map to construct probability maps of ALT for the region in four depth categories: 20–40, 40–60, 60–80, and >80 cm.

Acidic tundra typically has thaw depths of 20–60 cm while nonacidic surfaces are likely to have greater thicknesses. Sharp local contrasts in ALT are due to soil moisture differences between basins and uplands. The lowest ALT values are in the higher elevations of the foothills. Wet tundra areas in river valleys of the foothills are more likely to have deeper thaw penetration. Shiklomaov and Nelson (2002) show that in the coastal plain of northern Alaska spatial variability of ALT is related primarily to low-frequency topographic variations associated with partially drained or drained thaw lake basins, while in the foothills of the Brooks Range variability occurs at much higher spatial frequencies. Interannual climatic variability was also shown in the latter study to cause significant variation in ALT.

A **talik**, or unfrozen zone, may lie between the base of the seasonal active layer and the permafrost table. Taliks may also occur beneath lakes and river channels and may completely penetrate the permafrost if it is shallow. The water content of the talik soil plays a major role in the development of icings (see Chapter 6, p.283).

A progressive increase in ALT has been observed across Northern Hemisphere (Luo *et al.*, 2016), except Northern Alaska and West Siberia located in continuous permafrost zones

(Shiklomanov *et al.*, 2012). For example, Greenland shows a nonuniform increase in ALT since late 1990s. Scandinavian CALM sites show small positive trends in ALT. Active-layer thickening has been progressive in the Abisko area of Arctic Sweden since 1970s and has accelerated since 1995, resulting in the disappearance of permafrost in several mire landscapes (e.g., Callaghan *et al.*, 2010). In the Russian European North, permafrost degradation has been so significant that permafrost patches 10–15-m thick in Vorkuta had thawed completely. An increase in ALT has been observed in East Siberia and the Russian Far East.

Between 2002 and 2012, in 10 sites of five alpine ecosystems over the Beiluhe area of central Qinghai–Tibet Plateau, the average onset of spring thawing at 50-cm depth advanced by at least 16 days in all but the barren alpine, and the duration of thaw increased by at least 14 days for all but the desert grassland and barren ecosystems (Wu *et al.*, 2015). The average increase in ALT was 4.26 cm a^{-1} and the average increase in permafrost temperatures at 6-m and 10-m depths were, respectively, 0.13 °C and 0.14 °C. With an observed increasing trend in precipitation but not in the MAAT, it seems in the Beiluhe area, the ALT increase was primarily due to an increase in the summer rainfall and the higher permafrost temperature was the combined effects of increasing rainfall and the asymmetrical seasonal changes in soil temperatures

For North American CALM sites, a progressive increase of ALT is evident only in the Alaskan Interior. Thaw penetration into the ice-rich base of active layers at the North Slope of Alaska is accompanied by the loss of ice, manifested as ground subsidence up to 11 cm over 2001–2008. Similarly for the Russian European North, increases in ALT have caused ground subsidence of 20 cm in 1998–2007 (Mazhitova *et al.*, 2008). Thawing permafrost invigorates surface water and groundwater exchange (Walvoord and Kurylyk, 2016), while thicker active layers enable transmitting more water from upland to streams and lakes downstream. On the other hand, upland could become drier as water sources from glaciers or snowpack in the upland dwindle.

5.5 Remote sensing

The freeze–thaw status of the near-surface soil can be determined using satellite remote sensing. SAR gives data on the timing, duration, and spatial progression of near-surface freeze–thaw in autumn and spring, for example. Freezing results in a large increase in the dielectric of soil and vegetation, which causes a large decrease in L-band (15–30-cm wavelength) and C-band (3.75–7.5-cm wavelength) radar backscatter (~3 DB). Passive microwave radiation (PMR) data offer similar information at lower spatial resolution. Frozen soils relative to unfrozen soils exhibit (1) lower thermal temperatures, (2) higher emissivity, and (3) lower brightness temperatures. The PMR algorithm for frozen soils is

$$\frac{\partial}{\partial f} T_\mathrm{B}(f) \leq P_{SG} \tag{5.4}$$

and

$$T_{\mathrm{B}(37\,\mathrm{V})} \leq P_D, \tag{5.5}$$

where the spectral gradient is in K Ghz^{-1} and $T_\mathrm{B\,(37\,V)}$ is in K. P_{SG} and P_D are the cutoff spectral gradient and brightness temperature, respectively. Based on these equations, surfaces can be classified as frozen, dry (and hot), wet (and cool), and mixed (Zuerndorfer and England, 1992). A frozen surface has low brightness temperature (37 GHz) and a relatively low negative spectral gradient. Zhang and Armstrong (2001) and Zhang et al. (2003c) analyzed soil freeze–thaw status over the contiguous United States and southern Canada in winter 1997–1998. They used a negative spectral gradient and a threshold value of P_{37} = 258.2 K. They found that almost 80% of the time, the near-surface soil was frozen before snow accumulated on the ground. They applied the validated frozen soil algorithm to investigate near-surface soil freeze/thaw status from 1978 through 2003 over the Northern Hemisphere. The long-term average maximum area extent of seasonally frozen ground, including the active layer over permafrost, is approximately 50.5% of the landmass in the Northern Hemisphere, making it the most extensive component of the cryosphere (Figure 5.4). Preliminary results indicate that the extent of seasonally frozen ground has decreased about 15–20% during the past few decades.

Hachem et al. (2008) use the Land Surface Temperature measured by MODIS on Terra and Aqua for 2000–2005 to determine mean monthly and annual surface temperatures as well as freezing- and thawing-degree day totals over Labrador-Ungava. The 0.25 ratio line of ΣTDD/ΣFDD is shown to correspond closely here with the mean July 10 °C isotherm and the boundary between continuous and discontinuous permafrost. For this region, the 0.50 ratio line of ΣTDD/ΣFDD and the −5 °C MAAT isotherm delimit the southern boundary of discontinuous permafrost. Sporadic permafrost can occur as far south as the 0 °C MAAT isotherm.

Zheng et al. (2019) solely used satellite data to drive process-based simulation of soil freeze–thaw processes using the Geomorphology-Based Eco-Hydrological Model (GBEHM) modified to be fully driven with a combination of MODIS, TRMM, and AIRX3STD satellite products for the upper Yellow River Basin (~2.54 × 10^5 km^2 in area) in northeast TP. The model results are validated against field observations of freezing and thawing front depths (D_ft) and soil temperature (T_soil) at 54 stations, and frozen-ground types at 22 boreholes. GBEHM simulated reasonable D_ft (R^2 = 0.69; mean bias = −0.03 m) and T_soil (station averaged R^2 and mean bias range between 0.90–0.96 and −0.51 to −0.14 °C, respectively). The frozen-ground types are also (in general) accurately identified by the satellite-based approach, except for a few boreholes located near the permafrost boundary regions. Additionally, they also demonstrate the importance of considering dynamic soil water content in frozen soil simulation.

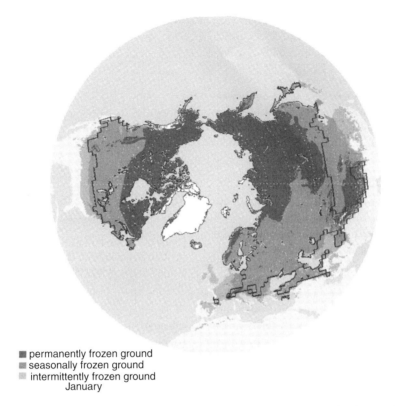

■ permanently frozen ground
■ seasonally frozen ground
▨ intermittently frozen ground
          January

Figure 5.4 The distribution of permafrost and seasonally frozen ground in the Northern Hemisphere based on passive microwave data, 1987–2003 (from Zhang *et al.*, 2003b) [Source: Zhang, T.-J. *et al.* 2003b. Distribution of seasonally and perennially frozen ground in the Northern Hemisphere. In M. Phillips, S. M. Springman and L. U. Arenson (eds). *Permafrost,* Vol. 2, *Proceedings of the 8th International Conference on Permafrost.* A.A. Balkema, Lisse, Netherlands, pp. 1291, fig. 1]

Ground-based remote sensing has begun to be widely used to determine thermal conditions and ice content. Four geophysical techniques in use involve: electrical resistivity (ER), electromagnetic induction (EMI), ground-penetrating radar (GPR), and refraction seismicity. Harris *et al.* (2009) provide a detailed review. ER in moist porous rock and soil increases markedly at the freezing point and in fine-grained soil it increases exponentially until most of the pore water freezes. One-dimensional profiles and 2-D tomography surveys have been applied to determine the presence/absence of ice, and to map permafrost structures. EMI measures the electrical conductivity in either the frequency domain or the time domain. It has been applied to assessing permafrost in rock glaciers, for example. GPR has been applied to mapping the summer active layer and ice-rich permafrost. Refraction seismics, which differentiate ice and water, are used to map the permafrost table. To resolve ambiguities, a combination of methods may be used: refraction seismics and ER tomography, or GPR and ER tomography.

5.6   Ground ice

There are four major types of ground ice: pore (or interstitial) ice, segregated ice, intrusive ice, and vein (or wedge) ice (Mackay, 1972). Pore ice provides the bonding that holds soil grains together. Segregated ice forms layers or lenses in the soil varying in thickness from a few millimeters to tens of meters. It forms by the migration and subsequent freezing of pore water. Intrusive ice forms by the intrusion of water, usually under pressure, into the frozen zone. It occurs as sill ice and dome-shaped pingos (see further). A special form occurs in lava tubes and caves (see Box 5.1). Vein ice forms when water drains into fissures formed by thermal contraction of the ground surface and freezes. Repeated events lead to the growth of deep ice wedges containing vertically orientated sheets of foliated ice. Ground ice may form at the same time as the surrounding sediments are deposited (syngenetic), or it may form later (epigenetic). Epigenetic ice wedges rarely exceed 4 m in depth. Syngenetic ice wedges are thickest and deepest on alluvial surfaces in central Siberia where there are high ice content sediments known as "yedoma" (Vasil'chuk and Vasil'chuk, 1997). They show from radio-carbon dating that ice wedges in northern Siberia grew at between 1.0 and 2.7 m per 1,000 years during the late Pleistocene (see Figure 5.5).

Massive ground ice has been extensively described in Siberia (Astakov, 1986), and the western Canadian Arctic (Mackay and Dallimore, 1992). Pollard and Couture (2008) report on massive ground ice in the Eureka Sound lowland of Ellesmere Island and Axel Heiberg Island (80° N). Here, horizontally layered massive ice 2–10 m thick is conformably overlain by 1–7 m of Holocene marine sediments. Ice contents increase beneath the active layer to 60–99% ice around 1.0–1.3 m and high ice contents persist to 8–9-m depth. Of 189 exposures of ground ice, 88% are in retrogressive thaw slump (RTS) headwalls; 70–90 of these have been active since 1995. As among the most dynamic landforms of thermokarst in permafrost areas, RTSs is a slope failure attributed to the thawing of ice-rich permafrost that leads to a debris flow in front of the slope.

Huang et al. (2020) applied DeepLabv3+, a deep learning algorithm for semantic segmentation, to Planet CubeSat images of high spatial and temporal resolutions, to automatically delineate 220 RTSs within an area of 5,200 km² in the TP with an average precision of 0.541. They find that (1) most of the RTSs in TP are small (areas <8 ha and perimeters <2,000 m) and (2) RTSs preferentially develop at locations with gentle slopes (four to eight degrees), and in areas lower than the surroundings (the mean topographic position index is −0.17) and receiving less solar radiation (i.e., north-facing slopes).

A recent increase in RTS activity has been observed in the Arctic polar desert where the mean annual ground temperatures are −16.5 °C, vegetation coverage is sparse and there is a lack of observations of RTS. For Banks Island of Canada (70,000 km²), Lewkowicz and Way (2019) found a 60-fold increase in numbers of RTS between 1984 and 2015, primarily following four particularly warm summers, resulting in increased turbidity in almost 300 lakes affected by RTS outflows. Modeled RTS initiation rates increased by an order of magnitude between 1906–1985 and 2006–2015, and are projected under Representative Concentration Pathway (RCP4.5) to rise to >10,000 per decade after 2075, which implies that ice-rich continuous permafrost terrain can be highly vulnerable to changing summer climate. Figure 5.5c shows a RTS comprises of a headscarp of thawing ice-rich sediments or massive ice, an overlying headwall composed of the active layer and low ice-content permafrost, and a bowl downslope filled with mud and debris derived from meltwater and soil from the collapse of the undercut headwall.

Jones *et al.* (2019) investigated $\sim$ 30-year record of annual RTS observations within the Eureka Sound Lowlands, Ellesmere, and Axel Heiberg Islands (Figure 5.5d). Record summer warmth in 2011/2012 promoted increasing active slumps from 100 or less per annum to over 200 regionally and promoting RTS initiation in previously unaffected terrain. Field mapping and remote sensing observations of 12 RTSs over 2011–2018 provide a mean headwall retreat rate of 6.2 m yr^{-1} and for some RTSs up to 26.7 m yr^{-1}. By correlating headwall retreat of RTS with thawing degree-days, annual precipitation, and terrain factors, the results indicate the sensitivity of cold permafrost in the high Arctic to climate-driven thermokarst initiation, but RTS dynamics could decouple from climate for certain RTS where terrain factors take on a greater role controlling the headwall retreat. Further, thermokarst development in a high Arctic permafrost are sensitive to changes in summer temperature.

Segal *et al.* (2016) assessed slump size, density, and growth rates in four ice-rich regions of western Canadian Arctic: Jesse Moraine, Tuktoyaktuk Coastlands, Bluenose Moraine, and Peel Plateau. They observed increases in the area impacted by slumps (up to +407%), average slump sizes (0.31–1.82 ha), and slump growth rates (169–465 m^2 yr^{-1}) which showed that thermokarst activity is rapidly accelerating in ice-rich morainal landscapes in the western Canadian Arctic, where slumping has become a dominant driver of geomorphic change, strongly influenced by topography, ground ice conditions, and Quaternary history. Observed increases and variation in slump activities occurred together with higher air temperature and precipitation suggests that increased precipitation has been an important driver of change. Further, the most rapid intensification of slump activity that occurred in the coldest, Jesse Moraine on Banks Island indicates that ice-cored landscapes in cold permafrost environments are highly vulnerable to climate change.

The "ice content" is defined as the weight of ice to dry soil, expressed as a percentage. Fine-grained soils may have ice contents of over 50%, while low ice content soils have values below 50%. Ice volumes determined in the upper 5 m of permafrost in the Arctic lowlands of western Canada range from 35 to 60%; most of this is pore or segregated ice. Pollard and French (1980) present a profile on Richards Island in the Mackenzie delta showing a maximum of just over 60% at 1.2-m depth diminishing to ~45% from 7 to 12 m depth.

Cryofacies can be defined based on volumetric ice content and ice crystal size. The use of the term "facies" is derived from sedimentology; a facies is a unit that exhibits lithological or structural characteristics that enable it to be distinguished from other units. Murton and French (1994) illustrate types of cryofacies for ice-rich sediments in the Pleistocene

Figure 5.5 Photographs of forms of ground ice

(a) Yedoma exposure of massive ground ice at the famous "Duvannuy Yar" location on the lower Koluma River in northern Yakutia, Russia. Dr. N. Shiklomanov: http://nsidc.org/frozengro/climate.ht ml (b) Syngenetic ice wedges in an 8-m high road cut about 8-km north of Fairbanks, AK in Late Pleistocene silt deposits, 2006. Courtesy Dr. Vladimir Romanovsky, University of Alaska. (c) A polycyclic coastal retrogressive thaw slump in southwest Banks Island (71.717° N, 124.127° W). Headscarp is thawing ice-rich permafrost (averaging 85% ice by volume), while overlying headwall is the former stabilized mudflow comprising the active layer and ice-poor permafrost. Undercutting of the headwall by ablation of the ground ice results in soil collapse that temporarily covers the ice (taken from Lewkowicz and Way, 2019). (d) A 2017 aerial photograph of a retrogressive thaw slump located at the Eureka Sound Lowlands area on west-central Ellesmere Island (taken from Jones *et al.*, 2019)

(c)   (d)

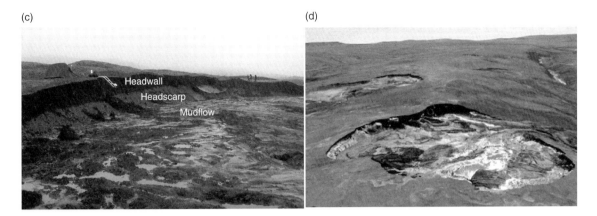

Figure 5.5 (cont.)

Mackenzie delta. The types range from pure ice (100% volumetric ice content), sand- and aggregate-poor ice (75–99%), sand- and aggregate-rich ice (50–75%), ice-rich sand, mud, or diamicton (25–50%), to ice-poor sand, mud, gravel, diamicton, or peat (<25%).

5.7 Permafrost models

Geographical permafrost modeling began in the United States with the work of Nelson and Outcalt (1983, 1987). The "frost index" method originally proposed by Nelson and Outcalt (1983) involves a straightforward ratio of freezing and thawing degree-day sums:

$$F = \left[\frac{\sum FDD}{\sum TDD} \right]^{0.5} \tag{5.6}$$

where ΣFDD is the sum of freezing degree-days and ΣTDD is the sum of thawing degree-days. They also introduced a seasonally weighted snow thickness to account for the snow cover's thermal conductivity. The procedure can be refined by employing the Stefan equation (see p. 233) and appropriate soil properties to calculate the depth of frost penetration; overestimates obtained with the Stefan procedure cancel one another when the ratio is formed. The frost-number approach to permafrost mapping was revised and extended by Nelson and Outcalt (1987), and physical justification for placement of zonal boundaries was derived. A surface frost number that takes account of snow cover effects is

$$F_* = \frac{FDD_*^{0.5}}{[FDD_*^{0.5} + TDD^{0.5}]} \tag{5.7}$$

Box 5.1 Ice in lava tubes and ice caves

A special form of ground ice is occasionally present in underground lava tubes and cave systems. On the Continental Divide, west of Albuquerque, New Mexico, there are more than 100 lava tubes and crevices with accumulations of perennial ice at elevations between 2,000 and 2,500 m in the El Malpais National Monument (35° N, 108° W). Navajo ice cave in Cibola County has ice stalagmites and an ice pond. An ice core obtained by Dickfoss *et al.* (1997) from Candelaria Ice Cave at the base of Bandera Crater (2,393 m asl) revealed a date of about 1,800–3,000 yr BP in the lowest meter of ice from 3.4 to 4.5-m depth. There is then a 1,500–2,000-year hiatus with the upper 3 m of ice dating from around AD 1600 to 1850. Accumulation in the ice pond was moderate (0.05 m a^{-1}) from 1924 to 1936, then decreased until 1947, resumed moderate accumulation and then slowed 1956–1970, began to increase in the early 1970s and peaked at 0.09 m a^{-1} in 1986–1991, since when there has been ablation to 1996. Temperature measured in the ice cave on three June days ranged between −1.8 °C and −2.7 °C and the relative humidity was 100%.

Ice caves occur throughout the United States – in California, Idaho, New Mexico (Halliday, 1934) – and are widespread in the limestone Alps of Bavaria, northern Austria (Dachstein, Totes Gebirge), and the Swiss Jura. Speleologists have mapped many of these extensive cave systems and reports may be found in the speleological and karst literature (Deutschen Höhlen- und Karstforscher, 2005). von Saar (1956) reports extensively on the Rieseneishöhle in the Dachstein citing climatic measurements inside and outside the cave during July 1928–September 1929. Others occur in the Tatra Mountains of Slovakia at Dobšinská (970 m altitude), where the ice surface area is ~9,800 m^2 and the volume is 110,000 m^3 (Thaler, 2008) and on Mt. Fuji, Japan at 1,120 m altitude there is a small lave tube ice cave (Ohata *et al.*, 1994).

Typically, there is a dynamic air circulation where in winter cold air and snow are sucked into the lower entrances promoting the growth of cave ice. When the cave slopes down from the entrance, cold air flows in during winter, as it is colder than the air in the cave. In spring and summer, the denser cold air is not able to leave the cave and is trapped (Figure 5.6). Static ice caves also occur where there is no air circulation but only a shaft through which winter cold air sinks into a cavity below and remains trapped there. In these caves, the ice usually forms from snow that becomes firn or that melts and refreezes to form congelation ice. Ice stalagmites and ice-covered walls form by the seepage of water through the karstic limestone and its freezing. The latent heat released by freezing is removed as the air is replaced by colder, outside air (Harrington, 1934).

The Schellenberger ice cave (1339/26) is located at an altitude of 1,570 m in Untersburg, Bavaria. It has an ice surface area of 60,000 m^2 and a thickness of up to 30 m. It was identified on a map of Bavaria in 1826 and explored in the 1870s. In 1925, it opened as the only tourist cave in the Berchesgaden Alps (Vonderthann, 2005). In the Tennengebirge Plateau at Werfen, near Salzburg, Austria (47.5°N, 13.2° E) at 1,650 – 1,775 m altitude, is one of the world's most famous caves with underground ice formations extending through a distance of ~0.7 km in a total cave system that extends 42 km (Spelolägische Institut, 1926; Spötl, 2007). There is a total ice-covered area of 10,000 m^2. It was discovered in 1879, when an ice lake and ice stalagmite were reported. In 1920–1921 extensive surveys were carried out (Oedl, 1922; Hauser and Oedl,

Box 5.1 (cont.)

1926). Abel (1955) reports on the ice conditions in the Himyrhalle, with an increase in ice thickness from 1922 to 1953 of 13.5 m and a total thickness of 26 m. Large 2–15-m-high ice stalagmites, suspended ice curtains up to 1.5 m long, and ice walls are common. The cave temperatures measured in March–April 1921 ranged from −2 °C to + 2° C (Hauser and Oedl, 1926). The cave slopes upward from the entrance and cold outside air penetrates in winter as warmer cave air rises through chimneys in the rock entraining fresh cold air through the entrance. Thaler (2008) analyzes data from the Eisriesenwelt system. He shows that the temperature at the entrance (where there is a closed door) tracks the external temperature in winter but from late April through early November it remains nearly constant just above freezing. Inside the cave the temperature averages ~0° C from July through October (1996–2007). Oberleitner *et al.* (2009) have made energy and mass balance measurements on the ice and find the mass balance to be negative with an ice surface lowering of 4 cm a^{-1}.

Measurements in the Bavarian Alps indicate overall decreases in ice volume as a result of global warming (Wisshak *et al.*, 2005). In the caves of the Schönberg system, in the Totes Gebirge, Austria, Wimmer (2007) reports ice shrinkage during 1970–2002, but since 2002 open shafts (due to melting ice masses) have allowed the downward penetration of cold air in winter and new ice buildup. In Candelaria Cave, NM, the rapid ice growth during 1986–1991 coincided with the wettest spell in 2000 years while the subsequent decrease occurred during a drought period (Dickfoss *et al.*, 1997).

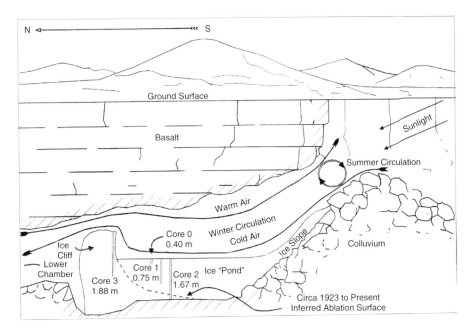

Figure 5.6 Schematic diagram of the air circulation in Candelaria Ice Cave, NM, in summer and winter (from Dickfoss *et al.*, 1997) [Courtesy of New Mexico Bureau of Mines and Mineral Resources]

The southern boundary of possible permafrost is delimited by the line where FDD$_*$ = TDD (i.e., $F_* = 0.5$). They also show that the discontinuous/continuous permafrost boundary is specified by $F_* = 0.666$. Nelson (1986) applied the methodology to map permafrost boundaries in central Canada and showed that it can predict the occurrence of permafrost in peatlands that lie outside the zonal limits based on climatic criteria alone.

Riseborough (2007) studied the effects of climatic variability on an equilibrium permafrost-climate model for the temperature at the top of permafrost (TTOP). He showed that stationary interannual variability introduces an error in the TTOP obtained with the equilibrium model that is higher where permafrost temperature is close to 0 °C. With a warming trend, the equilibrium model prediction tracked the changing TTOP until permafrost temperatures reached 0 °C, after which it produced significant error. Errors of up to 1 °C were due to the temperature gradient through the developing talik, and depended on the warming rate, and the thickness of the talik. The error was found to be largest when the permafrost table was about 4-m below the surface.

The first GCM simulations were off-line using temperatures taken from runs of the GFDL, GISS, and UKMO models for CO_2-doubling (Anisimov and Nelson, 1990, 1996). The GCM was not used to simulate permafrost change and thus there were no feedbacks. Similar analyzes were carried out subsequently using the results of transient GCM simulations (Anisimov and Nelson, 1997). Anisimov and Reneva (2005) show that the reduction in near-surface permafrost area by AD 2050 ranges from 19% to 34% according to four GCMs. The incorporation of permafrost directly within GCM land surface schemes was first undertaken by Lawrence and Slater (2005; Lawrence et al., 2008) using the NCAR Community Climate System Model (CCSM)-3 (see Section 10.5 of Chapter 10). The CCSM has a 5-layer snow model over a 10 layer, 3.4-m-deep soil model that treats thermal and hydrologic frozen soil processes. The Climate Model Intercomparison Project (CMIP)-3 used in the IPCC Assessment Report 4 in 2007 had the following models treating frozen ground: CNRM (France); MIROC and MRI (Japan); CSIRO (Australia); Canadian Climate Center, BCCR (Norway); BCC-CM1 (China); INM (Russia); GFDL GISS, and NCAR CCSM (USA); and UKMO/Hadley Centre.

Riseborough et al. (2008) categorize models according to temporal, thermal, and spatial criteria, and their approach to defining the relationship between climate, site surface conditions, and permafrost status. The most significant recent advances include the expanding application of permafrost thermal models within spatial models, and the application of transient numerical thermal models within spatial models,

5.8   Geomorphological features associated with permafrost

A wide range of meso- to microscale surface landforms occur in terrain that is underlain by permafrost. The mesoscale forms include thermal contraction-crack polygonal ground (patterned ground) that is from 15 to 40 m across in unconsolidated sediments and 5 to 15 m in bedrock. They form with ground surface temperatures of about −15 °C to −25 °C (French, 2007). Air temperature decreases of ~1.8 °C d^{-1} over a 4- day interval appear favorable for cracking (Mackay, 1993), but less than half of the fissures in a given area crack annually. The cracks may be filled with ice, forming ice wedges, or mineral soil.

There are several types of perennial frost mound that occur in areas of continuous and discontinuous permafrost (Mackay, 1986a). In peaty organic material, palsa mounds may form. The term is Finnish meaning peat-bog hummock with an ice core. A palsa is 1–8 m high and generally 5–25 m in diameter, comprising alternating layers of segregated ice and peat or mineral soil. A pingo is a perennial frost mound with a core of massive ice that is primarily produced through the injection of water. Pingo is an Inuktitut term from the Mackenzie Delta area (Mackay, 1962). Pingos may be a few meters to 50 m high, conical in shape, and up to 300 m in diameter. Hydrostatic (or closed system) pingos form by the doming of frozen ground through pore-water expulsion due to permafrost growth (aggradation) in the closed talik formed beneath drained lake bottoms that are underlain by saturated sediments. Repeated injections of water into the overlying frozen ground, followed by freezing, lead to the formation of massive ice in the pingo core and progressive doming of the feature. They are widespread in the Tuktoyaktuk Peninsula area of northwest Canada where the largest are over a thousand years old. Initially they may grow at a rate of about 1.5 m a^{-1}. Mackay (1986b) believe that only about 50 pingos were actively growing in the Mackenzie delta area. Hydaulic (or open system) pingos develop where groundwater under artesian pressure reaches the surface. They occur on valley slopes in central Alaska and the Yukon (Holmes *et al.*, 1968).

Needle ice – vertical ice crystals that grow upwards in the direction of heat loss – may form in the topsoil. They are a few millimeters to several centimeters long and cause microscale heaving of the soil particles. It occurs just below the lower limit of permafrost according to Lawler (1989). Figure 5.7 illustrates its altitudinal versus latitudinal occurrence. It may occur on a diurnal basis year-round in middle latitudes, but mainly in spring and autumn poleward of 60° N (Lawler, 1988).

Rock glaciers are tongues of frozen debris and boulders, containing interstitial ice, that deform downslope under gravity. They occur below talus slopes and below glacier end moraines in most alpine mountain ranges (Barsch, 1988). They move at a few centimeters to a few meters per year at rates determined by surface slope, composition and internal structure, thickness of the ice horizon, and ground temperature. Current research is devoted to better understanding of the deformation and creep of ice-rich permafrost bodies and also to inferring the lower altitudinal limits of past and present permafrost.

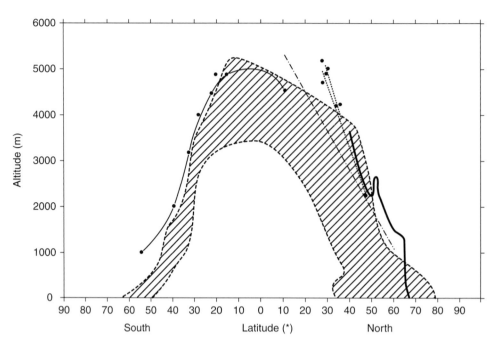

Figure 5.7 The latitudinal and altitudinal occurrence of needle ice (hatched); the solid, chained, and dotted lines indicate the lower limits of permafrost according to various sources (from Lawler, 1989; Weather 44(10), p. 407. Fig. 2) [Courtesy Royal Meteorological Society]

Cryoturbation is a term used to characterize soil movements due to freeze–thaw processes. It includes frost heave and thaw settlement. Frost heave occurs on an annual basis during the autumnal freezing of moisture in the active layer. Measured values vary from 1 to 32 cm (French, 2007, p. 145), but most observations are in the 2–5 cm range. A distinction is drawn between primary frost heave where there is a sharp boundary between the frozen and unfrozen layers and secondary frost heave where there is a thin intermediate layer of partially frozen soil in which ice lenses develop. Pore pressure in freezing soils sets up cryostatic capillary suction where the excess upward flux of water generates extra heave and ice lens growth beyond that due to expansion on freezing (Fowler and Krantz, 1994) (cf. segregation ice, p.238). The cryostatic suction increases as the unfrozen water content decreases. Clays exhibit the greatest tendency to form ice lenses because they have a greater increase in suction with decreasing water content. Silts show the greatest tendency to frost heave since they have moderately large suction, hydraulic conductivity, and permeation rates. Sorting of fine and coarse particles can occur through freezing and thawing from above. Sorting may occur by uplift of particles due to frost heave, by the preferential migration of fine particles ahead of a freezing plane, and by mechanical sorting when larger material migrates downward under gravity. Sorted and non-sorted patterned ground may occur as circles and polygons on level ground, and as elongated stone stripes on slopes. Most patterned ground has dimensions of 1–2 m with a relief of 10–30 cm.

Thermokarst terrain is another permafrost-related landscape form. When ground ice in permafrost regions melts it causes the ground to subside or slump on slopes. The hollows that so form are termed thermokarst terrain. The term "thermokarst" was introduced by M.M. Yermolayev in the USSR in 1932. When ice wedge polygons thaw and form depressions about 1-m deep, slumping begins and the polygon cores develop into conical mounds known in Yakutia as baydyarakhs. These gradually slide and slump into the hollows and the presence of water accelerates the process. Flat-bottomed circular or oval-shaped depressions – termed alases – develop with steep sides and a thaw lake in the center. In the lowlands of central Yakutia these have a depth of 3–40 m and a diameter from 100 m to 15 km (Soloviev, 1962). Taliks form beneath the thaw lakes. Around Yakutsk up to 40% of the land surface is affected by thermokarst formations, most of them dating from the Holocene thermal maximum (Czudek and Demek, 1970) (see Chapter 9, p.408).

5.9   Changes in permafrost and soil freezing

The oldest known permafrost was discovered near Dawson City, Yukon in 2008 (Froese *et al.*, 2008). An ice wedge within a few meters of the surface was dated to 740,000 ± 60,000 yr BP. Thus, this relict ice survived several glacial/interglacial cycles, including marine isotope stages 11 and 5e, considered to be longer and warmer than the Holocene. In the Pechora lowlands, Henriksen *et al.* (2003) report that sediment cores indicate that buried glacier ice has survived for about 80,000 years following glaciation that occurred about 90 ka. The dead ice began to degrade only around 13 ka. French (2008) notes that Russian geologists concluded in the late 1980s that the massive ground-ice bodies in western Siberia were of glacial origin.

A combined ice sheet – permafrost model study by Tarasov and Peltier (2007) suggests that, compared to present day, permafrost was thicker at the Last Glacial Maximum for regions that are still occupied by surface ice, but it was thinner for the areas of the present-day permafrost zone that are currently ice free. This illustrates the strong thermal insulation that is provided by ice sheets. Their results also show that in the central Canadian Arctic at present the permafrost lower boundary is as much as 250 m less than the equilibrium value. This implies deepening of the permafrost lower boundary while the upper layers are thawing due to climatic warming.

Romanovsky *et al.* (2008a) point out that permafrost that was present in Europe, northern Kazakhstan, and western Siberia during the LGM has disappeared, although in the Pechora River basin it is still present at depths of 200 m and more. In north central Siberia, eastern Siberia, and the Russian Far East there was no widespread thaw during the Holocene Thermal Maximum. Currently, there is thawing in the region of Vorkuta (67.5° N, 64° E). Permafrost patches 10–15-m thick have thawed completely and the southern boundary of permafrost shifted several tens of kilometers northwards. A calculation for Yakutsk (62.1° N, 129.8° E) using observed air temperatures from 1834 to 53 and 1887 to 2003 shows that ground temperature rose most rapidly in the second half of the nineteenth century. There were cold intervals in the 1940s and especially the 1960s to 20 m depth, but at 50 m temperatures continued to rise at a slower rate. Along a transect from 55° N to the Arctic Ocean, 122–138° E, Romanovsky *et al.* (2007b) analyzed air temperatures and ground temperatures at 1.6 m for 18 stations from 1956 to 1990. The average trend for MAGT at 1.6 m was 0.26 °C/10 yr compared with 0.29 °C/10 yr for air temperatures. The most significant trends were at latitudes 55–65° N. In the Pechora River basin (68.2° N, 54.5° E), Malkova (2008) finds small changes in permafrost temperatures. Trends are 2 to 10 times smaller than the corresponding changes in MAAT and between 1984 and 2007 were from 0.003 °C to 0.02 °C a^{-1}. In the northern Tien Shan mountains of Central Asia, Marchenko *et al.* (2007) measured permafrost temperatures at 10–14-m depth that rose by 0.3–0.6 °C during 1974–2004, and the ALT increased by 23%.

Temperatures of continuous and discontinuous permafrost in the Northern Hemisphere have increased with different rates in most regions since the early 1980s, with generally greater increase for colder than warmer permafrost. In 1975–2005, permafrost in the Russian European North has degraded significantly, with complete thawing of some warm permafrost, the southern boundary of discontinuous (continuous) permafrost moved north by up to 80 (50) km, surface subsidence associated with degradation of ice-rich permafrost occurred at various locations, and ALTs increased by a few centimeters to tens of centimeters in recent two to three decades. In northern North America, there were large interannual variations in changes in ALT but few significant trends. The thickness of seasonally frozen ground in some parts of Europe and Asia has also decreased, in places by more than 30 cm from 1930 to 2000 (Vaughan *et al.*, 2013).

In the Qinghai Plateau (Kekexeli Wildland area), the lower limit of permafrost has risen by about 70 m over the last three decades (Wu *et al.*, 2001) and the extent of seasonal thawing has extended over large areas of permafrost terrain. Soil temperature measurements to 12 m depth were made from 1995 to 2007 at 10 sites along the Qinghai–Tibet Highway from Kunlun Pass to Anduo in the south by Wu and Zhang (2010). The mean, spatially averaged ALT for 1995–2007 was 2.41 m with a range of 1.32–4.57 m. ALT showed little change from 1956 to 1983 and a sharp increase of ~ 39 cm from 1983 to 2005. The magnitude of the increase was greater in the region of warm permafrost than in the cold permafrost region and was primarily caused by an increase in summer air temperature.

Permafrost temperatures in Alaska increased dramatically in the last quarter of the twentieth century (Osterkamp, 2008). The permafrost surface warmed 3–4 °C on the Arctic Coastal Plain, 1–2 °C in the Brooks Range, and 0.3–1 °C south of the Yukon River. In Interior Alaska the increase in the late 1980s–1990s was due to greater snow cover. At Barrow, AK, about half of the recent warming was due to an increase in air temperature and half due to snow cover effects. However, there was no increase in active layer depths on the

Arctic Coastal Plain. This seems attributable to penetration of thaw into the ice-rich transient layer at the TTOP that led to thaw subsidence. Streletsky *et al.* (2008) show that over five years in the Alaska coastal plain and foothills this subsidence amounted to 12–13 cm. In the outer Mackenzie Delta, the ground is currently more than 2.5 °C warmer than in 1970 and ground temperatures have increased in the uplands of the delta by approximately 1.5 °C (Burn and Kokelj, 2009). The impact of climate change on permafrost is evident in the thickness of the active layer, which increased on average by 8 cm at 12 tundra sites on northern Richards Island from 1983 to 2008. Smith *et al.* (2010) show that permafrost has generally been warming in the western Arctic since the 1970s and in parts of eastern Canada since the early 1990s. The increases are generally greater north of the treeline and the magnitude of the change was less in warmer (>−2 °C) than in colder permafrost.

Using a global permafrost temperature dataset from the GTN-P, Biskaborn *et al.* (2019) evaluated temperature change across permafrost regions of Northern Hemisphere (Figure 5.8a). Between 2007 and 2016, they found permafrost temperature near the depth of zero annual amplitude in the continuous permafrost zone warmed by 0.39 ± 0.15 °C, discontinuous permafrost by 0.20 ± 0.10 °C, mountain permafrost by 0.19 ± 0.05 °C, and permafrost in Antarctica by 0.37 ± 0.10 °C (Figure 5.8b). Globally, the observed permafrost temperature trend of 0.29 ± 0.12 °C over 2007–2016 follows the amplification of Arctic warming. However, in the discontinuous zone, permafrost warming was attributed to thicker snow depth since air temperature remained statistically unchanged.

The Permafrost and Climate in Europe (PACE) program sites in Europe (and forerunners) give results of permafrost temperature measurements over varying intervals (Harris, 2009). At Jannsonhaugen, Svalbard, temperatures at 10 m depth warmed about 2 °C between 1998 and 2015. In the Upper Engadine, Swiss Alps at Murtel-Corvatsch (2,670 m), they rose ~1 °C between 1987 and 2015 at 11.6 m. They then cooled due to low winter snowfall and subsequently fluctuated, rising again in subsequent years of 2008 to 2015 (Figure 5.8c). Other shorter temperature measurements fluctuated between cooling and warming, but the latter is more dominant in recent years.

Data on permafrost thawing due to climate change are hard to obtain by direct observations of permafrost depth at scales larger than local. Lyon *et al.* (2009) suggest using streamflow recession analysis based on a long-term streamflow record and illustrate the method for the subarctic Abiskojokken catchment in northern Sweden. They estimate that permafrost in the catchment may be thawing at an average rate of about 0.9 cm a^{-1} during the past 90 years. The calculation is in good agreement with direct observations of permafrost thawing rates in the region from 0.7 to 1.3 cm a^{-1} over the past 30 years.

The relationships between winter air temperature, precipitation, and soil freezing have been examined by Henry (2008) for 31 sites in Canada, using 40 years of weather station data. The sites ranged from the temperate zone to the high Arctic. Annual soil freezing days were found to decline with increasing mean winter air temperature despite decreases in snow depth and cover; this is in contrast with the idea that a shallower snowpack may lead to increased soil freezing. Reduced precipitation only increased annual soil freezing days at the

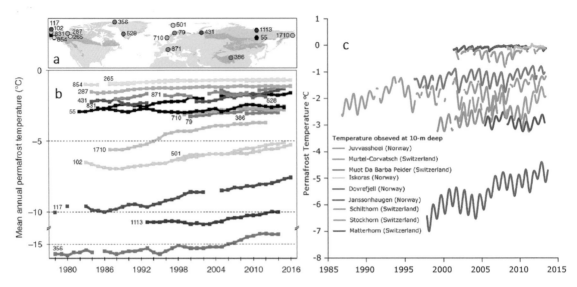

Figure 5.8 Long permafrost temperature records for selected sites. (a) Location of boreholes with long time-series data. Depth of measurements is according to the Global Terrestrial Network for Permafrost ID: 24.4 m (ID 356), 20 m (ID 55, 79, 102, 117, 501, 710, 831, 1113, and 1710), 18 m (ID 386), 16.75 m (ID 871), 15 m (ID 854), 12 m (ID 287), 10 m (ID 265, 431), and 5 m (ID 528). The light blue area represents the continuous permafrost zone (>90% coverage) and the light purple area represents the discontinuous permafrost zones (<90% coverage). (b) Mean annual permafrost temperatures at Swiss alpine sites at 10 m depth over time. The colors indicate the location of boreholes in permafrost zones derived from the International Permafrost Association (IPA) map: Iskoras; 591 m asl, SE-facing; Dovrefjell; Janssonhaugen: 275 m asl (fine-grained sandstone), S-facing; Juvvasshoe: 1,894 m asl (Gneiss), NW-facing; Schilthorn: 2,970 m asl, N-facing (rock slope with thick fine debris); Muot da Barba Peider: 2,900 m asl, NW-facing (debris slope); Murtèl Corvatsch: 2,670 m asl, NW-facing (rock glacier); Matterhorn: 3,295 m asl (vertical in rock ridge/Hörnligrat) [Courtesy: Swiss Permafrost Monitoring Network (PERMOS), University of Zurich] (A black and white version of this figure will appear in some formats. For the color version, please refer to the plate section.) [www.eea .europa.eu/data-and-maps/figures/observed-permafrost-temperatures-from-selected-4]

warmest sites. Annual soil freeze–thaw cycles increased in both warm and dry winter conditions. No relation has been found between a surface air temperature increase of 1–1.5 °C over 50 years and surface ground temperature in winter–spring at 8 sites in north-western Canada due apparently to the masking effects of freeze–thaw processes and latent heat (Woodbury et al., 2009) and the insulation by snow cover (Sokratov and Barry, 2002; Zhang, 2005). For the midwestern United States there are significant regional differences in soil temperature trends. In northwest Indiana, north-central Illinois, and southeast Minnesota), Sinha et al. (2010) show trends in extreme and mean seasonal soil temperature from 1967 onward that indicate an increase of 10-cm soil temperatures and a reduction in soil frost days. Slow permafrost degradations are observed in frozen fine-grained earth material with substantial amount of unfrozen water. By numerical experiments, Nicolsky and

Romanosky (2018) demonstrated that substantial unfrozen water content present in fine-grained material at below 0 °C temperature can significantly slow down the thawing rate and hence the rate of permafrost degradation to climate warming. However, the effect is highly nonlinear, with larger differences in the rates of thawing between fine- and coarse-grained materials under low than high heat fluxes coming into permafrost.

An important aspect of the thawing of permafrost is the release of carbon dioxide (CO_2) and methane (CH_4) trapped in the frozen soil because permafrost warming can potentially amplify global climate change for when frozen ground thaws, it releases soil organic carbon (SOC) to the atmosphere. The amount of SOC stored in the circum-Arctic permafrost regions is estimated at 1,460–1,600 Pg, with ~800 Pg SOC currently frozen in permafrost (Hugelius et al., 2014). About 45% of carbon eroded from melting permafrost due to climate warming is released to oceans, while the rest is oxidized by sunlight into CO_2, and by the microbial breakdown of organic carbon accelerated by climate warming, releasing CO_2 and methane to the atmosphere, fueling climatic change that causes permafrost to waste away, a positive feedback. This feedback can accelerate climate change, but the magnitude and timing of greenhouse gas emission from these regions and their impact on climate change remains uncertain partly due to poorly understood permafrost carbon dynamics. Schuur et al. (2015) suggest a gradual and prolonged release of greenhouse gas emissions from permafrost in a warming climate.

O'Connor et al. (2010) point out that whether release of frozen carbon happens as CO_2 or CH_4 is determined by whether decomposition proceeds aerobically or anaerobically, which generally depends on whether the thawing permafrost is water saturated or not. Zimov et al. (2006) estimate that frozen loess (termed **yedoma** in Siberia) that was deposited during glacial times covers more than one million km^2 of northern Siberia and central Alaska to an average depth of ~25 m and has an average carbon content of 2–5%. Tarnocai (2009) states that the average carbon content of cryoturbated mineral soils in the permafrost zone is ~49–61 kg m^{-2}, while that of peat land soils in the sub-Arctic is in the range 43–144 kg m^{-2}. The carbon reservoir in frozen yedoma is estimated to be ~500 Gt (50×10^3 Tg), nearly equal to that in vegetation (650 Gt), but another ~400 Gt of carbon are stored in non-yedoma permafrost (Zimov et al., 2006). Carbon that is beneath the widespread thermokarst lakes in yedoma terrain is decomposed anaerobically by microbes, yielding methane that bubbles to the surface. Hence, there is a significant greenhouse-gas warming potential currently locked up in permafrost. Another major methane source is the sediments of the Laptev, East Siberian, and Chukchi Sea shelves, an area of 2.1×10^6 km^2, three times that of the terrestrial Siberian wetlands. Shakhova et al. (2010) report that the annual average temperature of the bottom sea water is -1.8 °C to 1 °C, giving substantial potential for thawing of the frozen sediments. Most of the shelf region is supersaturated with CH_4 in the near-bottom waters, with >50% of the surface water supersaturated. The median background summer-time supersaturation was 880% and 8,300% in hotspots. In winter, both the bottom- and sur-face-water-dissolved CH_4 concentrations were 5–10 times higher than in summer. The total CH_4 flux for the period of open water, made up of the background plus hotspots, is estimated to be 2.19 Tg CH_4. This can be compared with total global emissions estimated to be about 500–600 Tg a^{-1} (0.5–0.6 Gt a^{-1}).

Tibetan Plateau (TP), also known as the third pole of Earth, contains the largest area of alpine permafrost in the world, covering ~42% of the entire TP, ~1.30×10^6 km^2 in area, and contains ~37.21 Pg perennially frozen SOC at the baseline period (2006–2015). TP has been subjected to serious permafrost degradation due to drastic climate warming in past decades (Ran and Cheng, 2018), including deepening of ALT, northward migration of the lower altitudinal limit of permafrost, and a continuous decline in the area of cold permafrost. The ALT in TP can be greater than 3 m. The thawing of permafrost carbon in response to future climate warming remains largely unknown over TP. Wang *et al.* (2020) investigated the spatiotemporal dynamics of SOC over the high-altitude TP. With continuous warming, they project the active layer to further deepen resulting in ~1.86 ± 0.49 Pg and ~3.80 ± 0.76 Pg permafrost carbon thawing by 2100 under RCP4.5 and RCP8.5 climate scenarios, respectively. This could largely offset the regional carbon sink and even potentially turn the region into a net carbon source. The findings highlight the importance of deep permafrost thawing generally ignored in current Earth system models.

5.10   Arctic infrastructure

According to Hjort *et al.* (2018), degradation of near-surface permafrost in the Northern Hemisphere under projected climatic change can put one-third of pan-Arctic engineering infrastructure, and 45% of the hydrocarbon extraction fields of the Russian Arctic in regions where thaw-related ground instability can cause severe damage to the built environment by 2050. Under projected climate change impact, nearly four million people and 70% of current infrastructure in the permafrost domain of NH are in areas with high potential for thawing of near-surface permafrost. For example, 470 km of the Qinghai–Tibet Railway and 280 km of the world's northernmost, Obskaya–Bovanenkovo railway, and currently more than 1,200 settlements (40 with population more than 5,000) are in the zone where permafrost thaw is likely, which could lead to ground subsidence and the loss of structural bearing capacity, causing severely damaged infrastructure. There are also 1,590 km of the Eastern Siberia–Pacific Ocean oil pipeline, 1,260 km of major gas pipelines originating in the Yamal-Nenets region, and 550 km of the Trans-Alaska Pipeline System located in areas where near-surface permafrost thaw may occur by 2050. Damage to pipelines and industrial facilities may lead to major ecosystem disruption, such as large-scale oil spills or disruption of critical energy delivery. Given uncertainties embedded in the projections of ground temperature and annual thaw depth, the extent of the Arctic infrastructure at risk could potentially be higher than estimated. On the other hand, reducing greenhouse gas emissions and stabilizing atmospheric concentrations, achieving a scenario consistent with the Paris Agreement could stabilize risks to the Arctic infrastructure after 2050s.

REVIEW QUESTIONS

(1) Which of the followings is the correction definition of permafrost: It is a frozen ground in which temperature remains below (a) $-30\,°C$ for 5 or more years, (b) $-15\,°C$ for 10 or more years, (c) $-40\,°C$ for 3 or more years, (d) $0\,°C$ for 2 or more years, and (e) $0\,°C$ permanently?

(2) Which of the following factors are necessary for the formation of permafrost: (a) annual precipitation ≥600 mm, (b) predominantly windy condition, (c) predominantly low air humidity, (d) mean annual air temperature (MAAT) $\leq-1\,°C$, (e) predominantly clayey soil materials, (f) a thin vegetative cover, (g) soil gaining latent heat from Air, and (h) soil losing latent heat to air?

(3) In the Northern Hemisphere, what is the MAAT that approximately represents the respective southern limits of continuous and discontinuous permafrost? Also, what is the respective percent of area underlain by frozen ground for these two types of permafrost in NH?

(4) List three climatic and/or physical factors that will influence the thickness of a permafrost developed over a long time, for example, thousands of year. Also, list three mesoscale geomorphologic features found in a continuous permafrost environment. Do you expect the degradation of permafrost due to global warming by 2100 in the Arctic will likely be extensive or limited even if air temperature changes will be as projected, about 3–8 °C by 2100 and why? (Sections 5.2 and 5.3)

(5) Name two types of satellite data, one of high and another of low resolutions, commonly used to map the occurrence of frozen ground in the upper surface soil layer. What are the expected changes in the following properties of soils under frozen and unfrozen states measured by the SAR or passive microwave sensors: (a) thermal temperatures, (b) dielectric constant, (c) emissivity, and (d) brightness temperatures?

(6) Compute the surface Frost Number F_+ (Equation 5.1) of Alert, Nunavuk (Latitude $\approx81°$ N) based on the mean monthly temperature of Alert, the tip of Canadian high Arctic. On the basis of F_+ computed, what type of permafrost will Alert have?

Mon	Jan	Feb	Mar	Apr	May	Jun	Jul	Aug	Sep	Oct	Nov	Dec
T_{mon}	-32.4	-33.4	-32.4	-24.4	-11.8	-0.8	3.3	0.8	-9.2	-19.4	-26.4	-30.1
Days	31	28	31	30	31	30	31	31	30	31	30	31

(7) Compute Z (Equation 5.2), the thickness of the active layer in meters of a permafrost composed of predominantly silt and clay at the end of a thawing season of 4-month duration when the average seasonal air temperature is 7 °C, if the water content is about 30%, and dry soil density is about 1,400 kg/m^3. Should you use the K_t (thermal conductivity) of unfrozen or frozen silt and clay shown below? Compute the rate of descent dz/dt (Equation 5.3) of the thawing front in m per month.

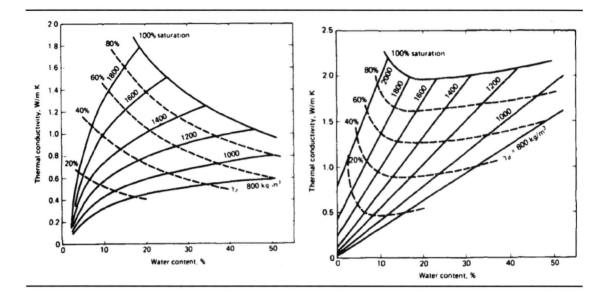

(8) Among a wide range of meso- to microscale surface landforms that occur in terrain underlain by permafrost, explain processes such as thermal contraction-crack polygonal ground in unconsolidated sediments/bedrock, and cracks filled with ice wedges or others, that lead to the formation of two common geomorphological features, pingo and palsa (Finnish word).

(9) Despite of the extent and rapidity of observed changes to permafrost due to climate warming, such as the significant and rapid permafrost thaw from the top in recent years, and the projected warming of 2.8–7.0 °C in the Arctic by 2100, why is the projected degradation in ice-rich permafrost in response to strong high-latitude warming over the twenty-first century remains highly uncertain?

(10) What is the approximate amount of soil organic carbon (SOC) stored in the circum-Arctic permafrost regions, and is the melting of permafrost due to climate warming a positive or a negative feedback process, given CO_2 will be released from carbon eroded from melting permafrost?

(11) For highly dynamic landforms in ice-rich permafrost areas attributed to the thawing of ice-rich permafrost, name three climatic factors that affect the formation and dynamics of retrogressive thaw slumps (RTSs) such as the retreat of headwall, and three terrain factors that can also exert significance influence on RTS dynamics in high Arctic?

(12) How can the degradation of near-surface permafrost in the NH under climate warming put pan-Arctic engineering infrastructure, hydrocarbon extraction fields, northern railways such as the Qinghai–Tibet Railway and various northern settlements at risk, such as damage to the built environment and major ecosystem disruption?

6   Freshwater ice

6.1   History

Engineering studies of freshwater ice began in the mid-nineteenth century in Eastern Europe. The flooding of Buda and Pest in 1838 led to studies of ice conditions on the River Danube during the winters of 1847/1848 and 1848/1849 by Arenstein (1849). Ashton (1986) and Barnes (1906) note that there were many nineteenth-century studies of ice formation and ice jams. Ireland (1792) mentions "ground ice" rising up from the bottom of the River Thames and there were other eighteenth-century references to it in France and Germany. Farquharson (1835, 1841) reports on anchor ice (ground-gru) observed in Lincolnshire, England, and proposed a theory of radiational cooling of rocks and vegetation in the river bed. Barnes (1906) published a study of frazil and anchor ice formation based on earlier literature and observations on the St. Lawrence River in Canada. Frazil is a French-Canadian term first used in 1831; anchor ice was originally termed ground ice (in Germany). Dunble (1860) studied the effects of lake ice on a 4-km-long railway bridge over Rice Lake, Ontario. Adams (1992) reports that Dunble (1860, p. 423) performed an experiment to demonstrate that "with the same change in temperature, the expansion and contraction of ice are equal."

Observations of lake and river freeze-up and breakup have a long history. von Cholnoky (1909) made a comprehensive study of ice conditions on Lake Balaton, Hungary. Adams (1992) notes that J. B. Tyrell gave a presidential address to the Canadian Institute on lake ice (Tyrell, 1910) discussing the freeze-up and breakup processes, ice types, and the shoreline effects of wind-driven ice shove during spring breakup. Observations of ice breakup started in Finland on the River Tornionjok in 1693, on lakes Kallavesi and Näsijärvi in 1833 and 1836, respectively, while dates of ice drift in Estonian rivers were combined in a chronology beginning in 1706 (Yoo and d'Odorico, 2002). Systematic work on Finnish lakes was undertaken by Simojoki (1940), updated by Palecki and Barry (1986). Field studies of polar lakes in Alaska, Greenland, and Antarctica began in the 1950s–1960s (Vincent *et al.*, 2008a).

For many years, Canada maintained an extensive network of ~240 observing sites from the 1940s to the 1990s, but almost all of these have since terminated (Brown, 1999). In order to monitor lake ice freeze-up and breakup over Great Slave Lake and Great Bear Lake in Canada, SSM/I passive microwave data has been employed. However, the 12.5-km resolution of SSM/I 85 GHz brightness temperature data limits this approach mainly to large lakes. In 1995, the Canadian Ice Service began a program of monitoring ice extent on small lakes using high-resolution satellite imagery. The amount of ice on each lake is determined weekly by visual inspection of Advanced Very High-Resolution Radiometer (AVHRR) and RADARSAT images. The program started in November 1995 with 34 lakes and increased to 118 lakes by 1998.

Icings, encountered during travels along the north coast of Siberia in 1820–1824, were reported by F. Wrangel according to Carey (1973). Also, von Middendorf studied icings in eastern Siberia in 1843–1844. Engineering studies began in the USSR in the 1920s–1930s associated with road and railway construction in Siberia, and in North America in connection with northern highways in the 1940s.

6.2   Lake ice

Lake ice formation is dependent on the unique density characteristic of fresh water, which reaches a maximum density at 3.98 °C. As a water body cools in the autumn it becomes isothermal at 3.98 °C. Further cooling of the surface allows a less dense layer to develop, the skin of which becomes supercooled (~−0.03 °C to −0.1 °C) (Devik, 1949). Ice forms on nuclei in the supercooled water or on a physical boundary (lake shore, rock, turbine intake, etc.). Frazil ice or sheet ice forms depending on the water motion. **Frazil ice** comprises randomly oriented needlelike structures or thin (25–100 μm), flat, circular platelets of ice 1–4 mm in diameter (Martin, 1981) formed in supercooled, turbulent water. Adjacent to the shore, sheet ice grows. Michel (1971) illustrates the three pathways to forming an ice cover (Figure 6.1). Frazil particles may adhere together forming slush, or snow may form snow slush, that aggregates into slush balls. Through collisions these grow into ice pancakes with upturned rims and eventually these amalgamate in ice floes and an ice sheet. In still water an ice sheet may form directly.

Once an ice cover is formed, it thickens as heat is lost to the atmosphere. In this case, the growth of ice thickness (h_i) can be estimated from accumulated freezing degree days (FDD) – the amount by which the mean daily temperature is below 0 °C accumulated over the winter. Based on 24 station years of data in the Russian Arctic (Lebedev, 1938), FDD are calculated with reference to a base temperature of 0 °C for each day of the winter:

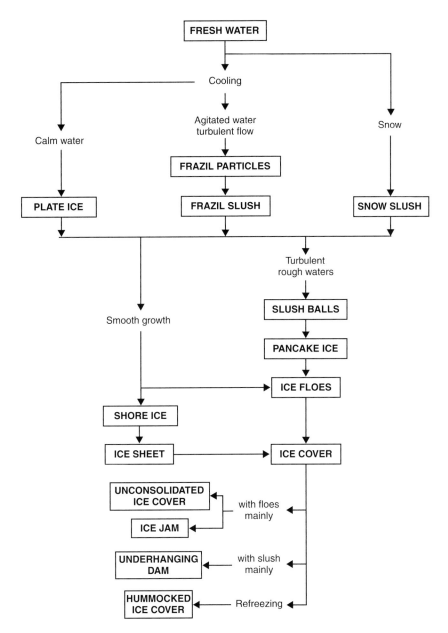

Figure 6.1 Pathways to the formation of a lake ice cover (from Michel, 1971) [A video showing the Yosemite frazil ice that happened in March–April can be found in www.youtube.com/watch_popup?v=9V9p4mFEYXc&vq=medium#t=15]

$$h_i = 1.33\text{FDD}^{0.58}. \tag{6.1}$$

Michel (1971) modifies the equation:

$$h_i = a\text{FDD}^{0.5}, \tag{6.2}$$

where a is a coefficient that depends on the characteristics of the water body (see below). Figure 6.2 shows minimum and maximum ice thickness curves (cm) for accumulated FDD (°C).

Snow accumulation on lake ice depresses the ice surface below the water level, causing the snow to become saturated and leading to the formation of slush; this freezes from the top down, forming white snow ice (in contrast to the black lake-water ice). White ice tends to be thicker near to shore and black ice in the center of a lake, with the total thickness showing limited spatial variation. Based on measurements in central Sweden, Bengtsson (1986) suggests that three observation points are sufficient to determine the mean lake ice thickness within 5 cm.

Maximum lake ice thicknesses reach up to 2 m in the high latitudes of northern Canada, Alaska, and Russia (Billelo, 1980). Gronskaya (2000) shows that maximum thickness on lakes in the Russian Far East and eastern Siberia increases from ~100 cm at latitude 45° N to

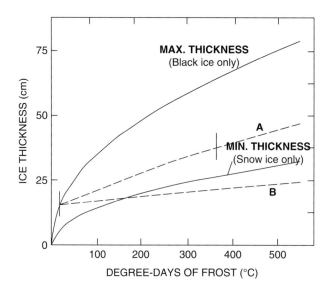

Figure 6.2 Curves of ice thickness (cm) versus freezing degree day (FDD in °C) (after Michel, 1971). The maximum thickness represents the Stefan relation for black ice; the minimum curve is for white (snow) ice when a fractured ice cover can be immediately flooded. Curve A shows the growth of white ice following a 30-cm snowfall on a fractured 15-cm cover of black ice. Curve B shows the growth by accretion of black ice under the same conditions when the original 15-cm cover cannot be flooded. [Source: B. Michel, Winter regime on rivers and lakes, U.S. Army CRREL Monograph III-B1a, 1971, p.79, fig. 65]

180 cm at 65° N and in European Russia from ~40 cm at 50° N to 60 cm at 65° N. In western Siberia, there is a steeper increase from ~100 cm at 65° N to ~200 cm at 75° N. Thickness decreases ~7 cm per 1 °C increase in air temperature according to Williams *et al.* (2004). Ice thickness also depends considerably on snow cover. A lake-ice model applied to Churchill, Manitoba, shows that 25–30 cm of snow cover leads to an ice thickness of ~140 cm compared with ~180 cm for a snow-free case (Duguay *et al.*, 2003). Caine (2002) reported observations of ice thickness in late March on Green Lake 4 at 3,580 m in the Colorado Front Range. A thinning of 2 cm a^{-1} since 1982 was noted and this correlated ($r = -0.44$) with precipitation during the preceding October to March.

The thermal regime of most lakes classifies them as **dimictic** – that is, they mix from top to bottom twice each year, in spring and autumn. In winter they are ice covered. In summer they are thermally stratified, with density differences separating the warm surface waters (the epilimnion), from the colder bottom waters (the hypolimnion) (Figure 6.3). In high latitudes, however, the ice-free period may be too short for the water to warm to 3.98 °C, so the water column remains unstable until refreezing takes place. Such lakes are called cold monomictic (Vincent *et al.*, 2008b).

When a lake is ice covered, the water will gradually warm during the winter (if there is no inflow). This warming is attributable to the energy released by the bottom sediments from that stored during the summer season. The heat flux can be calculated from heat conduction analysis if the thermal diffusivity of the sediments is known (Ashton, 1980).

In the McMurdo Dry Valleys of Antarctica (77.5° S, 162° E), there are perennially ice-covered lakes with ice thicknesses of 3–6 m being typical (Vincent *et al.*, 2008b). These include Lake Fryxell, Lake Hoare, and Lake Bonney in the Taylor Valley. There are also a few similar lakes in the High Arctic (Ellesmere Island and Greenland). In summer, the ice around the shores of these lakes may melt out forming a moat. The ice thickness is a function of the heat conduction out of the ice and the latent heat release at the ice/water interface (McKay *et al.*, 1985). The release of latent heat at the base of the ice is controlled by ice surface ablation of ~30 cm a^{-1}. The water beneath the ice in the Dry Valley lakes is highly stratified with a shallow fresh water layer overlying highly saline bottom water – a pattern known as **meromictic**. This water stratification is also observed in epishelf lakes where an ice shelf dams up a lake occupying a fiord. In summer, low-density meltwater flows over the more dense sea water below. White Smoke Lake is an epishelf lake in the Bunger Hills Oasis of Wilkes Land (66° S, 100° E) dammed by the Shackleton ice shelf; it has perennial ice 1.8–2.8-m thick (Doran *et al.*, 2000). The numerous freshwater lakes in the Vestfold Hills of East Antarctica (68.5° S, 78.2° E) lose all or most of their ice for a short interval in late summer, while the ice on the saline ones becomes thinner.

Permafrost thaw (or thermokarst) lakes (see Chapter 5) are widespread in the circumpolar Arctic. The coastal plain of Alaska has countless 2–3 m-deep lakes and 0.5 m-deep ponds. In the Mackenzie Delta (13,135 km^2), Emmerton *et al.*, (2007) counted 45,000 lakes >0.14 ha and 4,500 smaller ones representing 25% of the delta surface. Around Barrow, Alaska, the ponds are frozen solid from late September until mid-June (Vincent *et al.*, 2008b). Whether

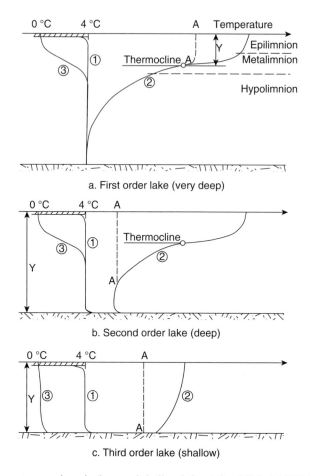

Figure 6.3 Temperature regimes in deep and shallow lakes (after Michel, 1971) [Source: B. Michel, Winter regime on rivers and lakes, US Army CRREL Monograph III-B1a, 1971, p. 38, fig. 30]

lakes freeze to the bottom or not can be mapped using synthetic aperture radar (SAR) imagery, with numerical ice growth modeling (Jeffries *et al.*, 1995) as the radar return differentiates the two.

The four standard observed ice condition dates are: first permanent ice (FPI), complete freeze over (CFO), first deterioration of ice (FDI), and water clear of ice (WCI). Usually CFO and WCI are the two primary indices. Ice data for 748 lakes and rivers were assembled by the Lake Ice Analysis Group (LIAG) and are available through the National Snow and Ice Data Center (Magnuson *et al.*, 2000a). The Global Lake and River Ice Phenology Database records span 1844–2005 and 170 of them are >50 years in length (http://nsidc .org/cgi-bin/catalog/adv_search.pl). The term "phenology" in the data set title refers to the seasonal cycle of freezing and thawing of lake and river ice.

The interannual variability in the dates of freeze-up and breakup is examined by Kratz *et al.* (2000) for 184 Northern Hemisphere lakes. They found that lakes with a short ice cover season tend to show more variability in freeze and thaw dates than lakes with a long ice season. Two hypotheses were offered to account for this difference: (1) that the rate of temperature change is larger in colder than in warmer climates giving a shorter transition interval for lake freeze-up and breakup; (2) among-year weather variability is less in early autumn and late spring, when cold climate lakes freeze and thaw than when warm climate lakes freeze and thaw. They demonstrated that the among-year variability in freeze-up is 1.5 times greater than in the thaw date. From hypothesis (1) the rate of change of air temperature may be less in autumn than in spring, or from hypothesis (2) among-year weather variability is greater in autumn than in spring. However, the data were not available to choose between the two hypotheses. The variability in freeze-up, breakup, and ice cover duration was also found to be greater in 1971–1990 than in 1951–1970.

Williams *et al.* (2004) assess statistical correlations between lake ice cover for 143 North American lakes and climatic, geographical, and bathymetric variables. They find significant correlations between ice duration and latitude, but morphometry is less important. In line with earlier work of Palecki and Barry (1986) they find that a 1 °C temperature increase leads to a delay of 5 days in freeze-up and 6 days earlier ice-out conditions. In further analyses of data for 128 lakes, they find that the best regression for freeze-up date is given by a logarithmic transform, whereas for breakup a linear regression is best. The best predictors of ice conditions are mean air temperature and latitude, followed by elevation. Lake surface area and depth have little effect. Assel and Herche (2000) determine that ice-on dates in North America (Finland and Russia) are 2.3 (2.6) days earlier per degree of latitude increase from about 40–65° N and ice-out dates are 1.2 (2.0) days later. In Sweden, ice cover data from 196 lakes between 55.7° N and 68.4° N show that the relationship between the timing of lake ice breakup and air temperature follows an arc cosine function (Weyhenmeyer *et al.*, 2004). Breakup occurs around Julian Day (JD) 90 in the south and JD150 at 65–67° N.

The relationship between freeze-up and seasonal weather was studied by Williams (1965) in Ontario. He showed that in early autumn, the water temperature declines in parallel with the air temperature. Then, when the water column is isothermal at 3.98 °C, a surface temperature inversion forms as the surface cools by heat loss through radiation and convection. The surface layer where there is a temperature inversion may only be ~0.4 m deep. Once the surface has cooled to 0 °C, further cooling removes latent heat and leads to ice formation. Weather variables play a determining role in cooling the surface to the freezing point. In temperate climates, such as Western Europe, lakes may freeze and thaw repeatedly during the autumn and, in mild winters, throughout the winter season. Adrian and Hintze (2000) illustrate this for Müggelsee in Berlin. During the winters of 1976/1977 to 1997/1998 there were eight seasons when the ice cover formed (>80% of the lake) and broke up more than twice. Between 1987/1988 and 1994/1995 the lake only froze once for more than four weeks.

The relationship of lake ice breakup and climatic conditions is more complicated than in freeze-up. As well as air temperature, as a proxy for incoming solar radiation, breakup also depends on the thickness of snow cover, runoff into the lake, and wind speed. Hodgkins *et al.* (2002) assembled and analyzed ice-out dates from 29 lakes in New England (Maine, New Hampshire, and Massachusetts) with 64 to 163 years of record. Ice-out dates have become significantly earlier that between 1850 and 2000 by 9 days in northern and mountainous areas of New England and by 16 days in more southerly locations.

Lake ice models

A 1-D energy balance model of lake ice growth is described by Liston and Hall (1995) that treats lake-ice freeze-up, breakup, total ice thickness, and ice type (Figure 6.4). There are four submodels: (1) describes the evolution of lake water temperatures and the lake stratification; (2) a snow submodel describes the depth and density of the snow cover as it accumulates on the ice, metamorphoses, and melts; (3) a lake ice submodel forms ice by two mechanisms – clear ice grows at the ice–water interface as a result of thermal gradients in the ice, and snow ice forms through the freezing of water-saturated snow (slush) from the upwelling of water due to the snow overburden, from snowmelt, and from rain on snow events; (4) a surface energy balance submodel calculates the surface temperature and energy available for freezing/melting. The model is forced by daily atmospheric data of precipitation, wind speed, and air temperature.

The thermal energy balance at the ice/water interface has the form:

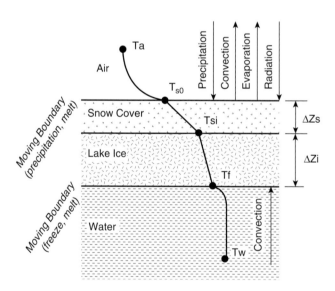

Figure 6.4 Outline of energy budget model for lake ice (from Liston and Hall, 1995b, Ann. Glaciol 21, p. 388, Fig. 1)

$$\rho_i L_i \frac{dz_i}{dt} = \frac{(T_i - T_s)(z_i + z_s + z_m + z_w)^{-1} - h_w(T_w - T_f)}{(k_i \; k_s \; k_m \; k_w)}, \quad (6.3)$$

where ρ_i = ice density, L_i = latent heat of fusion, h_w = convective transfer coefficient, T_s = lake surface water temperature, T_f = water freezing temperature, z = depth of each layer of thermal conductivity k, where the subscripts i, s, m, and w denote individual layers of ice, snow, snow/water mix, and water, respectively, and dz_i/dt is the velocity of the moving ice boundary. The model is validated for St. Mary's Lake, MT and Great Slave Lake, NWT, Canada.

Fang *et al.* (1996) developed a freeze-up algorithm that uses a full heat budget equation to estimate surface cooling, quantifies the effect of forced convective (wind) mixing and the latent heat removed by ice formation at daily time step. The vertical resolution is finer near the water surface where temperature gradients before freeze-up are largest. The algorithm was combined with a year-round temperature model for Minnesota and was tested satisfactorily against observations for Ryan Lake, MN (0.06 km^2 area) and eight other Minnesota lakes (1–38 km^2 in area) for multiple years. The 1-D, vertical unsteady diffusive heat transport equation for lake water is:

$$\frac{\partial T_w}{\partial t} = \frac{1}{A} \frac{\partial}{\partial z} \frac{(K_z A \partial T)}{(\partial z)} + H_w / pC_p \quad (6.4)$$

where ∂T_w = water temperature, t = time. A = horizontal area as a function of depth z, K_z = the vertical turbulent heat diffusion coefficient, pC_p = heat capacity per unit volume, and H_w = an internal heat source due to radiation absorption in the water column.

Vavrus *et al.* (1996) develop a Lake Ice Model Numerical Operational Simulator (LIMNOS), patterned after the thermodynamic sea-ice model of Parkinson and Washington (1979). The model treats the bulk diffusive vertical energy transfer through the snow and ice layers, with an energy balance required at each of the vertical interfaces of air, snow, ice, and water. The ice model applies the "zero-layer" parameterization of Semtner (1976) where the ice has zero heat capacity and therefore adjusts its temperature instantaneously to atmospheric forcing. Ice is assumed to form as a 1-cm slab across the lake. When the ice is snow free, part of the incident solar radiation penetrates the lake causing warming the base of the ice. LIMNOS has two lake layers with a mixing layer depth determined by the turbulent kinetic energy due to wind stirring and surface buoyancy. Snow ice formation due to wetting is calculated. The model, validated for three Wisconsin lakes, is shown to simulate the annual ice-on and ice-off dates of Lake Mendota with a median absolute error of only 2 d and 4 d, respectively.

Walsh *et al.* (1998) modify LIMNOS to run globally on a 0.5° by 0.5° latitude–longitude grid using average monthly climate data. First they simulate the ice phenology for lakes of 5- and 20-m mean depths across the Northern Hemisphere to demonstrate the effects of lake depth, latitude, and elevation on ice phenology. Lake depth (>100 m), large surface area, and snow cover are shown to be responsible for discrepancies in the calculated dates. For zonal means they show a linear increase in lake ice duration from 100 days at 40° N to 330 days at 75° N. Then they simulate the ice phenology of 30 lakes across the Northern Hemisphere that have long-term ice records and show that the ice-on date and duration are more accurately depicted than ice-off date. Lewis (2010) uses the Benson and Magnuson (2000) hemispheric database and finds a median ice cover duration of about 120 days at 43° N increasing curvilinearly to about 180 days at 63° N, with a ±30-day range.

Two models – PROBE and LIMNOS – are compared for Lake Pääjärvi at 60° N in Finland and Lake Mendota, WI, at 43° N by Elo and Vavrus (2000). Both models use temperature, relative humidity, cloudiness, and wind speed to calculate the surface energy exchange; PROBE also uses wind direction. Snowfall data are needed to determine the snow cover on the lake ice. PROBE uses a 3-hr time step and LIMNOS takes account of the daily temperature variation. In PROBE, a hypsometric curve is used to specify the horizontal area with depth. Turbulence in the lake is simulated via a kinetic energy/dissipation submodel. LIMNOS calculated snow and ice separately for both lakes and slush ice was determined. The growth and melting of ice on Lake Pääjärvi in the PROBE model was calculated with degree-days and local parameters; for Lake Mendota snow cover was calculated separately. The simulations were run for 1961–1990. Using LIMNOS for Lake Mendota, the predicted date of ice formation differed by no more than 3 days from the observed date of December 22 in 20/29 years. For PROBE, the calculated date was December 17 and the difference averaged 5 days. Observed melt occurred on average on 3rd April and the simulation with PROBE gave April 1. For Lake Pääjärvi, the observed average date of ice formation was December 6; the simulated dates were November 30 with PROBE and December 13 with LIMNOS. Ice melt occurred on May 3 on average and the simulated dates were April 29 with LIMNOS and May 3 with PROBE. Overall, the simulation results were encouraging.

The Canadian Lake Ice Model (CLIMo) is used by Menard et al. (2002) to model ice cover at Back Bay on Great Slave Lake, NWT, for 1960–2000 and compare the results with shore-based observations and SSM/I-derived ice dates. CLIMo is based on the 1-D landfast sea-ice model of Flato and Brown (1996). It solves the temperature profile through the snow and ice cover using the 1-D unsteady heat conductivity equation to compute the surface energy balance. The input climate variables are: mean daily air temperature, wind speed, relative humidity, cloud cover, and snow depth. Other parameters are number of ice cover layers, mixing depth, the site latitude, and the time step. Calculated freeze-up (breakup) dates were within 6 (4) days of shore-based observations. Pour et al. (2017) improved the retrieval of lake ice thickness of the Great Slave Lake and Baker Lake using the Moderate Resolution Imaging Spectroradiometer (MODIS) lake ice surface temperature and a heat balance equation. Compared to in situ measurements from the Canadian Ice Service, ice thickness estimates is improved when snow depth predicted by CLIMo is also used rather than an empirical relationship between snow depth and ice thickness. Over the study period (2002–2014), the mean bias error and the root-mean-square error are reduced from −0.42 to 0.07 m and 0.58 to 0.17 m, respectively. However, this approach is limited to ice thickness of less than 1.7 m.

Remote sensing

Optical imagery is primarily used to detect breakup and freeze-up of freshwater ice cover (Cooley and Pavelsky 2016; Muhammad *et al.*, 2016). However, misclassification from optical images could occur under certain ground or atmospheric conditions, for example, turbid water and thin cloud could be misclassified as snow/ice, and under thin river ice or topographic shadows, river ice could be misclassified as water. Dolan *et al.* (2019) used optical satellite data to detect long-term changes in ice covers of pan-Arctic rivers and lakes of Alaska, for rivers wider than 90 m using Landsat imagery, and for rivers wider than 150 m and lakes larger than 1 km^2, MODIS imagery. From 34 years of Landsat data, they show that rivers in northern Alaska experienced about 8 months of ice cover, compared to 6 months in southern Alaska. Based on twenty years of MODIS data, they found 8.9% of river reaches in Alaska experienced earlier breakup and 8.9% later freeze-up, while for 4241 Alaskan lakes, they found 10.4% of the lakes experienced earlier breakup and 2.2% later freeze-up.

Airborne radar and space-borne SAR imagery (Ku-, X-, C-, and L-band) have been used to determine whether lakes have a bedfast ice (BI) or floating ice (FI), or other form of ice foes. In late-winter, when the maximum ice thickness exceeds the maximum water depth, lakes freeze to the bottom, resulting in BI, and when it is less than the maximum water depth, lakes have FI. Engman *et al.* (2018) used ERS-1/2, RADARSAT-2, Envisat, and Sentinel-1 SAR imagery for seven lake-rich regions in the Arctic Alaska to analyze lake ice regime extents and dynamics over a 25-year period (1992–2016) by an interactive, intensity threshold classification method applied to C-band SAR data of different polarizations and incidence angles. They only detected significant declines in BI regimes in the Fish Creek area with 3% of lakes switching from BI to FI during this 25-year period. However, winter warming is expected to cause future Arctic lake ice to shift regimes.

Detection of ice cover by passive microwave data is possible on large lakes such as Great Slave Lake (61.7° N, 114° W) and Great Bear Lake (66° N, 121° W) in Canada since April 1992 (Walker and Davey, 1993). DMSP SSM/I data are acquired with a focus on ice freeze-up and breakup. It is possible to discriminate between areas of ice cover and open water using 85 GHz data. QuikSCAT data, validated by AVHRR Polar Pathfinder climate data, were used to monitor ice phenology on Great Bear Lake and Great Slave Lake by Howell *et al.* (2009). For 2000–2006, the average melt onset date on Great Slave Lake occurred on JD123, the average WCI date was on JD164, and the average freeze onset date was on JD330. On Great Bear Lake, the average melt onset date occurred on JD139, the average WCI date was JD191, and the average freeze onset date was JD321. Standard deviations were 4–5 days for melt onset and 7 days for freeze onset. Ice cover remained at least five weeks longer on Great Bear Lake than on Great Slave Lake. On Great Bear Lake, melt onset took place first in the eastern arm; open water occurred first in the southeastern and western arms, while freeze onset appeared first in the northern arm and along the shorelines. On Great Slave Lake, melt onset began first in the central basin; freeze onset occurred first within the east arm, closely followed by the north and west arms, and then finally in the center of the main basin.

Wynne *et al.* (1998) determined lake ice breakup dates from 1980 to 1994 for 81 lakes and reservoirs in the US upper Midwest and portions of Canada (60° N, 105° W to 40° N, 85° W) using images from the visible band of the *GOES-VISSR*. The analyzed breakup dates agreed closely with available ground observations. The pattern of breakup was accounted for by latitude and snow depth. The pooled records showed a significant trend toward earlier breakup over the 15 years.

6.3   Changes in lake ice cover

Northern North America is occupied by countless number of lakes of a wide range of sizes, forming an important part of the cryosphere. The timing of lake ice breakup and freeze-up is a useful indicator of climate variability and change, which is of particular relevance in the environmentally sensitive Arctic, where changes in the lake ice regimes could result in major ecosystem changes, such as shifts in species.

For a 1 °C rise in average air temperature: ice-onset date occurs ~5 days later, and ice-out date occurs ~6 days earlier (Williams *et al.*, 2004). Based on a global data base of 39 lake and river ice records (Benson and Magnuson, 2000), Magnuson *et al.* (2000b) find that over the period 1846 to 1995, there has been a 5.7-day per century delay in freeze-up and 6.3-day advance in break up (corresponding to a warming of +1.2 °C) (see Figure 6.5). A study of Canadian lake-ice cover from 1951 to 2000 by Duguay *et al.* (2006) found a shortening of the lake-ice season over much of Canada mainly attributable to earlier breakup. Latifovic and Poulio (2007) use AVHRR data to extend existing *in situ* measurements for 36 Canadian lakes and to develop records for six lakes in Canada's far north. Trend analysis of the combined *in situ* and AVHRR record (~1950–2004) shows earlier breakup (average 0.18 days yr^{-1}) and later freeze-up (average 0.12 days yr^{-1}) for the majority of lakes analyzed. Trends for the 20-year record in the far north show earlier breakup (average 0.99 days yr^{-1}) and later freeze-up (average 0.76 days yr^{-1}).

Given many Arctic lakes in the Canadian Arctic Archipelago are ice covered more than 10 months per year, warmer temperatures could result in ice regime shifts. Within the polar-desert environment are some secluded, small local warmer areas (polar oases) with longer growing seasons and greater biological productivity/diversity. Surdu *et al.* (2016) documented ice regimes of 11 lakes between 4 and 542 km^2 in areas and located in both polar-desert and polar-oasis environments. They investigated the recent ice cover of these lakes over 1997–2011 using RADARSAT-1/2 ScanSAR Wide Swath, ASAR Wide Swath, and Landsat data. All lakes show earlier melt onset, experienced earlier summer ice minimum and water-clear-of-ice (WCI) dates (except the Lower Murray Lake), with 9–24 days (2–20 days) earlier WCI dates for polar oases (polar-desert) lakes. Some lakes may be transitioning from a perennial/multiyear to a seasonal ice regime.

Brown and Claudy (2011) used CLIMo to simulate lake ice phenology across the North American Arctic from 1961 to 2100 using two climate scenarios produced by the Canadian

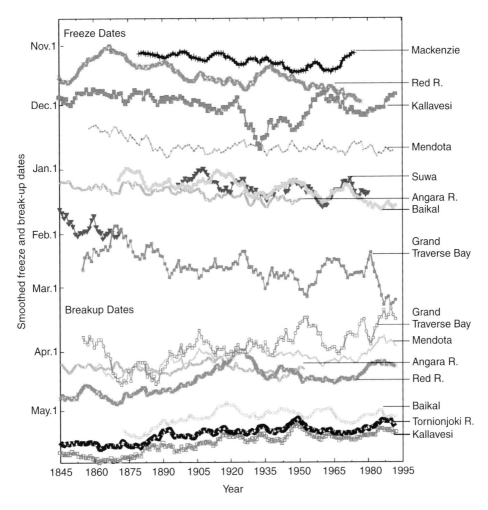

Figure 6.5 Trends in lake and river ice freeze-up and breakup from selected sites around the Northern Hemisphere, 1846–1995 (from Magnuson *et al.*, 2000b) [Source: Science 289(5485), 2000, p. 1744, fig. 1. Courtesy American Association for the Advancement of Science]

Regional Climate Model (CRCM) and validated the 1961–1990 results using 15 locations across the Canadian Arctic. Compared to the 1961–1990 data, projected ice cover in 2041–2070 suggest a projected shift in breakup and freeze-up dates ranging from 10 to 25 days earlier (breakup) and 0 to 15 days later (freeze-up), reduction in ice cover durations of 10–25 days for shallow lakes (3 and 10 m) and 10–30-day reduction for deep lakes (30 m). More extreme reductions of up to 60 days were simulated in coastal regions. The mean maximum ice thickness is projected to decrease by 10–60 cm with no snow cover and 5–50 cm with snow cover on the ice.

Assel *et al.* (2003) showed that the annual maximum ice cover for 1963–2001 was largest from 1977–1982 relative to 1963–1976 or 1983–2001. The 1998–2001 period saw the lowest ice on record since the last century for four of the five Great Lakes. Wang *et al.* (2010) document the severe ice conditions of winter 2008–2009, following a decade of mainly little ice (1997–1998 to 2007–2008). The ice extent in 2008–2009 was 166,000 km^2, similar to levels in the early 1980s and the winter of 2002–2003. The large variability is clearly linked to anomalous atmospheric circulation patterns, particularly the Pacific North America pattern, La Niña, and the Arctic Oscillation. Over the entire record from 1972 there is no overall trend. Jensen *et al.* (2007) analyzed trends in ice phenology and duration for 65 water bodies across the Great Lakes region (Minnesota, Wisconsin, Michigan, Ontario, and New York) during 1975–2004. Average rates of change in the dates of freeze-up (3.3 d per decade) and breakup (22.1 d per decade) were 5.8 and 3.3 times more rapid, respectively, than historical rates (1846–1995) for Northern Hemisphere water bodies. Differences in trends of ice duration were due to difference in elevation and the local rate of change in either air temperature or the number of days with snow cover.

For 29 lakes in New England, Hodgkins *et al.* (2002) provide between 64 and 163 years of ice-out dates. Between 1850 and 2000 the advance in the date is 8 days in the north and mountains and 16 days in southern New England. For Mirror Lake, NH (43.9° N, 71.7° W, 213 m asl), Likens (2000) find no significant change in the ice-on date for 1968–1998, but an earlier ice-out date during April of 0.36 d a^{-1}, correlating with an increase of 0.07 °C a^{-1} in April since 1963.

Observations of the freezing of the upper part of Lake Constance date back to AD 875 (Solow, 1991). It is shown that there were eight freezing events in the fifteenth century and seven in the sixteenth compared with five in the fourteenth, four in the eighteenth, two in the eleventh, thirteenth, seventeenth, and nineteenth centuries but only one each in the twelfth and twentieth centuries; the last freeze was in 1963. These occurrences clearly reflect the Little Ice Age (see Chapter 9, p. 408). Records for 11 lakes on the Swiss Plateau show that the freezing frequency varied from 13% to 75% over last 105 years (Franssen and Scherrer, 2007). However, there has been a significant reduction in ice cover over the last two decades and several lakes have not frozen since the 1980s. For many lakes in the Eastern Alps for 1895–1950, Eckel (1955) shows an altitude effect on the freeze-up and breakup dates, respectively, of 4 days earlier and 8 days later per 100 m increase in altitude. Lej da San, near St. Moritz, Switzerland, at 1,768 m altitude has observations of breakup date since 1832. Livingstone (1997) shows earlier thawing by 7.6 days/100 years with shifts in the mean date around 1857 and 1932. The timing is a function of the local and regional air temperature over 4–8 weeks in April.

On Lake Suwa, Japan, freeze dates became later over the 550-year record since AD 1443 by about 2 days per 100 years (Magnuson *et al.*, 2000b). Later freezing for relatively unbroken time windows ranged from 3.2 days per 100 years (1443 to 1592) to 20.5 days per 100 years (1897 to 1993). Lake Suwa was ice covered for 240 out of 243 winters from AD 1443 to 1686 but only for 261 out of 298 winters from AD 1687 to 1985.

In Finland, records from Lake Kallavesi (62.9° N, 27.7° E) began in 1833 and from Lake Näsijärvi (61.5° N, 23.7° E) in 1836. Both show a shortening of the ice cover over the entire record by 23 days with greater change in the breakup (Kuusisto and Elo, 2000), and a strong shift toward later freeze-up dates from 1910s to 1940s.

Ice cover on Lake Baikal has been monitored over 1896–1996. Todd and Mackay (2003) found that freeze-up and breakup trends of Lake Baikal are correlated with regional temperature changes and at decadal to multidecadal time scales the timing of the events is linked to the Scandinavian anticyclone and the Arctic Oscillation of surface air pressure.

6.4   River ice

River ice types

Thermal ice formed out of slow stream flow and calm water is usually solid and largely free of air bubbles. Frazil ice formed in turbulent flows of supercooled water tends to be embedded with spherical and irregularly bounded air bubbles. Snow ice that contains small and spherical air bubbles forms either from snow falling into cool water (≈ 0 °C) or when snow on an ice cover gets wet, say, due to rainfall, and then freezes. Consolidated ice that forms during freeze-up is a thick, porous, and rough-surfaced accumulation of ice floes or a variety of ice types.

Freeze-up

More than one-third of Earth's landmass is drained by rivers that seasonally freeze over (Yang *et al.*, 2020). River waters cool in the autumn due to convective heat loss to the colder atmosphere. Eventually, water temperatures reach 0 °C but a slight degree of supercooling is required for ice to form. Supercooled water (−0.01 °C) in turbulent motion leads to the formation of **frazil ice** – fine, needle-like structures or thin (1–100 μm), flat, circular plates of ice (1–4 mm in diameter) suspended in water (Martin, 1981). The word frazil is derived from Old French for cinders. The initial mechanism of frazil nucleation has been much debated (Ashton, 1980). The process appears to involve mass exchange whereby small ice crystals nucleate in the cold surface air and fall onto the water surface where nucleation occurs by crystal multiplication (Osterkamp, 1975). The frazil builds up into **pancakes** whose shape and rim are due to repeated collisions (see Figure 6.6).

These may amalgamate into a sheet of ice. In slow-flowing water near the banks, ice particles form a continuous layer of skim ice on the water surface. The resulting shore-fast ice effectively prevents further supercooling of the water underneath, and therefore, subsequent ice growth is primarily thermal in nature. Initially it grows laterally as heat is lost through the ice and into the banks, accompanied by thermally induced thickening (Hicks, 2008). Frazil particles readily freeze to each other and this causes them to flocculate, forming "frazil

Figure 6.6 Pancake ice floes [Courtesy of Dr. Faye Hicks, University of Alberta, Edmonton]

slush." Ultimately, buoyancy overcomes the ability of the fluid turbulence to keep slush balls in suspension, and they float to the water surface. The surface of the floating slush freezes to form a crust, creating pancake ice floes. These may freeze together and as the surface concentrations of ice floes increase to 80–90%, "bridging" may occur. This involves a congestion of ice floes and a subsequent cessation of their movement at a site along the river. Typical bridging locations are at tight bends and where the channel becomes constricted. The incoming ice floes may accumulate on the water surface, causing an upstream progression of the ice front by "juxtaposition." However, if flow velocities are sufficient, ice floes coming to the ice front could be swept under the ice cover and then deposited on the underside. This process is known as "hydraulic thickening." The flow speed needed to transport ice beneath an initial cover depends on the riverbed roughness, floe shape, and porosity, and the ice cover characteristics. Critical velocities for such ice transport are ~0.6–1.3 m s^{-1}. In steep reaches, frazil may continue to accumulate beneath a downstream ice sheet throughout the winter giving rise to a "hanging dam." These tend to form where there is a deep pool below rapids or where a river enters a lake. An accumulation of floes may build into an **ice jam. Anchor ice** forms in supercooled water when frazil freezes on to vegetation, gravel, and boulders on the riverbed. When the water temperature rises it may float to the surface.

When incoming ice pans come to a stop to form a juxtaposition cover, the interstitial water freezes, resulting in a continuous layer of solid ice. When frazil growth ceases, there is freezing of pore water in the frazil slush. Further cooling leads to thermal growth of the ice, forming long vertical crystals – columnar ice. Once formed, ice covers thicken via heat transfer between water and atmosphere, by flooding and refreezing of the surface, or by deposition of ice beneath the surface. When a snow cover is present in sufficient amounts, it may depress the top of the floating ice below the phreatic level. This occurs due to water seeping through cracks in the ice cover, which then saturates the lower portion of the snow cover. This subsequently freezes forming a new layer on top of the original ice surface, termed **snow ice**.

Semiempirically, the net surface heat flux from the water to the atmosphere (Q) can be expressed by

$$Q = K(T_w - T_a), \tag{6.5}$$

where K = heat transfer coefficient typically in the range 15–30 W m^{-2} K^{-1}
T_w = water temperature
T_a = air temperature.

This approach is generally more practical than using a full energy budget equation.

Maximum ice thickness is calculated via a form of the Stefan equation (Michel, 1971)

$$h_i = \alpha(\text{FDD})^{0.5}, \tag{6.6}$$

where h_i is ice thickness (mm), α is a coefficient that accounts for conditions of exposure and surface insulation (mm °C$^{-0.5}$ d$^{-0.5}$), and FDD is the accumulated degree-days below freezing (°C day) from the onset of freeze-up. The Stefan equation gives erroneous results for low values of h_i. Values of α range as follows:

Windy lake without snow cover	27 mm °C$^{-0.5}$ d$^{-0.5}$
Average lake with snow	17–24 mm °C$^{-0.5}$ d$^{-0.5}$
Average river with snow	14–17 mm °C$^{-0.5}$ d$^{-0.5}$
Sheltered small fast flowing river	7–14 mm °C$^{-0.5}$ d$^{-0.5}$

(Michel, 1971)

The average maximum thickness of ice on rivers in Canada ranges from ~0.3 m in the south to 1.7 m in the Arctic. Along the Ob' river the thickness increases from 0.6 m in the south to 1.6 m in the north (Vuglinsky, 2002b). Corresponding figures are 0.75 and 2.1 m on the Yenisei and 0.65 and 2.5 m on the Lena river. In Mongolia the thickness is between 1.0 and 1.8 m and shallow rivers are frozen to their bed for 5–6 months (Punsalmaa and Nyamsuren, 2002).

The SAR data acquired in the microwave range offers the potential to estimate river ice thickness. Mermoz *et al.* (2014) developed a river ice thickness retrieval model based on the polarimetric entropy (H), a measure of statistical disorder, or a function of ratios related to the relative power of surface and volumetric scattering mechanisms of the backscattered radar signal sensitive to river ice thickness (see Figure 6.7). The model was used to retrieve

Figure 6.7 Estimated river ice thickness for a 4 × 2.8-km section located at (45.8° N, 72.4° W) of the Saint-François River, Quebec, showing growth of the ice cover during the winter. The total power is in gray levels (taken from Mermoz *et al.*, 2014)

river ice thickness from C-band polarimetric SAR (PolSAR) SAR data acquired by RADARSAT-2 in the winter of 2009 over the Saint-François River and the Koksoak River of Quebec, and the Mackenzie River of Canada. Field campaigns were carried out to obtain river ice thickness data at 70 sites categorized under thermal, frazil, snow, and/or consolidated river ice types to account for the effect of the ice structure on the backscatter signals. By a leave-one-out cross-validation method, the accuracy of the river ice thickness retrieved from SAR data using the retrieval model based on H alone was encouraging, with a mean producer's accuracy of 99.1%, a mean user's accuracy of 90.8%, and the Kappa coefficient (Landis and Koch, 1977) of 0.97.

River ice floats with about 90% of its thickness submerged, reducing the area of active flow. The ice also resists the water flow. As a consequence, the water level with an ice cover is about 30% higher than for the same discharge without ice present. As the freeze-up front passes a location on the Mackenzie River, the water level increases ~0.5 m. There is an increase in river slope with ice cover compared with open water and this ratio is of the order of 1.6–2 for some rivers in Sweden (Ashton, 1980). Beltaos and Prowse (2009) emphasize that the rise in river stage caused by an ice cover is fundamental to ice-related hydrologic impacts; these include floods caused by freeze-up and breakup ice jams, low winter flows caused by water storage during freeze-up, and sharp waves generated by ice-jam releases. Ice thickness and strength, both controlled by weather conditions, also play major roles. Rises in water level associated with ice-jam formation and release can be extremely rapid – 0.5–0.8 m per minute has been reported – creating a severe hazard to local communities.

Ice jams

Ice jams usually accumulate with the toe of the jam against a solid ice cover; the head will backup upstream until it reaches a backwater at a rate determined by the ice supply, the river flow, the strength of accumulated ice blocks, and the channel characteristics (Hicks and

Figure 6.8 Ice jam with Jie Che conducting breakup monitoring at Hay River, NWT, 2005 [Courtesy Robyn Andirshak, University of Alberta, Edmonton]

Beltaos, 2008) (Figure 6.8). A classification of ice jams has been proposed by the IAHR Working Group on River Ice Hydraulics (1986; Beltaos, 1995, p. 99). It uses four criteria: the dominant formation process, season, spatial extent, and state of evolution. The formation processes are: congestion (surface jam), transport and deposition (hanging dam), submergence–frontal progression (narrow channel jam), shoving or collapse (wide channel jam), and anchor ice accretion (ice dam). Season is either freeze-up or breakup. Spatial extent is categorized vertically (floating or grounded) and horizontally (partial or complete). State of evolution is evolving or steady (nonequilibrium or equilibrium). Beltaos (2008a) notes that wide-channel jams, which form by collapse and shoving of ice floe accumulations, are just thick enough to withstand the longitudinal external forces applied on them. Narrow-channel jams (mostly formed during freeze-up) have a thickness controlled by the hydraulic conditions at their upstream end. The thickness is just sufficient for the net buoyancy of the ice to withstand submergence by the hydrodynamic forces and overturning moments that develop at the head of the jam. Water flows through voids in ice jams and Beltaos (2008) shows that the seepage flow through the jam should vary as the square root of the water surface slope multiplied by a seepage coefficient that ranges from 1.0 to 2.5 m s^{-1}.

There appear to be three phases in the deformation of a floating ice field: (1) consolidation, (2) deformation of the consolidated layer by underturning and rafting of ice floes, and (3) continued deformation of the jumbled field of floes and thickening of the rubble field

(Hopkins and Tukhuri, 1999; Beltaos, 2010). The rubble tends to behave as a Mohr–Coulomb material (see Note 6.1). The most common sites of ice jams are sharp bends in the river, abrupt reductions in slope, and channel constrictions that reduce flow velocity (Beltaos, 2007). An "equilibrium" jam may develop in a reach with nearly constant ice thickness and flow depth. Downstream the slope of the water surface steepens rapidly as the water level profile adjusts to the lower stage prevailing at the downstream toe. Here, the thickness of the ice jam increases and it may become grounded. Ice-jam thicknesses could reach 5–10 m on large northern rivers. Major ice jams that occasionally form in the lower Peace River of Canada could generate extensive flooding, which helps replenish the perched basins of Peace-Athabasca Delta, a Ramsar wetland of international importance and a UNESCO World Heritage Site (Beltos, 2018).

Ice-jam roughness was measured for freeze-up jams comprising loose slush, dense frozen slush, and solid-ice blocks by Nezhikhovskiy (1964) who showed that the dimensionless Manning roughness coefficient (n) increases with increasing jam thickness, solidity, surface roughness, and channel sinuosity (see Note 6.2). Beltaos (2001) determined a composite breakup jam roughness coefficient

$$n_o \approx (0.063 \text{ to } 0.076)h^{\frac{1}{2}} y^{\frac{-1}{3}},$$
(6.7)

where h and y represent laterally averaged jam thickness and flow depth, respectively; h/y is in the common range of 0.3–1.0. Composite roughness calculation is far more complex for very thin or thick ice jams.

The increase in water level height associated with ice jams based on 20 years of data on two reaches of the Lena River in Siberia is about 50–100% over the height for the same discharge in open water conditions (Ashton, 1980). River ice development has been investigated using web cameras over winters 2000/2001–2002/2003 at the confluence of the Allegheny River and Oil Creek in Oil City, PA suffered from frequent ice-jam flooding. Hourly images were classified in terms of stationary ice cover, frazil ice, brash ice, or open lead formation in an ice cover and the percentage of the channel width in the image covered by each ice category was recorded (Vuyovish et al., 2009). At high flows, significant amounts of frazil ice are generated on both the Allegheny River and Oil Creek, which tend to deposit in a dredged reach downstream of the confluence, forming a freeze-up jam. At low flows, a stationary ice cover will form in pool sections, reducing the amount of frazil ice generation. The FDD at the beginning of the stationary ice cover period was ~250 °C days in 2001 and 2003. The Cold Regions Research and Engineering Laboratory (CRREL) reports that 43% of New Hampshire ice jams have occurred in March and April, when the rivers begin to break up. The 47% of jams that occur in January and February could be either freeze-up or breakup ice jams. A database of ~18,000 ice jam events in Alaska, southern Canada, and the contiguous United States is maintained by CRREL (www.crrel.usace.army.mil/icejams).

Breakup

River ice breakup comprises several distinct phases: onset, drive, ice-jam formation, and wash (Beltaos, 2008b). While these are sequential at a given location, several phases may be occurring simultaneously along a particular reach of the river. The first sustained movement of the winter ice cover defines the onset of breakup (Beltaos, 2008b, p. 169). Drive refers to the transport of ice blocks by the river. Wash refers to the final clearance of ice presenting significant hydraulic resistance. The mean duration of breakup on the Mackenzie River in Canada averages 6 days, increasing from 4 days in the south to 12 days at the delta, with a mean maximum duration of 16 days (de Rham *et al.*, 2008). The drive phase averages 2 days in length and the wash phase 4 days.

The breakup of river ice on a reach may involve thermal processes where the ice gradually deteriorates and essentially melts *in situ*. Initially, the surface albedo of the ice cover decreases and absorption of solar radiation increases. The ice deterioration accelerates as the albedo decreases. Once leads (openings) develop in the ice, absorbed solar radiation warms the water and this heat melts the underside of the ice. Basal melt occurs when the transfer of heat from the water to the ice exceeds the rate of heat conduction into the ice. The most rapid melt takes place when the thermal gradient is small or the ice is isothermal at 0 °C. The flow of water beneath the ice promotes a more rapid ice melt. The water–ice heat transfer was found to vary beneath a solid ice cover from 10 to 30 W m^{-2} with water temperatures of only 0.003 °C to 0.1 °C and flow velocities of 0.4 to 0.9 m s^{-1} (Marsh and Prowse, 1987). The rate of heat transfer may be enhanced if the underside of the ice is rippled.

An empirical estimate of the rate of ice thinning was proposed by Bilello (1980) from surveys in Alaska and northern Canada. The ice cover thickness (z, in meters) is given by

$$z = z_i - k \, \Sigma\text{TTD}, \tag{6.8}$$

where z_i is the initial thickness prior to breakup, TTD = thawing degree days with respect to a base temperature of −5 °C, and k is an empirical coefficient ranging from 0.002 to 0.01 m(°C day)$^{-1}$.

In mid-latitudes, winter thaws are common. Rain on snow events lead to increases in river discharge and stage. Hence, breakup can occur while the ice cover is still competent giving rise to ice jams (Beltaos, 2008b), which may occur more than once per season. Breakup may occur suddenly due to the passage of a dynamic breakup front generally caused by a sudden increase in stream flow runoff due to upstream snowmelt, or an ice-jam release (Hicks, 2008). The downstream-propagating water wave due to ice-jam release can be meters high and is known as a jam release wave (or "jave") (Beltaos and Prowse, 2009). Javes can propagate at up to 10 m s^{-1}, with amplitudes of up to 4 m.

Hinge cracks may form parallel to the banks as the ice in the center of the stream is lifted and the border ice is depressed and flooded. They are located about 12 times ice thicknesses away from each bank. Transverse cracks may also form in the ice cover. In the Thames

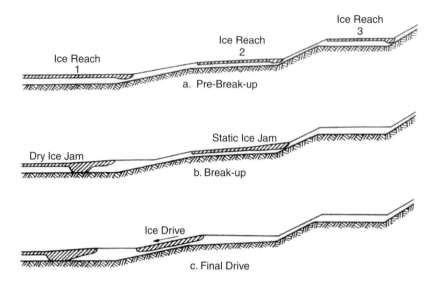

Figure 6.9 The three phases of river ice breakup (From Michel, 1971) [Source: B. Michel, Winter regime on rivers and lakes, US Army CRREL Monograph III-B1a, 1971, p. 84, Fig. 69]

River, Ontario, in 1982 and 1984, these had a spacing of about 300 m in ice covers about 30-cm thick (Beltaos, 2008b, p. 154). The rise in water level lifts the ice cover and it breaks into pieces and is carried downstream in an ice drive. Figure 6.9 illustrates the three phases of river ice breakup. The failure modes of an ice cover during breakup include vertical and horizontal flexure, tensile splitting, and crushing which depend on the ice strength and its spatial and temporal variation.

The breakup severity is a function of ice cover strength and integrity balanced by hydro-meteorological conditions which control streamflow discharge and water level. A less severe, thermal breakup occurs when the ice cover has substantially deteriorated by solar radiation and erosion from below to allow it to break up with little increase in discharge. In contrast, a mechanical breakup could occur when streamflow discharge and water level increase rapidly, causing the ice cover to break its bond with channel banks while it is still strong and competent, which tends to occur when a colder-than-normal period is followed by the rapid melt of a greater-than-normal snowpack and subsequent runoff. Optical imagery of MODIS may also be used to differentiate thermal and mechanical breakup events. The recent rising breakup water level trends in the Yukon River suggest that breakup severity is increasing for the greater energy inputs from higher winter and spring air temperatures, leading to an earlier onset and more rapid snowmelt events, resulting in higher peak flows in some regions (Janowicz and Hinzman, 2017). Similarly, weakening ice resistance and earlier breakup timing are detected near the Mackenzie Delta, and increasing trends in the magnitude and earlier onset of upstream discharge in the Lena, Ob, and Yenisey Rivers, which is expected to affect the breakup timing of these rivers of Siberia (Cooley and Pavelsky, 2016).

Besides some observed river ice break-ups such as described earlier, broad-scale analyses of climate warming on river-ice breakup severity have yet to be undertaken. Whether shifts toward shorter river-ice duration will produce more or less severe ice jams in breakup events remains in question, largely because precipitation which could control the driving (snowmelt runoff) and resisting (ice thickness, strength, composition) forces that affect breakup severity. On the other hand, a more thermal than dynamic breakup likely means reduced river-ice severity and ice-jam flooding.

Pavelsky and Smith (2004) analyzed 10 years of spring breakup along 1,600–3,300 km lengths of the Lena, Ob, Yenisey, and Mackenzie Rivers using MODIS and AVHRR. They show that at the watershed scale, spatial patterns in breakup seem to be primarily governed by latitude, timing of the spring flood wave, and location of confluences with major tributaries. However, channel-scale factors (slope, width, and radius of curvature) known to influence ice breakup at the reach scale, do not appear to be major factors at the watershed scale. The timing of breakup at eleven points 160 km apart, upstream from the mouth of each river, shows only a 15–20-day range on the Lena and Ob but a 40-day range on the Yenisei. Breakup patterns on the Mackenzie River vary between locations. A spatially integrated breakup date shows that the largest interannual variability occurs on the Ob' and the least on the Lena. Vuglinsky (2002b) shows that the duration of complete ice cover increases eastward from 150–200 days, on average, on rivers in the Ob basin, to 160–220 days in the Yenisei basin, and 180–230 days in the Lena basin. On Canadian Rivers, the mean ice cover duration (1970–2001) ranges from 65 days on the Thames River at Thamesville, Ontario to 121 days on the St. John River near East Florenceville, NB, to 174 days on the Athabaska River below Fort McMurray, Alberta, to 249 days on the Back River above the Hermann River, Nunavut (Milburn, 2008).

Projected changes in 0 °C conditions on four largest Arctic rivers (Lena, Mackenzie, Ob, and Yenlsey) for 2050s and 2080s with reference to 1979–2008 exhibit progressively earlier timing of the 0 °C isotherm over their entire length, and a tendency to greater warming in a downstream (primarily south to north) direction. Such a reduction in the current climatic gradient will likely lead to a more thermal than dynamic breakup characterized by reduced ice action and ice-jam flooding (Prowse *et al.*, 2010).

In Canada, Chen and She (2019) found river ice breakup dates to be more consistent among different data sources than freeze-up dates. Using the Water Survey of Canada HYDAT database, they constructed breakup dates from 1950 to 2016, which show an overall earlier breakup trend across terrestrial ecozones of Canada. In particular, stations of breakup dates occurring over 3 days per decade earlier are mainly located in southern parts of the Pacific and Western Mountains, and Central Plains where pronounced warming trends were also detected (Figure 6.10a). Mixed trends around the Atlantic and Great Lakes – St. Lawrence regions are evident where warming was not as pronounced, and stations showing later breakup dates of over 3 days per decade on the average are mainly located here. Earlier river ice breakup, later

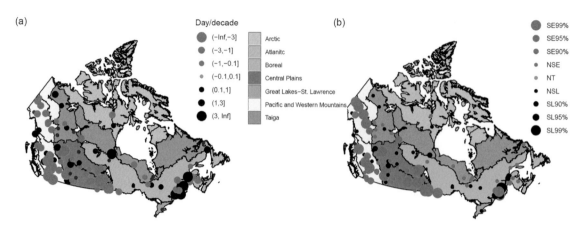

Figure 6.10 (a) Average rate of changes in river breakup timing, and (b) trends in breakup timing over each terrestrial ecozone of Canada from 1950 to 2016. Grey and black dots indicate earlier and later breakup, respectively. SE90% and SL90% indicate significant earlier and later trend at 90% confidence level (the absolute values of Gaussian frequency factor Z are between 1.645 and 1.96); for SE95% and SL95%, the absolute values of Z are between 1.96 and 2.576; the absolute values of Z for SE99% and SL99% are over 2.576, respectively (taken from Chen and She, 2019)

freeze-up and shorter river ice season were also found in Southern British Columbia rivers (Doyle and Ball 2008), the Mackenzie River (Goulding *et al.*, 2009), and northwestern Canada (Janowicz, 2010). Furthermore, the magnitude of change in breakup dates (both earlier and later) are larger in lower than higher latitudes partly because the climate of lower latitudes are warmer. Statistical significant trends in breakup timing at 90%, 95%, and 99% significant levels over Canada from 1950 to 2016 are shown in Figure 6.10b, where grey and black dots indicate earlier and later breakup, respectively.

River ice breakup can trigger severe spring ice jams which in turn trigger flooding events. An ice jam roughly 25 km long on the Athabasca River that flooded the downtown Fort McMurray of Alberta forced the evacuation of 13,000 people in late April of 2020. In Yukon annually the earliest flooding events typically occur in late April or May and are triggered by ice jams, where the spring breakup and associated flooding is usually followed by the snowmelt freshet several weeks later, but due to climate warming, there is an increasing frequency of overlapping river ice breakup events and freshet peaks in Yukon. In 2011 and 2015 the combined breakup and freshet peaks occurred on May 23 and 15, respectively. Furthermore, freeze-up of the Yukon River at Whitehorse has been delayed by approximately 30 days since 1902 (Janowicz, 2010), while recent breakup dates have advanced. Prior to 1989, only two April break-ups had been observed, but after 1989, eight April break-ups have been observed. Both the ice cover and breakup periods have become shorter in recent years (Janowicz and Hinzman, 2017).

River ice models

River ice models in general aim to simulate some of the following processes: cooling of water, ice generation, ice cover formation, thickening, ice transport, ice shove, erosion and deposition, melting, and breakup. These processes are illustrated in Figure 6.11. An early report treating autumn freeze-up and spring breakup was prepared by Shulyakovskii (1966). It is beyond the scope of this text to detail the various numerical models of ice jams. Petryk (1995) lists 12 different 1-D models – some of them proprietary – and provides a short summary of each of them. Since then new formulations have appeared in the public domain including 2-D models. Hicks *et al.* (2007) report on incorporating ice processes into the River 1-D hydrodynamic model. A study for the Peace River indicates that it adequately simulates water temperature and ice front progression; ice-jam formation and release components have also been added to the model. The model is found to simulate well the release wave speed and its peak magnitude. Wojtowicz *et al.* (2009) have developed the River 2-D model to simulate water and ice conditions in an 80-km reach of the lower Athabaska River. Processes to be treated include cooling/supercooling, frazil production, border ice formation, surface ice transport, transport and rise, bridging, and frontal progression. An overview of river ice process models is provided by Shen (2010).

The inclusion of such variables as water surface width, channel curvature, freeze-up stage, and ice competence in predictive schemes for river breakup is essential to enable accurate forecasts. Present major knowledge gaps concern the dynamic interaction of moving ice with the river flow and with the stationary ice cover, and the characteristics of the underside of the ice. Another factor pointed out by Beltaos (2008b) is the frictional resistance that develops at interfaces between moving and grounded rubble, typically located near the sides of the river.

The fact that most rivers in northern North America and Eurasia flow northwards toward the Arctic Ocean has important implications for the river ice regime during the breakup process. The ice is generally thinner and air temperatures rise faster in the south leading to earlier river breakup and increased streamflow. The ice thickness on the upper Peace and Athabaska Rivers is around 0.6 m, compared with 1 m on the lower Peace River and 1.5 m in the Mackenzie River delta (Hicks and Beltaos, 2008). Hence, the downstream movement of ice floes will encounter undeteriorated ice cover in the northern reaches of the rivers promoting ice jams as illustrated for the Ob River gulf in Figure 6.11. Also, low river gradients (<0.1 m km^{-1}) in the northern reaches do not generate the driving force for breakup by snowmelt runoff. Instead, waves triggered by the release of upstream ice jams are an effective mechanism for dynamic breakup (Beltaos, 2007; Hicks and Beltaos, 2008). The wave can dislodge intact ice cover, set it in motion, and break it up.

Breakup and ice jams in the Mackenzie delta are examined by Beltaos and Carter (2009). The low gradients, giving rise to low flows in the delta, and the thick ice cover tend to favor thermal breakup processes and this was the case in 2007. In 2008, relatively fast runoff led to rising river levels in the lower Mackenzie and Peel Rivers. A 13 km jam was present in the Peel River on May 19, but this had shrunk by May 21. On this date there was an 8 km jam on

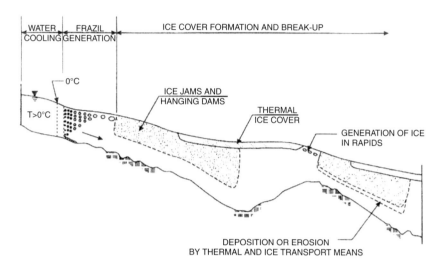

Figure 6.11 The ice conditions and processes to be simulated in a river ice-jam model (after Petryk, 1995) [Source: Petryk, 1995, in S. Beltaos (Ed), River ice jams. Water Resource Publications, Highlands Ranch, CO. p. 151, figure 5.2]

Middle Channel 44-km downstream of Point Separation and a 12 km jam on East Channel. By May 22, the head of the ice jam in Middle Channel was at Point Separation with ice delivery from upstream. The East Channel jam released on May 23 and reformed downstream. The Middle Channel jam did the same on the next day, and finally released on May 30 after shortening through attrition. The ice thicknesses in this jam, estimated from shear walls left on the riverbanks after the release of the jam, indicate a range from 3.3 to 5.3 m.

Goulding *et al.* (2009) analyze hydroclimatic conditions controlling breakup in the Mackenzie Delta over the period 1974–2006, with an emphasis on extreme flood events. They quantify both the upstream driving force, based on the spring discharge hydrograph at Arctic Red River, and the downstream resistance force, describing the competence of the downstream ice cover. The contribution of each to the severity and timing of breakup is determined. The severity of peak breakup stage is most influenced by upstream discharge and the balance between upstream and downstream melt; timing is related to delta ice conditions and the rise of the spring hydrograph. The highest (lowest) breakup events are characterized by rapid (protracted) upstream melt and by lower (higher) intensity of melt in the delta. Years of high peak stage are shown to coincide with a large temperature gradient between the Mackenzie basin and the delta.

The mass of ice that is discharged from a river system during breakup is significantly less than that contained in the pre-breakup ice cover as a result of ice melt in the river and stranding of floes. The ice volume produced by the clearance process can be expressed (Prowse, 1995):

$$V_i = LWH_t(1 - c), \tag{6.9}$$

where L = river length contributing ice, W = ice cover width, H_t = pre-breakup ice thickness, and c = an ice loss coefficient (smallest on shallow steep rivers, largest on long rivers with many tributaries).

6.5   Ice-jam floods

Ice-jam floods (IJFs) can be more severe than open water flood events because of much higher water depths caused by ice jams. They could seriously affect riverine communities in the NH where about 60% of the rivers experience significant seasonal effects of river ice. Unfortunately, they are difficult to predict because they could occur suddenly, but they are common in many northern countries such as in Canada, the USA, Russia, Poland, and China, especially in eastern North America (Beltos *et al.*, 2012). From analyzing ice-jam data of the CRREL, found more frequent IJFs observed in some northern states of the USA, especially in New York and Montana where more than 1,400 ice events were reported by 2005. From the Canadian Disaster Database, they found 44 out of 83 IJFs occurred in New Brunswick, Ontario, and Quebec of eastern Canada. In recent years, under climate warming impact, flood damages are expected to increase significantly (Winsemius et al., 2016), and in south-eastern and western Canada (Janowicz, 2010), Yukon, British Columbia, and Alberta, IJFs are shifting toward earlier dates, whereas Atlantic Canada is experiencing delayed timing in IJF events, generally with larger shifts in regulated rivers (+10 to −12 days per decade) than unregulated rivers (±7 days per decade), where trends in IJFs are also correlated with the basin size, with smaller basins showing larger variabilities. However, less sensitivity to anthropogenic factors is found in regulated rivers.

The Ice Jam Database (https://rsgisias.crrel.usace.army.mil/icejam/) developed at the CRREL of Hanover, NH in 1992 contains over 20,000 records of both historic and current ice jams that have occurred across the USA since 1780. It was intended to provide timely ice jam information, and to assist decision-makers in long-term planning to alleviate damages caused by ice-jam flooding, to significantly reduce the loss of life, the economic and environmental impacts of the floods. In Northern parts of the USA, ice jams can rapidly lead to flooding damages, leaving communities little time to prepare. Annual ice jams damages in the United States are estimated at $138 M (US 2015 dollars, White *et al.*, 2007). Recorded parameters include location (river name, state, city, year, coordinates), jam type, damages, gage stations, Hydrologic Unit Codes (HUCs), descriptions and relevant publications and photographs. To allow for a more complete picture of an ice-jam flooding event, a new timeline dataset consists of quantitative recording of stage variations during the ice jam including initiation, peak, and changes in flood stage status have been collected since 2014, which will be useful for emergency management and better responses during an event (Carr *et al.*, 2015).

6.6   Trends in river ice cover

Breakup on major rivers in European Russia (upper Volga, Oka, and Don) and western Siberia (upper Ob and Irtysh) have been shown to have advanced by an average 7–10 days/100 years, during ~1893–1985 (Soldatova, 1993) although some rivers in central and eastern Siberia (middle to lower Yenisey and upper Lena) had the opposite trend (later breakup dates). A study of records over 54–71 years from nine major Russian Arctic/sub-Arctic rivers has been carried out by Smith (2000). Compared with the longer-term and broad regional studies of Soldatova (1993), several opposing temporal trends were found for river-ice freeze-up. In the case of breakup, trend analysis failed to identify any statistically significant shifts in timing. Ginsburg and Soldatova (1997) analyzed trends over 160–286 years on the ten largest rivers in Russia. They found delays of 2–8 days per century in freeze-up and similar advances in breakup dates. Vuglinsky (2006) compared ice conditions on Russian rivers for the periods 1950–1979 and 1980–2000. He found later/earlier freeze-up/breakup dates by 5–7 days on the average in the rivers of northwest European Russia. For Siberian basins, freeze-up occurred 2–3 days later, and breakup 3–5 days earlier, on average. Changes in the maximum river-ice thickness between the same two periods, showed reductions of 2–5 cm on the Northern Dvina and Pechora, 4–6 cm on the lower Ob, Yenisei, and Lena, 8–12 cm on the middle Yenisei and Lena, and 7–9 cm on the upper Yenisei (Vuglinsky, 2006).

   Advances in long-term breakup dates have also been documented for rivers in northern Sweden/Finland and the eastern Baltic Sea/Scandinavian region according to Kuusisto and Elo (2000) and Prowse and Bonsal (2004). For 1709–1998 the trend toward earlier breakup was 13 days for the Tornio River in Lapland and 15 days for the Daugava River at Riga, Latvia (Kuusisto and Elo, 2000). In the Baltic countries, data from 17 rivers spanning 60–77 years show that breakup date has shifted earlier by 2.8–6.3 days per 10 years (Klavins *et al.*, 2009). Studies of the Tanana River in Alaska for 1917–2000 (Sagarin and Micheli, 2001) and the Yukon River at Dawson, northwestern Canada for 1896–1998 (Jasek, 1999) indicate that the average date of breakup has also advanced by approximately 5 days per century. For western and eastern parts of Canada, Zhang *et al.* (2001) observed a major spatial distinction, with the former showing trends toward earlier breakup over records from 1950 to 1998. Lacroix *et al.* (2005) also found a trend toward earlier breakup dates especially in west and southwest Canada. Breakup advanced by 1–2 days per decade with more rapid change toward the end of the twentieth century. A record of ice bridges across the St. Lawrence River at Quebec City for 1620–1910 shows that winters in the seventeenth to eighteenth centuries were warmer than in the nineteenth century (Houle *et al.*, 2007). During the period 1800–1910, winters were 2.4–4 °C colder than 1971–2000 with winter severity culminating between 1850 and 1900. In central Maine, the ice thickness on February 28 on the Piscataquis River decreased by 23 cm (45%) from 1912 to 2001 and the ice-out date advanced, by 0.21 days a^{-1} during 1931–2002 (Huntington *et al.*, 2003). The total number of days with ice-affected flow in nine unregulated rivers in northern New England decreased on average by 20 days from 1936 to 2000 (0.31 days a^{-1}) (Hodgkins *et al.*, 2005), with most of the decrease occurring after 1960. Twelve of 16 rivers studied had earlier last dates of ice-affected flow in spring, with the average last date advancing by 11 days from 1936 to 2000 (0.17 days a^{-1}).

On the Yellow River in China the location of earliest breakup has shifted downriver during 1950–2001 (JIang *et al.*, 2008). The duration of the river ice cover has decreased by 38 days at Bayangaole (~40° N, 107° E) during 1968–2001, and on the lower reaches of the river it has decreased by 12 days. In Mongolia, freeze-up is occurring 3–15 days later than in 1945–1955 and breakup is 5–20 days earlier (Punsalmaa and Nyamsuren, 2002). The autumn freeze-up is more delayed in the western part of the country (e.g., the Khovd River).

To determine the regional characteristics of mid-winter breakup events and their historical trends, Prowse *et al.* (2002) analyzed river ice conditions in the "temperate region" of North America. The southern edge was chosen as the 400 FDD isoline and the northern edge corresponded to an ice thickness of ~50 cm. An increasing trend in the number of mid-winter events through the twentieth century at the northern boundary was observed, particularly for the western region. The Atlantic and Central sub-regions have also experienced an increasing frequency of such events in recent decades.

Using 400,000 clear-sky Landsat images from 1984 to 2018, it was found that on average, the global extent of ice cover has declined by 2.5 percentage points, and the ice cover duration has declined by almost a week in the past three and half decades (Yang *et al.*, 2020; Derouin, 2020). Yang *et al.* (2020) also developed a climate model based on temperature and season, and calibrated and validated the model with observed river ice data. Driven under Representative Concentration Pathway (RCP) 8.5 climate scenarios, the model projects a mean decrease in seasonal ice duration by 6.10 ± 0.08 days per °C increase in the global mean surface air temperature. Conversely, the average river ice duration is projected to decline by 16.7 days for 2080–2100 compared with 2009–2029 globally, whereas under RCP4.5 it is projected to decline on average by 7.3 days. If rivers that never experience ice are not included, the projected decline in river ice cover duration will probably be over a month under RCP8.5 scenarios. Their results show that, globally, river ice is measurably declining and will continue to decline linearly with projected increases in surface air temperature toward the end of the twenty-first century.

6.7   Icings

Icings (or Aufeis from German) are sheet-like masses of ice that form on the ground surface or in fluvial channels when water seeps from the ground, a spring, or a river, onto the land or an ice surface during periods of subfreezing temperatures and accumulates in successive layers of ice (Figure 6.12). The term was first proposed by Muller (1947). The Russian term "naryn" is used for ground icings. River icings (naled in Russian) develop after a seasonal ice cover has formed. This mechanism prevails in high-gradient alpine streams as they freeze solid. Water flowing below the ice cover may be forced up onto the river ice by channel

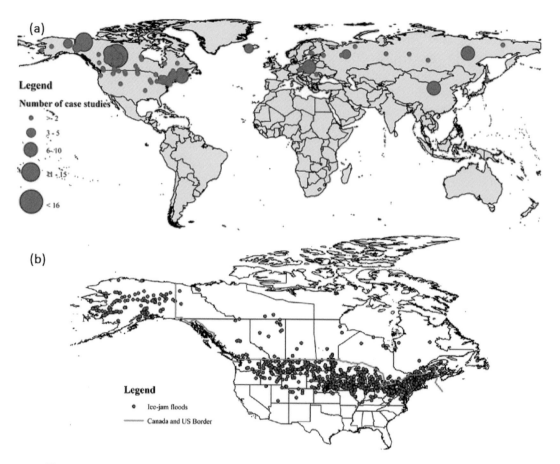

Figure 6.12 (a) Global geographic distribution and occurrence of case studies on IJF, and (b) reported IJF in North America (taken from Rokaya *et al.*, 2019)

restrictions or ice dams, building up further ice layers as it spreads out and freezes (Carey, 1973). Groundwater discharge is blocked by ice, perturbing the steady-state condition, and causing a small incremental rise in the local water table until discharge occurs along the bank at the top of the previously formed ice. The initiation of icing growth in a stream is due to the weight of snow on the initial ice cover, which increases the potential water level in the ice cover (Kane, 1981). If the hydrostatic head is higher than the surface of the icing, active growth can occur in early winter. Successive ice layers can lead to ice accumulations that are several meters thick. River icings are most common in flat braided streams channels where there is a change of gradient. The rate of freezing and the downstream extent of the icing are determined by the meteorological conditions. The surface of a river icing is broad and flat, occasionally with shallow terraces formed by successive overflows. The ice is in layers, or laminar; it is clear or white if there are trapped air bubbles and may have a faint yellowish or

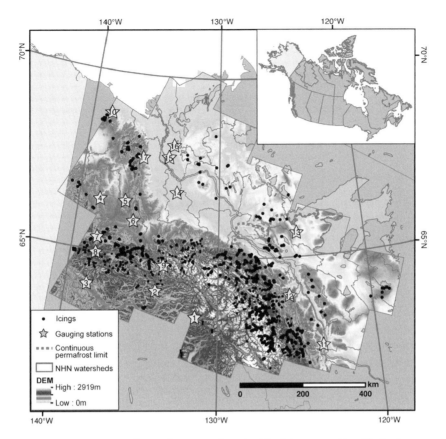

Figure 6.13 Icing occurrence in northwestern Canada. The icings were identified using a semi-automated approach and a dense stack of Landsat imageries. Active gauging stations and level-3 watersheds are also shown (taken from Crites *et al.*, 2020)

tan coloration from minerals in the source water. The size of a river icing depends on the magnitude of the water source and the topography around the river channel. River icings disintegrate through river channels that cut unto them and widen by thawing and collapse of the sides.

Ground icings usually occur where there is a break of slope or on a floodplain away from the river (Carey, 1973). Overall, they appear to be more frequent on south-facing slopes. Water appears on the ground surface from a seepage and freezes when the ground temperature drops below 0 °C.

They may thicken continuously or intermittently throughout the winter. Frost mounds may be associated with ground icings. They largely waste away *in situ* through melt on all sides.

Spring icings form in a variety of topographic situations – at the base of south-facing slopes, or from sources along the beds of large Siberian rivers crossing the permafrost zone. They generally continue to grow throughout the winter season. Spring icings are generally larger than ground icings. An exceptional case is the Ulakhan taryn in the Moma River (a tributary of the Indigirka River) in eastern Siberia; it grows to 25 km long, 5–8 km wide and up to 4 m thick in places (Chekotillo et al., 1960; Carey, 1973). In spring, these icings melt away on all sides but may also be breached by the channels from the spring water. A report on icings along 1,600 km of the Alaska Highway was given by Thomson (1966). He noted that a flow of only 26 l per minute can form an icing 25 cm thick covering 0.4 ha within a month.

Icings are widespread east of the Colville River on the North Slope of Alaska in the continuous permafrost area, but are less abundant to the west according to Harden et al. (1977). They occur where stream channels are wide and often braided. Their distribution is related to changes in stream gradient and to the occurrence of springs. Large icings, such as occur on the Kongakut, Sagavanirktok and Canning rivers, may survive the summer melt season. Icings in central Yakutia on the Anmangynda River and Ulakhan-Taryn River occupy ~6 km^2 and 0.77 km^2, respectively. Their thickness is about 2–3 m and lengths 3–7 km (Savko, 1973). According to Sokolov (1973) there are some 10,000 icings in northeastern Siberia; many are <0.1 km^2. East of the Lena River, in the Verkhoyano-Chukotskaya mountains, however, areas of separate gigantic icings are measured in dozen of square kilometers according to Tolstikhin (1968). In some areas they occupy up to 3% of a river basin. Other regions where they occur include the river valleys of the mountain ranges of Chersky, Selennyakh, Suntar-Khayata, the Buordakhskii Massif, and the Stanovoe Uplans (Romanovskii et al., 1978). The total area in northern Russia is estimated to reach 128,000 km^2 according to Kotlyakov (1997, vol. 2). They are predominantly river icings:

Type	Area (km^2)	Volume (km^3)
River	101,165	240,23
Groundwater	26,736	252

The regional distribution in the former Soviet Union (fSU) is illustrated in Table 6.1.

Alekseyev et al. (1973) provide detailed regional accounts of icings in Siberia. In the Ulakhan-Taryn valley of central Yakutia, where the January temperature averages around −40 °C, icings began to form during 1964–1965 to 1966–1967 anywhere from early December to early February (Gavrilova, 1973). Growth of ≥5 cm per day occurs with an air temperature of −36 °C to −25 °C. Icings are usually annual and thawing takes place mainly during May–June, although some icings are perennial. The icing in the Moma Basin, northeast of the Chersky Range, is up to 76–112 km^2 in area, although in some years it has virtually melted by August (Romanovskii et al., 1978). Osokin (1973) provides data on the altitudinal distribution of naleds in eastern Siberia. In the mountains of eastern Yakutia they are generally developed between 500 and 1,000 m altitude, except on the north slope of the Suntar-

Table 6.1 The spatial characteristics of icings in the former Soviet Union (from Kotlyalov, 1997)

	Rivers		Number	Groundwater	
	Area (km^2)	Volume (km^3)		Area (km^2)	Volume (km^2)
TOTAL fSU	101,165	24,023	60,450	26,736	252
TOTAL Arctic drainages	84,335	22,648	38,290	22,366	245
Kola	385	0.17	140	1.4	0.01
E.EurRus	1,030	0.44	180	27.4	0.04
Urals	480	1.58	430	216	0.58
W. Siberia	4,040	2.06	1,700	252	0.37
E. Siberia	26,150	1,100	13,530	3,840	10.60
Sayan-Altai	8,990	342	1,770	2,210	3.24
Baikal	12,820	521	1,880	4,720	6.45
NE Asia	26,740	1,100	8,350	11,000	24.00

Khayata range where they are at 1,100–1,300 m. In southern Yakutia they are between 600 and 1,200 m, in the Trans-Baikal up to 1,200–1,250 m, in eastern Sayan above 1,200 m, and in the Tian Shan between 2,500 and 4,000 m.

In a 618,400 km^2 area of northwestern Canada (Figure 6.13), Crites *et al.* (2020) investigated the distribution of icings from 573 Landsat images (1985–2017) and their relations with permafrost types, winter air temperature and baseflow conditions. Using hydrometric data they estimated the winter baseflow contribution to the total annual discharge of 17 rivers in the study area. In catchments with continuous permafrost, the 1,400 mapped icings occur preferentially at the foothills of heavily faulted karstic mountainous regions, where winter baseflow and its contribution to the annual discharge was lower than in discontinuous permafrost but showed an increasing trend over the 1970–2016 period. As icing conditions are sensitive to degrading permafrost and projected increase in groundwater discharge, a northward shift in the continuous permafrost boundary would likely reduce icing occurrences or perhaps a shift in their distribution if northerly regions become more conducive to icing development. As icings store on average 18% of winter baseflow contribution to total annual discharge of rivers, with an increasing storage trend along a latitudinal gradient, the change in icings conditions would affect local rivers hydrologically.

Seepage taliks are discussed by Romanovskii *et al.* (1978). There are two types: one where it is closed and extends into the seasonally thawed layer that freezes in winter, and another where an open or closed seepage talik forms an uninterrupted strip that increases downward through the section. In the latter case the dimensions of the icing depend strongly on the winter conditions. In a severe winter with little snow much of the groundwater issues at the surface forming large icings.

Sokolov (1973) determined that it takes about 210 days for the area of an icing to reach 0.2 km^2 with volumes >4 million m^3. Initially, the area increases faster than the volume, but later in the winter season the reverse occurs. Based on data for 310 naleds, the maximal volume (V) at the end of winter was found to be:

$$V = 0.96A^{1.09} \tag{6.10}$$

where V is in thousand m^3, and A = area in thousand m^2.

Early scientific work by Podyakonov (1903) developed a descriptive formula for the variables involved in icing formation (I). This was later modified as follows:

$$I = \frac{F}{B} \cdot \frac{A}{Q} \cdot \frac{w}{z} \cdot \frac{X}{T} \cdot \frac{1}{V} \tag{6.11}$$

where F = frost intensity (magnitude of negative air temperature); B = area of river basin above the reach under consideration; A = cross-sectional area of the open stream channel; w = stream width; z = stream depth; X = depth of soil freezing on the river banks; T = local water temperature in the river; and V = cross-sectional area of the unfrozen valley alluvium.

Sumgin (1941) published a conceptual model of a ground icing where a talik (see Chapter 5, p. 234) is confined by underlying permafrost and seasonal freezing from above in autumn. Pressure builds up in the talik as the upper portion freezes and the volume increases. There is flexure in the overlying frozen ground and a frost mound (bugry in Russian) develops. This may eventually fracture releasing the pressurized water on to the surface where it spreads out and forms a ground icing. Ice accumulates until the water is exhausted or temperatures rise above freezing. Petrov (1930) made field observations on the Amur-Yakutsk highway and performed laboratory experiments showing that high pressures could occur in groundwater subject to frost penetration into the ground. Chekotillo *et al.* (1960; transl. 1965) published a survey of current knowledge updating a report published 20 years earlier through the Permafrost Institute of Yakutsk.

Note 6.1 Mohr–Coulomb material

Mohr–Coulomb theory describes the response of brittle materials such as rubble piles, to shear stress as well as normal stress (Wikipedia: http://en.wikipedia.org/wiki/Mohr-Coulomb_theory). Coulomb's friction hypothesis is used to determine the combination of shear and normal stress that will cause a fracture of the material. Mohr's circle is used to determine which principal stresses will produce this combination of shear and normal stress, and the angle of the plane in which this will occur. The Mohr–Coulomb failure criterion represents the linear envelope that is obtained from a plot of the shear strength of a material versus the applied normal stress. This relation is expressed as:

$$\tau = \sigma \tan \varphi + c,$$

where τ is the shear strength, σ is the normal stress, c is the intercept of the failure envelope with the τ axis, and φ is the slope of the failure envelope. The quantity c is often called the cohesion and the angle φ is called the angle of internal friction. For ice rubble fields during their final phase of development $\varphi \sim 56$–$58°$, much higher than ordinary granular materials ($\sim 45°$) (Beltaos, 2010).

Note 6.2 The Manning coefficient

The Manning equation specifies open channel water flow

$$V = \frac{k}{n} R_h^{0.667} S^{0.5},$$

where V is the cross-sectional average velocity (m s^{-1}), k is a conversion constant equal to 1.0 for SI units, R_h is the hydraulic radius (m) $= A/P$ where A is the cross-sectional area of flow (m^2), P is the wetted perimeter (m), S is the slope of the water surface (m m^{-1}), n is the Manning coefficient – an empirical dimensionless coefficient that depends on surface roughness and channel sinuosity.

REVIEW QUESTIONS

(1) Name two climatic factors that contribute to the formation of lake ice of certain thickness. How are white and black ice formed in lakes? According to the Stefan relation for black ice (curve of maximum ice thickness shown in Figure 6.2), how thick in cm will the black ice grow out of accretion under a cumulative freezing degree day (FDD) of 400, and the corresponding white ice thickness under the same FDD? Why lake ice thickness increases with latitude, for example, in Russian Far East and eastern Siberia, lake ice thickness increases from ~100 cm at latitude 45° N to 180 cm at 65° N?

(2) Why does lake ice float on top during winter, given the density of water is maximum at 4 °C? Conversely, why does the warm water warmed by the sun during summer floats on top of colder water beneath? Due to the density differences between surface water (epilimnion) and near bottom water (hypolimnion), can we expect much vertical mixing in lakes during summer? How's about mixing in spring and autumn?

(3) Several 1-D energy balance models of lake ice growth have been developed and applied to lakes in Europe and North America. Briefly explain the four sub-models of the 1-D energy balance model of Liston and Hall (1995a) and the thermal energy balance at the ice-water interface shown in Equation 6.3. Name five climate variables typically needed to run a 1-D energy balance lake ice model.

(4) For large lakes such as the Great Slave and Great Bear lakes of Canada, how can passive microwave DMSP SSM/I data be used to monitor the lake ice cover, especially regarding ice freeze-up and breakup dates? Name two other types of satellite images used to study breakup dates.

(5) Explain how lake ice characteristics such as the timing of lake ice breakup, freeze-up, and ice cover durations are useful indicators of climate variability and change, particularly to environmentally sensitive, Arctic lakes? For example, the long-term average maximum ice covers of Lake Erie vary from 95% in severe winters to only 14% in mild winters. Also explain how large lake ice variability is linked to climate anomalies such as Pacific North American Pattern and Arctic Oscillation.

(6) What is the range of projected shifts in breakup and freeze-up dates in lakes across the North American Arctic between 1961–1990 and 2041–2070 simulated by the lake ice model CLIMo, and the change in water-clear-of-ice (WCI) dates of lakes in the Canadian Arctic Archipelago over 1997–2011 detected using RADARSAT-1/2 ScanSAR Wide Swath, ASAR Wide Swath, and Landsat data?

(7) What is a river ice jam, typical northern river sites that are prone to ice jamming, and how could ice jam cause a higher flood risk?

(8) Under the impact of climate warming, what are two key types of river ice breakup, and how will they differ in nature and severity resulted from different relative influence of meteorological and hydraulic factors?

(9) Name key heat fluxes occurring at the water–ice and ice–air interfaces, and two prerequisites for the formation of frazil ice.

(10) Will ice-jam floods (IJF) common in many northern countries such as in Canada, the USA, and Russia likely become more severe under climate change impact and why, since less severe, thermal breakup instead of mechanical breakup, could occur more often under a warmer climate? Further, will IJF likely be more severe in regulated or unregulated rivers?

(11) Briefly explain the mechanisms behind the formation of icings in high-gradient alpine streams and in flat braided streams, and the effects of terrestrial factors such as continuous/discontinuous permafrost boundaries, and winter baseflow on the occurrences of river icings, and a shift in their distributions.

Part II

The marine cryosphere

7 Sea ice

7.1 History

The earliest account of sea ice is due to Pytheas, a Greek sailor who encountered it southeast of Iceland in 325 BC (Sturm and Massom, 2010). Later encounters were made by Celtic monks in the northwest North Atlantic in AD 550 and 800 (Weeks, 1998). In the seventeenth to nineteenth centuries, whalers and sealers operated in Arctic waters of the North Atlantic, Barents Sea, and Greenland Sea and Scoresby (1820), a whaling captain, published a notable book on ice and ocean conditions in the Greenland Sea. A remarkable expedition was F. Nansen's drift across the Arctic Ocean in the Fram, 1893–1896; his observations of the vessel's motion led to V. W. Ekman's theory of the spiral of ocean currents with depth. The first book on the physics of sea ice was written by Malmgren (1927) based on observations made during the Norwegian Maud north polar expedition, 1918–1925.

Koch (1945) prepared an index of ice conditions off Iceland from AD 1150 and this analysis was updated by Wallevik and Sigurjónsson (1998) and Lassen and Thejll (2005). Ice edge positions were first documented in the Nordic seas in the 1550s (Vinje, 1999). Winter maximum ice extent in the Baltic Sea was tabulated from 1720 to 1956 by Betin and Preobrazhensky (1959), based on Speerschneider (1915, 1927), together with the opening date of the port of Riga from 1710, with less complete data from AD 1530. Sea ice terminology was initially developed by Scoresby (1820). Some of Scoresby's terms were carried over into Markham and Mill's (1901) contribution to the Antarctic manual used by the 1901 British Antarctic expedition. The Danish Meteorological Institute in Copenhagen charted sea ice for the months of April–August (or September) in the Arctic, mainly the North Atlantic sector, from 1898 (Kissler, 1934). The series spanned 1898–1939 and 1946–1950, with charts available from 1877 (Ryder, 1896). Finland began ice services for the Baltic Sea in the 1890s. The Soviet Union began regular summertime reconnaissance flights in the early 1930s mapping the sea ice conditions in the Siberian Arctic (Mahoney et al., 2008). The data set "Sea Ice Charts of the Russian Arctic in Gridded Format, 1933–2006" (with a gap from 1993 to 1996) is available at http://nsidc.org/cgi-bin/getmetadata.pl?id=g02176 (Arctic and Antarctic Research Institute, 2007).

In North America, similar airborne mapping only started in the 1950s. Major work on Arctic climate, oceanography, and sea ice was carried out by the Soviet North Pole (NP) drifting stations (Box 7.1). The NP-1 operated during 1937–1938 and the program resumed with NP-2 in 1950–1951. After a break until 1954, stations were manned until NP-31 in July 1991 and then the program was restarted in 2003 with NP-32. The Russian stations were mostly established on thick ice floes, while the United States installed camps on ice islands such as T-3 (Fletcher's ice island, named after J. O. Fletcher its discoverer) and the Arctic Research Laboratory Ice Stations (ARLIS) in the 1950s and 1960s. The first US station on sea ice was Station Alpha during the International Geophysical Year, 1957–1958 (Untersteiner and van der Hoeven, 2009). Another Soviet program carried out each spring, named SEVER (north), involved aircraft landings on the sea ice to measure snow depths and ice thicknesses in 1937, 1941, 1948–1952, and 1954–1993; up to 200 landings were made each year with 23 parameters measured (Romanov, 1995). Arctic-wide sea ice data for 1950–1994 are available in an atlas on CD-ROM (Arctic Climatology Project, 2000).

Remote sensing has revolutionized the study of sea ice since the mid-1960s. Massom (2009) provides a detailed overview of the different sensors and applications. Prior to the availability of imagery over the polar ice packs our knowledge was limited to observations from ships sailing along the ice margins or frozen into the ice, NP drifting stations, and expeditions such as W. Herbert's British trans-Arctic Expedition of 1968–1969 reported by Koerner (1970, 1973).

Box 7.1 North Pole drifting stations

A landmark program of Arctic sea ice, climate, and oceanographic research has been the Soviet–Russian operation of NP drifting stations. The first station NP-1 was set up in May 1937 by several ski-equipped aircraft landing on the ice near the NP. The planes flew from Moscow and made refueling stops en route, the last on Franz Josef Land. The expedition was led by Ivan Papanin. During 9 months NP-1 drifted 2,850 km. Measurements were made of ocean depth, water temperature, and meteorological conditions, and water samples were collected from different levels. By February 1938, the station had drifted out to the Greenland Sea and the icebreaker Yermak evacuated the camp. The program resumed after World War II with NP-2 in 1950–1951. Following a break until 1954, stations were manned until NP-31 in July 1991 and then the program was restarted in April 2003 with NP-32 to NP-40 in October 2013. The latest NP-2015 only drifted 714 km over April–August, 2015 because of fast thawing of the ice. Since it had been difficult to find a suitable ice floe to place a station camp in recent years, ice camps have been replaced with a drifting research vessel as a ice-resistant self-propelled platform (IRSPP) named "North Pole" in Dec of 2020.

The US–Canadian–Japanese Arctic Ice Dynamics Joint Experiment (AIDJEX) was organized and conducted by the University of Washington in the Beaufort Sea during 1970–1978 (Untersteiner *et al.*, 2007), see http://psc.apl.washington.edu/aidjex/. It led to early modeling of sea ice behavior (Pritchard, 1980). In the 1970s to 1980s, significant research on the physical properties of sea ice was undertaken at the US Army Cold Regions Research and Engineering Laboratory (CRREL) in Hanover, NH. Beginning in 1979, the Arctic Buoy Program was initiated by the University of Washington. In 1991 this became the International Arctic Buoy Program (IABP), which now involves eight nations. The Marginal Ice Zone Experiment (MIZEX) was conducted with aircraft and ships in the Greenland Sea during 1983, 1984, and 1987, and the Coordinated Eastern Arctic Experiment (CEAREX) in 1988–1989. Stress measurements were made on a multiyear floe whose deformation was analyzed. The data are available from http://nsidc.org/data/nsidc-0020 .html.

From 1982 to 2009, the German research vessel Polarstern completed 25 Arctic and Antarctic cruises. The ship travels to the Arctic during the boreal summer and spends her austral summers in the Antarctic, although several expeditions were also carried out during polar winters. Results are primarily reported in the Berichte zur Polar-und Meeresforschung. The Surface Heat Budget of the Arctic Ocean (SHEBA) experiment was performed from September 15, 1997 to October 31, 1998 when the Canadian icebreaker Des Groseillers was frozen into the ice of the Beaufort Sea and the ship drifted for 12 months. Landfast ice data for the Canadian Arctic have been assembled and analyzed by Brown and Cote (1992).

As part of the International Polar Year (IPY) activities, 2007–2008, the US National Science Foundation sponsored 34 projects under the Arctic Observing Network (AON) program. These projects have collected a wide range of Arctic Ocean and ice data available at http://aoncadis.ucar.edu/home.htm.

In the Antarctic, whalers and sealers noted ice edge locations in the 1920s and 1930s, but scientific interest in Southern Ocean ice was delayed until all-weather passive microwave data became available in 1973. Remote-sensing data on sea ice are discussed later. A Russian–US expedition carried out sea ice and oceanography investigations in the western Weddell Sea by drifting on an ice floe (71.4–65.6° S, ~52° E) from February through June 1992 (Myel'nikov, 1995). The Polarstern drifted in the same area from November 2004 to January 2005 (Hellner *et al.*, 2008).

Large-scale sea ice research facilities include: the environmental ice basins at the US Army CRREL in Hanover, NH, Helsinki University of Technology, and the HSVA Hamburg Ship Testing Ice Basin. The CRREL has a unique Ice Engineering Test Basin, which is a large refrigerated room where multiple ice sheets can be grown and tested. Opened in 1978, it is designed primarily for large-scale modeling of ice forces on structures such as drill platforms and bridge piers, and for tests using model icebreakers. There have been numerous field experiments on sea ice in the North American Arctic in connection with oil and gas exploration and production; an overview of Beaufort Sea studies is provided by Timco and Frederking (2009).

There are a number of books on sea ice covering most aspects of the subject, one of the first being by Zubov (1943); more recent ones are by Doronin and Kheisin (1977), Untersteiner (1986), Carsey (1992), Wadhams (2000), Leppäranta (2005), Thomas and Dieckmann (2010), Weeks (2010), Shokr and Sinha (2015), Lemieux *et al.* (2017), and Johannessen *et al.* (2020).

7.2   Sea ice climatology and characteristics

As millions of km^2 of frozen seawater floats on the ocean surface, sea ice forms and melts with the polar seasons, affecting both human activity and the biological habitat. In the Arctic, some sea ice persists over several years, while almost all Antarctic sea ice is "seasonal" because it melts away and reforms annually. Other than playing a vital role in the habitats of birds and marine life, sea ice also plays a crucial role in regulating polar climate, especially in the Arctic. Sea ice regulates exchanges of heat, moisture, and salinity in the polar regions by insulating the relatively warm ocean water from the cold polar atmosphere except cracks in the ice that allow exchange of heat and water vapor from ocean to atmosphere in winter.

Sea ice occupies on average about 7% of the global ocean, primarily in the Arctic and Antarctic. It undergoes large seasonal variations in extent (see Table 1.1) and plays a major role in the climate of high latitudes because of its high reflectivity (albedo) of incoming solar radiation and its insulating effect on the underlying ocean surface. Sea ice extent, usually defined as the area of ocean where there is at least 15% or more of sea ice concentration, tends to be of greater interest to the Arctic scientific community than other aspects of sea ice because satellites measure extent more accurately than other measurements, such as thickness. Arctic sea ice extent reaches its maximum extent on March each year, marking the beginning of the sea ice melt season until September of the same year. Sea ice typically covers about 14–16 million km^2 in late winter in the Arctic and 17–20 million km^2 in the Antarctic. On average, the seasonal decrease is much larger in the Antarctic, with only about 2–4 million km^2 remaining at summer's end, compared to about 7–9 million km^2 in the Arctic. In recent years, Arctic minima have been only 3.5–5 million km^2. In September 2007, only a little over 4 million km^2 remained following exceptional summer melt and a cumulative export of multiyear ice (MYI) through the Fram Strait. There was only slightly more ice remaining at the end of summers 2008, 2009, and 2010, but in 2012 it dropped to about 3.5 million km^2, the lowest record so far.

Arctic sea ice extent for March 2019 averaged 14.55×10^6 km^2, and together with 2011 is the seventh lowest extent in the 40-year satellite record. This is 0.88×10^6 km^2 below the 1981–2010 average and 0.26×10^6 km^2 above the lowest March average of 2017. The Bering Sea typically reaches its maximum ice extent in late March or early April but in 2019, the maximum occurred in late January and it was 34.5% below the 1981–2010 average maximum. Trends in ice extent are discussed in Section 7.7. In the Antarctic summer, the sea ice melts back to the coast in various locations between 20° E and 160° E.

Box 7.2 Sea ice off East Asia and in the Caspian and Aral seas

Sea ice in the Northern Hemisphere is farthest south off East Asia reaching 42° N off Hokkaido and 38–40° N in Laizhou Bay, Bohai Bay, and Liaodong Bay, China. The salinity in Bohai Bay is ~28–30 PSU (practical salinity units) and its average depth is 18 m. Ice forms in late November in Liaodong Bay and late December in Laizhou Bay and disappears in mid-March in Liaodong Bay and in late February in Liazhou and Bohai bays giving a duration of ~120 days in the north and northeast and 55 days in the southwest (Gong et al., 2007; Ning et al., 2009). Landfast ice in Bohai–Liaodong bays extends 1–4 km off the coast and has a thickness of 15–40 cm; drifting ice is 10–25-cm thick (Gu et al., 2005). According to Ning et al. (2009) in a light (heavy) ice year the thickness is 10–30 (30–100) cm. The freezing duration has decreased by 18 days from 1953/1954 to 2003/2004. The years of most severe ice conditions were in the mid-1950s, late 1960s, 1976/1977, and 2000/2001. Since 1990, ice conditions have mainly been light. Historically, there were very heavy ice conditions in 1935/1936, 1944/1945, and 1946/1947.

The Sea of Okhotsk is, generally, almost entirely covered with sea ice from late December to May (Ohshima et al., 2006). Maximum thicknesses range from about 40 cm around the Kuril Islands (45° N) to >100 cm at 58° N. Ship-based hydrographic observations show an average maximum sea ice thickness inferred from the spring salinity profiles at 350 stations during the past 80 years of 0.76 m (Ohshima and Riser, 2010). Salinity in the Sea of Okhotsk is lowered by the inflow of the Amur River. Ice formation is strongly dependent on growth in the coastal polynya areas in the northwestern parts of the sea. From there, ice is advected south and east by the wind and ocean currents. In the south, ice area is lost by rafting and ridging. The thickness is typically 0.7–1.2 m in the west and southwest and relatively small in the central and northeastern parts of the Sea of Okhotsk. South of 53° N, where a relatively large number of hydrographic observations have been made, the average thickness is estimated to have decreased from 0.91 m in the 1950s–1960s to 0.83 m for the 1970s–1980s, and 0.66 m for the 2000s. Sea ice thickness and volume fluctuations for 1987–1999 were analyzed by Kazutaka et al. (2001). Since 1995, sea ice extent and thickness have decreased. In 1996 and 1997 ice volume was only about 57% of that in 1988.

The northern Caspian Sea (48–49° N) is only 5–6-m deep and has a salinity of ≤ 1.2 PSU due to the Volga River inflow. Ice conditions were mapped from aerial surveys from 1927 to 2002 (Kouraev et al., 2004). Aerial surveys of the Aral Sea were made between 1950 and 1985. Updated records have been obtained from passive microwave data for 1988–2002 and TOPEX/Poseidon 13.6 GHz backscatter for 1992–2002. Maximum ice area in the Caspian Sea decreased from ~100,000 km^2 in the 1950s to 40–60,000 km^2 in the 1980s to 1990s. In 1999–2000 there was a low of 15,000 km^2. The duration of the ice season in the eastern part of the Caspian Sea ranges from 70 to 145 days, with typical values around 100–110 days. In the western part, the duration varies from 20 to 120 days, with a decreasing trend since the mid-1990s. In the Aral Sea there were maximum extents ranging from 14,000 to 44,000 km^2 in the 1950s to 1970s decreasing to <17,000 km^2 in the 1980s, partly as the sea shrank greatly in size and split into two parts, and also due to the much increased salinity that depressed the freezing temperature to −9 °C. Aral Sea has receded to three separate lakes since the end of the last century.

The seasonal cycle in Arctic sea ice extent is asymmetrical, with a more rapid retreat in spring and summer and a slower advance in autumn and winter. Eisenman (2010) shows that the asymmetry is a consequence of the distribution of continents; Arctic coastlines block southward ice extension in winter, but have little effect in summer. If we take the latitude of the Arctic sea ice edge, averaged zonally over locations where it is free to migrate, we find that the latitude of the zonal-mean sea ice edge during 1978–present has followed an approximately sinusoidal seasonal cycle, with a 2.5 month lag behind that of incoming solar radiation (Eisenman, 2010). Ice in the Northern Hemisphere extends far south in winter off eastern North America and East Asia as a result of the cold waters and cold air outbreaks from the eastern sides of the continents. Information on the sea ice off East Asia is given in Box 7.2, together with details of the ice conditions in the Caspian and Aral seas, the furthest south extent in western Eurasia (Granskog *et al.*, 2010).

Remote sensing

Sea ice in the Eurasian Arctic seas was routinely mapped by visual reconnaissance from aircraft flights in the Soviet Union starting in July 1933 and continuing until 1992 (Borodachev and Shilnikov, 2003). The coverage was initially only in late summer but by 1950 it was continuous throughout the year. From the 1950s on, 30–40 aircraft made 500–700 flights annually (Johannessen *et al.*, 2007). Side-Looking Airborne Radar (SLAR) mapping was used from the mid-1960s and in 1983 SLR was available from the Okean 01 series of satellites. Ice concentration and ice type were mapped at 10- to 30-day intervals. In 1940, the Canadian Department of Transport Marine Services began an "Ice Patrol" in the Gulf of St. Lawrence. Summer patrols in the Arctic began in 1957. Aerial ice reconnaissance data for the Canadian Arctic Archipelago are contained in atlases for the summer seasons of 1961–1978 (Lindsay, 1982). These give approximately six to ten charts for each year showing the existing fractional concentration of three ice types, and ice forms such as ridging. The first SLAR used for ice reconnaissance in Canada was installed in 1978; it had a 100-m resolution; SLAR measurements continue to be used along the eastern coast of Canada. Airborne Synthetic Aperture Radar (SAR) was introduced in 1990 with digital processing techniques and resolution in the range 5–30 m.

The use of satellite data from Very High Resolution Radiometer (VHRR) visible and infrared sensors began in 1966. In 1970, the National Oceanographic and Atmospheric Administration (NOAA) launched the first of a series of satellites with VHRR having improved resolution of 1 km. In 1978, the first satellite carrying the Advanced Very High Resolution Radiometer (AVHRR) was launched. This series continues to this day. In 1999, a classified US National Technical Means program called Medea began collecting 1-m--resolution imagery of sea ice at four sites around the Arctic basin, and two additional sites were added in 2005. This program continues during the summer season and a report of the Polar Research Board of the National Research Council recently examined the merits of releasing these images (Committee on Climate, Energy, and National Security, 2009). The report called for the priority release of images for 2007–2008 during the IPY, and for the Barrow region and Beaufort Sea, where there have been large changes in sea ice cover. These images are available at http://gfl.usgs.gov/ArcticSeaIce.shtml.

In December 1972, the National Aeronautics and Space Administration (NASA) launched the Electrically Scanning Microwave Radiometer (ESMR) on Nimbus 5. Until May 1977, this provided single channel horizontally polarized maps at a frequency of 19 GHz. Its ability to operate in darkness and through cloud cover yielded the first comprehensive maps of polar sea ice extent for 1973–1976 (Zwally *et al.*, 1983; Parkinson *et al.*, 1987). The brightness temperature data, gridded to 25 km (Parkinson *et al.*,1999) are available at http://nsidc.org/data/docs/daac/nsidc0077_esmr_tbs.gd.html

Figure 7.1 provides examples of late winter and late summer ice cover in the two hemispheres. The data are of ice concentration, which is the fraction, or percentage, of ocean area covered by sea ice. Ice extent is conventionally defined with reference to a limit of 15% ice

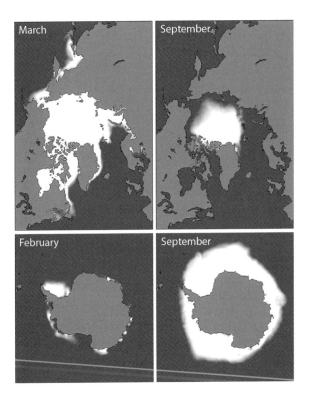

Figure 7.1 Sea ice climatologies: Arctic and Antarctic sea ice concentration climatology from 1981 to 2010, at the approximate seasonal maximum (March–September) and minimum (September–March) levels based on passive microwave (PM) satellite data. Image provided by National Snow and Ice Data Center, University of Colorado, Boulder (A black and white version of this figure will appear in some formats. For the color version, please refer to the plate section.)

concentration. Sea ice concentration can be estimated from passive microwave brightness temperature data because sea ice and water have differing passive microwave signatures. Water has a highly polarized signature within a certain frequency band (i.e. its brightness temperature in the vertical (V) polarization channel is higher than that in the horizontal (H) polarization), while sea ice does not. Most algorithms use some form of polarization ratio (PR) and a mixing diagram with brightness temperature "tie points" to estimate the concentration of sea ice within the field of view of the sensor. Tie points for ice and open water are the set of brightness temperature values that correspond to ice concentrations of 100 and 0%, respectively. This is elaborated later.

In October 1978, the Scanning Multichannel Microwave Radiometer (SMMR) was launched on Nimbus 7 and operated until August 1987. The instrument had three channels, two with dual polarization. Frequencies 18 and 37 GHz were used in various algorithms to derive sea ice concentrations for first-year ice (FYI) and MYI (Gloersen et al., 1993). The records continued with the Special Sensor Microwave Imager (SSM/I) on Defense Meteorological Satellite Program (DMSP) satellites. These instruments had four frequencies including 19 and 37 GHz. The spatial resolution of each channel is shown below.

Frequency (GHz)	Polarization	Resolution (km)	
		Along track	Along scan
19.35	V H	69	43
22.235	V	60	40
37.0	V H	37	28
85.5	V H	15	13

The NSIDC products are made available at 25-km resolution in polar stereographic and Equal-Area Scalable-Earth (EASE) grid formats (http://nsidc.org/data/ease/).

Sea ice can be discriminated in the microwave regime through differences in the emissive characteristics between ice and ocean; in general, sea ice is more emissive than open ocean. Use of combinations of frequencies allows more accurate discrimination between ice and ocean, as well as the ability to estimate fractional ice cover within regions of mixed ice and water. The NASA Team algorithm (Cavalieri et al., 1984) uses a PR and a gradient ratio (GR). The PR is

$$\mathrm{PR}_{[19V/H]} = \frac{T_B[19V] - T_B 19H}{T_B[19V] + T_B[19H]} \tag{7.1}$$

and the GR is

$$\mathrm{GR}_{[37V/19V]} = \frac{T_B[37V] - T_B[19V]}{T_B[37V] + T_B[19V]} \tag{7.2}$$

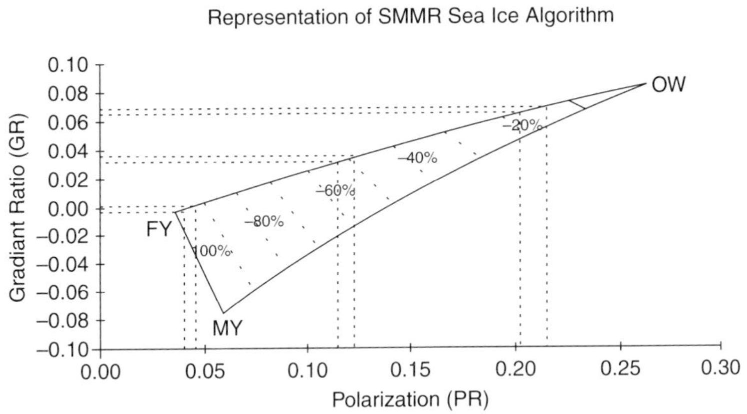

Figure 7.2 First-year ice (FYI), multiyear ice (MYI), and open water (OW) have typical values of PR and GR, as shown by site observations and airborne measurements. Values in between ice signatures and OW are interpreted as ice concentration (Gloersen *et al.*, 1993)

Where T_B is the brightness temperature, H and V are, respectively, horizontal and vertical polarizations.

The PR is small for ice and large for water while the GR is small for FYI but large for MYI. Figure 7.2 illustrates these differences (Cavalieri *et al.*, 1984). Combinations of PR and GR enable the brightness temperature (*TB*) signatures to be interpreted as ice type and there are some eight or so algorithms in use for this purpose. The algorithms adopted by NSIDC for its Sea Ice Index use the 19 GHz V, 19 GHz H, and 37 GHz V SSM/I channels, and the 18 GHz V, 18 GHz H, and 37 GHz V SMMR channels.

Another approach is used by the Bootstrap algorithm (Comiso, 1986), which employs linear combinations of 19 and 37 GHz frequencies at both horizontal and vertical polarizations to estimate fraction ice coverage. The NASA Team and Bootstrap, as well as other algorithms, require empirically derived "tie points," or coefficients for pure surface types (100% ice and 100% water). There are many uncertainties and limitations in using passive microwave data for sea ice detection. There are errors due to ambiguous emissivity signals, particularly from surface meltwater during summer and for thin ice.

Representation of SMMR sea ice algorithm

The available passive microwave frequencies can discriminate between at most three ice types, but often a region may have more than three unique microwave signatures. There can be erroneous ice retrievals over open water due to increased ocean surface emissivity from wind roughening. Atmospheric emission from cloud liquid water may be a factor in some conditions. Perhaps the major limitation is the low spatial resolution of passive microwave sensors, with footprints of 12–50 km. Thus, individual floes cannot be imaged and the ice edge location can be estimated to several kilometers' accuracy at best. However, passive microwave data are a valuable source of sea ice information because it is sunlight independent and is generally not affected by clouds and other atmospheric sources. Also, passive microwave sensors have wide swaths and sun-synchronous orbits that provide frequent coverage of the polar regions. Table 7.1 lists the primary and secondary remote-sensing instruments used to determine the principal sea ice characteristics. Passive microwave data provide a consistent and nearly complete daily record of sea ice conditions in both the Arctic and Antarctic since late 1978. More recent algorithms employ the higher frequency passive microwave channels on SSM/I and AMSR-E to obtain better spatial resolution and to resolve some of the surface ambiguities. These include the NASA Team 2 (Markus and Cavalieri, 2000) and the ARTIST (Spreen *et al.*, 2008). The NASA Team 2 algorithm is used for the AMSR-E standard sea ice product (Comiso *et al.*, 2003). Andersen *et al.* (2007) analyze sea ice concentration from the SSM/I for winter 2003–2004 in the central Arctic using seven different algorithms compared with 57 SAR scenes. They find that algorithms, using primarily 85 GHz information, consistently give the best agreement with observations. The 85 GHz information is more sensitive to atmospheric effects but these were shown to be secondary to the influence of the surface emissivity variability. Atmospheric errors are found to be important at low ice concentrations, while ice emissivity errors are important at high ice concentrations.

Recently, renewed attention has been given to improving weather filters for PMR data (Webster *et al.*, 2010). The filter improves estimates of sea ice concentration due to passing weather systems by (1) removing spurious ice over areas of open water, (2) increasing ice concentration estimates under clouds, and (3) decreasing ice estimates under relatively dry air. Over FYI, corrections to the estimates of ice concentration range from −10 to +30% during summer, while during winter and over MYI the corrections are of the order of ±10%.

Radars that measure the power of the return pulse scattered back to the antenna can be used to derive geophysical parameters of the illuminated surface, or volume, based on the scattering principles of microwave electromagnetic (EM) radiation. These instruments are known as scatterometers. The major instruments flown are the European Space Agency's (ESA) Earth Remote Sensing (ERS)-1 and -2 Active Microwave Instrument (C-band, 3 GHz, V), the first of which operated between 1992 and 1996 and the second of which has been operating since 1996, and the NASA QuikSCAT SeaWinds instrument (Ku band, 13.6 GHz, V, and H), flown from 1999 to the present. Scatterometry is useful for determining both ice extent (Allen and Long, 2006) and ice motion (Haarpaintner, 2006).

Radar altimeters and lidar altimeters are used to estimate ice thickness by measuring the freeboard – the height of the ice above the ocean surface, which is detected in leads and polynyas. The Geoscience Laser Altimeter System (GLAS) instrument on the Ice, Cloud, and Land Elevation Satellite (ICESat) was launched by NASA in January 2003 and operated until October 2009. It operated at frequencies of 532 and 1,064 nm. It had a 70-m footprint and a 170-m along-path spacing. The signal has a root-sum-square

Table 7.1 Satellite sensors used for sea ice research

Sensor type	Vis/IR	High res. vis/ IR	Passive microwave	SAR	Scatterometry	Radar altimeter	Laser altimeter
Primary recent, current and near-future satellites/ sensors	AVHRR MODIS, DMSP-OLS	Landsat-TM, Terra- ASTER Quick-Bird, SPOT 5, Formosat-2	DMSP-SSM/I, AMSR-E	Envisat, Radarsat-1 and 2, Terra SAR-X and –L	QuikScat	ERS-1 and 2, JERS, Envisat, Cryosat-2	IceSat GLAS
Ice extent	S		P	S	S	S	
Ice concentration	S	S	P	S	S		
Ice thickness	R		R	R		R	P
Ice motion	P	S	P	P	P		
Melt onset freeze-up	S	S	P	S	R		
Ice classification	S	S	P	P	S		

P = primary data source, S = secondary data source, R = research and development

(RSS) error of 0.2 cm. Lidar ranging may be either a discrete return, waveform recording, or photon counting. Data are currently available for fourteen intervals of about 35 days from February 2003 through March 2008 (http://nsidc.org/data/icesat/laser_op_periods .html). Sea ice thickness declines in the Arctic using ICESat data are reported by Kwok and Rothrock (2009).

In April 2010 ESA launched Cryosat-2, which has a Synthetic Aperture Radar (SAR)/ Interferometric Radar Altimeter (SIRAL). Over sea ice, coherently transmitted echoes are combined by synthetic aperture processing to reduce the surface footprint size. The altimeter's along-track footprint is divided into more than 60 separate beams with a resolution of around 250 m each, sufficient to differentiate ice floes from open water and often the leads between them.

Another active microwave sensor useful for sea ice studies is SAR. This is an imaging radar that synthesizes images from multiple looks during the satellite's motion in orbit to effectively create a large antenna and thus obtain much higher spatial resolution. The Canadian RADARSAT-1 sensor has been providing SAR coverage of sea ice since 1995. The RADARSAT-2 was launched in December 2007. However, as a purely commercial satellite, it is unlikely that data will be widely available to the science community. The resolution is high enough to capture small-scale ice motion and ice deformation events, allowing ice motion, ice age, ice volume, ice production, seasonal ice area to be estimated at fine spatial scales (Kwok et al., 1995; Kwok and Cunningham, 2002). The high resolution, all-sky capabilities are particularly useful for operational analysis of sea ice, and SAR imagery is widely used by operational sea ice centers such as the Canadian Ice Service and the US National Ice Center (NIC). However, the narrow swath of SAR sensors limits repeat coverage to every 3–6 days in many regions of the Arctic. In addition, SAR imagery of sea ice can be difficult to interpret and automated analysis has been largely unsuccessful.

Typical backscatter signatures from different ice types are shown in Table 7.2 for winter and summer in the Antarctic for C-band VV polarization.

Because of the change in emissivity of sea ice during melt, passive microwave imagery is useful for the determination of melt onset (Drobot and Anderson, 2001). Belchansky et al. (2004) estimated melt onset dates, freeze onset dates, and melt season duration over Arctic sea ice for 1979–2001, using passive microwave satellite imagery and surface air temperature data. Average melt duration varied from a 75-day minimum in 1987 to a 103-day maximum in 1989. On average, melt onset in annual ice began 10.6 days earlier than MYI, and freeze onset in MYI commenced 18.4 days earlier than annual ice. Ranges in melt duration were highest in peripheral seas, numbering 44 and 51 days, respectively, in the East Siberian and Chukchi seas.

Lukovich and Barber (2007) analyze spatial patterns of sea ice concentration anomalies derived from PMR data. They find that anomalies persist for 5–7 weeks in the Labrador Sea, 3–5 weeks in the Greenland Sea and around Svalbard, 4–7 weeks along the southern boundary in the Barents Sea, and 3 weeks in the southern Beaufort Sea. There are shorter

Table 7.2 Radar backscatter signatures over Antarctic sea ice (Drinkwater, 1998)

Month	Ice type	Mean backscatter (dB)
July 1992	Smooth FYI	−16
	Rough FYI	−10
	Second and MYI	−6
December 1992	New and young ice	−32 to −20
	Smooth FYI	−20 to −14
	Rough FYI	−14 to −11
	MYI and pancakes	−11 to −6
	Icebergs	−6 to 0

time scales in the Kara and Bering seas. The coherent regions of persistence appear to be linked with regions of high positive or negative meridional wind anomalies.

Surface methods to measure sea ice characteristics are described by Mahoney and Gearheard (2008). Eicken *et al.* (2009) provide a comprehensive account of field techniques for sea ice research, including ice thickness and roughness, snow cover, ice optics, strength, thermal, electrical, hydraulic, and biogeochemical properties. Shipboard observations are described in MANICE (Environment Canada, 2005) and by Worby (1999) under the Antarctic Sea Ice Processes and Climate (ASPeCt) program. In the latter program observations are made hourly within a 1-km radius of the ship. The recorded elements are concentration in tenths, 13 ice types and 10 thickness categories, topography, ridges per linear mile, snow depth and surface coverage, melt state, behavior of the ice (i.e., movement, developing or releasing pressure), ridge heights, and water temperature. Indigenous knowledge of sea ice (termed siku across the Arctic) is assembled by Krupnik *et al.* (2010) providing local terminologies and classifications.

Weekly, biweekly, or 10-day charts of ice conditions are produced routinely by national operational ice services in the Baltic countries, Canada, Denmark, Iceland, Japan, Norway, Russia, and the United States (see www.ipy-ice-portal.org/). Only the United States NIC produces biweekly ice charts for the ice covered areas of both hemispheres (Figure 7.3); the other services provide regional ice analyses. The NIC also provides analyses on a biweekly basis for the Alaskan waters and the Great Lakes, weekly for the Ross Sea during the austral summer navigation season, and Baffin Bay during the boreal summer. Analyses are also produced weekly for the Arctic Basin, Bering Sea and Cook Inlet, Chukchi Sea, Beaufort Sea, Barents Sea, Kara Sea, and the northern part of the East Greenland Sea.

Sea ice growth

Ice forms in the ocean when the surface cools to about −1.8 °C for average ocean salinity (34.5 PSU, or practical salinity units). Ice floats because it is less dense than water – about

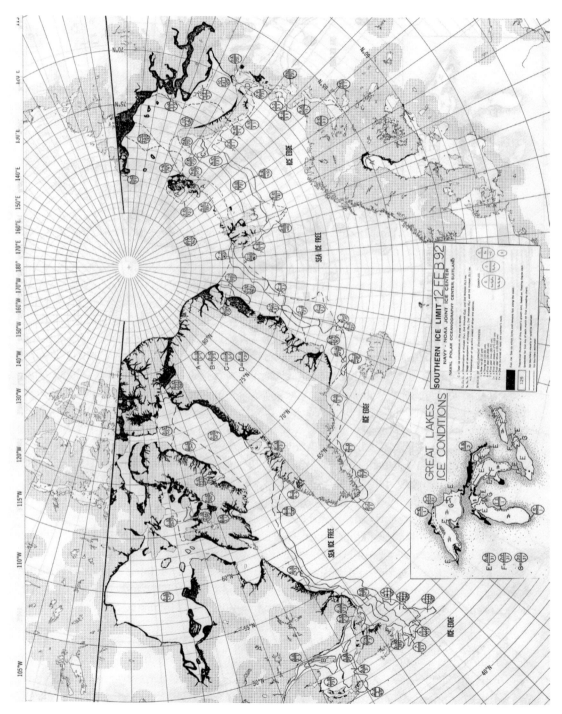

Figure 7.3 Sample National Ice Center (NIC), US ice chart: part of the Arctic analysis for December 20, 1994 showing egg code symbols (see Section 7.2)

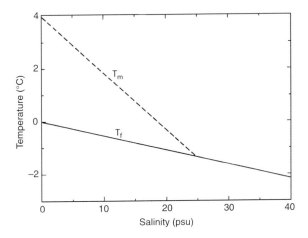

Figure 7.4 The dependence of the freezing point on ocean salinity (from Maykut, 1985). The lines for the temperature of maximum density (*Tm*) and the freezing temperature (*Tf*) meet at −1.33 °C and S = 24.695 PSU [Courtesy Applied Physics Laboratory, University of Washington, Seattle]

917 kg m^{-3} at 0 °C compared with 1,000 kg m^{-3}, respectively. Figure 7.4 illustrates the dependence of the freezing point on ocean salinity. For every 5 PSU increase in salinity, the freezing point decreases by 0.28 °C. The addition of salt to water lowers its temperature of maximum density – which is 3.98 °C for fresh water – and when the salinity exceeds 24.7 PSU, the temperature of maximum density disappears. Cooling of the water surface by the emission of infrared radiation, and contact with a cold air mass, makes the surface water denser and therefore sets up convection in the water column. However, the whole water column does not have to cool to freezing before ice can form, only the upper layer above the pycnocline, or level of density maximum. In the Arctic this is typically located at about 50–150-m depth (see Figure 7.5). The surface salinity in the Arctic is low (<30 PSU) off the Siberian river estuaries and 32–34 PSU elsewhere.

Sea ice has two phases: salt-free ice and liquid brine (Ackley, 1996). This is a result of the insolubility of salts in ice, as opposed to water, and the process of brine entrapment during ice growth. Sea ice growth is kinetic, occurring rapidly and episodically, triggered by a nucleation event. The local supercooling is used as a heat sink for the latent heat released in the phase transition from liquid to solid. At a growing ice interface, most of the salts are rejected. Brine, gas, and solid salts are usually trapped at sub-grain boundaries within a lattice of essentially pure ice (Timco and Weeks, 2010). First-year sea ice has a typical salinity in the range 4–6 ‰ (parts per thousand, or PSU). In the brine solution, the rate of heat removal is about ten times that at which salt can diffuse from a region of high concentration to a lower one. Thus, the solution becomes supercooled, because the salt

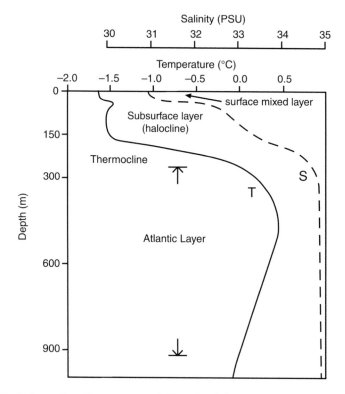

Figure 7.5 Typical profiles of temperature (°C) and salinity (PSU) in the Canadian Basin showing the vertical structure (modified after Melling and Lewis, 1982) [Courtesy of Elsevier: Deep-sea Research 29 (8A), 1982, p. 968, fig. 1]

diffusion needed to lower the equilibrium freezing point of the rest of the solution, cannot keep pace with the thermal cooling.

In calm conditions, the first sign of freezing of the sea is an oily appearance of the water caused by the formation of needle-like crystals. These crystals are pure ice, free of salt (Wadhams, 1998, 2008). Sea ice forms as a skim of crystals that are initially in the form of tiny discs, floating flat on the surface. They have a diameter <2–3 mm. Each disc has its c-axis vertical and grows outwards laterally. At a certain point, the growing crystals take on a hexagonal, stellar form, with long fragile arms stretching out over the water surface. These arms soon break off, leaving a mixture of discs and fragments. Turbulence in the water leads these fragments to breakup further into small crystals, which form a suspension of increasing density in the surface water. This ice type is called **frazil** or **grease ice**. In calm conditions, the frazil crystals soon freeze together to form a continuous thin sheet of transparent ice, called **nilas** (Figure 7.5). As the ice thickens, the nilas takes on a gray and finally a white appearance (white nilas). Nilas undergoes a quite different growth process, called **congelation** growth, in which water molecules freeze on to the bottom of the ice. This freezing process is easier for crystals with horizontal c-axes than for those with vertical c-axes (Weeks, 2010). The crystals with a horizontal c-axis grow at the expense of the others as the ice sheet grows thicker. Thus, the crystals near the top of a FYI sheet are small and randomly oriented, and then there is a transition to long vertical columnar crystals with horizontal c-axes. This columnar structure is a key identifier of **congelation ice** that has grown thermo-dynamically by freezing onto the base of an existing ice cover. The thickening of a sea ice sheet occurs by the addition or extension of ice platelets into the supercooled solution below. The platelets are dendrite-like crystals that are 1–3-mm thick and up to 100-mm across (Gow *et al.*, 1998). As the platelets penetrate into the solution, salt is rejected by the growing pure ice phase (Weeks and Ackley, 1986; Gow and Tucker, 1991). In McMurdo Sound in winter 2008, Gough *et al.* (2010) measured the evolution of sea ice as it grew from 0.88 m in late May to 2.08 m in late October, when the ice consisted of 0.12-m frazil ice, 0.88-m columnar ice, 0.40-m mixed columnar/platelet ice, and 0.68-m platelet ice.

More dense, salty regions descend as plumes and are replaced by less salty, upwelled water. This fluid fills the spaces between the platelets and is trapped as a brine inclusion as the ice platelets thicken and neck off the inclusion (Ackley, 1996). The salt concentration in brine pockets at $-10\,°C$ reaches 115 PSU. The spacing between brine inclusions represents the breakdown of the ice-crystal lattice spacing and is ~0.3–2 mm. Notz and Worster (2009) examine the roles of the initial fractionation of salt at the ice–ocean interface, brine diffusion, brine expulsion, gravity drainage, and flushing with surface meltwater. Analytical and numerical studies, as well as laboratory and field experiments, show that only gravity drainage and flushing contribute to any measurable net loss of salt (Untersteiner, 1968). In rapidly growing, young ice, when the ice is much warmer than the overlying air, some brine is forced upward and, along with rime crystals, forms "frost flowers" on the ice surface in patches about 3–4 cm in diameter.

In rough water, waves maintain the new ice as a dense suspension of frazil crystals. Because of particle orbits in the wave field, this frazil ice undergoes cyclical compression and during this process the crystals can freeze together to form small coherent cakes of slush (or shuga, see Figure 7.6a) which grow larger by accretion from the frazil ice and more solid through continued freezing between the crystals. This eventually turns into pancake ice (Figure 6.7) because collisions between the cakes force frazil ice onto the cake edges, then the water drains away leaving a raised rim of ice with the appearance of a pancake. At the sea ice margin, the pancakes are only a few cm in diameter, but they increase in diameter and thickness with increasing distance from the ice edge, and they may ultimately reach 3–5 m in diameter and 50–70-cm thickness. Away from the marginal ice zone (MIZ), where wave amplitudes are small, the pancakes begin to freeze together in groups and coalesce to form large floes, and finally a continuous sheet of FYI. At the time of consolidation, the pancakes are jumbled together and rafted over one another. The result is that the rafted ice is two to three times the ice thickness due to simple thermodynamic growth, and the edges of pancakes protrude upwards to give a highly irregular surface topography. In the Southern Ocean,

Figure 7.6 Pictures of various types of ice during the growth process. (a) Shuga and young FYI floes (rear) in the Laptev Sea, September 2005. (b) FYI broken up by wave action, Davis Strait, May 24, 1971. (c) Pressure ridge off Barrow, AK, March 16, 2006. Surveying ridge topography using GPS, as part of the NASA-funded "AMSRIce06" satellite data validation campaign for AMSR-E [Courtesy Dr. J. A. Maslanik, Aerospace Engineering, University of Colorado]. (d) Grounded pressure ridges in the Beaufort Sea, spring 1949. Photographer: Rear Admiral Harley D. Nygren, NOAA Corps (ret.) (www .photolib.noaa.gov/htmls/corp1014.htm)

pancake ice accounts for about 60% of the ice cover at maximum extent (Weeks and Ackley, 1986). Ice growth in leads and polynyas (see Section 7.4) occurs by horizontal accretion of frazil slush.

Wadhams (2000) considers that there are four major MIZs: in the Southern Ocean around Antarctica, the Bering Sea, the Greenland–Barents Sea, and the Labrador Sea–Baffin Bay system. The defining condition is an ice edge adjoining a rough open ocean with long, high waves. The Antarctic has the longest and most extensive MIZ ice forms continuously at the ice edge from April to September, as frazil and pancake ice, and the MIZ may exceed 250 km in width before the waves are sufficiently damped to permit consolidation of the pancakes.

Once an ice cover has formed, the ocean is isolated from the overlying cold air. The latent heat of freezing is transferred upward through the ice by thermal conduction except where leads open briefly enabling turbulent transfers. The rate of thickening of the ice is now determined by the temperature gradient in the ice and its thermal conductivity.

New ice is a technical term that refers to ice less than 10-cm thick. As the ice thickens, it enters the young ice stage, defined as ice that is 10–30-cm thick. FYI is thicker than 30 cm and in a single winter season may reach a thickness of 1.5–2 m. Ice forms first near the coast because a relatively small depth of water has to be cooled to the freezing point. Sea ice nomenclature is presented by WMO (2007) in a multilingual format. A sea ice glossary is available on line at NSIDC (http://nsidc.org/cgi-bin/words/topic.pl?sea%20ice).

Flat expanses of floating ice are called ice floes. They are classified according to size: giant: over 10-km across; vast: 2–10-km across; big: 500–2,000-m across; medium: 100–500-m across; and small: 20–100-m across. Wider areas of ice are termed ice fields.

In the stormy Southern Ocean, frazil ice formation can also be initiated by snowfall, rather than supercooling of the ocean surface. Slush is a floating mass formed initially from snow and water. Shuga is formed in agitated conditions by the accumulation of slush, or grease ice, into spongy pieces 5–10 cm in size. Antarctic sea ice also grows by the addition of snowfall. The weight of the snow depresses the initial ice cover and a slush layer forms as waves soak the snow on the ice. This slush freezes and adds a new layer of ice on the surface. About half of the ice cover in the Southern Ocean is estimated to have been flooded at some time in its history (Ackley, 1996).

Ice that survives a summer melt season becomes second-year ice, which rarely exceeds 2.5 m in thickness. However, it is difficult to differentiate such ice from older MYI even with on-ice measurements (Timco and Weeks, 2010). The thickness of MYI depends on both meteorological and ice dynamical processes. Old sea ice is largely fresh, since the ocean salt is expelled by the growing ice through a process called brine rejection. The ice–water interface moves downward in the form of parallel rows of cellular projections called dendrites. Brine that is rejected from the growing ice accumulates in the grooves between these rows of dendrites. As the dendrites advance, ice bridges develop across the narrow grooves that contain the rejected brine, leaving the brine trapped. The walls close in through freezing, until there remains a tiny cell of highly concentrated brine, concentrated enough to lower the freezing point to a level where the surrounding walls can close in no further. The cells persist as tiny inclusions that eventually drain out of the ice, by way of a network of brine drainage channels, which they create. Weeks and Lofgren (1967) also showed that the salt rejection mechanism becomes more effective as the ice growth slows down. As the ice sheet ages the brine concentration drops. The water from young sea ice may have a salinity of about 10 PSU, decreasing to 1–3 PSU in old ice. The resulting highly saline (and hence dense) water sinks down to the pycnocline.

First-year sea ice production in the Arctic occurs mainly by the ice growing outward from old ice that has survived the summer, as well as from the coasts and over the shelves where brines, released during sea ice formation, change seawater density. This seasonal signal is then communicated to the basin across the shelf slopes. Most of the freshwater signal from river runoff is confined to the coast in the form of buoyancy-driven coastal currents and to the upper water column, as determined via shelf–basin and atmosphere–ice–ocean exchanges. The vertical water column structure in the Arctic Ocean features a strong halocline at 50–200-m depth. The salinity increases from ~30 PSU at the surface to ~34.5 PSU in the Atlantic water below 200 m. The halocline is maintained by salt released during sea ice formation, which drains down due to its high density, and near the coasts by freshwater runoff.

In the Antarctic, sea ice production is largely in the open ocean. It depends significantly on its relatively thick snow cover, which controls three of the four modes of thermodynamic ice growth. These processes are: congelation, flooding and snow-to-ice conversion, and summer surface processes such as superimposed ice formation. Congelation involves frazil crystals, which form in open water areas and make a major contribution to the total ice mass. Under the influence of wind and wave action the frazil crystals coagulate, eventually consolidating into small circular pancakes of ice. These eventually freeze together to form larger floes or a consolidated ice cover. When the weight of the snow cover on the ice is sufficient, the ice surface may be depressed below sea level. This permits an influx of seawater through the permeable snow that saturates the lower layers of snow; these may subsequently refreeze to form "snow ice" Analysis of 173 cores taken on six voyages into the East Antarctic pack between 1991 and 1995 revealed that on average the pack comprised 47% frazil ice, 39% columnar ice, and 13% snow ice (www.aspect.aq/formation.html). Ackley *et al.* (1990) indicate that between 14% and 28% of the ice is flooded. The amount of open water in the East Antarctic pack ice decreases from almost 60% in December to little more than 10% in August, and the thinnest ice thickness category (0–0.2 m) shows a 30% seasonal change between December and March. In contrast, the amount of ice greater than 1.0 m shows very little seasonal variability. The dynamic processes of rafting and ridging are dominant mechanisms by which the ice thickens. Worby *et al.* (1998) give average sea ice thickness data for East Antarctica (60° S, 150° E) based on ship data from the ASPeCt program. They show the modal thickness changing from the thinnest category in March, to a maximum of 0.6–0.8 m in August and then back to open water in December when the ice-covered area is only 40%. Worby *et al.* (2008a) summarize all available data from 1980 to 2005. The long-term mean and standard deviation of total sea ice thickness (including ridges) is found to be

0.87 ± 0.91 m, which is 40% greater than the mean level ice thickness (0.62 m). The ice thickness distribution shows least variability in the western Weddell Sea, which contains up to 80% of the MYI, and is largely ice covered year-round. Mean sea ice thicknesses in the western Weddell Sea range from less than 1 m in the south to 1.5–2.5-m along the Antarctic Peninsula. There are similar thicknesses near the coast in the eastern Ross Sea. Most of the sea ice in the Indian Ocean and Pacific Ocean sectors ranges from almost 1 m near the coast to 0.3 to 0.5 m near the ice edge.

The mass balance of sea ice in the Antarctic was the focus of three projects during the IPY, 2007–2009. These used differing approaches to assess ice thickness and mass balance evolution. A network of twelve drifting buoys on sea ice and a ship were deployed in the Amundsen and Bellingshausen seas (80–120° W) in the Sea Ice Mass Balance (SIMBA) led by S. Ackley; helicopter-based radar and laser altimetry, and a remotely operated underwater vehicle were used in the Sea Ice Physics and Ecosystem eXperiment (SIPEX) between 120° and 130° E led by A. Worby; and a German expedition with the Polarstern operated in the Weddell Sea. The results show that the sea ice in the East Antarctic was more dynamic, swell affected, and heavily deformed in some areas compared to conditions off West Antarctica where the ice was more compact and homogenous. The dearth of oceanographic data from beneath winter sea ice in the Southern Ocean has recently been addressed by instrumenting elephant seals. Charrassin *et al.* (2008) obtained two temperature and salinity profiles daily, to an average depth of 566 m and a maximum depth of 2,000 m, collecting 8,200 profiles from south of 60° S. The salinity data can be used to estimate ice growth rates, which ranged from 3 cm d^{-1} in April 2004 at 65° S, 54° E to 1 cm d^{-1} from May to mid-August 2004 at 66.5° S, 84° E.

Landfast ice

Landfast ice is sea ice that is contiguous with the shore or the seabed and is immobile. It may be attached to an ice wall, ice front, shoals, or between grounded icebergs. Barry *et al.* (1979) list three criteria that can distinguish landfast ice from other forms of sea ice: (1) the ice remains relatively immobile near the shore for a specified time interval; (2) the ice extends from the coast as a continuous sheet; and (3) the ice is grounded or forms a continuous sheet which is bounded at the seaward edge by an intermittent or nearly continuous zone of grounded ridges (see Figure 7.6d). It develops first in sheltered bays and inlets, is generally shoreward of the 20-m isobath, and remains stable for much of the year. However, in western Baffin Bay, Jacobs *et al.* (1975) show that landfast ice extends out over water which is 180 m deep, 70 km offshore in Home Bay.

There appear to be two mechanisms that account for its spatial distribution (König Beatty and Holland, 2010). The initial landfast ice formation occurs in shallow water which allows faster local freezing due to the lack of deeper warmer water acting as a source of heat through convection. The second mechanism is the consolidation of pack ice that is transported by onshore winds. Once established, grounded ice ridges stabilize the landfast ice. Coastal

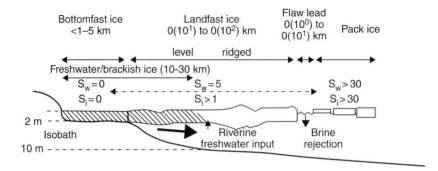

Figure 7.7 Schematic summary of the southern Laptev Sea ice cover and river processes (Sw, salinity of surface seawater, PSU; Si, bulk sea-ice salinity); the approximate width of the different zones is also indicated (Eicken *et al.*, 2005)

geometry (groups of islands, narrow passages, and concave coastlines) also enables the sea ice to remain landfast. In Antarctica, the location of annually recurring fast ice was found to be closely related to the distribution of icebergs, grounded in up to 500 m of water, that serve to anchor the ice sheet (Fraser *et al.*, 2010).

Thermodynamic processes largely determine the growth of fast ice and its thickness at the end of the season depends on the time interval that the ice cover has been stationary. In the northern Baltic Sea, Granskog *et al.* (2003) find that snow-ice layers contribute 24–32% of the total ice thickness. The snow in these layers contributes, on average, 18–21% of the total sea ice thickness (by mass). In the Laptev Sea, Eicken *et al.* (2005) report that stable-isotope data show that the landfast ice is composed of about 62% of river water that locks up 24% of the total annual discharge of the Lena and Yana Rivers (see Figure 7.7). In the late 1990s, the mean ice thickness amounted to 1.65 m; the older, core area of the landfast ice is around 2-m thick (Eicken *et al.*, 2005). Bottomfast ice was not as widespread as previously hypothesized, occupying only 250 km^2 of the Lena delta (from SAR data in 1996/1997 and 1998/1999). The floating landfast ice covers much of the southern Laptev Sea and in places extends more than 200-km out from the coast according to Timokhov (1994), whereas landfast ice in the Beaufort Sea is typically grounded at water depths of around 16–22 m by a line of grounded shear ridges or stamukhi (Barry *et al.*, 1979; Reimnitz *et al.*, 1994; Mahoney *et al.*, 2007a). For the Beaufort Sea, the landfast ice extent has been mapped for 1996–2004 based on RADARSAT-1 SAR imagery. The data are available at http://nsidc.org/data/docs/noaa/g02173_ak_landfast_and_leads/index.html#3. In the southern Laptev Sea, (modeled) net heat fluxes from the atmosphere and river flooding contribute 53 and 47%, respectively, to the melt of nearshore fast ice according to Bareiss and Görgen (2005).

The interannual variability of maximum fast ice thickness at four sites in the High Arctic over the period 1950–1989 was examined by Brown and Cote (1992). The insulating role of snow cover was found to explain 30–60% of the variance in maximum ice thickness values.

Other snow-related processes such as slushing and density variations were estimated to explain a further 15–30%. In contrast, the annual variation in air temperatures explained less than 4% of the variance. There were no signatures of global warming but recent ice thinning and thickening trends at Alert and Resolute were consistent with changes in the average depth of snow cover on the ice.

Melling (2002) analyzed data on ice thickness in the Canadian Arctic Archipelago from 123,700 drill holes collected in the 1970s; the sea area is $1.9 \times 10^6 \, \text{km}^2$. The ice is landfast for over half the year and the summer concentration is 7–9 tenths. The ice is a mixture of MYI, second year, and FYI with the latter subordinate except in the southeast. The average ice thickness in late winter is 3.4 m but regionally up to 5.5 m. The drift of ice is controlled by ice bridges that form across the channels. In the unusually warm summer of 1998, the ice plugs in two northwestern channels cleared for the first (known) time (Atkinson *et al.*, 2006) and this was repeated in 2007.

Snow depth

The annual maximum snow depth of the MYI region in the Arctic occurs in May and averages 34 cm (11-cm water equivalent) based on data from the Soviet NP Drifting Stations (Colony *et al.*, 1998; Warren *et al.*, 1998) (see Figure 7.8). The thickest snow cover is found north of Greenland and Ellesmere Island. Locally, it is up to a meter deep or more around ridges. Five to nine snowfall and wind events may occur during the winter season and after deposition the snow undergoes compaction, metamorphism, and wind erosion (Sturm, 2009). The ice cover is largely snow free during August. In the Antarctic snowfall is a major mechanism of sea ice growth, due to the large amounts that accumulate, as described earlier. The insulating effect of snow cover is important in the Antarctic, whereas in the Arctic the ice–albedo effect is dominant. Snow cover has a low thermal conductivity – $<0.1 \, \text{W m}^{-1} \, \text{K}^{-1}$ for depth hoar and $0.3 \, \text{W m}^{-1} \, \text{K}^{-1}$ for a snow slab – compared with 2.1–2.6 $\text{W m}^{-1} \, \text{K}^{-1}$ for sea ice.

Snow depth on sea ice can now be mapped by AMSR-E on NASA's Aqua satellite, calculated using the spectral GR of the 18.7- and 37-GHz vertical polarization channels (Comiso *et al.*, 2003). The GR becomes more negative as snow depth increases. The algorithm is appropriate for dry snow conditions and has an upper limit of 50 cm as a result of the limited penetration depth at 18.7 and 37 GHz. Worby *et al.* (2008b) find that in the rough sea ice of the SSIZ of East Antarctica in September–October 2003, the products underestimated the snow depth by a factor of 2.3.

Sea ice decay

At the end of winter, level pack ice in the Arctic has a snow cover of about 30–40-cm thickness, with locally around ice ridges a meter or more. During late May–June this snow cover begins to melt and is mostly gone by early July. The meltwater accumulates in melt

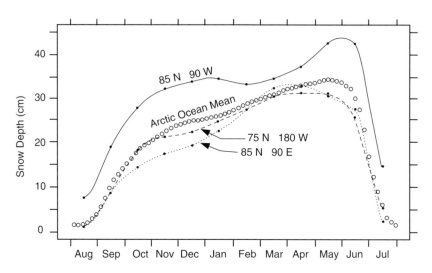

Figure 7.8 Annual cycle of mean snow depth (cm) from North Pole Drifting Station data on multiyear sea ice. The mean of all NP stations that sampled complete years (August–July) is shown, as well as the seasonal cycles at three locations (from Warren *et al.*, 1998) [Courtesy American Meteorological Society]

ponds formed on the irregular surface of the ice; these ponds are typically shallow on smooth FYI, but deeper on old hummocked ice that has survived at least one summer. Ten years' of Soviet drifting station data show a peak extent around 10 July of 25%, on average (Nazintsev, 1964). In the Beaufort/Chukchi Sea during summer 2004, Tschudi *et al.* (2008) found that the estimated pond coverage from MODIS increased rapidly during the first 20 days of melt from 10% on June 1 to 40% on July 1. Skyllingstad *et al.* (2009) show that the pond fractional area increases linearly with time during the melt season under both sunny and cloudy conditions. Aerial estimates during SHEBA suggest an increase from about 15% in late June to 24% in early August. Fetterer and Untersteiner (1998) find that local variability of pond cover is greatest at the beginning of the melt season, varying from 5 to 50% depending on ice type. Pond cover decreases with time on thick MYI, due to drainage, and it increases with time on thin ice (eventually leading to the disappearance of thin ice at the end of summer). Ponds gradually form an interlinked network and they may thaw through the ice due to the lower pond albedo (0.1–0.3), compared with 0.5 for bare, discolored ice, and 0.6 to 0.8 for a deteriorated surface or snow-covered ice (Tucker *et al.*, 1999). Pond reflectance, however, varies significantly with cloud cover, pond depth, physical properties of the ice bottom, amount of biogenic or particulate material, and ice type, according to Morassutti and LeDrew (1995) and Tucker *et al.* (1999). Ponds observed during the August 5 to September 30, 2005 Healy Oden Trans-Arctic Expedition had an average depth of 0.3 m (Perovich *et al.*, 2009), but during SHEBA maximum depths of 0.4–0.6 m were recorded (Skyllingstad *et al.* (2009). The pond water drains through thaw holes and cracks in

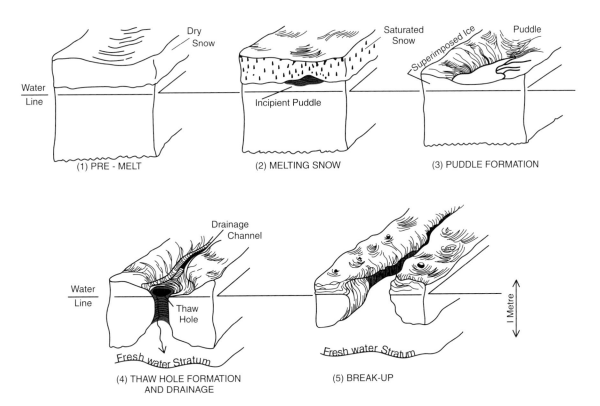

Figure 7.9 Landfast ice decay stages showing evolution of melt ponds (Jacobs *et al.*, 1975)

the ice (Jacobs *et al.*, 1975), which temporarily increases the albedo (see Figure 7.9). At Barrow and Wales in Alaska, a Seasonal Ice Zone Observing Network (SIZONet) has been set up to document the stages in landfast ice decay through a combination of community-based observations and geophysical measurements, such as coastal radar monitoring of ice stability and movement, SAR satellite imagery, and on-ice and airborne thickness measurements. Ice mass balance (IMB) data for the Chukchi Sea and Barrow are available from February 2000 (www.sizonet.org/data).

Ablation in ponds occurs at a rate 2.5 times that on bare ice as a result mainly of the low pond albedo and associated absorption of solar radiation (Untersteiner, 1961). The ice directly beneath the melt ponds is thinner and is absorbing more incoming radiation. This causes an enhanced rate of bottom melt so that the ice bottom develops a pattern of depressions that mirror the melt pond distribution on the upper surface. Some of the drained meltwater collects in these depressions to form under-ice melt pools, which refreeze in autumn and so partially smooth off the underside of the ice.

Melt ponds play a major role in determining the summer sea ice albedo (Barry, 1996). An albedo scheme that includes them has recently been developed for the ECHAM5 general

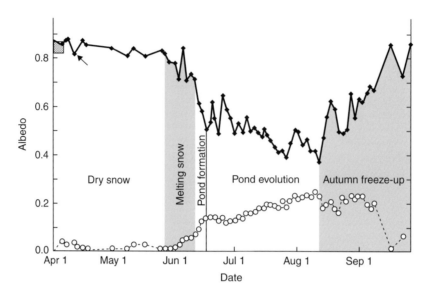

Figure 7.10 Seasonal course of wavelength-integrated surface albedo in the central Arctic from April to September 1998 over a 2-km line. The standard deviation is plotted in open circles [Source: Perovich *et al.*, 2002] [Courtesy American Geophysical Union]

circulation model by Pedersen *et al.* (2010). It includes important components such as albedo decay due to snow aging, bare sea ice albedo dependent on the ice thickness, and a melt pond albedo dependent on the melt pond depth. Four surface types are considered: snow-covered sea ice (a_s), bare sea ice (a_i), melt ponds (a_p), and open water (a_w). The total albedo is weighted according to the grid area's mean ice concentration. The sea ice albedo (ai_{ce}) is defined as:

$$a_{ice} = a_s f_s + a_i f_i + a_p f_p \tag{7.3}$$

where *f* denotes the fraction of the corresponding surface. Skyllingstad *et al.* (2009) show that melt pond coverage has a greater effect on surface albedo than pond depth.

   The surface albedo of the Arctic Basin declines from around 0.8 in May to 0.5 in mid-August based on calibrated estimates from DMSP satellite visible imagery over 10 summers (Robinson *et al.*, 1992). The SHEBA data for a 2-km line show an albedo that declines from 0.8 in April to 0.4 at the end of July (Figure 7.10). At this time there is significant spatial variability from 0.1 for dark, deep ponds to 0.65 for bare ice. Snowfalls occur in late August raising the surface albedo and insulating the new ice cover. Five centimeters of snow is sufficient to raise the albedo to 0.8. Grenfell and Perovich (2004) compare the different seasonal evolution of albedo at sites on sea ice, lagoon ice, and tundra near Barrow, AK, in 2000–2002.

Table 7.3 Representative all-wave solar albedos of surface types in the East Antarctic sea ice zone in spring (SON) and summer (DJF). Values in bold are derived from measurements (from Brandt et al., 2005) [Courtesy of the American Meteorological Society]

Ice type	Ice thickness (cm)	No snow		Thin snow (<3 cm)				Thick snow (>3 cm)			
				Clear		Cloudy		Clear		Cloudy	
		Clear	Cloudy	SON	DJF	SON	DJF	SON	DJF	SON	DJF
Open water	0	0.07	0.07								
Nilas	<10	0.14	0.16	0.42	0.39	0.45	0.42				
Young gray-white ice	15–30	0.32	0.34	0.64	0.59	0.68	0.64	0.76	0.70	0.81	0.76
First-year ice	30–70	0.41	0.45	0.74	0.69	0.79	0.74	0.81	0.75	0.87	0.82
First-year ice	>70	0.49	0.54	0.81	0.75	0.87	0.82	0.81	0.75	0.87	0.82

All-wavelength albedos in the Antarctic have been assembled by Brandt *et al.* (2005). Table 7.3 summarizes these values. However, in the Antarctic melt occurs mainly from the bottom and sides of the ice floes, which are in contact with the ocean.

Some of the meltwater works its way down through the ice along minor pores and channels, and in doing so drives out much of the remaining brine. This flushing process is the most efficient and rapid mechanism of brine drainage, and it operates to remove nearly all of the remaining brine in FYI. Finally, the ice may begin to fracture through a combination of wind-driven motion, tidal motions, and wave action (see Figure 7.6b).

There is a widespread distribution of sediment-laden ice in the Arctic. This seems to occur in relation to episodic storms that stir up sediment over the shallow shelves of the Beaufort Sea, off the Queen Elizabeth Islands, and the Laptev Sea. According to Darby (2003) most of the entrained sediment fits the criteria for suspension freezing in shallow water, but the presence of winter polynyas with offshore winds appears to be the critical factor for sea ice entrainment. Dirty ice has a lower albedo and so it decays more readily. The entire ocean floor is strewn with pebbles and rocks that were rafted by sea ice into the open ocean from shore (Schwarzacher and Hunkins, 1961).

Multiyear ice

Ice that survives two or more summer seasons of partial melt is called MYI. (However, from remote-sensing data only FYI and MYI can be distinguished; not second-year ice.) This old ice is much fresher and stronger than FYI and has a rougher surface. Typically, growth of MYI continues from year to year until the ice thickness reaches about 3 m, at which point summer melt matches winter growth and the thickness then oscillates through an annual cycle. In the Antarctic little ice survives the summer season. In the Arctic, however, sea ice commonly takes several years, either to make a circuit within the clockwise Beaufort Gyre surface current system (7–10 years) or else to be transported across the Arctic Basin and exported via the East Greenland Current (3–4 years) into the North Atlantic. Until the early 2000s over half of the ice in the Arctic was multiyear, but the record minimum sea ice extent of summer of 2002 resulted in the lowest area of surviving FYI up to that time (Kwok, 2004). Anomalous ice export over several years, and exceptional melt in 2005 and 2007, removed much of the MYI (Stroeve *et al.*, 2007). During 2000–2009, the extent of MYI declined at a rate of 1.5 million km^2 decade^{-1}, triple the rate of reduction during 1970–2000. In March 2009, the MYI extent was 3.0 million km^2 (Perovich *et al.*, 2009b) and in March 2019, MYI extent was less than 0.1 million km^2.

Zwally and Gloersen (2008) determine a local temporal minimum (LTM) of ice area that accounts for the non-simultaneity of the melt–freeze transition. Passive microwave data for 1979–2004 are analyzed for 25-km cells. The average ice area surviving the summer melt is found to be 2.6×10^6 km^2 (excluding ~0.7×10^5 km^2 above 84° N). This is about 45% less than the value determined for the total ice cover at the minimum extent in mid-September (3.8×10^6 km^2). The value of the LTM has decreased by 9.5% decade^{-1} similarly to the decline of the September minimum value. The timing of the LTM has become delayed from August 11 to August 19, indicating a later end of the melt season.

Johnston *et al.* (2009) report on extensive measurements of the thickness of MYI from the Canadian Arctic taken in the 1970s to 1980s and from Sverdrup Basin in 1978. The average MYI thickness in Sverdrup Basin, the Alaskan Beaufort, and the central Canadian Arctic ranged from 6 to 7 m. The average MYI thickness in the Canadian Beaufort was 7.2 m. The modes of MYI thickness in the Sverdrup Basin (9.2 m) and Canadian Beaufort Sea (10 m) are comparable, as are modes for the Alaskan Beaufort (7.9 m) and central Canadian Arctic (7.3 m). Maximum values were reported to be 40 m from the Canadian Beaufort Sea off Banks Island and 23 m in Sverdrup Basin. For the Eurasian sector of the Arctic Ocean, Eicken *et al.* (1995) reported an average MYI thickness of 2.86 m.

Mass balance

The mass balance of sea ice involves thermodynamic and dynamic processes, giving rise to growth/melt, advection, ridging, and the transformation of FYI into MYI. Building on earlier estimates of the average mass balance of Arctic sea ice, Koerner (1973) used measurements made during the 1968–1969 British trans-Arctic Expedition (Herbert, 1969), but it must be noted that the area of ice types has changed considerably and the ice thickness has decreased dramatically since 2004. Koerner (1973) shows that 1.1 m of ice must form annually of which 47% is the accumulation of FYI. About half of this grows in open water or below young ice and 20% is due to ridging. The mean ridge depth was ~13 m and the keel/sail ratio was ~4.5 for FYI ridges and 3.2 for MYI ridges. Ridging reduced the ice area by 3–4% annually. About 15% of the ice cover was exported annually representing the balance of FYI plus the balance of MYI of less than steady-state thickness.

Thomas and Rothrock (1993) calculated an ice balance for the Arctic Ocean using SMMR data for 1979–1985 and buoy-derived ice motion fields with Kalman filtering and smoothing. The evolution of the ice cover is interpreted in terms of advection, melt, growth, ridging, and aging of FYI into MYI. The 7-year average area change values are shown in Table 7.4.

Table 7.4 Area change values of FYI, MYI, and total ice in the Arctic Ocean for 1979–1985 (10^6 km^2 a^{-1}) (from Thomas and Rothrock, 1993)

Process	FYI	MYI	Total
Advection	−0.42	−0.42	−0.84
Ridging	−0.90	−0.09	−0.99
Growth	4.07	0	4.07
Melt	−1.88	−0.53	−2.41
Aging	−0.83	0.83	0
Net	0.03	−0.04	−0.01

Generally, the coastal regions of Alaska and Siberia, and the area just north of Fram Strait, are sources of FYI, with the rest of the Arctic Ocean acting as a sink via ridging and aging, which together equal the melt term. All of the Arctic Ocean except for the Beaufort and Chukchi seas is a source of MYI, with the Chukchi Sea being the only internal sink of MYI. Export through the Fram Strait accounts for 14% of the ice area annually. Given the recent dramatic changes in ice conditions in the Arctic (see Section 7.7), these mass balance estimates are clearly not representative of the present state of the sea ice. In the abnormally warm summer of 1998 during SHEBA, Eicken *et al.* (2001) measured an ablation of 0.9–1.2 m during May–August. Surface melt began around day 170 and ended about day 230. Bottom melt began around day 153 and ended around day 270.

Rothrock and Zhang (2005) examined the ice volume changes in the Arctic using daily air temperature and sea level pressures (SLP) for 1948–1999 to force an ice–ocean model. They found that the annual ice mass production and export are typically out of balance year to year by ±30%, but very nearly in balance over decades. The volume response to rising temperatures accounted for a reduction of over 25% in volume over the five decades. The total ice volume decreased by 36% between 1966 and 1999, as a result of the loss of 40% of the undeformed ice and 28% of the ridged ice. The central Arctic Ocean and the East Siberian Sea experienced the greatest decrease in ice volume up to 1999. In October 2012, the CryoSat satellite measured only 6,000 km^3 of Arctic sea ice, a record low, which bounced back by about 50% to 9,000 km^3 in October 2013 mostly out of the growth of MYI. However, despite of the bounce back, this sea ice volume is still low compared to historical averages of about 20,000 km^3 in the 1980s (www.cbc.ca/news/technology/arctic-sea-ice-volume-up-50-1.2465952).

The mass balance of Antarctic sea ice has received less attention until recently. Ackley (1979) analyzed the ice cycle in the Weddell Sea where the ice advances rapidly north-eastward in May and reaches about 55° S from August to November. Near the pack ice edge, ~38% of the ice was <1-m thick, 45% was older level floes, and 17% ridged ice. Snow cover ranged from 0.1 to 0.3-m depth. The total annual production is estimated to an average of about 3.1 m, 2.1 m for the winter and 6 months and 1 m for the summer. Surface-melt ablation is not seen on the Weddell Sea pack ice inside the summer ice edge. Wintertime Antarctic sea ice has been increasing marginally (~1% decade^{-1} as of 2016) but with large year-to-year variability since passive microwave satellite record has become available in 1979 to present, with trends that are statistically significant at the 95% significant level.

Ice symbology

Ice charts display information about ice conditions using an "egg" code. This is illustrated in Figure 7.11. The code refers to total and partial ice concentrations, the stage of development (ice type), and floe size.

7.3   Ice drift and ocean circulation

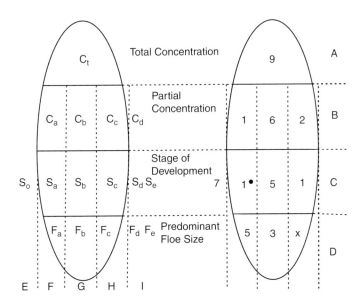

Figure 7.11 The International Egg Code showing (left) ice categories and (right) a numerical example [Source: Canadian Ice Service, Environment Canada] Row A: Total concentration expressed in tenths (in this example, 9/10). Row B: Partial concentration: breakdown of the total ice coverage expressed in tenths and graded by thickness. The thickest starting from the left and in this example, 1/10 is the thickest. Row C: Stage of development: the type of ice in each of the grades, determined by its age, that is 1/10 is medium FYI (1.), 6/10 is gray–white ice (5), and 2/10 is new ice (1). A trace of old ice is represented on the left side (outside the egg) by the number 7. Row D: Floe size: the form of the ice determined by its floe size for each section. In the example, big floes (5) for medium FYI (1.); small floes (3) for gray–white ice (5); and undetermined, unknown or no form floes (x) for new ice (1).

Column Description

Column	Description
E	Trace of ice thicker/older then S_a
F	Thickest
G	Second thickest
H	Third thickest
I	Additional groups

Pack ice is constantly in motion. On a daily basis about 80 percent of the ice drift is attributable to the wind and the rest to ocean currents and tides. In the schooner Tara's drift from the northern Laptev Sea to Fram Strait, from 24 April to 31 August 2007, a buoy network installed in a 400×400 km grid near $88°$ N drifted at 7–11 cm s^{-1} (Gascard et al., 2008). Häkkinen et al. (2008) observe that ice drift and wind stress both show gradual acceleration since the 1950s with significant positive trends in both winter and summer data. The cause of the observed trends is increasing storm activity over the Transpolar Drift Stream caused by a poleward shift of cyclone tracks. Lanfast ice is by definition immobile but a key issue is what ocean-atmosphere conditions may detach it from the shore. Mahoney *et al.* (2007b) calculate that the anchoring strength provided by grounded ridges off Barrow, AK, is 2–3 orders of magnitude greater than typical wind or water stresses. They conclude that decoupling processes, such as a sea level surge or thermal erosion of keels, must occur, in order to cause the landfast ice to detach. Another mechanism is the collision of a large pack ice floe with the landfast ice edge, detaching a portion of the landfast ice. During 1996–2004, the timing of breakup of landfast ice along the northern Alaska coast correlated strongly with daily mean air temperatures >0 °C (Mahoney *et al.*, 2007a) indicating that melting plays a significant role in destabilizing the landfast ice. Melting onset preceded spring breakup by an average of 18 days between 1997 and 2004.

Almost all of the recent information on ice drift kinematics in the Arctic comes from the IABP (see http://iabp.apl.washington.edu), which was started in 1979. Ice motion in the Arctic is generally clockwise in the western Arctic around the Beaufort due to the mean anticyclonic circulation in the atmosphere which transports ice from the Siberian coast into the North Atlantic (Serreze and Barrett 2010). In summer, there are temporary reversals when there is Arctic cyclone or easterly atmospheric circulation over the southern Beaufort Sea (Asplin *et al.*, 2009). The circulation patterns, direction and magnitude of ice drift of Arctic and Antarctic sea ice and decadal trends (%) in annual ice extent anomalies with reference to that for 1979–2012 are shown in Figure 7.12. Some of the ice supplied from the Laptev and East Siberian seas enters the Transpolar Drift Stream, which exits the Arctic via Fram Strait (Serreze *et al.*, 1989b). On average, about 15% of the ice mass in the Arctic Ocean is exported annually. especially when AO, NAO, and PNA are positive, for the anomalous anticyclonic circulation promotes more sea ice export through Fram Strait.

The local atmospheric circulation is teleconnected to the climate of remote regions through climate patterns, e.g., AO is characterized by low (high) SLP anomalies over the Arctic that lead to cyclonic (anticyclonic) atmospheric circulation anomalies, and a contracted (expanded) BG circulation (Kwok et al. 2013). These could minimize sea ice growth in winter (Hegyi and Taylor 2017). Positive AO with a stronger vortex is shown to increase the export of ice from the Arctic (Rigor and Wallace, 2004). There is also decadal-scale variability of the SLP over the Arctic Ocean that is likely to be associated with the AO and the closely related North Atlantic Oscillation (NAO) (Polyakov and Johnson, 2000). The NAO was negative (with higher pressures in the Icelandic Low) in the 1870–1900s and from 1960–1980, separated by an extended positive regime during 1900–1950, and again from the 1980s through 1997 (Portis *et al.*, 2001). During positive NAO, the enhanced north-south gradient in SLP over the North Atlantic would drive greater southward ice flux through Fram Strait (Armitage et al. 2018). From the surface forcing of meridional thermal gradients, NAO has affected the sea ice variability (Caian et al. 2018). Strong ENSO and NAO episodes were associated with anomalous sea-ice extent because strong SST anomalies and a deepened Icelandic Low lead to very strong northerly winds in the Labrador Sea. In the western (eastern) Arctic, the reaction of sea

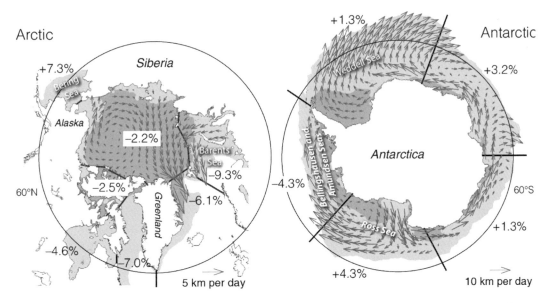

Figure 7.12 The mean circulation pattern of Arctic and Antarctic sea ice and decadal trends (%) in annual anomalies in ice extent (after removal of the seasonal cycle), in different sectors of the Arctic and Antarctic. Arrows show the average direction and magnitude of ice drift. The average sea ice cover for 1979–2012, from satellite observations, at maximum (minimum) extent is shown as gray shading (taken from FAQ 4.1, figure 1 of Vaughan *et al.*, 2013) [Source: Vaughan *et al.*, 2013]

ice to a positive AO is similar (opposite) to that of ENSO. The interaction between climate oscillations further complicates the teleconnections to sea ice: the NAO responses to ENSO in the central Pacific are mostly linear but nonlinear in the Eastern Pacific (Ding et al. 2017). This nonstationary interaction is affected by the Atlantic Multidecadal Oscillation (AMO), such that the negative ENSO-NAO interaction in late winter will only be significant when ENSO and AMO are in-phase (Zhang et al., 2018).

Based on passive microwave data in 1979–2007 the mean annual outflow at 81° N through Fram Strait was 700×10^3 km^2, with a maximum in 1994/95 of $1{,}002 \times 10^3$ km^2 (Kwok, 2009). Smedsrud *et al.* (2010) analyzed SAR-based ice velocities for 2004–2009 at 79° N and obtained an average of 880×10^3 km^2 compared with a long-term mean value for 1957–2010 of 770×10^3 km^2. They show an increasing outflow during the 5 years from 2004 to 2009. Kwok (2009) found no trend in the outflow but values were maximal when the NAO index was near its peak. The estimated annual volume flux of ice for 1991–1999 was 2,200 km^3 (~0.07 S_v; $S_v = 10^6$ m^3 s^{-1}). The net annual outflow via the passage between Svalbard and Franz Josef Land was 57×10^3 km^2. In 2007 there was an anomalous export of thick ice from the Lincoln Sea via Nares Strait (Wohlleben and Tivy, 2010). The strait is about 30-km wide and 500-km long. Normally, the southward flux of sea ice is obstructed between mid-February and mid-July when the ice in Nares Strait forms a stable ice arch at the Smith Sound. For the first time in the Canadian Ice Service ice chart records, which began in 1968, in 2007, this arch did not form. The duration of stoppage of ice movement through Nares Strait was only 58 days in 2007 compared to an average of 187 days. The ice export via Nares Strait averages ~5% of that through Fram Strait (42,000 km^2 over 5 years); however, in 2007 it was 87,000 km^2 (Kwok *et al.*, 2010). The corresponding volume fluxes were 141 km^3 (average) and 254 km^3 (2007) which is >10% of the Fram Strait outflow.

In compensation for the export of ice and water in the East Greenland Current, warm saline Atlantic Water enters the Arctic Ocean via the Norwegian Coastal Current where transit times from 60 °N are 4–6 years to the Kara Sea, 6–7 years to the Laptev Sea, 9–10 years to the NP, and 14–15 years to the Canada Basin (Dickson, 2009, fig. 42). The temperature and salinity of the waters flowing into the Norwegian Sea have recently been at their highest values for >100 years. At the eastern end of the inflow path, temperatures along the Russian Kola section of the Barents Sea (33.5° E) have equally never been greater in >100 years (Dickson, 2009).

In the Southern Ocean, data on ice motion began with buoys air-dropped into the Weddell Sea in 1979. Since 1994, the International Programme for Antarctic Buoys (IPAB) has coordinated the acquisition of data from drifting buoys in the Antarctic. Emery *et al.* (1997) used 85.5 GHz SSM/I data for 1988–1994 to map sea ice motion in both polar oceans. Schmitt *et al.* (2005) have produced an atlas combining these data sources and presenting monthly charts for March–November from 1979 to 1997.

The overall increase in Antarctic sea ice extent since 1970s is the sum of opposing trends in different sectors of the Southern Ocean (Comiso *et al.*, 2011), where advances and retreats of the ice edge have been attributed to trends in ice drift driven by wind, and large-scale intensification in surface winds associated with the circumpolar lows around Antarctica (e.g., Haumann *et al.*, 2014; Zhang, 2014).

With more than three decades of passive microwave data from SMMR, SSM/I, and AMSR-E sensors, Kwok *et al.* (2017) analyzed the mean and trends of the ice drift largely driven by wind, the variability in the location of the drift patterns, and ice export from the Weddell and Ross seas. From assessing the drift estimates in the passive microwave data using ice drift from high-resolution SAR imagery, they estimated uncertainties in ice drift data retrieved from successive satellite brightness temperature fields of 37 and 85 GHz to be about 3–4 km day^{-1}. Figure 7.13 shows the mean June–November ice drift of the Southern Ocean in 1982–2015 at a 200-km grid estimated from ERA-Interim SLP fields. There are three wind-driven, ice drift patterns within the Antarctic sea ice zone that are controlled by the location and depth of three atmospheric low centers, represented by red crosses in the monthly mean fields based on the SLP data of the ERA-Interim (http://apps.ecmwf.int/datasets/data/interim-full-daily/levty pe=sfc/) reanalysis data. Using a linear model that relates ice drift to wind, they estimated the drift speeds of sea ice in the Southern Ocean are ~1.4% that of the geostrophic wind, which is about 50% higher than results obtained in the Arctic, which suggests ice covers in the Southern Ocean is thinner, weaker, and less compact. The three distinct cyclonic, wind-driven drift patterns are centered over the Amundsen, Riiser-Larsen, and Davis seas. The first two drift patterns are dominant in all winter, while the third drift pattern is only visible in Figure 7.13 between August and November. In particular, circumpolar trends in ice edge is linked to trends

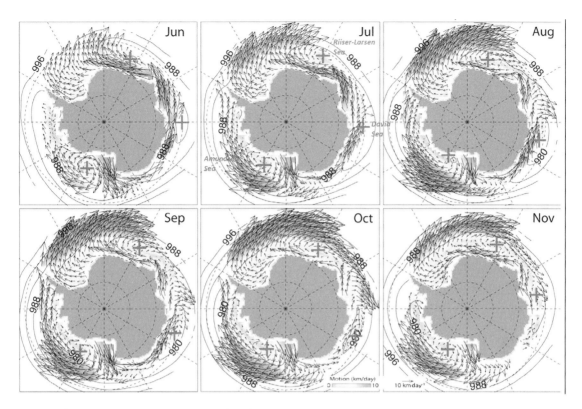

Figure 7.13 Mean monthly ice drift of Antarctica over 1982–2015 (June–November) with contours of isobars at 4-hPa intervals estimated from ERA-Interim SLP fields; drift estimates presented at a 200-km grid; and red crosses represent centers of the three atmospheric lows in the monthly mean fields (modified from Kwok *et al.*, 2017) (A black and white version of this figure will appear in some formats. For the color version, please refer to the plate section.)

in meridional winds, and to on-ice/off-ice trends in zonal winds. Anomalies in large-scale ice drift is also linked to the Southern Oscillation (SO) and the Southern Annular Mode (SAM) variability. The average sea ice export at flux gates that parallel the 1,000-m isobath in the Ross and Weddell seas are 0.75×10^6 km^2 (range of 0.35×10^6 km^2 to 1.36×10^6 km^2) and 0.32×10^6 km^2 (range of 0.02×10^6 km^2 to of 0.70×10^6 km^2), respectively.

From June through November, ice along the coasts between 175° E and 0° longitude, and in the Amundsen–Bellingshausen Sea, moves mostly from east to west. In the Ross Sea it moves northward and then turns eastward, while in the Weddell Sea it moves northwestward and then turns northeastward in the Drake Passage. There is also northward motion between 90° and 150° E. The pack ice in the Southern Ocean undergoes cyclical periods of convergence and divergence under the influence of winds and ocean currents. North of the Antarctic Divergence (~65° S latitude), the pack generally moves from west to east in the Antarctic Circumpolar Current at about 15 km d^{-1}, with a net northward component of drift.

Ice dynamics is based on five stresses: wind stress, water stress, internal ice stress due to ice interactions, Coriolis force, and the stress from the tilt of the sea surface. The Coriolis force and the tilt term are an order of magnitude less than the other three terms. The air and water stresses have a variable turning angle averaging about 25° in the Arctic and −25° in the Antarctic (Hibler, 2004), depending on the density stratification in the atmospheric and oceanic boundary layers. The wind stress, which drives the sea ice through frictional drag, is integrated over a large area – it has been estimated that concentrated pack ice responds to wind fields integrated over a distance of 400-km upwind. Internal ice stress is highly variable depending on ice conditions. It can be negligible when the ice cover is not compact and there are "free-drift" conditions, but it can be the largest force when there is thick, compact ice cover. The force due to ice resistance to deformation involves the relationship between stress and strain rate, which is termed the rheology (Flato, 2004). Feltham *et al.* (2002) report on a method to derive a geophysical sea-ice rheology.

Early work assumed that stress is linearly dependent on strain rate as in a linear viscous fluid (Campbell, 1965). Pritchard *et al.* (1977) used an elastic–plastic rheology where the stress is linearly dependent on strain up to a yield strength where failure occurs. Hibler (1979) developed a viscous–plastic model with an elliptical yield curve; the pre-yield stress states are linearly related to the strain rate. His approach has been widely adopted in ice–ocean models. The standard model treats sea ice as a visco-plastic material that flows plastically under typical stress conditions, but behaves as a linear viscous fluid, where strain rates are small and the ice becomes nearly rigid. Based on measurements from stress sensors around the SHEBA camp, however, Weiss *et al.* (2006) show that winter and/or perennial sea ice do not behave as a viscous material, even at large scales; the normal flow rule is not obeyed, and stresses are highly intermittent and spatially poorly correlated. Rather, brittle fracture and frictional sliding govern inelastic deformation over all spatial and temporal scales.

Timco and Weeks (2010) report that the tensile strength of FYI loaded in the horizontal direction ranges between 0.2 and 0.8 MPa. The few measurements for Old Ice give a range from 0.5 to 1.5 MPa. Tensile strength is important in the failure process for local and mesoscale failures. Shear strength values for FYI range from 400 to 700 kPa for granular ice and 550 to 900 kPa for columnar sea ice. There are no corresponding data for MYI. Shear strength is important for the failure mode for local failures. Typical values of the compressive strength of FYI range from 0.5 MPa to over 5 MPa. Measurements on undeformed second-year ice in the Canadian Arctic (Sinha, 1985) gave values ranging from 7 to 15 MPa. The compressive strength is important for ice crushing failures on structures.

7.4   Sea ice models

Sea ice models typically feature processes of ice thermodynamics and dynamics although early studies only treated thermodynamic ice growth and decay. The steady state Stefan (1890) relationship can be written:

$$\rho_I L H = k_i (T_m - T_a) H \tag{7.4}$$

where L = latent heat of fusion, T_m = melting point of the ice, T_a = upper boundary temperature of the ice, H = ice thickness, k_i = ice conductivity, ρ_I = ice density, and t = time. Ice growth/melt at the underside is a result of the difference between the upward ocean heat flux and the heat conducted away from the ocean/ice interface into the ice. Timco and Weeks (2010) show that

$$H = (2k_i / \rho_I L)^{0.5} [(T_b - T_a) t]^{0.5} \tag{7.5}$$

where T_b is the temperature at the bottom of the ice. This equation does not take into account snow cover, ocean heat flux, or wind and for these effects a coefficient α is included:

$$H = 0.035 \alpha \Upsilon \left[\sum (T_b - T_a) t \right]^{0.5} \tag{7.6}$$

H is in meters, T is in °C, and t is in days. For the Canadian Beaufort Sea, a best fit to ice thickness measurements is found with $\alpha \sim 0.75$. The sum of the number of freezing degree-days $\Sigma(T_b - T_a)t$ is incorporated. By the end of winter, level FYI typically reaches a thickness of about 2 m.

The first 1-D model of sea ice thermodynamics was developed by Maykut and Untersteiner (1971) and the model was applied to a large ocean area by Parkinson and Washington (1979). The model had four layers – ice, snow, ocean, and atmosphere – and 200-km horizontal resolution. The incorporation of detailed thermodynamic processes includes the presence of snow on the sea ice, leads and polynyas, melt ponds, the effect of internal brine-pocket melting on surface ablation, the storage of sensible and latent heat inside the snow–ice system, and the transformation of snow into slush ice when the snow–ice inter- face sinks below the waterline due to the weight of snow. An intermediate 1-D thermodynamic sea ice model developed by Ebert and Curry (1993) includes leads and a surface albedo parameterization that interacts strongly with the state of the surface, and explicitly includes meltwater ponds (see Figure 7.14). Four important positive feedback loops were identified: (1) the surface albedo feedback, (2) the conduction feedback, (3) the lead–solar flux feedback, and (4) the lead fraction feedback. The destabilizing effects of these positive feedbacks were mitigated by two strong negative feedbacks: (1) the outgoing longwave flux feedback and (2) the turbulent flux feedback.

	Koerner, 1973	Maykut, 1982	Holland et al., 1997	
Growth (cm)	110	130	107 bottom	7 lateral
Melt	60	94*	21 top 30 bottom	21 lateral
Net	50	36*	42	

* includes effects of melt ponds, lateral melting, and pressure ridge keels

Elbert and Curry: One-dimensional Thermodynamic Sea Ice Model

(a)

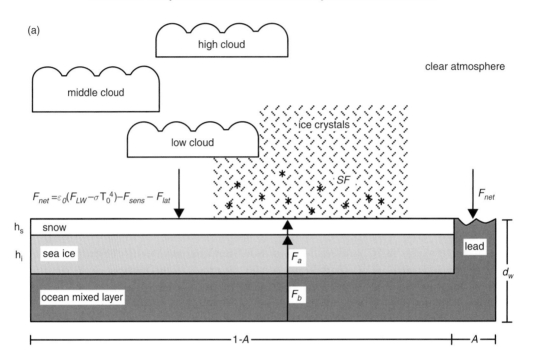

(b)

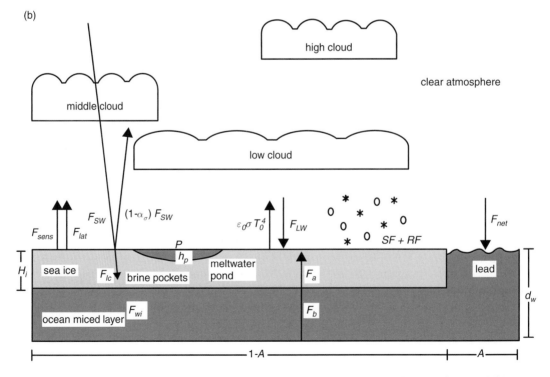

Figure 7.14 One-dimensional thermodynamic sea ice model; conditions for (a) winter and (b) summer (from Ebert and Curry, 1993) [Courtesy of American Geophysical Union]

A review of 1-D and 2-D thermodynamic models and observations is given by Steele and Flato (2000). They compare the annual sea ice growth and melt calculated by Maykut (1982) and Holland *et al.* (1997) with Koerner's (1973) observations, which are 130 and 94 cm, 107 and 51 cm, and 110 and 60 cm, respectively.

Modeling sea ice in either a stand-alone model or a GCM involves solution of the following equations (Hibler, 1979; Flato, 2004):

(1) for momentum, to obtain the ice velocity fields;
(2) for thermodynamic processes to obtain net ice growth/melt; and
(3) conservation equations including deformation and transport of ice, plus the thermo-dynamic sources and sinks.
(4) an equation of state.

Conservation equations are needed for ice area (*A*, concentration) and ice volume (*h*, thickness).

$$\partial h \partial t = -\nabla \cdot (uh) + Sh \tag{7.7}$$

$$\partial A \partial t = -\nabla \cdot (uA) + SA \tag{7.8}$$

where u is the ice velocity vector and S_h and S_A are source terms for mean ice thickness and concentration, respectively; $\nabla \cdot$ is the divergence operator. The second equation must also have the constraint that $A \le 1$. The ice strength is parameterized only in terms of h and A (Hibler, 1979).

Ice dynamics have been extensively treated by Hibler (1979, 2004). He couples the dynamics to the ice thickness characteristics by allowing the ice interaction to become stronger as the ice becomes thicker and/or contains a lower area percentage of thin ice. The dynamics in turn causes high/low oceanic heat losses in regions of ice divergence/convergence. The ice is considered to interact in a plastic manner with the plastic strength depending on the ice thickness and concentration. These in turn evolve according to continuity equations that include changes in ice mass and percent of open water due to advection, ice deformation, and thermodynamic effects. Anisotropic dynamic behavior of sea ice has also been investigated (Coon *et al.*, 1998; Hibler and Schulson, 2000), though such approaches are computationally intensive and currently are not commonly used in models. The standard model treats sea ice as a visco-plastic material that flows plastically under typical stress conditions but behaves as a linear viscous fluid where strain rates are small and the ice becomes nearly rigid. The standard viscous–plastic model has poor dynamic response to forcing on a daily time scale. Models do not generally account for high-frequency (sub-daily) inertial and tidal effects on dynamics, though research has shown that such effects can be important in the evolution of the ice cover (Heil and Hibler, 2002).

The thermodynamics and dynamics are coupled through the ice thickness distribution (Thorndike *et al.*, 1975). Essentially, deformation leads to pressure ridging and the formation of open water areas, while thermodynamic processes act to ablate ridges and remove open water by ice formation in winter and create thinner ice/open water in summer. Thus, deformation acts to spread out the thickness distribution by promoting thick and thin ice categories while thermodynamic processes work toward a central ice thickness value (Hibler, 2004).

An atmospheric GCM was coupled to a global 1-degree, 20-level ocean GCM with dynamic and thermodynamic sea ice by Washington and Meehl (1996) and run with increasing atmospheric CO_2. The Coupled Model Intercomparison Project (CMIP) allows a comparison of predicted Arctic sea ice (Meehl *et al.*, 1997). Of the 12 models, only seven include sea ice motion and only four of these have a prognostic solution to the momentum equation. Apart from errors and approximations in the sea ice representation, the models also suffer from errors in the atmospheric and oceanic forcing fields. While the Northern Hemisphere ice extent in winter is well simulated overall, the ice thickness does not capture the proper spatial distribution with thicker ice toward North America and Greenland and thinner ice in the Eurasian basin. The simulations for the Southern Hemisphere (SH) show a wider range of extents and thickness.

Johnson *et al.* (2007) examine the simulated sea ice concentration from nine ice–ocean numerical models in the Arctic Ocean Model Intercomparison Project (AOMIP). The models have similar characteristics in winter (100% cover is produced), and most models reproduce an observed minimum in sea ice concentration for September 1990. Martin and Gerdes (2007) make a comparison of sea ice drift results from different AOMIP sea ice–ocean coupled models and observations for 1979–2001. The models are capable of reproducing realistic drift pattern variability. However, one class of models has a realistic mode at drift speeds around 3 cm s^{-1} and a short tail toward higher speeds. Another class shows unrealistically a more even frequency distribution with large probability of drift speeds of 10–20 cm s^{-1}. Reasons for these differences lie in discrepancies of wind stress forcing as well as sea ice model characteristics and sea ice–ocean coupling. Hunke and Holland (2007) underscore the sensitivity of Arctic sea ice and ocean to small changes in forcing parameters. A comparison of three sets of forcing data, all variants of NCEP forcing, give significant differences in ice thickness and ocean circulation using a global, coupled, sea ice–ocean model.

An assessment of coupled climate models with respect to the development of Arctic sea ice thickness during the twentieth century is made by Gerdes and Koeberle (2007). Model behavior is compared with results from an ocean–sea ice model using the AOMIP atmospheric forcing for the period 1948–2000. The hindcast exhibits virtually no trend in Arctic ice volume over its integration period 1948–2000. Most of the coupled climate models show a negative trend over the twentieth century that accelerates toward the end of that century.

A study of GCMs used for the IPCC Fourth Assessment Report shows that while they produce reasonably similar ice extents in the Arctic, their equilibrium ice thickness values have a wide range due to differences in down-welling infrared radiation (Eisenman *et al.*, 2007). Holland *et al.* (2006) found that for some scenarios of future CO_2 concentrations the sea ice cover responds non-linearly with large decreases in extent within only 5–10 years, indicating that the current observed linear trends may not hold in the future. Stroeve *et al.* (2007) showed that the IPCC models substantially underestimate the observed decline in

Arctic sea ice extent compared to observations over the past 50 years. Hence their application in future scenarios should be treated with caution. Projections of CMIP5 climate models in the reductions of Arctic sea ice extent by the end of the twenty-first century range widely, from 43% (RCP2.6) to 94% (RCP8.5) in September (IPCC, 2019).

7.5   Leads, polynyas, and pressure ridges

Ice motion produces many important changes to the appearance and development of sea ice. The two most obvious features are leads (linear openings, 10 m to several kilometers wide and kilometers to tens of kilometers long) and pressure ridges. A large-scale divergent wind field, created by an appropriate pressure pattern, can create a divergent stress over a large field of sea ice. Since ice has little strength under tension (see p 328), divergence can open up cracks, which widen to form leads. Along the Siberian shelf the ice motion is commonly directed offshore forming large and persistent polynyas (a Russian term for irregular shaped open water areas) between the landfast ice and the moving pack ice.

Pressure ridges form due to the compressive effect of convergent ice motion, which piles up ice blocks into a linear ridge with a sail (Figure 7.6c) and a keel. Parmerter and Coon (1973) found the maximum height, crack location, and required force as functions of the mechanical and geometrical properties of the ice sheet. Hopkins and Thorndike (2002) analyze the causes of the orientation, location, and density of linear kinematic features (leads and ridges) in the Arctic and show that they are attributable primarily to wind forcing rather than to the configuration of the Arctic basin.

Leads refreeze in less than a day in winter because of the large temperature difference between the atmospheric boundary layer (typically $-30\ °C$) and the ocean surface ($-1.8\ °C$) and the high rate of heat loss as a result of infrared emission. They may also close due to convergent motion of the ice. The heat loss from a newly opened lead can exceed $1{,}000\ \mathrm{W\ m^{-2}}$ and the lead steams with Arctic sea smoke (steam fog), or frost smoke (ice fog) from the condensation/crystallization of the evaporated surface water in the cold air. Schnell et al. (1989) and Andreas et al. (1990) report cases where wide (>10 km) leads produced plumes that penetrated the Arctic inversion and extended up to 4-km altitude. The surface of the lead rapidly cools and, within hours, new ice (nilas) forms, if the surface is calm, and this cuts off the evaporation, but not the transfer of sensible heat to the atmosphere. During LEADEX in April 1992, Ruffieux et al. (1995) observed that when a 1-km-wide lead was covered with about 10 cm of ice, the sensible heat flux increased to about $170\ \mathrm{W\ m^{-2}}$, downwind of the lead. Over a 36-hour period, the average net surface heat flux was $-75\ \mathrm{W\ m^{-2}}$ over the pack ice, $-130\ \mathrm{W\ m^{-2}}$ over the lead, and $-250\ \mathrm{W\ m^{-2}}$ over the open water.

Leads occupy 0–5% of the central Arctic in winter with 10–20% in the MIZ. In an analysis of five winters of DMSP optical imagery for the western Arctic (90° W–150° E), Miles and Barry (1991, 1998) show that densities of large leads (~200–300-m wide) are observed to be highest in early winter, decreasing by 20% from November through April. The lead density averaged over all grid points for 1979–1985 was 9.9×10^{-3} km^{-1}. The measurements are in kilometers of lead length divided by area (km^2) which results in units of km^{-1}. The highest densities (14×10^{-3} km^{-1}) are observed in the central Canada Basin, and lowest in the East Siberian Sea. There is limited interannual variability in the positions of maximum and minimum densities. Preferred lead orientations are identified as generally north–south in the Beaufort Sea sector and east–west in the East Siberian Sea sector, with transitional orientations in the intermediate area. The mean distributions of lead density and orientation are observed to be associated with large-scale mean fields of ice divergence and shear, respectively. The preeminent geometric feature of the lead distribution is a characteristic rectilinear pattern, with a crossing angle of about 30°, in accordance with theory. The spatial and temporal distribution of recurring leads off the north coast of Alaska has been determined by Eicken *et al.* (2009) for December–June, 1993–2004, based on visible/infrared AVHRR data (http://nsidc.org/data/docs/noaa/g02173aklandfastandleads/index.html#3)

The analysis shows that for December–April the areal fraction was 1.9% and the number density was 0.6×10^{-3} km^{-2}; the corresponding values for May–June were 7.6% and 2.3×10^{-3} km^{-2}, respectively.

Polynyas – areas of open water within the ice pack – can be broadly divided into sensible and latent-heat polynyas. Sensible heat polynyas are large offshore openings in blankets of sea ice that releases large amount of heat stored below the ocean surface to the frigid atmosphere, resulting in loss of surface buoyancy which may lead to convective overturning, modify the ocean interior, with significant implications for large-scale ocean circulations (Pedro *et al.*, 2016). The heat release from offshore polynyas can affect the regional climate and possibly global climate patterns through atmospheric teleconnections (Weijer *et al.*, 2017). In coastal regions of Antarctica, latent-heat polynyas are a major source of the world's bottom waters, a warm upwelling ocean water that influences the process of thermohaline circulation (Figure 7.17). Polynyas are a source of heat and moisture to the atmosphere, sources of ocean bottom water, and important resources for marine life such as seals and penguins, by providing access between the ocean and atmosphere for them. Because they persist for long time, and because overturning ocean water brings nutrients to the surface, phytoplankton thrive in polynyas. During the summer, Antarctic polynyas are one of the most biologically productive regions in the world's oceans.

Polynyas tend to recur in the same location from one winter to the next. Figure 7.15 shows the Maud Polynya in the Lazarev Sea to the east of the Weddell Sea and the Antarctic Peninsula. Polynyas may range in size from 10 to 10^5 km^2. Barber and Massom (2007) provide summary tables of the physical characteristics of many Arctic and Antarctic polynyas (see also Hannah *et al.*, 2009). Figure 7.16 shows a map of the distribution of 17 polynya regions in the Arctic (Preußer *et al.*, 2016).

In the latent heat polynya (Figure 7.17), frazil ice is continually forming in streamers at the downwelling zones of Langmuir roll circulations in the ocean and is moved downwind to accumulate at the downstream edge. The open water loses sensible heat to the atmosphere and

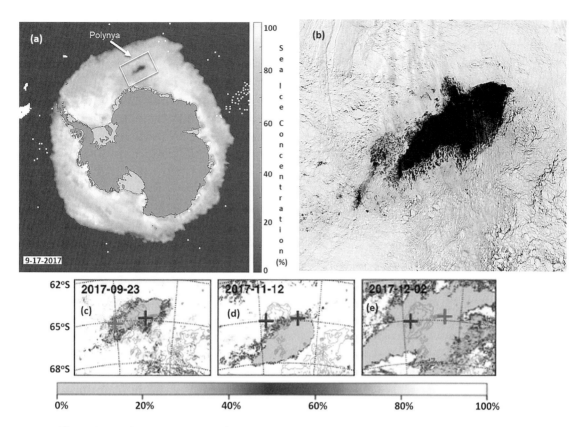

Figure 7.15 The Maud Polynya in the Lazarev Sea to the east of the Weddell Sea and the Antarctic Peninsula captured (a) on September 17, 2017 from a DMSP SSM/I image (http://mallemaroking.org/maud-polynya-2017/), (b) September 25, 2017 MODIS image of NASA Worldview (https://worldview.earthdata.nasa.gov/). AMSR2-ASI PM images of sea ice concentration (SIC) during the 2017 Maud Rise polynya of (c) September 23, (d) November 12, and (e) December 2, 2017 (adapted from Campbell *et al.*, 2019) (A black and white version of this figure will appear in some formats. For the color version, please refer to the plate section.)

radiates strongly in the infrared. The heat required to maintain the open water is supplied by upwelling warm water or by the latent heat of fusion released as new ice forms. Cold, dense, brine-rich water associated with sea ice formation accumulates over the shelf and eventually flows down the shelf slope to form deep water. The open water is a major source of latent heat transfer to the boundary layer. Smith *et al.* (1990) show that the sea–air heat flux ranges between 150 and 700 W m^{-2}, and the mean ice production rate is 0.1–0.3 m d^{-1}. The mean sensible heat flux from the Dundas Island polynya in Penny Strait in March 1980 was 204 W m^{-2} out of a total daily heat loss of 330 W m^{-2} (den Hartog *et al.*, 1983). Preußer *et al.* (2016) used an energy balance model driven with 13 sets of ice-surface temperatures retrieved from winter MODIS thermal infrared satellite data (2002/2003 to 2014/2015;

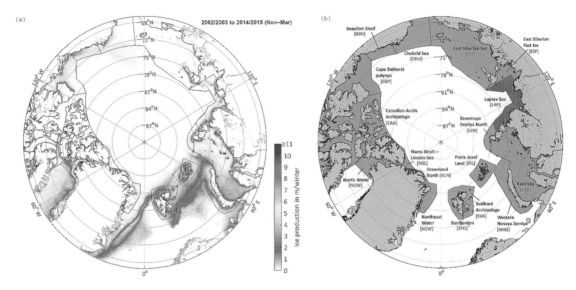

Figure 7.16 (a) Average Arctic cumulative ice production (m/winter) north of 68° N; and (b) applied polynya masks (blue) enclose the typical location of in each of 17 polynya regions between 2002–2003 and 2014–2015 (taken from Preußer *et al.*, 2016) (A black and white version of this figure will appear in some formats. For the color version, please refer to the plate section.)

November to March), and ECMWF ERA-Interim atmospheric reanalysis data to characterize circumpolar polynya dynamics and ice production (Figure 7.16a). They computed daily thin-ice thickness (TIT) distributions (<0.2 m) of 17 coastal polynya regions over the entire Arctic basin, from which they estimated polynya area and total thermodynamic ice production for the 17 polynya regions which together cover an average thin-ice area of $227 \pm 36 \times 10^3$ km^2 in winter, resulting in an average total winter-accumulated ice production of about 1811 ± 293 km^3, whereby the Kara Sea (15%), the North Water (NOW) (15%), western Novaya Zemlya (20%), and scattered polynyas in the Canadian Arctic Archipelago (all combined 12%) are the main contributors. Their results are distinctly different from earlier studies on pan-Arctic polynya characteristics because of the use of high-resolution MODIS data. They detected positive trends in ice production over the 13 seasons in the eastern Arctic and the NOW polynya.

Sensible heat polynyas are reported from the Sea of Okhotsk (Afultis and Martin, 1987) and Whaler's Bay north of Svalbard (Smith *et al.*, 1990), but are generally less common, although both processes may operate together as in the NOW of Baffin Bay (Steffen, 1985). A large (200,000 km^2) polynya formed In the Weddell Sea ice over the Maud Rise during 1974–1976 was identified from ESMR passive microwave data. Holland (2001) explains it as the result of a cyclonic eddy shed from the Maud Rise that then interacted with ocean thermodynamic processes. It would have required a heat flux of ~100 W m^{-2} for its survival (Morales Maqueda *et al.*, 2004).

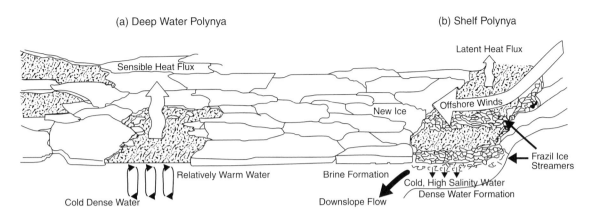

Figure 7.17 Schematic illustration of the physical processes taking place in (a) deep water and (b) a shelf polynya. Modified after Morales Maqueda *et al.* (2004)

Lately, Campbell *et al.* (2019) identified the strengthening of SAM, Weddell Low, cyclonic wind stress curl, and winter storm activity as positive multidecadal trends in polynya-favorable conditions in the Southern Ocean. In the Antarctic, the largest polynyas observed in the Weddell Sea region near the Maud Rise seamount and in the Cosmonaut Sea offshore of East Antarctica in the last few decades had recently recurred in 2016 and 2017 (Swart *et al.*, 2018). Kurtakoti *et al.* (2018) attributed this to topography–flow interactions at the Maud Rise that enhances upward heat fluxes and generates eddies that transmit divergent strain to the ice cover.

Instead of the traditional view of rapid escape of deep-ocean heat through convective mixing, by profiling float observations collected in 2016/2017 in the Weddell Sea region of the Southern Ocean, Campbell *et al.* (2019) demonstrated that polynyas observed near the Maud Rise seamount since 1976 were initiated and modulated by the passage of severe storms, and the intense heat loss drove deep overturning within them due to reduced haline stratification in the upper ocean related to persistent wind-driven upwelling. They show that past Weddell polynyas probably developed under similar anomalous climatic conditions associated with a mode of SH climate variability, predicted to strengthen under the effect of climate change.

Pease (1987) developed a steady-state model of a wind-driven polynya which showed good agreement with observations from the Bering Sea. Observed widths were 10–20 km and steady-state conditions were established within 24–36 hours. Polynya width is almost inversely proportional to the surface-to-air temperature difference, so that the total heat loss from the surface is almost independent of air temperature but is proportional to the wind speed.

Terra Nova Bay polynya in Antarctica is maintained during the winter season by 25–40 m s^{-1} katabatic winds that are channeled by glacial valleys and flow off the ice sheet over the adjacent sea ice (Bromwich and Kurtz, 1984). The NOW in northern Baffin Bay has received considerable attention since the 1960s (Dunbar, 1969; Müller *et al.*, 1977). It may extend over 50,000–80,000 km^2. Schneider and Budeus (1997) show that in winter, strong northerly winds push newly formed sea-ice out of the northern NOW area. An ice bridge regularly forms in Smith Sound between November and March, but it has formed later and broken up earlier in the 1990s compared with the 1980s (Barber *et al.*, 2001b). Since during winter the air–sea temperature difference can be up to 30 °C, new ice forms rapidly balancing the ice export. The NOW typically has some 90% ice cover in winter including a substantial proportion of FYI floes over 30-cm thick (Steffen, 1986). Barber *et al.* (2001a) note that localized sensible heat effects occur in autumn, winter, and spring along the Greenland side of the polynya, supporting the findings of Steffen (1985). There is a 5 °C temperature difference between the coast of Ellesmere Island and the coast of Greenland as a result of warm advection in the boundary layer. In summer, winds are weak and the northward flowing coastal current constitutes the dominant forcing of the NOW summer polynya, forming in the southern part of the area. During this season the polynya gradually increases its size toward the north since the air–sea heat budget is positive and no new ice-formation occurs.

On the 25–30-m deep Laptev Sea shelf, the surface water has a salinity of ~10 PSU due to river inflow, while the bottom water is at 32 PSU (Höllemann *et al.*, 2010). The seasonal amplitude of salinity and inferred net sea-ice production in winter during the 1960s–1990s is found to strengthen (weaken) when the AO is positive (negative) because wind-driven advection moves more (less) ice away from the coasts (Dmitrenko *et al.*, 2009). There is enhanced (diminished) polynya formation in the flaw lead, between the landfast and the pack ice, and more (less) brine release in the shelf waters during the respective AO modes. The polynya is 10–100-km wide and up to 2,000 km in length. Bareiss and Görgen (2005) show that from November to June 1979/1980–2001/2002, the mean area of the West New Siberian polynya in the southeast Laptev Sea averaged 4,000 km^2 and had a mean duration of 14 days, while the Annabar–Lena polynya averaged 3,000 km^2 and had a mean duration of 22 days. The mean cumulative areas of the two were $1,713 \times 10^3$ km^3 and $1,152 \times 10^3$ km^2, respectively, associated with a mean frequency of 12.4 polynya events in all investigated regions. In fast-ice areas exposed to surface flooding from rivers, coastal polynyas develop, on average, after 4 weeks. During January–April 2008, the Laptev polynya generated 1.8% of the total ice volume in the Arctic according to Rabenstein (2010).

Arctic polynyas provide a significant marine ecosystem; they are a source of plankton, krill, and fish, and large colonies of arctic birds breed nearby. Many marine mammals – seals, walruses, narwhal, whales, and polar bears – depend on them as feeding grounds and over-wintering areas. In the Antarctic, polynyas support plankton, krill, squid, fish, seals, and whales.

The coastal polynya area around Antarctica during JJAS (wintertime) 1992–2008 is estimated from SSM/I data to be 245,000 km^2 (Kern, 2009). The polynyas along East Antarctica (60°–160°E) comprise about 40% of the total; the most persistent are located along the Lars-Christensen Coast (LCC), Prydz Bay, the western Davis Sea, Mertz Glacier, and in the Ross Sea along the Ross Ice Shelf, and in Terra Nova Bay. The polynya at the LCC is observed on 110 ± 5 days during winters 1992–2008 and covered an average area of 2,400 km^2 on more than 90 days.

Ice in refrozen leads is the weakest part of the ice cover and is the first part to be crushed into piles of broken ice blocks, when the wind is convergent. Such linear deformation features are called pressure ridges, the above-water part being the sail and the much more extensive, below-water part the keel. In the Arctic, most keels are about 10–25-m deep; the deepest keel on record had a draft of 47 m (Lyon, 1961). Maximum pile-up height is ~15 m with a tendency for that height to increase with ice thickness (Timco and Barker, 2002). Ridged ice in the Arctic makes a major contribution to the overall mass of sea ice; probably about 40% on average and more than 60% in coastal regions. The typical ridge/sail ratio is ~3–4:1 (Tucker, 1989). The spacing of pressure ridges follows a lognormal distribution (Key and McLaren, 1989). The mean thickness of pressure ridges according to Johnston et al. (2009) is, on average, 9.9 m (± 4.7 m) and the most massive pressure ridge had a mean thickness of 24.7 m (Kovacs, 1975).

7.6   Ice thickness

Ice thickness is determined directly by drilling holes in the ice or by IMB buoys equipped with acoustic range-finder sounders and a thermistor string for internal temperatures (Richter-Menge et al., 2006). In addition, there are differential airborne EM induction and laser altimeter measurements, upward looking sonars on submarines or moored to the ocean floor, which record the ice draft below the sea surface, and airborne or satellite radar or laser altimeters that measure the ice surface height (freeboard) above the water surface. A detailed account of all existing methods, their advantages and limitations is given by Haas and Druckenmiller (2009).

The relationship between ice draft and ice thickness is determined from

$$\text{Thickness} = \text{draft}(R + 1)/R \qquad (7.9)$$

where $R = \dfrac{\rho_i h_i + \rho_s h_s}{h_i(\rho_w - \rho_i) + h_s(\rho_w - \rho_i)}$

ρ_i = mean ice density, ρ_s = mean snow density, ρ_w = mean density of sea water, h_i = mean ice thickness, and h_s = mean snow thickness. $R = 5.686$ for a snow thickness of 25 cm and 7.700 for a snow thickness of 5 cm (Wadhams et al., 1992).

The processes of ice growth, melt, advection and ridging all affect the frequency distribution of sea ice thickness. The mechanical terms involve the divergence of ice mass on the ocean surface, which forms leads; the advection of ice mass parcels from one location to another; and a random redistribution term that includes thermal processes like lateral growth and melt of ice, and ridging and piling-up of ice which results from inelastic collisions and ice sheet deformation.

Thorndike (1992) first described the relationship between the ice thickness probability density function and the growth rate of sea ice in a stochastic differential equation:

$$\frac{dg}{dt} = \frac{\partial(fg)}{\partial h} \tag{7.10}$$

where

$$f \equiv \frac{dh}{dt}$$

and g defines the ice thickness distribution. Rothrock (1986) defines $g(h)$ as the fraction of R (an area defined on the ocean surface) with ice thickness between h and $h+ dh$. Thus, h is a function of location on the ocean surface, $h(x,y)$, where $h(x,y)$ may be treated as a stochastic process (a random variable). Dedrick (2002) develops techniques for estimating the ice thickness distribution from digital sea ice charts accessed by a Geographic Information Systems (GIS).

In the Antarctic, two decades of data compiled by the SCAR ASPeCt) program, totaling over 23,000 observations, give a mean thickness of all ice as 0.87 ± 0.91 m compared with a level ice thickness of 0.62 m. Over the years, the ASPeCt data archive has been used in various studies on the Antarctic sea ice, for example, Hutchings *et al.* (2012). North/south and east/west transects revealed lag distances over which sea ice thickness decorrelates to be of the order of 100–300 km (Worby *et al.*, 2008a). Using data from ICESat for 2003–2009, Yi *et al.* (2010) measure ice thicknesses in the Weddell Sea. During winter (October–November), sea ice grows to its seasonal maximum both in area and thickness with the mean thicknesses of 2.1–2.2. In summer, the mean thicknesses are 1.6–2.1 m in the western Weddell Sea where ice persists.

In the Arctic ice thickness has been primarily mapped from upward-looking sonar (ULS) on submarines (Wadhams and Amanatidis, 2007). The earliest data were obtained by the USS Nautilus in August 1958 and compared with measurements from the USS Queenfish along the same track at the same time of year in 1970 (McLaren, 1989). Nautilus recorded generally more severe ice conditions within the Canada Basin than did Queenfish; overall mean drafts were 3.08 and 2.39 m, respectively. The thickest ice is found north of the Queen Elizabeth Islands and northern Greenland. Here, during the 1960s to 1980s, the thickness reached 6–7 m (Bourke and Garret, 1987; Bourke and McLaren, 1992). The ICESat measures ice height above the freeboard and enables ice thickness to be determined. Kwok and Cunningham (2008) obtained estimates for the Arctic for October–November (ON) 2005 and 2006 and February–March (FM) 2006, and March–April (MA) 2007. The mean thickness was 2.46 m in FM 2006, with a snow thickness of 40 cm, and 2.37 m in MA 2007. There was a higher MYI cover in ON 2005 of 37% versus 31% in ON 2006. Rabenstein et al. (2010) describe ice thickness and surface properties of different sea-ice regimes within the Arctic Trans Polar Drift from Summers 2001, 2004, and 2007. Lindsay (2010) reports on a new unified ice thickness data set.

A novel approach to determining basin-scale ice thickness is proposed by Wadhams and Dobie (2010). Small-amplitude, long period, infragravity waves in the ocean (compare p. 373) can be used to measure ice thickness by determining their travel time between measurement sites. The waves travel at different speeds in ice and open water, with the difference being a sensitive function of ice thickness. Measurements made near the NP show that the travel time of 15 s waves is reduced by around 7 hours for a typical 2 m ice thickness. Their results show that measurements are feasible for the region between Fram Strait and the central Arctic, where a relatively direct deep-water path exists from the North Atlantic source of the waves (thought to be generated by storm waves intersecting the coast of northwest Africa).

7.7   Trends in sea ice extent and concentration from paleo and NSIDC data

Under the impact of climate warming, Arctic sea ice has been decreasing and it appears to be heading in an unrecoverable direction, the melting season has lengthened (Stroeve and Notz, 2018) and sea ice has become younger and thinner (Lindsay and Schweiger, 2015). Given the Arctic sea ice is shrinking at an unprecedented rate, and reliable satellite data for the frozen north only began in 1978, Walsh *et al.* (2016) recently published a new Arctic sea ice database called SIBT1850, "Gridded Monthly Sea Ice Extent and Concentration, 1850 Onward," which serves as a benchmark in putting the current rate and extent of retreat of the Arctic sea ice in a better perspective (http://seaiceatlas.snap.uaf.edu) Walsh et al. gathered historical sea ice information dated between 1850 and 1978, which include Arctic sea ice charts from the Danish Meteorological Institute, compilations by US Navy oceanographers, whaling ship log books, a company's ice charts of waters around Alaska and Scandinavian ice edge data. Information from these historical sources were digitized and synthesized so that they are compatible with one another, and missing data were gap filled using an analog method. The SIBT1850 data in Figure 7.18a shows that since 1850, monthly Arctic sea ice extent for January to December had minor fluctuations but with no detectable trends until about 1970s, when negative trends began and has persisted and/or accelerated over the last four decades, especially in the summer (June–September). The recent global scale climate warming detected is the obvious driving force behind the current decline in Arctic sea ice.

In another paleo Arctic sea ice data study to address the question of whether or not declining trends for the last several decades are anomalous, Kinnard *et al.* (2011) used a network of high-resolution terrestrial proxies (Arctic ice cores, tree-ring and lake sediments) from the circum-Arctic region to reconstruct past extents of summer sea ice for the past 1,450 years. Despite of large uncertainties associated with the reconstructed, paleo

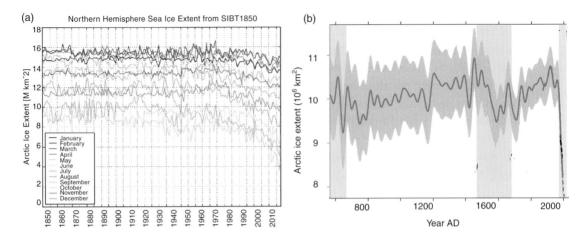

Figure 7.18 (a) The Arctic sea ice extent in millions of km^2 present at any concentration plotted for each month, from 1850 to 2013 (taken from Walsch et al., 2016); and (b) Reconstructed Arctic sea ice extent over the last 1,450 years (taken from Kinnard et al., 2011). (A black and white version of this figure will appear in some formats. For the color version, please refer to the plate section.)

Arctic sea ice data, especially before the sixteenth century, it is obvious the duration and magnitude of the current decline in sea ice is unprecedented for the entire past 1,450 years (Figure 7.18b). This resembles another ice hockey stick similar to the Arctic summer temperature hockey stick of Kaufman *et al.* (2009) and the warming of Atlantic water into the Arctic hockey stick (Spielhagen, 2011), which together with anthropogenic warming, are likely the main driving factor behind the decline of sea ice extent on multidecadal timescales, and may result from nonlinear feedbacks between sea ice and the Atlantic meridional overturning circulation.

One of the longest records of sea ice conditions is that compiled by Lauge Koch (1945) for East Greenland ice off the coasts of Iceland based on reports dating back to the early colonization of Iceland that were collected by several Icelandic authors, including P. Thoroddsen in 1917. The time series of the number of weeks that sea ice is observed during October–September near the Icelandic coasts was extended by the Icelandic Meteorological Office to 1983, and further updated to 1990 by Wallevik and Sigursjóhnsen (1998). There are two periods of frequent occurrence of ice (around AD 1300 and AD 1550–1900), separated by two to three centuries with nearly ice-free waters. The first maximum about AD 1300 coincides with a period of more severe climate in Europe. The second maximum is coincident with the Little Ice Age in Europe and is followed by an abrupt decrease in the first decades of the twentieth century. Wallevik and Sigursjóhnsen (1998) recalculated the Koch series using seven different algorithms. These included counts for four 3-month seasons. Lassen and Thejll (2005) analyze the data from AD 1500 and claim to find a correlation with the Gleissberg solar cycle (~88 years). Ogilvie (1984) does not consider the data prior to AD 1600 to be reliable enough for quantitative evaluation. She identified further sources showing that sea ice was common off the north coast in the 1590s whereas the Koch graph suggests little or none. However, there is evidence for a mild period between 1640 and 1670, and severe decades in the 1630s, 1690s, 1740s, and 1750s. Ogilvie and Jonsson (2001) also compiled a sea ice index, based on ice occurrence off the northwest, north, east, and south coasts, for 1601–1850 from contemporaneous sources. They show light ice conditions from about 1640 to 1680 and a large increase during 1780–1840. Sea ice years recurred during 1864–1872 and the 1880s, but the incidence decreased sharply after 1903 until the late 1960s. Speerschneider (1931) and Koch (1945) provide a related record of the Storis drift along the West Greenland coast from 1820 to 1930. This characterizes the northern extent of ice from the East Greenland Current that has been carried around Cape Farewell.

Decadal to centennial variability of maximum sea ice extent has been reconstructed for the western Nordic Seas for AD 1200–1997 by Macias Fauria et $al.$ (2009) by combining a regional tree-ring chronology from timberline areas in Fennoscandia and δ^{18} O from the Lomonosovfonna ice core in Svalbard. The twentieth century has sustained the lowest sea ice extent values since AD 1200; low sea ice extent also occurred in the mid-seventeenth and mid-eighteenth centuries, early fifteenth and late thirteenth centuries, but these periods were all less persistent than in the twentieth century. Largest sea ice extent values occurred from the seventeenth to the nineteenth centuries, during the Little Ice Age. Low-frequency variability centered at 70–90 year and 7–32 year frequency bands was probably linked in part to the AO/NAO. For the Barents Sea, Vinje (1999) compiled August records from 1580 to 2002, with sparse coverage from 1680 to 1740. He reports that the August edge was located around 76° N in 1640 and 1800, around 78–79° N from 1680 to 1780, and shifted north of 78° N again after 1930 (Vinje, 1999). There is a strong correlation of ice extent with the July–August temperature series for central England from 1695 and for the Northern Hemisphere from 1860. Divine and Dick (2006) use ice observations for the Nordic Seas from April through August to construct time series of ice edge position anomalies spanning the period 1750–2002. They found evidence of oscillations in ice cover with periods of about 60 to 80 years and 20 to 30 years, superimposed on a continuous negative trend. The lower frequency oscillations are more prominent in the Greenland Sea, while higher frequency oscillations are dominant in the Barents Sea. Vinje (2001) analyzed changes in April and August ice extent during 1864–1998. In April, the extent of ice in the Nordic Seas has decreased by ~33% (from 2300 to 1600 × 10^3 km^2) since 1864, with a much larger reduction in the western sector than the eastern sector. Nearly half of this reduction occurred between 1864 and 1900. Since 1920, the ice extent in August in the eastern sector has been more than halved. Using 85 GHz SSM/I data, Kern et $al.$ (2010) show that there is a 2 months' longer ice-free season in the Irminger Sea (west of Iceland) in the 2000s compared with the 1990s, and reductions in ice area between 1992–1999 and 2000–2008 by 17% in winter and 45% in summer. In the Barents Sea the corresponding reductions were 20% and 54%, respectively.

For eastern Canada, Hill et $al.$ (2002) compile an historical record of sea ice extent on the Scotian Shelf and in the Gulf of St. Lawrence from the early 1800s to 1962, extending back an earlier record from 1963 to 2000. The ice extent east of Cabot Strait over the Scotian Shelf increased from low values in the early 1800s to around 40,000 km^2 during 1850–1880, dipped to half of this in the first decade of the twentieth century, and then reached 60,000 km^2 in the 1920s with a record of 120,000 km^2 in 1923, before declining to ≤20,000 km^2 in the 1950s. Lowest values of ~10,000 km^2 were in the 1970s and 2000. In May, ice extended east of Cabot Strait 49% of the time from 1963 to 1997, and remarkably this was identical to the frequency from 1844 to 1962.

The historical record of sea ice extent in the Arctic dates back to 1870. Kinnard *et al.* (2008) show that the seasonal sea ice has gradually expanded over that time, particularly during the last three decades. In a separate reconstruction using historical observations and a coupled climate model simulation, Brönnimann *et al.* (2008) show that sea ice concentrations in late summer began to decrease sharply after about 1970.

Falkingham *et al.* (2002) analyze sea ice in the Canadian Arctic from 1969 to 2001 which includes the eastern and western Arctic, Hudson Bay, and the Labrador Sea. They use the total accumulated coverage (TAC) from the 17 weekly ice charts of June 25 to October 15. There is considerable variation in trends. For Hudson Bay the TAC declined 40% between 1971 and 2001 (13% decade^{-1}). For 1969–2001, the corresponding decadal values for both the eastern and western Arctic were 5%. However, in the Western Arctic Waterway (Amundsen, Coronation and Queen Maud Gulf) it was 11% decade^{-1}. In contrast, in Lancaster Sound there was no trend over the 30 years, yet in the northern Labrador Sea the TAC declined a remarkable 72% (24% decade^{-1}).

With reference to the 1979–2000 mean, Figure 7.19 shows the September Arctic sea ice concentration anomaly maps of 2004, 2007, 2010, and 2012 derived from the NSIDC Sea Ice Index have been declining, in particular north of Greenland where the concentration has

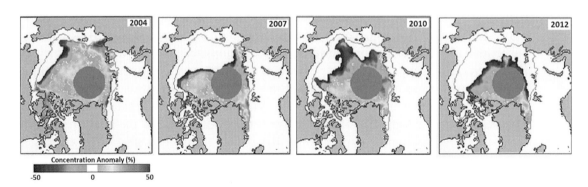

Figure 7.19 September Arctic sea ice concentration anomaly maps of 2004, 2007, 2010, and 2012 derived from the NSIDC Sea Ice Index. Each image shows the concentration anomaly (see color key) and the 1979–2000 mean September ice edge (pink line). Every year, the ice edge is north of its mean position off the coasts of Alaska and Siberia (Image of NSIDC, Boulder (https://nsidc.org/cryosphere/sotc/sea_ice .html) (A black and white version of this figure will appear in some formats. For the color version, please refer to the plate section.)

been decreasing so that since 2010, most anomalies have been negative. The Arctic sea ice extent has been shrinking, and to its lowest record in 2012, to less than half of the 1979–2000 mean. Using sea ice index data of NSIDC of 1979–2010 (http://nsidc.org/data/seaice_index/ n_plot.html), the trend in Northern Hemisphere and SH sea ice concentration was $-3.5 \pm 0.7\%$ decade^{-1} and $1.4 \pm 1.2\%$ decade^{-1}, with mean values for 1979–2000 of 12.2 and 13.8 million km^2, respectively.

Parkinson and Cavalieri (2008) show that annual averages for the Arctic for 1979–2006 (updated to 2008 by Comiso, 2010) have an overall negative trend of $-3.7 \pm 0.4\%$ decade^{-1}; negative trends of ice extent are also observed for each of the four seasons, and for each of the 12 months. For September 1979–2009, the trend is -11.9% decade^{-1} (Stroeve, 2010). For the yearly averages, 1979–2007, the largest area decreases occur in the Kara and Barents seas, with linear least squares slopes of $-7.4 \pm 2.0\%$ decade^{-1}, followed by Baffin Bay/Labrador Sea, with a slope of $-9.0 \pm 2.3\%$ decade^{-1}, the Greenland Sea, with a slope of $-9.3 \pm 1.9\%$ decade^{-1}, and Hudson Bay, with a slope of $-5.3 \pm 1.1\%$ decade^{-1}. The largest decreases have occurred for July through October (Deser and Teng, 2008). Eisenman (2010) analyzed the northward retreat of ice edge latitude (the point with ice-covered ocean to the north and ice-free ocean to the south) during 1978–2010 and found nearly identical rates in March and September. The annual mean trend is 8 km a^{-1}, giving a northward shift of 250 km over the 31-year period.

Rodrigues (2009) analyzed the length of the ice-free season (LIFS) and a variable designated by the inverse sea ice index (ISII); the LIFS at a certain point in a particular year is defined as the number of days between the clearance of the ice and the appearance of ice in that point in that year; the ISII measures the absence of sea ice throughout the year, which varies between zero (perennial ice cover) and one (open water all year-round). Between 1979 and 2006, the spatially averaged ice-free season in the Arctic lengthened by 1.1 days year^{-1} (from 119 days in the late 1970s to 148 days in 2006), but it increased to 5.5 days year^{-1} during 2001–2007. In 2007 and 2008, the average ice-free seasons in the Arctic were 168 and 158 days long, respectively. The ISII reached a maximum of 0.50 in 2007, while the minimum (0.40) was registered in 1982.

Trends in sea ice thickness

Kwok and Rothrock (2009) examined sea ice thickness records from 42 years of submarine records (1958 to 2000) and 5 years of ICESat records (2003 to 2008) and estimated that the mean Arctic sea ice thickness had declined by 1.75 m from 3.64 m in 1980 to 1.89 m in 2008. Laxon *et al.* (2013) compared sea ice volume between 2003–2008 and 2010–2012 using data from ICESat, the Pan-Arctic Ice-Ocean Modeling and Assimilation System (PIOMAS) and the ESA CryoSat-2 mission. They found that sea ice volume declined by 4,291 km^3 at the end of summer, and 1,479 km^3 at the end of winter, respectively.

Using maps that show Arctic sea ice thickness at the end of melt season for 2004–2008 from ICESat, and 2011–2014 from CryoSat-2, Kwok and Cunningham (2015) estimated declining

Table 7.5 Regional sea ice thickness (m) estimates of the Arctic from submarine (1958–1976 and 1993–1997), ICESat (2003–2007), and CryoSat-2 (2011–2017) data (adapted from Kwok, 2018)

Data Type	Chukchi Cap	Beaufort Sea	Canada Basin	North Pole	Nansen Basin	Eastern Arctic
1958–1976 Submarine	1.59	1.9	3.4	3.75	3.8	3.2
1993–1997 Submarine	0.95	0.95	2.05	2.2	2.02	1.25
2003–2007 ICESat	0.7	0.95	1.7	1.8	2.0	1.23
2011–2017 CryoSat-2	0.6	0.65	1.4	0.9	1.5	0.85

trends in sea ice volume of 402 km^3 year^{-1} in winter and 760 km^3 year^{-1} in summer, which works out to be higher than the estimates of Laxon *et al.* (2013). In a later study using ICESat and CryoSat-2 data, Kwok (2018) reported lower losses in sea ice volume at 287 km^3 year^{-1} in winter and 513 km^3 year^{-1} in fall over 2003–2017. He also estimated six regional sea ice thicknesses that cover about 38% of the Arctic Ocean using submarine (1958–1976 and 1993–1997) and satellite (2003–2007 and 2011–2017) data (Table 7.5). Overall, between the submarine (1958–1976) and the CryoSat-2 periods (2011–2017), the average thickness of the Arctic Ocean near the end of the melt season, over six regions, decreased by 2.0 m or about 66% over six decades. From analyzing about six decades of data from submarine sonars (1958–1997), and ICESat and CryoSat-2 satellites (2003–2017), Kwok (2018) show in a study area $\sim$ 38% of the Arctic Ocean, that the average Arctic sea ice thickness near the end of the melt season had an overall thinning of $\sim$ 1.7 m from a maximum estimated thickness of 3.64 m in 1980 to about $\sim$ 2 m in 2017. From the 1999 to 2017 record of QuikSCAT and ASCAT scatterometers, he estimated the Arctic has lost more than 50% (2×10^6 km^2) of its MYI, which now covers less than one-third of the Arctic Ocean. As a result, the Arctic Ocean is expected to be mainly controlled by seasonal ice, and will be more sensitive to climate forcing.

The ESA CryoSat-2 radar altimeter launched in 2010 (Laxon *et al.*, 2013) observed a mean sea ice thickness of 2.14 m in the Arctic Basin in April 2018, which is marginally lower than the average, 2.19 m measured over the 2010–2018 CryoSat-2 record ranging from 2.03- to 2.29-m thick (Perovich *et al.*, 2018). For the winter of 2017/2018, the mean Arctic sea ice thickness anomaly shows that thicknesses were above average in the East Siberian Sea shelf areas, with thinner ice in Fram Strait, Beaufort Sea, and Bering Strait. The recent launch of NASA's ICESat-2, a laser-based altimeter, will enhance large-scale measurements of sea ice thickness in the next several years.

Arctic sea ice age

The Arctic sea ice covers the largest area of the Arctic (Arctic sea ice maximum) at the end of the winter cold season, which usually begins in September and ends in March. The Arctic sea ice has been recovering less in the winter in recent years, meaning the sea ice is more likely to melt off over the summer, possibly because the underlying Arctic Ocean has been warmer. Sea ice age provides an early assessment of the Arctic Ocean most susceptible to melting out during the coming summer. The Arctic sea ice cover continues to become younger and thinner since younger sea ice tends to be thinner than older ice. Figure 7.20 shows Arctic sea ice age for 1984, 1995, 2005, 2015, and 2019, and Arctic sea ice age time series of mid-April as a percentage of the Arctic Ocean coverage from 1984 to 2019, which shows the almost complete loss of 4+-year-old ice in recent years. By 2019, almost all of the sea ice older than 4 years, which once made up about 30% of the Arctic sea ice, is gone. As of mid-April 2019, the 4+-year-old ice made up only 1.2% of the ice cover (Figure 7.20). On the other hand, 3 to 4-year-old ice increased slightly, increasing from 1.1% in 2018 to 6.1% in 2019. If the 3-year-old ice survives the summer melt season, it will somewhat replenish the 4+-year-old category going into the 2020 winter, but there has not been much replenishment lately (Source: https://nsidc.org/arcticseaicenews/).

Rigor and Wallace (2004) found that the age of sea ice explains more than half of the variance in summer sea-ice extent. Seasonal ice – which melts and refreezes every year – comprised about 70% of Arctic sea ice in winter, up from 40 to 50% in the 1980s and 1990s.

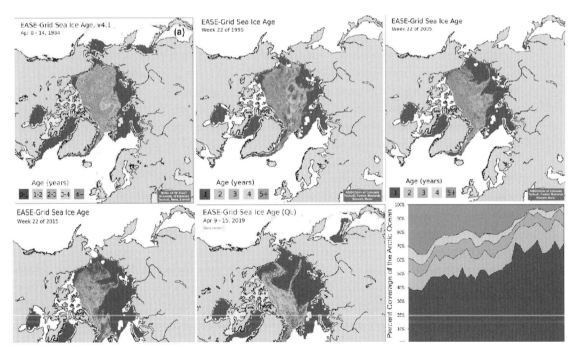

Figure 7.20 Arctic sea ice age for (a) April 8-14, 1984, Week 22 of 1995 (b), 2005 (c), and 2015 (d), and April 9-15, 2019 (e), and (f) Arctic sea ice age time series of mid-April as a percentage of Arctic Ocean coverage from 1984 to 2019, which shows the almost complete loss of 4+ year old ice in recent years. Note that the age time series is for ice within the Arctic Ocean only (Source: https://nsidc.org/arcticseaicenews/). (A black and white version of this figure will appear in some formats. For the color version, please refer to the plate section.)

The decreases in second and MYI have accelerated in recent years. Scatterometer data from the QuikSCAT satellite suggests a recent precipitous decrease in the perennial ice extent, for example, showing a 23% loss between March 2005 and March 2007 (Nghiem *et al.*, 2007). Wang *et al.* (2009) show that strong meridional wind anomalies drove more sea ice out of the Arctic Ocean from the western to the eastern Arctic and into the northern Atlantic during the summers of 1995, 1999, 2002, 2005, and 2007. This pattern reflects the Arctic atmospheric dipole anomaly (DA) of SLP; the wind anomaly blows from the western to the eastern Arctic during the +DA phase, accelerating the TransPolar Drift Stream, and vice versa during −DA. Ogi *et al.* (2010) extend Wang et al.'s study and show that the combined effect of winter and summer wind forcing accounts for 50% of the variance of the interannual change in September Arctic sea ice extent and it also accounts for about one-third of the downward linear trend of ice extent since 1979.

In September 2007, ice extent in the Arctic declined to a record daily low of 4.13 million km^2 (4.3 million km^2 for the monthly average) compared with 5.9 million km^2 in 2003 (Figure 7.21), and 7.88 million km^2 in 1996. The minimum value for September 2008 was slightly higher but still only 4.6 million km^2 (Perovich and Richter-Menge, 2009).

The Northwest Passage was open in 2007, 2008, and 2009 and the Northern Sea Route, north of Siberia, in 2008 and 2009, and have been navigated by Western commercial and passenger vessels in recent years. The summer of 2007 with the second lowest minimum sea ice recorded so far was relatively cloud free in the Beaufort Sea but calculations of the extra incoming solar radiation for June–August suggest that this factor was not the main determinant of the ice loss (Schweiger *et al.*, 2008). Nevertheless, Perovich *et al.* (2008) demonstrate that there was an extraordinarily large amount of bottom melting of the ice in the Beaufort Sea in summer 2007 and calculations indicate that solar heating of the upper ocean was the primary source of heat for the observed melting. The positive anomaly in solar heat input was due to an almost doubling of the area fraction of open water in 2007 compared to climatology.

During 1979–2005 there was an increase of 17% (amounting to 2.9×10^{15} MJ) in total heat input into the Arctic Ocean (Perovich *et al.*, 2007). This is enough heat to melt 9.3×10^{12} m^2 of ice. In the region of the Beaufort–Chukchi–East Siberian Sea the heat input from 1979 to 2005 increased by 69% (see also Carmack and Melling, 2011).

Based on passive microwave SMMR and SSM/I data collected since 1979, the annual average extent of ice in the Arctic has decreased by about 3.8% decade^{-1}. The decline in extent at the end of summer (in late September) has been even greater at 11% decade^{-1}, reaching a record minimum in 2012 of 3.39 million km^2, 44 % below the 1981–2010 average, and 16% below the previous 2007 record (Vaughan *et al.*, 2013). Since 2002, a new record has been set five times (2007, 2012, 2011, 2016, and 2019) and several other years have experienced near-record lows. Minimum extents from 2007, 2016, and 2019 are all statistically tied for the second lowest. As of October 2019, the 13 lowest September ice extents over the satellite record have all occurred since 2007. Arctic sea ice extent for April 2019 averaged 13.45 million km^2, which was 1.24 million km^2 below the 1981 to 2010 long-term average

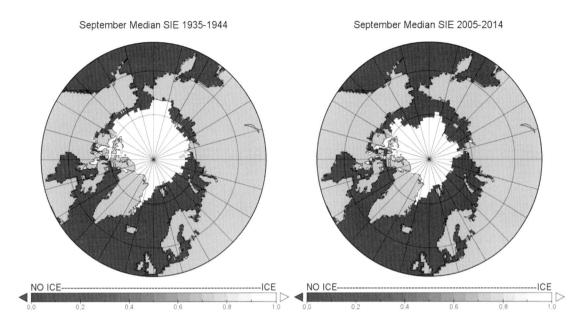

Figure 7.21 Median September Arctic sea ice extent for the lowest decade of the pre-satellite era (a) 1935–1944, and for the lowest decade of the satellite era (b) 2005–2014 [Source: https://neven1.typepad.com /blog/2016/01/september-arctic-sea-ice-extent-1935-2014.html]

extent and 230,000 km^2 below the previous record low set in April 2016 (https://nsidc.org /arcticseaicenews/). The Arctic ice volume is also decreasing. Changes in the relative amounts of perennial and seasonal ice are contributing to the reduction in ice volume. Since 1979 about 17% of perennial and seasonal sea ice decade^{-1} in the Arctic has been lost through melting, and about 40% since 1999.

In the midst of its significant decline, annually Arctic sea ice extent is also highly variable. Olonscheck *et al.* (2019) have proposed that this strong variability is closely related to fluctuations in the air temperature above the Arctic Ocean driven by atmospheric heat transport into the Arctic from lower latitudes. They argue that factors such as the ice–albedo feedback, cloud and water vapor feedbacks, and oceanic heat transported into the Arctic together explain only 25% of the year-to-year sea ice extent variations. Most of the sea ice variations are directly caused by mid-atmospheric temperature conditions shown in both observed data and climate models' simulations (Figure 7.22e). It seems that year-to-year fluctuations in sea ice extent are easier to understand than previously thought.

In contrast to the Arctic sea ice, sea ice extent in the Antarctic has not changed much since 1979, but has marginally increased with a trend of 0.8 ± 0.7% decade^{-1} over 1979–2019 (Figure 7.23). The mean sea ice values for 1981–2010 are 10.4 million km^2 for SH.

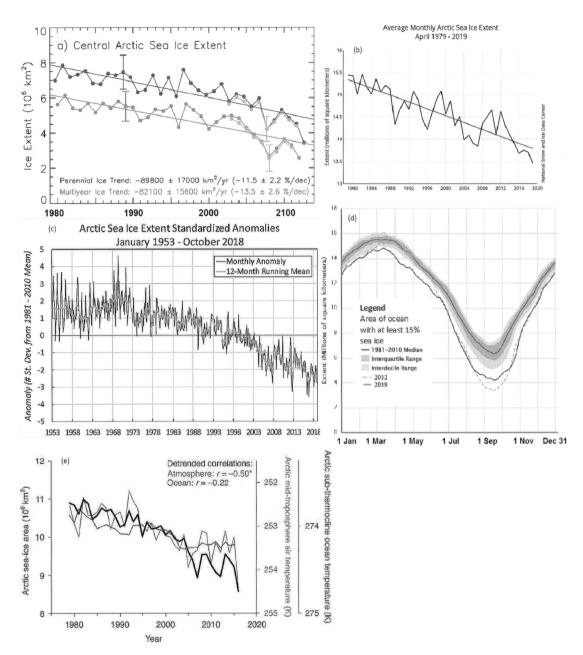

Figure 7.22 (a) Annual perennial (blue) and multi-year (green) sea ice extent in the Central Arctic from 1979 to 2012 derived from passive microwave (PM) data (Comiso, 2012). Perennial ice values are derived from summer minimum ice extent, while the multi-year ice values are averages of DJF. The gold lines (after 2002) are from AMSR-E data. Uncertainties in observations are indicated by error bars (Vaughan et al., 2013). (b) April Arctic sea ice extent for 1979 to 2019 showing a decline of 2.64 %/decade (https://nsidc.org/arcticseaicenews/). (c) Monthly sea ice extent anomalies for the NH over 1953-2018. Jan. 1953 to Dec. 1979 data were obtained from the UK Hadley Centre. For Jan. 1979 to present, data are derived from PM data (https://nsidc.org/cryosphere/sotc/sea_ice.html). (d) Arctic sea ice extent, the 1981–2010 median, the interquartile and interdecile ranges, the 2012 record low, and 2019 (https://nsidc.org/arctic seaicenews/charctic-interactive-sea-ice-graph/). (e) Time series of Arctic sea-ice area, mid-troposphere air temperature and sub-thermocline ocean temperature from 1979 to 2016. The correlation is significant at 99.9% marked with an asterisk. The detrended significant correlation of Arctic sea-ice area to 60–90 °N 2 m air temperature is r = −0.69 (from Olgonscheck et al., 2019).

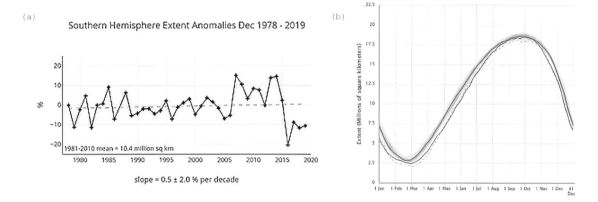

Figure 7.23 (a) Southern Hemisphere (SH) sea ice extent anomalies, 1978–2019, with a trend at $0.8 \pm 0.7\%$ decade^{-1}. (b) Antarctic sea ice extent, the 1981–2010 median, the interquartile and interdecile ranges, the 2017 record low, and 2019. The mean sea ice values for 1981–2010 are 10.4 million km^2 for SH (http:// nsidc.org/data/seaice_index/compare_trends)

Based on passive microwave SMMR and SSM/I data collected in 1979 and 2018, Figure 7.24 shows that between 1979 and 2018, the minimum Arctic sea ice extent in September (Figures 7.24c and 7.24d) has decreased more substantially compared to the maximum Arctic sea ice extent in March (Figures 7.24a and 7.24b). This means that the sea ice has melt off more over the summer in recent years even though the March maximum Arctic sea ice extent in 2018 has not decreased much compared to 1979.

Between the maximum Antarctic sea ice extent of September, 1979 and 2018, the changes are minimal with respect to the 1981–2010 median maximum Antarctic sea ice extent represented by the pink outline (Figure 7.25). Even for the minimum Antarctic sea ice extent of 2018, only between the Getz and Ross ice shelves showing some obvious retreat compared to the 1981–2010 median.

Sea ice volume export through the Fram Strait represents an important freshwater input to the North Atlantic, which in turn could modulate the intensity of the thermohaline circulation. It also contributes significantly to variations in Arctic IMB. Satellite observations of Fram Strait ice-area export also show an increase over the last 4 years, with ~37% increase in winter 2007–2008 (Smedsrud *et al.*, 2008). In 2007 there was also a large flux of ice out of the Arctic via Nares Strait due to the absence of the usual ice arches across the channel. Kwok *et al.* (2010) show that the area and volume outflows of 87×10^3 km^2 and 254 km^3, respectively, were more than twice the averages for 1997–2009 and represented about 10% of the outflow of ice through Fram Strait. Using CryoSat-2 sea ice thickness retrievals and three different ice drift products, Ricker *et al.* (2018) estimated that between 2010 and 2017, monthly winter sea ice volume export varies between 21 and 540 km^3 through the Fram

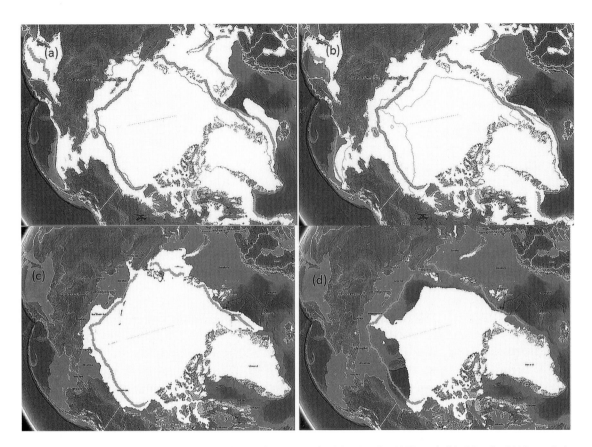

Figure 7.24 The Maximum Arctic sea ice extent in (a) March, 1979 and (b) March, 2018, and the minimum Arctic sea ice extent in (c) September, 1980, and (d) September, 2018 (Images of the Google Earth retrieved from https://nsidc.org/data/seaice_index/archives)

Strait. They found that annual to interannual ice volume export variability is mainly driven by ice drift variability, which in turn are driven by large-scale variability in the AO and NAO. The seasonal cycle in sea ice export is also driven by the mean thickness of exported sea ice. Over 50% of the variability in Arctic winter MYI volume changes can be explained by the variations in ice volume export through the Fram Strait. As the first mode of wintertime SLP variability for regions north of 20° N, the positive (negative) AO is characterized by low (high) SLP anomalies over the Arctic that lead to the cyclonic (anticyclonic) atmospheric circulation anomalies (Armitage et al., 2018), the east (west) origin of the Transpolar Drift Stream, and a contracted (expanded) Beaufort gyre circulation (Kwok et al., 2013), which in turn minimize sea ice growth in winter (Hegyi and Taylor, 2017). The enhanced north-south dipole in SLP over the North Atlantic centered on the Iceland Low and Azores High in the positive NAO years has also been shown to drive greater southward ice flux through the

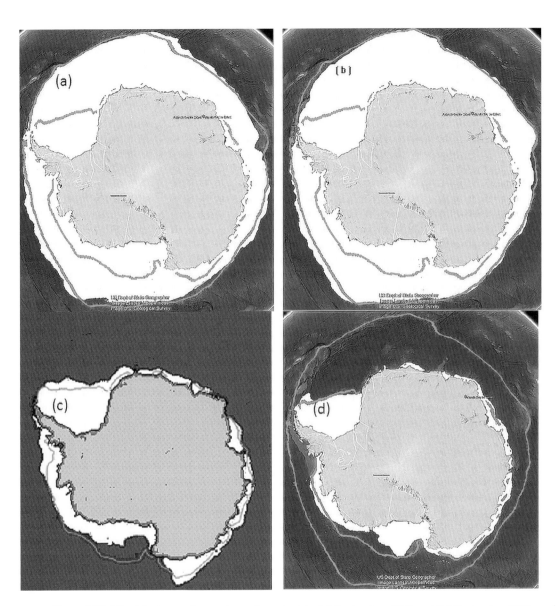

Figure 7.25 The Maximum Antarctic sea ice extent in (a) September 1979 and (b) September 2018, and the minimum Antarctic sea ice extent in (c) March 1979, and (d) March 2018 (Images of Google Earth retrieved from https://nsidc.org/data/seaice_index/archives)

Fram Strait. From the persistent surface forcing of quasi-stationary meridional thermal gradients, the NAO pattern has affected the sea ice variability at interannual time scales (Caian *et al.*, 2018).

Atmospheric poleward energy flux has declined since 1990, but advection of oceanic heat has recently increased. Woodgate *et al.* (2010) determine the Bering Strait volume and heat transports from 1991 to 2007. In 2007, both annual mean transport and temperatures were at record-length highs. The heat flux in 2007 was $5–6 \times 10^{20}$ Ja^{-1}, twice the 2001 heat flux. This amount is comparable to the annual shortwave radiative flux into the Chukchi Sea, and enough to melt one-third of the 2007 seasonal Arctic sea-ice loss. Between 1979 and 2001, the duration of the melt season in the Arctic has increased by 5.4 days decade^{-1} in the central Arctic, 9 days decade^{-1} in the Beaufort Sea, and 16.9 days decade^{-1} in the Barents Sea (Stroeve *et al.*, 2006). Since 1979, the length of the melt season of Arctic sea ice has grown by about 37 days. Arctic sea ice now starts melting 11 days earlier and it starts refreezing 26 days later than it used to, on average (NASA, 2016).

Min *et al.* (2008) demonstrate that human influence on the changes in Arctic ice extent can be robustly detected since the early 1990s. Stroeve *et al.* (2007) show that the timing of the decrease in 2007 is well ahead of IPCC model simulations for this century. Holland *et al.* (2006) find that in Community Climate System Model simulations ice retreat accelerates as thinning increases the open water formation efficiency for a given melt rate and the ice–albedo feedback increases shortwave absorption. The retreat is abrupt when ocean heat transport to the Arctic is rapidly increasing. Over 70% of the sea ice cover in spring 2008 consisted of young, fairly thin ice – an even more extreme situation than in spring 2007.

Tietsch *et al.* (2010) report on simulations with ECHAM5 coupled to the Max Planck Ocean Model for A1B1 scenarios. They find that for four removals of ice in July at 20-year intervals, there is always a recovery within 2 years, because the ocean cools through the relatively thin ice cover. The Arctic winter has a stabilizing effect and there appears to be no tipping point for persistent Arctic ice cover decline, in agreement with Eisenman and Wettlaufer (2009).

Drobot *et al.* (2008) projected a September 2008 low of 4.4 million km^2 using a complex linear regression model, close to the actual value of 4.67 million km^2 (Serreze and Stroeve, 2015). Ogi et al. (2008) show that the preconditioning by events in prior years, as represented by an index of May MYI, and current atmospheric conditions, as represented by an index of July–August–September SLP anomalies over the Arctic basin, account for ~60% of the year-to-year variance of September sea-ice extent since 1979.

The enormous loss of ice area in summer 2007 is attributed by Zhang *et al.* (2008) to preconditioning, anomalous winds, and ice–albedo feedback. The oldest and thickest ice within the MYI pack has been replaced in recent years by thinner FYI, thus preconditioning it to other factors. The atmospheric circulation associated with a highly amplified Pacific North America pattern strengthened the transpolar drift of sea ice, causing more ice to move out of the Pacific sector and the central Arctic Ocean. Thin ice and open water then allowed more surface solar heating due to a much reduced surface albedo, leading to amplified ice melting. The Arctic Ocean lost an additional 10% of its total ice mass with 70% due directly to the amplified melting and 30% to the unusual ice advection. Kauker *et al.* (2009) analyze the adjoint of a coupled ocean–sea ice model and find that four factors determined the 2007 ice minimum: May and June wind conditions, September 2-m air temperature, and March ice thickness accounted for 86% of the ice reduction. On the other hand, a reduced cloud cover or the inflow of more warm Pacific Water through Bering Strait have only minor effects on the 2007 ice cover. This is in contrast with the findings of Woodgate *et al.* (2010).

Despite these trends, Howell *et al.* (2008a) find that, from 1968 to 2006, MYI conditions in the western regions of the Northwest Passage remained relatively stable because the M'Clintock Channel and Franklin regions operated as a drain-trap mechanism for MYI. In addition to the Queen Elizabeth Islands region, the western Parry Channel and the M'Clintock Channel are also regions where a considerable amount of MYI forms in situ and combined with dynamic imports contributes to heavy MYI conditions. Multiyear sea ice increases occurred from 2000 to 2004 because of dynamic import and FYI being promoted to MYI, but this replenishment virtually stopped from 2005 to 2007, coincident with longer melt seasons. (Howell *et al.*, 2008b). In summer 2007, the Northwest Passage opened up for the first time.

Ice in the Eurasian Arctic has generally decreased since 1933 based on Soviet and Russia ice charts (Mahoney *et al.*, 2008). The retreat has not been continuous, however, with the data showing two periods of retreat separated by a partial recovery between the mid-1950s and mid-1980s. The charts, combined with air temperature records suggest that the retreat in recent years is pan-Arctic-wide and year-round in some regions, whereas the early to mid-twentieth-century retreat was confined to summer and autumn in the Russian Arctic. Rodrigues (2008) confirms the retreat during summers 1979–2007 in all regions of the Russian Arctic and in the Barents Sea in winter months.

Markus *et al.*, (2009) analyzed trends in melt onset and freeze-up using PMR data from 1979 to present. Melt trends are toward earlier melt ranging from −1.0 d decade^{-1} for the Bering Sea to −7.3 d decade^{-1} for the East Greenland Sea. Except for the Sea of Okhotsk all areas also show a trend toward later autumn freeze onset. The Chukchi/Beaufort seas and Laptev/East Siberian seas have the strongest trends with 7 d decade^{-1}. For the entire Arctic, the melt season length has increased by about 20 days over the last 30 years. The largest trends of over 10 d decade^{-1} are in Hudson Bay, the East Greenland Sea, the Laptev/East Siberian seas, and the Chukchi/Beaufort seas. For landfast ice along the Beaufort–Chukchi Sea coasts of Northern Alaska, Mahoney *et al.* (2007b) find that breakup is 21 days later during 1996–2004 than 1973–1977 (Barry *et al.*, 1979) but only 6 days later in the Chukchi Sea sector.

Rothrock *et al.* (1999) showed changes in Arctic ice thickness by comparing submarine sonar ice draft data from 1958 through 1976 to measurements from the 1990s. The results show that there was thinning at every point of comparison between 1993 and 1997 with similar data acquired between 1958 and 1976; the mean ice draft at the end of the melt season decreased by about 1.3 m (40%) in most of the deep water portion of the Arctic Ocean, from 3.1 m in 1958–1976 to 1.8 m in the 1990s. The decrease is greater in the central and eastern Arctic. In a further study, Rothrock *et al.* (2001) examine digitally recorded draft data from

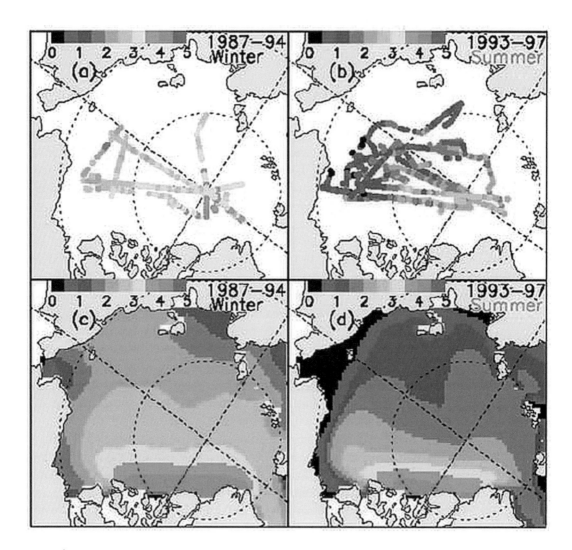

Figure 7.26 Composite of mean draft for (a) winter and (b) summer cruise tracks. Model mean draft for (c) period of winter cruises and (d) summer cruises (from Rothrock *et al.*, 2003) [Courtesy American Geophysical Union]

eight cruises spanning the years 1987 to 1997 and find a decrease of about 1 m over the 11-year span (see winter and summer cruise tracks shown in Figure 7.26). Yu *et al.* (2003) examine differences in the sea ice thickness distribution function between 1958–1970 and 1993–1997. Substantial losses occurred in ice thicker than 2 m, with an increase in the amount of 1–2-m ice. The volume of ice less than 4-m thick remained nearly the same and the total volume decreased about 32%. Part of the change is likely to be caused by increased ice area export through Fram Strait in the late 1980s and early 1990s, but a substantial shift in the peak thickness suggests that changes in thermal forcing were also a major factor in the observed thinning. Airborne EM-inductive measurements by Haas *et al.* (2008) show that between August–September 1991 and 2001 the modal thickness of large numbers of individual ice floes in the region of the NP decreased from 2.50 to 1.95 m (22%). This continued to 2007 when the reduction was 53%. The thinning was mainly due to a regime shift from predominantly MYI and second-year ice (Kwok *et al.*, 2009). The MYI volume has declined by >40% since 2005. Seasonal ice has now become the dominant ice type in the Arctic Ocean, having surpassed the MYI area and volume in the winter season. Kwok and Rothrock (2009) extend the analysis of 42 years' of submarine records (1958–2000) described by Rothrock *et al.* (1999, 2008) with data from ICESat (2003–2008). They find that declassified submarine sonar measurements (covering ~38% of the Arctic Ocean) give an overall mean winter thickness of 3.64 m in 1980 that can be compared to 1.89 m during winter 2008, a 1.75-m reduction. Prior to 1997, ice extent in the data release area of declassified submarine sonar measurements was >90% during the summer minimum, compared with <55% during the record setting low value in 2007.

Farrell *et al.* (2009) use data from the GLAS, on the Ice, Cloud and Land Elevation Satellite (ICESat) to analyze sea ice freeboard in the Arctic Ocean up to 86 °N. Using a new method for sea surface height retrieval, they construct a time series of ice freeboard spanning 5 years between March 2003 and 2008. The autumn (October–November) and winter (February–March) data illustrate the seasonal and interannual variations in freeboard, but the autumn 2007 and winter 2008 spatially averaged freeboards are below the seasonal means. During 2003–2008, mean freeboard has declined at a rate of < -1.8 cm a^{-1} during autumn and < -1.6 cm a^{-1} during the winter. They emphasize that it is unclear whether the results represent a long-term, downward trend or are a part of natural variability.

Haas *et al.* (2010) report on airborne EM ice thickness surveys in April 2009. They found that the modal thicknesses of old ice had changed little since 2007. North of Ellesmere Island and Greenland, thickness distributions showed broad modes between 3.1- and 4.5-m thick, with long, exponential tails representing large fractions of pressure ridges frequently thicker than 10 m. The thinnest ice, with mean thicknesses between 1.69 and 1.88 m, occurred in the Beaufort and Chukchi seas.

The CRREL has installed a network of ice-tethered IMB buoys in different parts of the Arctic, complemented by a few sea-floor moorings with ice profiling sonar (IPS). These mass balance buoys use thermistors to measure the ice temperature and above-ice and below-ice acoustic sounders to measure the positions of the surface and bottom within 5 mm. Results are now being analyzed.

The satellite record for the Antarctic shows a slight increase in the ice extent and in the length of the ice season (Figure 7.11), but historical data reconstructed from the location of whaling ships suggest a southward shift in the ice edge beginning in the late 1950s (de la Mare, 1997, 2009). Observations from whaling factory ships during 1930–1931, 1932–1933,

1933–1934, 1934–1935, and 1935–1936, together with Discovery and other expedition data were assembled into charts of ice edge by Mackintosh and Herdman (1940), which were re-worked by de la Mare. The estimated ice extent during October–March 1931/1932–1955/1956 was 1.9–2.8° latitude north of its position during 1971/1972–1986/1987 with an average difference of 2.4° latitude. The shift was largest in the Weddell Sea with reductions across the Indian Ocean sector. The reduction also appears to be corroborated by other direct ice measurements at Signy Island and proxy evidence from ice cores in Antarctica. Methane sulphonic acid (MSA) in ice cores from Law Dome is correlated with sea ice extent and indicates an abrupt decline in the 1950s (Curran *et al.*, 2003). Kukla and Gavin (1981) found that the monthly ice extents of 1973–1980 in Navy–NOAA Joint Ice Center satellite data analyses were significantly less than those in US and Russian atlases. The US Navy Oceanographic Atlas of the Polar Seas (1957) used all available past records including Navy expeditions between 1939 and 1957. The Russian "Atlas Antarktiki" (1966) was based on data from the 1930s and Russian, Japanese, and US expeditions between 1947 and 1962. The maximum monthly extent in September is 20.8 million km^2 in the Russian atlas, 20.1 million km^2 in the US atlas, and 18.5 million km^2 in the satellite data; the corresponding minimum values in March are 4.2, 4.5, and 4.7 million km^2, respectively. However, in all other months the satellite era values are lower than the atlas values. Cotté and Guinet (2007) use the whaling ship data for November to February 1931 to 1987 (with no data during 1941–1945 or 1961–1971). They find a difference between the whaling-derived mean ice edge before 1960 and a satellite-derived mean ice edge for 1973–1987. The mean latitudinal difference over the four summer months is 2.4°, which is identical to the change found by de la Mare (2009). In December, the difference was up to 3.5°. The reduction of the sea ice extent occurred in the 1960s, mainly in the Weddell Sea sector where the change ranged from 3° to 7.9° latitude. Using a climate model, Goosse *et al.* (2009b) simulated a decreased extent by 0.5×10^6 km^2 between the early 1960s and early 1980s, and also the observed slight increase over 1980–2000.

REVIEW QUESTIONS

(1) Define sea ice extent, and why sea ice extent is inevitably larger than sea ice area.
(2) Name three types of satellite data that are commonly used to estimate sea ice conditions and recent changes to sea ice in the Arctic and Antarctic.
(3) Based on passive microwave SMMR and SSM/I data of DMSP satellites collected since 1979, what is the average decreasing annual trend in Arctic perennial sea ice extent in percent $decade^{-1}$, and decreasing trend in summer (in late September) in percent $decade^{-1}$? What was the lowest summer sea ice extent ever recorded in million km^2 and when did it happen?
(4) Why cryospheric changes, such as the decline of Arctic sea ice in recent years, are likely the most obvious physical evidence of global warming?
(5) Why have changes in Antarctic sea ice receive less attention than changes in Arctic sea

ice? Explain the contrast between changes to Antarctic and Arctic sea ice in recent years, such as the average recent trend in Antarctic sea ice observed.

(6) What information can be obtained from paleoclimate record such as terrestrial proxies (Arctic ice cores, tree-ring, and lake sediments) of the circum-Arctic region to reconstruct past extents of summer sea ice in the Arctic?

(7) How have large-scale climate anomalies such as NOA and AO contributed to the southward sea ice flux through the Fram Strait?

(8) Explain how FYI and MYI are discriminated through differences in their emissive characteristics represented by polarization ratio (PR) and gradient ratio (GR) estimated from brightness temperature of passive microwave data at 19 V, 19 H, and 37 V?

(9) What are the key mechanisms behind the formation of leads and polynyas in the Arctic and Antarctic in terms of ice drift and local melting? Why are they a major source of brine and how have they affected the regional exchange of sensible heat, latent, and longwave radiation fluxes in the Arctic and Antarctic?

(10) Using Arctic sea ice thickness estimated from ICESat and CryoSat-2 data, what are the estimated declining trends in sea ice volume in winter and in summer ($km^3 \, year^{-1}$) in recent years by Kwok and Cunningham (2015) and Kwok (2018), and the average decrease in Arctic sea ice thickness in the last 6 decades between submarine (1958–1976) and CryoSat-2 (2011–2017) data?

(11) Since the Arctic sea ice has been getting younger and thinner, what has happened in summers to most of the Arctic sea ice recovered in recent past winters? How much has 4+-year-old ice shown in Figure 7.20 lost in percentage from 1984 to 2019?

(12) Explain the global climate impacts due to disappearing Arctic sea ice, such as the ice–albedo effect, release of methane from continental shelves of the Arctic Ocean to the atmosphere, and possible increase of atmospheric water vapor.

(13) Explain possible ecological implications of diminished Arctic sea ice, and new commercial opportunities, such as creating new shipping lanes and increased access to natural resources in the Arctic region.

8   Ice shelves and icebergs

8.1   History

The first observations of icebergs were probably made by Inuit hunters in the Arctic and then by early mariners, including Irish monks and Vikings. Martin Frobisher's expeditions to Baffin Island in the 1570s–1580s certainly witnessed them and whalers and sealers in Baffin Bay and the Greenland Sea frequently sheltered in their lee from storms and sea ice. Documentation of icebergs in the northwest Atlantic began in 1914 by the International Ice Patrol after the loss of the RMS *Titanic*, and over 1,500 lives, due to a collision with an iceberg in April 1912. The First International Conference for the Safety of Life at Sea established the Ice Patrol, operated by the US Coast Guard, in 1913. It conducts surveys of the icebergs that drift south of 48° N off Newfoundland. Initially this was from cutters, and then airborne reconnaissance flights started in 1946 using first visual observations; airborne radar studies began in 1957 and in 1983 Side-Looking Airborne Radar (SLAR) was deployed. After 1991 (1995) radar remote sensing made use of data from ERS-1 (ERS-2), and RADARSAT's synthetic aperture radar (SAR). A major concern is the hazard to drilling platforms off the coast of Newfoundland.

In the western South Atlantic, there are Spanish records of iceberg sightings during the second half of the eighteenth century, as recorded from logbooks (del Rosario Prieto *et al*., 2004). A total of five sightings have been identified, two of isolated bergs and three outbreaks (1770 and 1794). In June 1770, there were sightings around 49.7° S near Cape Horn and in January 1794 some 2,000 icebergs were observed between 52.6°–50 .7° S and 43.8° W. These complement the data of Burrows (1976) who tabulated 35 years with sightings north of 60° S in the southwest Pacific. There were major irruptions in 1852–1859, 1891–1898, and 1904–1912, but also in the 1770s, 1860s, 1920s, and 1930s. Icebergs were seen near New Zealand in 1855, reaching 40° S, 170° W, and many bergs were present between Chatham Island and Antipodes Island in 1892, the northernmost at 42.3° S.

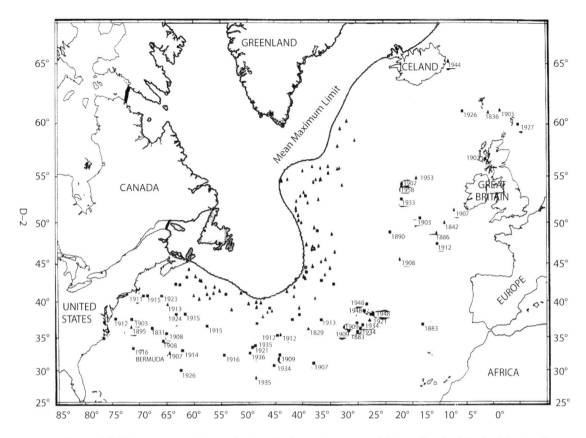

Figure 8.1 The mean maximum iceberg limit and extreme sightings in the North Atlantic Ocean (Ketchum and Hildebrand, 1977) [Source: Courtesy US Coast Guard]

In the North Atlantic, records of extreme iceberg sightings extend to Bermuda 32° N, 64° W in 1907 and 1914, 29° S, 49° W in 1935, 37° N, 18° W in 1903, 52° N, 12° W in 1907, and 60° N, 0° W in 1927 (see Figure 8.1)

Ice shelves were first encountered by Antarctic expeditions in the nineteenth century. James C. Ross's expedition (1841–1843) discovered the Ross Sea, and the Ross Ice Shelf. In 1911–1912, Wilhelm Filchner explored the eastern end of the shelf that bears his name in the Weddell Sea; the Ronne – the western part of the shelf – was first photographed from the air in 1947. The British Arctic Expedition of 1875–1876 first documented the ice shelf fringing northern Ellesmere Island. Research on this ice shelf by the Canadian Defense Research Board began in the 1960s. In the Antarctic, Richard Byrd established a series of "Little America" bases, located on the Ross Ice Shelf between 1929 and 1947 and used them to support airborne surveys of the continent. The last base operated during the International Geophysical Year, 1957–1958.

8.2   Ice shelves

An ice shelf is a sheet of very thick ice, with a nearly level surface that is attached to the land, but most of which is floating on the ocean. It is bounded on its seaward side by a steep ice cliff that can be up to 250 m high. Ice shelves form primarily when continental ice streams flow into the ocean and form a floating tongue or platform. The boundary between the floating ice shelf and the grounded ice (resting on bedrock) that feeds it is called the grounding line. The Ross Ice Shelf advances into the sea at a rate of between 1.5 and 3 m a day. Ice shelves flow through gravity-driven spreading of the ice floating on the ocean. A second mechanism of ice shelf formation is the growth of multiyear landfast ice up to about 10 m thickness. This process is observed in northern Ellesmere Island (Lemmen *et al.*, 1988). Modern-day ice shelves range in thickness from 1 to 2 km at the grounding line to a few hundred meters, or less, at the ice shelf front. The density contrast between glacial ice (917 kg m^{-3}) and seawater (1,025 kg m^{-3}) means that only about 11% of the floating ice is above the ocean surface. However, the presence of lower density firn and snow on the ice shelf means that a somewhat greater percentage of the ice thickness may be above water. Unlike the Antarctic, ice shelves in the Arctic are scarce and small, where the largest ice shelf is the Ward Hunt Ice Shelf of Nunavut, Canada with an area of 224 km^2 (Copland, 2017).

Antarctica

In the Antarctic, ice shelves comprise about 44% of the coastline with an aggregate area of 1.5 million km^2 (about 11% of the continent). The largest are the Ross Ice Shelf (0.47 million km^2), 800 km across, and the Filchner–Ronne Ice Shelf (or FRIS) (0.42 million km^2) in the Weddell Sea, separated into eastern (Filchner) and western (Ronne) parts by Berkner Island. The Ross Ice Shelf thins from about 800 m at its landward edge to about 300 m at its seaward edge (Figure 8.2) (Bamber and Bentley, 1994, fig. 2b). The FRIS has an average thickness of around 700 m (Nicholls *et al.*, 2009). Ice shelves are almost continuous from longitude 30° W eastward to 40° E – the Brunt (26° W), Riiser-Larsen (16° W), and Fimbul (0° longitude) shelves are the largest. In East Antarctica, the largest are the Amery (71° E), West (85° E), and Shackleton (100° E) shelves (Scambos *et al.*, 2007). Amery is the third largest ice shelf in Antarctica, and it is a key drainage channel for the eastern Antarctica. Suyetova (1966) gives an aggregate length of coastal ice shelves as 13,660 km, with a further 11,100 km of ice wall, and 2,860 km of outlet glaciers. Figure 4.8b illustrates locations and areas of the Antarctic shelves.

Antarctic Peninsula

The ice shelves along the Antarctic Peninsula (AP) have displayed significant retreat over the last few decades (Scambos *et al.*, 2003). Cook and Vaughan (2010) note that in recent decades

Figure 8.2 The edge of the Ross Ice Shelf, December 1996 [Courtesy Mike van Woert, Michael Van Woert, NOAA NESDIS, ORA] www.photolib.noaa.gov/htmls/corp2399.htm

7 out of 12 ice shelves round the Peninsula have either retreated significantly or have been almost entirely lost. The total area of ice shelf lost, from the earliest available records – from the 1950s or 1960s to 2009 – is 28,117 km^2 or 18% of the original area. The largest ice losses were on the Larsen A, B, and C, and the Wilkins ice shelves, which together accounted for 84% of the total.

The northernmost shelf on the east coast of the peninsula – the Larsen Ice Shelf – is a dramatic example of this process. It comprised three separate elements: Larsen A – a small shelf in the northernmost embayment – began a gradual retreat in the late 1940s that ended dramatically in January 1995, when almost 2,000 km^2 of ice disintegrated during a storm (Rott *et al.*, 1996). Larsen B (3,250 km^2 of ice, 220 m thick) to the south – collapsed and broke up between January 31 and March 7, 2002 (Figure 8.3a). Larsen C, which has a mean thickness of 289 m based on radar altimetry from ERS-1 and airborne radio echo-sounding data (Griggs and Bamber, 2009), had a rift of 175 km in length that completely broke off from the land on July 12, 2017 (Figure 8.3b). Geological evidence indicates that the former Larsen A shelf had previously broken up and reformed only about 4,000 years ago, although the former Larsen B had been stable for at least 12,000 years. The rapid breakup has been attributed to the effects of liquid water; meltwater ponds formed on the surface of the shelf during the 24-hour-long days, then drained down into cracks and, by acting like a multitude of wedges, levered the shelf into pieces (Scambos *et al.*, 2000).

Recent abrupt and catastrophic large-scale disintegrations of ice shelves along the AP could indirectly contribute to sea-level rise from the Antarctic Ice Sheet where 74% of its coastal margin is surrounded by ice shelves and floating outlet glaciers (Bindschadler *et al.*, 2011). Disintegration of ice shelves has been attributed to several processes such as surface meltwater ponding and crevasse enlargement by enhanced regional warming and

Figure 8.3 (a) Collapse of the Larsen B ice shelf on the east side of the Antarctic Peninsula, March 7, 2002, shown in a MODIS image. http://upload.wikimedia.org/wikipedia/commons/2/22/Larsen_B_Collapse .jpg (b) A rift of 175 km in length that occurred at the Antarctica's Larsen C ice shelf, with the iceberg completely separated from the land on July 12, 2017 www.esa.int/spaceinimages/Images/2017/08/Larse n_C_rift_from_the_air [Courtesy: European Space Agency]

weakening of ice material due to warm meltwater percolation (Scambos *et al.*, 2003), enhanced basal melting from ocean warming and feedback between ice shelf retreat and greater warm-water ingress (Jacobs *et al.*, 2011), brine infiltration, localized outer-margin fracturing by bending stresses induced by buoyancy forces (Scambos *et al.*, 2009), and the propagation of structural weaknesses due to crevasses and rifts (Khazendar *et al.*, 2007). Besides these factors that trigger ice shelf disintegration events, Massom *et al.* (2018) show that increased open-water duration (Figure 8.4a) and reduction of regional sea ice concentration (Figures 8.4b and c) substantially increase the ocean wave energy reaching the ice shelf fronts, resulting in increased absence of a protective sea ice buffer, allowing storm-generated ocean swell to increase flexure of outer ice shelf margins, weakening them to a point of calving, and sudden wide-scale disintegration.

When the Larsen B Ice Shelf collapsed in 2002, disintegration occurred during 4.5 months of continuous open-sea exposure in the Bellingshausen Sea (mid-November 2001 to late March 2002). The mean austral summer (December–February) sea ice concentration offshore in 2001/2002 was only 14.7%, which was associated with sustained strong and warm northerly/ northwesterly airflow across the AP. The warm winds simultaneously drove the sea ice away from the Larsen B Ice Shelf front (northwestern Weddell Sea) and caused extensive surface melt-pond coverage. Daily maximum ocean-wave hindcast data offshore from the Larsen and Wilkins (box areas L and W in Figure 8.4a) show peak periods of 12 s and 16 s, and peak significant wave heights in the periods before and during the disintegrations.

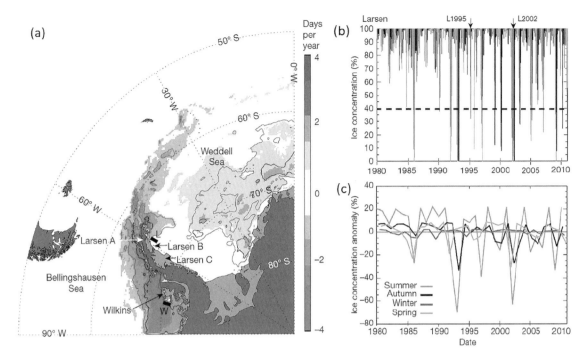

Figure 8.4 (a) The passive microwave derived shorter annual sea ice season duration for the Weddell and Bellingshausen seas for 1979/1980 to 2009/2010 offshore from the Larsen ice and Wilkins shelves is shown by red indicating fewer days of coverage. The black/blue contours delimit significance at $P<0.01$ and 0.10, respectively; (b) Time series showing reduction in daily sea ice coverage for region L offshore from Larsen ice shelves, and temporal coincidence with disintegration events where arrows denote the approximate onset timings of the major disintegration events of Larsen A in 1995 and Larsen B in 2002. The plots are summer in red, autumn in black, winter in dark blue, and spring in light blue; (c) Plots of mean seasonal sea ice concentration anomalies off the Larsen A and B from region L (taken from Masson *et al.*, 2018) (A black and white version of this figure will appear in some formats. For the color version, please refer to the plate section.)

From examining disintegration events of three selves, Larsen A of 1995, Larsen B of 2002, and Wilkins of 2008 and 2009, they proposed a conceptual model of key prerequisites for disintegration, which are extensive flooding and hydrofracture, reduced sea ice in fronts of ice shelves, outer-margin fracturing and rifting, and initial calving from outer ice shelf margins. They found wave-induced flexure is particularly effective in outermost ice shelf regions thinned by bottom crevassing.

A giant iceberg of 5,800 km^2 in area, known as A-68, broken off from the Larsen C ice shelf of the AP on July 17, 2017, leaving the Larsen C ice shelf, the fourth largest ice shelf in Antarctica of about 50,000 km^2 in area more than 12% smaller than before it broke off and adrift in the Weddell Sea (Figure 8.3). It was the third major calving of ice shelf in Antarctica following the collapse of the more northerly Larsen A ice shelf in 1995 and Larsen B in 2002. While Larsen A and Larsen B ice shelves could have disintegrated partly because of climate change impact, for example, thinning of ice shelves by warmer ocean water, the calving of A-68 was not linked to such processes, given recent data from the Scripps Institute of Oceanography show thickening of most of the shelf. Ice shelves are more likely to break up as they extend further out into the ocean. The progress of the rift, and the loss of the iceberg, can now be monitored from radar images acquired by the European Space Agency's Sentinel-1 mission.

The stress from bending of an ice shelf as a consequence of hydrostatic forces in the seawater increasing faster that inside the ice because of the difference in density of ice and seawater induces cracks which deepen as water fills the cracks, and eventually calving occurs. Massom et al. (2018) found that loss of sea ice means that ice shelves will be more exposed to ocean wave energy, leading to a faster disintegration.

Off the southwest of the AP, the Wilkins ice shelf occupies much of Wilkins Sound, located between the concave western coastline of Alexander Island and Charcot Island and Latady Island to the west. It was about 150 km by 110 km and received most of its sustenance from in situ accumulation (Vaughan et al., 1993). It began retreating in the 1990s (Lucchita and Rosanova, 1998). The shelf then had a total area of 17,400 km^2. Events in 1998 and the early 2000s reduced that to 13,680 km^2. By late February–early March 2008, the area of stable shelf had shrunk to ~10,300 km^2. A narrow strip of shelf ice that was protecting several thousand more kilometers of the ice shelf broke up on April 5, 2009, removing about 330 km^2 of ice, and enabling icebergs to start calving off the exposed shelf (Figure 8.5). Scambos et al. (2009) show from remote-sensing data that the breakup events of February 28 to March 6, May 27 to May 31, and June 28 to mid-July 2008, occurred mainly through a distinctive type of shelf calving, which they term "disintegration." This is characterized by repeated rapid fracturing that creates narrow ice-edge-parallel blocks, with subsequent block toppling and

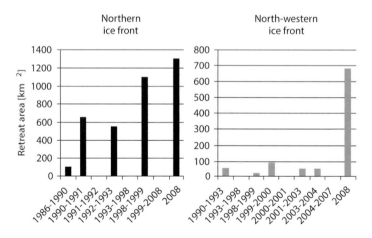

Figure 8.5 Wilkins ice shelf area loss and breakup events from 1990 to 2008. (Braun et al., 2009) [Source: *The cryosphere (EGU), 3, p. 47, Figure 4(h)*] Matthias.braun@uni-bonn.de

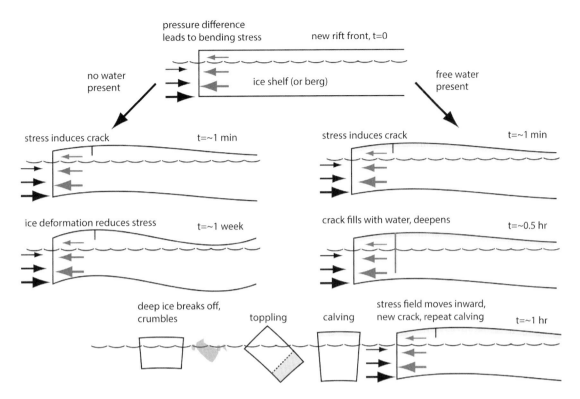

pressure difference
leads to bending stress   new rift front, t=0

no water   free water
present   ice shelf (or berg)   present

stress induces crack   t=~1 min   stress induces crack   t=~1 min

ice deformation reduces stress   t=~1 week   crack fills with water, deepens   t=~0.5 hr

deep ice breaks off,   stress field moves inward,
crumbles   toppling   calving   new crack, repeat calving   t=~1 hr

Figure 8.6 Diagram of forces at an ice plate margin and the effect of available water (from fig. 6 of Earth and Planetary Science Letters 280, Scambos *et al.*, 2009)

fragmentation forming an expanding iceberg and ice rubble mass. Ice plate bending stresses at the ice front, arising from buoyancy forces, can lead to runaway calving when free water is available (Figure 8.6).

The AP has experienced unprecedented warming in the past 50 years of ~2.5–3.0 °C, and several ice shelves have retreated in the past 30 years. Six of these shelves have collapsed completely: the Prince Gustav Channel, Larsen Inlet, Larsen A, almost all of Larsen B, Wordie, Muller, and Jones ice shelves. There appears to be a critical temperature threshold above which ice shelves cannot be sustained. Morris and Vaughan, 2003) show that ice shelves where temperatures are between the −5 °C and −9 °C mean annual isotherm have shown little change in terminus position, while shelves that exceed this isotherm have undergone dramatic retreat or complete collapse. However, Cook and Vaughan (2010) point out that not all ice shelf retreats are the same. There are a variety of responses to the various forcings: protracted retreat, rapid collapse, and possibly staged collapse. The rate of retreat does not appear to be related in any simple fashion to the rate of change of climate; rather, it is modulated by the ice shelf configuration and the conditions of mass balance.

Ellesmere Island

Ellesmere Island, Nunavut, was once bounded on the Arctic Ocean side by a single giant ice shelf that covered almost 10,000 km^2 and formed around 5,500 cal yr BP, based on a hiatus in driftwood deposition (England *et al.*, 2008). Today, the remnants cover less than 10% of the original area (Jeffries, 2002) (see Figure 8.7). They are composed of multiyear landfast sea ice sustained by the basal accretion of brackish seawater and intermittent years of net snow accumulation. The five remaining shelves are attached to the north coast of Ellesmere Island and lie north of 82° N. The Ward Hunt Ice Shelf (83° N, 74° W) is a 443 km^2 remnant of a much larger feature that extended along the northern coast of Ellesmere Island at the beginning of the twentieth century (Crary, 1960). The original ice shelf contracted 90% during 1906–1982 by calving from its northern edge (Vincent *et al.*, 2001). The Ward Hunt shelf, the largest remaining section, lost 600 km^2 of ice in a massive calving in 1961–1962. In 1981, its thickness was consistently 45–60 m (Narod *et al.*, 1988). Since then, the remnant ice shelves, including the Ward Hunt Ice Shelf, remained relatively stable until 2000. There has been an acceleration of the breakup of the Ellesmere ice shelves since 2000. The Ward Hunt ice shelf broke into two over the period 2000 to 2002, with additional fissuring and further ice island calving (Mueller *et al.*, 2003). The Ayles Ice Shelf broke off from the coast on August 13, 2005, forming a giant ice island 37 m thick and around 66 km^2 in size. In summer 2008, there was complete breakaway of the 50 km^2 Markham ice shelf, the Serson shelf lost

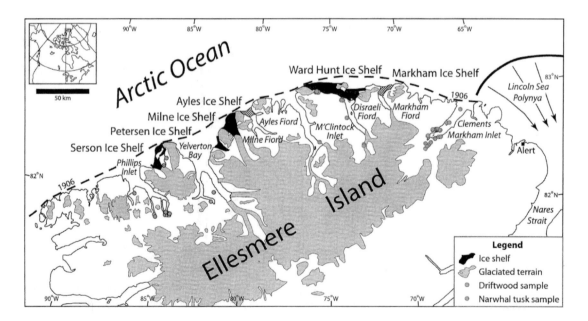

Figure 8.7 Map of Ellesmere Island ice shelves (England *et al.*, 2008) [Source: *Geophysical Research Letters*, 35: L19502, figure 1. American Geophysical Union]

two sections totaling an estimated 122 km^2 and a further 22 km^2 of ice split off from the Ward Hunt shelf. A total of 23% of the pre-2008 ice shelf area broke off during 2008 leaving an area of only 720 km^2 (Mueller *et al.*, 2008). The last fully intact ice shelf of Canada, the Milne Ice Shelf on the northwest coast of Ellesmere Island has broken-up in early August, 2020, which reduced the shelf's size from 187 to 106 km^2 (https://newsroom.carleton.ca/2020/nunavuts-milne-ice-shelf-suddenly-collapses/).

8.3   Ice streams

The FRIS is fed by eight large glaciers, from east to west.

They are the Bailey, Slessor, Recovery, Support Force, Foundation, Institute, Rutford, and Evans. Their thickness is generally in the range 1,000–2,000 m. Much of the bed is weak (basal shear stresses 4–20 kPa) according to Joughin *et al.* (2006), but in contrast to the Ross Ice Shelf ice streams, it is heterogeneous, with "sticky" spots providing resistance to flow. Ice streams entering the Ross Ice Shelf from the interior of Antarctica are large features about 500 km long, 20 to 100 km wide and up to 2,000 m thick. There are five major ice streams entering from the West Antarctic Siple Coast and a number of glaciers from the Transantarctic Mountains of East Antarctica. They typically have basal shear stresses <4 kPa and move at a rate of 1 to 2 m per day (approaching 3.5 m d^{-1} at the calving front), sliding over a bed of sediment saturated with liquid water. However, if the bed is cold enough for the water in it to freeze, the loss of lubrication causes the ice stream to slow and eventually stop. The ice streams are separated by slow-flowing ice ridges that are frozen to the bed; they are characterized by low slopes compared with the flanking ice.

Ice-flow velocity measurements from SAR data are used by Joughin and Tulaczyk (2002) to reassess the mass balance of the Ross ice streams. They find strong evidence for ice-sheet growth (+26.8 Gt yr^{-1}), in contrast to earlier estimates indicating a mass deficit (−20.9 Gt yr^{-1}). Average thickening is equal to ~25% of the accumulation rate, with most of this growth occurring on the Kamb Ice Stream (formerly C). Ice Streams D (Bindschadler) and E (MacAyeal) and the combined outflow of the Mercer Ice Stream (formerly A) and Whillans Ice Stream (formerly B) are in balance. This is consistent with radar altimetry observations (see Box 8.1) that show little thickening or thinning over their catchments. Ice stream F is significantly positive, but its contribution to the overall total mass balance is small relative to that of the other ice streams. The stagnant Kamb Ice Stream (C) has a strongly positive mass balance because of its negligible outflow, and it is the major contributor to the overall positive mass balance for the region. The ice stream flow is irregular over the last millennium as revealed by processed AVHRR images of the ice shelf surface (Fahnestock *et al.*, 2000). Whillans Ice Stream must have stopped its rapid flow about 850 calendar years ago and restarted about 400 years later and MacAyeal Ice Stream (formerly E) either stopped or slowed significantly between 800 and 700 years ago, restarting about 100 years later according to Hulbe *et al.* (2005).

Box 8.1 Ice shelf altimetry

Ice shelf thickness is an important boundary condition for ice-sheet and sub-ice-shelf cavity modeling, required near the grounding line to estimate the ice flux used for determining the ice-sheet mass balance, which is limited by the accuracy of the ice thickness. To undertake altimetry studies of ice shelves, we have first to remove several artifacts from the data. The primary ones are tidal effects and the inverse barometer effect. Tides in the southwestern part of the Ronne Ice Shelf in the Weddell Sea have a range of 7 m. This variation has to be filtered out using tidal models, which are reasonably accurate. The second effect is due to variations in atmospheric pressure on sea level: high (low) atmospheric pressure lowers (raises) sea-level height. The range is about 20–30 cm.

Comparison of ICESat altimetry over Greenland and Antarctica with European Remote-sensing Satellite-2 (ERS-2) and Envisat radar altimetry shows that the Geoscience Laser Altimeter System (GLAS) precision varies as a function of surface slope from 14 to 59 cm, and the radar precision varies from 59 cm to 3.7 m for ERS-2 and from 28 cm to 2.06 m for Envisat (Brenner *et al.*, 2007). Envisat elevation retrievals when compared with ICESat results over regions with less than 0.1° surface slopes show a mean difference of 9 ± 5 cm for Greenland and −40 ± 98 cm for Antarctica. ERS-2 elevation retrievals over these same low surface slopes differ from ICESat results by −56 ± 72 cm over Greenland and 1.12 ± 1.16 m over Antarctica. The ice sheet return from radars is a composite of surface and subsurface volume scattering and the penetration is up to 4.7 m in cold dry regions. Over sloping surfaces, usually the largest error in the elevations retrieved from the radar altimeter is due to the large footprint of the beam. The largest error source in calculating the elevation from each ICESat measurement is the precision of the pointing knowledge of the laser beam. There is no penetration by the laser beam (Brenner *et al.*, 2007).

Pritchard *et al.* (2012) used satellite laser altimetry and modeling of the surface firn layer to reveal the circum-Antarctic pattern of ice shelf thinning through increased basal melt, which they believe is the primary control of Antarctic ice-sheet loss, via a reduced buttressing of the adjacent ice sheet leading to accelerated glacier flow. The highest thinning rates occur where warm water at depth can access thick ice shelves via submarine troughs crossing the continental shelf. Their measurements of the Ross Ice Shelf reveal coherent patterns of ice shelf elevation change rate ($\Delta h/\Delta t$) at the scale of cm yr^{-1} (Figure 8.8). Griggs and Bamber (2011) retrieved the ice thickness for all Antarctic ice shelves using radar altimeter data from geodetic phases of ERS-1 during 1994–1995 supplemented by ICESat data for regions south of the ERS-1 latitudinal limit. Surface elevations derived from these data are converted to ice thicknesses using a modeled firn-density correction to reduce the error in ice thickness. Comparison to airborne data shows good agreement, with biases ranging from −13.0 m for areas where the hydrostatic equilibrium assumption breaks down to 53.4 m in regions where marine ice may be present.

Using data from radar altimeters of ERS-1, ERS-2, Envisat, and CryoSat-2, Adusumilli *et al.* (2018) constructed 23-year (1994–2016) of AP ice shelf height change time series. Combining these data with output from atmospheric and firn models, they partitioned the total height-change into contributions from surface and basal mass balance, firn state, and ice

Box 8.1   (cont.)

dynamics. On the Bellingshausen coast of the AP, ice shelves lost 84 ± 34 Gt a^{-1} to basal melting, and 50 ± 7 Gt a^{-1} to surface mass balance and ice dynamics. Net basal melting on the Weddell coast was 51 ± 71 Gt a^{-1}. Recent height increases over major AP ice shelves are driven by changes in firn state.

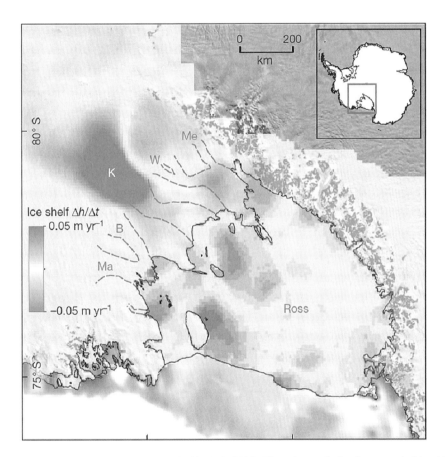

Figure 8.8 Surface $\Delta h / \Delta t$ of Ross Ice Shelf, 2003–2008. The color scale for the grounded ice $\Delta h / \Delta t$ signal is ± 30 cm yr^{-1}. The continental shelf is shown in gray, landward of the continental-shelf break. Labels Me, W, K, B, and Ma denote the ice streams Mercer, Whillans, Kamb, Bindschadler, and MacAyeal, respectively. Dashed gray lines show the lateral ice stream margins. Black lines show ice shelf boundaries (mapped between 1993 and 2003). The inset shows the location of the figure (green box) overlaid on the outline of Antarctica (taken from Pritchard *et al.*, 2012) (A black and white version of this figure will appear in some formats. For the color version, please refer to the plate section.)

8.4   Conditions beneath ice shelves

Under ice shelves, melting and freezing due to the gradient in the pressure-dependent seawater freezing point drive an overturning circulation that is referred to as the "ice pump" (Lewis and Perkin, 1986). The seawater freezing point decreases by about 0.75 °C for each additional kilometer of depth. Maximum melt rates occur where the thermal forcing (the difference between the *in situ* seawater temperature and the pressure-dependent freezing point) is largest, near grounding lines, defined as the point where the ice starts to float (Fricker *et al.*, 2009). In a "closed" ice pump, the freshwater fluxes associated with melting and freezing are equivalent. Water that refreezes at a higher temperature near the ice shelf front returns to the grounding line and supplies the latent heat necessary to melt ice there. Holland *et al.* (2008) find a quadratic relationship between total ice shelf basal melting and ocean warming. Freezing occurs when meltwater freshened water masses travel northward along the base of the ice shelf.

Freezing of seawater at the base of the Ross Ice Shelf was first detected by Zotikov *et al.* (1980). The bottom 6 m of a 416-m core at J-9 (82.4° S, 168.6° W, northeast of Crary Ice Rise), were composed of sea ice. Marine ice forms when frazil ice crystals accumulate at the base of an ice shelf, as part of the thermohaline circulation in the underlying cavity, and then become consolidated into layers (Khazendar *et al.*, 2009).

Within an ice shelf, the heterogeneous material that is composed of marine ice, sea ice, and firn is referred to as an ice mélange. There have been several suggestions that an ice mélange acts as a binding material that slows or halts rift enlargement or, on a much larger scale, holds extensive segments of an ice shelf together. Meteoric ice may have temperatures between −21 °C and −15 °C, while those of the ice mélange are −11 °C to −7 °C due to their different sources. This affects their rheological properties causing the meteoric ice to be almost twice as stiff as the marine ice.

MacAyeal (1984) provides a numerical simulation of the circulation beneath the Ross Ice Shelf. He suggests that vertically well-mixed conditions predominate in the southeastern part of the sub-ice-shelf cavity where the water column is shallow. Here basal melting is expected to be ~0.05–0.5 m a^{-1} and it will drive a thermohaline circulation where high salinity shelf water (at −1.8 °C), formed by winter sea-ice production in the open Ross Sea, flows along the seabed toward the tidal mixing fronts under the ice shelf; meltwater (at −2.2 °C), produced in the well-mixed region, flows out of the sub-ice-shelf cavity along the ice shelf bottom. Cavities beneath the ice shelves account for about 40% of the area of the Antarctic continental shelf and so play a major role in ice-ocean interactions (Nicholls *et al.*, 2009).

Ice shelf morphology plays a critical role in linking subsurface heat sources to the ice because the basal slope strongly influences the properties of buoyancy-driven flow near the base of the ice shelf (Little *et al.*, 2009). Observations suggest that the freezing of marine ice appears to be concentrated along western boundaries of ice shelves where there is northward flow of meltwater freshened water masses, with intensified melting in the eastern sectors where there is a southward heat flow. Little *et al.* (2008) attribute intensified freezing in the west to shoaling in the ice shelf topography. They note that topography may constrain oceanic circulation and thus basal melt–freeze patterns through its influence on the potential vorticity field. However, melting and freezing induce local circulations that can modify locations of heat transport to the ice shelf. They investigated the influence of buoyancy fluxes on locations of melting and freezing under different bathymetric conditions. Decoupled simulations show that flow in the interior is governed by large-scale topographic gradients, while recirculation plumes dominate near buoyancy fluxes. In coupled simulations, which allow freshwater and heat fluxes to migrate, strong cyclonic flow near the southern boundary (forced by melt-induced upwelling) drives inflow and melting to the east. Recirculation is less evident in the upper water column, as shoaling of meltwater-freshened layers dissipates the dynamic influence of buoyancy forcing, yet freezing remains intensified in the west. The flow throughout the cavity is relatively insensitive to bathymetry, but stratification, the slope of the ice shelf, and strong, meridionally distributed buoyancy fluxes weaken its influence on ice accretion/melt.

A suggestion by Sergienko *et al.* (2008) is that large-amplitude ocean waves could excite vibrational motions that propagate as flexural-gravity waves throughout the ice shelf interior. The possible effects of these ocean wave-induced motions on stress distributions around ice shelf rifts, for example, are being modeled. Bromirski *et al.* (2009) postulate that the breakup events of the Wilkins Ice Shelf in austral winter 2008 coincided with the estimated arrival time of infra-gravity waves from the North Pacific. Infra-gravity waves are a type of long-period ocean wave generated when ocean swell strikes continental coastlines.

8.5   Ice shelf buttressing

Ice shelves play a key role in buttressing the glaciers that flow into the shelves as was first pointed out by Mercer (1978) and Thomas (1979). Evidence for this role is provided by De Angelis and Skvarca (2003), who show major perturbations on former tributary glaciers (Boydell, Sjögren, Edgeworth, Bombardier, and Drygalski) that fed sections of the Larsen Ice Shelf on the AP before its collapse. Scambos *et al.* (2004) also find a two- to six-fold increase in centerline speed of four glaciers flowing into the now-collapsed section of the Larsen B Ice Shelf. The surface of Hektoria Glacier lowered by up to 38 ± 6 m in a 6-month period beginning 1 year after the break up in March 2002. Rignot *et al.* (2004) found that the mass loss associated with the flow acceleration exceeded 27 km^3 of ice per year. Changes in both summer melt percolation and in the stress field due to shelf removal appear to play a major role in glacier dynamics.

Gagliardini *et al.* (2010) examine the coupling of ice shelf melting and buttressing. They show, using a model incorporating grounding-line dynamics, that melting acts directly on the magnitude of the buttressing force by modifying both the area experiencing lateral resistance and the ice shelf velocity. Hence, the decrease of back stress imposed by the ice shelf is the prevailing cause of dynamical thinning inland. The distribution of melting is found to be a key parameter in determining forces at the grounding line and it is shown to be possible for an increase in the global melting to lead to a grounding line advance and growth of the grounded ice-sheet if the melting is not concentrated near the grounding line.

The thinning of outlet glaciers and ice streams in the Antarctica and Greenland has been attributed to an acceleration of the ground ice flow, which lead to an increase in the discharge of ice through the grounding line (Haseloff and Sergienko, 2018). There is a balance between the surface mass located upstream the grounding line and the ice discharge through the grounding line, whose location depends on the accumulation rate, bed elevation, and basal shear stress. For unconfined marine ice sheets, the bed slope is a major factor for determining if the grounding line is stable, that is, returns to equilibrium under small perturbations, or not. On the other hand, the buttressing of an ice shelf holds back the glaciers that feed them, leading to a reduction in ice flux through grounding line. The importance of buttressing is shown that due to the disintegration of the Larsen B ice shelf over time, the seaward velocities of Hektoria, Crane and Jorum, and Green Glaciers increase significantly (Scambos *et al.*, 2004). Basal melt rates near Bawden Ice Rise, a major pinning point of Larsen C Ice Shelf, showed large increases, which will lead to substantial loss of buttressing if sustained (Adusumilli *et al.*, 2018).

However, based on simulations of numerical ice-sheet models of varying complexity, Schannwell *et al.* (2018) show that the removal of the Larsen C ice shelf will not significantly affect up-stream tributary glaciers and its effect to global sea-level rise is marginal (<2.5 mm by 2100, <4.2 mm by 2300). This is because despite of its large size, Larsen C does not provide strong buttressing forces to upstream basins and its collapse is not expected to lead to large additional discharge from its tributary glaciers. In contrast, inland glaciers in response to a collapse of the George VI Ice Shelf may cause the global sea level to rise up to 8 mm by 2100 and 22 mm by 2300 partly because it provides both strong buttressing and mostly marine-based outlet glaciers on sloping bedrock topography expected to be favorable for marine ice sheet instability.

8.6   Icebergs

An iceberg forms when a mass of ice splits off the tongue of an ice shelf or a floating glacier or glaciers that have disintegrated. They are floating pieces of ice of various shapes, and commonly found near Antarctica and in the North Atlantic Ocean near Greenland. Tabular icebergs usually come from the calving of ice shelves. Factors leading to disintegration of ice shelves are such as bending stresses as a consequence of buoyancy forces; regional warming which leads to surface meltwater ponding and hence crevasse enlargement; and weakening and thinning of ice strength by the infiltration of warm water infiltration or brine infiltration.

Calving processes

Ice shelves move seaward at between 0.3 and 2.5 km a^{-1} and the seaward front experiences stresses from currents beneath the ice shelf, as well as from tides, storm waves, and ocean swell during the summer. Ocean swell causes the floating tongue to oscillate vertically until it fractures. Collisions with existing large icebergs are a further source of stress. The shelf normally possesses cracks and crevasses along which fractures occur, causing a piece of the ice to break off forming an iceberg. Rifts develop behind the ice shelf's front and periodically these rupture completely to form large tabular icebergs. This process is termed *calving*.

Kristensen (1983) identifies five mechanisms responsible for calving. These are: creep failure due to spreading of the ice tongue/shelf setting up lateral stresses; Reeh-type calving (Reeh, 1968) where there is fracture due to long-term creep at a distance from the ice edge roughly equivalent to the ice thickness; hinge-line calving where ice fractures at the grounding line due to storm waves, or other perturbations, generating very large icebergs; vibrational calving where the ice shelf is deformed due to long-period ocean waves (swell, storm surges, or tsunami) and fractures possibly along crevasses; and collisions between an iceberg and the ice shelf front such as occurred in 1967 at Troltunga ice tongue, Dronning Maud Land when a 5,000 km^2 iceberg was formed (Swithinbank, 1969). Figure 8.9 illustrates four calving mechanisms after van der Veen (2002). He points out that major calving events from Antarctic ice shelves are associated with large rifts that extend through the entire thickness of the shelf. Satellite mapping shows that rifts are common features; some reach several hundred kilometers in length and several kilometers wide (Lazzara *et al.*, 1999). They appear to form as a result of extensional stresses in the ice and crack nucleation initiated by some ice-weakening event upstream.

Reeh (1968) notes that at the base of the floating ice the hydrostatic forces are balanced, but they become increasingly unbalanced above the ocean level toward the ice surface. These unbalanced forces set up a normal tensile force that curves the top of the ice face toward the water. The downward movement sets up a buoyancy force acting on a section of ice on the landward side of the ice face. The effective shear stress in the ice reaches a maximum value at a distance back from the ice face equal to the ice thickness. The maximum shear stress is ~100–300 kPa.

Benn *et al.* (2007) suggest that the first-order control on calving is the strain rate due to spatial variations in velocity, which determine the location and depth of surface crevasses. Superimposed on this are second-order processes that include fracture propagation in response to local stress imbalances close to the glacier front, undercutting of the glacier terminus by melting at or below the waterline, and bending at the grounding line of an ice tongue. Calving of submarine platforms or ice toes is a third-order process. A key question is whether calving is the cause or the consequence of ice flow acceleration. The former view is exemplified by Meier and Post (1987) and the latter by van der Veen (2002).

Based on a statistical analysis of data on changes in terminus position, ice speed, ice thickness, and water depth from 12 Alaskan tidewater glaciers, Brown *et al.* (1982)

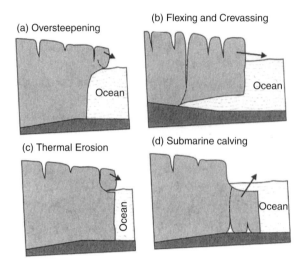

Figure 8.9 Four mechanisms for the production of icebergs from a grounded terminus (from van der Veen, 2002, p. 99, fig. 1). (a) Oversteepening of the ice cliff owing to faster ice flow at the surface of the glacier; (b) flexing that allows crevasses upglacier to penetrate the full ice thickness; (c) thermal erosion at the waterline; and (d) calving of submarine platforms or "toes" [Courtesy of *Progress in Physical Geography* 2002, 26, p. 99, figure 1. Sage Publications: http://ppg.sagepub.com/]

proposed a water-depth relation for calving rate from grounded glaciers. For 22 tidewater glaciers in Alaska, Greenland, and Svalbard, Pelto and Warren (1991) derived a water depth (Dw) – calving rate (Uc) relationship as follows: $Uc = 70 + 8.33\ Dw$ (m a^{-1}). However, Haresign (2004) found that the relationship between calving rate and water depth varies between regions and between tidewater glaciers and lacustrine glaciers. A problem noted by van de Veen (2002) is that the water-depth model applies only to annually averaged calving rates, whereas calving is a discrete process with icebergs detaching periodically from the ice front. When seasonal rates are considered, the water-depth model breaks down (Sikonia, 1982). A further observation that casts doubt on the water-depth model is that, during its rapid retreat, the speed of Columbia Glacier increased almost as much as did the calving rate. Why the increased calving would be linearly proportional to the increase in glacier speed is unclear. Other problems arise: from late1984 to early 1989, the calving rate on the Columbia Glacier remained more or less constant, yet the terminus continued retreating into water 100 m deeper. Subsequently, the calving rate almost doubled while the terminus remained grounded at about the same depth. Hanson and Hooke (2000) consider calving as a multivariate problem with water depth, longitudinal strain rate, and temperature as the three key factors controlling the rate of calving. Calving glaciers in southern Patagonia indicate that large (small) thinning rates occur on glaciers that are retreating rapidly (slowly) according to Naruse *et al.*

(1995). Patagonian glaciers also confirm the contrasting behavior of tidewater and lacustrine glaciers (Warren and Aniya, 1999). For comparable water depths, calving rates on glaciers calving into lakes are an order of magnitude smaller than those on tidewater glaciers. The reason for this is unclear although tidal effects and ocean wave action must be involved, as well as the difference in density between fresh water and salt water, subaqueous melt rates, frontal over-steepening and longitudinal strain rates.

van der Veen (1996) points out that, irrespective of the nature of the calving process, the ice thickness at the terminus of retreating glaciers always remains near the flotation thickness. If the terminus becomes sufficiently thin, the snout breaks off to maintain a thickness close to the flotation value. van der Veen notes that the calving front tends to be located where the height of the terminal ice cliff above buoyancy, Ho, is about 50 m. However, Benn et al. (2007) emphasize that the height-above-buoyancy model does not allow for the formation of ice shelves, because the model "cuts off" the glacier terminus before flotation can occur. They state that fracture propagation preconditions the location, magnitude, and timing of calving events.

Burgess et al. (2005) state that the annual discharge of ice calved from tidewater glaciers can be calculated as

$$Q_{total} = Q_{flux} - Q_{vloss} \tag{8.1}$$

where Q_{flux} is the annual volume of ice discharged at the tidewater terminal and $Q_{v\,loss}$ is the volume loss due to changes in the terminus positions of tidewater outlet glaciers. To determine Q_{total}, we need to know the cross-sectional area and average cross-sectional velocity of the ice at the terminus of each outlet glacier and the history of terminus advance or retreat.

The dominant cause of mass loss for the Antarctic ice sheet has been attributed to Iceberg calving. By iceberg tracking, Jacobs et al. (1992) estimated the total calving flux of Antarctic was $2,016 \pm 672$ Gt yr^{-1}, which agrees with the mean of estimates of 1970s and 1980s. However, Depoorter et al. (2013) estimated a total calving flux of $1,321 \pm 144$ Gt yr^{-1} (34% less than past estimates) and a total basal mass balance of $-1,454 \pm 174$ Gt yr^{-1}. They estimated ice shelf thickness based on altimetry data of ERS-1 supplemented by ICESat data for latitudes south of the ERS-1 limit, and corrected the elevation data to elevation rates from ERS-1 and ICESat to fit ice surface velocity data of InSAR (synthetic-aperture radar interferometry) and ice penetrating radar (IPR) data. Using these satellite measurements, they estimated the mass balance for all ice shelves in Antarctica, calving flux, and grounding-line flux. They concluded that about half of the ice-sheet surface mass gain is lost through oceanic erosion before reaching the ice front.

Iceberg data

Orheim (1987) estimated that the mean annual volume of icebergs coming from Antarctica is between 750 and 3,000 km^3 corresponding to a mass of ~6.4–25 × 10^{14} kg a^{-1}. Budd et al.

(1967, 1971) estimate that the Amery ice shelf has an annual output of 31 km^3, while the Filchner ice shelf has an output of 100 km^3. These shelves are basically the floating extension of glaciers that lose icebergs flow into the sea to maintain equilibrium, balancing the input of snow upstream. In 1963, a large area of the Amery Ice Shelf broke off and this single event discharged 870 km^3 (6×10^{14} kg) of ice into the ocean. On September 25, 2019, the Amery Ice Shelf produced its biggest iceberg since 1963 called D28 that covers 1,636 km^2 in area and weighs about 315 billion-ton. The breaking off of D28 from the Amery Ice Shelf is likely not linked to climate change impact, but not so with the B-49 iceberg at the Pine Island Glacier, which broke off on February 9, 2020 possibly because of melting caused by warm water eating away at the ice that surrounds the frozen continent.

Suyetova (1966) estimates that 62% of the total annual discharge of water (1,180 km^3) into the Southern Ocean is from the ice shelves, 22% from outlet glaciers and 16% from ice walls (see Chapter 8, Section 8.2). The highest concentrations of icebergs in the Pacific sector are found in the Amundsen and Bellingshausen seas (Glasby, 1990). Average coastal concentrations detected by radar range from 5 to 9 bergs per 1,000 km^2 and up to 25 per 1,000 km^2 in the western Ross Sea.

Bauer (1955) estimated that East Greenland produced 1.08×10^{14} kg a^{-1} (125 km^3) (50% of Greenland's total calving loss of 2.15–2.25×10^{14} kg a^{-1} (~250 km^3), while the West Coast south of Melville Bay generated 0.81×10^{14} kg a^{-1} (38%) and Melville Bay 8%. The glacier systems producing 80% of the icebergs on the East Coast are: (1) Storstrommen, L. Bistrops, and Soraner glaciers; (2) De Geers and Jaette glaciers; and (3) Daugaard–Jenssen glacier. On the West Coast the major sources are: Jakobshavn, Rinks, in Melville Bay the Steenstrup, Dietrichson, Nansen, Kong Oscar, and Gade, and in Kane Basin the Humboldt glacier (Robe, 1980).

In Baffin Bay, iceberg counts diminish greatly in the southward drift current. Andrews (2000) notes that the number of icebergs per 1,000 km^2 during 1963–1972 varied from a high of 15–60 per 1,000 km^2 in the extreme north of Baffin Bay, to between 0 and 2 per 1,000 km^2 east of Hudson Strait. Miller and Hotzel (1984) estimated the number of icebergs crossing east–west transects on the Labrador Shelf at ~6–15 icebergs per kilometer per year with total numbers of icebergs crossing the transects varying from 1,400 to 3,000. Iceberg numbers decreased from 25 to 35 over the marginal trough in the west to 5–9 over the shelf slope in the east. In the waters from Hudson Strait (67° N) to the Grand Banks (48° N), the annual number of icebergs decreases linearly from around 4,000 to 300 according to Ebbesmeyer *et al.* (1980). Afultis (1987) tabulates the monthly number of icebergs south of 48° N as documented by the International Iceberg Patrol. For 1900–1912, there was an average of 452 icebergs per year, with over 1,000 in 1909 and 1912. For 1913–1945 surface patrol vessels counted an annual average of 435 bergs, with over 1,300 in 1929 and over 1,000 in 1945. Visual aircraft reconnaissance for 1946–1982 counted an average of 273 bergs, with zero in 1966, and aircraft SLAR and RADARSAT for 1983–2007 counted an average of 983, with 2,200 in 1984. Figure 8.9 illustrates the annual variation of icebergs south of 48° N for 1912 to 2018, showing extreme numbers in the decade of the 1990s. Years with <300 icebergs are

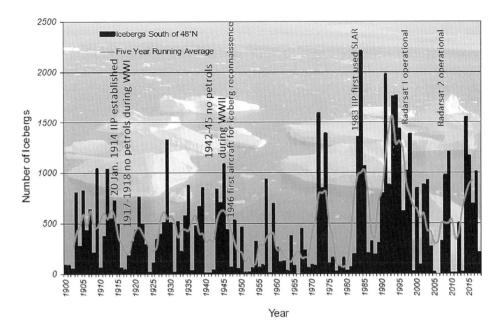

Figure 8.10 The annual number of icebergs south of 48° N during 1912–2018 Note the dates when different observing procedures were initiated (modified from U.S. Coast Guard, 2018)

considered light, 300–600 intermediate, 600–900 as heavy and >900 as extreme. As shown in Figure 8.10, the International Ice Petrol (IIP) was established in 1914. The extreme decadal iceberg limits for 1946–1955, 1956–1965, and 1966–1975 were all at latitude 40° N, 48–50° W according to Robe (1980, figure 18). A new Normalized Season Severity Index (NSSI) was proposed by Futch and Murphy (2003) by combining three normalized indices: the area enclosed by the Limit of All Known Ice (LAKI), the length of the iceberg season (LOS), and the number of icebergs south of 48° N. Values of the NSSI ranged from 0 to 5.8 during 1975 to 2009.

Abramov (1996) presents an atlas of icebergs in the Russian Arctic seas, based on shipboard and aerial reconnaissance observations. The data span 1899 to 1992 in the Barents Sea, ~1930 to 1992 in the Arctic Basin, the Kara Sea and the Franz Josef Land straits, and 1950 to 1992 in the Laptev Sea, East Siberian Sea, and Chukchi Sea. Monthly charts of the mean and maximum number of icebergs and the probability of their occurrence are provided. Shipboard observations for the Barents Sea show a mean size of 64 × 46 m and a sail height of 11 m (Abramov, 1992). These originate on the Spitzbergen banks between Hopen and BjØrnaya and between 30° and 60° E. The total number of icebergs observed by aerial reconnaissance was: ~29,000 in the Franz Josef Land Straits (1936–1992), ~17,900 in the Laptev Sea (1950–1991), 11,170 in the Barents Sea (1937–1992), 9,100 in the Kara Sea (1936–1993), and 6,900 in the Arctic Basin (1937–1992). Probabilities peak at >60% in July–

September east of Severnaya Zemlya and in the Franz Josef Land straits. The southern boundary of icebergs in the Barents Sea fluctuated between 74° N and 80° N in the 1930s–1950s and then was around 75° N during 1960–1990. In the western Kara Sea, the southern boundary was between 72 and 76° N and in the western Laptev Sea around 76° N from 1950 to 1990, The annual total iceberg volume is ~ 25,000 km^3, and the iceberg flux from the Eurasian archipelagos is estimated at 6.3 km^3 a^{-1}, of which about two-thirds comes from Franz Josef Land and Novaya Zemlya, and a quarter from Svalbard.

Physical characteristics

There are six size categories for icebergs according to the International Ice Patrol. The smallest ones are called growlers (<5 m long, ~1 m high). The next larger size is a bergy bit, which is 5–14 m long and 1–5 m high. The remaining four size categories are: small (15–60 m long and 5–15 m high), medium (60–120 m long and 16–45 m high), large (120–210 m long and 46–75 m high) and very large (>210 m long and >75 m high). Tabular icebergs are flat sheets of floating ice formed from ice shelves with a length/height ratio of ≥5:1; occasionally, their length may exceed 100 km (Figure 8.11b). They are most common, and form in much larger sizes off Antarctica. Non-tabular icebergs can take a variety of shapes, from pinnacles (the most common) to cube-like blocks, dry-dock U-form, domes, or completely irregular. Icebergs have an average width to length ratio of about 1:1.6 (Bigg *et al.*, 1997). The height to length ratio is approximated by

$$H = 0.402\ L^{0.89} \text{ (Hotzel and Miller, 1983).}$$

The iceberg size distribution has two distinct parts. One part is a slightly skewed distribution of parent icebergs and the second part shows an approximately exponential increase in

Figure 8.11 (a) Section of a large tabular iceberg B-15A off the Ross Ice Shelf in the Southern Ocean, 2001. http://amrc.ssec.wisc.edu/gallery1.html [Courtesy of Josh Landis, National Science Foundation]; (b) Icebergs of a variety of shapes when they break apart [Courtesy: Ted Scambos, NSIDC]

frequency with decreasing length in the bergy bit and growler size range (Crocker and Cammaert, 1994).

Icebergs have a wide range of sizes. In the Antarctic, the US National Ice Center identifies an average size of 1,500 km^2, with most between 150 and 2,000 km^2. The world's record iceberg B-15, which broke off the Ross Ice Shelf in March 2000 measured 10,800 km^2 (Ballantyne and Long, 2002) (Figure 8.12). One part, iceberg B15A broke up further when it grounded on a 215 m shoal off Cape Adare in October 2005 (Martin *et al.*, 2010). In October 1998, the iceberg A-38 broke off the FRIS. It had a size of roughly 150 by 50 km and was thus larger than Delaware. It later broke up into three parts. A similar-sized calving in May 2000 created iceberg A-43 measuring ~167 by 32 km. Remnants of it may have reached the South Island of New Zealand in November 2006. For Antarctic waters the size classes used by Jacka and Giles (2007) are: 25–100, 100–200, 200–400, 400–800, 800–1,600, 1,600–3,200, and >3,200 m width, for ship-based observations. Near the Antarctic coast the icebergs drift westward in the easterly winds, at 80–90° E they turn northward influenced by the bathymetry, and then move eastward in the Circumpolar Current. There are increased concentrations in all size classes from 50 to 90–100° E. From 50° to 80° E there is an approximately equal concentration of icebergs in the three classes 25–100, 100–200, and

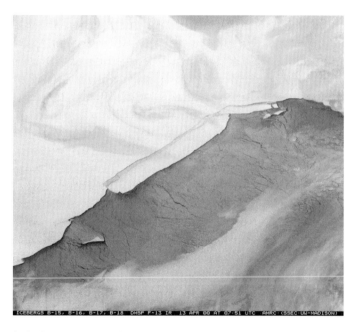

Figure 8.12 The iceberg B-15 breaking off the Ross Ice Shelf on April 13, 2000, shown together with B-16, B-17, and B-18. This infrared image was acquired by the Defense Meteorological Satellite Program F-13 satellite http://earthobservatory.nasa.gov/IOTD/view.php?id=552 [Courtesy Earth Observatory, NASA]

>200 m width. Eastward of 80–90° E there is a decrease in concentration with increasing size category (Jacka and Giles, 2007).

Volume (*V*) estimates for icebergs have been based on scaling of the length (*L*), width (*W*), and height (*H*) of the portion above water. For Antarctic icebergs, Shil'nikov (1965) gives 4.9 *LWH* for tabular bergs, 4.1 *LWH* for domed ones, and 2.5 *LWH* for pinnacles, where the volume is in m^3. For Greenland icebergs, Farmer and Robe (1977) estimated *V* = 3.35 *LWH*, using photogrammetry, and the iceberg mass (metric tons) as 3.01 *LWH*.

Iceberg studies off West Greenland between 1975 and 1978 are reported by Mangor and Zorn (1983). The mean iceberg mass observed in Disko Bay was in the range 5–11 $\times$ 10^9 kg and the maximum was 32 $\times$ 10^9 kg; the corresponding values for the area north and west of the Store Hellefiske Bank were 2 $\times$ 10^9 kg and 15 $\times$ 10^9 kg, respectively. The average iceberg depths off the Store Hellefiske bank were 80–125 m with a maximum of 187 m. Off Scoresby Sund in East Greenland iceberg drafts of up to ~300 m are reported.

Dowdeswell *et al.* (1992) measured the sizes, frequencies, and freeboards of 1,900 icebergs within the Scoresby Sund fjord system of East Greenland from the *Polarstern*, using ship X-band radar (corrected for beam spreading) and sextant. They found that 69% of icebergs were <200 m in width; only five were >1 km in length. Iceberg concentration (maximum 0.6 bergs km^{-2}) declines with distance from the major iceberg sources. The modal iceberg keel depth, calculated from freeboard measurements, is 400–500 m in the inner fjords, but less in the outer fjords, reflecting shallower bathymetry. During the penultimate Saalian glaciation (Marine Isotope Stage 6) there is evidence of iceberg scouring on the Lomonosov Ridge (85–87° N, 150° E), implying keels 1,000 m deep (Kristoffersen *et al.*, 2004), while iceberg plow marks on the Iceland–Faeroe Ridge indicate keel depths of 820 m in the (Saalian) glacial North Atlantic (Kuijpers and Werner, 2007) (see Chapter 9, Section 9.5).

Temperature measurements were made on a 1,000 $\times$ 300-m iceberg, 10–18 m high, by von Drygalski (1983) on the ice-beset German expedition ship *Gauss* (66° S, 90° E) from April 1902 to January 1903. Boreholes were drilled to 30 m depth and readings made in the air and ice once to four times a month. In August 1902 with an average air temperature of −22 °C on the *Gauss*, the 30 m ice temperature was −10.4 °C, and in December 1902 with an average air temperature of −1.1 °C the 30 m ice temperature reached −9.6 °C; the mean annual air temperature on the *Gauss* was −11.5 °C. Orheim (1980) indicates bulk temperatures for Antarctic icebergs at around 70° S in the Weddell Sea of ~−17 °C with temperatures in the upper 10 m about 6 °C higher than in the ice shelf. The refreezing of percolated meltwater is mainly responsible for raising internal temperatures at shallow depths. Icebergs that had reached the West Wind Drift had snow temperatures around 0 °C. Scambos *et al.* (2008) measured a temperature of −15 °C at 11 m on iceberg A-22 located to the northwest of Orcadas Island in March 2006. This temperature is significantly below the mean annual temperature for the Peninsula region, but higher than the mean annual temperatures near the Ronne Ice Shelf front (−17 °C to −25 °C). The surface firn temperature was −6.4 °C and at 1.5 m depth it reached −1.2 °C. Diemand (1983) considers that the temperature should remain essentially unchanged from its original value as an iceberg drifts from Baffin Bay to

Newfoundland. The ice temperature is an important control on crushing strength as this value drops off rapidly at temperatures above $-15\,°C$ (Butkovich, 1954).

Icebergs contain air bubbles that account for 2–8% of Greenland and Antarctic icebergs by volume (Scholander and Nutt, 1960; Robe, 1980).

Shelf icebergs have a density corresponding to that of firn, about $450\,kg\,m^{-3}$, at the surface increasing to 860–$890\,kg\,m^{-3}$ at about 60 m depth, whereas icebergs from outlet glaciers have nearly uniform densities of 880–$910\,kg\,m^{-3}$ (Crary *et al.*, 1962; Matsuo and Miyake, 1966). As a consequence, the ratio of the submerged to total volume of shelf icebergs is about 0.83 compared with 0.88 for all others (Robe, 1980).

Iceberg deterioration

During June–November in the Labrador Sea, a small, non-tabular iceberg takes five days to disintegrate completely, while medium to large bergs have a life expectancy of the order of weeks (Venkatesh and El-Tahan, 1988).

There are three major causes of iceberg deterioration – breaking, wave-induced melt at the water line, and melting at the top, bottom and sides (Huppert, 1980). Melting is generally very slow in view of the low surface-to-volume ratio of typical icebergs – 0.01–$0.4\,m^2\,m^{-3}$ – and the bulk of the iceberg below water (Robe, 1980). According to Savage (2001), the dominant mechanism for iceberg deterioration is wave erosion. Waves can erode a notch in an iceberg, with a vertical extent <15–$20\,m$, after which calving and/or fracture can occur. Icebergs have relatively high permeability at the waterline so that seawater can penetrate to form brine layers. Other important processes are wave-induced calving, and forced convection in the water. Incident solar radiation plays a relatively minor role. Scambos *et al.* (2008) study calving mechanisms during iceberg drift northward in the western Weddell Sea. The mechanisms identified are (1) rift calving, (2) edge wasting, and (3) rapid disintegration (see Figure 8.6). (1) exploits preexisting large fractures, (2) produces numerous small edge-parallel slivers, and (3) is associated with surface lakes and firn-pit ponding: the firn/ice transition in icebergs occurs at about 40–60 m depth. Mechanism (3) is identified as a probable factor in the disintegration of the Larsen B ice shelf.

Ocean waves are the primary cause of the breakup of tabular icebergs (Kristensen, 1983). Resonant bending of the iceberg is determined by the geometry and structure of the iceberg and by the period and amplitude of the waves (Wadhams, 2000, pp. 261–5). Accelerometers and tilt meters mounted on two Antarctic icebergs show that icebergs act to filter out short wave periods (Kristensen *et al.*, 1982). The dominant peaks were at 16 s for heave (vertical movement), 10 s for surge (fore/aft movement), and 11 s for roll (angular motion about the long axis). The strain spectrum has peaks at 10, 15, and 50 s showing that the iceberg is selectively extracting wave energy. The strains corresponding to those peaks are much larger than expected (typical of those for much thinner sea ice floes) indicating that resonance is occurring. This leads to fatigue and breakup of large tabular icebergs to dimensions less than 1.5–1.0 km, after which wave flexure becomes unimportant. Icebergs that are unstable and roll over develop a dome-shaped, whereas stable icebergs form a wave-cut notch up to twice the wave amplitude above the mean free surface and $1/K$ below that level, where K is the wave number – the number of wavelengths per unit distance (Robe, 1980).

Ivana *et al.* (2007) describes various deterioration mechanisms built in the iceberg deterioration component of an iceberg forecasting model developed for the Canadian Ice Service, which consist of melting due to solar radiation, buoyant convection, and forced convection, wave erosion and calving. They found that wave height, wind velocity and water temperature play significant roles in iceberg deterioration, and calving intervals this model predicted agree well with observations.

Iceberg motion

Iceberg motion is determined by the effects of six forces: wind drag, water drag, the force due to the wave "radiation stress" (excess momentum flux associated with the wave motion in a wave train), the horizontal pressure gradient force exerted by the water on the volume that the iceberg displaces, an effective force associated with the added mass of the iceberg, and forces due to interactions with sea ice (Savage, 2001). The wave radiation force is at least twice the air drag force according to Smith (1993). Nevertheless, the major factors that influence iceberg drift are the wind velocity and water currents. The air drag law depends upon the square of the relative wind velocity. The air drag force acts in the direction of the relative wind, so long as the iceberg shape is symmetrical about the two planes parallel to the flow direction. The wave radiation term depends upon the way in which the waves are diffracted and dissipated by the iceberg. Regarding the pressure gradient force, Bigg *et al.* (1997) argue that the material derivative dV_w / dt, where V_w is the water velocity, is the principal factor needed to reproduce realistic iceberg distributions in a dynamical model. They find that the basic force balance in iceberg motion is between water drag and water advection. These two forces contribute approximately $70 \pm 15\%$ of the total forcing of iceberg motion. Bigg *et al.* argue that the ocean current in which they are embedded basically advects icebergs. Icebergs are subject, however, to other forces that produce an offset from pure advection in the iceberg's motion making the water drag significant. The Coriolis force and the air drag generally make up roughly 15% each of the remainder of the force balance.

The water drag term actually has three components: (1) form drag, (2) frictional drag due to the viscous stress acting at the underwater surface of the iceberg, and (3) inertial drag due to the acceleration of the water relative to the iceberg. The steady-state form drag coefficient is a function of the iceberg's Reynold's number, $Re = LU/v$, where L is a characteristic length and U is the relative velocity of the iceberg, and v is the kinematic viscosity (Robe, 1980). For a nearly square 150 m iceberg, 100 m deep, with a relative drift speed of 0.05–0.12 m s^{-1}, the Reynolds number ranged from 1.8×10^6 to 1.4×10^7 (Russell *et al.*, 1978). Based on radar tracking of 33 icebergs off Saglek, Labrador, Soulis (1975) determined that the mean drift speed relative to the current was 2.5 times and the mean direction was 25° to the right of the wind and ~1° to the right of the current, but with great variability. The best correlation time lag for both wind and current was of the order of 210 minutes.

Drift speeds range from around 0.5 m s^{-1} off Dronning Maud Land (10° E) in the East Wind Drift to 0.1 m s^{-1} in the Weddell Sea gyre (Vinje, 1980). The drift direction follows the ocean currents and West Wind Drift toward South Georgia and the South Sandwich Islands (Kristensen, 1983). Once they enter the waters north of the Antarctic Convergence located between about 50° S in the Indian and South Atlantic oceans and 60° S in the southeast Pacific Ocean, the higher water temperatures (3–7 °C) hasten their decay. Vinje (1980) determined from satellite-tracked icebergs that they take from 1 to 5 years to move into the west Wind Drift from leaving the Antarctic coast between about 50° E and the AP. The drift trajectories reflect the integrated current in the upper 200–300 m of the ocean. The maximum northern extent of Antarctic icebergs is in the southwest Atlantic Ocean, off the east coast of New Zealand, and off South Africa (see p. 344).

Icebergs are common in the northwestern Atlantic, around Greenland, and off Svalbard. Icebergs from East Greenland travel mainly south in the East Greenland Current until they reach Cape Farewell at 60° S. They then turn northwestward and travel northward in the West Greenland Current toward Melville Bay or turn westward in Davis Strait before moving southward. Approximately 40,000 medium- to large-sized icebergs calve annually from Greenland's outlet glaciers, mostly from West Greenland (Robe, 1980). The Jakobshavn Glacier in West Greenland is responsible for producing at least one-tenth of the icebergs calving into the sea from the entire Greenland Ice Sheet. The icebergs circulate Baffin Bay in a counterclockwise direction and then travel south in the Labrador Current. About 1 to 2% (400–800) of those make it as far south as latitude 48° N off Newfoundland. The average drift speed of icebergs off Newfoundland is around 0.2 m s^{-1} (0.7 km h^{-1}). The maximum numbers normally occur during April–June and the numbers are at a minimum during November–January. There is a large interannual variation in the numbers of icebergs off Newfoundland. Marko *et al.* (1994) show that the spring ice extent off Labrador is the critical parameter in the interannual variability of iceberg numbers. More extensive sea ice decreases the initial ice mass needed for icebergs to survive south of 48° N. Labrador spring ice extent was found to be closely correlated with midwinter Davis Strait ice extent. Downstream iceberg numbers are relatively insensitive to iceberg production rates and to fluctuations in southerly iceberg fluxes from areas of northern Baffin Bay.

When the colossal iceberg A68 detached itself from Antarctica's Larsen C ice shelf in 2017, at 5,800 km^2 in size, it was one of the biggest icebergs ever recorded. As iceberg A68 drifted north from the AP, a small piece broke off, becoming A68a, which continued its slow drift through the water for over two years until it hit the Antarctic Circumpolar Current circling the Antarctic, and was propelled on the fast track, northeast route traveled by other chunks of ice that break off from the AP. After parking itself to the southeast of the ecologically sensitive South Georgia island in late 2020, it started to disintegrate into several icebergs and countless bits of floating ice.

8.7   Ice islands

The calving of ~8000 km^2 of ice shelves as numerous, large tabular, ice islands from the northern coast of Ellesmere Island over the twentieth century were first detected in the 1940s (Copland and Muller, 2017). After calving, these ice islands could either drift west and remain in the Arctic Ocean for up to several decades, or they enter the interior islands of the Canadian Arctic Archipelago and disintegrate relatively rapidly; or they occasionally drift along the east coast of Canada, reaching as far south as Labrador. Even though ice islands could serve as mobile military platforms for US or Soviet, or as mobile climate stations for measuring oceanographic and atmospheric properties over their drift paths, or their disintegration patterns are related to the effects of climate change, they could pose a hazard to offshore oil platforms and exploration, and a risk to shipping lanes off the Newfoundland Ocean.

   In the Arctic, the Ward Hunt, Milne, and other ice shelves along the northern coast of Ellesmere Island have repeatedly calved off forming *ice islands*, first reported by Koenig *et al.* (1952). T-1 was discovered by aerial reconnaissance in August 1946; T-2 and T-3 in July 1950. The first landing on the most famous ice island T-3 (or Fletcher's ice island, after J. Fletcher who first identified it from the air) was made in April 1952 when a manned station was installed. This continued to be operated for the next 27 years (Jeffries, 1992). T-3 had a length of 80 km and was 8 km wide at its narrowest part. It exited the Arctic Ocean via Fram Strait in 1984, Arctic Research Laboratory Ice Station (ARLIS)-II operated without interruption during 1961–1965. Five Soviet drift stations NP-6, NP-19, NP-22, NP-23, and NP-24, were situated on ice islands. These ice islands circulate in the clockwise Beaufort Gyre for a number of years before eventually exiting the Arctic, via the Transpolar Drift Stream and East Greenland Current, and occasionally via Nares Strait. Fragments of WH-5 drifted from north of Ellesmere Island in 1963 to the Grand Banks in 1964, for example (Nutt, 1966). Newell (1993) reviewed sightings of ice islands and exceptionally large icebergs in the waters off eastern Canada. For the area south of 55° N, the frequency of sightings and the maximum reported lengths were greater during the first half of this century than during the period 1950–1990. He cites a maximum length of ~550 m and an observed mass of 25–30 $\times$ 10^9 kg reported between 1973 and 1979, but icebergs 7 and 12 km in length were reported in Hudson Strait in summer 1928 and another ~13 km long in June 1934 in Davis Straight. About 11 months after a massive 251 km^2 ice island broke off from the Petermann Glacier on August 5, 2010 over the northwestern coast of Greenland, it was caught up in ocean currents off the coast of Labrador, Canada, slowly broke up and melted away on its journey of over 3,000 kilometers. (https://earthobservatory.nasa.gov/images/51264/ice-island-off-labrador).

REVIEW QUESTIONS

(1) Name the classic tragic accident involving an iceberg that happened in the northwest Atlantic leading to the establishment of the International Ice Patrol in 1914. The accident was featured in numerous films, TV movies and notable TV episodes for over a century?

(2) Explain the four calving mechanisms for the production of icebergs from a floating glacier or ice shelf.

(3) Define an ice shelf. Who discovered the world's largest, Ross Ice Shelf in the Antarctic, the size of this ice shelf in area (km^2), length (km), and thickness (m), and the rate it is advancing to the sea in m/day?

(4) According to Massom *et al.* (2018), what other factors (see Figure 8.4) could also trigger ice shelf disintegration events, besides processes such surface meltwater ponding, crevasse enlargement, weakening and thinning of ice, increased basal melt, localized outer-margin fracturing, and the propagation of structural weaknesses?

(5) Explain the key contributing factors behind the dramatic breaking off of Larsen A on January 1995, Larsen B on January–March 2002, and Larsen C on July 2017. What is the approximate total amount of ice lost in km^2 from the Antarctic Peninsula since the 1950s?

(6) After icebergs have broken off from ice shelves, what are major causes of iceberg deterioration? Why ice melting tends to play a relatively minor role in iceberg deterioration? Note that about 90% of an iceberg is submerged under water.

(7) Radar altimeters of satellites have been employed to study ice shelf altimetry. Name three such satellites, and the approximate range of ice shelf elevation change rate ($\Delta h/\Delta t$) of the Ross Ice Shelf over 2003–2008 (see Figure 8.8) estimated by Pritchard *et al.* (2012) using radar altimeter data and modeling of the surface firn layer.

(8) How is the ice shelf thinning or loss of the Antarctic controlled by a combination of factors such as iceberg calving, the basal melt, reduced buttressing of adjacent ice sheet, and ground ice flow through the grounding line?

(9) Why is buttressing – holding back the glaciers that feed them – important in the disintegration of the Larsen B but not the Larsen C ice shelves?

(10) By iceberg tracking, what was the total calving flux of Antarctic estimated by Jacobs *et al.* (1992) estimated in Gt yr^{-1}, and that by Depoorter *et al.* (2013) based on altimetry data of ERS-1 supplemented by ICESat data for latitudes south of the ERS-1 limit? Which of the two estimates is likely more representative?

(11) What are the six size categories for icebergs according to the International Ice Patrol? What is the most common shape of icebergs formed from ice shelves, and the average width to length ratio of such icebergs? How large was the world's largest recorded B-15 Iceberg in terms of surface area in km^2?

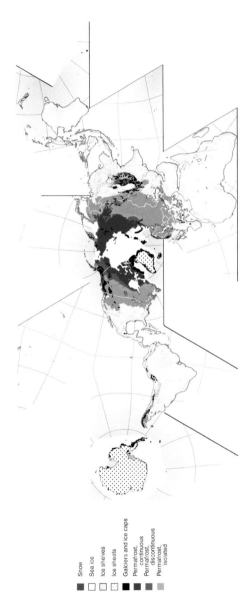

Snow
Sea ice
Ice shelves
Ice sheets
Galciers and ice caps
Permafrost, continuous
Permafrost, discontinuous
Permafrost, isolated

Figure 1.1 The global distribution of the components of the cryosphere (from Hugo Ahlenius, courtesy UNEP/GRID-Arendal, Norway)

(http://upload.wikimedia.org/wikipedia/commons/b/ba/Cryosphere_Fuller_Projection.png).

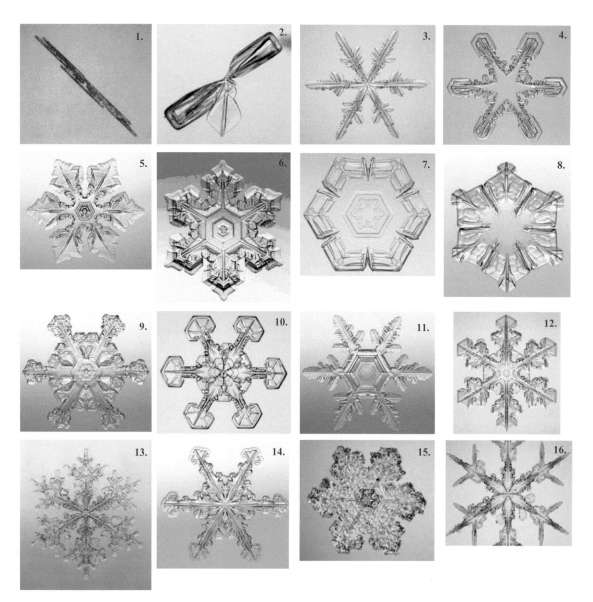

Figure 2.2 Examples of snowflakes classified according to Magono and Lee (1966): 1. Needle, 2. Sheath, 3. Stellar crystal, 4. Stellar crystal with sector-like ends, 5. Stellar crystal with plates at ends, 6. Crystal with broad branches, 7. Plate, 8. Plate with simple extension, 9. Plate with sector-like ends, 10. Rimed plate with sector-like ends, 11. Hexagonal plate with dendritic extensions, 12. Plate with dendritic extensions, 13. Dendritic crystal, 14. Dendritic crystal with sector-like ends, 15. Rimed stellar crystal with plates at ends, and 16. Stellar crystal with dendrites.

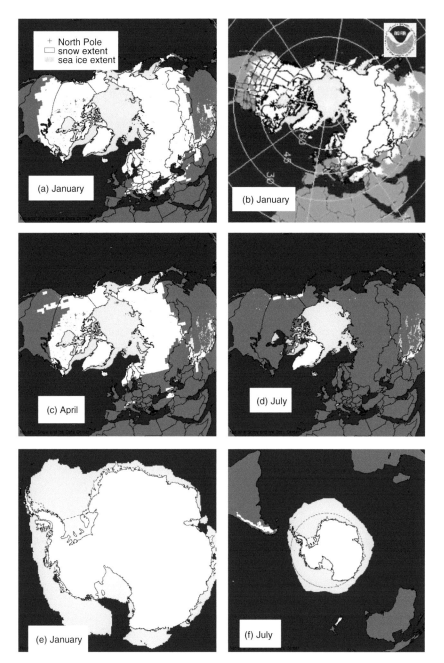

Figure 2.5 Seasonal variation in the mean monthly snow and sea ice cover extent for January, April, and July over Northern Hemisphere using data of NSIDC over 1967–2005 for snow and 1979–2005 for ice; for January and July over Antarctic/Southern Hemisphere over 1987–2002 for snow and 1979–2003 for ice (Maurer, 2007) by Lambert Azimuthal Equal-Area (http://nsidc.org/data/atlas) projection; January 31, 2008 snow and ice chart of NH adapted from NOAA-AVHRR image of NOAA (http://wattsupwiththat.com/2008/02/09/jan08-northern-hemisphere-snow-cover-largest-since-1966/).

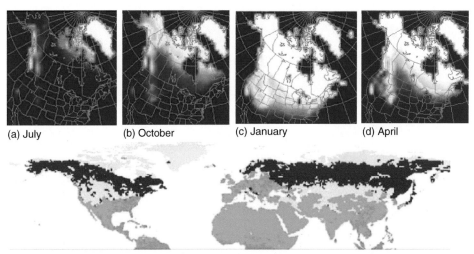

(a) July   (b) October   (c) January   (d) April

(e) Snow-free forests (dark green), unforested areas with snow cover (gray), and forests
with snow cover (red) are shown

Figure 2.6 Seasonal variation in the mean monthly snow cover extent for (a) July, (b) October, (c) January, and (d) April over North America computed from snow charts derived from weekly visible satellite images of NOAA-AVHRR over 1972–1993 (www.tor.ec.gc.ca/CRYSYS/cry-edu.htm); (e) Northern Hemisphere snow and forest covers for January, 2005 computed from the NSIDC Equal-Area Scalable Earth Grid (EASE-Grid) snow cover product (Armstrong and Brodzik, 2005) and the University of Maryland global land cover classification (Hansen *et al.*, 2000) (taken from Rutter *et al.*, 2009).

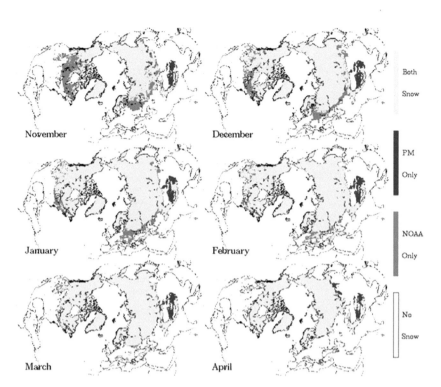

Figure 2.12 Monthly (November–April) snow cover extent climatology for Northern Hemisphere derived from long-term snow cover data of NOAA and passive microwave over 1978–2005 (taken from Armstrong *et al.*, 2006).

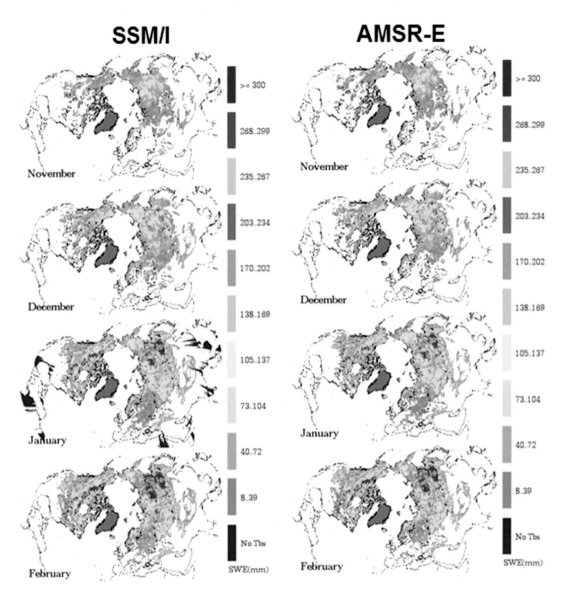

Figure 2.14 Northern Hemisphere monthly average snow water equivalent (SWE) derived from SSM/I (left) and AMSR-E (right), November 2003–February 2004 (Courtesy of Brodzik, M. J., NSIDC/CIRES).

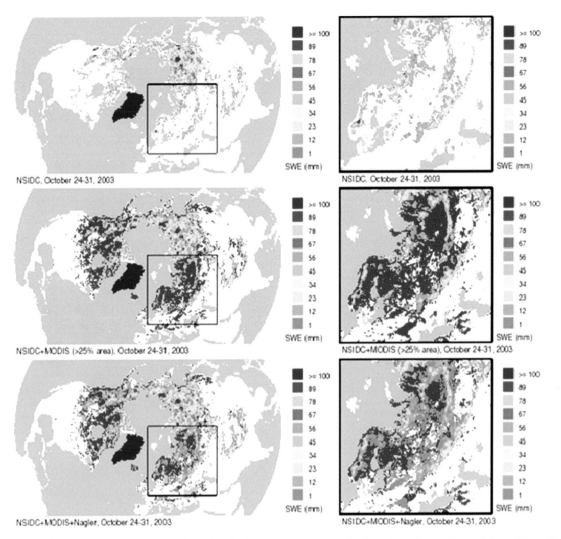

Figure 2.15 (a) shows Northern hemisphere snow water equivalent (SWE) map derived from NSIDC Global Monthly EASE-Grid SWE Climatology, with enlarged area in Northern Europe and Western Russia for October 24–31, 2003; (b) shows additional snow-covered area (red), determined from at least 25% of component MODIS CMG pixels indicating snow cover; (c) shows improvement in shallow snow SWE estimates by combining 89 GHz data with MODIS snow-covered area (select day w/ max diff 37 − 85 GHz) (Courtesy of Brodzik, M. J., NSIDC/CIRES).

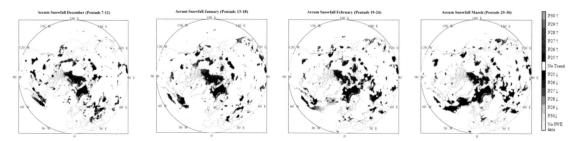

Figure 2.21 Spatial distributions of grids of the Northern Hemisphere GlobSnow data set with significant SWE (5-day averages) trends at $p < 0.05$ significant level over December–March of 1988–2017 using the Mann–Kendall test, with more statistically significant negative than positive SWE trends especially across Canada, the high Arctic, and Europe, while scatted positive trends in Russia.

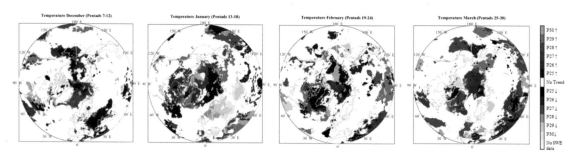

Figure 2.22 The spatial distribution of grids with statistically significant ($p < 0.05$) temperature Pentad (5-day averages) trends by the Mann–Kendall test in the Northern Hemisphere based on the 1988–2017 ERA-Interim reanalysis temperature data.

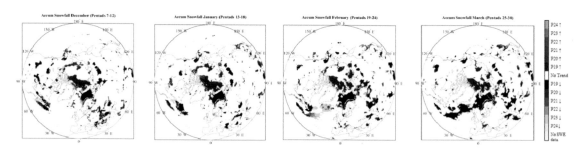

Figure 2.23 Spatial distributions of grids of the Northern Hemisphere with significant Spearman's rank correlation between SWE and air temperature over 1988–2017 at $p < 0.05$ significant level.

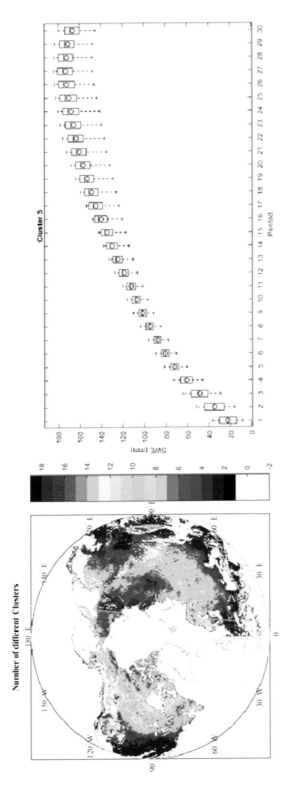

Figure 2.24 (a) Number of different clusters derived from the SOM and K-means clustering analysis that each grid in Northern Hemisphere is assigned over 1988–2017, and an example of (b) the SWE in boxplots for all grids of NH assigned to Cluster #5.

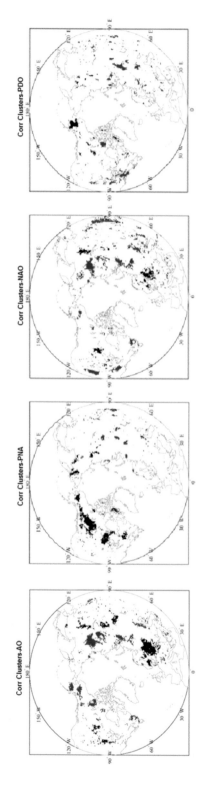

Figure 2.25 The spatial distribution of grids with significant ($p < 0.05$) positive (red) and negative (black) Spearman's rank correlation between cluster areas, AO, PNA, NAO, and PDO, respectively.

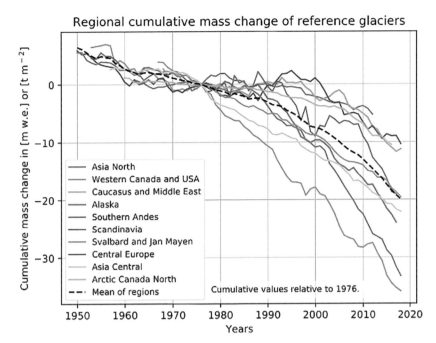

Figure 3.10 Cumulative mass change relative to 1976 given in m w.e. (water equivalent) for regional and global means based on data from a set of global reference glaciers with more than 30 continued observation years for 1950–2018. The data are compiled by the World Glacier Monitoring Service (WGMS) in annual calls for data from a scientific collaboration network in more than 40 countries worldwide. Regional values are arithmetic averages while global values are one single value averaged for each region with glaciers to avoid a bias to well-observed regions. Values before 1960 and in 2018 need to be taken with caution due to the limited sample size (https://wgms.ch/global-glacier-state/).

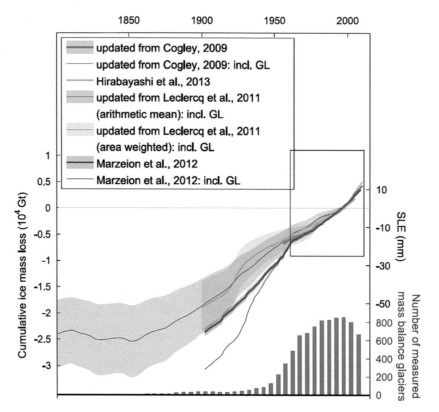

Figure 3.11 Global cumulative glacier mass change for 1801–2010 set to zero mean over 1986–2005. Estimates are based on glacier length variations (updated from Leclercq *et al.*, 2011), from area-weighted extrapolations of individual directly and geodetically measured glacier mass budgets (updated from Cogley, 2009), and from modeling with atmospheric variables as input (Marzeion *et al.*, 2012; Hirabayashi *et al.*, 2013). Uncertainties are based on comprehensive error analyses in Cogley (2009) and Marzeion *et al.* (2012) and on assumptions about the representativeness of the sampled glaciers in Leclercq *et al.* (2011), including Greenland (GL). Uncertainties are shown only for the Cogley and Marzeion curves excluding GL. The blue bars show the number of measured single-glacier mass balances per pentad in the updated Cogley (2009) time series (adapted from figure 4.12 of Vaughan *et al.*, 2013).

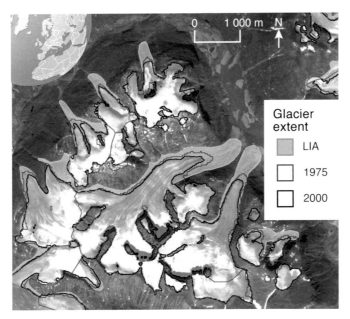

Figure 3.13 Glacier shrinkage since the Little Ice Age in the Cumberland Peninsula, Baffin Island [Source: F. Svoboda, University of Zurich] http://maps.grida.no/go/graphic/glacier-shrinking-on-cumberland-peninsula-baffin-island-canadian-arctic. Cartographer/designer Hugo Ahlenius, UNEP/GRID-Arendal.

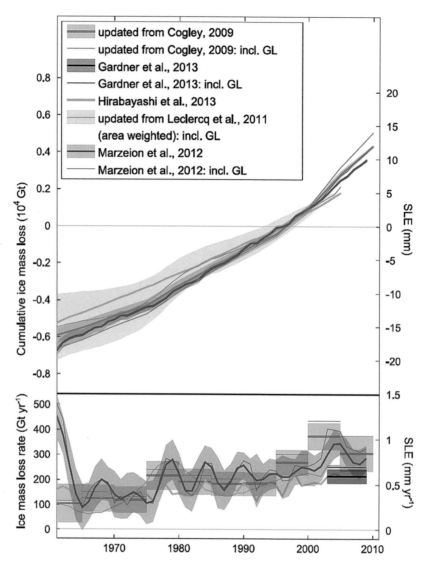

Figure 3.18 Global cumulative glacier mass change for 1961–2010 set to zero mean over 1986–2005. As in Figure 3.11, estimates are based on glacier length variations, area-weighted extrapolations of individual directly and geodetically measured glacier mass budgets, and from modelling with atmospheric variables as input (adapted from Vaughan *et al.*, 2013).

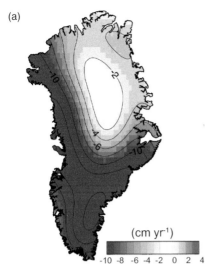

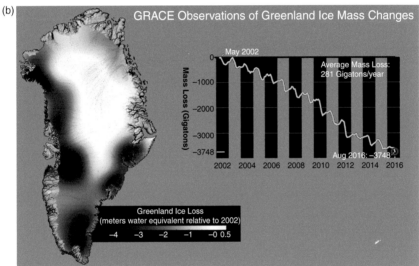

Figure 4.5 (a) Temporal evolution of mass ice loss of GIS estimated from GRACE time-variable gravity, shown in centimeters of water per year for 2003–2012, with color coded red (loss) (taken from Vaughan *et al.*, 2013), (b) GRACE observations of Greenland ice mass changes in Gt from May 2002 to August 2016. Orange and red shades showing areas with lost ice mass mainly concentrate in lower-elevation and coastal areas which experienced up to 4 m of ice mass loss (expressed in meter w.e.; dark red). Higher-elevation areas represented in white near the center of Greenland experienced minimal change in ice mass since 2002. (c) The Northern and Southern Hemisphere cryosphere in polar projection where the former shows the sea ice cover during the minimum summer extent (September 13, 2012). The yellow line is the average location of the ice edge (15% ice concentration) for the yearly minima from 1979 to 2012. Areas of continuous permafrost are shown in dark pink, discontinuous permafrost in light pink. The green line along the southern border of the map shows the maximum snow extent while the black line across North America, Europe, and Asia shows the 50% contour for frequency of snow occurrence. The Greenland Ice Sheet (blue/gray) and locations of glaciers (small gold circles) are also shown. The yellow line shows the average ice edge (15% ice concentration) during maximum extent of the sea ice cover from 1979 to 2012. Glacier locations were derived from the Randolph Glacier Inventory (Arendt *et al.*, 2012) (figure taken from Vaughan *et al.*, 2013).

(c)

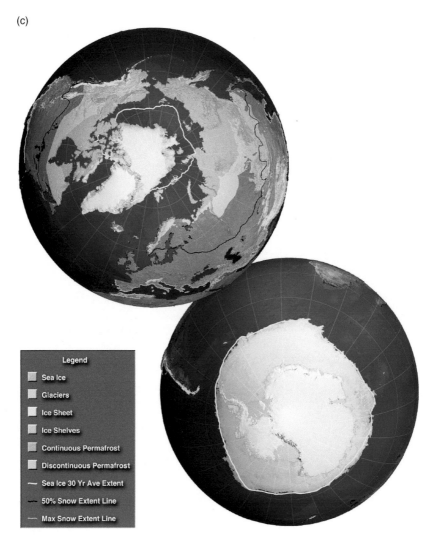

Legend

▢ Sea Ice
▢ Glaciers
▢ Ice Sheet
▢ Ice Shelves
▢ Continuous Permafrost
▢ Discontinuous Permafrost
— Sea Ice 30 Yr Ave Extent
— 50% Snow Extent Line
— Max Snow Extent Line

Figure 4.5 (cont.)

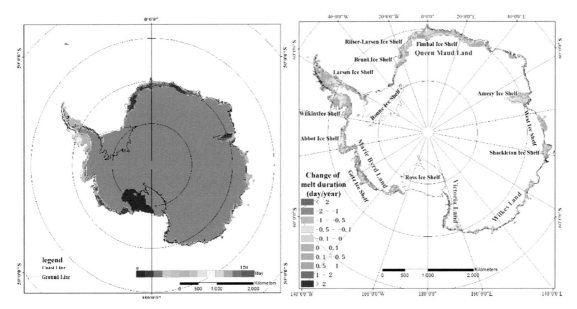

Figure 4.8 (a) Average annual melt duration (days), and (b) distribution of the rate of change of the melt duration (days) in Antarctica from 1978 to 2014 (taken from Liang *et al.*, 2019).

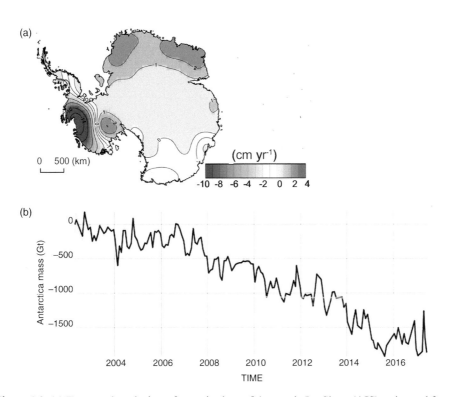

Figure 4.9 (a) Temporal evolution of mass ice loss of Antarctic Ice Sheet (AIS) estimated from GRACE time-variable gravity, shown in centimeters of water per year for 2003–2012, with color-coded red (loss) to blue (gain) (taken from Vaughan *et al.*, 2013). (b) The cumulative average loss of ice mass in Gt of the AIS from the GRACE gravity observations of 2002–2017.

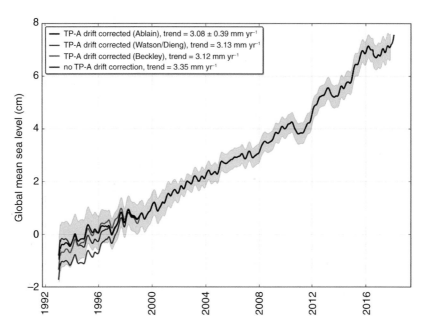

Figure 4.11 Evolution of ensemble mean global mean sea level (GMSL) time series (average of the six GMSL products from AVISO/CNES, SL_cci/ESA, University of Colorado, CSIRO, NASA/GSFC, and NOAA). On the black, red, and green curves, the TOPEX-A drift correction is applied, respectively, based on Ablain *et al.* (2017), Watson *et al.* (2015) and Dieng *et al.* (2017), and Beckley *et al.* (2017). Annual signal removed and 6-month smoothing applied; GIA correction also applied. Uncertainties (90% confidence interval) of correlated errors over a 1-year period are superimposed for each individual measurement (shaded area) (taken from Cazenave *et al.*, 2018).

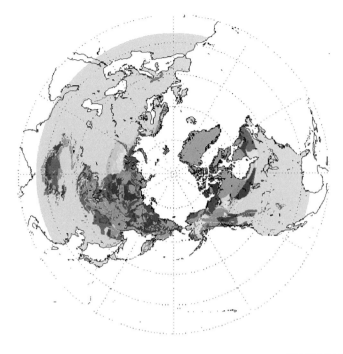

Legend for EASE-Grid Permafrost and Ground Ice Map

Permafrost Extent (percent of area)	Ground Ice Content (visible ice in the upper 10–20m of the ground; percent by volume)				
	Lowlands, highlands, and intra- and intermontane depressions characterized by thick overburden cover (>5–10 m)			Mountains, highlands, ridges, and plateaus characterized by thin overburden cover (<5–10 m) and exposed bedrock	
	High (> 20%)	Medium (10–20%)	Low (0–10%)	High to medium (>10%)	Low (0–10%)
Continuous (90–100%)	Ch	Cm	CL	Ch	CL
Discontinuous (50–90%)	Dh	Dm	DL	Dh	DL
Sporadic (10–50%)	Sh	Sm	SL	Sh	SL
Isolated Patches (0–10%)	Ih	Im	IL	Ih	IL

Variations in the extent of permafrost are shown by the different colors; variations in the amount of ground ice are shown by the different intensities of color. Letter codes assist in determining to which basic permafrost and ground ice class any particular unit belongs. Letter codes are defined in the documentation that accompanies the data files.

Ice caps and glaciers

Figure 5.1 Map of the distribution of frozen ground in the Northern Hemisphere based on Brown *et al.* (1997) [courtesy of NSIDC, Boulder, Colorado] http://nsidc.org/data/docs/fgdc/ggd318_map_circumarctic/index.html#format (spatial coverage map and legend).

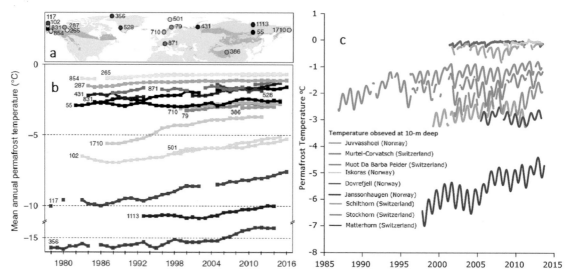

Figure 5.8 Long permafrost temperature records for selected sites. (a) Location of boreholes with long time-series data. Depth of measurements is according to the Global Terrestrial Network for Permafrost ID: 24.4 m (ID 356), 20 m (ID 55, 79, 102, 117, 501, 710, 831, 1113, and 1710), 18 m (ID 386), 16.75 m (ID 871), 15 m (ID 854), 12 m (ID 287), 10 m (ID 265, 431), and 5 m (ID 528). The light blue area represents the continuous permafrost zone (>90% coverage) and the light purple area represents the discontinuous permafrost zones (<90% coverage). (b) Mean annual permafrost temperatures at Swiss alpine sites at 10 m depth over time. The colors indicate the location of boreholes in permafrost zones derived from the International Permafrost Association (IPA) map: Iskoras; 591 m asl, SE-facing; Dovrefjell; Janssonhaugen: 275 m asl (fine-grained sandstone), S-facing; Juvvasshoe: 1,894 m asl (Gneiss), NW-facing; Schilthorn: 2,970 m asl, N-facing (rock slope with thick fine debris); Muot da Barba Peider: 2,900 m asl, NW-facing (debris slope); Murtèl Corvatsch: 2,670 m asl, NW-facing (rock glacier); Matterhorn: 3,295 m asl (vertical in rock ridge/Hörnligrat) [Courtesy: Swiss Permafrost Monitoring Network (PERMOS), University of Zurich]. [www.eea.europa.eu/data-and-maps/figures/observed-permafrost-temperatures-from-selected-4].

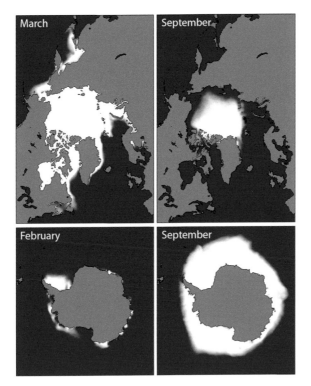

Figure 7.1 Sea ice climatologies: Arctic and Antarctic sea ice concentration climatology from 1981 to 2010, at the approximate seasonal maximum (March–September) and minimum (September–March) levels based on passive microwave (PM) satellite data. Image provided by National Snow and Ice Data Center, University of Colorado, Boulder.

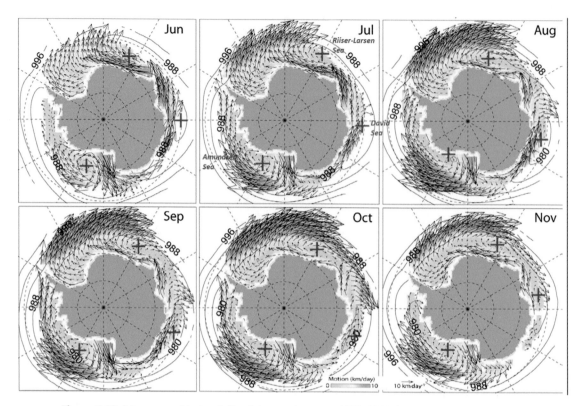

Figure 7.13 Mean monthly ice drift of Antarctica over 1982–2015 (June–November) with contours of isobars at 4-hPa intervals estimated from ERA-Interim SLP fields; drift estimates presented at a 200-km grid; and red crosses represent centers of the three atmospheric lows in the monthly mean fields (modified from Kwok *et al.*, 2017).

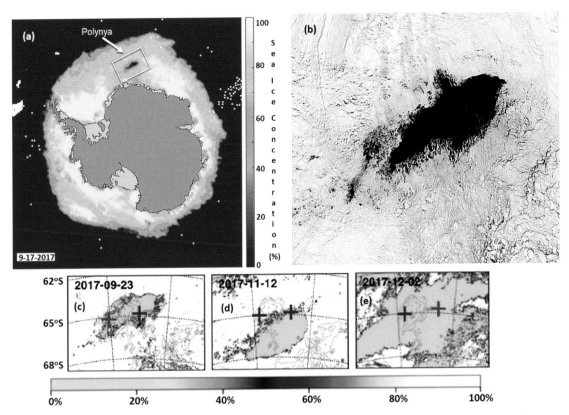

Figure 7.15 The Maud Polynya in the Lazarev Sea to the east of the Weddell Sea and the Antarctic Peninsula captured (a) on September 17, 2017 from a DMSP SSM/I image (http://mallemaroking.org/maud-polynya-2017/), (b) September 25, 2017 MODIS image of NASA Worldview (https://worldview.earthdata.nasa.gov/). AMSR2-ASI PM images of sea ice concentration (SIC) during the 2017 Maud Rise polynya of (c) September 23, (d) November 12, and (e) December 2, 2017 (adapted from Campbell *et al.*, 2019).

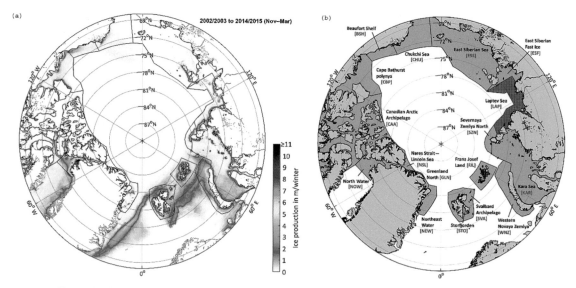

Figure 7.16 (a) Average Arctic cumulative ice production (m/winter) north of 68° N; and (b) applied polynya masks (blue) enclose the typical location of in each of 17 polynya regions between 2002–2003 and 2014–2015 (taken from Preußer *et al.*, 2016).

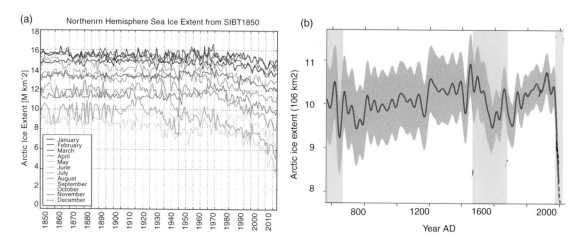

Figure 7.18 (a) The Arctic sea ice extent in millions of km² present at any concentration plotted for each month, from 1850 to 2013 (taken from Walsch et al., 2016); and (b) Reconstructed Arctic sea ice extent over the last 1,450 years (taken from Kinnard et al., 2011).

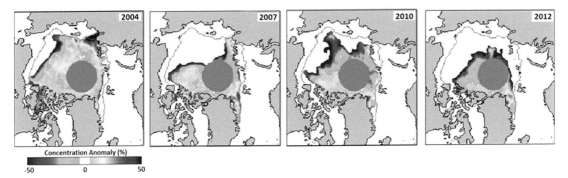

Figure 7.19 September Arctic sea ice concentration anomaly maps of 2004, 2007, 2010, and 2012 derived from the NSIDC Sea Ice Index. Each image shows the concentration anomaly (see color key) and the 1979–2000 mean September ice edge (pink line). Every year, the ice edge is north of its mean position off the coasts of Alaska and Siberia (Image of NSIDC, Boulder (https://nsidc.org/cryosphere/sotc/sea_ice.html).

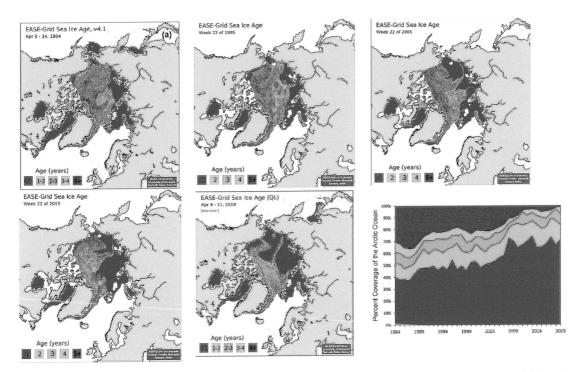

Figure 7.20 Arctic sea ice age for (a) April 8-14, 1984, Week 22 of 1995 (b), 2005 (c), and 2015 (d), and April 9–15, 2019 (e), and (f) Arctic sea ice age time series of mid-April as a percentage of Arctic Ocean coverage from 1984 to 2019, which shows the almost complete loss of 4+ year old ice in recent years. Note that the age time series is for ice within the Arctic Ocean only (Source: https://nsidc.org/arcticseaicenews/).

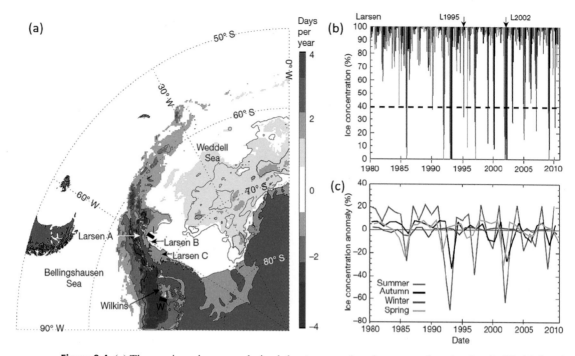

Figure 8.4 (a) The passive microwave derived shorter annual sea ice season duration for the Weddell and Bellingshausen seas for 1979/1980 to 2009/2010 offshore from the Larsen ice and Wilkins shelves is shown by red indicating fewer days of coverage. The black/blue contours delimit significance at $P<0.01$ and 0.10, respectively; (b) Time series showing reduction in daily sea ice coverage for region L offshore from Larsen ice shelves, and temporal coincidence with disintegration events where arrows denote the approximate onset timings of the major disintegration events of Larsen A in 1995 and Larsen B in 2002. The plots are summer in red, autumn in black, winter in dark blue, and spring in light blue; (c) Plots of mean seasonal sea ice concentration anomalies off the Larsen A and B from region L (taken from Masson *et al.*, 2018).

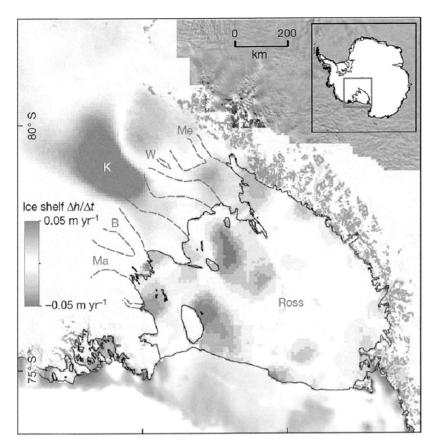

Figure 8.8 Surface $\Delta h/\Delta t$ of Ross Ice Shelf, 2003–2008. The color scale for the grounded ice $\Delta h/\Delta t$ signal is ± 30 cm yr^{-1}. The continental shelf is shown in gray, landward of the continental-shelf break. Labels Me, W, K, B, and Ma denote the ice streams Mercer, Whillans, Kamb, Bindschadler, and MacAyeal, respectively. Dashed gray lines show the lateral ice stream margins. Black lines show ice shelf boundaries (mapped between 1993 and 2003). The inset shows the location of the figure (green box) overlaid on the outline of Antarctica (taken from Pritchard *et al.*, 2012).

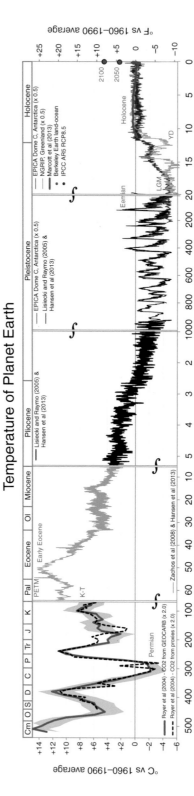

Figure 9.1 The temperature of planet Earth over the last 500 Ma on a semilogarithmic plot. [Source: Wikipedia http://en.wikipedia.org/ wiki/Geologic_temperature_record] By Glen Fergus – Own work; data sources are cited below, CC BY-SA 3.0, https://commons.wiki media.org/w/index.php?curid=31736468.

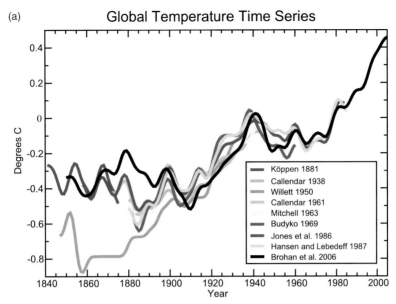

Figure 10.1 (a) Published records of surface temperature change over large regions. Köppen (1881) tropics and temperate latitudes using land air temperature. Callendar (1938) global using land stations. Willett (1950) global using land stations. Callendar (1961) 60° N to 60° S using land stations. Mitchell (1963) global using land stations. Budyko (1969) Northern Hemisphere using land stations and ship reports. Jones *et al.* (1986a, b) global using land stations. Hansen and Lebedeff (1987) global using land stations. Brohan *et al.* (2006) global using land air temperature and sea surface temperature data is the longest of the currently updated global temperature time series. All time series were smoothed using a 13-point filter. The Brohan *et al.* (2006) time series are anomalies from the 1961 to 1990 mean (°C). Each of the other time series was originally presented as anomalies from the mean temperature of a specific and differing base period (from Le Treut *et al.*, 2007). (b) Global temperature anomaly of 1960-2014 with reference to the 1951–1980 mean temperature where more red areas in recent years indicates global warming and the area inside the black box shows warming amplified in the Arctic, particularly within the past 25 years (Wendisch *et al.*, 2017). (c) Mean 2010–2019 global temperatures compared to a baseline 1951–1978 average [Source: Goddard Institute for Space Studies NASA's Goddard Space Flight Center, https://en.wikipedia.org/wiki/Instrumental_temperature_record] [Source: IPCC 2007, Figure 1.3].

(b)

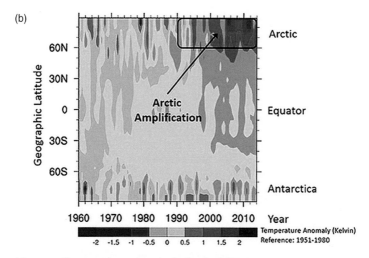

(c) Temperature change in the last 50 years

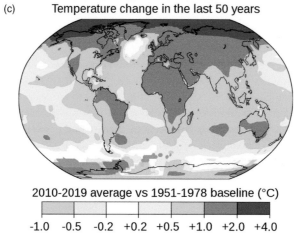

2010-2019 average vs 1951-1978 baseline (°C)

-1.0 -0.5 -0.2 +0.2 +0.5 +1.0 +2.0 +4.0

Figure 10.1 (cont.)

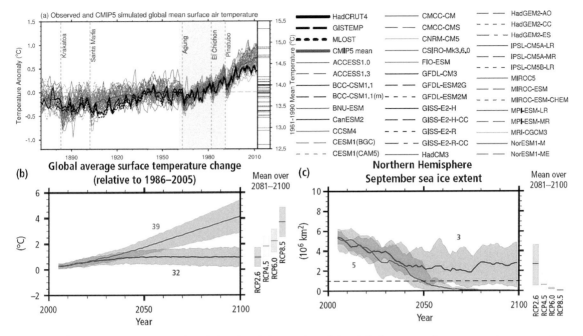

Figure 10.2 (a) Observed and simulated time series of the global mean surface temperature anomalies, differences from the 1961 to 1990 time-mean of each individual time series (colors), the CMIP5 multi-model mean (thick red), and the observations (thick black: Jones *et al.*, 1999; Flato *et al.*, 2013); and selected GCMs of CMIP5 shown in (a); (b) Global annual change in mean surface temperature for 2006–2100 relative to 1986–2005 from CMIP5 experiments; and (c) Change in NH September sea-ice extent (5 year running mean). The dashed line represents nearly ice-free conditions, when the September sea-ice extent is less than 10^6 km^2 for at least five consecutive years (taken from IPCC, 2014).

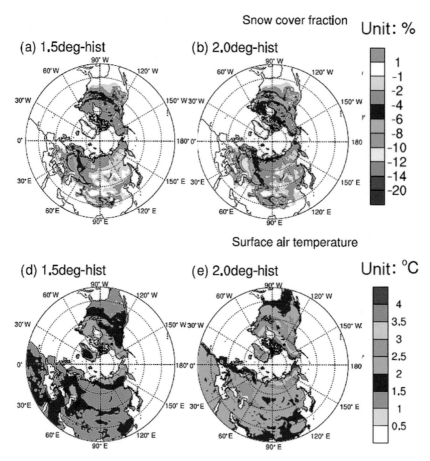

Figure 10.3 Snow cover fraction (SCF) differences in percent (%) between 2071–2100 and 1971–2000 ensemble mean from CESM-LE over Northern Hemisphere land area for (a) 1.5 °C scenarios and (b) 2.0 °C scenarios with reference to "Hist", (c) the land surface air temperature (LSAT) differences corresponding to (a), and (d) LSAT differences correspond to (b), respectively. "Hist" is the ensemble mean for 1971–2000 from CESM-LE. (taken from Wang et al., 2018).

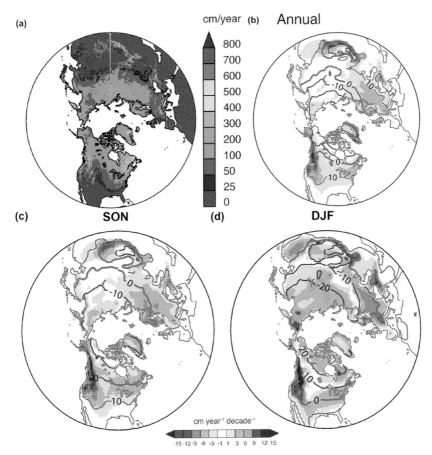

Figure 10.4 (a) The 1951–2005 average annual snowfall (cm yr^{-1}) derived from the University of Delaware Global Surface Air Temperature and Precipitation Climatology, version 2.01. Multimodel ensemble (MME) trends in snowfall (2006–2100; cm yr^{-1} decade^{-1}) for (b) annual, (c) September–November (SON), and (d) December–February (DJF), with contours of the 2-m temperature at 10°C intervals from the MME for 1986–2005 and hatching denotes regions with statistically significant trends (p ≤ 0.01) (taken from Krasting et al., 2013).

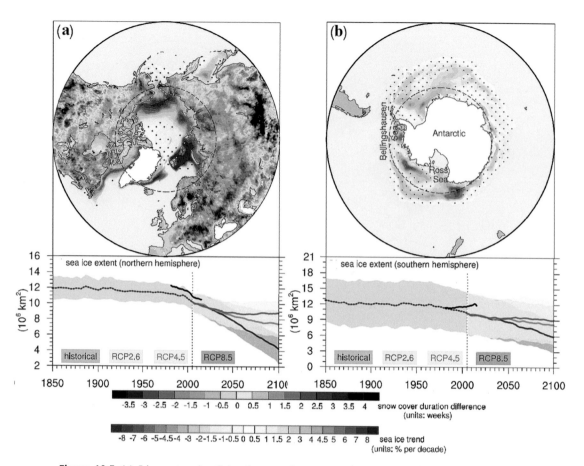

Figure 10.5 (a) Linear trends of Arctic annual-mean sea ice concentration for 1982–2016, in % per decade, alongside the difference in climatological snow cover duration (in weeks) between 2006–2015 and 1981–1990, and (b) linear trends of Antarctic annual-mean sea ice concentration for 1982–2016, in % per decade. The comparable 5-year running mean time series of annual-mean sea ice extent in the northern and southern hemispheres are shown below (a) and (b), where black, green, blue, orange, and red curves indicate observations, CMIP5 historical simulation, RCP2.6, RCP4.5, and RCP8.5 projections of 1850 to 2100, respectively; shading indicates ± standard deviation of multimodels. (a) and (b) are taken from NOAA/NSIDC Climate Data Record of Passive Microwave Sea Ice Concentration, Version 3 (https://nsidc.org/data/g02202). Snow cover duration in (a) was derived from a blend of four independent datasets, each covering the 1981–2015 period (Brown *et al.*, 2003; Takala *et al.*, 2011; Brun *et al.*, 2012; Reichle *et al.*, 2017). The models from CMIP5 include bcc-csm1-1, bcc-csm1-1-m, CanESM2, CCSM4, CESM1-CAM5, CNRM-CM5, GISS-E2-H, IPSL-CM5A-LR, IPSL-CM5A-MR, MIROC5, MIROC-ESM, MIROC-ESM-CHEM, MPI-ESM-LR, MPI-ESM-MR, MRI-CGCM3, NorESM1-M, NorESM1-ME.

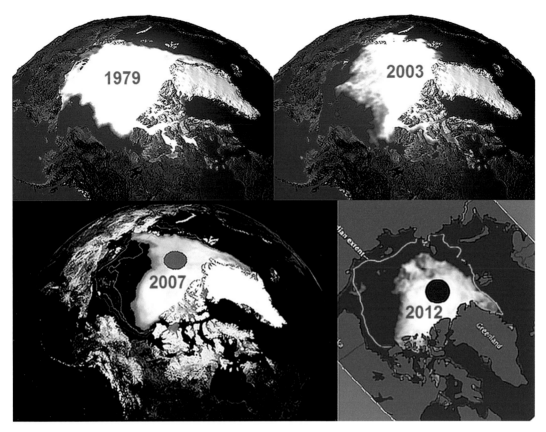

Figure 10.6 Arctic perennial sea ice has been decreasing at a rate of 2.7 (2.1 to 3.3)% per decade, with larger decreases in summer of 7.4 (5.0 to 9.8)% per decade. The first, second, third, and fourth images show the minimum sea ice concentration for 1979, 2003, 2007, and 2012, respectively. The third image of 2007 shows a drastic shrinkage of sea ice, while the fourth image shows that in the late summer of 2012, Arctic sea ice extent has dropped to its lowest level ever recorded, 4.10 million km^2. Images by the DMSP Special Sensor Microwave Imager SSMI (https://www.ncdc.noaa.gov/ssmi/data-access).

Figure 11.1 (a) The Hoover Dam of United States built in 1930s for flood control, water supply, hydropower, navigation, and water regulation (PictureNet/Corbis/Getty Images); (b) Hydropower generated at the Kananaskis Lake fed by snowmelt water from Mountains of the Banff National Park, Canada (picture taken by Kai Ernn Gan).

Part III

The cryosphere past and future

9   The cryosphere in the past

9.1   Introduction

The Earth has undergone enormous changes in its snow and ice cover and temperature during geological time (Figure 9.1). There have been at least six major Ice Ages when large parts of the Earth's surface are covered by glaciers and extensive ice sheets, as well as periods when there has probably been no ice, like the Cretaceous; however, there is no strict quantitative definition. Continental ice sheets grow during Ice Ages (glacial periods) and dwindle during interglacial periods throughout the Quaternary Period in the past 2.6 million years (https://www.ncdc.noaa.gov/abrupt-climate-change/Glacial-Interglacial %20Cycles). The latest glacial period occurred between about 120 ka and 11 ka ago. Since then, Earth has been in the interglacial period called the Holocene. Glacial (Interglacial) periods tend to occur during periods of less (more) intense summer solar radiation in the NH due to variations in Earth's orbit of a frequency of about 100,000 years. This "eccentricity" cycle of the solar radiation time series is weaker than the "precession of the equinoxes" cycles lasting about 23,000 years (Lisiecki and Raymo, 2005). Interestingly, solar radiation changes in the high northern (southern) latitudes are in (out of) phase with temperature variations in Antarctica. This means that the growth or decline of ice sheets has an important influence on the global climate. Warming at the end of glacial periods tends to happen more abruptly than the increase in solar insolation possibly because of positive feedbacks such as the ice-albedo feedback, and the feedback involving atmospheric CO_2. Historical CO_2 trapped in ice core bubbles shows that the amount of atmospheric CO_2 decreased during glacial periods (Bereiter et al. 2015) partly because more CO_2 is stored in the deep ocean, which weakens the atmosphere's greenhouse effect, resulting in lower temperatures. In contrast, warming at the end of the glacial periods liberates CO_2 from the ocean, enhancing the atmosphere's greenhouse effect and contributing to more warming.

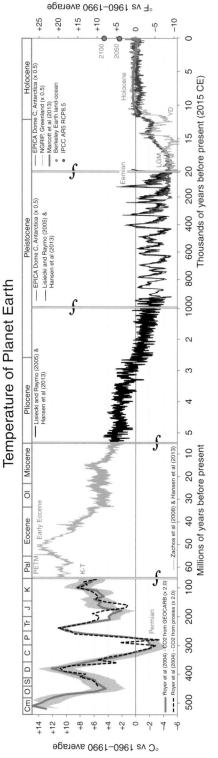

Figure 9.1 The temperature of planet Earth over the last 500 Ma on a semilogarithmic plot. [Source: Wikipedia http://en.wikipedia.org/wiki/Geologic_temperature_record] By Glen Fergus – Own work; data sources are cited below, CC BY-SA 3.0, https://commons.wikimedia.org/w/index.php?curid=31736468 (A black and white version of this figure will appear in some formats. For the color version, please refer to the plate section.)

The major ice ages were the Huronian/Makganyene glaciations in the NeoArchean–Paleoproterozoc (~2,700 to 2,200 million years ago, Ma), the Sturtian, Marinoan, and Ediacaran glaciations of the Neoproterozoic (~730–700 Ma, 665–635 Ma, and 635–542 Ma), the late Ordovician glaciation (460 Ma), the late Devonian glaciation (360 Ma), the Permo-Carboniferous glaciation (320–250 Ma), and the Late Cenozoic glaciation (beginning ~40 Ma). Glaciations appear to account for ~28% of Paleozoic time.

The interval 850–635 Ma is known as the Cryogenian period, which witnessed the two largest known glacial ages – the Sturtian and Marinoan. During approximately 730–540 Ma, cooling due to a dimmer early sun (~0.94 of today's value about 750 Ma) and lower CO_2 levels, may have produced a Snowball Earth in which permanent ice covered the entire globe. Tajika (2003) discounts the faint young sun and emphasizes the fluctuations in CO_2 levels caused by changes in the temperature-dependent rate of silicate weathering. The other Ice Ages were concentrated on the continents when they were located in high southern or northern latitudes as a result of continental drift (Jastrow and Rampino, 2008). Harris (2002) attributes the major temperature fluctuations (>25 °C) over the last 350 Ma to changes in land–sea distribution due to plate tectonics. Second-order controls are associated with changes in large-scale ocean currents and the thermohaline circulation (15–25 °C fluctuations), and third-order controls with the Milankovitch cycles affecting the Earth's orbital eccentricity and axial tilt (~10 °C fluctuations). Etkin (2010) shows how global temperature anomalies and CO_2 concentration in the atmosphere occupy a distinct region of state space over the last 420,000 years with chaotic temporal behavior but a high overall correlation. Temperature and CO_2 are in a closed loop, driven by orbitally induced solar radiation variations, until the Industrial Revolution when the temperature change begins to respond linearly to the anthropogenic addition of CO_2, shifting the state into a new domain.

The concept of a glaciation in the western Alps was first advanced by B. F. Kuhn (1787–1788) and J. Hutton (1795). Ignaz Venetz took this idea up in 1816, and in 1829 extended the concept to northern Germany. In 1861, he proposed the occurrence of four glaciations.

However, the idea of an Ice Age is attributed to Louis Agassiz who in 1840 published *Études sur les glaciers* (Carozzi, 1967), extending the views of J. G. von Charpentier and F. K. Schimper that erratic blocks in the Swiss Jura had been transported there by Alpine glaciers. He proposed that an ice sheet like that of Greenland had covered the Alps in the not too distant past.

In 1920, Milutin Milankovitch, a Serbian astronomer, proposed that glacial/interglacial cycles were caused by variations in the solar radiation received by the Earth as a result of changes in the eccentricity of the Earth's orbit around the Sun (~100,000 yr periodicity), the Earth's axial tilt (~41,000 yr), and the precession of the Earth's axis of rotation (~23,000 yr). Hays *et al.* (1976) found strong evidence to support the orbital hypothesis in the variations in oxygen isotope records in oceanic fossil foraminifera. The precessional and orbital-tilt periodicities force a lagged response in northern ice sheets, whereas the eccentricity operates mainly through internal feedbacks (ice–albedo and CO_2) that build up ice sheets gradually, but lead to their rapid decay (Ruddiman, 2006). Nevertheless, a comprehensive theory of Ice Age climatic variations is still elusive (Raymo and Huybers, 2008).

9.2   Snowball Earth and ice-free Cretaceous

There is widespread geological evidence from tillites, dropstones, and ice striae on rock surfaces, that several times during the Neoproterozoic era the Earth had a near-total ice cover – "Snowball Earth" (www.snowballearth.org/) – as originally hypothesized by Kirschvink (1992). He proposed a runaway albedo feedback, in which the world ocean was virtually covered by sea ice, but continental ice cover was thin and patchy because of the virtual elimination of the hydrologic cycle. The global mean temperature was about −50 °C and the equatorial temperature about −20 °C. During the Sturtian (716 Ma), Laurentia was at the Equator and ice was grounded below sea level. According to Macdonald et al. (2010), the glaciation was global in character. The several glaciations ended abruptly when subaerial volcanic out-gassing raised atmospheric CO_2 concentrations to about 350 times the modern level of 385 ppm (Hoffman et al., 1998), or about 330 times the 2019 global CO_2 level of 409.8 ± 0.1 ppm (https://www.climate.gov/sites/default/files/BAMS_SOTC_2019_co2_paleo_1000px .jp). Abbot and Pierrehumbert (2010) and Le Hir et al. (2010) suggest instead that continental and volcanic dust at the ice surface in the tropics may have significantly lowered the tropical surface albedo and encouraged deglaciation. Another possibility is that methane release from melting permafrost in low latitudes might have triggered the global meltdown (Kennedy et al., 2008; Shields, 2008). Allen and Etienne (2008) propose a modification of this view. Sedimentary rocks deposited during the cold intervals indicate that glaciers and ice streams continued to deliver large amounts of sediment to open oceans throughout the glacial cycles. The sediment-ary evidence, including wave-formed ripples, indicates that despite the severity of glaciation some oceans must have remained ice-free permitting free exchange with the atmosphere; the water cycle acted normally throughout the glacial epochs. They conclude that the Neoproterozoic era was probably characterized by many glacial advances, separated by inter-glacial periods; in some of the glaciations ice occupied low latitudes and was global in extent.

   Direct geological evidence of glaciation is lacking for the Mesozoic era (Triassic, Jurassic, and Cretaceous periods), 250–65 Ma. During most of the Cretaceous period, 140–65 Ma, temperatures were much higher than now. Tropical sea surface temperatures may have averaged around 37 °C and deep ocean temperatures were as much as 15 to 20 °C higher than today. This epoch probably represented an ice-free Earth (Barry, 2002b).

9.3   Phanerozoic glaciations

There appear to have been two Phanerozoic glacial modes. The late Ordovician (445–440 Ma) and late Devonian (375–368 Ma) glaciations typify short-duration, high atmospheric CO_2 events, characterized by cosmopolitan faunal distributions and episodes of catastrophic extinction. The Permo-Carboniferous (360–350 and 330–263 Ma) and Cenozoic glaciations typify long-duration, low atmospheric CO_2 events, characterized by abundant biogeographic differentiation and stable or rising biotic diversity (Raymond and Metz, 2004). The late Ordovician Ice Age probably lasted only about 0.5 Ma with glacier flow from the Sahara – then located over the South Pole – toward North Africa, and flow over Saudi Arabia. The late Devonian witnessed a reduction of CO_2 levels giving rise to global cooling and a harsh continental climate with a major mass extinction. The Devonian glaciation event occurred over a broad area of Pangea – a super-continent that existed from about 500 to 180 Ma – including much of Brazil, Bolivia, and sub-Saharan Africa.

In the mid-Carboniferous period (~320 Ma), ice covered Antarctica, Australia, southern Africa, the Indian subcontinent, Asia, and the Arabian Peninsula (Deynoux, 2004). The late Carboniferous–early Permian saw the south polar regions covered with deep layers of ice and glaciers covering much of the super-continent of Gondwana. The path of ice-center migration during the Paleozoic closely follows paleomagnetic wander paths according to Caputo and Crowell (1985). The record suggests that glacierization occurred when Gondwana was located in the south polar regions and disappeared when the pole was in oceanic or coastal regions. During the Permo-Carboniferous the northern continents were joined in the super-continent of Laurasia, extending from the subtropics to the Equator, and covered with tropical forests and desert dunes in arid areas. The North Pole was ocean so the glaciation may have been unipolar (Bleil and Thiede, 1990).

The history of global sea level provides another perspective on land ice volumes during the Paleozoic (540–250 Ma). Haq and Schuster (2008) review the evidence for fluctuations in sea level of the order of a few tens of meters to about 125 m. The nadir of Paleozoic sea level occurred in the early Permian. Glaciation has been attributed to between 28% and 38% of Paleozoic time so that other factors must also determine eustatic changes in sea level.

9.4 Late Cenozoic polar glaciations

The late Cenozoic witnessed the beginning of the present glacial era in the Antarctic at the Eocene/Oligocene boundary around 33–34 Ma (Goldner et al., 2014; Barker et al., 2007a), although alpine glaciation has been identified in Admiralty Bay, King George Island, and South Shetland Islands around 45–41 Ma by Birkenmajer et al. (2005). East Antarctica appears to have been fully glaciated in the earliest Oligocene, whereas there is evidence only from the late Oligocene (~25 Ma) in West Antarctica (Barker et al., 2007b). The onset of glaciation in Antarctica was probably related to the continuing breakup of Gondwana, the poleward drift of the Antarctic continent, and the opening of ocean passages around it (Kennett, 1977). The development of an ice sheet in East Antarctica about 33–34 Ma is attributed by Barker et al. (2007b) to the formation of the Antarctic Circumpolar Current in the Southern Ocean. Hay et al. (2005) point out that this current is driven by the year-round westerly winds whose constancy is determined by the permanent temperature gradient between the ice-covered continent and the ocean to the north. Hence, it is unclear whether the ocean structure led to glaciation or vice versa. Orbital forcing (low obliquity) is a further factor although DeConto et al. (2008) argue that the decrease of atmospheric CO_2 levels below 750 ppm and deep-ocean cooling of 4 °C played a critical role. CO_2 levels declined from between 1,500 and 1,000 ppm during 45–32 Ma to 750 ppm. During the early middle Miocene ~20 Ma, global surface temperatures were ~3 to 6 °C higher and sea level was 25–40 m above present, although pCO_2 appears to have been similar to modern levels (Tripati et al., 2009). Decreases in pCO_2 were apparently synchronous with major episodes of glacial expansion during the late Middle Miocene ~14 to 10 Ma. Between 14.1 and 13.9 Ma, there was a rapid cooling of about 8 °C in mean summer temperatures in the McMurdo Dry Valleys and tundra vegetation and insects were extinguished (Lewis et al., 2008). There is strong evidence for the presence of large ice sheets on both West and East Antarctica during the Miocene (Hambrey et al., 1989). Nash et al. (2007) identify a cold period with polar ice sheets during 13–10 Ma, followed by relative warmth from 9 to 6 Ma. In the Pliocene, 5–2 Ma, there was a dynamic ice margin in the Ross Sea sector with numerous interglacials. Recent work shows that there were intervals of collapse of the West Antarctic Ice Sheet (WAIS) with little or no marine ice from about 5 to 3 Ma (Pollard and DeConto, 2009). After 3 Ma, there were longer intervals with modern-to-glacial ice volumes, with ice-shelf or grounded-ice cover. In the late Pliocene (~3.3–2.4 Ma), decreases in pCO_2 were again synchronous with major episodes of glacial expansion according to Tripati et al. (2009). However, there was a dominant ~40-ka obliquity cyclicity identified in 40 sedimentary cycles during the Pliocene, up to 1.8 Ma (Nash et al., 2009). Brief super-interglacial collapses of the WAIS occurred, including a well-dated event at 1.07 Ma (Marine Isotope Stage 31). This was followed after 800 ka by colder conditions with extensive ice sheets (Nash et al., 2007).

Traditionally, the onset of glaciation in northern high latitudes has been regarded as occurring much later than in the Antarctic. This would imply an extended period of unipolar (Antarctic) glaciation. The earliest recorded glaciation in the Northern Hemisphere is in the Late Miocene. This involved a significant buildup of ice on southern Greenland around 7 Ma (Larsen et al., 1994). Recently, however, ice-rafted detritus has been identified as early as the Eocene, ~45 Ma, in the Arctic Ocean (Moran et al., 2006; St. John, 2008). According to Stickley et al. (2009), episodic sea-ice formation in marginal shelf areas of the Arctic started around 47.5 million years ago, about a million years earlier than estimates based on ice-rafted debris (IRD) evidence only. This appears to have been followed half a million years later by the onset of seasonal sea-ice formation in offshore areas of the central Arctic. Eldrett et al. (2007) report extensive IRD in sediments of late Eocene to early Oligocene age (between 38 and 30 Ma) from the Norwegian–Greenland Sea. The sediment rafting was by glacial ice, rather than sea ice, and points to East Greenland as the likely source. These data suggest the existence of at least mountain glaciers on Greenland about 20 million years earlier than previously documented. DeConto et al. (2008) calculate that for Northern Hemisphere glacierization to occur, CO_2 concentration needs to be around 280 ppm and this level was only reached about 25 Ma. Concentrations have remained around that level subsequently. The sedimentary record reveals cooling of the Arctic that was synchronous with the expansion of Antarctic ice around 14 Ma and Greenland ice about 3.2 Ma. During the Miocene there appears to have been perennial sea ice since 14 Ma (Darby, 2008). Overall, the glacial evidence supports arguments for bipolar symmetry in climate change, although the development of major ice sheets in northern high latitudes is still ~14 Ma later than in the Antarctic. Edgar *et al.* (2007) demonstrate using marine sediment records for ~41.6 Ma that the implied global ice volume can easily be accommodated in Antarctica without major Northern Hemisphere ice sheets. The threshold for Antarctic glaciation was reached about 34 Ma while that for the Arctic was only reached about 20 Ma, because of the differences in land–ocean distribution in the two hemispheres.

In the late Pliocene, there was glaciation in Alaska/Yukon around 3.5 Ma, the Eurasian Arctic and northeast Asia at 2.75 Ma, and major glaciation of the North American continent at 2.73 Ma (Haug *et al.*, 2005; Maslin *et al.*, 2006). The latter onset is attributed to a warm summer North Pacific Ocean acting as a moisture source. Lewis *et al.* (2010) argue that uplift of the North American Cordillera in the Late Miocene may have played an important role in priming the climate for the intensification of glaciation in the Late Pliocene. Dud-Rodkin *et al.* (2004) date glaciation onset in northwestern Canada and east-central Alaska to 2.54 Ma. Harris (2001) attributes the accumulation of land ice in Alaska, Iceland, and Greenland to the time when the Arctic Ocean underwent tectonic isolation from the North Atlantic and Pacific oceans. The Arctic Ocean froze during a cold event ~2.58 Ma as the western Arctic cooled rapidly. About this time Baffin Island and Labrador became ice sheet centers (Harris, 2005). Nevertheless, there were also extensive warm intervals in the Arctic during the Pliocene when summer sea ice was probably much reduced (Polyak *et al.*, 2009). Greenland was probably ice free in a warm interval around 2.4 Ma (Alley *et al.*, 2010).

For the Mid-Pliocene (~3 Ma), simulations with an atmosphere–ocean GCM coupled with the GLIMMER ice sheet model show that, compared with pre-industrial conditions, ice-sheet feedbacks and lower topography contributed 42%, increased CO_2, 35%, and vegetation changes 23% of the Arctic temperature change (Lunt *et al.*, 2009). Lunt *et al.* (2008) also tested different hypotheses for the glaciation of Greenland around 3 Ma. They find that neither the closure of the Panama Isthmus, nor tectonic uplift, nor the cessation of a permanent El Nino state (Huybers and Molnar, 2007), is sufficient to generate extensive glacierization of Greenland. However, the reduction of atmospheric CO_2 from a mid-Pliocene level of 400 ppm, to a Quaternary level of 280 ppm, is sufficient when coupled with the orbital characteristics of 115 ka BP. Nevertheless, Sarnthein *et al.* (2009) and Ruddiman (2010) report an absence of evidence for CO_2 variations of that magnitude. The tipping point for Northern Hemisphere glaciation occurred in association with a severe deterioration of climate in three steps between 3.2 and 2.7 Ma according to Sarnthein *et al.* (2009). Both models and paleoceanographic records indicate clear linkages between the onset of glaciation and three steps toward the final closure of the Panama seaway about 2.9 Ma. The closing of the seaways led to greater steric height of the North Pacific and this doubled the low-salinity throughflow from the Bering Strait across the Arctic Ocean to the East Greenland Current. Lower sea surface salinity favored increased sea ice in the Arctic Ocean, promoting albedo feedback and the buildup of continental ice sheets. The closure of the seaway led to an enhanced Atlantic meridional overturning circulation (MOC) that had counteracting effects on climate – increasing poleward heat and moisture transport, and thereby increasing precipitation over East Greenland and over Europe. Moreover, the Arctic throughflow helped offset the poleward heat transport over the North Atlantic.

Using the NCAR Community Atmosphere Model version 3 (CAM3) with a slab ocean model, Vizcaino *et al.* (2010) analyze the ELA changes over the Northern Hemisphere associated with a variety of possible forcings around 2.75 Ma and conclude that atmospheric CO_2 levels and the strength of the cold tongues in the eastern equatorial oceans are the main determinants of Northern Hemisphere glaciations. The exact timing of glacierization, however, was paced by changes in the Earth's orbital eccentricity and obliquity forcing according to Hays *et al.* (1976). However, Wunsch (2004) demonstrated that the fraction of the variance in deep-sea and ice-core records attributable to orbital changes never exceeds 20%. Moreover, most paleoclimatic records show that the 100 ka energy (in only 7 glacial cycles) is indistinguishable from a broadband stochastic process. Meyers and Hinnov (2010) show that development of the Northern Hemisphere ice sheets is paralleled by an overall amplification of both deterministic and stochastic climate energy. Progression from a more stochastic early Pliocene to a strongly deterministic late Pleistocene is primarily accomplished during two transitory phases of Northern Hemisphere ice-sheet growth. They report high amplitude 100 ka and 400 ka isotopic cycles during their respective theoretical eccentricity nodes (~2.75 Ma and ~3.5 Ma), coincident with active Northern Hemisphere ice buildup. Herbert *et al.* (2010) argue that the inception of a strong CO_2–greenhouse gas feedback (operating particularly in high-latitude oceans) and amplification of orbital forcing at ~2.7 Ma, connected the variations of Northern Hemisphere ice sheets with global ocean temperatures. Tropical SST records show unusually intense coolings at ~2.5, 2.1, and 1.7 Ma and tropical SST leads glacial cycles by about 2 to 5 ky.

9.5   The Quaternary

The Quaternary period includes the Pleistocene glaciations of the last 2.6 Ma and the postglacial Holocene, as determined by the International Commission on Stratigraphy in 2009. There is still dispute over terminology and some prefer to define the Pleistocene epoch and the Quaternary beginning at 1.8 Ma.

The glacial cycles and tropical SSTs show a 41,000-year signature, related to the axial tilt (or obliquity) of the Earth, from ~2.7 Ma until about 0.8 Ma (Raymo *et al.*, 2006) when they shift to a 100,000-year signature associated with the orbital eccentricity of the Earth about the Sun (Raymo and Huybers, 2008; Berger and Loutre, 2010; Herbert *et al.*, 2010), although note Wunsch's (2004) caveat mentioned earlier. The cause of this shift is unknown although it has been attributed to a nonlinear response to small changes in external boundary conditions. Clark and Pollard (1998) attribute the transition to a change from an all soft-bedded Laurentide Ice Sheet to a mixed hard–soft-bedded one through glacial erosion of a thick regolith and the resulting exposure of unweathered crystalline bedrock. Before the transition, a deforming sediment layer maintains a thin ice sheet, which responds linearly to the dominant 23 and 41 ka orbital forcings. Progressive removal of the sediment layer eventually causes a transition to thicker ice sheets whose dominant timescale of change (~100 ka) reflects nonlinear deglaciation processes. Bintanja and van de Wal (2008) propose that the development of a 100,000-year cycle in the late Pleistocene is due to an increase in the ability of North American ice sheets to persist through periodic maxima of solar radiation, itself a result of increasing volumes of glacier ice. Lourens et $al.$ (2010) argue that the dominant 41 ka component in $\delta^{18}O$ lags obliquity by ~6.5 ka in both the late Pliocene (2.56–2.4 Ma) and late Pleistocene. Maximum ice volume growth occurred in phase with obliquity minima, which invokes low total summer energy reducing ice-sheet ablation. The late Pliocene and late Pleistocene $\delta^{18}O$ records reveal significant power at ~28 ka, which appears to be bound to the major glacial terminations. They suggest that this beat likely reflects the sum frequency of the 41 ka prime and its multiples of 82 and 123 ka. The late Pliocene deglaciations lack a distinct precession (23–19 ka) signal, thus excluding Northern Hemisphere summer insolation as the major trigger. The origin, phase, and geometry of the late Neogene glacial cycles are primarily determined by the linear (41 ka) and nonlinear (28, 82, and 123 ka) response mechanisms of the ice sheets to the obliquity forcing. Throughout the last 800,000 years, ice sheets have taken about 90,000 years to grow and only 10,000 years to decay. For the last six glacial/interglacial cycles the mean interval between terminations is 102 ka with a range of ~85 to ~120 ka (Paillard, 2001).

The European Project for Ice Coring in Antarctica (EPICA) ice core from Dome C (76° S, 123° E) has enabled scientists to reconstruct atmospheric CO_2 concentrations for the last 800,000 years. Recently, Bereiter et $al.$ (2015) detected an analytical artifact in the ice core data, which increases over the deepest 200 m and reaches 10.1 ± 2.4 ppm in the oldest/deepest part. Their corrected ice core data partially resolves the issue with a different correlation between CO_2 and Antarctic temperatures found in this oldest part of the data. The ice core data shows that the most severe glacial intervals were MIS 2 (35.6–11.6 ka BP), 12 (451–425 ka BP), and 16 (651–621 ka BP) when CO_2 levels fell to 180–200 ppm (Masson-Delmotte et $al.$, 2010); in all eight glacial cycles there was a 25-fold increase in eolian dust levels (Lambert et $al.$, 2008). Between 740,000 and 430,000 years ago, interglacials occupied a considerably larger proportion of each glacial/interglacial cycle, but were not as warm as the subsequent five interglacials – Marine Isotope Stages (MIS) 11, 9, 7, 5, and 1 (EPICA, 2004; Tzedakis, 2009). The MIS 11 between 423 and 362 ka (an analogue to the Holocene with respect to orbital forcing) was an unusually long and warm interglacial when global sea level was +20 m in contrast to the preceding MIS 12 when it fell to −140 m. The Greenland Ice Sheet apparently melted completely during MIS 11 (Alley et $al.$, 2010). These changes occurred during a time when the Earth's orbit was nearly circular and precessional changes were small around 400 ka (Paillard, 2001). The warmer interglacials also had higher CO_2 levels (260–285 ppm) and higher sea levels than the earlier cooler ones (MIS 13, 15, 17, and 19) (Tzedakis *et al.*, 2009). The sea level during the last (Eemian) interglacial around 125 ka has traditionally been put at about 4–6 m above present. However, work by Kopp *et al.* (2009) and Clark and Huybers (2009) suggests that it was at least 6.6 m, and possibly as high as 8–9 m above present. The maximum contribution from Greenland was probably 3.4 m and the thermosteric plus mountain-glacier and ice-cap contribution was probably no more than 1 m. The Antarctic contribution would mainly have come from the WAIS, which holds at least 3.3 m of sea-level equivalent. Hence, the finding of Kopp *et al.* (2009) implies that most of the WAIS must have melted during the peak interglacial time.

On the continents there is generally evidence (from interglacial soils) for only four or five glacial cycles, due to subsequent erosion or burial of morainal features, but in deep Antarctic ice cores and in marine sedimentary records, many more are indicated. Glacial episodes in the northern Alps are named according to the Penck–Bruckner scheme as the Biber (<2.47 Ma), Donau (>780 ka), Günz (<780 ka), Haslach (<780 to >380 ka), Mindel (>380 ka), Riss (MIS 8–6), and Würm (MIS 4–2). However, the status of the Biber and Donau remains uncertain. Schlüchter (1988) provides evidence of at least 15 separate glaciations of the Swiss Alpine foreland during the Quaternary. Sibrava (2010) recommends discarding the classical Penck–Bruckner scheme and dates the earliest glaciation, which only affected the Italian Alps, to the Matuyama (between 2.4 and 0.7 Ma). Muttoni *et al.* (2003) date the first major Alpine glaciation to MIS 22 (0.87 Ma). In northern Europe, the youngest three glacials (documented by glacial sediments /moraines/tills) are: Elster (MIS 10), Saale (MIS 8, 7, 6), and Weichsel (MIS 4, 3, 2). The corresponding names for the last four glaciations in North America are the Nebraskan (680–620 ka), Kansan (455–380/300 ka), Illinoian (300–130 ka), and Wisconsinan (110–12 ka). However, the terms Nebraskan and Kansan have subsequently been abandoned (see later). The penultimate ice age (MIS 6, about 188–130 ka) may have produced the most extensive ice in Greenland (Wilken and Meinert, 2006) with evidence from East Greenland (Alley *et al.*, 2009), whereas some 400,000 years ago during the warm interval of MIS 11, coniferous forest indicates a nearly ice-free Greenland (De Vernal and Hillaire-Marcel, 2008). In the last interglacial (MIS 5e) around 130 ka the Greenland Ice Sheet covered a smaller area than now, but its extent is poorly constrained (Alley *et al.*, 2010). Subpolar, seasonally open water was present in the area north of Greenland at that time and most of the Arctic Ocean may have been free of summer ice at that time (Polyak *et al.*, 2009). The winter limit of sea ice did not extend south of the Bering Strait and was probably located at least 800 km north of its historical limits.

The last glacial cycle began about 115,000 years ago and reached a first maximum around 75,000 years ago. Sea level was about 100 m below present around 65 ka and 130 m below present at the Last Glacial Maximum (LGM) around 25–18 ka due to the massive buildup of ice in North America (the Laurentide and Cordilleran ice sheets) and in Fenno-Scandinavia. The Laurentide Ice Sheet reached a maximum thickness of around 3,000 m. At its maximum extent it reached latitude 37° N and covered an area of more than 13 million km². The Greenland Ice Sheet covered about 40% more area and had an estimated 42% greater volume than at present (Alley *et al.*, 2009). In northwest Greenland, the Greenland ice merged with the Innuitian Ice Sheet of the Canadian Arctic Archipelago. The total ice area at the LGM was about 40 million km², and the total volume of global ice was about 50 million km³ (Shum *et al.*, 2008). Land and sea ice together covered about 30% of the Earth's surface. Growth of the ice sheets to their maximum positions occurred between 33.0 and 26.5 ka in response to climate forcing from decreases in Northern Hemisphere summer insolation, tropical Pacific sea surface temperatures, and atmospheric CO_2 according to Clark *et al.* (2009). The LGM peaked between 26.5 and 19–20 ka.

Global ice sheets have been reconstructed using the geodynamical models ICE4G (Peltier, 1994) and ICE5G (Peltier, 2004). The two reconstructions differ in the spatial extent, height, and volume of the ice sheets at the LGM. In ICE5G, the Laurentide Ice Sheet contains significantly more volume than ICE4G, with the Keewatin Dome 2–3 km higher over a broad area of central Canada. Also, in ICE4G the Fennoscandian Ice Sheet extends farther east into northwestern Siberia.

The lowest temperatures of the LGM in 71 ice core records are found around 22 ka in both hemispheres (Shakun and Carlson, 2010). Temperatures were lowered by about 5.8 ± 1.4 °C, globally, based on a modeling study by von Deimling *et al.* (2006), and about 28 °C over the ice sheets. The global annual mean sea surface temperature was lowered by 1.9 ± 1.8 °C, according to a new analysis by MARGO Project Members (2009), with a cooling of 2.4 ± 2.2 °C for the entire Atlantic Ocean and 2.9 ± 1.3 °C for the tropical Atlantic 15° N–15° S.

In northern Asia, there was only local ice cover due to the extreme dryness and blocking effect on the atmospheric circulation of the Fenno-Scandinavian Ice Sheet. During the LGM, for example, glaciers in the Polar Urals were not much larger than today (Mangerud *et al.*, 2008). The Quaternary Environments of the Eurasian North (QUEEN) program showed that the largest ice sheet existed during the penultimate (Saalian) glaciation ~140 ka (Thiede *et al.*, 2001). The maximum extent of glaciation during the last ice age occurred during the Early and Mid-Weichselian; around 100–90 ka an ice sheet advanced southward from the Kara Sea into northwestern Russia and dammed up a large lake in the Pechora lowland. After an extensive deglaciation, regrowth of ice occurred during the Middle Weichselian. During 65–70 ka an ice sheet emanating from the Barents Sea shelf expanded onto the mainland and blocked the river system that flowed toward the Arctic Ocean creating a large ice-dammed lake – the White Sea Lake (Larsen *et al.*, 2006). This maximum ice sheet extent is identified from prominent end moraines across northwestern Russia. About 55–45 ka the Kara Sea ice sheet again expanded southward; it was independent of the Scandinavian Ice Sheet and the Barents Sea remained ice-free.

This glaciation was succeeded by a *c.* 20 ka-long ice-free and periglacial period before the Scandinavian Ice Sheet invaded from the west, and joined with the Barents Sea Ice Sheet in the northernmost part of northwestern Russia. The Svalbard archipelago was covered by an ice sheet that was centered on the floor of the Barents Sea to the east (Elverhøi *et al.*, 2002). The Barents Ice Sheet coalesced with the Scandinavian Ice Sheet forming a continuous ice cover that extended across the Barents Sea shelf, Novaya Zemlya, and east to the Kara Sea. The southern limit of the ice sheet must have been somewhat north of the Arctic coast, on the Kara Sea shelf. The eastern margin is thought to have been located west of Severnaya Zemlya and the Taimyr Peninsula (Thiede *et al.*, 2001).

During the LGM permafrost occupied all of Siberia, and most of Europe and Central Asia (Lisitsyna and Romaovskii, 1998). Sea level lowering resulted in the exposure of a vast continental shelf in the Arctic Ocean with the formation of "ice complex" sediments with high ice content in massive ice deposits.

The Scandinavian Ice Sheet developed about 117–105 ka in northern Sweden and the mountains of Norway. In the Rondane of central Norway, it reached its maximum extent about 100–90 ka ago during the early Weichselian; the next largest phase there was during early Middle Weichselian about 70–60 ka ago. (Dahl *et al.*, 2005). In MIS 4 virtually all of Sweden was glaciated (Lundquist, 2004). This largely melted away during MIS 3 (65–25 ka). The third largest ice extent in the Rondane occurred during the Late Weichselian maximum *c.* 20 ka ago (Dahl *et al.*, 2005; Donner, 2005). It is estimated to have covered about 6,600,000 km² and attained a thickness of up to 3,000 m around 18–20 ka. It then began shrinking and the retreat following the Younger Dryas (YD) proceeded evenly without readvances. About 12.6 ka BP an ice–dammed lake (the Baltic Ice Lake) formed between southern Sweden and Poland, and extended across the Baltic to Finland. This drained around 10.3 ka BP, and was replaced by the Yoldia Sea. The glacier ice had almost disappeared by 8 ka.

Evidence from the British Isles for the early Quaternary is sketchy. At 0.8 Ma a Scottish-centered ice sheet extended into the North Sea and around 0.55 Ma ice reached the western edge of the continental shelf, according to Boulton *et al.* (2002). Scotland was probably glaciated during the early Devensian (MIS 4) but the extent is unknown. It was mostly ice free during the Middle Devensian (MIS 3) and then reglaciated after 30–25 ka with centers in the Scottish Highlands and Southern Uplands and other centers in the Outer Hebrides, Lake District, Wales, and southeast Ireland. Scottish and Scandinavian ice was confluent in the North Sea between about 26 ka and 22.8 ka BP (Boulton *et al.*, 2002).

In North America magnetostratigraphy indicates that Early Pleistocene glaciations in the lower and upper Matuyama chron of reversed polarity (2.4 to 0.71 Ma) were characterized by eastern and western ice masses separated by a 2,000-km wide north–south ice-free corridor down the center of the continent (Barendregt and Irving, 1998; Barendregt and Dud-Rodkin, 2004). Accordingly, the area covered by ice and hence the ice volume, was considerably less in the first 2 Ma of the late Cenozoic than it was in the last 0.7 Ma (the Brunhes normal polarity chron). Roy *et al.* (2004) identify seven pre-Illinoian glaciations in the north-central United States. The oldest date from the Matuyama chron, another set from the polarity transition 1.3–0.8 Ma, and a third set from the Brunhes (<0.8 Ma) In the Wisconsinan interval (~110–12 ka) there were three major ice sheets – the Cordilleran in the west, the Laurentide covering most of Canada and the northern United States, and the Innuitian in the Canadian Arctic Archipelago.

Simulations incorporating orbital effects and trace gases using Earth system Models of Intermediate Complexity (EMICs) show that the inclusion of vegetation–albedo feedback effects leads to greater ice sheet buildup in North America than in Eurasia, in line with observational evidence (Mysak, 2008). The general picture is that Wisconsinan glacierization in North America reached its maximum extent around 70 ka. The final Wisconsinan episode, known as the Tioga, began about 30 ka and reached its greatest extent around 21 ka. Ice covered most of Canada, the Upper Midwest, and New England, as well as parts of Montana and Washington. There was separate Cordilleran (Fraser) ice in the western mountain ranges and local (Pinedale ~8 ka) glaciations in the central Rocky Mountains. Winsborrow *et al.* (2004) identify 34 major ice streams in the Laurentide Ice Sheet, mainly in the northwest quadrant. These are larger than contemporary ice streams in the Antarctic and also show a wider range of dimensions. Large ice streams at the northwest margin of the Laurentide Ice Sheet, equivalent in size to the Hudson Strait Ice Stream, underwent major changes during deglaciation (~21–9.5 cal ka), resulting in intermittent delivery of icebergs into the Arctic Ocean (Stokes *et al.*, 2009).

The growth of the Laurentide Ice Sheet resulted in a split jet stream that temporarily favored augmented precipitation and growth of the Innuitian Ice Sheet, which advanced in the mountain sectors as recently as 19 ^{14}C ka BP (England *et al.*, 2006). The western islands were occupied by local island-based ice caps that coalesced, constituting the southwest extremity of the ice sheet.

In Tibet the maximum glaciation, named the Kunlun glaciation, is dated to 0.8–0.6 Ma and overall was about 2.4 times more extensive than the modern glaciation (Shi *et al.*, 2008b). This ratio increased to 12:1 in the Tanggula Range. The Guliya ice cap (35.2° N, 81.5° E, 6,200 m asl) developed during this interval according to a basal ice core date obtained by Thompson *et al.* (1997). Precipitation in the central and eastern parts of the plateau was 1.8–3.2 times greater than today but temperatures at the ELA were 1–2 °C higher than present. During the penultimate glaciation the ice cover on the Tibetan Plateau around Golog Mountain in southeast Qinghai was about 70% of that during the Maximum Glaciation; the corresponding figure for the western Kunlun is 83%. For the last glaciation the values are, respectively, 48% and 75% (Shi *et al.*, 2008b). The south slopes of Mount Tomur in the central Tian Shan record the Penultimate Glaciation; the Terang Glacier was about 80 km long and the Muzart Glacier 180 km long. In the Last Glaciation ice here was about 1.6 times its present extent. Corresponding figures for Mt. Golog are 44 times and for the western Kunlun 1.4 times. The early part of the last glacial phase (75–58 ka) was somewhat more extensive than the final phase (21 ka cal BP). The glacial area during the LGM averaged 7.5 times that at present; the ratio was only 2.2 times in the western Kunlun and 3.3 times in the Tanggula Range, but increased to 8.7 times in the Qilian Shan and 41 times in the Hengduan Mountains.

The late Pleistocene history of Antarctic glaciation is poorly known in terms of ice extent and the chronology of advances and retreats (Ingólfsson, 2004). The West and East Antarctic ice sheets (EAISs) do not appear to have behaved synchronously. Sediment cores show that the West Antarctic was ice free at times between ~14 and ~3.5 Ma (Fox, 2008). The EAIS did not apparently reach the edge of the continental shelf during the LGM, whereas the WAIS did so. The sea ice in winter appears to have extended to the Antarctic Convergence and in summer to the present winter limits, based on the use of diatoms as a proxy (Gersonde and Zielinski, 2000). Scherer *et al.* (1998) provide sedimentary evidence (diatoms) that during the late Pleistocene there was a partial or complete collapse of the Ross embayment of the WAIS. The most likely candidate for the time of WAIS collapse is MIS 11, ~400 ka, which was an unusually long interglacial period. The last retreat of the WAIS from ~14.5 ka to the present is shown by the observed retreat of grounding lines in the Ross Sea and the rapid sea-level rise at that time (Clark *et al.*, 2009). The LGM configuration of the Antarctic Peninsula Ice Sheet is reconstructed from geomorphological evidence by Bentley *et al.* (2009). The ice sheet expanded with several hundred meters of thickening and with radial flow away from the positions of two present-day ice domes in the southern part of the Peninsula. The ice sheet probably merged with expanded grounded ice from the WAIS in the Weddell Sea.

Mountain glaciers were considerably more extensive during the last glacial cycle and reached elevations 1,000 m or more below those at present. Already in 1914, Machatschek had tabulated snow line depressions in Europe, Asia, and North America showing maximum depressions of 1,300–1,400 m in the Allgau and Salzkammergut of Austria and western Caucasus, and minimum values of ~600 m on the east side of the Sierra Nevada and western Tien Shan. Nine decades later, Porter (2000) showed depressions ranging from 440 to 1,400 m at 12 tropical sites in Africa, the Americas (to 10° S latitude), and Pacific islands, but most are in the range 800–1,000 m. Regionally, in the southern tropical Andes (8–22° S), an average lowering of 920 ± 250 m has been reported. The timing of the advances is in some regions synchronous with the ice sheet variations but in other regions is different (Thackray *et al.*, 2008). These latter regions include the ice caps of northeastern Russia – discussed earlier, the tropical Andes (dominant advances *c.* 25–22 ka, *c.* 15 ka, and *c.* 13–10.5 ka), Alaska, the coastal Olympic Mountains and the eastern flank of the Cascade Mountains (MIS 4 and 3 rather than MIS 2). Gillespie and Molnar (1995) concluded that in many of the mountain ranges around the world, alpine glaciers reached their maximum extents between 45 and 30 ka. They attribute dates of 70–55 ka for maximum glaciation in Japan. In the Sierra Nevada, the records strongly suggest that the largest glaciers advanced early in the last glaciation, perhaps around 100–70 ka. The differences are a result of changes in atmospheric circulation and precipitation patterns, partly resulting from the presence of the major ice sheets and different coastal configurations.

The rapid climatic shifts observed in the Northern Hemisphere during the last ice age have been shown to be most likely noise induced, rather than driven by some hidden periodicity.

Ditlevsen and Ditlevsen (2009) show that the waiting times depend on the climate state based on annual layer counts over 60 ka in the North Greenland Ice Core Project (NGRIP) core. The mean waiting time is ~800 yr during the warm interstadials and ~1,600 yr during the cold stadials. The residence time in a given state indicates how stable that state is to perturbations.

Barker *et al.* (2009) note that in contrast to the abrupt temperature changes observed in the Northern Hemisphere in late glacial time, fluctuations over Antarctica were more gradual and approximately out of phase with their northern counterparts. Ice-core evidence from Dome C, Antarctica (75.1° S, 123.3° E, 3,233 m asl) identifies a direct relationship between the extent of warming across Antarctica and the duration of cold, stadial conditions over Greenland (EPICA Members, 2006). There was an abrupt onset of warming in the Southern Ocean about 18,000 years ago. This can be attributed to the bipolar temperature seesaw in the Atlantic Ocean (Barker *et al.*, 2009). There was a switch-off in the Atlantic meridional circulation in response to the increasing injection into the surface ocean of meltwater from the decaying northern ice sheets. The mechanism of glaciation terminations has remained uncertain up to now. Wolff *et al.* (2009) propose that the initial process involves a warming in Antarctica. Such warmings, known as Antarctic Isotopic Maxima, generally begin to reverse with the onset of a warm Dansgaard–Oeschger (D–O) event in the Northern Hemisphere due to the bipolar seesaw. However, in the early stages of a termination, Antarctic warming is not followed by abrupt warming in the north. The lack of an Antarctic climate reversal enables southern warming and the associated atmospheric CO_2 rise to reach a point at which full deglaciation becomes inevitable.

The initial retreat of the Northern Hemisphere ice sheets was about 21,000–19,000 years ago (Severinghaus, 2009), but substantial decay occurred around 14.5-k calendar years ago, at the start of the Bölling–Allerød warm interval (Alley and Clark, 1999). In Fram Strait (81° N, 2° E), perennial sea-ice cover prevailed for most of the LGM, but warming about 14,800 years ago was briefly associated with ice-free conditions in summer (Müller *et al.*, 2009). Empirical orthogonal function analysis of 71 ice core records spanning 19–11 ka indicates that two modes explain 72% of deglacial climate variability. The EOF1 (61% of the variance) shows a globally near-uniform pattern, with its principal component strongly correlated with changes in atmospheric CO2; EOF2 exhibits a bipolar seesaw pattern between the hemispheres, with its principal component resembling changes in Atlantic MOC (Shakun and Carlson, 2010).

Postglacial warming was interrupted by the abrupt YD cold interval around 12.8–11.5-k calendar years ago, that lasted ~1,300 years. Recently, this event has been attributed to multiple cometary airbursts that impacted at least North America at the onset of the YD triggering massive environmental changes, abrupt mega-faunal extinctions, and the disappearance of the Clovis culture (Kennett *et al.*, 2009), but the evidence for the proposed impacts remains controversial. In contrast, others suggest that the YD was caused by a significant reduction of the North Atlantic thermohaline circulation in response to a sudden influx of fresh water. The source of this water remains unresolved. One possibility is the release of water eastward from the ice-dammed glacial Lake Agassiz in central North

America during the deglaciation (Carlson *et al.*, 2007; Lewis and Teller, 2007). However, Murton *et al.* (2010) show that at ~12.9 cal kyr BP a corridor was opened from Lake Agassiz to the Arctic Ocean along the Mackenzie River valley. It remains to be established that this inflow could suppress the North Atlantic thermohaline circulation. Bradley and England (2008) hypothesize that extremely thick multiyear sea ice ("paleocrystic ice"), formed in a smaller Arctic Ocean during glacial sea level lowering during MIS 2, played a key role. Accumulation of the (limited) snowfall as firn would have been accompanied by minimal ablation and, in the absence of a net export of sea ice, these surface conditions would have resulted in the growth of sea ice of exceptional thickness (Walker and Wadhams, 1979). The opening of Bering Strait by 11.5 ^{14}C ka BP would have provided an important flow of Pacific water into the Arctic Basin, and the increased volume of warm Atlantic water entering the Arctic Ocean as the Barents Sea Shelf became deglaciated accompanied this. Hence, more dynamic atmospheric and oceanic circulations forced the export of paleocrystic ice into the critical region of North Atlantic Deepwater formation in the Greenland Sea. Thermal and salinity stratification of near-surface waters was responsible for an abrupt reduction in deep water formation in the Greenland Sea, and this was an important factor in triggering the YD anomaly. For 50-m-thick ice, the calculated freshwater discharge through Fram Strait is equivalent to ~10.2 Sv, almost double that from the Lake Agassiz/Ojibway system during the cold event at 8.2 cal ka BP. The vast proglacial lakes dammed up by the Laurentide Ice Sheet must have played a major role in its rapid decay through calving into them, rather than just surface ablation of the southern lobes. A further hypothesis to explain the YD event has recently been proposed by Broecker *et al.* (2010). They suggest that in the context of the last three glacial terminations, cold reversals equivalent to the YD appear to be integral to the global switch from glacial to interglacial climate. A one-time catastrophe is not required, although a catastrophic flood could have served to pre-trigger the YD. Shakun and Carlson (2010) show that the magnitude of the YD climate anomaly (cooler/drier) increases with latitude in the Northern Hemisphere, from 2 °C in mid-latitudes to 5 °C in high latitudes, with the opposite pattern (0–2 °C warmer/wetter) in the Southern Hemisphere, reflecting a bipolar seesaw response.

In much of Antarctica, there was also a cold event – the Antarctic Cold Reversal – which preceded the YD by about 1,000 years (Blunier *et al.*, 1997). However, the reversal appears to have been synchronous with the YD in Taylor Dome, inland of the Dry Valleys (Steig *et al.*, 1998). The findings appear to show spatial inhomogeneity in climate changes over Antarctica.

The YD appears to be the last in a series of Heinrich events (H0–H6, increasing with age) that occurred during the last 70,000 years; H1–H5 have a spacing of about 7,000 years. They are associated with major periods of IRD, mainly from Hudson Bay, being deposited in the North Atlantic Ocean. The picture is one of a thermally oscillating ice sheet with periodic surges occurring in the Hudson Bay region (MacAyeal, 1993). Heinrich events occur during cold intervals. They are the culmination of successively colder D–O oscillations, each spanning ~1,500 years; an H event occurs during the cold phase of a D–O oscillation, followed by rapid warming (Bond *et al.*, 1993). There are conflicting interpretations as to the causes. One possibility is that the ice sheet surges are triggered by a D–O cooling phase. Thomas *et al.* (2009) show from an analysis of the North-GRIP ice core in Greenland that long-range transport of dust from East Asia changed first, followed by snow accumulation, moisture source conditions, and finally the atmospheric temperature in Greenland. The sequence of events shows that changes in atmospheric and oceanic sources and circulation preceded the DO warming by several years. The abrupt climate changes (~10-year timescale) are linked to the reduction or elimination of North Atlantic deep water formation and associated changes in the oceanic MOC, as suggested above for the YD event. Alvarez-Solas *et al.* (2010) propose instead that massive iceberg discharges are triggered by warming of the subsurface water in the North Atlantic, a result of altered ocean circulation. This warm water erodes the ice shelves fringing the Laurentide Ice Sheet, removing the impediment to ice-stream discharge that has entrained basal debris. Certainly ocean–ice sheet interaction must be a major element in any explanation of the IRD events.

It remains controversial as to whether glaciers in the Andes, New Zealand, and Greenland responded to the YD event (Davis *et al.*, 2009). However, glaciers in Alaska, Baffin Island, British Columbia, Washington state, Iceland, Scandinavia, and the Alps did witness YD advances.

The Fenno-Scandinavian Ice Sheet finally disappeared about 9 ka BP and remnants of the Laurentide Ice Sheet persisted until about 6 ka BP in Baffin Island and 5.7 ka BP in northern Labrador-Ungava (Carlson *et al.*, 2008) – within the "postglacial" Holocene.

The termination of glacial cycles in the Late Pleistocene has long remained problematic. The increase of summer insolation in the Northern Hemisphere due to orbital changes is an insufficient reason because such increases also occurred elsewhere in the glacial record without terminations taking place. Denton *et al.* (2010) propose a comprehensive hypothesis to account for the last four terminations. They argue that the collapse of the major Northern Hemisphere ice sheets created stadial conditions with expanded sea ice in the North Atlantic, due to the weakening of the oceanic MOC, which disrupted global patterns of ocean and atmospheric circulation. The westerlies in both hemispheres were displaced southward and produced ocean upwelling and warming that together accounted for much of the termination in the Southern Ocean and Antarctica. The last termination was associated with two southern warming pulses; the first coincided with the Heinrich 1 stadial (18–15 ka) and the second with the YD. The upwellings raised atmospheric CO_2 levels above the threshold needed for interglacial conditions, thus terminating the glacial phase.

9.6 The Holocene

The postglacial Holocene epoch began around 11 ka ago. A timescale based on multi-parameter annual layer counting in the NorthGRIP ice core in Greenland gives an age of 11,700 cal yr before AD 2000 for the base of the Holocene (Walker *et al.*, 2009). During the Boreal phase there was a sudden cooling event around 8.2 ka that lasted about 150 years in the Northern Hemisphere. It may have been linked to the sudden discharge of icebergs from the Laurentide ice sheet (Wiersma and Jongma, 2010) associated with the final drainage of meltwater from Glacial Lake Agassiz in North America. Yu *et al.* (2010) identify a major drainage event at 9,300 cal yr BP from Lake Superior through the Lake Huron–North Bay–Ottawa River–St. Lawrence River valleys. Conditions then ameliorated and there was a thermal maximum, or Hypsithermal, around 6,000–5,000 years ago. By comparing δ18O from Greenland Ice Sheet ice cores with δ18O in ice cores from small marginal ice caps, a new temperature history reveals a pronounced Holocene climatic optimum in Greenland coinciding with maximum thinning near the ice sheet margins (Vinther *et al.*, 2009). In the North Atlantic the Holocene thermal maximum (or Hypsithermal) lagged behind that in the High Arctic as a result of ocean conditions, with the discharge of glacial meltwater from the remains of the Laurentide Ice Sheet slowing the warming. The middle Holocene was relatively warm off East Greenland but in the Norwegian Sea ice-rafting peaked in the mid-Holocene, 6.5–3.7 ka (Risebrobakken *et al.*, 2003). In Scandinavia, there were early Holocene glacier readvances around 11,200, 10,500, 10,100, 9,700, 9,200, and 8,400–8,000 cal yr BP according to Nesje (2009). Norwegian glaciers appear to have melted away at least once during the early/mid-Holocene; glaciers were most contracted between 6,600 and 6,000 cal yr BP.

Most mountain ranges in the Northern Hemisphere saw maximum glacier recession during the early Holocene, with some glaciers disappearing, according to Davis *et al.* (2009). Iceland, for example, became mostly ice free after 8 ka during the Holocene thermal maximum. Norway is an exception, with abrupt Late Glacial and early Holocene glacier fluctuations between ~11.2 and 8.2 ka. Also, in the Himalaya, Karakoram, and Tibet there were substantive glacier advances throughout most of the Holocene. In contrast, some alpine areas in the Southern Hemisphere saw glaciers reach their maximum postglacial extent during the early to middle Holocene. This is the case in most areas of South America except for the dry subtropical Andes.

The Hypsithermal was followed by a cooling that led to four minor Neoglaciations; the best known of these is the Little Ice Age (LIA) dated in Europe around AD 1550–1850 (Grove, 2004). The term LIA was proposed by Matthes (1939) to refer to the various glacier advances during the late Holocene, but has subsequently been reserved for the most substantial glacier advances of the last millennium. Davis *et al.* (2009) assert that alpine glaciers in many parts of the world reformed and/or advanced during Neoglaciation, reaching their maximum Holocene extents during the LIA. Matthews and Briffa (2005) argue that "Little Ice Age" glacierization occurred over about 650 years and can be defined most precisely in the European Alps (AD 1300–1950) when extended glaciers were larger than before or since. "Little Ice Age" climate is defined as a shorter time interval of about 300 years (c. AD 1400–1700) when Northern Hemisphere summer temperatures (land areas north of 20° N) fell significantly below the AD 1961–1990 mean (Mann *et al.*, 2009). However, "Little Ice Age" glacierization was highly dependent on winter precipitation. The LIA saw widespread glacier advances in mountain regions, and snowlines were about 100 m lower than in the late twentieth century. Glaciers in northern Sweden probably reached their maximum LIA extent between the seventeenth and the beginning of the eighteenth centuries, whereas most Norwegian glaciers attained their maximum extent during the mid-eighteenth century (Nesje, 2009). Glaciers in the Alps attained their LIA maximum extents in the fourteenth, seventeenth, and nineteenth centuries, with most reaching their greatest LIA extent in the final advance about AD 1850/1860 (Ivy-Ochs *et al.*, 2009). Temperatures were around 1 °C lower than today with multidecadal fluctuations. Miller *et al.* (2010) find evidence of ice cap growth in northeastern Arctic Canada between AD 1250 and 1300 and around AD 1450, with the ice remaining in an expanded state until the last few decades. In southern Tibet, maritime glaciers were 30% larger during the LIA than at present, while continental glaciers in the western part of the plateau were <10% larger (Shi *et al.*, 2008b). Since the LIA maximum, glaciers in most mountain regions of the world have lost about 25–33% of their area (Barry, 2006) (see Chapter 3), especially in the last two decades (SROCC, 2019). Sea ice conditions in the North Atlantic during the LIA are discussed in Chapter 7 (Section 7.7) and below.

The timing of Southern Hemisphere glacier variations during the Holocene shows complex relationships with those in the Northern Hemisphere. Schaefer *et al.* (2009) report that sometimes glaciers in New Zealand were larger than at present when those in the Alps were smaller, but at other times both appear to have advanced simultaneously. There is a notable interhemispheric disparity in the timing of the maximum ice extent. Glaciers at Mount Cook were more extensive about 6,500 years ago than at any subsequent time. In contrast, most Northern Hemisphere glaciers reached their greatest Holocene extents during the LIA (1300 to 1860 AD). Several glacier advances beyond the positions of the nineteenth century termini occurred in New Zealand during northern warm periods characterized by diminished Northern Hemisphere glaciers, such as between about 1500 to 900 BC, during 200 BC to 300 AD, and during the Medieval Warm Period 800–1300 AD. Coherency between the records at Mount Cook and the Northern Hemisphere was greatest during 300–700 AD, and broad similarities were apparent during 1200–1850 AD (the northern LIA), with multiple glacier advances followed by a general termination beginning in the mid to late nineteenth century.

Masiokas *et al.* (2009) review the evidence of the last millennium for extratropical South America (17–55° S). They find that most records indicate that dates of maximum glacier expansion range from the sixteenth to the nineteenth centuries (the LIA), but with considerable variability in the extent and timing of events. Glaciers from the North Patagonian Icefield consistently showed that the LIA maximum extent there occurred sometime during the nineteenth century. In contrast, for other glaciers in southern Patagonia the LIA maxima have been generally identified one to three centuries earlier. The number and extent of glaciers increased significantly in the Patagonian region. In spite of the occurrence of several readvances over the past ~100 years, most areas of the extratropical Andes have experienced a general pattern of recent glacier recession and significant ice mass losses.

In the tropical Andes (10° N–16° S), the first glacial advance is dated around AD 1200–1350 (Jomelli *et al.*, 2009). The maximum glacial extent (MGE) – the furthest down-valley extent recorded synchronously by the majority of glaciers – occurred around AD 1630–1680 in Bolivia and Peru (the outer tropics) and around AD 1730 in Ecuador, Colombia, and Venezuela (the inner tropics). Subsequently, glaciers retreated more or less continuously during the eighteenth and nineteenth centuries. In the outer Tropics of South America, minor glacial advances occurred around 1730, 1760, 1800, 1850, and 1870. In the inner Tropics, synchronous minor advances occurred around 1760, 1820, and 1880. Between the MGE and the early twentieth century, Andean glaciers lost about 30% of their total length.

In the early Holocene (10–8 ka), the distribution of bowhead whale bones indicates at least periodically ice-free summers along the length of the Northwest Passage (NWP) and the same pattern is repeated from about 500–1250 AD (Polyak *et al.*, 2009). Vare *et al.* (2009) use a sea ice biomarker chemical in a sediment core from Barrow Strait in the Canadian Arctic Archipelago to reconstruct Holocene sea ice history. They find spring sea ice occurrence was lowest during the early–mid Holocene (10.0–6.0 cal kyr BP).

During 6.0–4.0 cal kyr BP spring sea ice occurrence showed a small increase. Between 4.0 and 3.0 cal kyr BP, sea ice increased abruptly to above the median. Elevated spring sea ice occurrences continued from 3.0 to 0.4 cal kyr BP, although they were more variable. Within this fourth interval, there is evidence for slightly lower (higher) spring sea ice occurrence during the Medieval Warm Period (LIA), respectively. An analysis of dinoflagellate records in a 7,700-year record of sediment cores along the main axis of the NWP by Ledu *et al.* (2010) found extensive sea ice cover in the eastern sector – Lancaster Sound – for most of the Holocene, but less ice than now from 10.8 to ~8.5 cal kyr BP in the rest of the NWP. In the central part of the NWP (Barrow Strait), there were millennial-scale sea ice fluctuations from 8.5 to 6.0 cal kyr BP. After 6.0 cal kyr BP, there was a decrease of the sea ice cover in the central part of the NWP with respect to modern conditions. In the westernmost part of the NWP (Dease Strait south of Victoria Island), there was little change from 6.0 to 4.0 cal kyr BP, followed by a slight increase in sea ice. Maximum sea ice cover was reached at 1,500 and 1,000 cal yr BP.

In the Northern Hemisphere, the sea-ice area and volume as simulated by an Earth System Climate Model (Sedláček and Mysak, 2009) during the LIA are larger than the present-day area and volume. The wind-driven changes in sea-ice area are about twice as large as those due to radiative forcing. For the sea-ice volume, changes due to wind and radiative forcing are of similar magnitude. Before 1850, the simulations suggest that volcanic activity was mainly responsible for the thermodynamically produced area and volume changes, while after 1900 the slow greenhouse gas increase was the main driver of sea-ice changes.

Sea levels rose sharply by about 30 m to −80 m asl around 14.5 ka BP. Carlson *et al.* (2008) estimate that between 11 and 9 ka BP, the Laurentide Ice Sheet contributed a further 15 ± 1.8 m of sea-level rise and during the subsequent phase of rapid ice retreat 9–8.5 ka BP, it contributed another 6.6 ± 0.8 m. Between 8.5 and 6.8 ka BP, a further 9.2 ± 1.1 m of sea-level rise was added from the Laurentide Ice Sheet. Major changes in the land surface occurred in the Bering Sea, the Arafura Sea north of Australia, and the North Sea–English Channel, where former lowlands were inundated by rising sea levels with major implications for human migrations. The rapid sea-level changes are also considered to account for the Biblical (Noah) and numerous other indigenous "flood histories."

REVIEW QUESTIONS

(1) Name six major Ice Ages over the geological time starting from the NeoArchean–Paleoproterozoc (~2,700 to 2,200 million years ago, Ma) to the Holecene (12,000 years to present). Approximately during what period (Ma) was the earth believed to be entirely covered with ice, the Snowball Earth?

(2) The three geologic events attributed to major temperature fluctuations are the Plate tectonics, large-scale ocean currents and thermohaline circulations, and Milankovitch cycles. Explain how glacial/interglacial cycles identified in paleo ice core data collected from, say, the Vostok Station of Antarctica, supported the orbital hypothesis of the Milankovitch cycles, such as the eccentricity of the Earth's orbit around the Sun has a periodicity of ~100,000 years.

(3) In your opinion, which of these hypotheses provides more plausible explanation(s) that the global glaciation period between 730 and 540 Ma ended abruptly: major volcanic eruptions that lead to a drastic increase in atmospheric CO_2 (350 times the modern level), or significantly lower tropical surface albedo, or methane released from melting permafrost in low latitudes, or some oceans remained ice free permitting free exchange with the atmosphere?

(4) What were the characteristics of the Phanerozoic glaciations modes in the late Ordovician (445–440 Ma), the late Devonian (375–368 Ma), the Permo-carboniferous (360–263 Ma), and the Cenozoic (beginning ~40 Ma) glaciations, with respect to atmospheric CO_2, biogeographic changes, and continental drifts?

(5) What evidence has been identified about the full glaciations of the East Antarctica in the earliest Oligocene (37–24 Ma), West and East Antarctica during the Miocene (24–5 Ma), and which hypothesis regarding the formation of the ice sheets, such as the breaking up of Gondwana, ocean passages or ocean current such as the Antarctic Circumpolar Current, and orbital forcing is most convincing to you and why?

(6) Approximately when did the glaciations in the Northern Hemisphere begin, and what evidence has been found that supports or goes against the traditional belief that glaciations in Greenland and the Arctic began in the late Miocene, much later than the glaciations of the Antarctic?

(7) For the Quaternary period, among various speculations on the cause of glacial cycles of the Pleistocene glaciations of the last 2 Ma that shifted from a 41,000-year to a 100,000-year signature, do you agree with the argument that a progressive removal of a deforming sediment layer, which maintained thin ice sheets responding to dominant 23 and 41 ka orbital forcings, had caused the transition to thicker ice sheets of dominant timescale of change of ~100 ka?

(8) What were the range of CO_2 levels obtained for the most severe glacial intervals of marine isotope stages (MIS) 2, 12, and 16 from ice cores of the European Project Ice Coring in Antarctica (EPICA) in Dome C, to the warmer interglacial intervals, say, MIS 11, 9, 7, 5, and 1? Are the current, much higher global CO_2 levels (>400 ppm) mainly caused by the burning of fossil fuels worldwide since the industrial revolution began in about 1760?

(9) Name several ice sheets in North America and Scandinavia during the Quaternary periods and the postglacial Holocene, of which some have been reconstructed by geodynamical models ICE4G and ICE5G, and how they have retreated over postglacial warming or deglaciation periods, say ~21–9.6 ka.

(10) Among a series of Heinrich events that interrupted the postglacial warming, what could have caused the abrupt Younger Dryas cold interval that resulted in massive environmental changes to occur and lasted ~1,300 years?

(11) In the postglacial Holocene epoch, among the four minor Neoglaciations, what were the characteristics of the Little Ice Age (LIA) dated around 1550–1850, besides the most substantial glacier advances over the last millennium observed in the European Alps and in mountain regions worldwide, and disparity in the maximum ice extent between Southern and Northern hemispheres?

10 The future cryosphere: Impacts of global warming

10.1 Introduction

There have been published records of surface temperature change over large regions since the late 1800s (Figure 10.1), but only in recent decades that there have been many studies showing that the Earth is experiencing unprecedented climate warming on a global scale. Under continued climate warming, a critical issue is the contributions to sea-level rise (SLR) through the melting of mountain glaciers and possible disintegration of parts of the West Antarctic ice sheet, added to ocean thermal expansion. Meltwater from Greenland, together with ice export from the Arctic Ocean, affects circulation in the North Atlantic through the associated energy and water fluxes, and these changes further modify atmospheric storm tracks. Snow-covered surfaces in the northern continents reflect 70% or more of the incoming solar radiation, so as snow and ice cover melts because of climate warming, the reduction in surface albedo to 15–25% due to exposed ground leads to increased absorption of solar radiation. This, in turn, results in more melting, leading to further warming or a positive feedback. A similar ice–albedo feedback arises from the reduction of sea-ice cover exposing an ocean surface with an albedo of only 5%. Recent research has provided better quantitative understanding of the contribution of snow/ice–albedo feedback to the pronounced warming signal in high-latitude regions (Landy *et al.*, 2014; Kashiwase *et al.*, 2017). Greenhouse gases (GHGs), such as CO_2 and CH_4 (methane) released by thawing permafrost to the atmosphere, will further exacerbate global warming, but again this has not been determined quantitatively.

Climate models indicate that global warming will be more pronounced in high latitudes, particularly the Arctic. Paleo temperature data, derived from the Vostok ice core in East Antarctica, indicate that variations of temperature in the polar regions over the last 160,000 years are double those of global average temperature. Major changes have been observed in

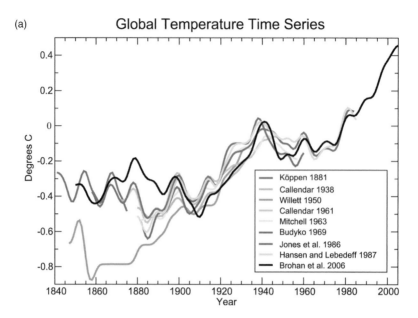

(a)

Figure 10.1 (a) Published records of surface temperature change over large regions. Köppen (1881) tropics and temperate latitudes using land air temperature. Callendar (1938) global using land stations. Willett (1950) global using land stations. Callendar (1961) 60° N to 60° S using land stations. Mitchell (1963) global using land stations. Budyko (1969) Northern Hemisphere using land stations and ship reports. Jones *et al.* (1986a, b) global using land stations. Hansen and Lebedeff (1987) global using land stations. Brohan *et al.* (2006) global using land air temperature and sea surface temperature data is the longest of the currently updated global temperature time series. All time series were smoothed using a 13-point filter. The Brohan *et al.* (2006) time series are anomalies from the 1961 to 1990 mean (°C). Each of the other time series was originally presented as anomalies from the mean temperature of a specific and differing base period (from Le Treut *et al.*, 2007). (b) Global temperature anomaly of 1960–2014 with reference to the 1951–1980 mean temperature where more red areas in recent years indicates global warming and the area inside the black box shows warming amplified in the Arctic, particularly within the past 25 years (Wendisch *et al.*, 2017). (c) Mean 2010–2019 global temperatures compared to a baseline 1951–1978 average [Source: Goddard Institute for Space Studies NASA's Goddard Space Flight Center, https://en.wikipedia.org/wiki/Instrumental_temperature_record] For the color version, refer to the plate section. {Source: IPCC 2007, Figure 1.3] (A black and white version of this figure will appear in some formats. For the color version, please refer to the plate section.)

the Arctic, including the highest temperatures in the last 400 years, earlier melting of ice on lakes and rivers, and a sharp decline in the extent of summer Arctic sea ice (ASI).

Polar climate shows complex interactions and feedbacks between atmosphere, land, cryosphere, and ocean (Serreze and Barry, 2005). It exhibits interannual to multidecadal oscillations, because of teleconnections to large-scale climate anomalies such as the Arctic Oscillation (AO) (or Northern Annular Mode – NAM), North Atlantic Oscillation (NAO), Pacific-North American (PNA) pattern, and the Pacific Decadal Oscillation

(b)

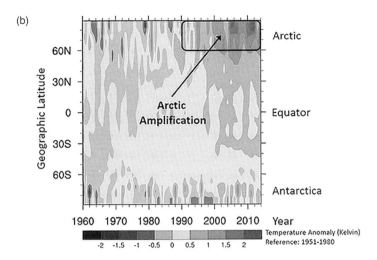

(c)

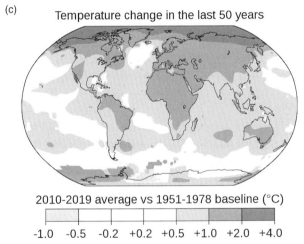

Figure 10.1 (cont.)

(PDO) (Barry and Carleton, 2001). For example, the positive phase of AO/NAO is associated with warmer, wetter winters in Siberia and colder, drier winters in western Greenland and northeastern Canada. The cold phase of PDO/PNA is associated with a wetter western Canada, and a deeper and eastward-shifted Aleutian Low advects warmer and moister air into Alaska. With these teleconnections, complex interactions that are not well understood or modeled, we expect that projections of future polar climate by global climate models will be associated with a large range of uncertainties (ACIA, 2005). However, some scientists have begun to conduct coupled atmosphere-ice-ocean Regional Climate Model experiments for the Arctic (Rinke *et al.*, 2003; Mikolajewicz, 2005).

Arctic Amplification (AA)

Arctic amplification is the climate change impact amplified in the Arctic, which has been warming at twice the global rate since 1980s (Assessment Report 5 of IPCC, 2014), a climate paradox (Maslanik et al., 2007a) which is likely related to several feedback mechanisms, namely, temperature, Planck, ice–albedo, lapse-rate, water vapor, and cloud feedbacks. ASI loss is likely one of the most important or obvious responses to AA since the loss of ASI allows strong heat flux from the ocean to escape to the atmosphere (Serreze et al., 2009), and because the albedo of seawater is much lower than sea ice, sea ice loss results in a substantial increase in the net incoming solar radiation, a positive feedback. Reconstructions of past climates from proxy records show that enhanced warming of the Arctic, which leads to the rapid reduction of ASI, has been a distinctive feature of the global climate system (Chapin III et al., 2005; Graversen et al., 2008; Mauritsen, 2016).

Various studies have investigated feedback mechanisms that are accelerators of Arctic warming, namely, (1) temperature feedback that results from the effect of rising temperatures on outgoing longwave radiation at the top of atmosphere (Pithan and Mauritsen, 2014); (2) Planck feedback due to the contribution from vertically uniform warming of the surface and troposphere; (3) lapse-rate feedback due to the contribution from tropospheric warming that deviates from the vertically uniform profile (Bintanja et al., 2011; Graversen et al., 2014); (4) ice–albedo feedback due to the increase in surface absorption of solar radiation when snow and ice retreat (Perovich et al., 2007; Serreze and Barry, 2011; Kashiware et al., 2017); (5) water vapor feedback due to the greenhouse effect of additional atmospheric water vapor (Held and Soden, 2000; Solomon et al., 2010; Dessler et al., 2013); (6) cloud feedback due to effect of clouds on the Earth's radiative balance (Wetherald and Manabe, 1988; Vavrus, 2004; Bony et al., 2015); and (7) sea ice change (Screen and Simmonds, 2010). Changes in atmospheric and oceanic heat transport also contribute to AA.

Arctic amplification has been found in past warm and glacial periods, as well as in historical observations and climate model experiments. Ice–albedo feedback has often been cited as the main contributor to AA. Pithan and Mauritsen (2014) analyzed climate model simulations from the Coupled Model Intercomparison Project Phase 5 (CMIP5) archive to quantify the contributions of various feedback mechanisms. They found that the largest contribution to AA comes from temperature feedback: as the surface warms, more energy is radiated back to space in low latitudes, compared with the Arctic.

On the other hand, these feedbacks themselves cannot fully explain the observed reduction of ASI given that AA is also found in simulations of climate models without considering changes in snow and ice cover (Curry et al., 1995; Graversen and Wang 2009; Graversen et al., 2014). ASI is shown to teleconnect to climate patterns such as the AO (Liu et al., 2004; Overland and Wang 2005), the NAO, the Pacific-North America Pattern (L'Heureux et al., 2008), and the El Nino Southern Oscillation (ENSO) (Hu et al., 2016), to a jet stream (Francis and Vavrus, 2015), and to ocean heat transport (Tomas et al., 2016). However, the observed correlation between climate patterns and the loss of ASI was not consistent

(Maslanik *et al.*, 2007a) and observations show lower correlations than simulations from various climate models that include many of the feedback mechanisms related to ASI (Johannessen *et al.*, 2004; Stroeve *et al.*, 2007; Comiso *et al.*, 2008; Onarheim *et al.*, 2018; Petty *et al.*, 2018).

The rapid rising air temperature and substantial decline of sea ice cover in the Arctic have led to significant changes in the Arctic basins, where the hydrologic processes have intensified, precipitation and discharge have increased, permafrost degraded, and even green vegetation have been observed. All these recent changes observed in the Arctic have been shown to be related to the significant decline in ASI and various related feedback mechanisms associated with temperature, water vapor, clouds, and surface albedo.

10.2   General observations

The basic process of global warming is attributed to the global atmospheric concentration of major GHGs:

(1) CO_2 has increased from a pre-industrial, 280 ppm to 414 ppm in 2019 (48%), and has been increasing at a rate of an average 2.1 ppm a^{-1} since 2005;
(2) Methane (CH_4) has increased from a pre-industrial, 715 ppb to 1,732 ppb in the early 1990s, and was 1,865 ppb in 2019 (161%);
(3) Nitrous oxide (N_2O) has increased from a pre-industrial, 265 ppb to 320 ppb in 2005.

Based on Antarctic ice cores, the current CO_2 concentration is far higher than it has been over the past 650,000 years. The CO_2 concentration typically ranged from about 180 to 280 ppm per glacial cycle. Other GHGs such as methane (CH_4)which is about 25 times more powerful than CO_2 as a GHGs (the global warming potential or GWP of CH_4 is 25), remained around 750 ppb from AD 1000 AD until the early 1800s. Like CO_2, it rose throughout the Industrial Revolution and is now around 1,875 ppb. There are other GHGs that are much more powerful than methane such as the chlorofluorocarbons (CFCs) but their overall effects are less significant because of much smaller quantities. In contrast, aerosols and dust particles have cooling effects. The warming and cooling effects of GHG and particles are assessed in terms of radiative forcing, such that about 3 Wm^{-2} will change the surface temperature by about 1 °C if nothing else changes (Houghton, 2009). The total anthropogenic radiative forcing for 2011 relative to 1750 is 2.3 Wm^{-2} (IPCC, 2014). A possible additional surface warming effect important in middle and high latitudes is that of ice crystal nuclei in stratiform clouds, according to Zeng *et al.* (2009), but the effect needs to be quantified.

Analyses of radiosonde and satellite data consistently show warming of the troposphere, the lower atmosphere. More than 90% of the excess energy absorbed by the climate system since the 1970s has been stored in the oceans as shown by global records of ocean heat content. Global SLR since the nineteenth century has been driven by warmer oceans (water expands), melting glaciers, sea ice and vece sheets, and changes in storage and usage of water on land. Global analyses show that specific humidity has increased over both the land and the oceans. Globally glaciers have been declining every year for last several decades. Spring snow cover has shrunk across the NH since the 1950s. Substantial losses in ASI have been observed since satellite records began in 1979, especially at September when sea ice will be the minimum at the end of the annual melt season. By contrast, the increase in Antarctic sea ice has been smaller.

Climate warming has not been uniform globally, but the global average annual temperature increase from 1850–1899 to 2001–2005 is 0.76 ± 0.18 °C. The two warmest years recorded are 2016 and 2019, and the five warmest years have all occurred since 2015 with nine of the 10 warmest years occurring since 2005, according to NOAA's National Centers for Environmental Information (NCEI). Records of 2016 and 2019 mean that 19 of the 20 warmest years in the instrumental record of global surface temperature since 1850 have occurred after 2000. The average global temperature in 2016 was 0.99 °C above the twentieth-century average. The 2010s are likely the warmest decade of the past 1,300 years in the Northern Hemisphere. The average Arctic temperatures have increased at almost twice the global average rate in the past 100 years. From 1900 to 2005, significantly increased precipitation has been observed in eastern parts of North and South America, northern Europe, and northern and central Asia. More intense and longer droughts have been observed over wider areas, particularly in the tropics and subtropics since 1970s, due to higher temperatures and decreased precipitation. Widespread changes in extreme temperatures have been observed over the last 50 years. However, Antarctic sea ice extent continues to show no statistically significant average trends, consistent with the lack of warming reflected in atmospheric temperatures averaged across Antarctica.

Our climate exhibits natural variability, which is partly caused by fluctuations in atmospheric and oceanic circulations, changes in solar irradiance, volcanic eruptions, and long-term changes in Earth's orbit relative to the Sun (called the Milankovitch cycles) (see Chapter 9). In other words, in the past, large polar ice sheets grew and dwindled over tens of thousands of years caused by natural variability of climate. However, the observed widespread warming of all the continents except Antarctica for the last several decades, massive losses of sea ice in the Arctic, retreat of mountain glaciers, thawing permafrost, and the earlier onset of spring snowmelt in the Northern Hemisphere, lead scientists to believe that the observed global climate warming is extremely unlikely to be caused by natural climate variability alone. Rather, it is mainly anthropogenic in nature. The IPCC concludes that it is very likely (greater than 90% chance) that most of the warming since the mid-twentieth century is due to continuing increases in GHG concentrations caused by human activity (Le Treut et al., 2007; Vaughan et al., 2013).

10.3   Recent warming and cryospheric changes

In Figure 10.2a on observed and simulated time series of the anomalies in annual and global mean surface temperature, vertical dashed gray lines represent times of major volcanic eruptions, single simulations for CMIP5 models (thin lines), CMIP5 multimodel mean (thick red line), different observations (thick black lines). Observational data are Hadley Centre/Climatic Research Unit gridded surface temperature data set 4 (HadCRUT4; Morice *et al.*, 2012), Goddard Institute for Space Studies Surface Temperature Analysis (GISTEMP; Hansen *et al.*, 2010) and Merged Land–Ocean Surface Temperature Analysis (MLOST; Vose *et al.*, 2012) and are merged surface temperature 2 m height over land and surface temperature over the ocean. Following the CMIP5 protocol (Taylor *et al.*, 2012), all simulations use specified historical forcings up to and including 2005 and use RCP4.5 after 2005.

In Figure 10.2b, observations are shown in black. Blue shading is the model time series for natural forcing simulations and pink shading is the combined natural and anthropogenic forcings. The dark blue and dark red lines are the ensemble means from the model simulations. All panels show the 5 to 95% intervals of the natural forcing simulations, and the natural and anthropogenic forcing simulations. For surface temperature the results are from Jones *et al.* (2013). The observed surface temperature is from Hadley Centre/Climatic Research Unit gridded surface temperature data set 4 (HadCRUT4). For land and ocean surface temperatures panel, solid green lines at bottom of panels indicate where data spatial coverage being examined is above 50% coverage and dashed green lines where coverage is below 50%.

Evidence of changes in the cryosphere in recent decades has been widely observed, namely, smaller glaciers, lesser and wetter snow, thawing permafrost, shorter freshwater ice covers, and dwindling ASI. A summary is given later, while details are provided in respective chapters:

(1) The duration of ice cover in rivers and lakes in high and middle latitudes of the Northern Hemisphere decreased by about 2 weeks over the twentieth century;
(2) Significant retreat of glaciers worldwide during the twentieth century, contributing to global SLR (Figures 3.17 and 3.18);
(3) Decrease of Arctic sea-ice extent and thickness was about 40% in late summer in recent decades, with the minimum sea ice concentration in 2012. Since 1979, the decrease was very likely about 3.5–4.1% per decade but for the summer sea ice minimum, the decrease was very likely about 9.4–13.6% per decade. In contrast, the annual mean Antarctic sea ice extent increased about 1.2–1.8% per decade between 1979 and 2012 (Figures 7.22–7.25);
(4) Reduction in spring snow-covered area has been about 10% since global observations began in the late 1960s, and a very high confidence that in 1967–2012, the extent of NH

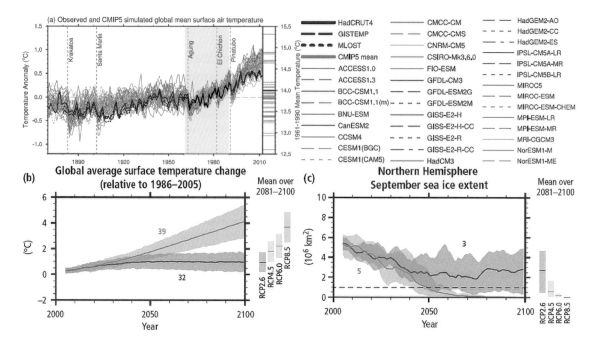

Figure 10.2 (a) Observed and simulated time series of the global mean surface temperature anomalies, differences from the 1961 to 1990 time-mean of each individual time series (colors), the CMIP5 multimodel mean (thick red), and the observations (thick black: Jones *et al.*, 1999; Flato *et al.*, 2013); and selected GCMs of CMIP5 shown in (a); (b) Global annual change in mean surface temperature for 2006–2100 relative to 1986–2005 from CMIP5 experiments; and (c) Change in NH September sea-ice extent (5 year running mean). The dashed line represents nearly ice-free conditions, when the September sea-ice extent is less than 10^6 km^2 for at least five consecutive years (taken from IPCC, 2014) (A black and white version of this figure will appear in some formats. For the color version, please refer to the plate section.)

snow cover has decreased since the mid-twentieth century by 1.6% per decade for March and April, and 11.7% per decade for June (Figures 2.20 and 2.21);

(5) Degradation of permafrost has been detected in parts of the polar and subpolar regions, and in response to increased surface temperature and changing snow cover, very likely permafrost temperatures have increased in most parts of the NH since early 1980s, with reductions in thickness and areal extent in some regions.

Global average sea level has risen by 1.8 (1.3–2.3) mm yr^{-1} during 1961 to 2003, and 3.6 mm yr^{-1} from 2006 to 2015. The total twentieth-century rise was ~0.17 (0.12–0.22) m, and ~0.22 (0.21–0.24) m since 1880, due to melting glaciers and ice sheets and ocean thermal expansion. Mass losses from Greenland and Antarctic ice likely contributed to SLR since the mid-1990s.

Could another ice age slow down or even reverse the global warming trend? Orbital variations make it very unlikely that the Earth would naturally enter another ice age for at least 30,000–50,000 years.

10.4 Climate projections

Based on a variety of assumptions regarding population growth, economic activity, and future emissions of GHGs, a wide range of climate scenarios called representative concentration pathways (RCPs) include time series of emissions and concentrations of the full suite of GHGs and aerosols and chemically active gases, as well as land use/land cover has been published (IPCC, 2013). "Representative" signifies that each RCP provides only one of many possible scenarios that would lead to specific radiative forcings, while "pathway" emphasizes both the long-term concentration levels and the trajectory taken over time to reach that outcome of interest. In contrast to SRES scenarios, radiative forcing trajectories in the RCPs are not associated with predefined storylines and can reflect various possible combinations of economic, technological, demographic, and policy developments (Moss et al., 2010). RCPs were used to develop climate projections in CMIP5.

The RCPs span a wide range of stabilization, mitigation/non-mitigation pathways with the resulting range of temperature changes larger than those projected under SRES scenarios, which do not consider mitigation options. The SRES scenarios span between just above 1.0 °C and 6.5 °C when considering the likely ranges of all scenarios, from B1 as the lowest and A1FI as the highest. RCP8.5 project the highest emissions with projected temperature changes ranging from 4.0 °C to 6.1 °C by 2100. The lowest RCP2.6 assumes significant mitigation, resulting in the global temperature change likely remaining below 2 °C. Similar temperature change projections by 2080–2099 are obtained under RCP8.5 and SRES A1FI, RCP6 and SRES B2, and RCP4.5 and SRES B1. However, there remain large differences in the transient trajectories, with rates of change slower or faster for the different pairs. These differences can be traced back to the interplay of the (negative) short-term effect of sulfate aerosols and the (positive) effect of long-lived GHGs.

RCP2.6: A stringent mitigation pathway where radiative forcing peaks at about 3 W m^{-2} and then declines to 2.6 W m^{-2} in 2100, with CO_2-eq concentration of 430–500 ppm (the corresponding Extended Concentration Pathway [ECP] assumes constant emissions after 2100).

RCP4.5: An intermediate stabilization pathway in which radiative forcing is limited at about 4.5 W m^{-2} in 2100, with CO_2-eq concentration of 500–650 ppm (the corresponding ECPs assumes constant concentrations after 2150).

RCP6.0: An intermediate stabilization pathway in which radiative forcing is limited at about 6.0 W m^{-2} in 2100, with CO_2-eq concentration of 651–850 ppm.

RCP8.5: One high pathway which leads to >8.5 W m^{-2} in 2100, with CO_2-eq concentration of 851–1,370 (the corresponding ECP assumes constant emissions after 2100 until 2150 and constant concentrations after 2250).

Shared socioeconomic pathways (SSPs) of the Coupled Model Intercomparison Project Phase 6 (CMIP6) were developed to complement the RCPs with varying socioeconomic challenges to adaptation and mitigation (O'Neill et al., 2014). Based on five narratives, the

SSPs describe alternative socioeconomic futures in the absence of climate policy intervention, comprising sustainable development (SSP1), regional rivalry (SSP3), inequality (SSP4), fossil-fueled development (SSP5), and a middle-of-the-road development (SSP2) (O'Neill *et al.*, 2017). SSP5-8.5 features much higher CO2 concentrations than RCP8.5 by 2100, but its the projected CH4 concentrations by 2100 are substantially lower than RCP8.5. Even though CH4 is about 25 times more powerful than CO2, the overall warming impact of CO2 is still substantially higher than CH4 because its concentration is much higher. Therefore, the projections of SSP-8.5 is worst than that of RCP8.5.

Sillman *et al.* (2013) analyzed climate scenario simulations of climate models participated in the CMIP5 ensemble for the twentieth and twenty-first centuries (Taylor *et al.*, 2012). The CMIP5 model output is available from the data archives of the Program for Climate Model Diagnosis and Intercomparison (PCMDI, www-pcmdi.llnl.gov) and the Earth System Grid (ESG) data distribution portal (www.earthsystemgrid.org). Based on the results of the RCPs of CMIP5 ensemble of 19 Global Climate Models, the global temperature is projected to increase by 1 °C (RCP2.6), 1.8 °C (RCP4.5), 2.3 °C (RCP6.0), and 3.7 °C (RCP8.5) by 2081–2100 with respect to the 1986–2005 period. Table 10.1 shows the difference between two

Table 10.1 Projected changes in daily minimum (TN) and maximum (TX) of near surface temperature (°C) and total annual precipitation on wet days >1 mm of precipitation (PRCPTOT in %) using different RCP emission scenarios in the Coupled Model Intercomparison Project Phase 5 (CMIP5) multimodel ensembles on global and regional scales over the twenty-first century relative to the 1981–2000 base period. Changes in daily minimum temperatures are more pronounced than changes on daily maximum temperatures.

	Temperature response Minimum[#] (TN) °C			Temperature response Maximum[#] (TX) °C			Precipitation response PRCPTOT* (%)		
	RCP2.6	RCP4.5	RCP8.5	RCP2.6	RCP4.5	RCP8.5	RCP2.6	RCP4.5	RCP8.5
Global	1.8	3.4	6.7	1.4	2.7	5.4	3.5	5.8	9.3
Arctic (60° N,180° E–90° N,180° W)	4.4	6.7	10.8	3.2	4.6	7.7	14	22	38
Antarctic (90° S,180° E–60° S,180° W)	2.4	3.1	4.9	1.2	2.2	3.4	15	35	60
Alaska (60° N,170° W–2° N,103° W)	3.3	6.2	12.8	2.1	2.7	5.3	1.1	1.5	2
Greenland (50° N,103° W-85° N,10° W)	2.6	4.8	10.1	1.5	2.7	5.3	0.7	1.1	1.8

periods, the 1981–2000 base period and 2081–2100 in terms of mean values from the 19-model ensemble for minimum and maximum temperature and precipitation.

As GHG concentrations rise, terrestrial Arctic is projected to warm between 4.4 °C and 10.8 °C in minimum temperature and between 3.2 °C to 7.7 °C in maximum temperature by 2100 (see Table 10.1) depending on the GHG emissions scenarios of RCP and the global climate model used (Sillman *et al.*, 2013). Climate models indicate that ground will warm considerably because of surface warming, leading to large-scale thawing of the near-surface permafrost (Lawrence *et al.*, 2008a; Zhang *et al.*, 2008b; Hjort *et al.*, 2018; Biskaborn *et al.*, 2019), snow extent and sea ice are also projected to decrease further in the Northern Hemisphere, and glaciers and ice caps are expected to continue to retreat (IPCC, 2014). As ground in terrestrial northern high-latitude undergoes considerable warming, we expect snow conditions to change substantially over the twenty-first century. Soil warming has been found to be mostly less than near-surface air warming due to the thermal damping of the warming signal by the heat capacity of soil.

The projected annual mean Arctic warming is almost double the projected global mean warming in the CMIP5 models, while the winter warming in the central Arctic is even higher than the global annual mean when averaged over the climate models. If global warming is to continue as projected according to the RCP climate scenarios, what will be the prospects for the cryosphere by the end of this century?

10.5   Projected changes to Northern Hemisphere snow cover

Climate model simulations indicate that there will be wintertime warming and increases in winter precipitation in mid- to high-latitude regions of the Northern Hemisphere (Meehl *et al.*, 2007). Based on the Community Climate System Model (CCSM)'s simulation of the twentieth and twenty-first centuries (SRES A1B scenario) climate, Lawrence and Slater (2010) found increased winter snowfall (+10–40%), decreased maximum snow depth (−5 ± 6 cm), and a shortened snow season (−14 ± 7 days in spring, +20 ± 9 days in autumn). They found that increasing snowfall counters the predominantly snowpack thinning influence of warmer winters and shorter snow seasons. In other words, CCSM projects both thinning and deepening snowpacks, depending on locations.

In general, a shorter snow season tends to warm the upper soil layers due to increased solar absorption; a shallower snowpack mitigates soil warming due to weaker winter insulation from the cold air, but a deeper snowpack has comparatively less impact due to the saturation of snow's insulating effect at deeper snow depths. Trends in snow depth and the length of snow season tend to be positively related, but they have opposite effects on the soil temperature. Therefore, because of these opposite effects, at the century timescale it is unclear whether changes to snow state can amplify or dampen soil warming. Lawrence and Slater (2010) found that snow state changes explain less than 25% of total soil temperature change by 2100.

According to the Clausius–Clapeyron relationship, climate warming increases the atmospheric moisture-holding capacity approximately by 7% per °C rise in temperature, resulting in more total precipitation. On the other hand, the fraction of precipitation falling as snowfall decreases and spring snowmelt tends to occur earlier with warming. Therefore, whether the amount of snowpack on ground will increase or decrease depends on the balance between these two competing factors.

Significant declines in Arctic (north of 60° N) May and June snow cover extent of −3.5% (±1.9%) and −13.4%, respectively, per decade between 1967 and 2018 (relative to the 1981–2010 mean) were determined from multiple datasets (Mudryk et al., 2017). The loss of spring snow extent is reflected in shorter snow cover durations estimated between −0.7 and −3.9 days decade^{-1}, depending on location and time period, but all spring snow cover duration trends estimated are negative, from surface observations (Bulygina et al., 2011; Brown et al., 2017), satellite data (Wang et al., 2013; Estilow et al., 2015), and model-based analyses (Liston and Hiemstra, 2011).

Snow water equivalent (SWE) increases in high latitudes of the Northern Hemisphere, especially in winter (DJF) such as northern Canada and Siberia, and decreases elsewhere (Figure 10.4). Under climate warming, SWE is expected to increase in high latitudes (Räisänen, 2008), broadly where the NDJFM mean temperature is ≤−20°C in the late twentieth century, where an increase in total precipitation generally dominates over reduced snowfall. Below this threshold, the winter precipitation and snowmelt are more sensitive to warming and so SWE will decrease. On the other hand, under warming the snow season is almost uniformly shortened because it should start later in the autumn and ends earlier in the spring. As expected, moving from early to the late twenty-first century, differences between model projections widen as their signal-to-noise ratio decreases. So the projections of SWE change are subjected to more uncertainties in the distant future, but their projections could be potentially improved by accounting adequately for biases in simulated winter temperature since the simulated SWE changes are strongly temperature dependent. Brown et al. (2017) detected negative trends in SWE over 1981–2016 from gridded products of satellite data and land surface models for both the Eurasian and North American sectors of the Arctic (Brown et al., 2017). They also estimated approximately 800,000 km^2 of spring snow cover lost per °C in warming.

Krasting et al. (2013) analyzed projections of the NH snowfall for 2006–2100 from simulations of 18 coupled atmosphere–ocean General Circulation Models (GCMs) of the CMIP5 under RCP4.5 climate scenarios. These GCMs generally simulated representative twentieth-century snowfall, but with a positive bias in many regions, such as western NA and central plateau of Siberia, compared with the 1951–2005 global average annual snowfall (cm yr^{-1}) map (Figure 10.3) derived from the University of Delaware Global Surface Air Temperature and Precipitation Climatology, version 2.01 (Willmott and Matsuura, 2009). The annual snowfall is projected to decrease across much of the NH during the twenty-first century, but projected to increase at higher latitudes (Räisänen, 2008). On a seasonal basis, the transition zone between negative and positive snowfall trends corresponds approximately to the −10 °C isotherm of the late twentieth-century mean surface air temperature, for example, positive trends prevail in

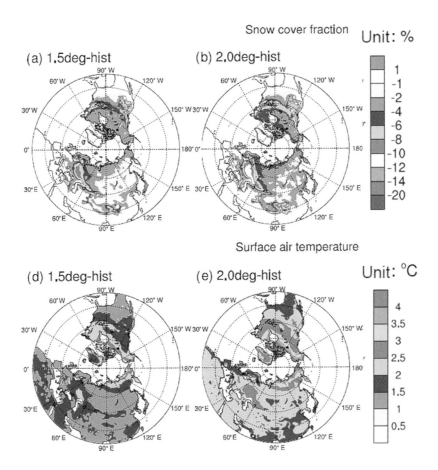

Figure 10.3 Snow cover fraction (SCF) differences in percent (%) between 2071–2100 and 1971–2000 ensemble mean from CESM-LE over Northern Hemisphere land area for (a) 1.5 °C scenarios and (b) 2.0 °C scenarios with reference to "Hist", (c) the land surface air temperature (LSAT) differences corresponding to (a), and (d) LSAT differences correspond to (b), respectively. "Hist" is the ensemble mean for 1971–2000 from CESM-LE. (taken from Wang et al., 2018). (A black and white version of this figure will appear in some formats. For the color version, please refer to the plate section.)

winter over large regions of Eurasia and North America, with projected redistributions of snowfall throughout the entire snow season. The decrease in the amount of precipitation falling as snow contribute to decreases in snowfall across most NH, while changes in total precipitation typically contribute to increases in snowfall, especially in DJF. The snowfall signal emerges more slowly than the temperature signal, implying that changes in snowfall are related to other factors besides regional warming.

Wang *et al.* (2018) project the ensemble mean % change in the annual snow area extent (SAE) and its standard deviation (SD) over NH between 1971–2000 and 2071–2100 divided by the mean of 1971–2000 for the Community Earth System Model (CESM). CESM's 1.5 °C

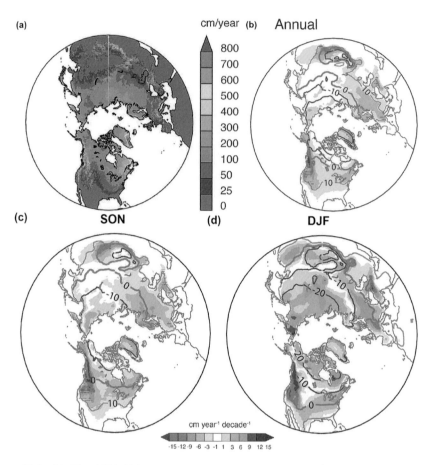

Figure 10.4 (a) The 1951–2005 average annual snowfall (cm yr^{-1}) derived from the University of Delaware Global Surface Air Temperature and Precipitation Climatology, version 2.01. Multimodel ensemble (MME) trends in snowfall (2006–2100; cm yr^{-1} decade^{-1}) for (b) annual, (c) September–November (SON), and (d) December–February (DJF), with contours of the 2-m temperature at 10°C intervals from the MME for 1986–2005 and hatching denotes regions with statistically significant trends (p ≤ 0.01) (taken from Krasting et al., 2013). (A black and white version of this figure will appear in some formats. For the color version, please refer to the plate section.)

and 2 °C projects designed to achieve the goal of the 2015 Paris Agreement, and CMIP5 RCP2.6 and RCP4.5 climate scenarios project the change in SAE to be −8.02 ± 0.78%, −10.92 ± 0.52%, −8.5 ± 5.58%, and −14.47 ± 5.71%, respectively. The reduction in SAE is strongly correlated negatively with the projected land surface air temperature (LSAT) change over NH. Figure 10.4 shows the 30-year annual mean SCF and LSAT between 2071–2100 and 1971–2000 for both the 1.5 and 2.0 °C scenarios, with the NH average SAE changes to be −1.69 × 10^6 km^2 in 1.5 °C and −2.36 × 10^6 km^2 in 2.0 °C. The largest magnitude of change could be above 10%, which occurs at mountain ranges in mid-latitudes, such as northern Canada, western America along the Rocky Mountains, and the western Tibetan Plateau. As expected, the ensemble mean for LSAT exhibits the largest warming over the Arctic for 2071–2100, projected to exceed 4 °C along the coastline of the Arctic Ocean, with less warming over the mid- and low latitudes (Figures 10.3c and d). The inconsistent spatial variations in LSAT and SCF suggest that LSAT is not the only factor in determining the SCF change.

For 2006–2100, the regression coefficients of SAE anomalies on the LSAT anomaly are $-1.37 \pm 0.56 \times 10^6$ km^2 °C^{-1} (1.5 °C), $-1.12 \pm 0.07 \times 10^6$ km^2 °C^{-1} (2°C), $-1.18 \pm 0.19 \times 10^6$ km^2 °C^{-1} (RCP2.6), and $-0.97 \pm 0.44 \times 10^6$ km^2 °C^{-1} (RCP4.5), respectively. The contribution of an increase in LSAT to the reduction of snow cover differs across seasons, with the greatest occurring in boreal autumn (49–55%). The spring SAE is projected to decrease by $-3.7\% \pm 1.1\%$ per decade within the CMIP5 ensemble over the twenty-first century, which exhibits variability from both model differences and internal climate variability. Internal variability in NH extratropical land warming trends and variability in winter precipitation can affect SCE trends (Thackeray et al., 2016). Climate model simulations indicate that there will be wintertime warming and increases in winter precipitation in mid- to high latitudes of the NH (Meehl et al., 2007).

Avalanches

Projecting changes in snow cover due to climate warming is important for many societal issues, including the risk of snow avalanche. Snow avalanches can be triggered naturally by meteorological factors such as snowfall load or infiltration of surface meltwater, rain-on-snow, or by the passage of people or skiers in avalanche terrain, by falling ice or rocks, or by explosives used for avalanche control (Schweizer et al., 2003). Changes in snow-cover characteristics could also induce changes in natural avalanche activity, which include changes in friction and flow regime (Naaim et al., 2013). A study of changes in avalanche timing for Aspen, Colorado, in AD 2030 and 2100 was conducted using the Snowmelt Runoff model, SNTHERM, and the output of five GCMs (Lazar and Williams, 2008). The occurrence of wet avalanches – defined as likely to occur when average daily temperature exceeds 0 °C – are predicted to occur between 2 and 19 days earlier than historical averages by AD 2030, and by 16 to 45 days earlier by AD 2100, with the wide range depending on the CO_2 emission scenarios used (low, medium, or high).

In the European Alps and Tatra Mountains, with less snow and higher air temperature, the numbers of snow avalanches and runout distance have decreased (Teich et al., 2012). Further, a decrease of avalanches with a powder part since the 1980s, a decrease of avalanche numbers below 2,000 m, an increase above (Eckert et al., 2013), and increase in the proportion of avalanches involving wet snow in December–February in recent decades have been observed (Naaim et al., 2016). Land use and land cover changes also contributed to changes in avalanches (García-Hernández et al., 2017). In summary, in particular in Europe, it is likely an increase in avalanche activity involving wet snow, and a decrease in the size and runout distance of snow avalanches over the past decades.

Global climate models mostly project an overall decrease in snow depth and snow cover duration at lower elevation, even though occasional occurrences of major snowfall events remain possible throughout most of the twenty-first century. Castebrunet *et al.* (2014) estimated an overall 20 and 30% decrease in mean and interannual variability of natural avalanche activity in the French Alps for 2020–2050 and 2070–2100, respectively, under A1B scenarios, compared to the reference period of 1960–1990, which are relatively high compared to changes in snow and meteorological variables. The decrease is projected to be more in spring and at low altitude, but more avalanche activity in winter at high altitude because conditions favorable to wet-snow avalanches is projected to occur earlier in the season. The projections for Northern Japan (Katsuyama *et al.*, 2017), and for North America are similar (Lazar and Williams, 2008). Avalanches involving wet snow are projected to occur more frequently during winter at all elevations due to surface melt or rain-on-snow (e.g., Castebrunet *et al.*, 2014, for the French Alps). On the other hand, the total number and runout distance of snow avalanches is projected to decrease in regions and elevations with significant reduction in snow cover (Mock *et al.*, 2017). On a whole, it is likely that there will be more changes in avalanches in mountain regions in the future, with generally less avalanche hazard at lower elevation, and mixed changes at higher elevations.

10.6 Projected changes in land ice

By combining observations from multiple missions (ICESat, GRACE, and CryoSat2), sea-level budget analyses, analysis of Arctic glacier and ice cap trends combined with statistical modeling, Bamber *et al.* (2018) presented annual and pentad (5-year mean) time series for the East, West Antarctic and Greenland Ice Sheets (GISs) and glaciers separately and combined. By averaging over pentads, they show a monotonic trend in land ice contribution to the oceans, increasing from 0.31 ± 0.35 mm of sea-level equivalent for 1992–1996 to 1.85 ± 0.13 for 2012–2016, which is lower than many estimates of GRACE-derived ocean mass change for the same periods.

There are many projections of loss of glacier ice area and volume during this century. For the Swiss Alps, Haeberli and Hohmann (2008) project a 75% reduction in glacier area by 2050. For glaciers in the Blackfoot–Jackson Glacier Basin of Glacier National Park, Montana, Hall and Fagre (2003) estimate that with CO_2–induced global warming, all remaining glaciers will disappear by the year 2030, despite predicted increases in precipitation. For the Himalaya, Qin (2002) estimated that by AD 2050 with a 2 °C warming, 35% of glaciers would disappear and runoff will increase, peaking between 2030 and 2050. Rees and Collins (2006) project that for a linear temperature increase of 0.06 °C a^{-1}, stream flow in the western (eastern) Himalaya will peak around AD 2050 (2070) with 150 (170)% of the initial flows. In the Eastern Alps, Weber *et al.* (2011) project that for the IPCC A1B scenario, by AD 2051–2060, the ice storage in the Inn Basin will be reduced by two-thirds of its current value. The contribution to river discharge due to ice melt at Vent in the Ötztal will shrink from its current 33% to only 1%, with snowmelt contributing 26% and rainfall 73%. During summer dry spells, former glacial streams will almost dry up completely.

For Vatnajökull, Hofsjökull, and Langjökull ice caps in Iceland, Björnsson *et al.* (2006) project their volume changes by coupling a mass balance model to a 3-D ice flow model. They show that by AD 2100, the first two ice caps will have decreased by about 70%, while Langjökull will have almost disappeared.

For the Andes, glacier behavior in the inner and outer tropics appears to be quite coherent throughout the region, despite different sensitivities to climatic forcing such as temperature, precipitation, and humidity (Vuille *et al.*, 2008). Climate model projections indicate a continued warming of the tropical troposphere throughout the twenty-first century, with a temperature increase that is enhanced at higher elevations. Based on different IPCC scenarios for AD 2050 and 2080, simulations with a tropical glacier–climate model indicate that glaciers will continue to retreat. Many smaller, low-elevation glaciers will disappear within a few decades. For the Cordillera Blanca, Peru, Juen *et al.* (2007) project reductions in glacier area of 49% (B1 scenario) to 75% (A2 scenario) by AD 2080, with major implications for water resources.

Schneeburger *et al.* (2003) apply a GCM with CO_2 doubling, a glacier mass-balance model (a temperature index melt model), and a glacier flow model to 11 small glaciers and the first two models only to six glaciers and six large, heavily glacierized areas in the Arctic. For the modeled glaciers, an average volume loss of 60% is predicted by AD 2050. Grant (2010) shows that by the end of this century summer temperatures in Novaya Zemlya will increase by between 3 °C and 5 °C (for the B2 and A2 SRES scenarios, respectively), resulting in a glacier retreat of between 26 and 44% of their original length.

For the GIS, Mernild *et al.* (2010) use Snow Model to simulate variations in the melt extent, surface water balance components, changes in surface mass balance, and freshwater influx to the ocean. The simulations are based on the A1B scenario for 1950–2080 with the HIRHAM4 regional climate model. There was a <90% increase in surface melt extent by AD 2080 with maximum changes in the southern part of the GIS; the greatest changes in the number of melt days occurred in the east (<50–70%). They report an average surface mass balance loss of 331 km^3 from 1950 to 2080. Surface freshwater runoff yielded a eustatic rise in sea level from 0.8 mm a^{-1} for 1950–1959 to 1.9 mm a^{-1} in 2070–2080. The accumulated freshwater runoff from surface melting would contribute 160-mm SLE from 1950 through 2080.

The GIS is projected to contribute 0.07 m (likely ranging 0.04–0.12 m) under RCP2.6, and 0.15 m (likely ranging 0.08–0.27 m) under RCP8.5 to SLR by 2100, while the AIS is projected to contribute 0.04 m (likely ranging 0.01–0.11 m) under RCP2.6, and 0.12 m (likely ranging 0.03–0.28 m) under RCP8.5 by 2100. Currently GIS is contributing more to SLR than the AIS, but that could change toward the end of the twenty-first century, for the divergence between GIS and AIS's relative contributions to SLR rise could increase beyond 2100 (IPCC, 2019). According to the SPM of AR5 (IPCC, 2013), the global mean SLR for 2081–2100 relative to 1986–2005 will likely be between 0.26 and 0.55 m under RCP2.6, and between 0.45 and 0.82 m under RCP8.5 scenarios, and between 0.52 and 0.98 m by 2100 under RCP8.5. Further, it is virtually certain that the global mean SLR will continue beyond 2100.

In the latest assessment for 122,800 glaciers and ice caps in the updated World Glacier Inventory, upscaled to 19 mountain regions, Radic and Hock (2011) determine a SLR by 2100 of 0.124 m, with the largest contribution from glaciers in Arctic Canada, Alaska, and Antarctica. Total glacier volume will decrease by 21.6%, but for the Alps the loss is projected to reach 75%.

10.7 Projected permafrost changes

Permafrost thaw can cause ground subsidence or landslides, destabilizes human infrastructure and Arctic coasts, and contribute to global climate warming through emissions of GHGs from microbial breakdown of previously frozen organic carbon in frozen ground. Anisimov (1989) and Anisimov and Nelson (1993) examined the effect of climate warming on permafrost in the Russian Arctic and for Russia, respectively, using paleo-reconstructions as analogues of future warm states. These were for the Holocene thermal maximum, the Eemian interglacial (125–122 ka), and the Pliocene thermal maximum (~4 Ma). The implied reductions in permafrost area were significantly greater than those obtained later using GCMs. For example, Anisimov and Nelson (1996) ran three GCMs (GFDL, GISS, and UKMO) with 2 °C warming to assess the effect on permafrost boundaries. For all permafrost zones, the reduction implied ranged from 25 to 28%, compared with 44% for the Holocene paleo-reconstruction. It is unclear what the reasons are for the large differences. Subsequently, Anisimov and Nelson (1997) ran model simulations with transient GHG concentrations for AD 2050. The models used were the GFDL89, ECHAM-1A, and the UKTR. The reductions for all permafrost zones ranged from 12 to 22%.

In another study, Anisimov et al. (1997) used the same transient models to simulate changes in active layer thickness (ALT) for AD 2050. Three soil types were considered – sand, silt, and peat, each with high (low) water contents of 0.35 (0.15), 0.30 (0.10), and 6.0 (3.5) kg kg^{-1}, respectively. The models differ substantially in the regional distribution of changes in ALT. The GFDL89 model projects rather small changes in ALT compared with the other models. Changes >30% are confined to the tundra regions of the Canadian Arctic Archipelago, the Russian Far East, and the Yamal/Gydan region of West Siberia. In the ECHAM-1A model, the largest relative increases in ALT are in the Russian Far East and western Canadian Arctic. The pattern of changes with both these models is similar for all soil types. The UKTR model predicts the most substantial warming and ALT changes are >30% throughout the continuous and extensive discontinuous permafrost zones and over the Tibetan Plateau.

Lawrence and Slater (2005) incorporate permafrost directly within a GCM land surface scheme using the NCAR CCSM. They show that by AD 2100, as little as 1.0 million km^2 of near-surface permafrost remains. Burn and Nelson (2006) critique this analysis pointing out that there is a 1.3–4.1 million km^2 area underestimate of the current permafrost extent by the CCSM and noting that the presence of excess ice, which would slow permafrost thawing, is not modeled. Moreover, the permafrost temperature in Alaska is overestimated by >5 °C in the CCSM. Lawrence and Slater (2006) concur that the extent to which permafrost degrades in response to strong high-latitude warming over the next 100 years remains highly uncertain. Lawrence *et al.* (2008) using the NCAR CCSM-3 perform explicit accounting of the thermal and hydrologic properties of soil organic matter and deepen the soil column to 50 m. The rate of near-surface permafrost degradation, in response to strong warming of <+7.5 °C over Arctic land areas from 1900 to 2100, is slower in the improved version of the model, particularly during the early twenty-first century, but is still sufficient to reduce the extent of near-surface permafrost by AD 2100 to only 15% of its extent in 1970–1989. Further CCSM runs for the SRES A1B scenario show 10–40% increases in snowfall and a snow season shorter by ~34 days in this century (Lawrence and Slater, 2010). The changes account for 10–30% of total soil warming at 1 m depth and ~16% of the simulated twenty-first century decline in near-surface permafrost extent. Zhang *et al.* (2008a, b) examine the transient response of permafrost in Canada for AD 1850–2090 for six climate scenarios in six GCMs. They found that although permafrost thaw from the top would be significant and respond quickly to climate warming, deep permafrost would persist for a long time. The predicted reduction in permafrost extent by AD 2100 was in the range 20–24%, despite air temperature changes of 2.8–7.0 °C, due to the slow response in the ground. The results also showed an exponential increase in the area with supra-permafrost taliks during this century, because of deeper summer thawing.

For the Swiss Alps, Haeberli and Hohmann (2008) project for AD 2050 a +2 °C/+10% change in winter temperature and precipitation and a +3 °C/−10% change in summer. These changes are shown to lead to deep warming of alpine permafrost (and a 75% decrease in glacier area, as noted earlier). However, the changes to ice-rich permafrost depend heavily on changes in snow cover, which are difficult to predict.

Major warming at ~10–20 m depth in permafrost have been recorded in many long-term monitoring sites in the Northern Hemisphere circumpolar permafrost region (AMAP, 2017), as much as 2–3 °C higher than 30 years ago. Between 2007 and 2016, the rising trend in permafrost temperatures was 0.39 °C ± 0.15 °C for colder continuous zone permafrost monitoring sites, 0.20 °C ± 0.10 °C for warmer discontinuous zone permafrost, giving a global average of 0.29 °C ± 0.12 °C across all polar and mountain permafrost (Biskaborn *et al.*, 2019). The increase in permafrost temperature in warmer sites is less due to the heat absorbed by the ice-to-water phase change, which may also a result in increased thickness in the active layer. However, there is only medium confidence that ALT across NH has increased because decadal trends vary across regions and sites (Shiklomanov *et al.*, 2012) and because mechanical probing of the active layer could underestimate the degradation of permafrost because the surface subsides when ground ice melts and drains (Streletskiy *et al.*, 2017).

According to projections of AR5 (IPCC, 2013), widespread permafrost thaw is projected for the twenty-first century that by 2100, near-surface (within 3–4 m) permafrost area could decrease by 24 ± 16% under RCP2.6, and 69 ± 20% under RCP8.5. Under RCP8.5, there could be a total release of tens to hundreds of Gt of permafrost carbon as CO_2 and methane by 2100, but under RCP2.6 scenarios, carbon emissions from the permafrost region will be considerably dampened.

In the Arctic, the extend and rapidity of recently observed landscape changes in coastal erosion, expansion of channel network, and degradation of frozen ground previously stable for thousands of years suggest that Arctic landscapes may be particularly sensitive to perturbations caused by global warming (Jorgenson et al., 2006). Regional warming and thawing of permafrost in Alaska and the Arctic could result in systemwide, rapid geomorphic responses (Hinzman et al., 2005), and release of large amount of carbon currently stored in permafrost due to permafrost thawing may exert a positive feedback on global warming (Schuur et al., 2008).

10.8   Projected changes in freshwater ice

There is evidence (but limited) that freshwater ice duration in lakes has been decreasing and the average seasonal ice cover shrinking. For 75 lakes located mostly in Scandinavia and northern United States, the most rapid changes were observed in the most recent period, with freeze-up occurring 1.6 days decade^{-1} later and breakup 1.9 days decade^{-1} earlier. In the North American Great Lakes, the average duration of ice cover declined 71% over 1973–2010 (IPCC, 2014). Even though river ice is known to have destructive impacts on hydraulic structures, navigation, and fish habitat, they can be beneficial such as replenishing ecosystems through ice-jam flooding (Weyhenmeyer et al., 2011).

As records of river ice freeze-up and breakup differ widely in length, and documented by different sources, it is challenging to conduct regional scale spatial and temporal analyses of river ice freeze-up and breakup. Possible drivers behind the shifts of freshwater ice pattern include air temperature, snowfall (Fu and Yao, 2015), timing of 0 °C isotherms (Šmejkalová et al., 2016), precipitation (Newton et al., 2017), runoff (Obyazov and Smakhtin, 2014), solar activity (Fu and Yao, 2015), and atmospheric CO_2 (Sharma et al., 2016). Among these, air temperature is probably the most important factor influencing freshwater ice (Bieniek et al., 2011; Newton et al., 2017), for ice breakup dates are often correlated with mean monthly air temperature for March and April (Fu and Yao, 2015). From analyzing lake ice breakup dates and climate data for 152 lakes across NH from 1951 to 2014, Lopez et al. (2019) detected 97% of study lakes exhibited earlier ice breakup trends and shorter ice seasons, with about one-half of ice breakup trends driven by spring air temperatures across NH. They also found abrupt changes in mean ice breakup for 53% of lakes with shift years identified between 1970 and 2002. Earlier ice breakup and shorter ice season will affect local economies, and lake ecosystems around the world.

Prowse et al. (2007) use a 3–7 °C rise in spring temperatures by AD 2100 to project a 15–35-day advance of river ice breakup over northern Canada. They cite unpublished work by B. R. Bonsal and others that shows a 20-day decrease, relative to 1961–1990, in the duration of river ice over most of Canada by AD 2050. Another effect of warming in winter will be the increase frequency of mid-winter thaws and a northward shift in their occurrence. Beltaos *et al.* (2006) project that the ice season in the Peace-Athabasca delta will be 2–4 weeks shorter by AD 2100. The increased occurrence of mid-winter thaws on the Peace River will translate into fewer dynamic river-ice breakup events. In high latitudes, projected increases in the frequency of mid-winter thaw events is expected to reduce the frequency of major ice jams during the spring breakup via diminished snowpacks (Beltaos *et al.*, 2006).

For Great Slave Lake, using the 1-D thermodynamic Canadian Lake Ice Model, Ménard *et al.* (2002) show that lake ice duration changes by 6 days for a 1 °C change in temperature. A 4 °C increase in MAT from the 1961 to 1990 average, projected for the CO_2 increase at this latitude, shortens the ice season by about 3 weeks. The model simulates only a 1-day change associated with a 25% increase in snow depth. However, changes in MAT have only a minor effect on ice thickness, whereas a 25% decrease in snow depth increases the maximum ice thickness by 17 cm.

For the largest lakes and reservoirs in Russia, Vuglinsky *et al.* (2002) use the historical records to determine the impacts of a 2 °C warming on ice cover phenology and thickness. They find that the central part of Lake Ladoga (61° N, 30.5° E), which for 1897–2000 had 121 days of ice cover with a mean maximum thickness of 61 cm, would have no stable ice cover. The shift to later ice-on dates ranges from 11 days for the southwestern part of Lake Baikal (54° N, 109° E), to 25 days for Lake Taimyr (74.5° N. 102.5° E), to 35 days for the western part of Lake Onega (61.5°, 35.7° E). The corresponding advance of breakup for these three lakes was 4, 15, and 5 days, respectively. Maximum ice thicknesses decreased by 40 cm (on 191 cm) for Lake Taimyr and 23 cm (on 66 cm) for Lake Onega. Among the smallest changes projected were for Lake Khanka (45.7° N, 132.3° E) in the Far East. There the ice season would be reduced by only 8 days and the maximum average thickness by 5 cm (on 95 cm).

In Canada, Chen and She (2019) found river ice breakup dates more consistent among different data sources than freeze-up dates. Using the Water Survey of Canada HYDAT database, they constructed a long record of breakup dates from 1950 to 2016 and show an overall trend of earlier breakup across Canada, particularly in the Pacific and Western Mountains, Central Plains, and the Arctic where pronounced warming trends were also detected. In Atlantic and Great Lakes–St. Lawrence, where warming was not as pronounced, mixed but significant trends in earlier breakup time were also detected. They also found the breakup timing to be more sensitive to spring warming in lower latitude regions of Canada. Earlier river ice breakup, later freeze-up and shorter river ice season were also found in Southern British Columbia rivers (Doyle and Ball, 2008), the Mackenzie River (Goulding *et al.*, 2009), and northwestern Canada (Janowicz, 2010).

River ice breakup can trigger severe spring ice jams which in turn trigger flooding events. In Yukon, annually the earliest flooding events typically occur in late April or May and are triggered by ice jams, where the spring breakup and associated flooding is usually followed by the snowmelt freshet several weeks later, but due to climate warming, there is an increasing frequency of overlapping river ice breakup events and freshet peaks in Yukon. In 2011 and 2015, the combined breakup and freshet peaks occurred on May 23 and 15, respectively. Furthermore, freeze-up of the Yukon River at Whitehorse has been delayed by approximately 30 days since 1902 (Janowicz, 2010), while recent breakup dates have advanced. Prior to 1989, only two April breakups had been observed, but after 1989, eight April breakups have been observed. Both the ice cover and breakup periods have become shorter (Janowicz and Hinzman, 2017).

For northern rivers, an almost universal trend toward earlier breakup dates during the last few decades of the twentieth century was noted but with considerable spatial variability in those for freeze-up. In the twentieth century, warming in spring and autumn has led to about 10–15 days earlier breakup and later freeze-up in many areas (IPCC, 2014). Besides earlier ice-off dates for lakes and rivers across NH, Schmidt et al. (2019) also found that the short-term behavior of ice-off time series to be affected by climate anomalies, particularly the NAO, Pacific-North American Pattern (PNA), and the El Niño-Southern Oscillation (ENSO). Further, the spatial pattern of ice-off dates associated with NAO, PNA, and ENSO matches the surface temperature patterns associated with NAO or PNA of high-frequency modes. Schmidt et al. (2019) conjecture that short-term weather variations play a stronger role than lower-frequency, ENSO, PDO, AMO in driving ice-off.

10.9   Projected sea ice changes

Liu and Curry (2010) use the NCAR CCSM-3 and the GFDL CM2.1 coupled models to analyze three scenarios for changes in Antarctic sea ice during this century. The losses amount to 0.4×10^5 km^2 decade^{-1} for scenario B1, 2×10^5 km^2 decade^{-1} for A1B, and 3.02×10^5 km^2 decade^{-1} for A2, with the greatest losses occurring in the Atlantic and Indian Ocean sectors. In addition, the rate of decline accelerates after the late 2060s for the A1B and A2 scenarios. The declines are attributed to increased heating from the ocean and atmosphere and increased liquid precipitation associated with an enhanced hydrological cycle.

Holland et al. (2010) assess ASI mass budgets for the twentieth century and project changes through the twenty-first century using 14 coupled global climate models. They show that large inter-model scatter in contemporary mass budgets is strongly related to variations in absorbed solar radiation, due mainly to differences in the surface albedo simulations. All models simulate a twenty-first century decrease in ice volume resulting from increased annual net melt, but the models vary considerably in the magnitude of ice volume loss and the relative roles of changing melt and growth in driving it. Models with thicker initial ice in the mid-twentieth century generally exhibit larger volume losses. Change in net ice melt is significantly related to changes in downwelling longwave and absorbed shortwave radiation. Eight of the models show the Arctic as being ice free in September by AD 2100 and some as early as 2050 for the A1B emissions scenario. From an analysis of individual runs of the CMIP3 models, selected where the ice area sensitivity to surface air temperature was 0.42° to 0.70° × 10^6 km^2 K^{-1}, Zhang (2010) finds that an ice-free summer Arctic Ocean may occur between AD 2037 and 2065, using a criterion of 80% sea ice area loss. He projects that the Arctic regional mean surface air temperature will likely increase by 8.5 ± 2.5 °C in winter and 3.7 ± 0.9 °C in summer by the end of this century.

Arzel et al. (2006) and Goosse et al. (2009a) show from model simulations that the variability of September sea ice extent in the Arctic of the twenty-first century first increases when the mean extent decreases from present-day values. A maximum of the variance is found when the mean September ice extent is around 3 million km^2. For lower extents, the variance declines with the mean extent. In contrast, around Antarctica the variance always decreases as the mean ice extent decreases, following roughly a square-root relationship. It appears that the land-locked Arctic Ocean limits the southward extension of the sea ice in winter, and thus reduce the amplitude of changes, while the summer ice edge is free to evolve in response to the various forcings.

Simulations for the Canadian Arctic Archipelago by Sou and Flato (2009) show little projected change in winter ice by AD 2041–2060 but a 45% decrease in summer ice. Ice thickness decreases by 17% in winter and by 36% in summer. They state that a completely ice-free Archipelago in summer is not likely by AD 2050. Projections for AD 2100 with CMIP3 models and the A1B scenario of the season with an open Northwest Passage increase from 2 to 4 months and from three to 6 months for the Northern Sea Route, according to Khon et al. (2010) .

Boé et al. (2009) analyze the simulated trends in past sea-ice cover in 18 state-of-art-climate models and find a direct relationship between the simulated evolution of September sea-ice cover over the twenty-first century and the magnitude of past trends in sea-ice cover. Under a scenario with medium future GHG emissions, they find that the Arctic Ocean will probably be ice-free in September before the end of the twenty-first century. Wang and Overland (2009; Kerr 2009) report that analysis of the best six sea ice models (with sophisticated sea ice physics packages) projects AD 2037 as the most likely date for the development of an open Arctic in summer. The median duration interval for the sea ice to be reduced from 4.6 to 1.0 million km^2 is 30 years (1 million km^2 allows for ice to persist north of Greenland). The first quartile of the distribution for the timing of September sea ice loss will be reached by AD 2028.

Due to global warming impact, the melting season of ASI has lengthened (Stroeve et al., 2017) and the ice cover has become younger and thinner (Lindsay and Schweiger, 2015; Kwok, 2018). Over 1979–2012, the annual ASI extent has continually decreased at 3.5–4.1% (0.45–0.51 million km^2) decade^{-1}, the perennial sea ice extent (summer minimum) decreased at 11.5 ± 2.1% (0.73–1.07 million km^2) decade^{-1} (IPCC, 2014) (Figure 10.2d), and the multiyear ice (survived ≥ two summers) at 13.5 ± 2.5% (0.66–0.98 million km^2) decade^{-1}. The ice–albedo feedback is a key driver in the evolution of summer ASI minimum, which could be partly driven by warm and moist air advection, increased downwelling longwave radiation due to heightened cloudiness and humidity (Kapsch et al., 2013). The annual surface melt period on Arctic perennial sea ice lengthened by 5.7 ± 0.9 days decade^{-1}, and between the East Siberian Sea and the western Beaufort Sea, the duration of ice-free conditions increased by almost 3 months. In contrast, the annual Antarctic sea ice extent had likely increased between 1.2 and 1.8% (0.13 to 0.20 million km^2) decade^{-1}. From 1980 to 2008, the average winter sea ice thickness within the Arctic Basin had likely decreased between 1.3 and 2.3 m. Over half of the observed ASI loss (Figure 10.5a) was due to rising concentrations of atmospheric GHGs, with the remainder due to natural climate variability (Stroeve *et al.*, 2012; Notz and Stroeve, 2016).

According to climate model simulations (Stroeve *et al.*, 2012) that include feedbacks leading to accelerated ice loss such as water-vapor feedbacks (Dessler *et al.*, 2013), lapse-rate feedbacks (Feldl *et al.*, 2017; 2020), and ice-albedo feedbacks (Kashiwase *et al.*, 2017), future reduction of ASI will be continuous and amplified (Pithan and Mauritsen, 2014), with ice-free summers occurring as early as the 2030s, and an ice-free year occurring as early as the 2050s (Onarheim *et al.*, 2018). The GCMs that most closely reproduce the observations project a nearly ice-free Arctic Ocean in September is likely for RCP8.5 before the mid-twenty-first century.

Compared with observations (Figure 10.6), climate models of the Coupled Model Intercomparison Project Phase 3 (CMIP3) (IPCC, 2007) have under-simulated the observed September Arctic ice extent (Figure 10.7). Stroeve *et al.* (2012) show that the newer CMIP5 climate models simulated Arctic ice extent trends that agree better with the observed 1979–2011 September trend for the Arctic. However, most of the ice extent trends simulated by CMIP5 models are still smaller than the observed downward trend.

It is very likely that the Antarctic sea ice extent increased between 1979 and 2017 at an annual-mean rate of $20.2 \pm 4.0 \times 10^3 \, km^2 \, yr^{-1}$ (Comiso *et al.*, 2017), but with strong negative departures in 2016 and 2017, and strong regional differences, with an overall increase in extent. The overall increase in Antarctic sea ice from rapid ice loss in the Amundsen and Bellingshausen seas outweighed by rapid ice gain in the Weddell and Ross seas (Figure 10.5b) is strongly regional and seasonal (Holland, 2014). The slow response of Antarctic sea ice cover to rising concentrations of GHGs compared to the Arctic could be due to anthropogenic warming delayed by the Southern Ocean circulation, which transports heat downwards into the deep ocean as shown by simulations of coupled climate models, along with differing cloud and lapse rate feedback (Armour *et al.*, 2016).

10.10 Projected glacier changes

Since mountain glaciers are sensitive indicators of global climate change, there have been many projections of the melting of glaciers across the world. Worldwide mountain glaciers have been retreating, with significant contributions to SLR, and a potential threat to people relying on water supply from glaciers. However, projections of glacier mass losses likely involve large uncertainties because of scales, limitations of surface mass balance glaciation models, effects of complex terrain characteristics, and local climate, and so forth.

Worldwide glaciers have been declining in length, area, volume, and mass and more rapidly since 1970s. Total mass loss from all glaciers globally, excluding those on the periphery of the two ice sheets, was estimated at $226 \pm 135 \, Gt \, yr^{-1}$ in 1971–2009, $275 \pm 135 \, Gt \, yr^{-1}$ in 1993–2009, and

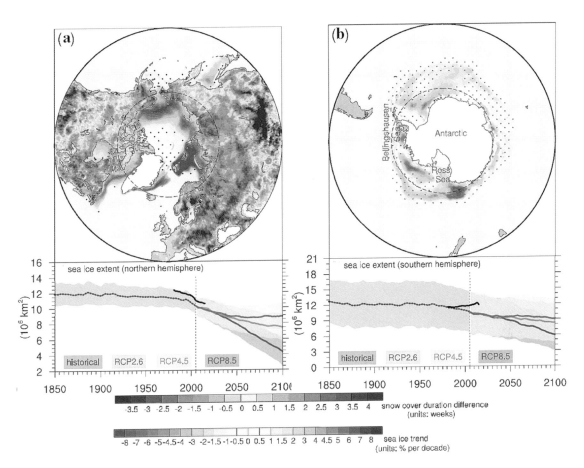

Figure 10.5 (a) Linear trends of Arctic annual-mean sea ice concentration for 1982−2016, in % per decade, alongside the difference in climatological snow cover duration (in weeks) between 2006–2015 and 1981–1990, and (b) linear trends of Antarctic annual-mean sea ice concentration for 1982−2016, in % per decade. The comparable 5-year running mean time series of annual-mean sea ice extent in the northern and southern hemispheres are shown below (a) and (b), where black, green, blue, orange, and red curves indicate observations, CMIP5 historical simulation, RCP2.6, RCP4.5, and RCP8.5 projections of 1850 to 2100, respectively; shading indicates ± standard deviation of multimodels. (a) and (b) are taken from NOAA/NSIDC Climate Data Record of Passive Microwave Sea Ice Concentration, Version 3 (https:// nsidc.org/data/g02202). Snow cover duration in (a) was derived from a blend of four independent datasets, each covering the 1981–2015 period (Brown et al., 2003; Takala et al., 2011; Brun et al., 2012; Reichle et al., 2017). The models from CMIP5 include bcc-csm1-1, bcc-csm1-1-m, CanESM2, CCSM4, CESM1-CAM5, CNRM-CM5, GISS-E2-H, IPSL-CM5A-LR, IPSL-CM5A-MR, MIROC5, MIROC-ESM, MIROC-ESM-CHEM, MPI-ESM-LR, MPI-ESM-MR, MRI-CGCM3, NorESM1-M, NorESM1-ME. (A black and white version of this figure will appear in some formats. For the color version, please refer to the plate section.)

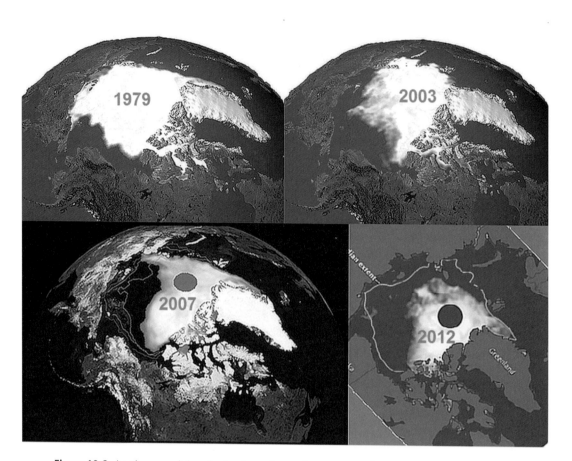

Figure 10.6 Arctic perennial sea ice has been decreasing at a rate of 2.7 (2.1 to 3.3)% per decade, with larger decreases in summer of 7.4 (5.0 to 9.8)% per decade. The first, second, third, and fourth images show the minimum sea ice concentration for 1979, 2003, 2007, and 2012, respectively. The third image of 2007 shows a drastic shrinkage of sea ice, while the fourth image shows that in the late summer of 2012, Arctic sea ice extent has dropped to its lowest level ever recorded, 4.10 million km^2. Images by the DMSP Special Sensor Microwave Imager SSMI (https://www.ncdc.noaa.gov/ssmi/data-access) (A black and white version of this figure will appear in some formats. For the color version, please refer to the plate section.)

301 $\pm$ 135 Gt yr^{-1} in 2005–2009 (Vaughan *et al.*, 2013). Between 2003 and 2009, more than 80% of the total ice lost was from glaciers in Alaska, Canadian Arctic, the periphery of the GIS, the Southern Andes, and the Asian Mountains.

Huss and Hock (2018) projected global glacier runoff and glacier-volume changes for 56 large-scale glacierized drainage basins selected from North and South America, Europe, and Asia. Based on three RCP emission scenarios of 14 GCMs, between 2010 and 2100, the total glacier volume in all the investigated basins is projected to decrease by 43 $\pm$ 14% (RCP2.6), 58 $\pm$ 13% (RCP4.5), and 74 $\pm$ 11% (RCP8.5). They found that by 2017, the maximum runoff has already occurred in 45% of the glacierized basins, for thereafter, the annual runoff is

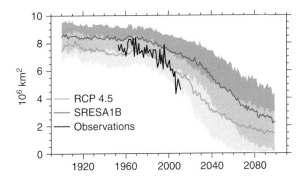

Figure 10.7 The observed September Arctic sea ice extent for 1952 to 2011 (bold black line) is compared against that for 1900–2100 simulated by CMIP3 models using the SRES A1B climate scenarios (the upper darker grey line represents the average results from all CMIP3 climate model runs with the surrounding shading showing the ±1 standard deviation of the different model runs) and from the CMIP5 archive using the RCP 4.5 scenarios (lower light grey line and surrounding shading). The shading between the two lines shows where the simulations from CMIP3 and CMIP5 overlap each other (http://nsidc.org/arctic seaicenews/2012/08/)

expected to decline. In the remaining basins, the modeled annual glacier runoff continues to rise until a maximum is reached, which tends to occur later in basins with larger glaciers and higher ice-cover fractions. By 2100, one-third of them might experience runoff decreases greater than 10% due to glacier mass loss in at least 1 month of the melt season, with the largest reductions in central Asia and the Andes. In an earlier report, using RCP2.6–8.5 climate scenarios, Huss (2012) projected the mean mass balance of Alpine glaciers in the European Alps to shrink to only 4–18% of their 2003 extent by the end of 2100. With moderate atmospheric warming, glacier mass balances are projected to decrease to -1.29 m w.e. a^{-1} by 2050 with a fast area loss. While mass balances of some Alpine glaciers may stabilize at drastically reduced glacier size, others could result in a run-away effect with mass balances lower than -2 m w.e. a^{-1}.

Radić *et al.* (2014) projected twenty-first century glacier mass changes in response to RCP4.5 and RCP8.5 climate scenarios of 14 global climate models of CMIP5. The multi-model mean projects a SLR of 155 ± 41 mm (RCP4.5) and 216 ± 44 mm (RCP8.5) over 2006–2100, reducing the current global glacier volume by 29 or 41%, with glaciers in the Canadian and Russian Arctic, Alaska, and glaciers peripheral to the Antarctic and GISs as the largest contributors to projected global glacier volume loss.

To study glaciers of western North America, Clark *et al.* (2015) developed a high-resolution regional glaciation model, coupling physics-based ice dynamics with a surface mass balance model, driven by four RCP climate scenarios of an ensemble of global climate models to project the melting of glaciers in western Canada. Their results indicate that by 2100, the volume of glacier ice in western Canada will shrink by $70 \pm 10\%$ relative to 2005. Few glaciers will remain in Interior and Rockies regions, while maritime glaciers in north-western British Columbia will remain but in a diminished state.

Over the past decade, ice loss from the GIS has increased because of both increased surface melting and the acceleration of ice flow and thinning of fast-flowing marine terminating outlet glaciers. Nick *et al.* (2013) applied a glacier flow model that includes a fully dynamic treatment of marine termini outlet glaciers to estimate the future dynamic contribution of four major marine-terminating outlet glaciers that collectively drain about 22% of the GIS. Using atmospheric and oceanic forcing from a mid-range (extreme) future warming scenario that predicts warming by 2.8 °C (4.5 °C) by 2100, they projected a contribution of 19–30 mm (29–49 mm) to SLR from these glaciers by 2200.

According to the Hindu Kush Himalaya Assessment report (2019), two-thirds of glaciers of Himalaya, the world's "Third Pole," where glaciers that feed 10 of the world's most important river systems (Ganges, Indus, Yellow, Mekong, and others) are critical water sources for about 250 million people in the mountains and to 1.65 billion others in the river valleys below, could melt by 2100 if global emissions are not sharply reduced. Even if the Paris climate agreement goal of limiting global warming to 1.5 °C is achieved, one-third of the glaciers would disappear by 2100.

Earlier studies also project major glacier melts in other parts of the world. For example, Haeberli and Hohmann (2008) project a 75% reduction in glacier area in the Swiss Alps by 2050. Based on the projected warming of the tropical troposphere with an enhanced temperature increase at higher elevations, and simulations by a tropical glacier–climate model, Vuille *et al.* (2008) project that tropical Andean glaciers will continue to retreat, with many smaller, low-elevation glaciers totally disappeared within a few decades.

From remotely sensed data on glaciers in the greater Himalaya collected between 2000 and 2008, Scherler *et al.* (2011) found variable retreat rates, and strong spatial variations in frontal changes and surface velocities in Himalayan glaciers under climate change impact. They highlight the importance of debris cover effect on the retreat of glaciers in the rugged central Himalaya where more than two-thirds of the monsoon-influenced glaciers observed are retreating, where debris-covered glaciers with stagnant low-gradient terminus regions typically have stable fronts. In contrast, more than one-half of glaciers investigated in the westerlies-influenced, Karakoram of the northwestern Himalaya are advancing or stable.

REVIEW QUESTIONS

(1) Name five key elements of the global cryosphere, for example, snow cover.
(2) What are obvious physical evidence that the cryosphere has been changing under the impact of global warming observed since the 1980s?
(3) Why is the global warming impact more pronounced in high latitudes, especially the Arctic, or why does global warming affect the cryosphere more than the hydrosphere?
(4) What is the common term indicating that climate change impact is amplified in the Arctic? Name three feedback mechanisms that have caused the Arctic to warm at about twice the global rate since 1980s.
(5) In terms of the radiative forcing in W m^{-2}, the CO_2-eq concentration in ppm, and the

projected temperature change in 2100, explain differences between several Representative Concentration Pathways climate projections, RCP2.6, RCP4.5, and RCP8.5 of CMIP5. Given the global CO_2 concentration is about 410 ppm in 2020, and CO_2 concentration is increasing at about 2 ppm yr^{-1}, in your opinion which of the above three RCPs is a more realistic projection for 2100?

(6) What is the difference in projected changes in daily minimum (TN) and maximum (TX) of near surface temperature (°C) and the total annual precipitation on wet days (%) based on the RCP 4.5 emission scenarios of (CMIP5) multimodel ensembles on global and the Arctic (60° N, 180° E–90° N, 180° W) over the twenty-first century relative to the 1981–2000 base period? Other than projected to be much warmer, why is the Arctic projected to be much wetter than the globe? Consider the Clausius–Clapeyron relationship.

(7) According to simulations of GCMs of CMIP5, how will the Northern Hemisphere snowpack projected to change with respect to winter snowfall, annual snowfall, maximum snow depth, and the duration of snow season, and latitudes? Even though GCMs generally simulated representative twentieth-century snowfall, what are the possible deficiencies of GCMs of CMIP5?

(8) What meteorological, human, and physical factors could trigger snow avalanches? With lower snow depths and higher air temperature because of climate warming, the numbers of snow avalanches and runout distance should generally decrease. However, under what snow conditions, altitude, and seasons could we expect the proportion of avalanches to increase?

(9) Name a few satellite missions that remotely observed the melting trends of land ice such as glaciers, Arctic sea ice, Greenland Ice Sheet (GIS), and Antarctic Ice sheet (AIS). According to the RCP2.6 and RCP8.5 scenarios of AR5 (IPCC, 2013), what likely will be the range of global mean sea-level rise (SLR) for 2081–2100 relative to 1986–2005?

(10) Why are Arctic and subarctic areas largely covered or affected by permafrost (frozen soil), which represent about 25% of emerged lands of the Northern Hemisphere, highly vulnerable to global warming? As permafrost thaws, what changes in hydrologic and thermal processes of the Arctic regions are expected in response to climate change, and how can such changes be predicted?

(11) What have been the observed changes in lake and river ice in response to climate warming impact, such as the timing river ice breakup, freeze-up, and river ice season, and how would such changes affect possible flooding caused by river ice jams during the breakup season? Besides climate warming, what large-scale climate anomalies have been observed to contribute to earlier ice-off dates for lakes and rivers across NH?

(12) Since 1979 when passive microwave mapping of sea ice became available, what has been the decreasing trends of annual and summer minimum Arctic sea ice extent, and the multiyear ice (survived ≥ two summers) in % and millions of km^2 decade^{-1} detected from passive microwave data (Figure 10.6)? In contrast, what was the annual mean

increasing trend of the Antarctic sea ice extent in $km^2\ yr^{-1}$ estimated between 1979 and 2017? What could have caused the slow response of Antarctic sea ice cover to rising concentrations of greenhouse gases compared to the Arctic?

(13) According to Vaughan *et al.* (2013), what was the total mass loss from all glaciers globally, excluding those on the periphery of the two ice sheets, in $Gt\ yr^{-1}$ in 1993–2009, and in 2005–2009, and predominantly from which regions? Based on RCP scenarios of multiple GCMs, what were the projected glacier-volume changes from North and South America, Europe and Asia between 2010 and 2100 under RCP2.6, RCP4.5, and RCP8.5? Will the glacier runoff continue to increase or will they decline after reaching the peak runoff due to glacier mass loss? Will the maximum glacier runoff tend to occur earlier or later in basins with larger glaciers and higher ice-cover fractions, and why?

Part IV

Applications

11 Applications of snow and ice research

In this chapter, we briefly review the principal applications of research on snow and ice phenomena and provide references to further readings. Each main component of the cryosphere is treated separately.

11.1 Snowfall

A heavy snowfall season means good skiing (Lind and Sanders, 2004, describe the physics of skiing), snowshoeing, tobogganing, and beautiful winter scenery. It also means a lot of snow clearance on our side walks and car parks, expensive snow plowing to clear our roads, terrible traffic conditions, driving hazards, and occasionally, flooding during spring snow-melt seasons. Snowfall is a primary factor in disrupting transportation on highways, in cities, and at airports. In January 2008, for example, heavy snows in southeastern China caused widespread disruption to road and rail traffic, 17 deaths, and the collapse of 3,635 houses in Anhui province under the weight of snow. On November 17–19, 2014, over 1.65 m of snow fell over some areas near Lake Erie and east of Buffalo, resulted in 13 fatalities, hundreds of major roof collapses and structural failures, over a thousand stranded motorists, and scattered food and gas shortages due to impassable roads. The amount of water that would yield 10 mm of rain can produce 5–10 cm or more of snow. Even 5 cm of snow is enough to create disruptions to traffic. This is particularly true in places where snowfall is uncommon, but heavy falls can also occur there, such as Atlanta, Seattle, London, Canberra, Christchurch, and Vancouver. On July 25, 2011, Christchurch was gripped by up to 30 cm of snowfall in parts of the City for it had no provision to clear snow from roads.

Rooney (1967) and de Freitas (1975) investigated the effect of snowfalls on traffic disruption in North America. Rooney examined seven "dry snow" cities across the United States and a further ten with wet snow. He used a one (most severe disruption) to five (minor disruption) point scale and found that snowfall and wind speed were important controls. De Freitas used a one (most severe) to nine (minor) scale of disruption and showed that 24-hour snowfall, wind speed, and air temperature during the snowfall are the most important variables. Quebec City had the largest number of disruptions, followed by London (Ontario), Montreal, Regina, and Toronto. In places where snowfall is common, such as Chicago, Detroit, Montreal, Quebec City, Toronto, and Minneapolis, disruption by small snowfalls is rare, although it occurs with snowfalls >15 cm. Traffic accidents in Montreal are shown to be most closely related to snowfalls (Andreescu and Frost, 1998).

For the northeastern United States, Kocin and Uccelline (2004) developed a Northeast Snowfall Impact Scale (NESIS). This represents a measure of the integrated impact of a snowfall within and outside the Northeast, calibrated by 30 major storms that occurred from West Virginia northeast to Maine during 1950–2000, and concurrent snowfalls east of the Rocky Mountains. It scales snowfalls of 10, 25, 50, and 75 cm with the areas affected and their populations. The largest NESIS score was 12.5 for a snowstorm on March 12–14, 1993. The 25 cm snow area covered 550 million km and affected 60 million people. The Presidents' Day storm of February 15–18, 2003 and the January 22–24, 2016 has a NESIS score of 8.9 and 7.6, respectively. NESIS scores are categorized as (1) notable (1–2.499) to (4) crippling (6–9.999) and (5) extreme (>10). Out of 98 storms during 1888–2018, 30 were category 1, 27 category 2 (significant), 27 category 3 (major), 12 category 4, and only 2 category 5.

A massive snowstorm with strong winds and other conditions is known as a blizzard. A large number of heavy snowstorms, some of which were blizzards, occurred in the United States during the early and mid-1990s, and the March 12–15, 1993 "Storm of the century" was manifest as a blizzard in most of the affected eastern United States. Every airport from Halifax, Nova Scotia to Atlanta, Georgia was closed for some time because of the storm and 300 deaths were attributed to it. Daily snowfall maps for the United States and southern Canada are available at:

www.intellicast.com/Travel/Weather/Snow/Cover.aspx
and for the United States at www.nohrsc.nws.gov/nsa/

Large snowstorms can be quite dangerous: a 15 cm snowstorm will make some unplowed roads impassable, and it is possible for vehicles to get stuck in the snow. Snowstorms exceeding 30 cm, especially in generally warm climates, will cave in the roofs of buildings and cause the loss of power. Precautionary measures include salting the highways (~25–40 g m^{-2}) and plowing when snow depths reach 3–4 cm. In France, bad weather plans for snowfall include road clearance, blocking of vehicles right on the road, compulsory exit or route, motorway access control, compulsory diversion for bypassing built-up areas or specific points, movement of heavy goods vehicles in convoy, parking of heavy goods vehicles on motorway lanes or in motorway service centers, and rescue and assistance services for road users. On mountain passes in the United States, snow tires or tire chains may be obligatory.

Fences to prevent snow from drifting across highways have been erected in many US states, especially Wyoming. There are three main types of snow fence according to Pugh and Price (1954). A collecting fence is used upwind and adjacent to the highway in order to reduce the wind speed, thereby collecting and depositing the snow before it drifts onto the road. A solid guide fence is aligned at an angle to the prevailing wind direction in order to deflect the snow laterally. A blower fence is aligned at an angle to the wind vertically in order to accelerate the local flow and transport the snow elsewhere. A fence with a porosity of 50% and a bottom gap of about 30 cm has maximum efficiency. It forms a lee drift with a length 30 times the fence height and a windward length of 12 times the fence height (Perry and Symons, 1991). Tabler (1975) developed a regression equation for the snow slope over the main part of a drift using predictors of the ground slopes for upwind and downwind distances from the lip of a topographic trap.

11.2 Freezing precipitation

Ice storms with freezing rain can cause severe problems by depositing a glaze layer –"black ice" – that disrupts traffic and brings down power lines. The US National Weather Service defines an ice storm as one that results in the accumulation of at least 0.6 cm of ice on exposed surfaces. From 1982 to 1994, there was an average of 16 ice storms per year in the United States. Highway operations are affected by loss of traction, loss of stability/maneuverability, lane obstruction, impaired mobility, and loss of visibility. Also, road damage, loss of life, property damage, loss of communications/power, and operational delays occur. Prediction of the threat is needed 24–48 hours ahead of time. A notorious example was the January 5–9, 1998 ice storm that shut down much of southern Quebec, Maine, New Hampshire, and parts of New York, and Vermont. A wide area received 50–100 mm of freezing precipitation in three successive waves. It is estimated that 32,000 km of transmission lines and 96,000 km of distribution lines were brought down by the storm in Canada. Over 4 million households in Canada and in the United States lost power for weeks. Losses totaled $6 billion in Canada and $2 billion in the United States. In addition, a fifth of Canada's maple syrup-producing trees suffered severe damage (www.islandnet.com/~see/weather/almanac/arc2008/alm08jan.htm).

In late January 2009, ice storms affected the southeastern United States, particularly Arkansas and Kentucky. Most areas affected saw over 5 cm of ice accumulation. The storm caused 55 deaths and left more than 2 million people without power. On December 20–23, 2013, a crippling ice storm that hit central and eastern Canada, Central Great Plains, southern and northeastern United States resulted in 27 deaths, loss of power to over a million residents and over $200 million in damages. This storm was similar to the ice storm of 1998 and affected similar areas.

11.3 Avalanches

Avalanches are a major hazard in mountainous terrain. Some avalanches are released naturally but majority of avalanches are triggered by human activity, such as snowboarding, skiing, snowmobiling, hiking, and mountain climbing. Avalanches cause about 250 fatalities worldwide per year according to Meister (2002). They currently claim about 40 lives annually in North America, compared with only several fatalities in the 1950s. In the United States, about half of the victims are snowmobilers and the next largest category is backcountry skiers. Snow avalanches are a serious mountain hazard in Canada. Between 1980 and 2016, 428 people died in Canada in 301 avalanche accidents, with most fatalities occurred in western Canada (British Columbia: 72%; Alberta: 20%). Among virtually all avalanche fatalities in the last decade happened in winter backcountry, 48% were mountain snow-mobile riders, 31% backcountry skiers, 5% out-of-bounds skiers, and 16% in other winter backcountry activities. There were about 1,020 fatalities in the European Alps during the ten winters 1996–1997 to 2005–2006 (Schweizer, 2008) and in 2006 there were at least 86 fatalities, over half of them in the French Alps, due to unusual late-season snowstorms coupled with more backcountry skiing and snowboarding. From analyzing avalanche fatalities from the European Alps between 1937 and 2015, Techel et al. (2016) found consistently about 100 people lost their lives each year in the Alps in last four decades. However, the number of fatalities in controlled terrain (settlements and transportation corridors) has decreased significantly since the 1970s but fatalities in uncontrolled terrain (recreational accidents) almost doubled between 1960s and 1980s and remained relatively stable since then, despite a strong increase in the number of backcountry recreationists. It seems the proportion of fatalities in uncontrolled terrain had increased from 72% to 97%.

Avalanche specialists divide avalanches into three main zones: the starting zone with slopes of 30–50°, the track with slopes of 20–30°, and the run-out zone with slopes of less than 20° (Armstrong and Williams, 1986). An avalanche may be initiated when more dense snow is deposited on top of a less dense layer of snow, resulting in the formation of a slab that is not well bonded to the weaker snow layers below. Large avalanches can have a path length of up to 3 km and a volume of 10^5–10^6 m^3 of snow and are big enough to destroy a village or a forest (McClung, 2008). For example, in 1803 a snow avalanche destroyed the village of Àrreu of the Valls d'Àneu mountain territory in the Catalan Pyrenees, knocking down houses and killing its people (Oller et al., 2020). The destructive potential of an avalanche depends on its size, which is typically classified into either four or five size categories. On average, the ratio of slab length to thickness for dry snow avalanches is ~100, as is the ratio of slab width to thickness; the median ratio of width to length is ~1.2 for confined and unconfined avalanches (McClung, 2009).

Several countries have avalanche warning services (e.g., Switzerland, Canada) and about 20 countries are members of the International Commission for Alpine Rescue located in Switzerland. In the United States, the US Department of Agriculture Forestry Service ended its snow and avalanche research in the 1980s under budgetary pressure, but it continues to house and fund a number of backcountry avalanche forecast centers across the country. The Forest Service partners with many different private and public funding sources to run the avalanche centers, which typically provide daily avalanche advisories and a variety of avalanche education for the public. There are currently 15 National Forecast avalanche centers in the United States, the state-run Colorado Avalanche Information Center, six local avalanche centers (https://avalanche.org/us-avalanche-centers/), and a Canadian Avalanche Center. For example, the Northwest Weather and Avalanche Center provides warnings and data on avalanches in the northwestern United States (www.nwac.us). The Forest Service National Avalanche Center in Ketchum, ID and Bozeman, MT, manages the military artillery for avalanche control, works on various avalanche projects affecting national forests, provides program guidance for the avalanche centers, and serves as a national point of contact for all issues relating to avalanches for the Forest Service, as well as transferring new technology to the regional avalanche centers and the avalanche community as a whole. Snow avalanche hazards and mitigation in the United States are reviewed in a US National Academy (1990) report. United States and Canada have worked to come to consensus on a common avalanche danger scale for North America. This scale was implemented for the winter of 2010/2011, as shown in Table 11.1. A similar avalanche risk table of five levels was adopted in Europe in April 1993 and updated in May 2003.

Risk assessment is increasingly being used in avalanche studies (Keylock, 1997). Risk is defined as the product of three terms: encounter probability, exposure, and vulnerability. Encounter probability is the chance that in a given time interval an avalanche reaches a particular point on the ground. Exposure is the probability that people, vehicles, or buildings

Table 11.1 Danger scale for public avalanche warnings in North America (Statham *et al.*, 2010)

Level	Conditions	Avalanche likelihood	Avalanche size and distribution
1 Low	Generally safe avalanche conditions	Natural/human-triggered avalanches unlikely.	Small avalanches in isolated areas/extreme terrain
2 Moderate	Heightened avalanche conditions on specific terrain	Natural avalanches unlikely. Human-triggered avalanches *possible.*	Small avalanches in specific areas/large avalanches in isolated areas
3 Considerable	Dangerous avalanche conditions	Natural avalanches possible. Human-triggered avalanches *likely.*	Widespread small avalanches or large avalanches in specific areas or very large avalanches in isolated areas
4 High	Very dangerous avalanche conditions	Natural avalanches likely; human-triggered avalanches *very likely.*	Large avalanches in many areas/very large avalanches in specific areas
5 Extreme	Avoid all avalanche terrain	Natural and human-triggered avalanches *certain*	Large to very large avalanches in many areas

are at the specified point when an avalanche occurs. Vulnerability is the degree of damage (loss of life, damage to property) that the avalanche causes at a specific location. Each term has a value between zero and unity.

Vulnerability depends on the impact pressure of the avalanche and the strength of the object that is exposed. Encounter probability depends on the interaction of the avalanche and the terrain. Exposure is independent of the avalanche. Avalanche risk levels are usually classified into 5 possible levels with a risk level 2 or higher potentially involving fatalities.

Statham *et al.* (2017) formulated an avalanche hazard conceptual model for qualitative avalanche hazard assessments in a risk-based, avalanche problem framework defined in terms of type, location, likelihood, and size: (1) what type of avalanche problem(s) exists? (2) where are these problems located? (3) how likely will an avalanche occur? and (4) how big will be the avalanche? Combining likelihood of avalanche(s) with destructive avalanche size gives an estimate of avalanche hazard, a qualitative counterpart to the frequency–magnitude matrices of avalanche hazard (CAA, 2016). The conceptual model illustrates key components of avalanche hazard and structures them into a systematic workflow for hazard and risk assessment.

Avalanche control methods include the use of artillery shells fired in to the starting zones, or the discharge by remote control of canisters set into the starting zones that dislodge an unstable snowpack (McClung and Schaerer, 2006).

11.4   Ice avalanches

Since 1700, more than 22 catastrophic events have resulted from ice avalanches that have caused outburst floods. The floods, known in Perú as aluviónes, come with little or no warning and are composed of liquid mud that generally transports large boulders and blocks of ice. The floods have destroyed a number of towns, and many lives have been lost. One of the hardest hit areas has been the Río Santa valley in northern Peru. Of these catastrophes, the most serious were the aluviónes that destroyed part of the city of Huaraz in 1725 and 1941. In addition, two destructive, high-speed avalanches from the summit area of Huascarán Norte (6,655 m asl) in 1962 and 1970 destroyed several villages, killing more than 40,000 inhabitants. Reports of these events include those by Morales Arnao (1966), Lliboutry (1975), Plafker and Ericksen (1978), and Hofmann *et al.* (1983).

Hugel *et al.* (2008) report on rock and ice avalanches triggered by seismic activity on volcanoes in Alaska. Kotlyakov *et al.* (2004) and Hugel *et al.* (2005) describe the massive failure of the Kolka Glacier in the northern Caucasus in 2002, following the collapse onto the glacier surface of 10–20 million m^3 of rock and ice. The avalanche almost completely entrained the Kolka glacier (~100 million m^3), traveled down valley for 20 km, stopped at the entrance of the Karmadon gorge, and was finally succeeded by a mudflow which continued for another 15 km. The event caused about 140 deaths and massive destruction. Tibet was struck by two huge glacier avalanches of similar scales on the Aru mountain range in July 21 and September 24, 2016, respectively. The first avalanche in July shifted 65 million m^3 of snow and ice, killing nine herders and hundreds of animals. The second avalanche was smaller, caused no loss of life, but still stretched over several kilometers. The first collapse might have been caused by a phenomenon called surging when ice flows from top to bottom of a glacier at velocities up to 100 times faster than normal (www.sciencealert.com/two-mysterious-ice-avalanches-have-struck-tibet).

11.5   Winter sports industry

A ski resort can be considered to have reliable snow if, in 7 out of 10 winters, a snow cover of at least 30 to 50 cm is available on at least 100 days between December 1 and April 15 (Bürki *et al.*, 2003). It is projected that by AD 2030 half of Switzerland's 230 resorts will not have enough regular snow to sustain skiing. That represents an economic disaster for resort owners, an environmental disaster as lack of winter snow changes water and weather patterns and, of course, a recreational disaster for winter-sports enthusiasts. Artificial snow is made by simulating the same conditions needed for natural snow. While the ambient temperature does not have to be at freezing (depending on humidity), it has to be close. Also, the warmer it is, the more expensive it is to make snow. According to Elsasser and Messerli (2001), 85% of all Swiss ski areas had sufficient snow cover in the late twentieth century. However, Swiss Alps could see a 50% reduction in the depth of snow cover and no snow at elevations lower than 1,200 m by 2100, if greenhouse gas emissions aren't curbed. However, declines in snow depth could be limited to 30% if global temperature rise is kept to no more than 2 °C above preindustrial levels (Marty *et al.*, 2017). A 300-m rise of the snow line, however, would reduce this to about 63%. As a consequence, skiers will expect more artificial snow, go on winter holidays less often, and concentrate on ski areas at higher altitudes. United States' $20 billion ski industry is concerned, particularly in the eastern United States. The annual snow mass in the western United States has decreased about 40% since early 1980s, and over United States, the average snow season was shortened by 34 days (Zhang *et al.*, 2018). By 2050s, home values near ski resorts could drop by at least 15% due to warmer winters, and as much as 55% at lower elevation ski areas, such as in Utah, Nevada, and parts of California (Butsic *et al.*, 2009). From Europe to North America, a lack of snow has delayed the start of ski season in such iconic destinations as Zermatt of Switzerland, Breckenridge of Colorado, and Whistler of British Columbia. Winter sports, which are extremely vulnerable to global warming, are a crucial part of Canada's economy, culture, and identity. According to the David Suzuki Foundation (Bruce, 2009), global warming could cripple the winter sport industry of Canada estimated at about $5 billion annually.

11.6   Water resources

Snowfall is the major source of water in western North America. In most western cordilleras, it accounts for 50–65% of annual precipitation according to Serreze *et al.* (1999). The percentage of annual precipitation represented by snowfall is highest for the Sierra Nevada (67%), northwestern Wyoming (64%), Colorado (63%), and Idaho/western Montana (62%), representing high SWE precipitation ratios and winter-half-year precipitation maxima. Lower percentages are found in the Pacific Northwest (50%) and Arizona/New Mexico (39%), where the seasonal distribution is different and temperatures are higher. In the Canadian Prairies, over 30% of the total precipitation comes as snowfall. The ratio increases to over 50% in Northwest Territories and in the high Arctic, the ratio can be as high as 90%. The Canadian Rockies can receive up to several meters of snowfall per year. It is noteworthy that the shallow snow cover of the Canadian Prairies generates as much as 80% of the annual surface runoff (Gray and Landine, 1988). For the hydrologic cycle, global warming has important consequences particularly in northern regions with annual runoff dominated by melting of snow or ice because with higher surface temperature, less winter precipitation falls as snow and the spring snowmelt will occur earlier. Even if precipitation volume and intensity remain unchanged, both effects lead to a shift in peak river runoff to winter and early spring, away from summer and autumn when the water supply demand is the highest. Where storage water capacities are lacking, much of the winter/spring runoff will discharge to the oceans, which could severely impact the future water supply of more than one-sixth of the Earth's population relying on glaciers and seasonal snow packs for their annual water supply, as are already observed in some regions (Barnett *et al.*, 2005; Kerkhoven and Gan, 2011).

Snowfall for water resources in the United States is assessed by the National Operational Hydrologic Remote Sensing Center (NOHRSC) of the National Weather Service, by the US Department of Agriculture's National Resource Conservation Service (NRCS) and by state agencies that maintain snow courses. For Canada, daily snowfall and precipitation measurements are handled by Environment Canada's National Climate Data and Information Archive.

11.7 Hydropower

Hydropower relies on a water supply from rivers that originates from rainfall and snowfall. Hydropower (P) is the extraction of energy from falling water (Q) of density ρ over a certain hydraulic head (H), driving electric turbines and generators, which operate at a certain rate of efficiency (e). $P = \rho e Q H$. Run-of-the-river hydropower is the generation of hydroelectricity depending on the natural flow and elevation drop of a river and so it is usually developed on a river with a consistent and steady flow. Hydropower stations on rivers with large seasonal fluctuations require a reservoir to impound excess water during wet or snowmelt seasons for uninterrupted operation during the dry season. The construction of reservoirs results in flooding large tracts of land which may have significant negative impacts on the environment.

The world generated 4,200 terawatt hours (TWh) of hydropower in 2018, the highest ever contribution from a renewable energy source, as worldwide installed hydropower capacity consistently increased from about 150 GW in 1950s to almost 1,300 GW in 2018. Hydropower generates about 17% of the world's electricity. With an installed hydropower capacity of 352 GW, China produced the most amount of hydropower, followed by Brazil, Canada, and the United States. In 2018, about 720 TWh of hydropower was generated in North America. In Canada, the hydropower capacity was about 81 GW in 2017, which amounts to about two-thirds of its renewable energy, and hydropower accounts for about 60% of Canada's electricity generation. Quebec alone has 61 hydropower plants, which together have about 34.5 GW of total installed capacity. Norway generates over 99% of its electricity from hydropower. The contribution from snowfall is the greatest in Canada, India, and Scandinavia. In the United States, only about 7% of the total energy production is from hydropower; the states of Washington, California, and New York are the most important producers. The operation and management of complex hydropower systems that may consist of networks of reservoirs connected in series or in parallel, or both, can be optimized to maximize the system revenue and minimize generation costs by using system analysis techniques (Labadie, 2004; Goor *et al.*, 2011).

The key hydrologic parameters associated with hydropower production are snowfall (especially the snow elevation level); changes in volume, timing, and density of the snowpack; and snowmelt and runoff (Aspen Environmental Group and Cubed, 2005). Climate changes that reduce overall water availability or change the timing of that availability have the potential to affect adversely the production of hydroelectricity. Changes in snowfall and snowmelt in mountain watersheds, or areas with significant snowmelt runoff, are expected to lead to important changes in water availability (Gleick, 1998). Temperature increases will have three effects: (1) increase in the ratio of rain to snowfall in cold months; (2) a decrease of the overall duration of snowpack; and (3) an earlier onset in the spring snowmelt, and increase in the rate and intensity of snowmelt. As a result, average winter runoff and average peak runoff both increase, peak runoff occurs earlier in the year, and there is a faster and more intense loss of warm-season soil moisture. Earlier snowmelt has major implications for reservoir storage capacity and hydropower generation. There will be reduced flows in summer and autumn, partly due to enhanced evaporation losses caused by higher temperature and earlier onset of spring snowmelt. Lower storage at the end of the summer would reduce the ability of the system to meet present hydropower output during the winter months.

11.8   Snowmelt floods

Until frozen ground thaws in spring, any melting snow (or rainfall) cannot readily penetrate into the ground, and the resulting surface runoff can lead to flooding in river basins predominantly covered with snowpack in winters and dominated by snowmelt runoff in the spring. Runoff during rain-on-snow events when rain falls on existing snowpacks has been associated with mass-wasting of hill slopes, damage to riparian zones, downstream flooding, property damage and loss of life. In the Upper Mississippi River watershed, snowmelt floods in March are the most frequent cause of floods. In 1997, floods attributed to melting snow in the Dakotas and Minnesota caused damage that exceeded US\$3 billion dollars. The 1997 flood of the Red River near Winnipeg, Manitoba with a peak discharge of 3,900 m³ s⁻¹ was categorized as a 100-year flood and 9,000 people were evacuated. The 2009 Red River spring flood of 3,630 m³ s⁻¹ was the second highest of the river in Manitoba since the official records began in 1912. Heavy snowmelt, frozen soil, and ice jams in rivers are the most common causes of flooding in Canada (Beltos, 2018). Under global warming impact, the higher possibility of midwinter thaws and heavy rainfall events could increase the risk of sudden ice breakup, leading to ice-jam flooding which can be further exacerbated if the ground is still frozen and unable to soak up rainwater, and major snowmelt is combined with heavy rainfall. The 2009 floods on the Red River in North Dakota exemplified this situation.

Todhunter (2007) shows that the Grand Forks flooding of the Red River resulted from the principal flood-producing factors occurring at either historic or extreme levels. Above normal autumn precipitation increased the soil moisture storage and reduced the spring soil moisture storage potential. A frozen soil layer developed that reduced the soil infiltration capacity to zero. Record snowfall totals and snow cover depths occurred across the basin and a severe, late spring blizzard delayed the snowmelt and replenished the snow cover to record levels for early April. This was followed by a sudden transition to an extreme late season thaw. The presence of river ice contributed to backwater effects and affected the timing of tributary inflows to the main stem of the Red River.

Data collected by the Airborne Snow Survey flights conducted by the NOHRSC of NOAA-NWS are used by National River Forecast Centers to predict likely areas of snow-melt flooding.

11.9 Freshwater ice

Before the invention of refrigeration, ice blocks were cut from frozen lakes and rivers and placed in ice houses, insulated with straw, to keep produce cold in the summer. Their origin can be traced back to the seventh century BC in China and 1700 BC in northwest Iran. They were introduced to Britain in 1660 and ice was imported from Scandinavia until the 1950s. Trade in ice was a major part of the early economy of New England in the United States from where it was shipped to the southern states. In a modern-day equivalent, a microbrewery in Narsaq, Greenland has begun producing beer using water melted from the Greenland ice sheet.

Ice on lakes and rivers has important effects in three areas: disruption of shipping, ice-jam floods, and frazil ice blocking the intakes of turbines at hydropower installations (Murphy, 1909). Disruption to shipping by freshwater ice is common on the Great Lakes of North

America, rivers in the American Midwest and in the Yukon-Northwest Territories, and rivers in Eurasia. In Russia, the period when rivers are completely frozen varies from 70 days a year in the west of the country to as much as 250 days in northern Siberia. Ice-cover thickness in the rivers of the Siberian Far East can attain 200 cm or more, while it may be only 50–60 cm in the rivers of European Russia if the winter is mild (Vuglinsky, 2002). Ice-jam breakup and related floods account annually for about $300 million of damage in North America, globally one of the most prone regions to ice-jam flooding (IJF), particularly in eastern NA (Kontour *et al.*, 2018). However, these approximations could be seriously underestimated given single IJF events could result in hundreds of millions of dollars in damages, such as the 1996 flood in the Susquehanna River Basin of USA with an estimated damage of $800 million (Rokaya *et al.*, 2018). In recent years, mid-winter breakup events have also been increasingly reported in temperate and maritime regions of NA. In Canada, trend analyses show IJFs occurring earlier in south-eastern and western Canada but later in Atlantic Canada. Shifts in magnitude of ice-jam floods in Canada varies from $+3.5\%$ to -5% yr^{-1} in unregulated rivers and $\pm 3.5\%$ yr^{-1} in regulated rivers which are more dominated by negative trends. The trends in IJF are also correlated with size of the basins in unregulated rivers with small basins showing larger variabilities.

Ice charts for the Great Lakes have been produced since 1973. Charts show ice extent and concentration three times weekly during the ice season. The plot of the average ice concentration (IC) of the Great Lakes for winters of 1973–2020 (Figure 11.1) shows the average February–March IC of Lake Superior (39%), Michigan (21.6%), Huron (42.7%), Erie (49%), and Ontario (12.3%), which is the lowest. The corresponding annual maximum ice concentrations (AMICs) are Lake Superior (94.7%), Michigan (55.8%), Huron (95.7%), Eric (94.3%), and Ontario (39.8%). A temporal trend analysis in the AMICs indicates there have been several ice-cover regimes over the 47 winters between 1973 and 2019. Ice cover in all the Great Lakes combined has been lower in recent years, especially in 2020, but there is no long-term linear trend from 1973 to 2020 because of high annual variabilities. For example, the AMICs of the Great Lakes of 2019 were among the highest, but AMICs in 2020 were among the lowest.

One American and two Canadian large icebreakers are stationed in the Great Lakes. Climate change is having an effect on the duration of the shipping season. Overall, despite of some spikes in water levels due to increased rainfall and winter snowmelt, the water level of the Great Lakes have been falling since the 1980s, due to higher evaporation loss from climate warming impact. Projections of the future water levels of the Great Lakes vary over a wide range, with water levels falling between 0.23 and 2.5 m over the twenty-first century. Possible future lower lake levels could restrict vessel cargos, increasing the number of trips, and the cost of moving cargo, but warmer lake temperatures could extend the navigation season due to declining ice cover (Millerd, 2011). In future, the Montreal–Lake Ontario section that remained open for 279 days in 2002–2006 could remain open longer than 280–290 days.

(a)   (b)

Figure 11.1 (a) The Hoover Dam of United States built in 1930s for flood control, water supply, hydropower, navigation, and water regulation (PictureNet/Corbis/Getty Images); (b) Hydropower generated at the Kananaskis Lake fed by snowmelt water from Mountains of the Banff National Park, Canada (picture taken by Kai Ernn Gan) (A black and white version of this figure will appear in some formats. For the color version, please refer to the plate section.)

Ice-jam floods are a major problem for many parts of Canada, accounting for about 35% of all floods (Beltaos, 2008b). Nova Scotia, New Brunswick, and Newfoundland each experience one or two major ice-related flood events per decade.

Spring breakups in the Mackenzie, the largest river of Canada, begins in April in its southern tributaries, generally works its way northward, and is completed in about 6 weeks. The breakup process of the ice-covered Mackenzie is often triggered by the spring snowmelt runoff of its major tributary, the Liard River. In late April 2020, Fort McMurray in northern Alberta was hit by a major ice-jam flood on the Athabasca River, the southern tributary of the Mackenzie. After a very cold winter of 2019–2020 in Alberta, the Athabasca River that flows northward and eastward developed a thick ice cover. Ice jams can be a threat for north-flowing rivers with headwaters typically located in warmer climate zones, where the spring warm-up occurs earlier than in downstream farther north. So meltwater can encounter still frozen waterways as it moves downstream. As the River approaches Fort McMurray, the topography becomes flatter and the river widens, which slow the current, resulting in ice floes slowing down and piling up. The Clearwater River that feeds into the Athabasca from the east in the city further slows the flow of Athabasca, which together led to a massive ice jam over 24 km long developed in and near Fort McMurray. Almost 13,000 people were evacuated from homes for a week.

Before the construction of embankments and control structures along the St. Lawrence River, Montreal was regularly affected by ice-jam floods. The north-flowing rivers of southeastern Quebec province frequently experience ice-jam floods. Ice-jam floods are created also by rivers in southern Ontario flowing into the Great Lakes. A similar situation exists at Hay River, NWT, where the Hay River enters Great Slave Lake.

11.10 Ice roads

Ice roads across frozen lakes are used in areas where construction of year-round roads is too expensive or impractical. The roads in winter can be built up with a system of holes in the ice to flood and thicken the route. After an ice road is plowed across a lake or along a river, the ice there grows much thicker than the surrounding ice, because the snow cover has been swept off – exposing the road directly to air temperatures well below freezing. Depending on the region, these seasonal links last up to several months.

Ice roads (snow-covered ground and frozen lakes) have a strong influence on the reliability and costs of transportation in some remote northern communities (Parlee and Furgal, 2012). They are used mainly by large trucks and tractor-trailer units supplying mining sites and remote communities with no other access, such as those connecting Yellowknife, capital of the Northwest Territories, and Echo Bay Mines on Great Bear Lake, and between Yellowknife and the diamond mines at Ekati, Diavik, and Snap lakes. Ice roads across Lake Ladoga, Russia, provided a winter supply route to Leningrad from November 1941 to January 1943 during the Second World War blockade of the city by the German Army. Ice roads are opened annually in northern Scandinavia, with varying ice thickness limits of 20–40 cm. The ice road between Inuvik and Tuktoyaktuk in the Northwest Territories of Canada provides almost level driving for several months of the year.

For these communities, the period of safe ice cover will affect the duration of use of ice roads over land and inland waterways, travel risks, time, and costs (Goldhar *et al.*, 2014). Rising temperatures are leading to a shortening of the ice road transport season and melting ice roads are creating transportation challenges. The opening dates for ice roads in northern Alaska have shifted 2 months later from early November (pre-1991) to January (recent years), dramatically decreasing the transportation duration of ice roads. Knowland *et al.* (2010) examine extreme dates for early and late opening of the winter road between Tulita and Norman Wells, NWT. The earliest dates (December 16–23) were associated with high pressure and anomalous cold in the preceding November, while the latest dates (January 18–27) had a strong Aleutian Low and El Niño conditions in November. In recent years, there have been significantly shorter ice road shipping seasons due to unusually warm early winters (Sturm *et al.*, 2017). A shorter operational season for ice roads due to warmer winter is projected (Mullan *et al.*, 2017).

11.11 Sea ice

Sea ice services can be grouped into four categories of needed information: for marine transportation and the use of ice as an operational platform; the role of ice as a marine hazard, as a coastal buffer from wave erosion, and as a local-regional climate regulator; its support of biodiversity and the marine food web; and cultural aspects of the "icescape" (Eicken *et al.*, 2009a). For the southern Beaufort Sea, Barnett's (1976) index of ice severity shows a linear correlation with the summer minimum ice extent, for example. The presence of sea ice leads to changes in wave regime and coastal erosion, and affects marine mammal and fish distributions. Attention is now being given to the role of Arctic sea ice in the lives of indigenous communities (Mahoney, 2010). The sea ice cover determines food supplies from the marine environment (fish, seals, and whales), clothing from skins, and travel possibilities over the ice. Because of its importance, a sea ice monitoring program has been established by six northern communities with a guidebook (Mahoney and Gearheard, 2008).

Sea ice is a shipping hazard and merchant vessels may require icebreaker support. The most extensive use of icebreakers is in the Russian Arctic along the estuaries of the major rivers, and the Northern Sea Route. The latter was officially opened to commercial exploitation in 1935, following the 1932 expedition of O. Y. Schmidt and trials in 1933 and 1934. Open water conditions are generally present near the coast from August to October. Armstrong (1952) provided an historical account of activities along the Northern Sea Route. Commercial navigation in the Siberian Arctic declined in the 1990s following the breakup of the Soviet Union. At present, more or less regular shipping is to be found only from Murmansk to Dudinka in the west and between Vladivostok and Pevek in the east. The Northern Sea Route was open in summers 2008–2009 and increased attention is being given to international shipping possibilities (Brigham *et al.*, 1999; Ragner, 2000; Østreng, 2006). In recent years, warming over the Arctic region is twice the global mean, known as Arctic amplification which occurs because the warmer atmosphere induces ice loss, leading to feedbacks that accelerate sea ice loss and higher net incoming solar radiation in the Arctic. According to climate model simulations, future reduction of Arctic sea ice will be continuous and amplified (Pithan and Mauritsen, 2014), with ice-free summers occurring as early as the 2030s, and an ice-free year occurring as early as the 2050s (Onarheim *et al.*, 2018).

11.12 Glaciers and ice sheets

In arid regions like northwest China and northern Chile, glaciers are significant sources of water for agriculture and domestic use. To cope with shrinking glaciers, China is building many reservoirs in Xinjiang to collect glacial runoff. In developed countries like Canada, Norway, and Switzerland glacial meltwater is a major component of hydropower. In British Columbia, generating stations in the Columbia River Basin, Mica, and the Kootenay Canal are all at least partially glacier fed. In the eastern and central Himalaya about 90% of snow and glacier melt occurs in about 2 months of the year during the summer monsoon. The River Ganga is partly fed by the meltwater from around 4,000 glaciers in the Himalaya and the River Indus receives meltwater from more than 3,300 glaciers, although the bulk of the discharge is from rain and snowmelt (Thayyen and Gergan, 2010). They show that the important role of glaciers in this precipitation dominant system is to augment runoff during the years of low summer discharge.

Glacier associated features also cause many natural hazards. These include glacier-dammed lake outburst floods (GLOFs) and glacier "surges." The occurrence of surging glaciers varies widely. Less than 1% of the world's glaciers exhibit surge-type behavior, but clusters are found in Alaska, Canada, the Andes, the Tien Shan, Pamir and Karakoram, Iceland, Greenland, and Svalbard according to Paterson (1994). About 13% of Svalbard's glaciers surge (Jiskoot *et al.*, 1998); long glaciers overlying shale or mudstone with steep surface slopes have the highest probabilities of surging. Figure 3.7 shows the surging, Wahlenbergbreen glacier in Svalbard, Norway seen in September 2013 (Sevestre, 2017). There are two main types of glacier lakes, moraine-dammed lakes formed from glacier retreated from a moraine, and ice-dammed lakes formed when drainage is blocked by a glacier that advances or becomes thicker. GLOFs occur when a moraine dam (that can be ice cored) that was supporting a glacial lake gives way releasing large volumes of water with debris into the valleys downstream (Ives, 1985). Thirty-five destructive GLOF events have been recorded in the Upper Indus River system in the past 200 years and there have been 12 since 1935 in the Tibetan part of the Himalaya. In Nepal there are 2,315 glacial lakes, out of which 20 are potentially dangerous (Mool *et al.*, 2001), and in Bhutan there are 2,674 lakes, out of which 24 are potentially dangerous. The event that occurred on August 4, 1985 from Dig Tsho (Langmoche) glacial lake in Nepal destroyed the nearly complete Namche Small Hydropower Plant, 14 bridges, and cultivated land, causing about $2 million in damages (Ives, 1986). In Peru alone, GLOFs were responsible for ~32,000 deaths in the twentieth century (Carey, 2005). A GLOF was observed in West Greenland on August 31, 2007 when a 0.5 km^2 ice-dammed lake blocked by the Russell Glacier east of Kangerlussuaq gave way and released some 29 million m^3 of water into the Watson River (Mernild *et al.*, 2008). The collapse came after 4 days with an average temperature of 9.5 °C. Another example of a glacial lake outburst flood at the Russel Glacier was observed on July 28, 2015 that released ~7.5 million m^3 of water in ~20 hours (Carrivick *et al.*, 2017). The threat from moraine-dammed GLOFs is typically greatest during glacier retreat, while ice-dammed GLOFs are highest during periods of glacier growth. Therefore, the former could increase as mountain glaciers continue to shrink worldwide. On the other hand, as moraine dams are usually destroyed in lake outbursts, the number of GLOFs will likely decrease over time as the capacity for storing glacial meltwater gradually is lost.

Glaciers and ice caps, although making up only about 4% of the total land ice area, may have provided about 60% of the total land ice contribution to sea-level rise (SLR) since 2000s (Meier *et al.*, 2007; IPCC, 2019). Current estimates indicate that the mass balance for the Antarctic ice sheet is in approximate equilibrium and it may represent only about 13% of the current contribution to SLR from land ice. In contrast, the Greenland Ice Sheet may be contributing about 25% of all glacier melt to SLR. The retreat of glaciers, ice caps, and ice sheets has been calculated to have contributed 0.69 mm a^{-1} to the rise of mean sea level for the period 1961–2003, whereas the observed rise due to thermal expansion was 0.42 mm a^{-1}. Between 1993 and 2003, however, the contribution to SLR increased for both sources to 1.19 and 1.60 mm a^{-1} (IPCC, 2007), and to 1.8 and 1.4 mm a^{-1} over 2006–2015 (IPCC, 2019), respectively. Trenberth (2009) pointed out that ice melt is more effective in SLR by 40–70 times than ocean thermal expansion when excess heat is deposited in the upper 700 m of the world ocean, and this increases to 90 times if the heating is below 700 m depth.

Sea level has been rising since early twentieth century, estimated at about 0.16 m over 1902–2015 (IPCC, 2019). Casenave *et al.* (2008) show that thermal expansion has slowed since 2003 but glacier melting and mass loss from the ice sheets have increased their contributions to SLR in recent years. Sea-level budgets for 1993–2003 and 2006–2015 are shown in Table 11.2. Fiedler and Conrad (2010) show that the effect of dams impounding water has depressed sea level by 30 mm since 1900 and reduced the rate of SLR by ~10%. The land ice component now accounts for ≈60% of the observed SLR, compared to about 43% over 1993–2003. A minor contribution, not included in Table 11.2, is attributable to the melting of floating sea ice, as a result of the 2.6% density difference between freshwater and seawater, This amounts to a 43 μm a^{-1} rise in mean sea level (1.5% of the total) during 1994–2004 according to Shepherd *et al.* (2010). SLR has accelerated from about 1.4 mm yr^{-1} over 1902–1990 to 2.8 mm yr^{-1} over 1993–2003 to 3.2 mm yr^{-1} over 2006–2015 (Casenave *et al.*, 2008; IPCC, 2019).

Glacier scenery provides significant economic value to the local area through tourism. In areas where glaciers are relatively crevasse-free, glaciers provide opportunities for skiing even in mid-summer, although in the Alps many of these are rapidly shrinking. Large vehicles with snow tracks transport visitors around the Athabasca Glacier, one of the eight glaciers fed by the Columbia icefield that lies between Banff and the southern end of Jasper National Park in the Rocky Mountains of Canada. Scenic overflights with glacier landings are major attractions in the Mount Cook region of New Zealand. Cruise ships visit the calving front of the Columbia Glacier in Alaska. Glacier tourism has a long history in the European Alps, the Caucasus, the northern Rocky Mountains, and Alaska. Glacier National Park in Montana was established in 1910, for example. More recently, glacier tourism has flourished also in Scandinavia, Iceland, the Andes, the Southern Alps, the Himalaya, Antarctica, and western China (Liu *et al.*, 2006). Glacier disappearance is very serious in areas such as Glacier National Park, which by the end of the twentieth century had lost three-quarters of the 150 glaciers it had in 1850 and is projected to have no glaciers at all by AD 2030.

Table 11.2 Rates of sea-level rise (SLR) for 1993–2003 (Cazenave *et al.*, 2008) and 2006–2015 (IPCC, 2019)

Source	1993–2003 (mm yr^{-1})	2006–2015 (mm yr^{-1})
Thermal expansion	1.6 ± 0.3	1.4 ± 0.1
Glaciers	0.8 ± 0.1	0.61 ± 0.08
Greenland	0.2 ± 0.04	0.77 ± 0.03
Antarctica	0.2 ± 0.17	0.43 ± 0.05
Total	2.8	3.21
Satellite altimetry	3.1 ± 0.4	3.58

In the midst of ongoing worldwide glacial retreat and significant mass loss, particularly in European Alps and New Zealand (Purdie, 2013), glacier tourism is threatened by surface morphology changes associated with glacier retreat and thinning: steepening ice slopes, increased debris cover, increase in the rockfall hazard, and more difficult access for guided walks on glaciers. On the other hand, enlarging proglacial lakes caused by glacier melt could increase tourism opportunities. Despite the challenges of maintaining glacier-related tourism as glaciers worldwide retreat, the response seems to be increasing interest in glacier tourism, a phenomenon defined as "tourists explicitly seeking vanishing landscapes." Increasingly, tourists appear drawn to locations where particular landscapes, species, or social heritages are disappearing (Lemelin *et al.*, 2012). However, even though glacier tourisms are big business in some countries, for example, earnings from international Tourism in New Zealand is only surpassed by the dairy industry, there has been only limited studies conducted to quantify the potential loss to tourism from glacier disappearance. Yuan *et al.* (2006) estimate that between 20% and 40% of the total 3.5 million domestic tourists to Lijiang (Yunnan, China) in 2004, would not go there in the absence of the Yulong Mountain glacier, resulting in an economic loss of $85–185 million. To maintain its glacier tourism in the midst of dramatic changes in glaciers, New Zealand introduces boat tours instead of walking tours on Tasman Glacier and mechanized access in the Fox and Franz Josef Glaciers (Purdie, 2013).

11.13 Icebergs

Icebergs have long been considered to be a potential water source for arid coastal regions of the world. However, extensive discussions at a special symposium decided that for a variety of reasons (wrapping and towing the iceberg, and discharging the ice at a port), this approach was not practicable (Husseiny, 1978).

With about 90% of their mass submerged below water and the shape of the underwater portion unknown (as implied by the common expression "tip of the iceberg"), icebergs remain a threat to shipping and to oil drilling platforms in the northwest Atlantic and Barents Sea. The Hibernia platform off Newfoundland (46.8° N, 48.8° W), installed in 1997, is the first and only iceberg-resistant offshore drilling structure in the world; it uses aircraft and helicopter surveillance and, when a berg threatens to collide with the platform, a ship is called in to tie polypropylene towropes around it and affix a wire towline to tow it out of the way. Hibernia has contributed billions to the economy of Newfoundland since 1997, but like all other oil producers, Hibernia's future is uncertain by a plunge in oil prices in 2020, with the consumption down substantially because of COVID-19.

The Canadian Ice Service (CIS), a division of Environment Canada, puts out ice reports using aerial observation, ship-to-shore reports, and readings on currents and water temperature from buoys to track the paths of icebergs on a daily basis that are of interest to fishing and cargo vessels, and marine oil and gas projects. As frequent mapping of large areas by plane can be costly, CIS has been using satellite images acquired by synthetic aperture radar (SAR) sensors of Radarsat-1, ENVISAT, and Radarsat-2, to complement aerial reconnaissance of icebergs. In addition to SAR sensors, the National Ice Center of the United States that provides global operational ice analyses, also use optical sensors such as MODIS to track large icebergs. The International Ice Patrol (IIP) of the US Coast Guard has monitored icebergs in the northwest Atlantic since the sinking of *RMS Titanic* in 1912 (see Chapter 8, Section 8.1). Despite of IIP issuing detailed daily information on limits of iceberg danger, and even though iceberg risk in 1912 was exceptional (Bigg and Wilton, 2014), icebergs still threaten ships over 100 years after Titanic.

11.14 Permafrost and ground ice

Frozen ground has major effects on structures – roads, railways, pipelines, sewage lines, airstrips, dams, and buildings – that are situated on it. Such structures are vulnerable to the shifting, or settlement of the ground caused by thawing of permafrost. The presence of a structure leads to changes in the ground thermal regime and typically causes melt of ice in the ground. The ground sinks to fill the space left by the ice. Because different patches of ground may thaw at different rates, melting ice can make the surface uneven leading to wavy road surfaces or rail lines and the displacement and the potential collapse of buildings. These geotechnical problems are addressed by permafrost engineers working in cold region environments; the detailed procedures employed are beyond the scope of this book; see, for example, Yershov (1998) and Senneset (2000). Smith and Riseborough (2010) use observations from the Norman Wells, NWT, pipeline corridor and thermal modeling with a soil profile consisting of 1 m of peat overlying a fine-grained mineral soil. The Right-of-Way (ROW) was cleared up to 25 m wide in winter 1 year prior to pipeline construction in winters 1984 and 1985. Between 1985 and 2007 thaw depths increased by more than 2 m beneath the

ROW near Norman Wells and more than 3 m in the warmer and thinner permafrost further south of Fort Simpson. Ground temperatures beneath the ROW at 4 m depth are up to 2 °C higher than those in the adjacent undisturbed terrain. A set of scenarios examine the effects of: ROW clearing alone; ROW clearing followed by vegetation recovery; ROW clearing combined with climate warming of 0.5 °C per decade; and ROW clearing followed by vegetation recovery combined with climate warming. Simulation results for warm thin permafrost (mean annual ground temperature above −1 °C; 20 m thick) indicate that the combined effects of ROW disturbance and climate warming are likely to result in permafrost degradation within 20–40 years, and ROW-disturbance effects may extend off-ROW under scenarios of climate warming. In colder and thicker permafrost (mean annual ground temperature below −1 °C; 50 m thick), the combined effect of climate warming and ROW disturbance will not likely lead to talik formation within 50 years, although seasonal thaw penetration will increase. The results of the simulations indicate that the effects of ROW disturbance outweigh those associated with climate warming in the initial 10 to 15 years following disturbance, although climate warming becomes important on longer time scales.

Structures can be elevated on concrete piles so that cold air can flow underneath and prevent the permafrost from thawing. A more general approach is to construct a gravel pad on the surface in order to maintain the thermal equilibrium of the permafrost. This method was used in the 1960s beneath the entire area of the townsite of Inuvik, NWT, Canada, for example (French, 2007).

The TransAlaska Pipeline crosses permafrost and the high temperature (>60 °C) of the oil in the pipeline would thaw the permafrost and cause the pipeline to sink and break. Engineers therefore built the pipeline above the ground in many places. In warm permafrost and other areas where heat might cause undesirable thawing, the supports contain two 2-inch "heat pipes;" these contain anhydrous ammonia, which vaporizes below ground, but rises and condenses above-ground, removing ground heat whenever the ground temperature exceeds the temperature of the air. Heat is transferred through the walls of the heat pipes to aluminum radiators on top of the pipes.

In 2006, China completed a railway across the Tibetan Plateau to Lhasa. Since much of the route was over permafrost, engineers used crushed rock embankments to insulate the ground, and built bridges to raise the train tracks above the permafrost (Zhang et al., 2008). The track extends for 630 km over permafrost – a large fraction of which is warm, ice-rich permafrost.

Coastline erosion is a major concern in the Arctic. On the Seward Peninsula in Alaska, the settlement of Shishmaref was built on permafrost along the Bering Sea coast. In the past, the ground stayed frozen and sea ice protected the shore from wave erosion. With global warming the summer sea ice that protected the coast has disappeared, and now waves actively erode the coast. Every year, the coastline recedes about seven meters and so Shishmaref is being relocated. Along a 100 km section of the Beaufort Sea coast of Alaska, annual erosion during 1955–2002 averaged 5.6 m yr^{-1} and the rate increased by 24% after 1970 (Jones et al., 2008). The permafrost in the upper 10–20 m comprises over 20%

ice by volume (Brown *et al.*, 1997). Retreat rates of up to 30–100 m yr^{-1} are currently being recorded (Lantuit and Pollard, 2008; Lantuit *et al.*, 2008). Major retreat events do not necessarily coincide with the occurrence of ocean storms, but the duration of sea ice-free conditions is important. A correlation between rapid coastline retreat and rising ground and water temperatures also suggests there is decreasing resistance of coastal bluffs to wave attack, and an increase in the rates of melting along coastal bluffs with permafrost. Section 5.10 also briefly discusses the impact of Arctic infrastructure under the degradation of near-surface permafrost.

Thermokarst development, the ablation of ice in permafrost due to increased ground temperature and the consolidation of the soil or bedrock results from loss of ice, are initiated or retarded by local or regional changes, and may cease before all of the excess ice has melted, or it can be counteracted by permafrost aggradation. Global warming initiated widespread thermokarst in the twentieth century, which can intensify and spread further over the twenty-first century (Murton, 2009).

11.15 Seasonal ground freezing

When water freezes, it expands. This can make the ground move, causing frost heave that raises the surface and anything on it. Frost heave can be strong enough to move and damage roads, bridges, and buildings. It tends to be especially strong where there is permafrost or deep, seasonally frozen ground. On a small scale, ground freeze–thaw processes lead to the formation of so-called periglacial terrain features – patterned ground (stone or mud circles and polygons), stone stripes on slopes, ice wedges, and pingos (mounds with an icy core) (see Chapter 5, Section 5.8). Descriptions of these and their mechanisms of formation may be found in Washburn (1980) and French (2007), for example. The typical freeze depth over the northern United States, excluding the Pacific Northwest, is about 90 cm or more (see http://nsidc.org/frozenground/whereis_fg.html).

Ground freezing has agricultural benefits in that it leads to soil breakdown and inhibits certain pests. However, it also causes cessation of most construction work delaying project completion.

REVIEW QUESTIONS

1. What are three main types of snow fence used to prevent snow from drifting across highways in many US states? Briefly explain how each of these fences is designed to reduce snow drift.
2. The albedo of fresh snow is between 0.7 and 0.9. Snow falling in cities gets dirty quickly. What effect would dirt have on the melting rate of snow? Suppose, dZ/dt is the rate of change of snowpack depth, L_m is the latent heat of fusion, and ρ_s is the snow density, then $\rho_s L_m dZ/dt = -R_n$. Using this equation, explain qualitatively, how you can quantify the

difference between the melt rates of clean and dirty snow. Note that R_n changes from fresh to dirty snow (see Equations 2.28, 2.29, 2.39).

3. How do ice storms/freezing rain disrupt traffic, damage properties, and result in the loss of power and communication? Provide your explanations with reference to a recent major ice storm.

4. Explain in terms of changes in runoff volume, timing, and amount of snowfall qualitatively how global warming will affect the generation of hydropower from river basins that are dependent on spring snowmelt and have fairly large seasonal runoff fluctuations. Suppose the mean monthly discharge of a river is given in the table below. Estimate the annual hydropower output of a run-on-river hydro system in kWh, given its monthly turbine combined with generator efficiency coefficient (e) is as given in the table.

Mon	January	February	March	April	May	June	July	August	September	October	November	December
$Q\,m^3\,s^{-1}$	1,500	1,080	1460	2000	2480	1900	981	483	609	787	995	1,240
e	0.80	0.80	0.85	0.85	0.85	0.85	0.80	0.75	0.75	0.77	0.80	0.80

Suppose the tailwater elevation H_{TL} (m) = 70 m + 0.005Q, the forebay elevation = 110 m, and the frictional head loss of the system is 5 m.

5. Before making a trip, from where and what information should backcountry skiers and snowboarders collect to assess the risk of an avalanche? Given about 100 people lost their lives each year in the Alps in last four decades, roughly what is the odd that a skier will encounter a fatal avalanche in the Alps?

6. How will global warming impact the winter spots industry in ski resorts with respect to reduction in snow cover, rise in snow lines in mountains, and home values near ski resorts?

7. Globally what is the approximate fraction of people relying on spring snowmelt, and how will their water supply be affected by changes to the hydrologic cycle in response to global warming?

8. Climate change tends to have negative impact to our environment. However, how has climate change affected the duration of the shipping season in the Montreal – Lake Ontario section of the St. Lawrence Seaway of the great lakes? For example, in 1982–1986, the average open period for this Seaway was 269 days; for 2002–2006 it increased by 10 days to 279 days. Will the shipping seasons over the great lakes lengthen under climate warming?

9. As Arctic sea ice continues to decline, discuss the new future opportunities in a more accessible Arctic and risks, such as oil and mineral extraction in the Arctic, the northern route shipping and transportation, and tourism at the Arctic.

10. In the midst of ongoing worldwide glacial retreat and significant mass loss, what are the challenges of glacier tourism across the world? Will you expect glacier tourism to disappear by 2100?

11. Which parts of North America are more prone to ice-jam flooding (IJF)? How have shifts

in magnitude of IJF in Canada varied between unregulated and regulated rivers, and size of river basins?

12. Besides releasing greenhouse gases to the atmosphere as permafrost thaws, what impacts will be expected from thawing frozen ground to the Arctic infrastructure, such as northern railway, pipelines, and foundations of northern buildings?

13. Using the power duration curve of a run-of-river hydro plant, estimate the annual usable power of the hydropower plant in kWh.

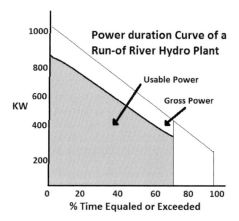

14. With reference to its observed annual hydrograph (1961–1990), explain qualitatively the projected climate change impact of 7 GCMs to the annual hydrograph of the Athabasca River Basin of Alberta for 2080s, such as the onset of spring snowmelt and summer runoff.

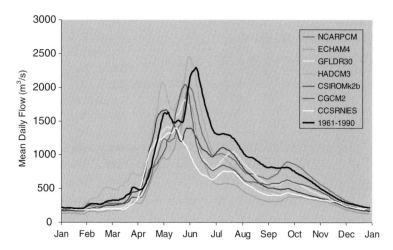

References

Abbot, D. S. and Pierrehumbert, T. T. (2010). Mudball: surface dust and Snowball Earth deglaciation. *Geophys. Res. Lett.*, **115**: D03104, doi: 10.1029/2009JD012007.

Abdalati, W. (2007). *Greenland Ice Sheet melt characteristics derived from passive microwave data*. Boulder, CO: National Snow and Ice Data Center. Digital media.

Abdalati, W. and Steffen, K. (1997). Snowmelt on the Greenland Ice Sheet as derived from passive microwave satellite data. *J. Climate*, **10**: 165–75.

Abdalati, W. and Steffen, K. (2001). Greenland Ice Sheet melt extent: 1979–1999. *J. Geophys. Res.*, **106** (D24): 33, 983–8.

Abel, G. (1955). Températures et formation de glace dans les grottes du Salzburg (Autriche), *Proc. 1st Int. Cong. Speleol.*, Paris, 1953. **2**: 321–4.

Ablain, M., *et al.* (2017). TOPEX-A Drift and Impacts on GMSL Time Series, https://meetings.aviso.altimetry.fr/fileadmin/user_upload/tx_ausyclsseminar/files/Poster_OSTST17_GMSL_Drift_TOPEX-A.pdf.

Abram, N. J., Wolff, E. W., and Curran, M. A. J. (2013). A review of sea ice proxy information from polar ice cores. *Quat. Sci. Rev.*, **79**: 168–83, doi: 10.1016/j.quascirev.2013.01.01.

Abramov, V. A. (1992). Russian iceberg observations in the Barents Sea, 1933–1990. *Polar Res.*, **11**: 93–7.

Abramov, V. A. (1996). *Atlas of Arctic icebergs*. Fair Lawn, NJ: Backbone Publishing Co. 70 pp.

ACIA (2005). *Arctic climate impact assessment*. New York: Cambridge University Press, 1042 pp.

Ackerman, S. A., *et al.* (1998). Discriminating clear sky from clouds with MODIS, *J. Geophys. Res.*, **103** (D24): 32, 141–57.

Ackerman, S. A., *et al.* (1995). Cirrus cloud properties derived from high spectral resolution infrared spectrometry during FIRE II .2. *Aircraft HIS results, J Atmos. Sci.*, **52**: 4246–63.

Ackley, S. F. (1979). Mass-balance aspects of Weddell sea pack ice. *J. Glaciol.*, **24** (90): 391–405.

Ackley, S. F. (1996). Sea ice. In George L., Trigg *et al.*, (eds.), *Encyclopedia of applied physics*. New York: VCH Publishers. **17**: pp. 81–103.

Ackley, S. F., Lange, M., and Wadhams, P. (1990). Snow cover effects on Antarctic sea ice thickness. In Ackley, S. F. and Weeks, W. F. (eds.), *Sea ice properties and processes*. CRREL Monograph 90–1. Hanover, NH: US Army Cold Regions Research and Engineering Laboratory Engin. Lab. pp. 225–9.

Adam, J. C., Hamlet, A. F., and Lettenmaier, D. P. (2009). Implications of global climate change for snowmelt hydrology in the twenty-first century. *Hydrol. Proc.*, **23**: 962–72.

Adams, P. (1992). J. B. Tyrell and D. H. Dunble on lake ice. *Arctic* **45**: 195–8.

Adrian, R. and Hintze, T. (2000). Effects of winter air temperature on the ice phenology of the Müggelsee (Berlin, Germany). *Verh. Int. Verein. Limnol.*, **27**: 2808–11.

Adusumilli, S., Fricker, H. A., Siegfried, M. R., Padman, L., Paolo, F. S., and Ligtenberg, S. R. M. (2018). Variable basal melt rates of Antarctic Peninsula ice shelves, 1994–2016, *Geophys. Res. Lett.*, **45**: 4086–95, doi: 10.1002/2017GL076652.

Afultis, M. A. (1987). Iceberg populations south of 48° N since 1900. *Report of the International Ice Patrol in the North Atlantic.* Bull. No. 74, CG 188–42. U.S. Dept. of Transportation, U.S. Coast Guard. pp. 63–7, 358.

Afultis, M. A. and Martin, S. (1987). Satellite passive microwave studies of the Sea of Okhotsk ice cover and its relation to oceanic processes, 1978–1982. *J. Geophys. Res.*, **92**: 13,013–28.

Agassiz, L. (1967). *Studies on glaciers* (trans. of Agassiz, L. 1840 *Études sur les glaciers*, trans., ed. Carozzi, A. V). New York: Hafner Publ. Co. 213 pp.

Ahlmann, H. W. (1924). Le niveau de glaciation comme fonction de l'accumulation d'humidité sous forme solide. *Geogr. Ann*, **6**: 223–72.

Ahlmann, H. W. (1935). Scientific results of the Norwegian-Swedish Spitsbergen Expedition in 1934. *Part V. Geogr. Annal.*, **17**: 167–218.

Ahlmann, H. W. (1948). Glaciological research on the North Atlantic coast. *Roy. Geogr. Soc., Res. Ser.*, no. 1. 83 pp.

Ahlmann, H. W. and Tryselius, O. (1929). Der Kårsa Gletscher in Swedisch Lappland. *Geogr. Annal.*, **11**: 1–32.

Aizen, V. B., *et al.* (2006). Glacier changes in the central and northern Tien Shan during the last 140 years based on surface and remote-sensing data. *Annals Glaciol.*, **43**: 202–13.

Alekseyev, V. R., *et al.* (eds.). (1973). Siberian naleds. USSR Academy of Sciences (1969). Draft Translation 399. Hanover, NH: US Army Cold Regions Research and Engineering Laboratory. 300 pp.

Alexeev, S. V., Alexeeva, L. P., and Kononov, A. M. (2008). Permafrost and cryopegs of the Anabar shield. In Kane, D. L. and Hinkel, K. M. (eds.), *Proceedings of the ninth international conference on permafrost*, Fairbanks, AK: University of Alaska, Institute of Northern Engineering. pp. 31–5.

Alford, D. and Armstrong, R. (2010). The role of glaciers in stream flow from the Nepal Himalaya. *The Cryosphere Discuss*, **4**: 469–94.

Allen, C., *et al.* (1997). Airborne radio echo sounding of outlet glaciers in Greenland, *Int. J. Re. Sens.* **18**(14): 3103–8.

Allen, J. R. and Long, D. G. (2006). Microwave observations of daily Antarctic sea-ice edge expansion and contraction rates. *IEEE Geosci. Remote Sensing Lett.*, **3**: 54–8.

Allen, P. A. and Etienne, J. L. (2008). Sedimentary challenge to Snowball Earth. *Nat. Geosci.*, **1**: 818–25.

Alley, R. B. (2000). *The two-mile time machine: ice cores, abrupt climate change, and our future.* Princeton, NJ: Princeton University Press. 229 pp.

Alley, R. B., *et al.* (2009). Past extent and status of the Greenland Ice Sheet. In Alley, R. B., Brigham-Grette, J., Miller, G. H., and Polyak, L., (eds.), *Past climate variability and change in the Arctic and at high latitudes.* U.S. Climate Change and Science Program, Synthesis and Assessment 1.2. Reston, VA: US Geological Survey. pp. 303–415.

Alley, R. B., *et al.* (2010). History of the Greenland Ice Sheet: Paleoclimatic insights. *Quat. Sci. Rev.*, **29** (15–16): 1728–56.

Alley, R. B. and Clark, P. U. (1999). The deglaciation of the Northern Hemisphere: a global perspective. *Ann. Rev. Earth Planet. Sci.*, **27**: 149–82.

Allison, I., Barry, R. G., and Goodison, B. E. (2001). *Climate and cryosphere (CliC) project science and co-ordination plan (Version 1).* Geneva: World Meteorological Organization. WMO/TD 1053, 96 pp.

Allison, I. and Kruss, P. (1977). Estimation of recent climatic change in Irian Jaya by numerical modeling of its tropical glaciers. *Arct. Alp. Res.*, **9**: 49–60.

Allison, I. and Peterson, J. A. (1976). Ice areas on Puncak Jaya – their extent and recent history. In Hope, G. S., Peterson, J. A., Radok U., and Allison I. (eds.), *The equatorial glaciers of New Guinea – results of the 1971–1973 Australian Universities' expeditions to Irian Jaya: survey, glaciology, meteorology, biology and paleoenvironments*. Rotterdam: A. A. Balkema. pp. 27–38.

Allix, A. (1924). Avalanches. *Geogr. Rev.*, **14** (4): 519–60.

Alvarez-Solas, J., *et al.* (2010). Links between ocean temperature and iceberg discharge during Heinrich Events. *Natur Geeosci.*, **3**: 122–6.

AMAP (2017). *Snow, water, ice and permafrost in the Arctic (SWIPA)*. Oslo, Norway: Arctic Council Secretariat.

Anacona, P. I., Mackintosh, A., and Norton, K. P. (2015). Hazardous processes and events from glacier and permafrost areas: lessons from the Chilean and Argentinean Andes. *Earth Surf. Proc. Land.*, **40** (1): 2–21, 37, doi: 10.1002/esp.3524.

Anandakrishnan, S., *et al.* (2007). Discovery of till deposition at the grounding line of Whillans Ice Stream. *Science*, **315**: 1835–8.

Anderson, E. A. (1976). A point energy and mass balance model of a snow cover, NOAA Technical Report NWS, **19**: 150 pp.

Andersen, S., *et al.* (2007). Intercomparison of passive microwave sea ice concentration retrievals over the hh-concentration Arctic sea ice. *J. Geophys. Res.*, **112**: C08004. doi: 10.1029/2006JC003543.

Anderson, M. R., Crane, R. G., and Barry, R. G. (1985). Characteristics of Arctic Ocean ice determined from SMMR data for 1979: case studies in the seasonal sea ice zone. *Adv. Space Res*,. **5** (6) G. Ohring and H. J. Bolle (eds.). Space Observations for Climate Studies: 257–61.

Anderson, R. K., *et al.* (2008). A millennial perspective on Arctic warming from 14 C in quartz and plants emerging from beneath ice caps. *Geophys. Res. Lett.*, **35**: L01502. doi: 10.1029/2007GL032057.

Andreadis, K. M. and Lettenmaier, D. P. (2006). Assimilating remotely sensed snow observations into a macroscale hydrology model. *Adv. Water Resour.*, **29**: 872–86.

Andreas, E. L., *et al.* (1990). Lidar-derived particle concentrations in plumes from Arctic leads. *Annals Glaciol.*, **14**: 9–12.

Andreas, E. L., Jordan, R. E., and Makshtas, A. P. (2005). Simulations of snow, ice, and near-surface atmospheric processes on Ice Station Weddell. *Bound.-Layer Met.*, **114**: 439–60.

Andrews, J. T. (2000). Icebergs and iceberg rafted detritus (IRD) in the North Atlantic: facts and assumptions. *Oceanography*, **13** (3): 100–8.

Andrews, J. T. and Miller, G. H. (1972): Quaternary history of northern Cumberland Peninsula, Baffin Island, N. W.T., Canada. Part IV: maps of the present glaciation limits and lowest equilibrium line altitude for north and south Baffin Island. *Arct. Alp. Res.*, **4**: 45–59.

Andreescu, M.-P. and Frost, D. B. (1998). Weather and traffic accidents in Montreal, Canada. *Clim. Res.*, **9**: 225–30.

Anisimov, O. A. (1989) Changing climate and permafrost distribution in the Soviet Arctic. *Phys. Geog.*, **10**(3): 285–93.

Anisimov, O. A. and Nelson, F. E. (1990). Application of mathematical models to investigate the interaction between the climate and permafrost. *Soviet Met. Hydrol.*, **1990** (10): 8–13.

Anisimov, O. A. and Nelson, F. E. (1996). Permafrost distribution in the Northern Hemisphere under scenarios of climatic change. *Global Planet. Change*, **14**: 59–72.

Anisimov, O. A. and Nelson. F. E. (1997). Permafrost zonation and climate change in the Northern Hemisphere: results from transient general circulation models. *Clim. Change*, **35**: 241–58.

Anisimov, O. A. and Reneva, S. (2006). Permafrost and changing climate: the Russian perspective. *Ambio*, **35** (4): 169–75.

Anisimov, O. A., Shiklomanov, N. I., and Nelson, F. E. (1997). Global warming and active layer thickness: results from transient general circulation models. *Global Planet. Change*, **15**: 61–77.

Anisimov, O. A., Shiklomanov, N. I., and Nelson, F. E. (2002). Variability of seasonal thaw depth in permafrost regions: A stochastic modelling approach. *Ecol. Modelling*, **153**: 217–27.

Aniya, M., *et al.* (1996). Inventory outlet glaciers of the Southern Patagonia Icefield, South America. *Photogram, Eng, Rem. Sensing*, **62**: 1361–9.

Anschutz, H., *et al.* (2009). Revisiting sites of the South Pole Queen Maud Land Traverses in East Antarctica: Accumulation data from shallow firn cores. *J. Geophys. Res.*, **114**: D24106, doi: 10.1029/2009JD012204.

Arctic and Antarctic Research Institute. (2007). *Sea ice charts of the Russian Arctic in gridded format, 1933–2006.* Edited and compiled by Smolyanitsky, V., *et al.*, Boulder, CO: National Snow and Ice Data Center. Digital media.

Arctic Climatology Project. (2000). *Environmental Working Group joint U.S.-Russian sea ice atlas.* F. Tanis and V. Smolyanitsky (eds.). Ann Arbor, MI: Environmental Research Institute of Michigan in association with the National Snow and Ice Data Center. CD-ROM.

Arctic Report Card (2009). www .arctic.noaa.gov/reportcard/

Arendt, A., *et al.* (2002). Rapid wastage of Alaska glaciers and their contribution to rising sea level. *Science*, **297** (5580): 382–6.

Arendt, A. A., *et al.* (2009). Validation of high-resolution GRACE mascon estimates of glacier mass changes in the St. Elias Mountains, Alaska, USA, using aircraft laser altimetry. *J. Glaciol.*, **54** (188): 778–87.

Arendt, A., *et al.* (2012). Randolph Glacier Inventory [v2.0]: A Dataset of Global Glacier Outlines. Global Land Ice Measurements from Space, Boulder Colorado, USA. Digital Media 32 pp. Available online at: http://www.glims.org/RGI/RGI_Tech_Report_V2.0.pdf

Arendt, A., *et al.* (2015). *Randolph Glacier Inventory – A Dataset of Global Glacier Outlines: Version 5.0.* Boulder, CO, USA: Global Land Ice Measurements from Space.

Arenstein, J. (1849). *Beobachtungen über die Eisverhältnisse der Donau.* 1847/48 und 1848/49. Wissenschaft, Vienna: Sitzungsbericht Akad. **5**: 331.

Armitage, T. W. K., Bacon, S., and Kwok, R. (2018). Arctic sea level and surface circulation response to the Arctic Oscillation. *Geophy. Res. Lett.*, **45**: 6576–84.

Armour, K. C. *et al.* (2016). Southern Ocean warming delayed by circumpolar upwelling and equatorward transport. *Nat. Geosci.*, **9** (7): 549, doi: 10.1038/Ngeo2731.

Armstrong, B. R. and Williams, K. (1986). *The avalanche book.* Golden, CO: Fulcrum. 240 pp.

Armstrong, R. (2001). *Historical Soviet daily snow depth version 2 (HSDSD).* Boulder, CO: National Snow and Ice Data Center. CD-ROM.

Armstrong, R. L., Brodzik, M. J., Savoie, M., and Knowles, K., (2006). Multi-Sensor snow mapping and global trends, WDC for Glaciology, Boulder, 30th Anniversary Workshop, University of Colorado, Boulder, October 25, 2006.

Armstrong, R., Alford, D., and Racoviteanu, A. (2009). Glaciers as indicators of climate change – the special case of the high elevation glaciers of the Nepal Himalaya. In *Water storage.*

A strategy for climate change adaptation in the Himalaya. Sustainable Mountain Development No. 56. Kathmanudu, Nepal: ICIMOD. pp. 16–18.

Armstrong, R. L. and Armstrong, B. R. (1987). Snow and avalanche climates of the western United States: A comparison of maritime, intermountain and continental conditions. Avalanche Formation, Movement and Effects. *IAHS Publ.*, **162**: 282–94.

Armstrong R. and Brodzik, M. J. (2002). Hemispheric-scale comparison and evaluation of passive microwave snow algorithms. *Annals Glaciol.*, **34**: 38–44.

Armstrong R., Brodzik, M. J., and Savoie, M. H. (2003). *Multi-sensor approach to mapping snow cover using data from NASA's EOS Aqua and Terra spacecraft (AMSR-E and MODIS).* Boulder, CO: National Snow and Ice Data Center (NSIDC), University of Colorado.

Armstrong, R. L., Brodzik, M. J., Knowles, K., and Savoie, M. (2005). *Global monthly EASE-Grid snow water equivalent climatology.* Boulder, CO: National Snow and Ice Data Center. Digital media.

Armstrong, R. L. and Brun E. (eds.) (2008). *Snow and climate: Physical processes, surface energy exchange and modeling.* Cambridge: Cambridge University Press. 222 pp.

Armstrong, R. L. (2010). The glaciers of the Hindu Kush–Himalayan region. Technical Paper. Kathmandu, Nepal: CIMOD. 20 pp.

Armstrong, R. L., and Brodzik, M. J. (2001). Recent Northern Hemisphere snow extent: a comparison of data derived from visible and microwave sensors. *Geophy. Res. Lett.*, **28** (19): 3673–6.

Armstrong, T. E. (1952). *The northern sea route: Soviet exploitation of the North East Passage.* Cambridge: Scott Polar Research Institute. Special publ. no. **1**, 20 pp.

Arzel, O., Fichefet, T., and Goosse, H. (2006). Sea ice evolution over the 20th and 21st centuries as simulated by current AOGCMs. *Ocean Modelling*, **12**: 401–15.

Ashton, G. D. (1980). Freshwater ice growth, motion, and decay. In Colbeck, S. C. (ed.), *Dynamics of snow and ice masses.* New York: Academic Press. pp. 261–304.

Aspen Environmental Group and Cubed, M. (2005). *Potential changes in hydropower production from global climate change in California and the western United States. CEC-700–2005–010.* California Energy Commission. 65 pp.

Asplin, M. G., Lukovich, J. V., and Barber, D. G. (2009). Atmospheric forcing of the Beaufort Sea ice gyre: Surface pressure climatology and sea ice motion. *J. Geophys. Res.*, **114** (D00D05): 9.

Assel, R. and Herche, L. (2000). Coherence of long-term lake ice records. *Verh. Int. Verein. Limnol.*, **27**: 2789–92.

Assel, R., Cronk, K., and Norton, D. (2003). Recent trends in Laurentian Great Lakes ice cover. *Clim. Change* **57**: 185–204.

Ashton, G. D. (ed.) (1986). *River and lake ice engineering.* Highlands Ranch, CO: Water Resources Publication. 486 pp.

Astakov, V. I. (1986). Geological conditions for the burial of Pleistocene glacier ice on the Yensisey. *Polar Geog. Geol.*, **10**: 286–95.

Atkinson, D. E., *et al.* (2006). Canadian cryospheric responses to an anomalous warm summer: synthesis of the climate change action fund project the state of the arctic cryosphere during the extreme warm summer of 1998. *Atmos.-Ocean*, **44**: 347–76.

Aubekerov, B. and Gorbunov, A. P. (1999). Quaternary permafrost and mountian glaciation in Kazakhstan. *Permafrost Periglac. Proc.*, **10**: 65–80.

Ávila, E. E., *et al.* (2009). Initial stages of the riming process on ice crystals, Geophys. *Res. Lett.*, **36**: L09808. doi:10.1029/2009GL037723.

Bader, H. (1961). The Greenland ice sheet, CRREL Mongr, I B2; Hannover, NH, US Army Cold Regions Research and Engineering Laboratory, 18 pp. https://erdc-library.erdc.dren.mil/jspui/bitstream/11681/2674/1/CRREL-Mono-1-B2.pdf

Bader, H., *et al.* (1939). Der Schnee und seine Metamorphose. *Beitr. Geologie der Schweiz, Geotechnische Serie-Hydrologie*, Issue **3**. Bern: Kümmerly and Frey, (In English as Snow and its metamorphosis. Snow, Ice and Permafrost research Establishment, SIPRE Translation No. 14, 1954, 313 pp.)

von Baer, K. E. (1838a). On the ground ice or frozen soil of Siberia. *J. Roy. Geog. Soc.*, **8**: 210–13.

von Baer, K. E. (1838b). Intelligence upon the frozen ground in Siberia. *J. Roy. Geog. Soc.*, **8**: 401–6.

Bahr, D. B. (1997a). Global distribution of glacier properties: A stochastic scaling paradigm. *Water Resour. Res.*, **33**: 1669–79.

Bahr, D. B. (1997b). Width and length scaling of glaciers. *J. Glaciol.*, **43** (145): 557–62.

Bahr, D. and Dyurgerov, M. B. (1999). Characteristic mass-balance scaling with valley glacier size. *J. Glaciol.*, **45** (149): 17–21.

Bahr, D. B., Dyurgerov, M., and Meier, M. F. (2009). Sea-level rise from glaciers and ice caps: A lower bound. *Geophys. Res. Lett.*, **36**: L03501. doi: 10.1029/2008GL036309.

Bahr, D. B., Meier, M. F., and Peckham, S. D. (1997). The physical basis of glacier volume – area scaling. *J. Geophys. Res.*, **102** (B9): 20, 355–62.

Bakkehøi, S. Domaas, U., and Lied, K. (1983). Calculation of snow avalanche run-out distance. *Annal. Glaciol.*, **4**: 24–9.

Ballantyne, J. and Long, D. G. (2002). A mulitdecadal study of the number of Antarctic icebergs using scatterometer data. Geoscience and Remote Sensing Symposium 2002, IGARSS '02, IEEE International, vol. 5: 3029–31. https://ieeexplore.ieee.org/stamp/stamp.jsp?arnumber=1237370

Bales, R. C., *et al.* (2009). Annual accumulation for Greenland updated using ice core data developed during 2000–2006 and analysis of daily coastal meteorological data. *J. Geophys. Res.*, **114** (D06115): 14.

Baldocchi, D. D., Matt, D. R., Hutchison B. A., and McMillen, R. T. (1984). Solar radiation within an oak-hickory forest: an evaluation of the extinction coefficients for several radiation components for fully-leafed and leafless periods. *Agric. For. Meteorol.*, **32**: 307–22.

Bamber, J. L. (1994). Ice sheet altimeter processing scheme. *Int. J. Remote Sens.*, **15**: 925–38.

Bamber, J. L., Alley, R. B., and Joughin, I. (2007). Rapid response of modern day ice sheets to external forcing. *Earth Planet. Sci. Lett.*, **257**: 1–13.

Bamber, J. L. and Bentley, C. R. (1994). Comparison of satellite-altimetry and ice-thickness measurements of the Ross Ice Shelf, Antarctica. *Ann. Glaciol.*, **20**: 357–64.

Bamber, J. L. and Kwok, R. (2004). Remote sensing techniques. In Bamber, J. L. and Payne, A. J. (eds.), *Mass balance of the cryosphere: Observations and modelling of contemporary and future changes.* Cambridge: Cambridge University Press. pp. 59–113.

Bamber, J. L. and Payne, A. J. (eds.). (2004). *Mass balance of the cryosphere: Observations and modelling of contemporary and future changes.* Cambridge: Cambridge University Press. 666 pp.

Bamber, J. L. and Rivera, A. (2007). A review of remote sensing methods for glacier mass balance determination. *Glob. Planet. Change*, **59**: 133–48.

Bamber, J. L., Vaughan, D. G., and Joughin, I. (2000). Widespread complex flow in the interior of the Antarctic Ice Sheet. *Science* **287** (5456): 1248–50.

Bamber, J. L., *et al.* (2009). Reassessment of the potential sea-level rise from a collapse of the West Antarctic Ice Sheet. *Science* **324**: 901–3.

Bamber, J. L., Westaway, R. M., Marzeion B., and Wouters, B. (2018). The land ice contribution to sea level during the satellite era, *Environ. Res. Lett.*, **13** (6): 063008, doi:10.1088/1748-9326/aac2f0.

Barber, D. G., *et al.* (2001a). Physical processes within the North Water (NOW) polynya, *Atmosphere – Ocean*, **39** (3): 163–6.

Barber, D. G., *et al.* (2001b). Sea-ice and meteorological conditions in Northern Baffin Bay and the North Water Polynya between 1979 and 1996. *Atmosphere – Ocean* **39** (3): 343–59.

Barber, D. G. and Massom, R. A. (2007). The role of sea ice in bipolar polynya processes. In W. O. Smith and D. G. Barber (eds.), *Polynyas: Windows into polar oceans.* New York: Elsevier. 474 pp.

Barclay, D. J., Wiles, G. C., and Calkin, P. E. (2009). Holocene glacier fluctuations in Alaska, *Quat. Sci. Rev.* **28**: 2034–48.

Bareiss, J. and Görgen, K. (2005). Spatial and temporal variability of sea ice in the Laptev Sea: and review of satellite passive-microwave data and model results, 1979 to 2002. *Glob. Planet. Change* **48**: 28–54.

Barendregt, R. W. and Dud-Rodkin, A. (2004). Chronology and extent of Late Cenozpic ice sheets in North America: A magnetostratigraphic assessment. In Ehlers, J. and Gibbard, P. L. (eds.), *Quaternary glaciations extent and chronology. Part 2. North America.* Amsterdam: Elsevier. pp. 1–8;

Barendregt, R. and Irving, E. (1998). Changes in the extent of North American ice sheets during the late Cenozoic. *Can. J. Earth Sci.* **35** (5): 504–9.

Barichivich J., Briffa K. R., and Myneni R. B., *et al.* (2013). Large-scale variations in the vegetation growing season and annual cycle of atmospheric CO_2 at high northern latitudes from 1950 to 2011. *Glob. Chang. Biol.*, **19**: 3167–83, doi: 10.1111/gcb.12283.

Barker, P. F., Dickmann, B., and Escutia, C. (2007b). Onset of Cenozoic Antarctic glaciation. *Deep-Sea Res. II*, **54**: 2293–307.

Barker, P. F., *et al.* (2007a). Onset and role of the Antarctic Circumpolar Current, Deep-sea Res. Part 2. *Topical Studies in Oceanography*, **54** (21–22): 2388–98.

Barker, S., *et al.* (2009). Interhemispheric Atlantic seesaw response during the last deglaciation. *Nature*, **457**: 1097–103.

Barletta, V. R., *et al.* (2013). Scatter of mass changes estimates at basin scale for Greenland and Antarctica. *The Cryosphere*, **7**: 1411–32, https://doi.org/10.5194/tc-7-1411-2013, 2013

Barnes, H. T. (1906). *Ice formation: with special reference to anchor-ice and frazil.* New York: J. Wiley and Sons. 260 pp.

Barnett, T. P., Adam, J. C., and Lettenmaier, D. P. (2005). Potential impacts of a warming climate on water availability in snow-dominated regions, *Nature*, **438**: 303–9.

Barnett, T. P., Pierce, D. W., Hidalgo, H. G., Bonfils, C., Santer, B. D., Das, T., Bala, G., Wood, A. W., Nozawa, T., Mirin, A. A., Cayan, D. R., and Dettinger, M. D. (2008). Human-induced changes in the hydrology of the western United States. *Science*, **319**: 1080–3.

Barrans, N. E. and Sharp, M. J. (2010). Sustained rapid shrinkage of Yukon glaciers since the 1957–1958 International Geophysical Year, *Geophys. Res. Lett.*, **37**: L07501. doi: 10.1029/2009GL042030.

Barry, R. G. (1966). Meteorological aspects of the glacial history of Labrador-Ungava with special reference to vapor transport, *Geogr. Bull. (Ottawa)* **8** (4): 319–40.

Barry, R. G. (1985). Snow and ice data. In Hecht, A. D. (ed.), *Paleoclimate analysis and modeling*. New York: J. Wiley and Sons, 259–90.

Barry, R. G. (1987). The cryosphere – neglected component of the climate system. In Radok, U. (ed.), *Towards understanding climate change*. Boulder, CO: Westview Press, 35–67.

Barry, R. G. (1989). The present climate of the Arctic Ocean and possible past and future states. In Herman, Y. (ed.), *The arctic seas: climatology, oceanography, biology and geology*. Van Nostrand: Reinhold Co. pp. 1–46.

Barry, R. G. (1991). Observational evidence of changes in global snow and ice cover. In Schlesinger, M. E. (ed.), *Greenhouse gas-induced climatic change: a critical appraisal of simulations and observations*. Amsterdam: Elsevier. pp. 329–45.

Barry, R. G. (1993). Canada's cold seas. In French, H. M. and Slaymaker, O. (eds.), *Canada's cold environments*. Montreal: McGill-Queen's University Press. pp. 29–61.

Barry, R. G. (1995a). Observing systems and data sets related to the cryosphere in Canada: A contribution to planning for the Global Climate Observing System. *Atmosphere-Ocean*, **33** (4): 771–807.

Barry, R. G. (1996). The parameterization of surface albedo for sea ice and its snow cover. *Progr. Phys. Geog.*, **20** (1): 61–77.

Barry, R. G. (1997). Cryospheric data for model validations: requirements and status. *Annals Glaciol.*, **26**: 371–5.

Barry, R. G. (2000). Data on the geographical distribution of sea ice. In Tanis, F. and Smolianitsky, V. (eds.), *Atlas climatology project environmental working group. Joint U.S.-Russian Atlas of Arctic Sea Ice*. NSIDC, Boulder, CO. CD-ROM.

Barry, R. G. (2002a). History of the World Data Center for Glaciology, Boulder, and the National Snow and Ice Data Center at the University of Colorado. Glaciol. Data Report GD-30. Twenty-fifth Anniversary. Monitoring an Evolving Cryosphere. NSIDC, Univ. of Colorado, Boulder, CO. pp. 1–7.

Barry, R. G. (2002b). The role of snow and ice in the global climate system: A review. *Polar Geog.*, **24** (3): 235–46.

Barry, R. G. (2003). Mountain cryospheric studies and the WCRP Climate and Cryosphere (CliC) Project. *J. Hydrology Special Issue: Mountain Hydrology and Water Resources* (eds., H. Lang and G. Kaser) **282** (1–4): 177–81.

Barry, R. G. (2006). The status of research on glaciers and global glacier recession: A review, *Progr. Phys. Geogr.*, **30** (3): 285–306.

Barry, R. G. (2008). *Mountain weather and climate*. 3rd edn. Cambridge: Cambridge University Press. 506 pp.

Barry, R. G., (2009). Snow cover. In Cuff, D. and Goudie, A. (eds.), *The Oxford companion to global change*. Oxford Reference Online. Oxford University Press. University of Glasgow. May 26, 2009. www.oxfordreference.com/view/10.1093/acref/9780195324884.001.0001/acref-9780195324884

Barry, R. G. and Carleton, A. M. (2001). *Synoptic and dynamic climatology*. London: Routledge. 620 pp.

Barry, R. G., Fallot, J.-M., and Armstrong, R. L. (1995). Twentieth-century variability in snow cover conditions and approaches to detecting and monitoring changes: Status and prospects. *Progr. Phys. Geog.*, **19** (4): 520–32.

Barry, R. G. and Maslanik, J. A. (1989). Arctic sea ice characteristics and associated atmosphere-ice interactions in summer inferred from SMMR data and drifting buoys: 1979 to 1985. *R. G. GeoJournal*, **18**: 35–44.

Barry, R. G., Moritz, R. E., and Rogers, J. C. (1979). The fast ice regimes of the Beaufort and Chukchi Sea coasts, Alaska. *Cold Regions Sci. Technol.*, **1**: 129–52.

Barry, R. G., *et al.* (1989). Characteristics of Arctic sea ice from remote sensing data and their relationship to atmospheric processes. *Annals Glaciol.*, **12**: 9–15.

Barry, R. G., *et al.* (1993). The Arctic sea-ice-climate system: Observations and modeling. *Rev. Geophys.*, **31**: 397–422.

Barry, R. G. and Serreze, M. C. (2000). Atmospheric components of the Arctic ocean freshwater balance and their interannual variability. In Lewis, E. L., *et al.* (eds.) *The freshwater budget of the arctic ocean.* Springer, Netherlands: Kluwer Academic Publ.: pp. 45–56

Barry R. G., Jania, J., and Birkenmajer, K. (2011). A. B. Dobrowolski – the first cryospheric scientist – and the subsequent development of cryospheric science. *Hist. Geo-Space Sci.*, **2**: 75–79

Barry, R. and Gan, T. Y. (2011). *Global Cryosphere, Past, Present and Future*, 472 pages, UK: Cambridge University Press, ISBN: 9780521769815 (Hardcover) & 9780521156851 (Paperback).

Barsch, D. (1988). Rock glaciers. In Clark, M. J. (ed.), *Advances in periglacial geomorphology.* Chichester: John Wiley and Sons. pp. 69–90.

Bartelt, P., Salm, B., and Gruber, U. (1999). Calculating dense-snow avalanche runout using a Voellmy-fluid model with active/passive longitudinal straining. *J. Glaciol.*, **45** (150): 242–54.

Bartelt, P. and Lehning, M. (2002). A physical SNOWPACK model for the Swiss avalanche warning services, Part I: Numerical model. *Cold Reg. Sci. Technol.*, **35** (3): 123–45.

Bartholomew, I., *et al.* (2010). Seasonal evolution of subglacial drainage and acceleration in a Greenland outlet glacier. *Nature Geosci.*, **3** (6): 408–11.

Bartlett, P. A., MacKay, M. D., and Verseghy, D. L. (2006). Modified snow algorithms in the Canadian Land Surface Scheme: Model runs and sensitivity analysis at three boreal forest stands. *Atmos. Ocean*, **44** (3): 207–22, doi: 10.3137/ao.440301.

Bassford, R. P., *et al.* (2006). Quantifying the mass balance of ice caps on Severnaya Zemlya, Russian High Arctic. I: Climate and mass balance of the Vavilov Ice Cap. *Arct. Antarct. Alpine Res.*, **38**: 1–12.

Batirov R. S., *et al.* (2003). *Avalanches of Uzbekistan.* Tashkent: SANIGMI. 119 pp.

Battle, W. R. B. and Lewis, W. V. (1951). Temperature observations in Bergschrunds and their relationship to cirque erosion. *J. Geol.*, **59** (6): 537–45.

Bauer, A. (1955). The balance of the Greenland ice sheet, *J. Glaciol.*, **2** (17): L456–62.

Bayr, K. J., Hall, D. K., and Kovalick, W. M. (1994). Observations on glaciers in the eastern Austrian Alps using satellite data. *Internat. J. Remote Sens.*, **15**: 1733–42.

Bazant, Z. P., Zi, G., and McClung, D. (2003). Size effect law and fracture mechanics of the triggering of dry snow slab avalanches, *J. Geophys. Res.*, **108** (B2): 2119. doi: 10.1029/2002JB001884.

Bazhev, A. (1997). Methods determining the internal infiltration accumulation of glaciers. In Kotltakov, V. M. (ed.), *34 Selected papers on main ideas of the Soviet glaciology, 1940s to 1980s,* Moscow: Glaciological Association, Institute of Geography, RAN. pp. 371–81.

Beaty, C. B. (1975). Sublimation or melting: Observations from the White Mountains, California and Nevada, USA, *J. Glaciol.*, **14**: 275–86.

Becker, A., *et al.* (2013). A description of the global land-surface precipitation data products of the Global Precipitation Climatology Centre with sample applications including centennial (trend) analysis from 1901–present, *Earth Syst. Sci. Data*, **5**: 71–99.

Beckley, B. D., *et al.* (2017). On the 'Cal-Mode' Correction to TOPEX Satellite Altimetry and Its Effect on the Global Mean Sea Level Time Series, *J Geophy Res-Oceans*, **122**: 8371–84, https://doi.org/10.1002/2017jc013090

Bedford, D. P. and Barry, R. G. (1994). Glacier trends in the Caucasus, 1960s to 1980s, *Phys. Geog.*, **15**: 414–24.

Bedford, D. and Douglass, A. (2008). Changing properties of snowpack in the Great Salt Lake Basin, western United States, from a 26-year SNOTEL record. *Prof. Geog.* **60**: 374–86.

Beedle, M. J. (2005). Climatic drivers of glacier mass balance in southeast Alaska in the second half of the twentieth century. M. A. thesis, Boulder, CO: University of Colorado, 172 pp.

Beedle, M. J., *et al.* (2008). Improving estimation of glacier volume change: a GLIMS case study of Bering Glacier System, Alaska. *The Cryosphere*, **2**: 33–51.

Belchansky, G. I., Douglas, D. C., and Platonov, N. G. (2004). Duration of the Arctic melt season: regional and interannual variability, 1979–2001. *J. Climate*, **17**: 67–80.

Bell, R., *et al.* (2007). Large subglacial lakes in East Antarctica at the onset of fast-flowing ice streams. *Nature*, **445**: 904–7.

Bell, R., *et al.* (2011). Widespread persistent thickening of the East Antarctic Ice Sheet by freezing from the base. *Science*, 331 (6024): 1592–5, doi: 10.1126/science.1200109.

Beltos, S. (2017). Frequency of ice-jam flooding of Peace-Athabasca Delta. *Can.*

J. Civ. Eng., NRC Res. Press, **45**: 71–5 dx.doi.org/10.1139/cjce-2017-0434

Beltaos S., Tang P., and Rowsell R. Ice jam modelling and field data collection for flood forecasting in the Saint John River, Canada, *Hydrological Processes*, **26**: 2535–45, doi: 10.1002/hyp.9293.

Beltaos, S. (ed.). (1995). *River ice jams.* Highlands Ranch, CO: Water Resources Publ., 390 pp.

Beltaos, S. (2001). Hydraulic roughness of breakup ice jams. *ASCE J. Hydraul. Eng.*, **127** (8): 650–6.

Beltaos, S. (2007). The role of waves in ice-jam flooding of the Peace–Athabasca delta, *Hydrol. Process.*, **21** (19): 2548–59.

Beltaos, S. (2008a). Progress in the study and management of river ice jams. *Cold Reg. Sci. Technol.*, **51**: 2–19.

Beltaos, S. (ed.). (2008b). *River ice breakup.* Highlands Ranch, CO: Water Resources Publ., 462 pp.

Beltaos, S. (2010). Internal strength properties of river ice jams. *Cold Reg. Sci. Technol.*, **62**: 83–91.

Beltaos, S. and Carter, T. (2009). Field studies of ice breakup and jamming in the Mackenzie delta. 15th Workshop on river ice. St. John's, Newfoundland. Committee on River Ice Processes and the Environment. pp. 266–83.

Beltaos, S. and Prowse, T. (2009). River-ice hydrology in a shrinking cryosphere. *Hydrol. Process.*, **23**: 122–44.

Beltaos, S., *et al.* (2006). Climatic effects on ice-jam flooding of the Peace-Athabaska delta. *Hydrol. Proc.*, **20** (19): 4031–50.

Bengtsson, L. (1986). Spatial variability of lake ice covers. *Geogr. Ann.*, **68A** (1–2): 113–21.

Beniston, M., Farinotti, D., Stoffel, M., Andreassen, L. M., Coppola, E., Eckert, N., *et al.* (2018). The European mountain cryosphere: a review of its current state, trends, and future challenges.

Cryosphere **12**: 759–94, doi: 10.5194/tc-12-759-2018.

Benn, D. I. and Lehmkuhl, F. (2000). Mass balance and equilibrium-line altitudes of glaciers in high mountain environments. *Quart. Int.*, **65–66**: 15–29.

Benn, D. I., Warren, C. R., and Mottram, R. H. (2007). Calving processes and the dynamics of calving glaciers. *Earth Sci. Rec.*, **82**: 143–79.

Benn, D., *et al.* (2009). Englacial drainage systems formed by hydrologically driven crevasse propagation. *J. Glaciol.*, **55** (191): 513–23.

Benson, B. and Magnuson, J. (2000), updated 2007. *Global lake and river ice phenology database*. Boulder, CO: National Snow and Ice Data Center/World Data Center for Glaciology. Digital media.

Benson, C. S. (1962). Stratigraphic studies in the snow and firn of Greenland ice sheet. US Army, Hanover, NH. CRREL Research Report 70, 93 pp.

Bentley, M. J., *et al.* (2006). Geomorphological evidence and cosmogenic 10Be/26Al exposure ages for the Last Glacial Maximum and deglaciation of the Antarctic Peninsula Ice Sheet. *Bull. Geol. Soc. Amer.*, **118**: 1149–59.

Bentley, M. J., *et al.* (2009). Mechanisms of Holocene palaeoenvironmental change in the Antarctic Peninsula region. *Holocene*, **19**: 51–69.

Bentley, W. A. and Humphries, W. J. (1931). *Snow crystals*. New York: McGraw-Hill (reprinted by Dover Publications, New York, 1964, 1973).

Benn, D. L. and Evans, D. J. A. (1998). *Glaciers and glaciation*. London: Arnold. 734 pp.

Bereiter, B., *et al.* (2015). Revision of the EPICA Dome C CO2 record from 800 to 600 kyr before present. *Geophys. Res. Lett.*, **42**: 542–9, doi: 10.1002/2014GL061957.

Berg, N. H. (1986). Blowing snow at a Colorado alpine site: Measurements and implications. *Arctic Alp. Res.*, **18**: 147–61.

Berger, A. and Loutre, M. F. (2010). Modeling the 100-kyr glacial–interglacial cycles. *Glob. Planet. Change*, **72** (4): 275–81.

Berger, C. L., *et al.* (2002). A climatology of northwest Missouri snowfall events: Long-term trends and interannual variability. *Phys. Geogr.*, **23**: 427–48.

Bergeron, V., Berger, C., and Betterton, M. D. (2006). Controlled irradiative growth of penitentes. *Phys. Rev. Lett.*, **96** (098502): 4.

Berghuijs, W. R., Woods, R. A., and Hrachowitz, M. (2014). A precipitation shift from snow towards rain leads to a decrease in streamflow. *Nat. Clim. Chang.* **4**: 583–6, doi: 10.1038/nclimate2246.

Bergström, S. (1995), The HBV model. In Singh, V. P. (ed.), *Computer models of watershed hydrology*. Highlands Ranch, CO: Water Resources Publications, pp. 443–76.

Berro, D. C., Mercalli, L., and Mortara, G. (2007). Evoluzions dei ghiacciai italiani nel periodo 2000–2007. *Nimbus*, **15** (3–4): 6–29.

Berthier E., *et al.* (2007). Remote sensing estimates of glacier mass balances in the Himachal Pradesh (Western Himalaya, India). *Rem. Sensing Environ.*, **108**: 327–38.

Berthier, E., *et al.* (2010). Contribution of Alaska glaciers to sea-level rise derived from satellite imagery. *Nature Geosci.*, **3**: 92–5.

Betin, V. V. and Preobazhensky, Y. V. (1959). Variations in the state of the ice on the Baltic Sea and in the Danish Sound. *Trudy Gos. Okean. Inst. (Moscow)*, **37**: 3–13. Translation 102, U.S. Navy Hydrographic Office, 1961.

Betterton, M. D. (2001). Theory of structure formation in snowfields motivated by penitentes, suncups, and dirt cone. *Phys. Rev. E.*, **63** (056129): 12.

Bhampri, R. and Bolch, T. (2009). Glacier mapping: A review with special reference to

the Indian Himalayas. *Progr. Phys. Geog.*,
33: 672–705.

Bhattacharya, I., *et al.* (2009). Surface melt
area variability of the Greenland ice sheet:
1979–2008. *Geophys. Res. Lett.*, **36**:
L20502, doi: 10.1029/2009GL039798.

Bhatt, U. S., *et al.* (2007). Examining glacier
mass balances with a hierarchical modeling
approach. *J. Computing Sci. Eng.*, **9**: 61–7.

Bianchi Janetti, E., *et al.* (2008). Regional
snow-depth estimates for avalanche
calculations using a two-dimensional
model with snow entrainment. *Annla
Glaciol.*, **49**: 63–70.

Bieniek, P. A., Bhatt, U. S., Rundquist, L. A.,
Lindsey, S. D., Zhang, X., and
Thoman, R. L. (2011). Large-scale climate
controls of interior Alaska river ice
breakup. *J. Clim.* **24**: 286–97. https://doi
.org/10.1175/2010JCLI3809.1

Biftu, G. F. and Gan, T. Y. (2001). Semi-
distributed, physically based, hydrological
modeling of the Paddle River Basin,
Alberta using remotely sensed data.
J. Hydrol., **244**: 137–56, doi: 10.1016/
S0022–1694(01)00333-X.

Bigg, G. R. (1999). An estimate of the flux of
iceberg calving from Greenland. *Arct.
Antarct. Alp. Res.*, **31**: 174–8.

Bigg, G. R., *et al.* (1997). Modelling the
dynamics and thermodynamics of icebergs.
Cold Regions Sci. Technol. **26**: 113–35.

Bigg, G. R. and Wilton, D. J. (2014). Iceberg
risk in the Titanic year of 1912: was it
exceptional? *Weather*, **69** (4): 100–4., RMS,
https://doi.org/10.1002/wea.2238C

Bilello, M. A. (1980). Maximum thickness and
subsequent decay of lake, river, and fast sea
ice in Canada and Alaska. Hanover, NH:
U.S. Army Cold Regions Research and
Engineering Laboratory, CRREL Report
80–6.

Bindschadler, R., *et al.* (1996). Surface
velocity and mass balance of ice streams
D and E, West Antarctica. *J. Glaciol.*, **42**
(142): 461–75.

Bindschadler, R., *et al.* (2008). The Landsat
image mosaic of Antarctica. *Rem. Sensing
Environ.*, **112** (12): 4214–26.

Bindschadler, R., *et al.* (2011). Getting around
Antarctica: new high-resolution mappings
of the grounded and freely-floating
boundaries of the Antarctic ice sheet
created for the International Polar Year.
Cryosphere, **5**: 569–88.

Bintanja, R. and van de Wal, R. S. W. (2008).
North American ice-sheet dynamics and
the onset of 100,000-year glacial cycles.
Nature, **454**: 869–72.

Bintanja, R., van der Linden, E. C., and
Hazeleger, W. (2011). Boundary layer
stability and Arctic climate change:
a feedback study using EC-Earth. *Climate
Dynamics*, **39**: 2659–73.

Birkenmajer, K., *et al.* (2005). First Cenozoic
glaciers in West Antarctica. *Polish Polar
Res.*, **26**: 3–12.

Birkeland, K. W. (1998). Terminology and
predominant processes associated with the
formation of weak layers of near-surface
crystals in the mountain snowpack. *Arct.
Alp. Res.*, **30** (2): 193–9.

Birkeland, K. W. (2001). Spatial patterns of
snow stability throughout a small mountain
range, *J. Glaciol.*, **47** (157): 176–86.

Bishop, M. and Barry, R. G., *et al.* (2004).
Global land ice measurements from
space (GLIMS): Remote sensing and
GIS investigations of the Earth's
cryosphere. *Geocarto Internat.*, **19**:
57–84.

Biskaborn, B. K., Smith, S. L., Noetzli, J.,
Matthes, H., Vieira, G.,
Streletskiy, D. A., Schoeneich, P.,
Romanovsky, V. E., Lewkowicz, A. G.,
Abramov, A., *et al.* (2019). Permafrost is
warming at a global scale, *Nat.
Commun.*, **10**: 264.

Bjørk, A. A., *et al.* (2012). An aerial view of 80
years of climate-related glacier fluctuations
in southeast Greenland, *Nat. Geosci.*, **5**:
427–32.

Björnsson, H. (2002). Subglacial lakes and jökulhlaups in Iceland. *Global Planet. Change*, **35**: 255–71.

Björnsson, H. (2009). *Jöklar á Íslandi (Glaciers in Iceland)*. Reyjavik: Opna, 478 pp.

Björnsson, H., *et al.* (2003). Surges of glaciers in Iceland. *Ann. Glaciol.*, **36**: 82–90.

Björnsson, H., *et al.* (2006). Climate change response of Vatnajökull, Hofsjökull and Langjökull ice caps, Iceland. European Conference on Impacts of Climate Change on Renewable Energy Sources Reykjavik, Iceland, June 5–9, 2006, 4 pp.

Bleil, U. and Thiede, J. (1990). The geological history of Cenozoic polar oceans: Arctic versus Antarctic – an Introduction. In Bleil, U. and Thiede, J. (eds.), *The geological history of Cenozoic polar oceans: Arctic versus Antarctic*. Dordrecht, Netherlands: Kluwer. pp. 1–8.

Bliss, A., Hock, R., and Cogley, J. G. (2013). A new inventory of mountain glaciers and ice caps for the Antarctic periphery. *Ann. Glaciol.*, **54**: 191–9.

Blunier, T., *et al.* (1997). Timing of the Antarctic Cold Reversal and the atmospheric CO2 increase with respect to the Younger Dryas event. *Geophys. Res. Lett.*, **24** (21): 2683–6.

Bockheim, J. G., *et al.* (2008). Distribution of permafrost types and buried ice in ice-free areas of Antarctica. In Kane, D. L. and Hinkel, K. M. (eds.), *Proceedings of the Ninth International Conference on Permafrost*, Fairbanks, AK: University of Alaska, Institute of Northern Engineering. pp. 125–30.

Boé, J., Hall, A., and Qu, X. (2009). September sea-ice cover in the Arctic Ocean projected to vanish by 2100. *Nature Geoscience* **2**: 341–3.

Bolch, T., Menounos, B., and Wheate, R. (2010). Landsat-based inventory of glaciers in western Canada, 1985–2005. *Remote Sens. Environ.* **114**: 127–37. doi: 10.1016/j.rse.2009.08.015.

Bolch, T. (2007). Climate change and glacier retreat in northern Tien Shan (Kazakhstan/Kyrgyzstan) using remote sensing data. *Global Planet. Change* **56**: 1–12.

Bolsenga. S. J. (1968). *River ice jams. A literature review. U.S. Lake Survey, Rep. 5–5*. Detroit, MI: Dept. of the Army, Corps of Engineers, Lake Survey District. 568 pp.

Bond, G., *et al.* (1993). Correlations between climate records from North Atlantic sediments and Greenland ice. *Nature*, **365**: 143–7.

Bony, S., et al. (2015). Clouds, circulation and climate sensitivity, *Nat. Geosci.*, **8**: 261–8, https://doi.org/10.1038/ngeo2398.

Boon, S., *et al.* (2010). Forty-seven years of research on the Devon Island Ice Cap, Arctic Canada. *Arctic*, **63**: 13–29.

Borodachev, B. E. and Shilnikov, V. I. (2003). *Istoriya L'dovoi Aviatsionnoi Razedki v Arktikei na Zamerzayushchikh Moryakh Rossii (1924–1993)* (The History of Aerial Ice Reconnaissance in the Arctic and Ice-covered Seas of Russia, 1924–1993). Gidrometeoizdat: St. Petersburg, 441 pp.

Borstad, C. P. and McClung, D. M. (2009). Sensitivity analyses in snow avalanche dynamics modeling and implications when modeling extreme events. *Canad. Geotech. J.* **46** (9): 1024–33.

Boulton, G. S., Peacock, J. D., and Sutherland, D. G. (2002). Quaternary. In Trewin, N. H. (ed.), *The geology of Scotland*. 4th edn., London: The Geological Society. pp. 409–30.

Bourke, R. H. and Garrett, R. P. (1987). Sea ice thickness distribution in the Arctic Ocean. *Cold. Regions Sci. Technol.* **13**: 259–80.

Bourke, R. H. and Mclaren, A. S. (1992). Contour mapping of Arctic Basin ice draft and roughness parameters. *J. Geophys. Res.*, **97**: 17, 715–28.

Bovis, M. J. (1977). Statistical forecasting of snow avalanches, San Juan Mountains, southern Colorado. *USA J. Glaciol.*, **18** (78): 87–99.

Bovis, M. J. and Mears, A. I. (1976). Statistical prediction of snow avalanche runout from terrain variables in Colorado. *Arct. Alp. Res.*, **8**: 115–20.

Box, J. E., *et al.* (2009). Greenland. Arctic Report Card 2009. www.arctic.noaa.gov /reportcard/

Bradley, R. S. and England, J. H. (2008). The Younger Dryas and the sea of ancient ice. *Quat. Res.*, **70** (1): 1–10.

Bradley, R. S., *et al.* (2009). Recent changes in freezing level heights in the Tropics with implications for the deglacierization of high mountain regions. *Geophys. Res. Lett.* **36**: L17701. doi: 10.1029/ 2009GL037712.

Braithwaite, R. J., Zhang, Y., and Raper, S. C. B. (2002). Temperature sensitivity of the mass balance of mountain glaciers and ice caps as a climatological characteristic. *Zeit. Gletscherk. Glazial.*, **38**: 35–61.

Brandt, R. E., *et al.* (2005). Surface albedo of the Antarctic sea ice zone. *J. Clim.*, **18**: 3606–22.

Bras, R. L. (1990). *Hydrology, An introduction to hydrologic science*. Reading, MA: Addison Wesley. 643 pp.

Braun, L. N., Weber, M., and Schulz, M. (2000). Consequences of climate change for runoff from Alpine regions. *Annals Glaciol.*, **31**: 19–25.

Braun, M., Humbert, A., and Moll, A. (2009). Changes of Wilkins Ice Shelf over the past 15 years and inferences on its stability. *The Cryosphere*, **3**: 41–56.

Brenner, A. C., DiMarzio, J. P., and Zwally, H. J. (2007). Precision and accuracy of satellite radar and laser altimeter data over the continental ice sheets. *IEEE Trans. Geosci. Remote Sens.*, **45**: 321–31.

Brigham, L. W., Grishchenk, V. D., and Kamesaki, K. (1999). The natural environment, ice navigation and ship Technology. In Østreng, W. (ed.), *The natural and societal challenges of the Northern Sea Route. A reference work*. Dordrecht: Kluwer Academic Publishers. pp. 48–120.

British Glaciological Society. (1949). Joint Meeting of the British Glaciological Society, the British Rheologists' Club and the Institute of Metals. *J. Glaciol.*, **1**: 231–40.

Brockamp, B. and Mothes, H. (1930). Seismische Untersuchungen aufdem Pasterzegletscher. *Zeit. Geophys.*, **6**: 482–500.

Broecker, W. S. (1997). Thermohaline circulation, the Achilles Heel of our climate system: Will man-made CO2 upset the current balance? *Science*, **278** (5343): 1582–8.

Broecker, W. S., *et al.* (2010). Putting the Younger Dryas cold event into context. *Quat. Sci. Rev.*, **29** (9–10): 1078–81.

Brohan P., *et al.* (2006). Uncertainty estimates in regional and global observed temperature changes: A new data set from 1850. *J. Geophys. Res.*, **111**: D12106, doi: 10.1029/2005JD006548.

Bromirski, P. D. , Sergienko, O. V. , and MacAyeal, D. R. (2009). Transoceanic infra-gravity waves impacting Antarctic ice shelves. *Geophys. Res. Lett.*, **37**: L02502, doi: 10.1029/2009GL041488.

Bromwich, D. H. and Kurtz, D. D. (1984). Katabatic wind forcing of the Terra Nova Bay polynya. *J. Geophys. Res.*, **89**: 3561–72.

Bromwich, D. H. and Parish, T. R. (1998). Chapter 4 of Meteorology of the Antarctic, Meteorology of the Southern Hemisphere. In Karoly, D. J. and Vincent, D. G. (eds.), *American Meteorological Society*, ISBN: 978-1-935704-10-2

Brönnimann, S., *et al.* (2008). Can we reconstruct Arctic sea ice back to 1900 with a hybrid approach? *Clim. Past Discuss.*, **4**: 955–79.

Bronselaer B., Winton, M., Griffies, S., Hurlin, W., Rodgers, K., Serigenko, O., Stouffer, R., and J. Russell, J. (2018). Change in future climate due to Antarctic meltwater, *Nature*, **564**: 53–58, doi: 10.1038/s41586-018-0712-z.

Brown, C. S., Meier, M. F., and Post, A. (1982). Calving speed of Alaska tidewater glaciers, with application to Columbia Glacier. U.S. Geol. Surv. Profess. Paper 1258-C, 13 pp.

Brown, J., Ferrians, Jr., O. J., Heginbottom, J. A., and Melnikov, E. S. (1997). Circum-arctic map of permafrost and ground-ice conditions, circum-pacific mMap series, *US Geological Survey*, ISBN 0-607-88745-1

Brown, J., *et al.* (1998). revised (2001). *Circum-Arctic map of permafrost and ground ice conditions*. Boulder, CO: National Snow and Ice Data Center/World Data Center for Glaciology. Digital Media.

Brown, J., Hinkel, K. M., and Nelson, F. E. (2000). The circumpolar active layer monitoring (CALM) program: research designs and initial results. *Polar Geogr.*, **24**: 165–258.

Brown, L. and Duguay, C. (2011). The fate of lake ice in the North American Arctic. *The Cryosphere*, **5** (4): 869–92, doi: 10.5194/tc-5-869-2011.

Brown, R. D. (1998). El Nino and North American snow cover. Proc. 55th Eastern Snow Conference, Jackson, N. H., June 4–6, pp. 165–72.

Brown, R. (Coordinating editor). (1999). *Canadian contributions to GCOS*. Freshwater ice.

Brown, R., Vikhamar-Schuler, D., Bulygina, O., Derksen, C., Loujus, K., Mudryk, L., *et al.* (2017). Arctic terrestrial snow cover. In *Snow, Water, Ice and Permafrost in the Arctic (SWIPA) 2017*, Oslo, Norway: Arctic Monitoring and Assessment Programme (AMAP). pp. 25–64.

Brown, R. D., Brasnett, B., and Robinson, D. (2003). Gridded North American monthly snow depth and snow water equivalent for GCM evaluation. *Atmosphere-Ocean*, **41** (1): 1–14, doi: 10.3137/ao.410101.

Brown, R. D. (2000). Northern Hemisphere snow cover variability and change, 1915–1997. *J. Climate*, **13**: 2339–55.

Brown, R. and Armstrong, R. L. (2008). Snow-cover data: measurement, products and sources. In Armstrong, R. L. and Brun, E. (eds.), *Snow and climate: physical processes, surface energy exchange and modeling*. Cambridge, UK: Cambridge University Press. pp. 181–216.

Brown, R. D. and Cote, P. (1992). Interannual variability of landfast ice thickness in the Canadian High Arctic, 1950–89. *Arctic*, **45**: 273–84.

Brown, R. D. and Mote, P.W. (2009). The response of Northern Hemisphere snow cover to a changing climate. *J. Climate*, **22**: 2124–45.

Brown, R. D., Walker, A., and Goodison, B. E. (2000). Seasonal snow cover monitoring in Canada – an assessment of Canadian contributions for global climate monitoring. Proc. 57th Eastern Snow Conference, Syracuse, NY, May 17–19, 2000: pp. 131–41.

Brown, R. D., *et al.* (2004). Climate variability and change – cryosphere. In Threats to water availability in Canada, NWRI scientific assessment, Report Series No. 3, Environment Canada, 128 pp.

Brown, R. D. and Robinson, D. A. (2011). Northern Hemisphere spring snow cover variability and change over 1922–2010 including an assessment of uncertainty. *Cryosphere*, **5**: 219–29, doi: 10.5194/tc-5-219-2011.

Bruce, I. (2009). *On thin ice: winter sports and climate change*, 54 pages, David Suzuki Foundation, ISBN 978–1-897375–24-2.

Brun, E., *et al.* (1992). A numerical model to simulate snow cover stratigraphy for operational avalanche forecasting. *J. Glaciol.*, **38**: 13–22.

Brun, E. *et al.* (2012). Simulation of northern Eurasian local snow depth, mass, and density using a detailed snowpack model and meteorological reanalyses. *Journal of Hydrometeorology*, **14** (1): 203–19, doi: 10.1175/jhm-d-12-012.1.

Brun, F., Berthier, E., Wagnon, P., Kääb, A., and Treichler, D. (2017). A spatially resolved estimate of High Mountain Asia glacier mass balances from 2000 to 2016. *Nat. Geosci.*, **10**: 668. Available at: https://doi.org/10.1038/ngeo2999

Brutsaert, W. (1982). *Evaporation into the atmosphere*, 299 pp., D. Reidel, Dordecht, Netherlands.

Budd, W. F., Dingle, R., and Radok, U. (1964). *Byrd snow drift project. meteorology dept.*, Australia: University of Melbourne, Publ. No. 6.

Budd, W. F., Jenssen, D., and Radok, U. (1971). Derived physical characteristics of the Antarctic ice sheet. ANARE interim report series, A(IV) Glaciology, 120, 178 pp.

Budd, W. F. and Jenssen, D. (1975). Numerical modeling of glacier systems. Proc. Moscow Sympos. on Snow and Ice in Mountainous Regions Internat. Assoc. Hydrol. Sci. Publ. No. 104: 257–91.

Budd, W. F., and McInnes, B. J. (1979). Periodic surging of the Antarctic ice sheet – an assessment by modelling. *Hydrol Sci. Bull.*, **24**: 95–104.

Budd, W. F., Smith, I. L., and Wishart, E. (1967). The Amery ice shelf. In Oura, H. (ed.), *Physics of snow and ice*. Proceedings of the international conference on low temperature science. I (1). Sapporo: Hokkaido University. pp. 447–67.

Budyko, M. I. (1969). The effect of solar radiation variations on the climate of the earth. *Tellus*, **21**: 611–19.

Bulygina, O. N., Razuvaev, V. N., and Korshunova, N. N., (2009), Changes in snow cover over Northern Eurasia in the last few decades, Environ. Res. Lett., **4** (045026): 6, doi: 10.1088/1748–9326/4/4/045026.

Bulygina, O., Groisman, P. Y., Razuvaev V. N., and Korshunova, N. N. (2011). Changes in snow cover characteristics over Northern Eurasia since 1966. *Environ Res Lett.*, 6 (4): doi: 10.1088/1748-9326/6/4/045204.

Burgess, D. O., *et al.* (2005). Flow dynamics and iceberg calving rates of the Devon ice cap, Nunavut, Canada. *J. Glaciol.*, **51**: 219–30.

Burgess, D. and Sharp, M. J. (2008). Recent changes in thickness of the Devon Island ice cap, Canada. *J. Geophys. Res.*, **113** (B7): B07204, doi: 10.1029/2007JB005238.

Burgess, E. W., *et al.* (2010). A spatially calibrated model of annual accumulation rate on the Greenland Ice Sheet (1958–2007). *J. Geophys. Res.*, **115**: F02004, doi: 10.1029/2009JF001293.

Bürki, R. , Elsasser, H. , and Abegg, B. (2003). Climate change and winter sports: Environmental and economic threats. 5th World Conference on Sport and Environment, Turin, December 2–3, 2003. IOC/UNEP.

Burn, C. R. and Nelson, F. E. (2006). Comment on "A projection of severe near-surface permafrost degradation during the 21st century" by David M. Lawrence and Andrew G. Slater. *Geophys. Res. Lett.*, **33**: L21503, doi: 10.1029/2006GL027077.

Burn, C. R. and Kokelj, S. V. (2009). The environment and permafrost of the Mackenzie Delta area. *Permafrost Periglac. Proc.*, **10**: 83–105.

Burnett, A. W., *et al.* (2003). Increasing Great Lake-effect snowfall during the Twentieth Century: A regional response to global warming? *J. Clim.*, **16**: 3535–41.

Burrows, C. J. (1976). Icebergs in the Southern Ocean. *New Zealand Geographer*, **32**: 127–38.

Buser, O. (1983). Avalanche forecast with the method of nearest neighbours: an interactive approach.*Cold Reg. Sci. Technol.*, **8**: 155–63.

Butkovich, T. R. (1954). Ultimate strength of ice. *SIPRE Res. Rep.*, **11** (US Army): 12.

Butsic, V., Hanak, E., and Valletta, R. (2009). Climate Change and Housing Prices: Hedonic Estimates for Ski Resorts in Western North America, Working Paper 2008–12, Federal Reserve Bank San Francisco.

Butt, M. (2009). Application of global snow model for the estimation of snow depth in the UK. *Meteorol. Atmos. Phy.*, **105**: 181–90.

Caian, M., Koenigk, T., Döscher, R., and Devasthale, A. (2018). An interannual link between Arctic sea-ice cover and the North Atlantic Oscillation. *Climate Dynamics*, **50**: 423–41.

CAA, 2016, Technical aspects of snow avalanche risk management—Resources and Guidelines for Avalanche Practitioners in Canada. In Campbell, C., Conger, S.,

Caine, N. (2002). Declining ice thickness on an alpine lake is generated by increased winter precipitation. *Clim. Change*, **54** (4): 463–70.

Callaghan, T. V., *et al.* (2010). A new climate era in the sub-Arctic: Accelerating climate changes and multiple impacts, *Geophysical Research Letters*, **37**: L14705, doi: 10.1029/2009GL042064.

Callaghan, T. V., *et al.* (2011a). Chapter 4. Changing snow cover and its impacts. In *Snow, water, ice and permafrost in the Arctic (SWIPA)*, 4:1-58 Oslo: Arctic Monitoring and Assessment Programme (AMAP), 538 pp.

Callaghan, T. V., *et al.* (2011b). The changing face of Arctic snow cover: A synthesis of observed and projected changes. *Ambio.*, **40** (Suppl 1): 17–31, doi: 10.1007/s13280-011-0212-y.

Campbell, E. C., *et al.* (2019). Antarctic offshore polynyas linked to Southern Hemisphere climate anomalies. *Nature*, **570**: 319–25, https://doi.org/10.1038/s41586-019–1294-0

Campbell, I. B. and Claridge, G. C. C. (2009). Antarctic permafrost soils. In Margesin, R. (ed.), *Permafrost soils*. Berlin: Springer Verlag. pp. 17–31.

Campbell, W. J. (1965). The wind-driven circulation of ice and water in a polar ocean. *J Geophys Res.*, **70**: 3279–01.

Caputo, M. V. and Crowell, J. C. (1985). Migration of glacial centers across Gondwana during Paleozoic Era Geol. *Soc. Amer. Bull.*, **96** (8): 1020–36.

Callendar, G. S. (1938). The artificial production of carbon dioxide and its influence on temperature. *Quart. J. Roy. Met. Soc.*, **64**: 223–40.

Callendar, G. S. (1961). Temperature fluctuations and trends over the earth. *Quart. J. Roy. Met. Soc.*, **87**: 1–12.

Carenzo, M., *et al.* (2009). Assessing the transferability and robustness of an enhanced temperature-index glacier-melt model. *J. Glaciol.*, **55** (190): 258–74.

Carey, K. L. (1973). *Icings developed from surface water and ground water. Cold Regions Sci, Eng. Monogr. III D3*. Hannover, NH: US Army Cold Regions Research and Engineering Laboratory. 65 pp.

Carey, M. (2005). Living and dying with glaciers: people's historical vulnerability to avalanches and outburst floods in Peru. *Global and Planetary Change*, **47**: 122–34.

Carmack, E. and Melling, H. (2011). Warmth from the deep. *Nature Geosci*, **4**: 7–8, https://doi.org/10.1038/ngeo1044.

Carlson, A. E., *et al.* (2007). Geochemical proxies of North American freshwater routing during the Younger Dryas cold event. *Proc. Nat. Acad. Sci.*, **104**: 6556–61.

Carlson, A. E., *et al.* (2008). Rapid early Holocene deglaciation of the Laurentide ice sheet. *Nature Geoscience*, **1**: 62–4.

Carozzi, A. V. (ed.). (1967). *Studies on glaciers.* Transl. of Agassiz, L. 1840 Etudes sur les glaciers, Neuchatel. New York: Hafner Publishing Co. 213 pp.

Carr, M. L., Gaughan, S. P., George, C. R., and Mason, J. G., (2015). CRREL's Ice Jam Database: Improvements and Updates, 18th Workshop on Hydraulics of Ice Covered Rivers Quebec City, Canada, August 18–20, 2015.

Carrivick, J. L., *et al.* (2012). Late-Holocene changes in character and behaviour of land-terminating glaciers on James Ross Island, Antarctica. *J. Glaciol.*, **58**: 1176–90.

Carrivick, J. L., *et al.* (2017). Ice-dammed lake drainage evolution at Russell Glacier, West Greenland. *Frontiers in Earth Science*, **5**: 100.

Carsey, F. D. (ed.). (1992). *Microwave remote sensing of sea ice.* Washington, DC: American Geophysical Union. 462 pp.

Carsey, F. D., *et al.* (1993). Status and future directions of remote sensing of sea. In: F.D. Carsey (ed.) Microwave Remote Sensing of Sea Ice. *Amer. Geophys. Union*, **26**: 443–6.

Casassa, G., *et al.* (2002). Current knowledge of the Southern Patagonia Icefield. In Casassa, G., Sepúlveda, F. V., and Sinclair, R. (eds.), *The Patagonian ice fields: a unique natural laboratory for environmental and climate change studies.* New York: Kluwer Academic/Plenum Publishers. pp. 67–83.

Castebrunet, H., *et al.* (2014). Projected changes of snow conditions and avalanche activity in a warming climate: the French Alps over the 2020–2050 and 2070–2100 periods. *The Cryosphere*, **8** (5): 1673–97, doi: 10.5194/tc-8-1673-2014.

Cavalieri, D. J., Gloersen, P., and Campbell, W. J. (1984). Determination of sea ice parameters with Nimbus 7 SMMR. *J. Geophys. Res.*, **89** (D4): 5355–69.

Cayan, D. R., *et al.* (2001). Changes in the onset of spring in the western United States. *Bull. Am. Met. Soc.*, **82**: 399–415.

Cazenave, A., *et al.* (2018). Global sea level budget 1993-present. *Earth System Science Data*, **10** (3), http://doi.org/10.5194/essd-10–1551-2018

Cazenave, A., Lombard. A., and Llovel, W. (2008). Present-day sea level rise: A synthesis, *Comptes Rendus Geosciences*, **340** (11): 761–70.

Chamberlin, T. C. (1894). Glacial studies in Greenland. *J. Geol.*, **2** (7): 649–66.

Chamberlin, T. C. (1897). Glacial studies in Greenland. X. The Bowdoin Glacier. *J. Geol.*, **5** (3): 229–40.

Chang, A. T. C., Foster, J. L., and Hall, D. K. (1987). Nimbus-7 derived global snow cover parameters. *Annals Glaciol.*, **9**: 39–44.

Chang, A. T. C., Foster, J. L., and Rango, A. (1991). Utilization of surface cover composition to improve the microwave determination of snow water equivalent in a mountain basin. *Int. J. Remote Sensing*, **12** (11): 2311–9.

Chang, A. T. C., *et al.* (1982). Snow water equivalent accumulation by microwave radiometry. *Cold Reg. Sci. Technol.*, **5** (3): 259–67.

Chang A. J. , *et al.* (1997). Snow parameters derived from microwave measurements during the BOREAS winter field campaign. *J. Geophys. Res.*, **102**: 29663–71.

Chapin, F. S., *et al.* (2005). Role of land-surface changes in Arctic summer warming. *Science*, **310** (5748): 657–60.

Chapman, W. L. and Walsh, J. E. (2007). Simulations of Arctic temperature and pressure by global coupled models. *J. Clim.*, **20**: 609–32, doi: 10.1175/JCLI4026.1.

Charrassin, J.-B., *et al.* (2008). Southern Ocean frontal structure and sea-ice formation rates revealed by elephant seals. *Proc. Nat. Acad. Sci.*, **105** (33): 11,634–9.

Chekotillo, A., Tsvid, A. A., and Makarov, V. N. (1960). *Naledy na territorii SSSR i bor'ba s nim* (Icings in the USSR snf their control). Blaoveshchensk: Amur. Knizhn. Izdat. 207 pp. (transl. 1965 for CRREL, US Army, Hanover, NH).

Chen, F., *et al.* (2014). Modeling seasonal snowpack evolution in the complex terrain and forested Colorado Headwaters region: A model intercomparison study. *J. Geophys. Res. Atmos.*, **119** (13): 795–13,819, doi: 10.1002/2014JD022167.

Chen, J. L., Wilson, C. R., and Tapley, B. D., (2013). Contribution of ice sheet and mountain glacier melt to recent sea level rise. *Nat. Geosci.*, **9**: 549–52, https://doi.org/10.1038/NGEO1829

Chen, J.-Y. and Funk, M. (1990). Mass balance of Rhonegletscher during 1982/83–1986/87. *J. Glaciol.*, **36** (123): 199–209.

Chen, J.-Y. and Ohmura, A. (1990). Estimation of Alpine glacier water resources and their change since the 1870s. Hydrology in Mountainous Regions. I – Uydrological Measurements; the Water Cycle (Proceedings of two Lausanne Symposia, August 1990). IAHS Publ. no. 193, pp. 127–35.

Chen, J. L., *et al.* (2009). Accelerated Antarctic ice loss from gravity measurements. *Nature Geosci.*, **2**: 859–62.

Chen, J. L., Wilson, C. R., and Tapley, B. D. (2006). Satellite gravity measurements confirm accelerated melting of Greenland ice sheet. *Science*, **313**: 1958–60.

Chen, Y. and She, Y. (2019). Temporal and Spatial Variations of River Ice Breakup Timing across Canada, CGU HS Committee on River Ice Processes and the Environment, 20th Workshop on the Hydraulics of Ice Covered Rivers Ottawa, Ontario, Canada, May 14–16, 2019.

Chen, J. M., Rich, P. M., Gower, S. T., Norman, J. M., and Plummer, S. (1997). Leaf area index of boreal forests: Theory, techniques and measurements. *J. Geophys. Res.*, **102**: 29,429–44.

Cheng, G. and Dramis, F. (1992). Distribution of mountain permafrost and climate. *Permafrost & Periglac. Proc.*, **3**: 83–91.

Cherry, J. E., *et al.* (2007). Development of the pan-Arctic snowfall reconstruction: new land-based solid precipitation estimates for 1940–99. *J. Hydromet.*, **8** (6): 1243–63.

Chinn, T. J. (1999). New Zealand glacier response to climate change of the past two decades. *Global Planet. Change*, **22** (1–4): 155–68.

Christoffersen, P., *et al.* (2018). Cascading lake drainage on the Greenland Ice Sheet triggered by tensile shock and fracture, *Nat. Commun.*, Vol. **9**, Article number:1064.

Choudhury, B. J., (1993), Reflectivities of selected land surfaces types at 19 and 37 GHz from SSM/I observations. *Remote Sens. Environ.*, **46**: 1–17.

Chudinova, S. M., Frauenfeld, O. W., Barry, R. G., Zhang, T.-J., and Sorokovikov, V. A. (2006). Relationship between air and soil temperature trends and periodicities in the permafrost regions of Russia. *J. Geophys. Res.*, **111** (F02008): 15.

Church *et al.* (2013). Sea level change, in climate change 2013, the physical science basis. In Stocker, T. F., *et al.* (eds.), *Contribution of WG I to AR5 of the intergovernmental panel on climate change*, Cambridge, UK: Cambridge University Press.

Ciracì, E., Velicogna, I., and Swenson, S. (2020). Continuity of the mass loss of the world's glaciers and ice caps from the GRACE and GRACE follow-on missions. *Geophy. Res. Lett.*, 47 (9): e2019GL086926. https://doi.org/10.1029/2019GL086926

Clair, T. A. and Ehrman, J. M. (1998). Using neural networks to assess the influence of changing seasonal climates in modifying discharge, dissolved organic carbon, and nitrogen export in eastern Canadian rivers. *Water Res. Res.*, **34** (3): 447–55.

Clark, M. P., Serreze, M. C., and Barry, R. G. (1996). Characteristics of Arctic Ocean climate based on COADS data, 1980–1993. *Geophys. Res. Lett.*, **23** (15): 1953–6.

Clark, P. U. and Huybers, P. (2009). Global change: Interglacial and future sea level. *Nature*, **462**: 856–7.

Clark, P. U. and Pollard, D. (1998). Origin of the middle Pleistocene transition by ice sheet erosion of regolith. *Paleoceanog.*, **13**: 1–9.

Clark, P. U., *et al.* (2009). The last glacial maximum. *Science*, **325** (5941): 710–14.

Clarke, G. K. C. (1991). Length, width and slope influences on glacier surging. *J. Glaciol.*, **36**: 236–46.

Clarke, G. K. C. (2003). Hydraulics of subglacial outburst floods: new insights from the Spring-Hutter formulation. *J. Glaciol.*, **49** (165): 299–313.

Clarke, G. K. C. (2005). Subglacial processes. *Annu. Rev. Earth Planet. Sci.*, **33**: 247–76.

Clarke, G. K. C. et al. (2015). Projected deglaciation of western Canada in the twenty-first century. *Nat. Geosci.*, **8** (5): 372–7, doi: 10.1038/ngeo2407.

Clifford, D. (2010). Global estimates of snow water equivalent from passive microwave instruments: history, challenges and future developments. *Int. J.Rem. Sensing*, **31** (14): 3707–26.

Cline, D. K. , *et al.* (2007). Overview of the Second Cold Land Processes Experiment (CLPX-II). IEEE Proc. International Geoscience and Remote Sensing Symposium, Barcelona.

Cogley, J. G. (2009). Geodetic and direct mass-balance measurements: comparison and joint analysis. *Ann. Glaciol.*, **50**: 96–100.

Colbeck, S. C. (1983). Theory of metamorphism of dry snow. *J. Geophys. Res.*, **88**: 5475–82.

Colbeck, S. C. (1997). *A review of sintering in seasonal snow. CRREL Report 97–10.* Hanover, NH: US Army Cold Regions Research & Engineering Laboratory. 17 pp.

Collins, D. N. (2006). Climatic variation and runoff in mountain basins with differing proportions of glacier cover. *Nordic Hydrol.*, **37**: 315–26.

Cogley, J. G. (2005). Mass and energy balances of glaciers and ice sheets. In Anderson, M. G. (ed.), *Encyclopedia of hydrological sciences*, vol. **4**. New York: J. Wiley and Sons. pp. 2555–74.

Cogley, J. G., (2008). Measured rates of glacier shrinkage. *Geophys. Res. Abstracts*, **10**: EGU2008-A-11595.

Cogley, J. G. (2009a). Geodetic and direct mass-balance measurements: comparison and joint analysis. *Annals Glaciol.*, **50**: 96–100.

Cogley, J. G. (2009b). A more complete version of the World Glacier Inventory. *Annals Glaciol.*, **50** (53): 32–8.

Collins, D. N. (2008). Climatic warming, glacier recession and runoff from Alpine basins after the Little Ice Age maximum. *Ann. Glaciol.*, **48**: 119–24.

Colony, R., Radionov, V., and Tanis, F. I. (1998). Measurements of precipitation and snow pack at the Russian North Pole drifting stations. *Polar Record*, **34**: 3–14.

Comiso, J. C. (1986). Characteristics of Arctic winter sea ice from satellite multispectral microwave observation. *J. Geophys, Res.*, **91** (Cl): 975–94.

Comiso, J. C. (2010). Variability and trends of the global sea ice cover. In Thomas, D. N. and Dieckmann, G. S. (eds.), *Sea ice.* 2nd edn. Chichester: Wiley-Blackwell. pp. 205–46.

Comiso, J. C. (2012). Large decadal decline in the Arctic multiyear ice cover. *J. Clim.*, **25**: 1176–93.

Comiso, J. C., Kwok, R., Martin, S., and Gordon, A. L. (2011). Variability and trends in sea ice extent and ice production in the Ross Sea. *J Geophys Res.*, **116** (C04):021. https://doi.org/10.1029/2010JC006391

Comiso, J. C., *et al.* (2017). Positive trend in the Antarctic sea ice cover and associated changes in surface temperature. *J. of Climate*, **30** (6): 2251–67, doi: 10.1175/Jcli-D-16-0408.1.

Comiso, J. C., Parkinson, C. L., Gersten, R., and Stock, L. (2008). Accelerated decline in the Arctic Sea ice cover. *Geophys. Res. Letters*, **35**: L01703. https://doi.org/10.1029/2007GL031972

Comiso, J. C., Cavalieri, D. J., and Markus, T. (2003). Sea ice concentration, ice temperature, and snow depth using AMSR-E data. *IEEE Trans. Geosci. Remote Sensing*, **42**: 243–52.

Comiso, J. C., Cavalieri, D., Parkinson, C., and Gloersen, P. (1997). Passive microwave algorithms for sea ice concentrations: A comparison of two techniques. *Remote Sensing Environ.*, **60** (3): 357–84.

Committee on Climate, Energy, and National Security. (2009). *Scientific value of Arctic sea ice imagery derived products.* Washington, DC: National Research Council. 48 pp.

Connolly, R., *et al.* (2019). Northern hemisphere snow-cover trends (1967–2018): a comparison between climate models and observations. *Geoscience*, **9**: 135, doi: 10.3390/geosciences9030135.

Conway, H. and Abrahamson, J. (1984). Snow-slope stability – a probabilistic approach. *J. Glaciol.*, **34** (117): 170–7.

Conway, H., *et al.* (1999). Past and future grounding-line retreat of the West Antarctic ice sheet. *Science*, **286**: 280–3.

Cook, A. J., *et al.* (2005). Retreating glacier fronts on the Antarctic Peninsula over the past half-century. *Science*, **308**: 541–4.

Cook, A. J. and Vaughan, D. G. (2010). Overview of areal changes of the ice shelves on the Antarctic Peninsula over the past 50 years. *Cryosphere*, **4**: 77–98.

Cooley, S. and Pavelsky, T. (2016). Spatial and temporal patterns in Arctic river ice breakup revealed by automated ice detection from MODIS imagery. *Remote Sensing of Environment*, **175**: 310–22, http://dx.doi.org/10.1016/j.rse.2016.01.004

Coon, M. D., *et al.* (1998). The architecture of anisotropic elastic-plastic sea ice mechanics constitutive law. *J. Geophys. Res.*, **103** (C10): 21, 915–25.

Copland, L. and Mueller, D. (2017). *Arctic ice shelves and ice islands*. Dordrecht: Springer. DOI: 10.1007/978-94-024-1101-0_11.

Copland, L., Sharp, M. J., and Dowdeswell, J. A. (2003). The distribution and flow characteristics of surge-type glaciers in the Canadian High Arctic. *Ann. Glaciol.*, **36L**: 73–81.

Costard, F., *et al.* (2007). Impact of the global warming on the fluvial thermal erosion over the Lena River in central Siberia. *Geophys. Res. Lett.*, **34**: L14501, doi: 10.1029é2007GL030212.

Cotté, C. and Guinet, C. (2007). Historical whaling records reveal major regional retreat of Antarctic sea ice. *Deep Sea Res.*, **54**: 243–52.

Cox, J. (2005). The snow/snow water equivalent ratio and its predictability across Canada. MSc thesis, Montreal: McGill University. 102 pp.

Crary, A. P. (1958). Arctic ice island and ice shelf studies. *Part I. Arctic*, **11**: 2–42.

Crary, A. P. (1960). Arctic ice islands and ice shelf studies. *Part. II. Arctic*, **13**: 32–50.

Crary, A. P., *et al.* (1962). Glaciological regions of the Ross Ice Shelf. *J. Geophys. Res.*, **67**: 2791–807.

Crocker, G. B. and Cammaert, A. B. (1994). Measurements of bergy bit and growler populations off Canada's East Coast. In: Proc. of IAHR Ice Symposium, Trondheim, Norway, August 23–26, 1994, Vol 1: 167–76.

Crites, H., Kokelj, S. V., and Lacelle, D. (2020). Icings and groundwater conditions in permafrost catchments of northwestern Canada. *Sci Rep*, **10**: 3283, https://doi.org/10.1038/s41598-020–60322-w

Cullen, *et al.* (2013). A century of ice retreat on Kilimanjaro: The mapping reloaded. *Cryosphere*, **7**: 419–31.

Curran, M. A. J., *et al.* (2003). Ice core evidence for sea ice decline since the 1950s. *Science 302*, **295**: 1890–2.

Curry, J. A., Schramm, J. L., and Ebert, E. E. (1995). Sea ice-albedo climate feedback mechanism. *Journal of Climate*, **8**: 240–7.

Cuffey, K. M. and Paterson, W. S. B. (2010). *The physics of glaciers*. 4th edn. Burlington, MA: Butterworth-Heinemann/Elsevier. 704 pp.

Czudek, T. and Demek, J. (1970). Thermokarst in Siberia and its influence on the development of lowland relief. *Quat. Res.*, **1**: 103–20.

Dadic, R., *et al.* (2010). Wind influence on snow depth distribution and accumulation over glaciers. *J. Geophys. Res.*, **115**: F01012. doi: 10.1029/2009JF001261.

Dahl, S. O., *et al.* (2005). Weichselian glaciation history in the Rondane 'dry valleys' of central Scandinavia. Geological Society of America. 2005 Salt Lake City Annual Meeting, paper 178–9.

Dahl-Jensen, D., *et al.* (2009). *The greenland ice sheet in a changing climate: snow, water, ice and permafrost in the arctic (SWIPA)*. Oslo: Arctic Monitoring and Assessment Programme (AMAP). 115 pp.

Dai, J. C., *et al.* (2009). Cold decade (AD 1810–1819) caused by Tambora (1815) and another (1809) stratospheric volcanic eruption. *Geophy. Res. Lett.*, 36: L22703, doi: 10.1029/2009GL040882.

Daly, S. F. (2008). Evolution of frazil ice. Proceedings of 19th IAHR International Symposium on Ice "Using New Technology to Understand Water-Ice Interaction." Jasek. M. (ed.) Vol. 1: 29–47.

Dansgaard, W., *et al.* (1969). One thousand centuries of climatic record from camp century on the greenland ice sheet. *Science*, **166** (3903): 377–80.

Darby, D. A. (2003). Sources of sediment found in sea ice from the western Arctic Ocean, new insights into processes of entrainment and drift patterns. *J. Geophys. Res.*, **108** (C8): 3257.

Darby, D. A. (2008). Arctic perennial ice cover over the last 14 million years. *Paleoceanog.*, **23**: PA1S07.

Darwin, C. (1839). Journal of researches into the natural history and geology of the countries visited during the voyage of H.M.S. Beagle round the world, under the Command of Capt. *Fitz Roy, R.N.* 2nd edn. London, UK: H. Colburn. p. 325. (http://darwin-online.org.uk/content/frameset?itemID=F20&viewtype=text&pageseq=1).

Das, S. B., *et al.* (2008). Fracture propagation to the base of the Greenland Ice Sheet during supraglacial lake drainage. *Science*, **320**: 778–81.

Davies, B. J. and Glasser, N. F. (2012). Accelerating shrinkage of Patagonian glaciers from the "Little Ice Age" (c. AD 1870) to 2011. *J. Glaciol.*, **58**: 1063–84.

Davis, P. T., Menounos, B., and Osborn, G. (2009). Introduction, Holocene and latest Pleistocene alpine glacier fluctuations: a global perspective. *Quat. Sci. Rev.*, **28** (21–22): 2021–33.

De Angelis, H. and Skvarca, P. (2003). Glacier surge after ice shelf collapse. *Science*, **299** (5612): 1560–2.

DeBeer, C. M., Wheater, H. S., Carey, S. K., and Chun, K. P. (2016). Recent, climatic, cryospheric, and hydrological changes over the interior of western Canada: A review and synthesis. *Hydrology and Earth System Sciences*, **20**: 1573–98.

DeConto, R. M., *et al.* (2008). Thresholds for Cenozoic bipolar glaciation. *Nature*, **455**: 652–6.

DeConto, R. M. and Pollard, D. (2016). Contribution of Antarctica to past and future sea-level rise. *Nature*, **531**: 591–7.

de Freitas, C. R. (1975). Estimation of the disruptive impacts of snowfalls in urban areas. *J. Appl. Met.*, **14**: 1166–73.

de la Mare, W. K. (1997). Abrupt mid-twentieth-century decline in Antarctic sea-ice extent from whaling records. *Nature*, **389**: 57–60.

de la Mare, W. K. (2009). Changes in Antarctic sea-ice extent from direct historical observations and whaling records. *Climatic Change*, **92**: 461–93.

de Quervain, M. R. (1950). Die Festigkeitseigenschaften der Schneedecke und ihre Messung. *Geofis. pura appl.*, **18**: 3–15.

de Quervain, M. and Meister, R. (1987). Fifty years of snow profiles on the Weissflujoch and relations to the surrounding avalanche activity (1936/37–1985–86). In Salm, B. and Guler, H. (eds.), *Avalanche formation, movement and effects. Proceedings of the Davos Symposium), IAHS Publ. no. 162*, Wallingford, UK: IAHS. pp. 161–81.

de Rham. L. P., Prowse, T. D., and Bonsal, B. R. (2008). Temporal variations in river-ice breakup over the Mackenzie River Basin, Canada. *J. Hydrol.*, **349**: 441–54.

de Scally, F. A. (1992). Influence of avalanche snow transport on snowmelt runoff. *J. Hydrol.*, **137**: 73–97.

Dedieu, J. F., *et al.* (2003). Glacier mass balance determination by remote sensing in the French Alps: Progress and limitation for time series monitoring. International Geoscience and Remote Sensing Symposium (IGARSS) '03, Proceedings 4: 2602–04.

Dedrick, K. R. (2002). Estimating sea ice thickness distributions and modeling their evolution in time. *Oceans '02 MTS/IEEE.*, **2**: 877–83.

Dee, D. P., *et al.* (2011). The ERA-Interim reanalysis: configuration and performance of the data assimilation system. *Q. J. R. Meteorol. Soc.*, **137**: 553–97, doi: 10.1002/qj.828.

del Rosario Prieto, A., García-Herrera, B. R., and Hernández Martin, E. (2004). Early records of icebergs in the South Atlantic Ocean from Spanish documentary sources. *Clim. Change*, **66**: 29–48.

Demuth, M. N., Munro, D. S., and Young, G. J. (eds.). (2006). *Peyto Glacier – One century of science*. Saskatoon, Saskatchewan: National Water Research Institute.

den Hartog, G., *et al.* (1983). An investigation of a polynya in the Canadian archipelago, Pt. 3. Surface heat flux. *J. Geophys. Res.*, **88**: 2911–16.

Denton, G. H., *et al.* (2010). The last glacial termination. *Science*, **328** (5986): 1652–6.

Depoorter, M. A., *et al.* (2013). Calving fluxes and basal melt rates of Antarctic ice shelves. *Nature*, **502** (7469): 89–92, doi: 10.1038/nature12567.

Derksen, C., *et al.* (2009). Northwest Territories and Nunavut snow characteristics from a Subarctic traverse: Implications for passive microwave remote sensing. *J. Hydromet.*, **10**: 448–63.

Derksen, C., *et al.* (2014). Physical properties of Arctic versus subarctic snow: Implications for high latitude passive microwave snow water equivalent retrievals. *J. Geophys. Res. Atmos.*, **119**: 7254–70, doi: 10.1002/2013JD021264.

Derksen, C., Walker, A., Goodison, B., and Strapp, J. W. (2005). Integrating in situ and multi-scale passive microwave data for estimation of sub-grid scale snow water equivalent distribution and variability. *IEEE Transactions on Geoscience and Remote Sensing*, **43** (5): 960–72.

Derouin, S. (2020). River ice is disappearing, Eos, 101, AGU, Published on February 18, 2020, https://doi.org/10.1029/2020EO140159.

Déry, S. J. and Brown, R. D. (2007). Recent Northern Hemisphere snow cover extent trends and implications for the snow-albedo feedback. *Geophys. Res. Lett.*,

34 (22): L22504, doi: 10.1029/ 2007GL031474.

Deser, C. and Teng, H.-Y. (2008). Recent trends in Arctic sea ice and the evolving role of atmospheric circulation forcing, 1979–2007. In De Weaver, E. T., Bitz, C., and Tremblay, L.-B. (eds.), *Arctic sea ice decline: Observations, projections, mechanisms, and implications*. Washington, DC: American Geophysical Union. pp. 7–26.

Desinov, L. V. and Konovalov, V. G. (2007). *Distancionny monitoring mnogoletnego regima oledenenia Pamira (Monitoring of multiannual glacial regime in the Pamir using remote sensing)*. Moscow: Inst. of Geography, RAS. Data Glaciol. Studies **103**: 129–34 (in Russian).

Dewalle, D. R. and Rango, A. (2008). *Principles of snow hydrology*. Cambridge: Cambridge University Press. 410 pp.

Dobhal, D. P., Gergan, J. T., and Thayyen, R. J. (2004). Recession and morphogeometrical changes of Dokriani glacier (1962–1995), Garhwal Himalaya, India. *Curr. Sci.*, **86** (5): 101–7.

Deutschen Höhlen- und Karstforscher. (2005). *Berchtesgadener Alpen. Karst und Höhle 2004/2005*. Munich: Verband Deutschen Höhlen- und Karstforscher. 237 pp.

De Vernal, A. and Hillaire-Marcel, C. (2008). Natural variability of Greenland climate, vegetation, and ice volume during the past million years. *Science*, **320** (5883): 1622–5.

de Woul, M. (2008). *Response of glaciers to climate change. Dissertations from the Department of Physical Geography and Quaternary Geology no. 13*. Stockholm University. 20 pp.

de Woul, M. and Hock, R. (2005). Static mass-balance sensitivity of Arctic glaciers and ice caps using a degree-day approach. *Ann. Glacio.*, **42**: 217–24.

Dessler, A. E., Schoeberl, M. R., Wang, T., Davis, S. M., and Rosenlof, K. H. (2013). Stratospheric water vapor feedback. *Proceed. of National Academy of Sciences of the United States of America*, **110**: 18087–91.

Devik, O. (1949). Freezing water and supercooling, anchor ice and frazil ice. *J. Glaciol*, **1** (6): 307–9.

Deynoux, M. (2004). *Earth's glacial record*. Cambridge: Cambridge University Press. 384 pp.

Diablobanquisa, *et al.* (2016). September Arctic sea ice extent: 1935–2014, Revista de Climatología.

Dickfoss, P. V., *et al.* (1997). History of ice at Candelaria Ice Cave, New Mexico. In Maybery, K. (comp.) *A natural History of El Malpais*. Socorro, NM: New Mexico Bureau of Mines and Mineral Resources. Bulletin No. 156. pp.91–112.

Dickfoss, P., Betancourt, J. L., and Thompson, L. (1997). History and paleoclimatic potential of Candelaria Ice Cave, west-central New Mexico. In Zidek, G. (ed.), *A Natural History of El Malpais*. New Mexico Bureau of Mines and Mineral Resources. Bulletin No. 156t, pp.91–112.

Dickinson, R. E., *et al.* (1991). Evapotranspiration models with canopy resistance for use in climate models. *A review. Agric. For. Meteorol.*, **54**: 373–88.

Dickson, R. R. (2009). The integrated Arctic Ocean Observing System (iAOOS) in 2008. A Report of the Arctic Ocean Sciences Board. 84 pp.

Diemand, D. (1983). Measurement of iceberg temperatures. *Icebrg Res.* (Scot Polar Res, Inst., Cambridge), No. **5**: 3–16.

Dieng, H. B., *et al.* (2017). New estimate of the current rate of sea level rise from a sea level budget approach. *Geophy. Res. Lett.*, **44**: 3744–51, https://doi.org/10.1002 /2017GL073308

Ding, S., W. Chen, J. Feng, and H.-F. Graf, (2017). Combined Impacts of PDO and

Two Types of La Niña on Climate Anomalies in Europe. Journal of Climate, 30, 3253–3278.

Ding, Y.-J. and Liu, S.-Y. (2006). The retreat of glaciers in response to recent climate warming in west China. *Ann. Glaciol.*, **43**: 97–106.

Dingman, S. L., *et al.* (1980). Climate, snow cover, microclimate anhydrology of the Arctic coastal plain. In Brown, J., *et al.* (eds.), *An arctic ecosystem: the coastal tundra at Barrow*, Alaska. Stroudsburg, PA: Dowden, Hutchinson and Ross. pp. 30–65.

Diolaiuti, G., *et al.* (2012). Evidence of climate change impact upon glaciers' recession within the Italian Alps—The case of Lombardy glaciers. *Theor. Appl. Climatol.*, **109**: 429–45.

Dirmeyer, P. A., *et al.* (2006). GSWP-2: Multimodel analysis and implications for our perception of the land surface. *Bull. Amer. Met. Soc.*, **87**: 1381–97.

Ditlevsen, P. D. and Ditlevsen, O. D. (2009). On the stochastic nature of the rapid climate shifts during the Last Ice Age. *J. Clim.*, **22**: 446–57.

Divine, D. V. and Dick, C. (2006). Historical variability of sea ice edge position in the Nordic Seas. *J. Geophys. Res.*, **111**: C01001. doi: 10.1029/2004JC002851.

Dmitrenko, I. A., *et al.* (2006). Seasonal variability of Atlantic water on the continental slope of the Laptev Sea during 2002–2004. *Earth Planet. Sci. Lett.*, **244**: 736–43.

Dmitrenko, I. A., *et al.* (2009). Sea-ice production over the Laptev Sea 244: shelf inferred from historical summer-to-winter hydrographic observations of 1960s-1990s. *Geophys. Res. Lett.*, **36**: L13605.

Dobhal, D. P. (2004). Retreating Himalayan glaciers – An overview. Proc: Receding glaciers in Indian Himalayan Region (IHR) – Environmental and Social Implications, pp. 26–38.

Dobrowolski, A. B. (1923). *Historja naturalna lodu (Natural history of ice)* (in Polish, French summary). Warsaw: Naklad H. Lindenfelda. 940 pp.

Dolan, W., Yang, X., Zhang, S., and Pavelsky, T. (2019). Working towards optical remote sensing of pan-Arctic river and lake ice, CGU HS Committee on River Ice Processes and the Environment, 20th Workshop on the Hydraulics of Ice Covered Rivers Ottawa, Ontario, Canada, May 14–16, 2019.

Dong, J., Walker, J. P., and Houser, P. R. (2005). Factors affecting remotely sensed snow water equivalent uncertainty. *Remote Sens. Environ.*, **97** (1): 68–82.

Donner, J. (2005). *The Quaternary history of Scandinavia*. Cambridge: Cambridge University Press. 212 pp.

Doran, P. T., *et al.* (2000). Sedimentology and geochemistry of a perennially ice-covered epishelf lake in Bunger Hills Oasis, East Antarctica. *Antarct. Sci.*, **12**: 131–40.

Doronin, Y. P. and Kheisin, D. E. (1977). *Sea ice*. Rotterdam: Balkema. 323 pp.

Dowdeswell, J. A. (1989). On the nature of Svalbard icebergs. *J. Glaciol.*, **35**: 224–34.

Dowdeswell, J. A., Whittington, R. J., and Hodgkins, R. (1992). The sizes, frequencies, and freeboards of East Greenland icebergs observed using ship radar and sextant. *J. Geophys. Res.*, **97** (C3): 3515–28.

Dowdeswell, J. A., Glazovsky, A. F., and Macheret, Y. Y. (1995). Ice divides and drainage basins on the ice caps of Franz Josef Land, Russian High Arctic, defined from Landsat, KFA-1000 ad ERS-1 SAR imagery. *Arct. Alp, Res*, **27**: 264–70.

Dowdeswell, J. A., *et al.* (2004). Form and flow of the Devon Island Ice Cap, Canadian Arctic. *J. Geophys. Res.*, **109**: F02002, doi: 10.1029/2003JF000095.

Dowdeswell, J. A. and Hagen, J. O. (2004). Arctic ice caps and glaciers. In Bamber, J. L. and Payne A. J. (eds.), *Mass* balance of the cryosphere. Ch. 14, Cambridge: Cambridge University Press, pp. 527–57.

Dowdeswell, J. A., *et al.* (2010). The glaciology of the Russian High Arctic from Landsat imagery. In Williams, R. S. Jr. and Ferrigno, J. G. (eds.), Satellite image atlas of glaciers. Glaciers of Asia. U.S. Geological Survey Profess. Paper, 1386-F, pp. 94–125.

Doyle, P. F. and Ball, J. F. (2008). Changing ice cover regime in southern British Columbia due to changing climate. In *Proceedings of the 19th IAHR International Symposium on Ice. Using New Technology to Understand Water-Ice Interaction* (Jasek M., Ed.). St Joseph Communications, Vancouver, Canada. pp. 51–61.

Dozier, J., Schneider, S. R, and McGinnis, D. F. Jr. (1981). Effect of grain size and snowpack water equivalence on visible and near-infrared satellite observations of snow. *Water Resour. Res.*, **17**: 1213–21.

Dozier, J., *et al.* (2009). Interpretation of snow properties from imaging spectrometry. *Remote Sens. Environ.*, **113**: S25–S37.

Drenkhan, F., *et al.* (2015) The changing water cycle: climatic and socioeconomic drivers of water-related changes in the Andes of Peru. *Wiley Interdiscip. Rev. Water*, **2** (6): 715–33, doi: 10.1002/wat2.1105.

Drewry, D. J., Jordan, S. R., and Jankowski, E. (1982). Measured properties of the Antarctic ice sheet: surface configuration, ice thickness, volume and bedrock characteristics. *Ann. Glaciol.*, **3**: 83–91.

Driedger, C. L. and Fountain, A. G. (1989). Glacier outburst floods at Mount Rainier, Washington State, USA. *Ann. Glaciol.*, **13**: 51–5.

Drinkwater, M. R. (1998). Active microwave remote sensing of observations of Weddell Sea ice. In Jeffries, M. (ed.), *Antarctic sea ice physical processes, interactions and variability. Antarctic Res. Ser. 74*, Washington, DC: American Geophysical Union. pp. 187–212.

Drobot, S. D. and Anderson, M. R. (2001). An improved method for determining snowmelt onset dates over Arctic sea ice using Scanning Multichannel Microwave Radiometer and Special Sensor Microwave/Imager data. *J. Geophys. Res.*, **106** (D20): 24,033–50.

Drobot, S. D., *et al.* (2008). Evolution of the 2007–2008 Arctic sea ice cover and prospects for a new record in 2008. *Geophys. Res. Lett.* **35** (L19501): 5.

Drygalski, E. von and Machatschek, F. (1942). *Encyclopaedie der. Erdkunde*. Gletscherkunde: ViennaL Franz Deuticke. 261 pp.

Dud-Rodkin, A., *et al.* (2004). Timing and extent of Plio-Pleistocene glaciations in north-western Canada and east-central Alaska. In Ehlers, J. and Gibbard, P. L. (eds.), *Quaternary glaciations -extent and chronology, Part II, North America*. New York: Elsevier. pp. 313–45.

Duguay, C. R., *et al.* (2003). Ice cover variability on shallow lakes at high latitudes: Model simulations and observations. *Hydrol. Proc.*, **17**: 3465–83.

Duguay, C. R., *et al.* (2006). Recent trends in Canadian lake ice cover. *Hydrol. Proc.*, **20**: 781–801.

Dunbar, M. (1969). The geographical position of the North Water. *Arctic*, **22**: 438–41.

Dunbar, M. and Greenway, K. R. (1956). *Arctic Canada from the air*. Ottawa: Queen's Printer. 541 pp.

Dunble. D. H. (1860). On the contraction and expansion of ice. *Canad. J. Industry, Sci., Art, n.s.* No. **29**: 418–25.

Dutra, E., *et al.* (2010). An improved snow scheme for the ECMWF land surface model: description and offline validation. *Journal of Hydrometeorology*, https://doi.org/10.1175/2010JHM1249.1

Dutrieux, P. *et al.* (2014). Strong sensitivity of Pine Island ice shelf melting to climatic variability. *Science*, **343**: 174–8, https://doi.org/10.1126/science.1244341, 2014.

Dye, D. G. (2002). Variability and trends in the annual snow-cover cycle in Northern Hemisphere land areas, 1972–2000. . *Hydrol. Proc.*, **16**: 3065–77.

Dyer, J. L. and Mote, T. L. (2006). Spatial variability and trends in observed snow depth over North America. *Geophys. Res. Lett.*, **33**: L16503, doi: 10.1029/ 2006GL027258.

Dyunin, A. K., *et al.* (1977). Strong snow-storms, their effect on snow cover and snow accumulation. *J. Glaciol.*, **19** (81): 441–9.

Dyurgerov, M. B. (2001). Mountain glaciers at the end of the twentieth century: global analysis in relation to climate and water cycle. *Polar Geog.*, **25**: 241–336.

Dyurgerov M. (2003). Mountain and subpolar glaciers show an increase in sensitivity to climate warming and intensification of the water cycle. *J. Hydrol.*, **282**: 164–76.

Dyurgerov, M. B. (2010). Reanalysis of glacier changes: from the IGY to the IPY, 1960–2008. *Data Glaciol. Stud.*, **108**: 1–116.

Dyurgerov, M. B. and Bahr, D. B. (1999). Correlations Between glacier properties – Finding appropriate parameters for glacier monitoring. *J. Glaciol.*, **45** (149): 9–16.

Dyurgerov, M. B. and Meier, M. F. (1999). Analysis of winter and summer glacier mass balances. *Geog. Ann.*, **81A**: 541–54.

Dyurgerov, M. B. and Meier, M. F. (2005). *Glaciers and the changing Earth system: a 2004 snapshot. Inst. Arct. Alp. Res. Occas. Pap. 58*. Boulder: University of Colorado. 117 pp.

Dyurgerov, M. B., Meier, M. F., and Bahr, D. B. (2009). A new index of glacier area change: A tool for glacier monitoring. *J. Glaciol.*, **55** (192): 710–16.

Ebbesmeyer, C. C., Okubo, A., and Helset, H. J. M. (1980). Description of iceberg probability between Baffin Bay and the Grand Bank using a stochastic model. *Deep-Sea Res.*, **27A**: 975–86.

Ebert, E. E. and Curry, J. A. (1993). An intermediate one-dimensional thermodynamic sea ice model for investigating ice-atmosphere interactions. *J. Geophys. Res.*, **98** (C6): 10,085–110.

Eckert, N., Baya, H., and Deschatres, M. (2010). Assessing the response of snow avalanche runout altitudes to climate fluctuations using hierarchical modeling: Application to 61 winters of data in France. *J. Climate*, **23**: 3157–80.

Eckert, N., *et al.* (2013). Temporal trends in avalanche activity in the French Alps and subregions: from occurrences and runout altitudes to unsteady return periods. *J. Glaciol.*, **59** (213): 93–114, doi: 10.3189/ 2013JoG12J091.

Edgar, K. M., *et al.* (2007). No extreme bipolar glaciation during the main Eocene calcite compensation shift. *Nature*, **448**: 908–11.

Eicken, H., *et al.* (1995). Thickness, structure, and properties of level summer multi-year ice in the Eurasian sector of the Arctic Ocean. *J. Geophys. Res.*, **100** (19): 22697–710.

Eicken, H., *et al.* (2005). Zonation of the Laptev Sea landfast ice cover and its importance in a frozen estuary. *Global Planet. Change*, **48**: 55–83.

Eicken, H., *et al.* (2009a). *Field techniques for sea ice research*. Fairbanks, AK: University of Alaska Press. 566 pp.

Eicken, H., *et al.* (2009b). *Recurring spring leads and landfast ice in the Beaufort and Chukchi Seas, 1993–2004*. Boulder, CO: National Snow and Ice Data Center. Digital media.

Eicken, H., Lovecraft, A. L., and Druckenmiller, M. J. (2009). Sea-ice system services: A framework to help identify and meet information needs relevant for Arctic observing networks. *Arctic*, **62**: 119–36.

Eicken, H., Tucker, W. B. III, and Perovich, D. K. (2001). Indirect measurements of the mass balance of summer Arctic sea ice with an electromagnetic induction technique. *J. Glaciol.*, **33**: 194–200.

Eisen, O., Harrison, W. D., and Raymond, C. F. (2001). The surges of Variegated Glacier, Alaska, U.S.A., and their connection to climate and mass balance. *J. Glaciol.*, **47** (158): 351–58.

Eisen, O., *et al.* (2008). Ground-based measurements of spatial and temporal variability of snow accumulation In East Antarctica. *Rev. Geophys.*, **46**: RG2001, doi: 10.1029/2006RG000218.

Eisenman, I. (2010). Geographic muting of changes in the Arctic sea ice cover. *Geophys. Res. Lett.*, **37**: L16501, doi: 10.1029/2010GL043741.

Eisenman, I., Untersteiner, N., and Wettlaufer. J. S. (2007). On the reliability of simulated Arctic sea ice in global climate models. *Geophys. Res. Lett.*, **34**: L10501, doi: 10.1029/2007GL029914.

Eisenman, I. and Wettlaufer, J. S. (2009). Nonlinear threshold behavior during the loss of Arctic sea ice. *Proc. Nat. Acad. Sci.*, **106**: 28–32.

Eckel, O. (1955). Statisches zur Vereisubg der Ostalpenseen. *Wetter u. Leben*, **7**: 49–57.

Elder, K. and Armstrong, B. (1987). A quantitative approach for verifying avalanche hazard ratings. In Salm, B. and Gubler, H. (eds.), *Avalanche formation, movement and effects*. Int. Assoc, Hydrol Sci., **162**: 593–601.

Eldrett, J. S., *et al.* (2007). Continental ice in greenland during the eocene and oligocene. *Nature*, **446**: 176–9.

Ellis, A. W. and Johnson, J. J. (2004). Hydroclimatic analysis of snowfall trends associated with the North American Great Lakes. *J. Hydrometeorol.*, **5**: 471–86.

Elo, A-R. and Vavrus, S. (2000). Ice modelling calculations comparison of the PROBE and LIMNOS models. *Verh. Int. Verien. Limnol.*, **27**: 2816–19.

Elsasser, H. and Messerli, P. (2001). The vulnerability of the snow industry in the Swiss Alps. *Mountain Res. Devel.*, **21** (4): 335–9.

Elsberg, D. H., *et al.* (2001) Quantifying the effects of climate and surface change on glacier mass balance, *J. Glaciol.*, **47**: 649–58.

Elverhøi, A., *et al.* (2002). The Eurasian Arctic during the last ice age. *Amer. Scientist*, **90**: 32–9.

Emery, W. J., Fowler, C. W., and Maslanik, J. A. (1997). Satellite derived maps of Arctic and Antarctic sea-ice motion: 1988–1994. *Geophys. Res. Lett.*, **24**: 897–900.

Emmerton, C. A., Lesack, L. F. W., and Marsh, P. (2007). Lake abundance, potential water storage, and habitat distribution in the Mackenzie River Delta, western Canadian Arctic. *Water Resour. Res.*, **43**: W05419, doi: 10.1029/2006WR005139.

Engell. M. C. (1910). Die Enstehung der Eisberge. *Zeit. f. Gletscherk.*, **5**: 122–32.

England, J. H., *et al.* (2006). The Innuitian Ice Sheet: configuration, dynamics and chronology. *Quart. Sci. Rev.*, **25**: 689–703.

England, J. H., *et al.* (2008). A millennial-scale record of Arctic Ocean sea ice variability and the demise of the Ellesmere Island ice shelves. *Geophys. Res. Lett.*, **35** (L19502): 5.

Engram. M., Arp, C. D., Jones, B. M., Ajadi, O. A., and Meyer, F. J. (2018). Analyzing floating and bedfast lake ice regimes across Arctic Alaska using 25 years of space-borne SAR imagery. *Remote Sensing of Environment*, **209**: 660–76, https://doi.org/10.1016/j.rse.2018.02.022

Ensminger, S. L., *et al.* (1999). Example of the dependence of ice motion on subglacial drainage system evolution, Matanuska Glacier, Alaska, United States, in Mickelson, D. M. and Attig, J. W. (eds.), Glacial processes: Past and present, Geol. Soc. Amer., Special Paper 337, pp. 11–22.

Environment Canada. (2005). MANICE. *Manual of standard procedures for observing and reporting ice conditions*. Ottawa:

Environment Canada, Canadian Ice Services.

EPICA Community Members. (2004). Eight glacial cycles from an Antarctic ice core. *Nature*, **429**: 623–8.

EPICA Community Members. (2006). One-to-one coupling of glacial climate variability in Greenland and Antarctica. *Nature*, **444**: 195–8.

Escher-Vetter, H. (1985). Energy balance calculations from five years meteorological records at Vernagtferner, Oetztal Alps. *Zeit. Gletscherk. Glazialgeol.*, **21**: 397–402.

Essery, R. (2013). Large-scale simulations of snow albedo masking by forests. *Geophys. Res. Lett.*, **40**: 5521–5, doi: 10.1002/grl.51008.

Essery, R. and Yang, Z.-L. (2001). An overview of models participating in the snow model intercomparison project (SnowMIP), 8th Scientific Assembly of IAMAS, Innsbruck, www.cnrm.meteo.fr /snowmip/

Essery, R., Long, Li and Pomeroy, J. W. (1999). A distributed model of blowing snow over complex terrain. *Hydrol. Proc.*, **13**: 2423–38.

Essery, R., *et al.* (2006). Boundary layer growth and advection of heat over snow and soil patches: Modelling and parametrization. *Hydrological Processes*, **20** (4): 953–67, doi: 10.1002/hyp.6122.

Essery, R., *et al.* (2009). SNOWMIP2, An evaluation of forest snow process simulations. *Bull. Amer. Met. Soc.*, **90**: 1120–35.

Essery, R., *et al.* (2009). SNOWMIP2, an Evaluation of Forest snow Process simulations. *Bulletin America Meteorological Society*, doi: 10.1175/2009BAMS2629.1.

Estilow, T. W., Young, A. H., and Robinson, D. A., (2015). A long-term Northern Hemisphere snow cover extent data record for climate studies and monitoring. *Earth Syst. Sci. Data*, **7**: 137–42, doi: 10.5194/essd-7-137-2015.

Ethan, C. , Campbell, E. C. , *et al.* (2019). Antarctic offshore polynyas linked to Southern Hemisphere climate anomalies. *Nature*, **570**: 319–25.

Etkin, B. (2010). A state space view of the ice ages – a new look at familiar data. *Clim. Change*, **100**: 403–6.

Evans, S. (1967). Progress report on radio echo sounding. *Polar Rec.*, **13** (85): 413–20.

Eyring, V., Bony, S., Meehl, G. A., Senior, C. A., Stevens, B., Stouffer, R. J., and Taylor, K. E., (2016). Overview of the Coupled Model Intercomparison Project Phase 6 (CMIP6) experimental design and organization. *Geosci. Model Dev.*, **9**: 1937–58, doi: 10.5194/gmd-9-1937-2016.

Fahnestock, M., *et al.* (1993). Greenland ice sheet surface properties and ice dynamics from ERS-1 SAR imagery. *Science*, **262** (5139): 1530–4.

Fahnestock, M. A., *et al.* (2000). A millennium of variable ice flow recorded in the Ross Ice Shelf, Antarctica. *J. Glaciol.*, **46** (155): 652–64.

Fahnestock, M., Scambos, T. , Haran, T. , and Bauer, R. (2006). *AWS data: characteristics of snow megadunes and their potential effect on ice core interpretation*. Boulder, CO: National Snow and Ice Data Center. Digital media.

Fahnestock, M. A., *et al.* (2000). Snow megadune fields on the East Antarctic Plateau: extreme atmosphere-ice interaction. *Geophys. Res. Lett.*, **27** (22): 3719–22.

Falkingham, J. C., Chagnon, R., and McCourt, S. (2002). Trends in sea ice in the Canadian Arctic. Ice in the environment, Vol. 1, Squire, V. and Langhorne, P. (eds.), Proc. 16th IAHR Internat. Sympos. on Ice, Int. Assoc. Hydraulic Eng. Rea., Dunedin, New Zealand. pp. 352–9.

Fallot, J.-M., Barry, R. G., and Hoogstrate, D. (1996). Variations of mean cold season temperature, precipitation and snow depths during the last 100 years in the Former Soviet Union (FSU). *Hydrol. Sci. J.*, **42** (3): 301–27.

Fang, X., Ellis, C. R., and Stefan, H. G. (1996). Simulation and observation of ice formation (freeze-over) in a lake. *Cold Reg. Sci. Technol.*, **24**: 129–45.

Farinotti, D., Huss, M., Bauder, A., and Funk, M. (2009). An estimate of the glacier ice volume in the Swiss Alps. *Glob. Planet. Change*, **68**: 225–31. doi: 10.1016/j.gloplacha.2009.05.004.

Farinotti, D., *et al.* (2015). Substantial glacier mass loss in the Tien Shan over the past 50 years. *Nature Geoscience*, doi: 10.1038/ngeo2513.

Farmer, C. J. Q., *et al.* (2010). Identification of snow cover regimes through spatial and temporal clustering of satellite microwave brightness temperatures. *Remote Sensing Environ*, **114**: 199–210.

Farmer, L. D. and Robe, R. Q. (1977). Photogrammetric determinations of iceberg volumes, photogram. *Eng. Remote Sensing*, **43**: 183–9.

Farquharson, J. (1835). On the ice formed, under peculiar circumstances, at the bottom of running water. *Phil. Trans. Roy Soc. London*, **125**: 329–43.

Farquharson, J. (1841). On ground Gru, or ice formed, under peculiar circumstances, at the bottom of running water. *Phil. Trans. Roy Soc. London*, 131: 37–9.

Farrell, S. L., *et al.* (2009). Five years of Arctic sea ice freeboard measurements from the Ice, Cloud and land Elevation Satellite. *J. Geophys. Res.*, **14**: C04008, doi: 10.1029/2008JC005074.

Feldl, N., Anderson, B., and Bordoni, S. (2017). Atmospheric eddies mediate lapse rate feedback and Arctic amplification. *J. Clim.*, 9213–24, AMS, https://doi.org/10.1175/JCLI-D-16-0706

Feldl, N., Po-Chedley, S., Singh, H. K. A., *et al.* (2020). Sea ice and atmospheric circulation shape the high-latitude lapse rate feedback. *Np.J. Clim. Atmos. Sci.*, 3, 41, https://doi.org/10.1038/s41612-020-00146-7

Feltham, D., Sammonds, P., and Hatton, D. (2002). Method of determining a geophysical-scale sea ice rheology from laboratory experiments. Ice in the environment. Proceedings, 16th IAHR International Symposium on Ice. Dunedin, New Zealand, pp. 94–9.

Ferraro, R., *et al.* (1994). Microwave measurements produce global climatic, hydrologic data. *EOS Trans., AGU*, **75** (30): 337–43.

Ferrians, O. J., Kachadoorian, R., and Green, G. W. (1969). Permafrost and related engineering problems in Alaska. USGS Prof. Paper 678, 37 pp.

Fetterer, F. and Untersteiner, N. (1998). Observations of melt ponds on Arctic sea ice. *J. Geophys. Res.*, **103** (C11): 24, 821–35.

Fiedler, J. W. and Conrad, C. P. (2010). Spatial variability of sea level rise due to water impoundment behind dams. Geophys. *Res. Lett.*, **37** (L12603): 6, doi: 10.1029/2010GL043462.

Fierz, C., *et al.* (2009). The international classification for seasonal snow on the ground. *IHP-VII Technical Documents in Hydrology No83, IACS Contribution No1*. Paris: UNESCO-IHP. 90 pp.

Finsterwalder, R. (1932). *Wissenschaftliche Ergebnisse der Alai-Pamir Expedition 1928. I. Geodatische, topographische und glaziologische Ergebnisse*. Berlin: D. Reimer.

Finsterwalder, S. (1897). *Der Vernagtferner. Seine Geschichte und seine Vermessung in den Jahren 1888 und 1889. Wissenschaft. Ergänzungshefte, Zeitschr*. Dtsch. Österreich. Alpenvereins **1**: 1–96 & 2 maps.

Fischer, A. (2010). Glaciers and climate change: Interpretation of 50 years of direct mass balance of Hintereisferner. *Global Planet. Change*, **71**: 13–26.

Fischer, A., *et al.* (2016). Future Challenges for Glacier Monitoring in Austria, in *Mountain Ice and Water, Developments in Earth Surface Processes*. Elsevier Science.

Fitzharris, B. B. and Schaerer, P. A. (1980). Frequency of major avalanche winters. *J. Glaciol.*, **26** (94): 43–52.

Fitzharris, B. B., *et al.* (12 lead authors and 15 contributing authors). (1996). The cryosphere: changes and their impacts, Ch 7. In Watson, R. T., Zinyowera, M. C., Moss, R. H., and Dokken, D. J. (eds.), *Climate change 1995: Impacts, Adaptations, and Mitigation of Climate Change: Scientific-Technical Analyses*, IPCC (WMO, UNEP): Cambridge University Press, pp. 241–65.

Flato, G. M. (2004). Sea-ice modelling. In Bamber, J. l. and Payne, A. J. (eds.), *Mass balance of the cryosphere: Observations and modelling of contemporary and future change*. Cambridge, UK: Cambridge University Press, pp. 367–90.

Flato, G., *et al.* (2013). Evaluation of Climate Models. In Stocker, T. F., *et al.* (eds.), *Climate Change 2013: The Physical Science Basis. Contribution of Working Group I to the Fifth Assessment Report of the Intergovernmental Panel on Climate Change*, Cambridge, NY: Cambridge University Press, pp. 741–882.

Flato, G. M. and Brown, R. D. (1996). Variability and climate sensitivity of landfast Arctic sea ice. *J. Geophys. Res.*, **101** (C11): 25, 767–77.

Föhn, P. M. B. (1987). The rutschblock as a practical tool for slope stability evaluation In Avalanche formation, movement and effects, IASH Publ. 162 (Symposium at Davos, 1986), pp. 223–8.

Föhn, P., *et al.* (1977). Evaluation and comparison of conventional and statistical methods of forecasting avalanche hazard. *J. Glaciol.*, **19** (81): 375–87.

Forbes, J. D. (1859). *Occasional papers on the theory of glaciers*. Edinburgh: A. and C. Black, 278 pp.

Forel, F. A. (1895). Les variations périodiques des glaciers. Discours préliminaire. *Extrait, Archives Sciences Phys. Nature*, **34**: 209–29.

Forsberg, R., *et al.* (2017). Greenland and Antarctica ice sheet mass changes and effects on global sea level. *Surv. Geophys.*, **38**: 89–104, https://doi.org/10.1007/s10712-016–9398-7, 2017

Foster, J. L., *et al.* (2008). Spring snow melt timing and changes over Arctic lands. *Polar Geog.*, **31**: 145–57.

Foster, J. L., *et al.* (2009). Seasonal snow extent and snow mass in South America using SMMR and SSM/I passive microwave data (1979–2006). *Remote Sensing Environ.*, **113**: 291–305.

Foster, G. L., Lunt, D. J., and Parrish, R. R. (2010). Mountain uplift and the glaciation of North America – a sensitivity study. *Clim. Past*, **6**: 707–17.

Fountain, A. and Vecchia, A. (1999). How many stakes are required to measured the mass balance of a glacier. *Geog. Ann.*, **81A**: 563–8.

Fowler, A. C. and Krantz, W. B. (1994). A generalized secondary frost heave model. *SIAM J. App.. Math.*, **54** (6): 1650–75.

Fox, D. (2008). Freeze-dried findings support a tale of two ancient climates. *Science*, **320**: 1152–4.

Francis, J. A. and Vavrus, S. J. (2015). Evidence for a wavier jet stream in response to rapid Arctic warming. *Environmental Research Letters*, **10**: 014005.

Frank, F. C. and Lee, R. (1966). Potential solar beam irradiation on slopes: tables for 30° to 50° latitude. U.S. Dept. of Agriculture. Forest Service. Research Paper RM-18, 116 pages.

Franssen, H. J. H. and Scherrer, S. C. (2007). Freezing of lakes on the Swiss Plateau in the period 1901–2006. *Int. J. Climatol.*, **28** (4): 421–33.

Fraser, A. D. *et al.* (2010). High-resolution East Antarctic landfast sea-ice extent and variability from 2000 to 2008. Paper 57A008, Proceedings, Tromso Sea Ice Symposium, Int. Glaciol. Soc.

Frauenfeld, O. W., Zhang, T.-J., Barry, R. G., and Gilichinsky, D. (2004). Interdecadal changes in seasonal freeze and thaw depths in Russia. *J. Geophys. Res.*, **109** (D05101): 1–12.

Frauenfeld, O. W., Zhang, T.-J., and McCreight, J. L. (2007). Northern Hemisphere freezing/ thawing index variations over the twentieth century. *Int. J. Climatol.*, **27**: 47–63.

Frei, A. and Robinson, D. A. (1995). Evaluation of snow extent and its variability in the Atmospheric Model Intercomparison Project. *J. Geophys. Res.*, **103** (D8): 8859–71.

Frei, A. and Robinson, D. A. (1999). Northern Hemisphere snow extent: regional variability 1972–1994. *Int. J. Climatol.*, **19**: 1535–60.

Frei, A., Miller, J. A., and Robinson, D. A. (2003). Improved simulations of snow extent in the second phase of the Atmospheric Model Intercomparison Project (AMIP-2). *J. Geophys. Res.*, **108** (D12): 4369, doi: 10.1029/2002JD003030.

Frei, A., Tedesco, M., Lee, S., Foster, J., Hall, D. K., Kelly, R., and Robinson, D. A. (2012). A review of global satellite-derived snow products. *Adv. Space Res.*, **50** (8): 1007–29.

French, H. M. (2007). *The periglcial environment.* 3rd edn. New York: Wiley, 458 pp.

French, H. (2008). Recent contributions to the study of past permafrost. *Permafrost Periglac. Process,* **19** (2): 179–94.

French, H. M. and Nelson, F. E. (2008). The permafrost legacy of Siemon W. Muller. In Kane, D. L. and Hinkel, K. M. (eds.), *Proceedings of the Ninth International Conference on Permafrost*, Fairbanks, AK: University of Alaska, Institute of Northern Engineering, pp. 475–80.

French, H. and Shur, Y. (2010). The principles of cryostratigraphy. *Earth-Sci. Rev.*, **101**: 190–206.

Frezotti, M., *et al.* (2002). Snow dunes and glazed surfaces in Antarctica: new field and remote-sensing data. *Annals Glaciol.*, **34**: 81–8.

Fricker, H. A., Coleman, R., Padman, L., Scambos, T. A., Bohlander, J., and Brunt, K. M. (2009). Mapping the grounding zone of the Amery Ice Shelf, East Antarctica using InSAR, MODIS and ICESat, *Antarctic Science*, **21** (5): 515–32.

Friedman, J. H. (1985). Classification and multiple regression through projection pursuit. Technical Report LCS012, Department of Statistics, Stanford University.

Friedman, J. H. and Stuetzle, W. (1981). Projection pursuit regression. *J. Amer. Stat. Assoc.*, **82**: 249–66.

Fritze, H., *et al.* (2011). Shifts in Western North American Snowmelt Runoff Regimes for the Recent Warm Decades. *J. Hydromet.*, https://doi.org/10.1175/2011JHM1360.1

Froese, D. G., *et al.* (2008). Ancient permafrost and a future, warmer Arctic. *Science*, **321**: 1648.

Fu, C. and Yao, H. (2015). Trends of ice breakup date in south-central Ontario. *J. Geophys. Res. Atmos.*, **120**: 9220–36.

Fujita, K. (2008). Effect of precipitation seasonality on climate sensitivity of glacier mass balance. *Earth Planet. Sci. Lett.*, **276**: 14–19.

Furbish, D. J. and Andrews, J. T. (1984). The use of hypsometry to indicate long-term stability and response of valley glaciers to changes in mass transfer. *J. Glaciol.*, **30**: 199–211.

Fyke, J., Sergienko, O., Lofverstrom, M., Price S., and Lenaerts, J. (2018). An overview of interactions and feedbacks between ice sheets and the Earth system. *Rev. Geophys.*, **56**: 361–408. https://doi.org/10.1029/2018RG000600

Gagliardini, O., *et al.* (2010). Coupling of ice-shelf melting and buttressing is a key process in ice-sheets dynamics. *Geophys. Res. Lett.*, **27** (L14501): 5.

Gan, T. Y. (1996). Passive microwave snow research at the Canadian High Arctic. *Canad. J. Remote Sensing*, **22**: 36–44.

Gan, T. Y., Barry, R., Gobena, A., and Rajagopalan, B., (2013). Changes in North American snowpacks for 1979–2007 detected from the snow water equivalent data of SMMR and SSM/I passive microwave and related climatic factors. *J. Geophysic. Research-Atm.*, **118** (14): 7682–97, https://doi.org/10.1002/jgrd.50507

Gan, T. Y., Gobena, A., and Wang, Q. (2007). Precipitation of southwestern Canada: Wavelet, scaling, multifractal analysis, and teleconnection to climate anomalies. *J. Geoph. Res.–Atm.*, **112**: D10110, http://doi.org/10.1029/2006JD007157

Gan, T. Y., Kalinga, O., and Singh, P. R., (2009). Comparison of snow water equivalent retrieved from SSM/I passive microwave data using artificial neural network, projection pursuit & nonlinear regressions. *Remote Sensing of Environment*, **25** (21): 4593–615, doi: 10.1016/j.rse.2009.01.004.

García-Hernández, C., *et al.* (2017). Reforestation and land use change as drivers for a decrease of avalanche damage in mid-latitude mountains (NW Spain). *Global and Planetary Change*, **153**: 35–50, doi: 10.1016/j.gloplacha.2017.05.001.

Gardelle, J., Arnaud, Y., and Berthier, E. (2010). Contrasted evolution of glacial lakes along the Hindu Kush Himalaya mountain range between 1990 and 2009. *Global Planet. Change*, **75**: 47–55.

Gardner, A. S., *et al.* (2013). A reconciled estimate of glacier contributions to sea level rise: 2003 to 2009. *Science*, **340**: 852–7, https://doi.org/10.1126/science.1234532

Gascard, J.-C., Hervé le Goff, J. F. , and Weber, M. (2008). Exploring Arctic transpolar drift during dramatic sea ice retreat. *Eos*, **89** (3): 21–2.

Gascard, J. C., Zhang, J., and Rafizadeh, M. (2019). Rapid decline of Arctic sea ice volume: Causes and consequences. *The Cryosphere Discuss.*, doi: 10.5194/tc-2019-2.

Gavrilova, M. K. (1973). Meteorological observations in Naled valley of Ulakhan-Taryn (Central Yakutia) In Alekseyev, V. R., *et al.* (eds.), *Siberian naleds. USSR Academy of Sciences (1969). Draft Translation 399*, Hanover, NH: US Army Cold Regions Research and Engineering Laboratory. pp. 136–57.

Ge, Y. and Going, G. (2009). North American snow depth and climate teleconnection patterns. *J. Clim.*, **22**: 217–33.

Gearheard, S., *et al.* (2006). "It's not that simple": A collaborative comparison of sea ice environments, their uses, observed changes, and adaptations in Barrow, Alaska, USA, and Clyde River, Nunavut, Canada. *Ambio*, **35** (4): 204–12.

General Secretariat of the Andean Community. (2007). *The end of snowy heights? Glaciers and climate change in the Andean community*. Peru, Lima: United Nations Programme for the Environment and Spanish International Cooperation Agency. 104 pp.

Gerdes, R. and Koeberle, C. (2007). Comparison of Arctic sea ice thickness variability in IPCC climate of the 20th Century experiments and in ocean–sea ice hindcasts. *J. Geophys. Res.*, **112** (C4): C04S13. 12 pp.

Gerrard, J. A. F., Perutz, M. F., and Roch, A. (1952). Measurement of the velocity distribution along a vertical line through a glacier. *Proc. Roy. Soc., London, A*, **213** (1115): 546–58.

Gersonde, R. and Zielinski, U. (2000). The reconstruction of late Quaternary Antarctic sea-ice distribution – the use of diatoms as a sea-ice proxy. *Palaeogeog., Palaeoclimatol., Palaeoecol.*, **162**: 263–86.

Gidrometeoizdat: The Avalanche Cadastre of the USSR, The Federal Service for Hydrometeorology and Environmental Monitoring of the USSR, Leningrad, USSR, 1–20, 1984–1991.

Gillan, B. J., Harper, J. T., and Moore, J. N. (2010). Timing of present and future snowmelt from high elevations in northwest Montana. *Water Resour. Res.*, **46** (1): W01507. http://dx.doi.org/10.1029/2009WR007861

Gillespie, A. and Molnar, P. (1995). Asynchronous maximum advances of mountain and continental glaciers. *Rev. Geophys.*, **33**: 311–64.

Ginsburg, B. M. and Soldatova, I. I. (1997). Long-term variability of ice phenomena dates on rivers as an indicator of climate variations in transitional seasons. *Soviet Met. Hydrol.*, **11**: 73–8.

Giovinetteo, M. B. (1964). Distribution of diagenetic snow facies in Antarctica and in Greenland. *Arctic*, **17**: 32–40.

Glasby, G. P. (ed.) (1990). *Antarctic sector of the Pacific*. Amsterdam: Elsevier. pp. 108–16.

Glazyrin, G. E., Kaminyanskii, G. M., and Pertziger, F. I. (1993). *Rezhim Lednika Abramova. (Regime of the Abramov glacier) (in Russian)*. St. Petersburg: Gidrometeoizdat. 228 pp.

Gleick, P. H. (1998). Water planning and management under climate change. *Water Resources Update*, **112**: 25–32.

Glen, J. W. (1952). Experiments on the deformation of ice. *J. Glaciol.*, **2** (12): 111–14.

Glen J. W. (1953). Rate of flow of polycrystalline ice. *Nature*, **172** (4381): 721–2.

Glen, J. W. (1958). The flow law of ice. A discussion of the assumptions made in glacier theory, their experimental foundations and consequences. Physics of the movement of the ice (Proc. Chamonix Symposium). *Bull. Int. Assoc. Sci. Hydrol.*, **47**: 71–83.

Global Climate Observing System (GCOS). (2004). *Implementation plan for global observing system for climate in support of the UNFCC*. Geneva: World Meteorological Organization. WMO/TD No. 1219 (GCOS-92).

Global Cryosphere Watch, (2015). Snow dataset inventory, http://globalcryospherewatch.org/reference/snowinventory.php, accessed on May 19, 2016.

Global Snow Laboratory (GSL) (January 2008). Northern Hemisphere snow cover: largest anomaly since 1966, http://wattsupwiththat.com/2008/02/09/jan08-northern-hemisphere-snow-cover-largest-since-1966/, accessed on July 22, 2009.

Gloersen, P., Campbell, W. J., Cavalieri, D. J., Comiso, J. C., Parkinson, C. L., and Zwally, H. J. (1993). *Arctic and Antarctic sea ice, 1978–1987: Satellite passive-microwave observations and analysis. NASA SP-511*. Washington, DC: NASA. 290 pp.

Gobena, A. K. and Gan, T. Y. (2006). Low-frequency variability in southwestern Canadian streamflow: links to large-scale climate anomalies. *Int. J. Climatol.*, **26**: 1843–69, doi: 10.1002/joc.1336.

Goita, K., Walker, A. E., and Goodison B. E. (2003). Algorithm development for the estimation of snow water equivalent in the boreal forest using passive microwave data. *Int. J. Remote Sensing*, **24**: 1097–102.

Goldhar, C., Bell T., and Wolf, T. (2014). Vulnerability to Freshwater Changes in the Inuit Settlement Region of Nunatsiavut, Labrador: A Case Study from Rigolet.

Arctic, **67** (1): 71–83, doi: 10.14430/arctic4365.

Golding, D. L. and Swanson, R. S. (1986). Snow distribution patterns in clearings and adjacent forest. *Wat. Resour. Res.*, **22**: 1931–40.

Goldner, A., Herold, N., and Huber, M. (2014). Antarctic glaciation caused ocean circulation changes at the Eocene–Oligocene transition. *Nature*, **511** (7511): 574–7, doi: 10.1038/nature13597.

Gong, D.-Y., Kim, S.-J., and Ho, C.-H. (2007). Arctic oscillation and ice severity in the Bohai Sea, East Asia. *Int. J. Climatol.*, **27**: 1287–302.

Goodison, B., *et al.* (2007). State and fate of the polar cryosphere, including variability of the Arctic hydrological cycle. *WMO Bull.*, **56** (4): 284–92.

Goodison, B. E., Barry, R. G., and Dozier, J. (eds.). (1987). *Large-Scale Effects of Seasonal Snow Cover, International Association of Hydrological Sciences, Publ. No. 166*. Wallingford, Oxfordshire, UK: IAHS Press, 425 pp.

Goodison B. E. and Walker, A. E. (1994). Canadian development and use of snow cover information from passive microwave satellite data. In Choudhury, B. J., Kerr, Y. H., Njoku, E. G., and Pampaloni, P. (eds.), *ESA/NASA International Workshop*, VSP, Utrecht, Netherlands. pp. 245–62.

Goodison, B. E., Walker, A. E., and Thirkettle, F. W. (1990). Determination of snow water equivalent on the Canadian Prairies using near real-time passive microwave data. In Kite, G. W. and Wankiewicz, A. (eds.), *Proceedings of the Workshop on Applications of Remote Sensing in Hydrology*, NHRI Symposium Series, Saskatoon, pp. 297–309.

Goodwin, I. D. (1990). Snow accumulation and surface topography in the kata batic zone of eastern Wilkes Land. *Antarctica. Antarct. Sci.*, **2** (3): 235–42.

Goor, Q., Kelman, R., and Tilmant, A. (2011). Optimal multipurpose multireservoir operation model with variable productivity of hydropower plants. *J. Water Res. Plann. Manage.*, **137** (3): 258–67.

Goosse, H., *et al.* (2009a). Increased variability of the Arctic summer ice extent in a warmer climate, Geophys. *Res. Lett.*, **36**: L23702, doi: 10.1029/2009GL040546.

Goosse, H., *et al.* (2009b). Consistent past half-century trends in the atmosphere, the sea ice and the ocean at high southern latitudes. *Clim. Dynam.*, **33** (7–8): 999–1016.

Gorbunov, A. P. (2009). Consistent ice and icings of Central Asia: geography and dynamics. In Braun, L. N., *et al.* (eds.), *Assessment of snow, glacier and water resources in Asia*. (Selected papers from the Workshop in Almaty, Kazakhstan, 2006). UNESCO-IHP and the German IHP/HWRP National Committee. Koblenz: IHP/HWRP Secretariat. pp. 145–50.

Gordon, M., Savelyev, S., and Taylor, P. A. (2009). Measurements of blowing snow. Part II: Mass and number density profiles and saltation height at Franklin Bay, NWT, Canada. *Cold Reg. Sci. Technol.*, **55**: 75–85.

Gough, A., *et al.* (2010). Sea ice on a supercooled ocean: field measurements of ice growth and structure in McMurdo Sound during winter 2009. Paper 57A098. Proceedings, Tromso Sea Ice Symposium. Int. Glaciol. Soc.

Gould, B., Haegeli, P., Jamieson, B., and Statham, G. (eds.), Canadian Avalanche Association, Revelstoke, BC, Canada.

Goulding, H. L., Prowse, T. D., and Beltaos, S. (2009). Spatial and temporal patterns of breakup and ice-jam flooding in the Mackenzie Delta, NWT. *Hydrol. Process. An Int. J.*, **23**: 2654–70.

Goulding, H. L., Prowse, T. D., and Bonsal, B. (2009). Hydroclimatic controls on the occurrence of breakup and ice-jam flooding in the Mackenzie Delta, NWT, Canada. *J. Hydrol.*, **379**: 251–67.

Gow, A. J. and Tucker, W. B. (1991). *Physical and dynamical properties of sea ice in the polar oceans. CRREL Monograph 91–1.* Hanover, NH: US Army Cold Regions Research and Engineering Laboratory.

Gow, A. J., *et al.* (1998). Physical and structural properties of landfast sea ice in McMurdo Sound, Antarctica. In Jeffries, M. O. (ed.), *Antarctic sea ice: Physical processes, interactions and variability*. Washington, DC: Amer. Geophys. Union. Antarct. Res. Ser., **74**, pp.69–88.

Granskog, M. A., Martma, T. A., and Vaikmäe, R. A. (2003). Development, structure and composition of land-fast sea ice in the northern Baltic Sea. *J. Glaciol.*, **49** (164): 139–48.

Granskog, M. A., Kaartokallio, H., and Kuosa, H. (2010). Sea ice in non-polar regions. In Thomas, D. N. and Dieckmann, G. S. (eds.), *Sea ice*. 2nd edn. Chichester: Wiley-Blackwell. pp. 531–77.

Grant, K. (2010). Changes in glacier extent since the Little Ice Age and links to 20th/21st century climatic variability on Novaya Zemlya, Russian Arctic. PhD Dissertation. University of Reading, UK: Department of Geography. 480 pp.

Grant, K. L., Stokes, C. R., and Evans, I. S. (2010). Identification and characteristics of surge-type glaciers on Novaya Zemlya. *Russian Arctic. J. Glaciol.*, **55** (194): 960–72.

Grassl, H. (1999). The cryosphere: An early indicator and global player. *Polar Res.*, **18**: 119–25.

Graversen, R. G., Mauritsen, T., Tjernström, M., Källen, E., and Svensson, G. (2008) Vertical structure of recent Arctic warming. *Nature*, 451: 53–6.

Graversen, R. G., Langen, P. L., and Mauritsen, T. (2014). Polar Amplification in CCSM4: Contributions from the Lapse Rate and Surface Albedo Feedbacks. *Journal of Climate*, **27**: 4433–50.

Graversen, R. G. and Wang, M. H. (2009). Polar amplification in a coupled climate model with locked albedo. *Climate Dynamics*, **33**: 629–43.

Gray, D. M. and Landine, P. G. (1988). An energy-budget snowmelt model for the Canadian Prairies. *Canad. J. Earth Sci.*, **25**: 1292–303.

Gray, D. M. and Male, D. H. (1981). *Handbook of snow: Principles, processes, management and use*. Toronto: Pergamon Press, 776 pp.

Grenfell, T. C. and Perovich, D. K. (2004). Seasonal and spatial evolution of albedo in a snow-ice-land-ocean environment. *J. Geophys. Res.*, **109** (C01001): 18.

Grenfell, T. C., *et al.* (2010). Expedition to the Russian Arctic to survey black carbon in snow. *Eos*, **90** (43): 386–7.

Greve, R. and Hutter, K. (1995). Polythermal three-dimensional modelling of the Greenland ice sheet with varied geothermal heat flux. *Annals Glaciol.*, **21**: 8–12.

Greve, R. and Blatter, H. (2009). *Dynamics of ice sheets and glaciers*. New York: Springer, 287 pp.

Grey, D. M. and Prowse, T. (1993). *Chapter 7 in Handbook of Hydrology*, Editor-in-Chief, Maidment, D., McGraw Hill, ISBN 0-07-039732-5.

Griggs, J. A. and Bamber, J. L. (2011). Antarctic ice-shelf thickness from satellite radar altimetry. *J. Glaciol.*, **57** (203): 485–98.

Griggs, J. and Bamber, J. L. (2009). Ice shelf thickness over Larsen C, Antarctica, derived from satellite Altimetry. *Geophys. Res. Lett.*, **36**: L19501, doi: 10.1029/2009GL039527.

Grinsted, A. (2013). An estimate of global glacier volume. *The Cryosphere*, 7: 141–51, doi: 10.5194/tc-7-141-2013, y Copernicus Publications.

Groh, A. and Horwath, M. (2016). The method of tailored sensitivity kernels for GRACE mass change estimates. *Geophys. Res. Abstract*, **18**: EGU2016–12065.

Gronskaya, T. P. (2000). Ice thickness in relation to climate forcing in Russia. *Verh, Int, Verin Limnol.*, **27**: 2800–2.

Grove, J. (ed.). (2004). *Little Ice Ages ancient and modern*. **2** vols., London: Routledge. 402 pp. and 406–718 pp.

Gruber, S. and Haeberli, W. (2009). Mountain permafrost. In Margesin, R. (ed.), *Permafrost soils*. Berlin: Springer Verlag. pp. 33–44.

Gu, N., *et al.* (2005). Study on spatial characteristics of sea ice reserves in Liaodong Bay of China. *J. Agric. Met.*, **61**: 105–11.

Gudmandsen, P. (1975). Layer echoes in polar ice sheets. *J. Glaciol.*, **15** (73): 95–101.

Haarpaintner, J. (2006). Arctic-wide operational sea ice drift from enhanced resolution QuikScat/SeaWinds scatterometry and its validation. *IEEE Trans. Geosci. Remote Sensing*, **42**: 1433–43.

Haas, C. and Druckenmiller, M. (2009). Ice thickness and roughness measurements. In Eicken, H., *et al.* (eds.), *Field techniques for sea ice research*. Fairbanks, AK: University of Alaska Press. pp. 49–116.

Haas, C., *et al.* (2008). Reduced ice thickness in Arctic Transpolar Drift favors rapid ice retreat. *Geophys. Res. Lett.*, **35**: L17501.

Haas, C., *et al.* (2010). Synoptic airborne thickness surveys reveal state of Arctic sea ice cover. *Geophys. Res. Lett.*, **37** (L09501): 5.

Hachem, S., Allard, M., and Duguay, C. (2008). A new permafrost map of Quebec–Labrador derived from near-surface temperature data of the Moderate Resolution Imaging Spectroradiometer (MODIS). In Kane, D. L. and Hinkel, K. M. (eds.) *Ninth International Conference on Permafrost, June 29–July 3, 2008, University of Alaska Fairbanks. Proceedings, Vol. 1*. Fairbanks, AK: University of Alaska, Fairbanks, Institute of Northern Engineering. pp. 591–6.

Haeberli, W. (1973). Die Basistemperatur der winterliche Schneedecke als moeglicher Indikator fuer die Verbreitung von Permafrost in denAlpen. *Zeit. Gletscherk. Glazialgeol.*, **9**: 221–7.

Haeberli, W. (1975). *Untersuchungen zur Verbreitung von Permafrost zwischen Flüellapass und Piz Grialetsch (Graubunden)*. Mitteil. Versuchsanstalt Wasserbau, Hydrologie u, Glaziologie, ETH, Zurich. **17**: 221 pp.

Haeberli, W. (1990). Glacier and permafrost signals of 20th-century warming. *Annals Glaciol.*, **14**: 99–101.

Haeberli, W. and Gruber, S. (2009). Global warming and mountain permafrost. In Margesin, R. (ed.), *Permafrost soils*. Berlin: Springer Verlag. pp. 205–18.

Haeberli, W. and Hohmann, R. (2008). Climate, glaciers and permafrost in the Swiss Alps 2050: Scenarios, consequences, and recommendations. In Kane, D. L. and Hinkel, K. M. (eds.), *Ninth International Conference on Permafrost, 29 June–3 July 2008, University of Alaska Fairbanks. Proceedings, Vol. 2*. Fairbanks, AK: University of Alaska. pp. 607–12.

Haeberli, W., Cihlar, J., and Barry, R. G. (2000). Glacier monitoring within the Global Climate Observing System. *Annals Glaciol.*, **31**: 241–6.

Haefeli, R, (1940). Zur Mechanik aussergewohnlicher Gletscherschwankungen, Schweiz. *Bauzeitung*, **115** (16).

Haegeli, P. (2019). Avalanches in Canada: Understanding and mitigating the risks, Annual Mountain Report, Alpine Club of Canada.

Hägeli, P. and McClung, D. M. (2003). Avalanche characteristics of a transitional snow climate – Columbia Mountains, British Columbia, Canada. *Cold Reg. Sci. Technol.*, **37** (3): 255–76.

Hagg, W., *et al.* (2005). A comparison of three methods of mass balance determination in the Tuyuksu Glacier Region, Tien Shan. *J. Glaciol.*, **50**: 505–10.

Hagg, W., *et al.* (2013). Glacier changes in the Big Naryn basin, Central Tian Shan. *Global Planet. Change*, **110**: 40–50, doi: 10.1016/j.gloplacha.2012.07.010.

Häkkinen, S., Proshutinsky, A., and Ashik, I. (2008). Sea ice drift in the Arctic since the 1950s. *Geophys. Res. Lett.*, **35** (L19704): 5.

Hall, D. K. and Martinec, J. (1985). *Remote sensing of snow and ice*. London: Chapman and Hall. 196 pp.

Hall, D. K. and Riggs, G. A. (2007). Accuracy assessment of the MODIS snow products. *Hydrol Processes*, **21**: 1534–47.

Hall, D. K., *et al.* (1995). Development of methods for mapping global snow cover using Moderate Resolution Imaging Spectroradiometer data. *Remote Sens. Environ.*, **54**: 127–40.

Hall, D. K., *et al.* (2009). Evaluation of surface and near-surface melt characteristics on the Greenland ice sheet using MODIS and QuikSCAT data. *J. Geophys. Res.*, **114**: F04006, doi: 10.1029/2009JF001287.

Hall, M. H. P. and Fagre, D. B. (2003). Modeled climate-induced glacier change in Glacier National Park, 1850–2100. *BioScience*, **53**: 131–40.

Halliday, M. D. (1954). Ice caves of the United States. *Nat, Speleol. Soc. Bull.*, **16**: 3–28.

Hallikainen, M. T. (1989). Microwave radiometry of snow. *Adv. Space Res.*, **9**: 267–75.

Hallikainen, M. T. and Jolma, P. A. (1992). Comparison of algorithms for retrieval of snow water equivalent from Nimbus-7 SMMR data in Finland. *IEEE Trans. Geosci. Remote Sensing*, **30**: 124–31.

Hamberg, A. (1910). Die Gletscher des Sarekgebirges und ihre Untersuchung. *Sveriges geolog. Undersök.*, **5**: 1–26.

Hambrey, M. J. and Alean, J. (2004). *Glaciers*. Cambridge: Cambridge University Press. 376 pp.

Hambrey, M. J., Larsen, B., and Ehrmann, W. U. (1989). Forty million years of Antarctic glacial history yielded by Leg 119 of the Ocean Drilling Program. *Polar Record*, **25**: 99–106.

Hamilton, R. A., *et al.* (1956). British North Greenland Expedition 1952–4. Scientific results. *Geog. J.*, **122**: 203–37.

Han, L., *et al.* (2019). A novel approach for cloud detection in scenes with snow/ice using high resolution Sentinel-2 images. *Atmosphere*, https://doi.org/10.3390/atmos10020044

Hansen et al., 2000, Global land cover classication at 1km spatial resolution using a classification tree approach, *International Journal of Remote Sensing* 21(6–7): 1331–1364, DOI: 10.1080/014311600210209.

Hanna, E., *et al.* (2005). Runoff and mass balance of the Greenland ice sheet: 1958–2003. *J. Geophys. Res.*, **110**: D13108, doi: 10.1029/2004JD005641.

Hannah, C. G., Dupont, F. and Dunphy, M. (2009). Polynyas and tidal currents in the Canadian Arctic Archipelago. *Arctic*, **62**: 83–95.

Hansen, J. E. and Lebedeff, S. (1987). Global trends of measured surface air temperature. *J. Geophys. Res.*, **92**: 13345–72.

Hansen, J., Ruedy, R., Sato, M., and Lo, K. (2010). Global surface temperature change. *Rev. Geophys.*, 48: RG4004, doi: 10.1029/2010RG000345.

Hanson, B. and Hooke, R. LeB. (2000). Glacier calving: a numerical model of forces in thecalving speed – water depth relation. *J. Glaciol.*, **46**: 188–96.

Hanson, C. L., Johnson, G. L., and Rango, A. (1999). Comparison of precipitation catch between nine measuring systems. *J. Hydrol. Engineering*, **4**: 70–5.

Haq, B. U. and Schutter, S. R. (2008). A chronology of Paleozoic sea-level changes. *Science*, **322** (5898): 64–8.

Haran, T., *et al.* (compilers). (2006), *MODIS mosaic of Antarctica (MOA) image map*.

Boulder, CO: National Snow and Ice Data Center. Digital media.

Harden, D., Barnes, P., and Reimnitz, E. (1977). Distribution and character of naleds in northeastern Alaska. *Arctic*, **30**: 28–40.

Hardy, J. P. and Hansen-Bristow, K. J. (1990). Temporal accumulation and ablation patterns in forests representing varying stages of growth. Proc. of the 58th Western Snow Conf. Sacramento, CA: 23–34.

Hardy, J. P., *et al.* (1997). Snow ablation modeling at the stand scale in a boreal jack pine forest. *J. Geophys. Res.*, **102** (D24): 29, 397–405.

Hardy, J. P., *et al.* (1998). Snow ablation modelling in a mature aspen stand of the boreal forest. *Hydrol. Process.*, **12**: 1763–78.

Haresign, E. C. (2004). Glacio-limnological interactions at lake-calving glaciers. Unpubl. PhD thesis, University of St Andrews, Scotland.

Harington, E. R. (1934). The origin of ice caves. *J. Geol.*, **42**: 433–6.

Harlan, R. L. and Nixon, J. F. (1978). Ground thermal regime. In Andersland, O. B. and Anderson, D. M. (eds.), *Geotechnical engineering for cold regions*. New York: McGraw-Hill. pp. 103–50.

Harris, C., *et al.* (2009). Permafrost and climate in Europe: monitoring and modelling thermal, geomorphological and geotechnical responses. *Earth-Sci. Rev.*, **92**: 117–71.

Harris, S. A. (2001). Sequence of glaciations and permafrost events. In Paepe, R. and Melnikov, V. (eds.), *Permafrost response on economic development, environmental security and natural resources*. Dordrecht: Kluwer. pp.227–52.

Harris, S. A. (2002). Global heat budget, plate tectonics and climatic change. *Geogr. Annal.*, **84A**: 1–9.

Harris, S. A. (2005). Thermal history of the Arctic Ocean environs adjacent to North America during the last 3.5 Ma and a possible mechanism for the cause of the cold events (major glaciations and permafrost events). *Progr. Phys. Geog.*, **29**: 218–37.

Harris, S., Brouchklov, A., and Guodong, C., 2018. *Geocryology: Characteristics and Use of Frozen Ground and Permafrost Landforms*. Leiden, The Netherlands: CRC Press, 765 pp, ISBN: 978-1-138–05416-5.

Harrison, W. D, *et al.* (2001). On the characterization of glacier response by a single time-scale. *J. Glaciol.*, **47** (159): 659–64.

Hartmann, D. L., *et al.* (2013). Observations: Atmosphere and Surface. In *Climate Change 2013: The Physical Science Basis. Contribution of Working Group I to the 5th Assessment Report of the Intergovernmental Panel on Climate Change* [Stocker, T.F., *et al.* editors]. Cambridge University Press, NY, USA.

Haseloff, M. and Sergienko, O. V. (2018). The effect of buttressing on grounding line dynamics. *Journal of Glaciology*, **64** (245): 417–31.

Hastenrath, S. (1981). *The glaciation of the Ecuadorian Andes*. Rotterdam: A.A. Balkema. 159 pp.

Hastenrath, S. (2008). *Recession of equatorial glaciers: a photo documentation*. Madison, WI: Sundog Publishing. 22 pp.

Hastenrath, S. (2009). Past glaciation in the tropics. *Quat. Sci. Rev.*, **28** (9–10): 790–8.

Hastenrath, S. (2010). Climatic forcing of glacier thinning on the mountains of equatorial East Africa. *Int. J. Climatol.*, **30**: 46–52.

Hattersley-Smith, G., *et al.* (1955). Northern Ellesmere Island, 1953 and 1954. Arctic 8: 2–16.

Haug, F., *et al.* (2005). North Pacific seasonality and the glaciation of North America 2.7 million years ago. *Nature*, **433**: 821–5.

Haumann, F. A., Notz, D., and Schmidt, H. (2014). Anthropogenic influence on recent circulation-driven Antarctic sea ice changes. *Geophys Res Lett.*, **41** (23): 8429–37, https://doi.org/10.1002/2014GL061659

Hauser, E. and Oedl, R. (1926). Eisbildung und meteorologische Beobachtungen. In *Die Eisriesenwelt in Tennengebirge (Salzburg)*. Speolog. Institut, Vienna. **6**, pp. 77–105.

Hay, J. E. and Fitzharris, B. B. (1988). The synoptic climatology of ablation on a New Zealand glacier. *J. Climat.*, **8**: 201–15.

Hay, W. W., Flögel, S., and Söding, E. (2005). Is the initiation of glaciation on Antarctica related to a change in the structure of the ocean? *Global Planet. Change*, **45**: 23–33.

Hays, J. D., Imbrie, J., and Shackleton, N. J. (1976). Variations in the Earth's orbit: Pacemaker of the Ice Ages. *Science*, **194**: 1121–31.

Headland, R. K. (2009). A chronology of Antarctic exploration. *A synopsis of events and activities from the earliest times until the International Polar Years, 2007–09*. London: B. Quaritch. 722 pp.

Hedstrom, N. and Pomeroy, J. W. (1998). Measurements and modelling of snow interception in the boreal forest. *Hydrol. Proc.*, **12**: 1611–25.

Hegyi, B. M. and Taylor, P. C. (2017). The regional influence of the Arctic Oscillation and Arctic Dipole on the wintertime Arctic surface radiation budget and sea ice growth. *Geophys. Res. Lett.*, **44**, 4341–50.

Heierli, J., Gumbsch, P., and Zaiser, M. (2008). Anticrack nucleation as triggering mechanism for slab avalanches. *Science*, **321** (5886): 240–3.

Heil, P. and Hibler, W. D. III. (2002). Modeling the high-frequency component of Arctic sea ice drift and deformation. *J. Phys. Oceanogr.*, **32**: 3039–57.

Heim, A. (1885). *Handbuch der Gletscherkunde*. Stuttgart: J. Engelhorn, 560 pp.

Held, I. M. and Soden, B. J. (2000). Water vapor feedback and global warming. *Annu. Rev. Energy Environ.*, **25**: 441–75.

Hellner, H. N., *et al.* (2008). The ISPOL drift experiment. *Deep-sea Res. (Topical studies in oceanology)*, **55** (8–9): 913–17.

Henderson, G. R. and Leathers, D. J. (2010). European snow cover extent variability and associations with atmospheric forcings. *Int. J. Climatol.*, **30** (10): 1443–51, doi: 10. 1002/ joc. 1990.

Henriksen, M., *et al.* (2003). Lake stratigraphy implies an 80,000 yr delayed melting of buried dead ice in northern Russia. *J. Quat. Sci.*, **18**: 663–79, doi: 10.1002/jqs.788.

Henry, H. A. L. (2008). Climate change and soil freezing dynamics: historical trends and projected changes. *Climatic Change*, **87**: 421–34.

Herbert, T. D., *et al.* (2010). Tropical ocean temperatures over the past 3.5 million years. *Science*, **328** (5985): 1530–4.

Herbert, W. (1969). *Across the top of the world. The British Trans-Arctic Expedition, Harlow*. Essex: Longmans. 209 pp.

Hernández-Henríquez, M. A., Déry, S. J., and Derksen, C. (2015). Polar amplification and elevation-dependence in trends of Northern Hemisphere snow cover extent, 1971–2014. *Environ. Res. Lett.*, **10**: 044010, doi: 10.1088/1748-9326/10/4/044010.

Heron, R. and Woo, M.-K. (1994). Decay of a High Arctic lake-ice cover: observations and modelling. *J. Glaciol.*, **40** (135): 283–92.

Hess, H. (1904). *Die Gletscher*. Braunschweig: F. Vieweg und Sohn. 426 pp.

Hess, H. (1935). Die Bewegung im innern des Gletschers. *Zeit. Gletscherkunde*, **23**: 1–35.

Hewitt, K. (2009). The Karakoram anomaly? Glacier expansion and the elevation effects, Karakoram Himalaya. *Mountain Res. Devel.*, **25L**: 332–40.

Hibler, W. D., III. (1979). A dynamic-thermodynamic sea ice model. *J. Phys. Oceanogr.*, **9**: 815–46.

Hibler, W. D., III. (2004). Modelling the dynamic response of sea ice. In Bamber, J. L. and Payne, A. J. (eds.), *Mass balance of the cryosphere: Observations and modelling of contemporary and future change*, Cambridge: Cambridge University Press. pp. 227–334.

Hibler, W.D., III and Flato, G. M. (1992). Sea ice models. In Trenberth, K. (ed.), *Climate System Modeling*, Cambridge: Cambridge University Press. pp. 413–36.

Hibler, W. D., III and Schulson, E. M. (2000). On modeling the anisotropic failure and flow of flawed sea ice. *J. Geophys. Res.*, **105** (C7): 17, 105–20.

Hicks, F. (2008). An overview of river ice problems: CRIPE07 guest editorial. *Cold Regions Sci. Technol.*, **55**: 175–85.

Hicks, F. and Beltaos., S. (2008). River ice. In Woo, M.-k. (ed.), *Cold region atmospheric and hydrologic studies The Mackenzie GEWEX experience, Vol. 2: Hydrologic processes*. Dordrecht: Springer-Verlag. pp. 281–305.

Hicks, F., Andrishak, R., and She, Y.-T. (2007). Modeling thermal and dynamic river ice processes. Current practices in cold regions engineering. Proceedings of the 13th International Conference on Cold Regions Engineering July 23–26, 2006, Orono, Maine. doi: 10.1061/40836(210)11.

Hill, B., Ruffman, A., and Drinkwater, K. (2002). Historical records of the incidence of sea ice on the Scotian Shelf and the Gulf of St. Lawrence, In: Ice in the environment, Vol. 1, Squire, V. and Langhorne, P. (eds.), Proc. 16th IAHR Internat. Sympos. on Ice, Int. Assoc. Hydraulic Eng. Res., Dunedin, New Zealand. pp. 313–20.

Hinkel, K. M. and Nelson, F. E. (2003). Spatial and temporal patterns of active layer thickness at Circumpolar Active Layer Monitoring (CALM) sites in northern Alaska,1995–2000. *J. Geophys. Res.*, **108** (D2): 8168, 13, doi: 10.1029/2001JD000927.

Hinzman, L. D., *et al.* (2005). Evidence and implications of recent climate change in northern Alaska and other Arctic regions. *Clim. Change*, **72** (3): 251–98.

Hirabayashi, Y., Döll, P., and Kanae, S. (2010). Global-scale modeling of glacier mass balances for water resources assessments: Glacier mass changes between 1948 and 2006. *J. Hydrol.*, **390**: 245–56.

Hirabayashi, Y., Zhang, Y., Watanabe, S., Koirala, S., and Kanae, S. (2013). Projection of glacier mass changes under a high-emission climate scenario using the global glacier model HYOGA2. *Hydrol. Res. Lett.*, **7**: 6–11.

Hirashima, H., *et al.* (2008). Avalanche forecasting in a heavy snowfall area using the SNOWPACK Model. *Cold Regions Sci. Technol.*, **51**: 191–203.

Hjort, J., Karjalainen, O., Aalto, J., Westermann, S., Romanovsky, V. E., Nelson, F. E., Etzelmüller, B. , and Luoto, M. (2018). Degrading Permafrost Puts Arctic Infrastructure at Risk by Mid-Century. *Nature Communications*, **9**: 5147.

Hobbs, W. (1910). The ice masses on and about the Antarctic continent. *Zeit. f. Gletscherk.*, **5**: 36–73, 87–122.

Hock, R. (2003). Temperature index melt modelling in mountain areas. *J. Hydrol.*, **282**: 104–15.

Hock, R. (2005). Glacier melt: a review of processes and their modeling. *Progr. Phys. Geog.*, **29**: 362–91.

Hock, R., *et al.* (2009). Mountain glaciers and ice caps around Antarctica make a large sea-level rise contribution. *Geophys. Res. Lett.*, **36**: L07501, doi: 10.1029/2008GL037020.

Hock, R., *et al.* (2019). High mountain areas. In Pörtner, H.-O., *et al.* (eds.), *IPCC Special Report on the Ocean and Cryosphere in a Changing Climate*.

Hock, R., Bliss, A., Marzeion, B., Giesen, R. H., Hirabayashi, Y., Huss, M., Radić, V., and Slangen, A. B. A. (2019). GlacierMIP – A model intercomparison of global-scale glacier mass-balance models &

projections. *J. of Glaciology*, **65** (251): 453–67, doi: 10.1017/jog.2019.22.

Hock, R. and Holmgren, B. (1996). Some aspects of energy balance and ablation of Storglaciären, northern Sweden. *Geogr. Ann.*, **78A**: 121–31.

Hodgkins, G. A., James, I. C., II, and Huntington, T. G. (2002). Historical changes in lake ice-out dates as indicators of climate change in New England, 1850–2000. *Int. J. Climatol.*, **22** (15): 1819–27.

Hodgkins, G., Dudley, R., and Huntington, T. (2005). Changes in the number and timing of days of ice-affected flow on northern New England rivers, 1930–2000. *Clim. Change*, **71**: 319–40.

Hoelzle, M. (1992).Permafrost occurrence from BTS measurements and climatic parameters in the eastern Swiss Alps. *Permafrost & Periglac. Proc.*, **3**: 143–7.

Hoelzle, M., *et al.* (2003). Secular glacier mass balances derived from cumulative glacier length changes. *Global and Planetary Change*, **36**: 295–306.

Hoelzle, M., *et al.* (2007). The application of glacier inventory data for estimating past climate change effects on mountain glaciers: A comparison between the European Alps and the Southern Alps of New Zealand. *Global Planet. Change*, **56**: 69–82.

Hoffman, M. J., Fountain, A. G., and Achuff, J. M. (2007). 20th-century variations in area of cirque glaciers and glacierets, Rocky Mountain National Park, Rocky Mountains, Colorado, USA. *Annals Glaciol.*, **46**: 349–54.

Hoffman, P. F., *et al.* (1998). A neoprotezoic snowball earth. *Science*, **281** (5381): 1342–6.

Hofmann, W. and Patzelt, G. (1983). *Die Berg- und Gletscherstürze von Huascaran, Cordillera Blanca, Peru.*

Hochgebirgsforschung 6. Innsbruck: Universitätsverlag Wagner. 110 pp.

Hodgkins, G. A., James, I. C., and Huntington, T. G. (2002). Historical changes in lake ice-out dates as indicators of climate change in New England, 1850–2000. *International Journal of Climatology*, **22**: 1819–27.

Høgda, K. A., Storvold, R., and Lauknes, T. R. (2010). SAR imaging of glaciers. In Pellikka, P. and Rees, W. G. (eds.), *Remote sensing of glaciers*, London: CRC Press, Taylor and Francis. pp. 153–78.

Holland, P. R. (2014). The seasonality of Antarctic sea ice trends. *Geophys. Res. Lett.*, **41**: 4230–37, doi: 10.1002/2014GL060172.

Holland, D. M. (2001). Explaining the Weddell Polynya – a large ocean eddy shed at Maud Rise. *Science*, **292** (5522): 1697–700.

Holland, D. M., *et al.* (2008). Acceleration of Jakobshavn Isbrae triggered by warm subsurface ocean waters. *Nature Geoscience*, **1**: 1–6, 46.

Holland, M. M., Curry, J. A., and Schramm, J. L. (1997). Modeling the thermodynamics of a sea ice thickness distribution. 2. Sea ice/ocean interactions. *J. Geohys. Res.*, **102**: 23, 93–107.

Holland, M. M., Bitz, C. M., and Tremblay, H. (2006). Future abrupt reductions in the summer Arctic sea ice. *Geophys. Res. Lett.*, **33**: L23503, doi: 10.1029/2006GL028024.

Holland, M. M., Serreze, M. C., and Stroeve, J. (2010). The sea ice mass budget of the Arctic and its future change as simulated by coupled climate models. *Clim. Dynam.*, **34**: 185–200.

Holland, P. R., Jenkins, A., and Holland, D. M. (2008). The response of ice-shelf basal melting to variation in ocean temperature. *J. Clim.*, **21**: 2558–72, doi: 10.1175/2007JCLI1909.

Höllemann, J., *et al.* (2010). Ocean-sea ice-atmosphere observations in the Laptev Sea polynya, Proceedings, Tromso Sea Ice Symposium. Int. Glaciol. Soc., Paper 57A122.

Holmes, G. W., Hopkins, D. M., and Foster, H. l. (1968). Pingos in central Alaska. U.S. Geol. Survey Bull., 1241-H, 40 pp.

Hood, E., Williams, M., and Cline, D. (1999). Sublimation from a seasonal snowpack at a continental, mid-latitude alpine site. *Hydrol. Proc.*, **13**: 1781–97.

Hooke, R. LeB. (1989). Englacial and subglacial hydrology: A review. *Arct. Alp. Res.*, **21**: 221–33.

Hooke, R. LeB. (2005). *Principles of glacier mechanics*. 2nd edn. Cambridge: Cambridge University Press. 248 pp.

Hope, G. S., Peterson J. A., Radok, U., and Allison, I. (1976). *The equatorial glaciers of New Guinea Rotterdam*. A./A. Balkema. 244 pp.

Hopkins, M. A. and Thorndike, A. S. (2002). Linear kinematic features in Arctic sea ice. In: Ice in the environment, Vol. 1, Squire, V. and Langhorne, P. (eds.), Proc. 16th IAHR Internat. Sympos. on Ice, Int. Assoc. Hydraulic Eng. Res., Dunedin, New Zealand. pp. 466–73.

Hopkins, M. A. and. Tuhkuri, J. (1999). Compression of floating ice fields. *J. Geophys. Res.*, **104** (C7): 15, 815–25.

Horwath, *et al.* (2016). ESA Climate Change Initiative (CCI) Sea Level Budget Closure (SLBC_cci) Sea Level Budget Closure Assessment Report D3.1, Version 1.0.

Hotzel, I. S. and Miller, J. D. (1983). Icebergs: their physical dimensions and the presentation and application of measured data. *Annals Glaciol.*, **4**: 116–23.

Houghton, J. (2009). *Global Warming*. 4th edn. Cambridge: Cambridge University Press. 438 pp.

Houle, D., Moore, J. D., and Provencher, J. (2007). Ice bridges on the St. Lawrence River as an index of winter severity, from 1620 to 1910. *J. Climate*, **20** (4): 757–64.

Howell, S. E. L., *et al.* (2008a). Multi-year sea-ice conditions in the western Canadian Arctic Archipelago region of the Northwest Passage: 1968–2006. *Atmos. – Ocean*, **46**: 229–42.

Howell, S. E. L., *et al.* (2008b). Changing sea ice melt parameters in the Canadian Arctic Archipelago: Implications for the future presence of multiyear ice. *J. Geophys. Res.*, **113** (C9): C09030, doi: 10.1029/2008JC004730.

Howell, S. E. L., *et al.* (2009). Variability in ice phenology on Great Bear Lake and Great Slave Lake, Northwest Territories, Canada, from SeaWinds/QuikSCAT: 2000– 2006. *Remote Sens. Environ.*, **113**: 816–34.

Hu, C., *et al.* (2016). Shifting El Niño inhibits summer Arctic warming and Arctic sea-ice melting over the Canada Basin. *Nat. Commun.*, 7: 11721, doi: 10.1038/ncomms11721

Huang, L., Luo, J., Lin, Z., Niu, F., and Liu, L. (2020). Using deep learning to map retrogressive thaw slumps in the Beiluhe region (Tibetan Plateau) from CubeSat images. *Remote Sensing of Environment*, **237**: 111534.

Hubbard, A., *et al.* (2000). Glacier mass-balance determined by remote sensing and high-resolution modelling. *J. Glaciol.*, **46** (154): 491–8.

Hugelius, G., *et al.* (2014). Estimated stocks of circumpolar permafrost carbon with quantified uncertainty ranges and identified data gaps. *Biogeosciences*, **11**: 6573–93.

Huggel, C., Caplan-Auerbach, J., and Wessels, R. (2008). Recent extreme avalanches: triggered by climate change? *Eos*, **89** (47): 469–70.

Hugel, C., *et al.* (2005). The (2002) rock/ice avalanche at Kolka/ Karmadon,Russian Caucasus: assessment of extraordinary avalanche formation. *Natural Hazards Earth System Sci.*, **5**: 173–87.

Huggel, C., *et al.* (2008). The 2005 Mt. Steller, Alaska, rock–ice avalanche: a large slope failure in cold permafrost. In Kane, D. L. and Hinkel, K. M. (eds.), Proceedings, the Ninth International Conference on Permafrost, Fairbanks, AK: University of Alaska, Institute of Northern Engineering. pp. 747–52.

Hughes, P. D. (2009). Twenty-first century glaciers and climate in the Prokletije Mountains, Albania. *Arct. Antarct. Alp. Res.*, **41**: 455–9.

Hughes, T. J. (1998). *Ice sheets.* New York: Oxford University Press. 343 pp.

Hulbe, C., Fahnestock, M., and Shuman, C. (2005). Ice streams stop and start: evidence from the Ross Ice Shelf, interpreted using numerical models of ice shelf flow. American Geophysical Union, Fall Meeting 2005, abstract C44A-02.

Hunke, E. C. and Holland, M. M. (2007). Global atmospheric forcing data for Arctic ice-ocean modelling. *J. Geophys. Res.*, **112**: C0451413.

Huntington, T. G., Hodgkins, G. A., and Dudley, R. W. (2003). Historical trend in river ice thickness and coherence in hydroclimatological trends in Maine. *Clim. Change*, **61**: 217–36.

Huntington, T. G., *et al.* (2004). Changes in the proportion of precipitation occurring as snow in New England (1949–2000). *J. Clim.*, **16**: 2626–36.

Huppert, H. E. (1980). The physical processes involved in the melting of icebergs. *Ann. Glaciol.*, **1**: 97–101.

Huss, M. (2012). Extrapolating glacier mass balance to the mountain-range scale: the European Alps 1900–2100. *Cryosph.*, **6**: 713–27, doi: 10.5194/tc-6-713-2012.

Huss, M., *et al.* (2008). Modelling runoff from highly glacierized alpine drainage basins in a changing climate. *Hydrol. Process.*, **22**: 3888–902.

Huss, M., *et al.* (2010). 100-year mass changes in the Swiss Alps linked to the Atlantic Multidecadal Oscillation. *Geophys. Res. Lett.*, **37** (10): L10501

Huss, M. and Farinotti, D. (2012). Distributed ice thickness and volume of all glaciers around the globe. *J. Geophy. Res. Atmos.*, **117** (F4): F04010, doi: 10.1029/ 2012JF002523.

Huss, M. and Hock, R. (2018). Global-scale hydrological response to future glacier massef loss. *Nature Climate Change*, https://doi.org/10.1038/s41558-017–0049-x

Husseiny, A. A. (ed.). (1978). Iceberg utilization. Proceedings of the First International Conference on Iceberg Utilization for Fresh water Production, Weather Modification, and Other Applications. Vol. 1. Elmsford, NY: Pergamon Press.

Hutchings, J. K., Heil, P., Steer, A., and Hibler, W. D., III (2012). Small-scale spatial variability of sea ice deformation in the western Weddell Sea during early summer. *J. Geophys. Res.*, **117**: C01002, doi: 10.1029/2011JC006961.

Hutton, J. (1795). *The theory of the Earth, with proofs and illustrations.* Vol. 2. London: Caddell and Davies. p. 218.

Huybers, P. and Molnar, P. (2007). Tropical cooling and the onset of North American glaciation. *Clim. Past*, **3**: 549–57.

Huybrechts, P. (1992). The Antarctic ice sheet and environmental change: a three-dimensional modelling study. *Berichte Polarforsch.*, **99**: 241.

Huybrechts, P., Payne, A. J., and EISMINT Intercomparison Group. (1996). The EISMINT benchmarks for testing ice-sheet models. *Ann. Glaciol.*, **23**: 1–12.

Huybrechts, P., *et al*. (2000). Balance velocities and measured properties of the Antarctic ice sheet from a new compilation of gridded data for modeling. *Ann. Glaciol.*, **30**: 52–60.

IAHR Working Group on River Ice Hydraulics (1986). River ice jams; a state of the art report. Proceedings International Ice Symposim, Iowa City, USA. III: 561–94

IceSat-2 (2018). Measuring the Height of Earth's Ice from Space, NASA. NP-2018–07-231-GSFC.

IMBIE (2018). Mass balance of the Antarctic Ice Sheet from 1992 to 2017. *Nature*, **558**: 219, https://doi.org/10.1038/s41586-018–0179-y

Immerzeel, W. W., van Beek, L. P. H., and Bierkens, M. F. P. (2010). Climate change will affect the Asian water towers. *Science*, **328**: 1382–5.

Ingólfsson, O. (2004). Quaternary glacial and climatic history of Antarctica. In Ehlers, J. and Gibbard, P. L. (eds.), *Quaternary glaciations – extent and chronology. Part III*. Dordecht, Netherlands: Elsevier. pp. 3–44.

Intergovernmental Panel on Climate Change (2007). *The Physical Science Basis, Contribution of Working Group I to the Fourth Assessment Report of the Intergovernmental Panel on Climate Change*, Solomon, S., Qin, D., Manning, M., Chen, Z., Marquis, M., Averyt, K. B., Tignor M., and Miller, H. L. (eds.), Cambridge, UK: Cambridge University Press.

Ireland, S. (1792). *Picturesque views of the River Thames, from its source in Gloucestershire to the Nore*. London: T. and J. Egerton. **2** vols. 209 and 258 pp.

IPCC (2019). Summary for Policymakers. In *IPCC Special Report on the Ocean and Cryosphere in a Changing Climate* [H.-O. Pörtner, D. Roberts, V. Masson-Delmotte, P. Zhai, M. Tignor, E. Poloczanska, K. Mintenbeck, M. Nicolai, A. Okem, J. Petzold, B. Rama, N. Weyer (eds.)], Springer Int. Publishing, Cham., Switzerland.

IPCC, (2013). SPM in Climate Change 2013: The Physical Science Basis. Contribution of Working Group I to the Fifth Assessment Report of the IPCC [Stocker, T. F., D. Qin, G.-K. Platter, M. Tignor, S. K. Allen, J. Boschung, A. Nauels, Y. Xia, V. Bex and P. M. Midgley (eds.)]. Cambridge University Press, United Kingdom.

IPCC, (2014). Climate Change 2014: Synthesis Report. Contribution of Working Groups I, II and III to the Fifth Assessment Report of the Intergovernmental Panel on Climate Change [Core Writing Team, R. K. Pachauri and L. A. Meyer (eds.)]. IPCC, Geneva, Switzerland, 151 pp.

IPCC, I. P. O. C., (2014). Fifth Assessment Report–AR5. Disponível em: www.ipcc.ch /report/ar5/. Acesso em, 20.

Ivana, K., et al., 2007, An Operational Iceberg Deterioration Model, Proceedings of the 16th (2007) International Offshore and Polar Engineering Conference, Paper 2007-JSC-409, 2007-07-01

Ives, J. D. (1985). Glacial lake outburst floods and risk engineering in the Himalaya, Occas. Paper No. 5, International Center for Integrated Mountain Development, Nepal: Kathmandu.

Ives, J. D. (1986). Glacial lake outburst floods and risk engineering in the Himalaya: a review of the Langmoche disaster, Khumbu Himal, August 4, 1985. Occas. Paper No, 10., International Center for Integrated Mountain Development, Nepal: Kathmandu:

Ives, J. D. (2007). *Skaftafell in Iceland: A thousand years of change*. Rwyjjavik: Ormstunga. 256 pp.

Ives, J. D., Shrestha, R. B., and Mool, P. K. (2010). *Formation of glacial lakes in the Hindu Kush-Himalayas and GLOF risk assessment*. Kathmandu, Nepal: International Centre for Integrated Mountain Development. 66 pp.

Ivy-Ochs, S., *et al.* (2009). Latest pleistocene and holocene glacier variations in the

European Alps. *Quat. Sci. Rev.*, doi: 10.1016/j.quascirev.2009.03.009.

Jacka, T. H. and Giles, A. B. (2007). Antarctic iceberg distribution and dissolution from ship-based observations. *J. Glaciol.*, **53** (182): 341–56.

Jacob, T., Wahr, J., Pfeffer, W. T., and Swenson, S. (2012). Recent contributions of glaciers and ice caps to sea level rise. *Nature*, **482**: 514–8. doi: 10.1038/nature10847.

Jacobs, J. D., Barry, R. G., and Weaver, R. L. (1975). Fast ice characteristics, with special reference to the eastern Canadian Arctic. *Polar Record*, **17**: 521–36.

Jacobs, S. S., Helmer, H. H., Doake, C. S. M., Jenkins, A., and Frolich, R. M. (1992). Melting of ice shelves and the mass balance of Antarctica. *J. Glaciol.*, **38**: 375–87.

Jacobs, S. S., et al., 2011, Stronger ocean circulation and increased melting under Pine Island Glacier ice shelf. Nature Geosci. 4, 519–523.

Jamard, A. L., Garcia, S., and Bélanger, L. (2002). L'enquête permanente sur les avalanches (EPA). *Statistique descriptive générale des événemnets et des sites. DESS Ingiénérie Mathématiques Option statistique.* Grenoble, France: Université Joseph Fourier. 111 pp [Available online at www.avalanches.fr/].

Jamieson, B., Campbell, C., and Jones, A. (2008). Verification of Canadian avalanche bulletins including spatial and temporal scale effects. *Cold Reg. Sci. Technol.*, **51**: 204–13.

Jamieson, B. and Geldsetzer, T. (1996). *Avalanche accidents in Canada, 1984–1996, Vol. 4.* Ottawa: National Research Council. 202 pp.

Janowicz, J. R. (2010). Observed trends in the river ice regimes of northwest Canada. *Hydrol. Res.*, **41**: 462–70, https://doi.org/10.2166/nh.2010.145

Janowicz, J. R. and Hinzman, L. (2017). In Sesser, A. L., Rockhill, A. P., Magness, D. R., Reid, D., DeLapp, J.,

Burton, P., Schroff, E., Barber, V., and Markon, C. (eds.), *Drivers of landscape change in the northwest boreal region of North America: Implications on policy and land management.* U.S. Geological Survey Circular.

Jansson, P., Hock, R., and Schneider, T. (2003). The concept of glacier storage: A review. *J. Hydrol.*, **282**: 116–29.

Jasek, M. J. (1999). 1998 breakup and flood on the Yukon River at Dawson – Did El Nin˜o and climate change play a role? In Shen, H. T. (ed.), *Ice in surface waters,* Rotterdam: Balkema. pp. 761–8.

Jastrow, R. and Rampino, M. (2008). *Origins of life in the universe.* Cambridge: Cambridge University Press. 444 pp.

Jeffries, M. O. (1992). Arctic ice shelves and ice islands: Origin, growth and disintegration, physical characteristics, structural-stratigraphic variability, and dynamics. *Rev. Geophys.*, **30**: 245–67.

Jeffries, M. O. (2002). Ellesmere Island ice shelves and ice islands. In Williams, R. S. and Ferrigno, J. G. (eds.), *Satellite image atlas of glaciers of the world: Glaciers of North America,* Washington, DC: United States Geological Survey. pp. J147–64.

Jeffries, M. O., Morris, K., and Liston, G. E. (1995). A method to determine lake depth and water availability on the North Slope of Alaska with spaceborne imaging radar and numerical ice growth modeling. *Arctic,* **48**: 367–74.

Jenkins, A., *et al.* (2010). Observations beneath Pine Island Glacier in West Antarctica and implications for its retreat. *Nature Geosci.*, **3**: 468–72.

Jenness, J. L. (1949). Permafrost in Canada. *Arctic,* **2**: 13–27.

Jensen, O. P., *et al.* (2007). Spatial analysis of ice phenology trends across the Laurentian Great Lakes region during a recent warming period. *Limnol. Oceanog.,* **52** (5): 2013–26.

Jenssen, D. (1977). A three-dimensional polar ice sheet model. *J. Glaciol.,* **18** (80): 373–89.

Jezek, K. C. (2003). Observing the Antarctic ice sheet using the Radarsat-1 synthetic aperture radar. *Polar Geog.*, **27**: 197–209.

Jezek, K. C. (1999). Glaciological properties of the Antarctic ice sheet from RADARSAT-1 synthetic aperture radar imagery. *Annals Glaciol.*, **29**: 286–90.

Jezek, K. C. (2008). *The RADARSAT-1 Antarctic mapping project. BPRC Rep. No. 22.* Columbus, OH: Byrd Polar Res. Center, Ohio State University. 64 pp.

Jiang, Y.-D., *et al.* (2008). Long-term changes in ice phenology of the Yellow River in the past decades. *J. Climate*, **21** (18): 4879–86.

Jin, J., *et al.* (1999). Comparative analyses of physically based snowmelt models for climate simulations. *J. Climate*, **12**: 2643–57.

Jing, Z., *et al.* (2006). Mass balance and recession of Urumqi glacier No. 1, Tien Shan, China, over the last 45 years. *Annals Glaciol.*, **43**: 214–7.

Jiskoot, H., Boyle, P., and Murray, T. (1998). The incidence of glacier surging in Svalbard: evidence from multivariate statistics. *Computers & Geosci.*, **24**: 387–99.

Johannessen, O. M., Bobylev, L., Shalina, E. V., and Sandven, S. (2020). *Sea Ice in the Arctic, Past, Present and Future*, 575 pp., Springer Polar Sciences, ISBN 978-3-030-21301-5.

Johannessen, O. M., *et al.* (2004). Arctic climate change: observed and modelled temperature and sea-ice variability. *Tellus*, **56A**: 328–41.

Johannessen, O. M., *et al.* (2007). *Remote sensing of sea ice in the Northern Sea Route: studies and applications.* Chichester, UK: Springer, Praxis Publishing. 472 pp.

Jóhannesson, T., Raymond, C. F., and Waddington, E. D. (1989). A simple method for determining the response time of glaciers. In Oerlemans, J. (ed.), *Glacier fluctuations and climate change*, Dordrecht: Kluwer. pp. 407–17.

Johanesson, T., *et al.* (1995). Degree-day glacier mass-balance modelling with applications to glaciers in Iceland, Norway and Greenland. *J. Glaciol.*, **41** (138): 345–58.

Johnson, B. C., Jamieson, J. B., and Stewart, E. R. (2004). Seismic measurement of fracture speed in a weak snowpack layer. *Cold Reg. Sci. Technol.*, **4**: 41–5.

Johnson, M., *et al.* (2007). A comparison of Arctic Ocean sea ice concentration among the coordinated AOMIP model experiments. *J. Geophys. Res.*, **112** (C04S11L): 16.

Johnston, M. B., Masterson, D., and Wright, B. (2009). Multi-year ice thickness: knowns and unknowns. Proceedings 20th POAC Conference, Paper POAC09–120. Lulea, Swededn: Lulea University of Technology.

Jomelli, V., *et al.* (2009). Fluctuations of glaciers in the tropical Andes over the last millennium and palaeoclimatic implications: A review. *Palaeogeog., Palaeoclimatol., Palaeoecol.*, **281**: 269–82.

Jones, A., *et al.* (eds.). (2010). *Soil atlas of the northern circumpolar region.* European Commission: Office for Official Publications of the European Communities. 144 pp.

Jones, G. S., Stott, P. A., and Christidis, N. (2013). Attribution of observed historical near-surface temperature variations to anthropogenic and natural causes using CMIP5 simulations. *J. Geophys. Res. Atmos.*, **118**: 4001–24, doi: 10.1002/jgrd.50239.

Jones, B. M., *et al.* (2008). Modern erosion rates and loss of coastal features and sites, Beaufort Sea coastline, Alaska. *Arctic*, **61**: 361–72.

Jones, H. G. (2008). From Commission to Association: the transition of the International Commission on Snow and Ice (ICSI) to the International Association of Cryospheric Sciences (IACS). *Annals Glaciol.*, **48**: 1–5.

Jones, M. K. W., Pollard, W. H., and Jones, B. M. (2019). Rapid initialization of retrogressive thaw slumps in the Canadian High Arctic and their response to climate and terrain factors. *Environmental Research Letters*, **14**: 055006.

Jones, P. D., Raper, S. C. B., and Wigley, T. M. L. (1986a). Southern Hemisphere surface air temperature variations: 1851–1984. *J. Appl. Meteorol.*, **25**: 1213–30.

Jones, P. D., *et al.* (1986b). Northern Hemisphere surface air temperature variations: 1851– 1984. *J. Clim. Appl. Meteorol.*, **25**: 161–79.

Jordan, R. (1991). *A one-dimensional temperature model for a snow cover: Technical documentation for SNTHERM 89. CRREL Special Report 91–16*. Hanover, NH: U.S. Army Cold Regions Research and Engineering Laboratory. 49 pp.

Jordan, R., Andreas, E., and Makshtas, A. (1999). Heat budget of snow covered sea ice at North Pole 4. *J. Geophysical Res.*, **104** (C4): 7785–806.

Jorgenson, M. T. and Kreig, R. (1988). A model for mapping permafrost distribution based on landscape component maps and climatic variables. In Sennesset, K. (ed.), *Permafrost. Proceedings of the fifth international conference on permafrost*. Trondheim: Tapir. Vol.1, pp. 176–82.

Jorgenson, M. T., Shur, Y. L., and Pullman, E. R. (2006), Abrupt increase in permafrost degradation in Arctic Alaska. *Geophys. Res. Lett.*, **33**: L02503, doi: 10.1029/2005GL024960.

Joughin, I., Abdalati, W., and Fahnestock, M. (2004). Large fluctuations in speed on Greenland's Jakobshavn Isbrae glacier. *Nature*, **432**: 608–10.

Joughin, I. and Tulaczyk, S. (2002). Positive mass balance of the Ross ice streams. *West Antarctica Science*, **295** (5554): 476–80.

Joughin, I., *et al.* (1999). Ice flow of Humboldt, Petermann, and Ryder Gletscher, northern Greenland. *J. Glaciol.*, **45** (150): 231–41.

Joughin, I., *et al.* (2006). Integrating satellite observations with modelling: basal shear stress of the Filcher-Ronne ice streams, Antarctica. *Phil. Trans. Roy. Soc., A*, **364**: 1795–814.

Joughin, I., *et al.* (2008). Continued evolution of Jakobshavn Isbrae following its rapid speedup. *J. Geophys. Res.*, **113** (F04006): 14.

Juen, I., Kaser, G., and Georges, C. (2007). Modeling observed and future runoff from a glacierized tropical catchment (Cordillera Blanca, Perú). *Global Planet. Change*, **59**: 37–48.

Juliussen, H., *et al.* (2010). NORPERM, the Norwegian Permafrost Database – a TSP NORWAY IPY legacy. *Earth Syst. Sci. Data*, **2**: 235–46.

Kääb, A. (2002). Monitoring high-mountain terrain deformation from digital aerial imagery and ASTER data. *J. Photogramm. Remote Sens.*, **57**: 39–52.

Kääb, A. (2008). Glacier volume changes using ASTER satellite stereo and ICESat GLAS laser altimetry: A test study on Edgeøya, eastern Svalbard. *IEEE Trans. Geosci, Remote Sensing*, **46** (10): 2823–30.

Kaenzig, R. (2015). Can glacial retreat lead to migration? A critical discussion of the impact of glacier shrinkage upon population mobility in the Bolivian Andes. *Population and Environment*, **36** (4): 480–96, doi: 10.1007/s11111-3 014–0226-z.

Kalinin, V. M. and Yakupov, V. S. (1994). Permafrost thickness along meridional profile from East Siberian Sea to Sea of Okhotsk. ICAM-94 Proceedings: Permafrost and Engineering Geology: 320–2.

Kamb, B. (2001). Basal zone of the West Antarctic ice streams and its role in lubrication of their rapid motion. In Alley, R. B. and Bindschadler, R. A. (eds.), *The West Antarctic Ice Sheet*, Washington, DC: Am. Geophys. Union. pp. 157–99

Kamb, B. and LaChapelle, E. (1964). Direct observation of the mechanism of glacier sliding over bedrock. *J. Glaciol.*, **5** (38): 159–72.

Kamniansky, G. M. and Pertziger, F. L. (1996). Optimization of mountain glacier mass balance measurements. *Zeit. Gletscherk. Glazial geol.*, **32**: 167–75.

Kanaev, L. A., Sezin, V. M., and Tsarev, B. K. (1987). Principles of avalanche danger forecast in the USSR. Proceedings of 2nd All-USSR Avalanche Meeting. Leningrad: Gidrometeoizdat, pp. 37–46.

Kane, D. (1981). Physical mechanics of aufeis growth. *Canad. J. Civil Engin.*, **8**: 186–95.

Kang, E.-S., *et al.* (2008). Glacial runoff and its modeling. In Shi, Y.-F. (ed.), *Glaciers and related environments in China*, Beijing: Science Press. pp. 261–316.

Kapnick, S. and Hall, A. (2012). Causes of recent changes in western North American snowpack. *Clim. Dyn.*, **38**: 1885–99, doi: 10.1007/s00382-011-1089-y.

Kapsch, M.-L., Graversen, R. G., and Tjernström, M. (2013). Springtime atmospheric energy transport and the control of Arctic summer sea-ice extent. *Nature Climate Change*, **3**: 744, doi: 10.1038/nclimate1884.

Kargel, J. S., Leonard, G., Bishop, M. P., Kääb, A., and Raup, B. H. (eds.) (2014). *Global land ice measurements from space*. Springer-Praxis.

Kaser, G., Fountain, A., and Jansson, P. (2003). A manual for monitoring the mass balance of mountain glaciers with particular attention to low latitude characteristics. *Technical documents in hydrology No. 59*. Paris: UNESCO. 137 pp.

Kaser, G., *et al.* (2004). Modern glacier retreat on Kilimanjaro as evidence of climate change: Observations and facts. *Int. J. Climatol.*, **24**: 329–39.

Kaser, G., *et al.* (2006). Mass balance of glaciers and ice caps: consensus estimates for 1961–2004. *Geophys. Res. Lett.*, **33** (19): L19501, doi: 10.1029/2006GL027511.

Kaser, G. and Osmaston, H. (2002). *Tropical glacviets*. Cambridge: Cambridge University Press. 207 pp.

Kashiwase, H., Ohshima, K. I., Nihashi, S., and Eicken, H. (2017). Evidence for ice-ocean albedo feedback in the Arctic Ocean shifting to a seasonal ice zone. *Scientific Reports*, **7**: 8170.

Kasser, P. (1973). Influence of changes in the glacierized area on summer run-off in the Porte du Scex drainage basin of the Rhône Symposium on the hydrology of glaciers. *Int. Assoc. Sci. Hydrol., Publ.*, **95**: 221–5.

Katsuyama, Y., Inatsu, M., Nakamura, K., and Matoba, S. (2017). Global warming response of snowpack at mountain range in northern Japan estimated using multiple dynamically downscaled data. *Cold Regions Science and Technology*, **136**: 62–71, doi: 10.1016/j.coldregions.2017.01.006.

Kattelmann, R. and Elder, K. (1991). Hydrologic characteristics and water balance of an alpine basin in the Sierra Nevada. *Water Resour. Res.*, **27**: 1553–62.

Kaufman, D., *et al.* (2009). Recent warming reverses long-term Arctic cooling. *Science*, **325** (5945): 1236–39, doi: 10.1126/science.1173983.

Kauker, F., *et al.* (2009). Adjoint analysis of the 2007 all time Arctic sea-ice minimum. *Geophys. Res. Lett.*, **36**: LL03707, doi: 10.1029/2008GL036323.

Kavanaugh, J. L., *et al.* (2009a). Dynamics and mass balance of Taylor Glacier, Antarctica: 1. Geometry and surface velocities *J. Geophys. Res.*, **114** (F4): F04010, doi: 10.1029/2009JF001309.

Kavanaugh, J. L., *et al.* (2009b). Dynamics and mass balance of Taylor Glacier, Antarctica: 3. State of mass balance. *J. Geophys. Res.*, **114** (F4): F04012.

Kay, J. E., Holland, M. M., and Jahn, A. (2011). Inter-annual to multi-decadal Arctic sea ice extent trends in a warming world. *Geophys. Res. Lett.*, **38**: L15708, doi: 10.1029/2011GL048008.

Kayastha, R. B. (2001). Study of glacier ablation in the Nepalese Himalayas by the energy balance model and positive degree-day method. PhD Thesis. Graduate School of Science, Nagoya University, 95 pp

Kazaryan, P. (2005). Lena river. In Nuttall, M. (ed.), *Encyclopedia of the Arctic*. London: Routledge. 2380 pp.

Kazutaka, T., Hiroyuki, E., and Fumihiko, N. (2001). Observation of sea ice in the Sea of Okhotsk by using the thin/thick ice detecting algorithm. *Seppyo*, **63**: 21–34.

Kendra, J. R., Sarabandi, S., and Ulaby, F. T. (1998). Radar measurements of snow: experiment and analysis. *IEEE Trans. Geosci. Remote Sens.*, **36** (3): 864–79.

Kennedy, M., Mrofka, D., and von der Borch, C. (2008). Snowball Earth termination by destabilization of equatorial permafrost methane clathrate. *Nature*, **453**: 642–5.

Kennett, D. J., *et al.* (2009). Nanodiamonds in the Younger Dryas boundary sediment layer. *Science*, **323**: 94.

Kennett, J. P. (1977). Cenozoic evolution of Antarctic glaciation, the circum-Antarctic ocean, and their impact on global palaeoceanography. *J. Geophys. Res.*, **82**: 3843–60.

Kerkhoven, E. and Gan, T. Y. (2011). Differences and sensitivities in potential hydrologic impact of climate change to regional-scale Athabasca and Fraser River basins of the leeward and windward sides of the Canadian Rocky Mountain respectively. *Clim. Chang.*, **106** (4): 583–607, https://doi.org/10.1007/s10584-010-9958-7

Kern, S. (2009). Wintertime Antarctic coastal polynya area: 1992–2008. *Geophys. Res. Lett.*, **36**: L14501, doi: 10.1029/2009GL038062.

Kern, S., Kaleshcke, L., and Spreen, G. (2010). Climatology of the Nordic (Irminger, Greenland, Barents, Kara and White/Pechora) Seas ice cover based on 85 GHz satellite microwave radiometry: 1992–2008. *Tellus*, **62A**: 411–34.

Kerr, R. A. (2009). Arctic summer sea ice could vanish soon but not suddenly. *Science*, **323** (5922): 1655.

Kershaw, G. P. and McCulloch, J. (2007). Midwinter snowpack variation across the Arctic treeline, Churchill, Manitoba, Canada. *Arct. Ant. Alp. Res.*, **39**: 9–15.

Ketchum, H. G. and Hildenbrand, R. N. (1977). Unusual iceberg sightings. Report of the International Ice Patrol Services in the North Atlantic Ocean. Appendix D. Bull. 63, CG-188–32. Dept. of Transportation, Coast Guard.

Key, J. R. and McLaren, A. S. (1989). Periodicities and keel spacing in the under-ice draft of the Canada Basin recorded by the USS Queenfish, August 1970. *Cold Regions Sci. Technol.*, **16**: 1–10.

Key, J., Drinkwater, M., and Ukito, J. (2007). A cryosphere theme report for the IGOS Partnership. Geneva: World Meteorological Organization, WMO/TD No. 1405, 100 pp.

Keylock, C. (1997). Snow avalanches. *Progr. Phys. Geog.*, **21**: 481–500.

Khalsa, S. J. S., Dyurgerov, M., Khromova, T., Raup, B., and Barry, R. G. (2004). Space-based mapping of glacier changes using ASTER and GIS tools. *IEEE Trans. Geosciences Remote Sensing*, **42** (10): 2177–83.

Khan, S. A., *et al.* (2010). Spread of ice mass loss into northwest Greenland observed by

GRACE and GPS. *Geophys. Res. Lett.*, **37**: L06501.

Khazendar, A., Rignot, E., and Larour, E. (2007). Larsen B Ice Shelf rheology preceding its disintegration inferred by a control method. *Geophys. Res. Lett.*, **34**: L19503, doi: 10.1029/2007GL030980.

Khazendar, A., Rignot, E., and Larour, E. (2009). Roles of marine ice, rheology, and fracture in the flow and stability of the Brunt/Stancomb-Wills Ice Shelf. *J. Geophys. Res.*, **114**: F04007, doi: 10.1029/2008JF001124.

Khon, V. C., *et al.* (2010). Perspectives of Northern Sea Route and Northwest Passage in the twenty-first century. *Clim. Change*, **100**: 757–68.

Kivinen, S. and Rasmus, S. (2015). Observed cold season changes in a Fennoscandian fell area over the past three decades. *Ambio*, **44**: 214–25, doi: 10.1007/s13280-014-0541-8.

Khromova, T. E., Dyurgerov, M. B., and Barry, R. G. (2003). Late-twentieth century changes in glacier extent in the Ak-Shirak Range, Central Asia, determined from historical data and ASTER imagery. *Geophys. Res. Lett.*, **30** (16): 1863, pp. HLS 2–1 to 2–5, doi: 10.1029/2003GL017233.

Khromova, T. E., Osipova, G. B., Tsvetkov, D. G., Dyurgerov, M. D., and Barry, R. G. (2006). Changes in glacier extent in the eastern Pamir, Central Asia, determined from historical data and ASTER imagery. *Remote Sensing of Environment*, **102**: 24–32.

Kieffer, H., Kargel, J., Barry, R. G., *et al.* (2000). New eyes in the sky measure glaciers and ice sheets. *EOS*, **81** (24): 265, 270–1.

Kienzle, S. W. (2008). A new temperature based method to separate rain and snow. *Hydrol. Process.*, **22** (26): 5067–85.

King, C. A. M. and Ives, J. D. (1956). Glaciological observations on some of the outlet glaciers of southwest Vatnajökull,

Iceland, 1954. *Pt II: Ogives. J. Glaciol.*, **2** (18): 563–9.

King, C. A. M. and Lewis, W. V. (1961). A tentative theory of ogive formation. *J. Glaciol.*, **3** (29): 915–39.

King, J. C., *et al.* (2008). Snow-atmosphere energy and mass balance. In Armstrong, R. L. and Brun, E. (eds.), *Snow and climate: physical processes, surface energy exchange and modeling*, Cambridge, UK: Cambridge University Press. pp. 70–124.

Kingdon-Ward, F. (1949). *Burma's icy mountains*. London: Jonathon Cape. 287 pp.

Kinnard, C., *et al.* (2008). A changing Arctic seasonal ice zone – Observations from 1870–2003 and possible oceanographic consequences. *Geophys. Res. Lett.*, **35** (L02507): 5.

Kinnard, C., *et. al.* (2011). Reconstructed changes in Arctic sea ice over the past 1,450 years. *Nature*, **479** (7374): 509–12, doi: 10.1038/nature10581.

Kirschvink, J. L. (1992). Late Proterozoic low-latitude global glaciation: The snowball Earth. In Schopf, J. W. and Klein, C. (eds.), *The Proterozoic biosphere: A multidisciplinary study*, Cambridge: Cambridge University Press. pp. 51–2.

Kissler, F. (1934). Eisgrenzen und Eisverschiebungen in der Arktis zwischen 50° W und 105° E in 34-jährigen Zeitraum 1898–1931. *Gerlands Beitr. Geophys.*, **42**: 12–55.

Klavins, M., Briede, A., and Rodinov, V. (2009). Long term changes in ice and discharge regime of rivers in the Baltic region in relation to climatic variability. *Clim. Change*, **95**: 485–98.

Klebelsberg, R. von. (1948/49). *Handbuch der Gletscherkunde und Glazialgeologie*, 2 vols. Vienna: Springer. 403 pp. and 602 pp.

Klein, A. G. and Kincaid, J. L. (2006). Retreat of glaciers on Puncak Jaya, Irian Jaya, determined from 2000 and 2002 IKONOS satellite images. *J. Glaciol.*, **52** (176): 65–79.

Klene, A. E., *et al.* (2001). The N-factor in natural landscapes: Variability of air and soil-surface temperatures, Kuparuk river basin, Alaska, USA. *Arct. Antarct. Alp. Res.*, **33**: 140–8.

Knight, P. G. (1999). *Glaciers*. London: Routledge. 261 pp.

Knowland, K. E., Gyakum, J. R., and Lin, C. A. (2010). A study of the meteorological conditions associated with anomalous early and late openings of a Northwest Territories winter road. *Arctic*, **63**: 227–39.

Koboltschnig, G. R., *et al.* (2009). Glaciermelt of a small basin contributing to runoff under the extreme climate conditions in the summer of 2003. *Hydrol. Proc.*, **23** (7): 1010–8.

Kobayashi, T. (1961). The growth of snow crystals at low supersaturations. *Phil. Mag.*, **6** (71): 1363–70.

Koch, J., Menounos, B., and Clague, J. J. (2009). Glacier change in Garibaldi Provincial Park, southern Coast Mountains, British Columbia, since the Little Ice Age. *Global Planet. Change*, **66** (3–4): 161–78.

Koch, L. (1945). The East Greenland ice. *Medd. Grønland (Coenhagen)*, **130** (3): 374 pp.

Kocin, P. J. and Uccellini, L. W. (2004). A snowfall impact scale derived from Northeast snowfall distributions. *Bull. Amer. Met. Soc.*, **85**: 177–94.

Koenig, S. L., Greenaway, E. R., and Dunbar, M. (1952). Arctic ice islands. *Arctic*, **5**: 68–95.

Koerner, R. M. (1970). Weather and ice observations of the British trans-Arctic expedition 1968–69. *Weather*, **25**: 218–28.

Koerner, R. M. (1973). The mass balance of the sea ice of the Arctic Ocean. *J. Glaciol.*, **12**: 173–85.

König, M., Winther. J.-G., and Isaksson, E. (2001). Measuring snow and glacier ice properties from satellite. *Rev. Geophys.*, **39**: 1–27.

König Beatty, C. and Holland, D. M. (2010). Modeling landfast sea ice by adding tensile strength. *J. Phys. Oceanog.*, **40**: 185–98.

Kohonen, T., Oja, E., Simula, O., and Kangas, J. (1996). Engineering application of the self-organizing map. *Proc. IEEE*, **84** (10): 1358–83.

Koivusalo, H. J. and Burges, S. (1996). Use of 1-dimensional snow cover model to analyze measured snow depth and snow temperature data from southern Finland, Water Resources Series, Tech. Rept. **150**. Seattle: University of Washington. 109 pp.

Kontar, Y. Y., *et al.* (2018). Advancing spring flood risk in the Arctic through interdisciplinary research and stakeholders collaboration, Chapter 25. In Beer, T., *et al.* (eds.), *Global change and future Earth: The geoscience perspective*, Cambridge: Cambridge University Press. 430 pp. ISBN-13 : 978-1107171596.

Kopp, P. E., *et al.* (2009). Probabilistic assessment of sea level during the last inter-glacial stage. *Nature*, **462**: 863–7.

Köppen, W. (1881). Über mehrjährige Perioden der Witterung – III. Mehrjährige Änderungen der Temperatur 1841 bis 1875 in den Tropen der nördlichen und südlichen gemässigten Zone, an den Jahresmitteln. untersucht. Zeitschrift der Österreichischen Gesellschaft für Meteorologie, Bd XVI, 141–50.

Korona, J., *et al.* (2009). SPIRIT. SPOT 5 stereoscopic survey of polar ice: Reference images and topographies during the fourth International Polar Year (2007–2009). ISPRS. *J. Photogramm. Remote Sens*, **64**: 204–12.

Kotlarski, S., *et al.* (2008). Representing glaciers in a regional climate model. *Clim. Dynam.*, **34**: 27–46.

Kotlarski, S., Jacob, D., Podzun, R., and Paul, F., 2008, Representing glaciers in a regional climate model, Climate Dynamics, **34** (1):27–46, DOI: 10.1007/s00382-009-0685-6

Kotlyakov, V. M. and Lebedeva, I. M. (1974). Nieve and ice penitentes, their way of formation and indicative significance. *Zeit. f. Gletscherk. Glazialgeol.*, **10**: 111–27.

Kotlyakov, V. M., Rototaeva, O. V., and Nosenko, G. (2004). The September 2002 Kolka glacier catastrophe In North Ossetia, Russian Federation: Evidence and analysis. *Mt. Res. Dev.*, **24**: 78–83.

Kotlyakov, V. M. (ed. In chief) (1997). *World Atlas of snow and ice resources.* Moscow: Institute of Geography, Russian Academy of Sciences. Vol. 1, Atlas, 392 pp.; Vol. 2, Snow and ice phenomena and processes, 372 pp.; Vol. 3, Legends and explanations for all the maps in English, 144 pp.

Kouraev, A. V., *et al.* (2004). Sea ice cover in the Caspian and Aral Seas from historical and satellite data. *J. Marine Systems*, **47**: L 89–100.

Kovacs, A. (1975). A study of multi-year pressure ridges and shore ice pile-up. *Calgary, Alberta: Arctic Petroleum Operators Association (APOA) Project*, **89**: 45.

Koyama, T. and Stroeve, J. (2019). Greenland monthly precipitation analysis from Arctic System Reanalysis (ASR):2000–2012. *Polar Sc.*, **19**: 1–12, doi.org/10.1016/j .polar.2018.09.001.

Krabill, W., *et al.* (2000). Greenland ice sheet: high-elevation balance and peripheral thinning. *Science*, **289**: 428–9.

Krabill, W. B., *et al.* (2004). Greenland ice sheet: increased coastal thinning. *Geophys. Res. Lett.*, **31**: L24402.

Krasting, J. P., *et al.* (2013). Future changes in northern hemisphere snowfall. *J. Climate*, AMS, 26: 7813–28, doi.org/10.1175/JCLI-D-12–00832.1.

Kratz, T. K., *et al.* (2000). Patterns in the interannual variability of lake freeze and thaw dates. *Verh. Int. Verein. Limnol.*, **27**: 2796–9.

Krawczynski, M. J., *et al.* (2009). Constraints on the lake volume required for hydro-fracture through ice sheets. *Geophys.*

Res. Lett., **36**: L10501, doi: 10.1029/ 2008GL036765.

Kristensen, M. (1983). Iceberg calving and deterioration in Antarctica. *Progress Phys. Geog.*, 7: 313–28.

Kristensen, M., Squire, V. A., and Moore, S. C. (1982), Tabular icebergs in ocean waves. *Nature*, **297** (5868): 669–71.

Kristoffersen, Y., *et al.* (2004). Seabed erosion on the Lomonosov Ridge, central Arctic Ocean: A tale of deep draft icebergs in the Eurasia Basin and the influence of Atlantic water inflow on iceberg motion? *Paleoceanog.*, **19**: PA3006, doi: 10.1029/ 2003PA000985.

Krupnik, I., *et al.* (eds.) (2010). *SIKU: Knowing our ice. Documenting Inuit sea ice knowledge and use.* New York: Springer. 300 pp.

Kudryavtsev, V. A., *et al.* (1974). *Fundamentals of frost forecasting in geological engineering investigations.* Draft Translation 606. Hanover, NH: US Army, Cold Regions Research and Engineering Laboratory. 489 pp.

Kuhn, B. F. 1787 (1788). Versuch ueber den Mechanismus der Gletscher. A. Hopfner's Magazine Naturkunde Helvetiens (Zurich). 1: 119–36 and 3, 427–3: Odell and Davies, pp. 343–51 and 384–93. (1956 Facsimile reprint: University of Illinois Press, Urbana).

Kuhn, M., Dreiseitl, E., Hofinger, S., Markl, G., Span, N., and Kaser, G. (1999). Measurements and models of the mass balance of Hinterreisferner. *Geografiska Annaler*, **81A**: 541–54.

Kuijpers, A. and Werner, F. (2007). Extremely deep-draft iceberg scouring in the glacial North Atlantic Ocean. *Geo-Mar. Lett.*, **27**: 383–9.

Kukla, G. and Gavin, J. (1981). Summer ice and carbon dioxide. *Science*, **214** (4520): 497–503.

Kulkarni, A. V., *et al.* (2007). Glacial retreat in Himalayas using Indian remote sensing satellite data. *Current Sci.*, **92**: 69–74.

Kulkarni, A. V. and Karyakarte, Y. (2014). Observed changes in Himalayan glaciers. *Current Sci.*, **106** (2): 237–44, www.jstor.org/stable/24099804.

Kunz, M., King, M. A., Mills, J. P., Miller, P. E., Fox, A. J., Vaughan, D. G., *et al.* (2012). Multi-decadal glacier surface lowering in the Antarctic Peninsula. *Geophys. Res. Lett.*, **39**: L19502, doi: 10.1029/2012GL052823.

Kurtakoti, P., Veneziani, M., Stössel, A., and Weijer, W. (2018). Preconditioning and formation of maud rise polynyas in a high-resolution earth system model. *J. Clim.*, **31**: 9659–78.

Kutuzov, S. and Shahgedanova, M. (2009). Glacier retreat and climatic variability in the eastern Terske – Alatoo, inner Tien Shan between the middle of the 19th century and beginning of the 21st century. *Global Planet. Change*, **69**: 59–70.

Kuusisto, E. and Elo, A. R. (2000). Lake and river ice variables as climate indicators in northern Europe. *Verh. Int. Ver. Limnol.*, **27**: 2761–4.

Kuz 'min, P. P. (1960). *Formirovanie snezhnogo pokrova i metody opredeleniya snegozapaso v.*Leningrad: Gidrometeoizdat (Transl. Snow cover and snow reserves. Jerusalem: Israel Program for Scientific Translation. 1963). 139 pp.

Kuzmichenok, V. A. (1989). Tekhnologiya i vozmozhnosty aerotopogropchicheskogo kar- togrphirovaniaizmeneniy lednikov (na primere oledenenya khrebta Akshiirak) (Methods and opportunities of the aero topographic cartography in context of glaciers changes (e.g. Akshiirak range glaciers)). Moscow: Inst. of Geography, RAS. Data Glaciol. Studies 67: 80–7 (in Russian).

Kwok, R. (2004). Annual cycles of multiyear sea ice coverage of the Arctic Ocean: 1999–2003. *J. Geophys. Res.*, **109**: C11004.

Kwok, R. (2009). Outflow of Arctic Ocean sea ice into the Greenland and Barents seas: 1979–2007. *J. Climate*, **27** (9): 2438–57.

Kwok, R., *et al.* (1995). Determination of the age distribution of sea ice from Lagrangian observations of ice motion. *IEEE Trans. Geosci. Remote Sensing*, **33**: 392–400.

Kwok, R., *et al.* (2007). Ice, Cloud, and land Elevation Satellite (ICESat) over Arctic sea ice: Retrieval of freeboard. *J. Geohys. Res.*, **112**: C12013, 19.

Kwok, R., *et al.* (2009). Thinning and volume loss of the Arctic Ocean sea ice cover: 2003–2008. *J. Geophys. Res.*, **114**: C07005.

Kwok, R., *et al.* (2010). Large sea ice outflow into the Nares Strait in 2007. *Geophys. Res. Lett.*, **37**: L03502, doi: 10.1029/2009GL041872.

Kwok, R. and Cunningham, G. F. (2002). Seasonal sea ice area and volume production of the Arctic Ocean: November 1996 through April 1997. *J. Geophys. Res.*, **107**: 8038, doi: 10.1029/2000JC000469.

Kwok, R. and Cunningham, G. F. (2008). ICESat over Arctic sea ice: Estimation of snow depth and ice thickness. *J. Geophys. Res.*, **113** (C08010): 17, 1025–30, doi: 10.1029/2008JC004753.

Kwok, R. and Cunningham, G. F. (2015). Variability of Arctic sea ice thickness and volume from CryoSat-2. *Phil. Trans. R. Soc. A*, **373**: 20140157.

Kwok, R. and Rothrock, D. A. (2009). Decline in Arctic sea ice thickness from submarine and ICESat records: 1958–2008. *Geophys. Res. Lett.*, **36**: L15501, doi: 10.1029/2009GL039035.

Kwok, R. (2018). Arctic sea ice thickness, volume, and multiyear ice coverage: losses and coupled variability (1958–2018). *Environ. Res. Lett.*, **13**: 105005.

Kwok, R., Pedersen, L. F., and Gudmandsen, P. (2010). Large sea-ice outflow into the Nares Strait in 2007. Paper 57A081. Proceedings, Tromo Sea Ice Symposium. Int. Glaciol. Soc.

Kwok, R., Pang, S. S., and Kacimi, S. (2017). Sea ice drift in the Southern Ocean: Regional patterns, variability, and trends. *Elem. Sci. Anth.*, **5**: 32, https://doi.org/10.1525/elementa.226

Kwok, R., Spreen, G., and Pang, G. (2013). Arctic sea ice circulation and drift speed: Decadal trends and ocean currents. *J. Geophys. Res Oceans*, **118**: 2408–25.

Labadie, J. W. (2004). Optimal operation of multireservoir systems: State-of-the-art review. *J Water Resour. Plan. Manage.*, **130** (2): 93–111.

LaChapelle, E. R. (1965). Avalanche forecasting – a modern synthesis. U.S. Forest Service, www.avalanche.org/~moonstone/forecasting/avalanche%20forecasting-a%20modern%20synthesis.htm

La Chapelle, E. (1966). Avalanche forecasting a modern synthesis. *International symposium on scientific aspects of snow and ice.* IASH, Publ. No. 69: 350–6.

Lacroix, M., *et al.* (2005). River ice trends in Canada, *Proc. 13th Workshop on the Hydraulics of Ice-covered Rivers, 2005.* Canadian Geophysical Union, Committee on River Ice Processes and the Environment. pp. 41–54.

Lacroix, P., *et al.* (2012). Monitoring of snow avalanches using a seismic array: Location, speed estimation, and relationships to meteorological variables. *J Geophys. Res.*, **1171**: F01034, doi: 10.1029/2011JF002106.

Laine, V. (2008). Antarctic ice sheet and sea ice regional albedo and temperature change, 1981–2000, from AVHRR Polar Pathfinder data. *Remote Sensing Environ.*, **112** (3): 646–67.

Lambrecht, A. and Kuhn, M. (2007). Glacier changes in the Austrian Alps during the last three decades derived from the new Austrian Glacier Inventory. *Annals Glaciol.*, **46**: 177–84.

Lambrecht, A. and Mayer, C. (2009). Temporal variability of the non-steady contribution from glaciers to water discharge in western Austria. *J. Hydrol.*, **376**: 353–61.

Lambert, F., *et al.* (2008). Dust-climate couplings over the past 800,000 years from the EPICA Dome C ice core. *Nature*, **452**: 616–19.

Landis, J. and Koch, G. (1977). The measurement categorical data. *Biometrics*, **33**: 159–74.

Landy, J., Ehn, J., Shields, M., and Barber, D. (2014). Surface and melt pond evolution on landfast first-year sea ice in the Canadian Arctic Archipelago. *J Geophys Res Oceans*, **119** (5): 3054–75.

Langlois, A., *et al.* (2009). Simulation of snow water equivalent using thermodynamic snow models in Quebec, Canada. *J. Hydrometeorology*, **10** (6), doi: 10.1175/2009JHM1154.1.

Langway, C. C. Jr. (2008). The history of early polar ice cores. *Cold Reg. Sci. Technol.*, **52**: 101–17.

Lantuit, H. and Pollard, W. H. (2008). Fifty years of coastal erosion and retrogressive thaw slump activity on Herschel Island, southern Beaufort Sea, Yukon Territory, Canada. *Geomorphology*, **95**: 84–102.

Lantuit, H., *et al.* (2008). Sensitivity of coastal erosion to ground ice contents: An Arctic-wide study based on the ACD classification of Arctic coasts. In Kane, D. L. and Hinkel, K. M. (eds.), *Proceedings of the Ninth International Conference on Permafrost*, Fairbanks, AK: University of Alaska. pp. 1025–30.

Larsen, E., *et al.* (2006). Late Pleistocene glacial and lake history of northwestern Russia. *Boreas*, **35**: 394–424.

Larsen, H. C., *et al.* (1994). Seven million years of glaciation in Greenland. *Science*, **264** (5161): 952–5.

Lassen, K. and Thejll, P. (2005). Multi-decadal variation of the East Greenland sea-ice extent, AD 1500–2000. *Sci. Rep. 05–02,*

Danish Meteorological Institute. Denmark: Copenhagen. 13 pp.

Laternser, M. and Schneebeli, M. (2003). Long-term snow climate trends of the Swiss Alps (1931–99). *Int. J. Climatol.*, **23** (7): 733–50.

Latifovic, R. and Poulio, D. (2007). Analysis of climate change impacts on lake ice phenology in Canada using the historical satellite data record. *Rem. Sens. Environ.*, **106**: 492–507.

Laumann, T. and Reeh, N. (1994). Sensitivity to climate change of the mass balance of glaciers in southern Norway. *J. Glaciol.*, **39** (133): 656–65.

Lawler, D. M. (1988). Environmental limits of needle ice: a global survey. *Arct. Alp. Res.*, **20**: 137–59.

Lawler, D. M. (1989). Some observations on needle ice. *Weather*, **44** (10): 406–9.

Lawrence, D. M. and Slater, A. G. (2005). A projection of severe near-surface permafrost degradation during the 21st century. *Geophys. Res. Lett.*, **32**: L24401, doi: 10.1029/2005GL025080.

Lawrence, D. M. and Slater, A. G. (2006). Reply to comment by C. R. Burn and F. E. Nelson on "A projection of near-surface permafrost degradation during the 21st century." *Geophys. Res. Lett.*, **33**: L21504, doi: 10.1029/2006GL027955. 7: 153–8.

Lawrence, D. M. and Slater, A. G. (2010). The contribution of snow condition trends to future ground climate. *Clim. Dyn.*, **34**: 969–81, doi: 10.1007/500382–009–0537–4.

Lawrence, D. M., *et al.* (2008). Accelerated Arctic land warming and permafrost degradation during rapid sea ice loss. *Geophys. Res. Lett.*, **35** (11): L11506, 1–5.

Lawrence, D. M., *et al.* (2008). Sensitivity of a model projection of near-surface permafrost degradation to soil column depth and inclusion of soil organic matter. *J. Geophys. Res.*, **113**: F02011, doi: 10.1029/2007JF000883.

Laxon, S. W., Giles, K. A., Ridout, A. L., Wingham, D. J., Willatt, R., Cullen, R., Kwok, R., Schweiger, A., Zhang, J., Haas, C., Hendricks, S., Krishfield, R., Kurtz, N., Farrell, S., and Davidson, M. (2013). CryoSat-2 estimates of Arctic sea ice thickness and volume. *Geophys. Res. Lett.*, **40**: 732–7, doi: 10.1002/grl.50193.

Lazar, B. and Williams, M. (2008). Climate change in western ski areas: Potential changes in the timing of wet avalanches and snow quality for the Aspen ski area in the years 2030 and 2100. *Cold Regions Sci. Tech.*, **51**: 219–28, doi: 10.1016/j.coldregions.2007.03.015.

Lazzara, M. A., *et al.* (1999). On the recent calving of icebergs from the Ross Ice Shelf. *Polar Geog.*, **23**: 201–12.

Lebedev, V. V. (1938). Rost l'do v arkticheskikh rekakh i moriakh v zavisimosti ot otritsatel'nykh temperatur vozdukha (The growth of Arctic river and sea ice in dependence on negative air temperatures). *Problemy Arktikii*, **5**: 9–25.

Le Brocq, A. M., *et al.* (2008). Subglacial topography inferred from ice surface terrain analysis reveals a large un-surveyed basin below sea level in East Antarctica. *Geophys. Res. Lett.*, **35**: L16503, 1–6, doi: 10.1029/2008GL034728.

Leclercq, P. W., Oerlemans, J., and Cogley, J. G. (2011). Estimating the glacier contribution to sea-level rise for the period 1800–2005. *Surv. Geophys.*, **32**: 519–35.

LeDoux, C. M., Hulbe, C. L., Forbes, M., Scambos, T., and Alley, K. (2017). Structural provinces of Ross Ice Shelf, Antarctica. *Annals of Glaciology*, **58** (75): 1–11, doi: 10.1017/aog.2017.24.

Ledu, D., *et al.* (2010). Holocene sea ice history and climate variability along the main axis of the Northwest Passage, Canadian Arctic. *Paleoceanogr*, **25**: PA2213, 21.

Legates, D. R. and Bogart, T. A. (2009). Estimating the proportion of monthly precipitation that falls in solid form. *J. Hydromet.*, **10** (5): 1299–306.

Legeais, J. F., *et al.* (2018). An improved and homogeneous altimeter sea level record from the ESA Climate Change Initiative. *Earth Syst. Sci. Data*, **10**: 281–301, https://doi.org/10.5194/essd-10-281-2018.

Legget, R. F. (1954). Permafrost research. *Arctic*, **7**: 153–8.

Legget, R. F. (1966). Permafrost in North America. In *Proceedings, Permafrost International Conference*. Washington, DC: National Research Council. pp. 2–7.

Le Hir, G., *et al.* (2010). Toward the snowball earth deglaciation. *Clim. Dynam.*, **35**: 285–97.

Lemelin, H., Dawson, J., and Stewart E. (2012). *Last chance tourism: Adapting tourism opportunities in a changing world*. London, UK: Routledge.

Lemieux, J.-F., Buehner, M., Pedersen, L. T., and Carrieres, T. (2017). *Sea ice analysis and forecasting towards an increased reliance on automated prediction systems*. Cambridge, UK: Cambridge University Press. 219 pp, ISBN: 9781108417426, 1108417426.

Lemke, P., *et al.* (2007). The cryosphere. In Climate change 2007: The physical science basis. Contribution of Working Group I to the Fourth Assessment Report of the Intergovernmental Panel on Climate Change [Solomon, S. D., *et al.* (eds.)]. Cambridge, UK: Cambridge University Press. pp. 337–83.

Lemmen, D. S., Evans, D. J. A., and England, J. (1988). Discussion of "Glaciers and the morphology and structure of the Milne Ice Shelf, Ellesmere Island, N.W.T., Canada" by Martin O. Jeffries. *Arct. Alp. Res.*, **20**: 366–71.

Lenhning, M., *et al.* (2002). A physical SNOWPACK model for the Swiss avalanche warning. Part II. Snow microstructure. *Cold Reg. Sci. Technol.*, **35**: 147–67.

Leppäranta, M. (2005). *The drift of sea ice*. Berlin: Springer. 266 pp.

Le Treut, H., Somerville, R., *et al.* (2007). Historical overview of climate change. In Solomon, S. D., *et al.* (eds.), *Climate Change 2007: The Physical Science Basis. Contribution of Working Group I to the Fourth Assessment Report of the Intergovernmental Panel on Climate Change*, Cambridge: Cambridge University Press. pp. 93–127.

Lewis, A. R., *et al.* (2008). Mid-Miocene cooling and the extinction of tundra in continental Antarctica. *Proc. Nat. Acad. Sci.*, **105**(21): 10, 676–80.

Lewis, E. L. and Perkin, R. G. (1986). Ice pumps and their rates. *J. Geophys. Res.*, **91**: 11,756–62.

Lewis, C. F. M. and Teller, J. T. (2007). North American late-Quaternary meltwater and Floods to the ocean: Evidence and impact – Introduction. *Palaeogeog., Palaeoclimatol., Palaeoecol.*, **246**: 1–7.

Lewis, W. M. Jr. (2010). Global primary production of lakes. *19th Baldi Memorial Lecture, Inland Waters*, **1**: 1, 1–28, doi: 10.5268/IW-1.1.384.

Lewis, W. V. (1949). Glacial movement by rotational slipping. *Geograf. Annal.*, **31**: 146–58.

Lewkowicz, A. G. and Way, R. G. (2019). Extremes of summer climate trigger thousands of thermokarst landslides in a High Arctic environment. *Nature Communications*, **10**: 1329.

L'Heureux, M. L., Kumar, A., Bell, G. D., Halpert, M. S., and Higgins, R. W. (2008). Role of the Pacific-North American (PNA) pattern in the 2007 Arctic sea ice decline. *Geophysical Research Letters*, **35**: L20701, doi:10.1029/2008GL035205.

Li, B., *et al.* (2006). Glacier change over the past 4 decades in the middle Chinese Tien Shan. *J. Glaciol.*, **52**: 425–32.

Liang, L., Li, X., and Zheng, F. (2019). Spatio-temporal analysis of ice sheet snowmelt in

Antarctica and Greenland using microwave radiometer data. *Remote Sensing*, **11**(16): 1838, https://doi.org/10.3390/rs11161838.

Lieven, H., *et al.* (2019). Snow depth variability in the Northern Hemisphere mountains observed from space. *Nat. Commun.*, **10**: 4629. https://doi.org/10.1038/s41467-019–12566-y

Likens, G. E. (2000). Along-term record of ice cover for Mirror Lake, New Hampshire: effects of global warming? *Verh. Int. Verein Limnol.*, **27**: 2765–9.

Lin, C.-H., *et al.* (2008). Glaciers and their distribution in China. In Shi, Y.-F. (ed.), *Glaciers and related environments in China*, Beijing: Science Press. pp. 16–94.

Lind, D. and Sanders, S. P. (2004). *The physics of skiing: Skiing at the triple point.* New York: Springer. 266 pp.

Lindenschmidt, Karl-Erich, *et al.* (2011). Characterizing river ice along the Lower Red River using RADARSAT-2 imagery. In *Proceedings of the 16th Workshop on River Ice*, Winnipeg, MB, September 18–22, 2011. pp. 198–213.

Lindsay, D. G. (ed.) (1982). Sea Ice Atlas of Arctic Canada, 1961–1968; Sea Ice Atlas of Arctic Canada, 1969–1974; Sea Ice Atlas of Arctic Canada 1975–1979, Energy, Mines and Resources, Canada. 213 pp., 219 pp., 139 pp.

Lindsay, R. (2010). New unified sea ice thickness climate data record. *EOS*, **91**(44): 405–6.

Lindsay, R. W., *et al.* (2009). Arctic sea ice retreat in 2007 follows thinning trend. *J. Clim.*, **22**: 165–75.

Lindsay, R. and Schweiger, A. (2015). Arctic sea ice thickness loss determined using subsurface, aircraft, and satellite observations. *The Cryosphere*, **9**: 269–83.

Lingle, C. S. and Fatland, D. R. (2003). Does englacial water storage drive temperate glacier surges? *Ann. Glaciol.*, **36**: 14–20.

Lisiecki, L. E. and Raymo, M. E. (2005). A Pliocene-Pleistocene stack of 57 globally distributed benthic δ18O records. *Paleoceanogr. Paleoclimatol*, **20**: PA1003. https://doi.org/10.1029/2004PA001071

Lisitsyna, O. M. and Romanovskii, N. N. (1998). Dynamics of permafrost in northern Eurasia during the last 20,000 years. In Proceedings of the Seventh International Permafrost Conference, Yellowknife, Canada, June 23–27, pp.675–81.

Liston, G. E. (2004). Representing subgrid snow cover heterogeneities in regional and global models. *J. Climate*, **17**: 1381–97.

Liston, G. E. and Elder, K. (2006). A distributed snow-evolution modeling system (SnowModel). *J. Hydromet.*, **7**: 1259–76.

Liston, G. E. and Hall, D. K. (1995a). An energy-balance model of lake-ice evolution. *J. Glaciol.*, **41** (138): 373–82.

Liston, G. E. and Hall, D. K. (1995b). Sensitivity of lake freeze-ip and breakup to climate change: A physically based modeling study. *Annals Glaciol.*, **21**: 387–93.

Liston, G. E. and Hiemstra, C. A. (2008). A simple data assimilation system for complex snow distributions (SnowAssim). *J. Hydromet.*, **9**: 989–1004.

Liston, G. E. and Hiemstra, C. (2011). The changing cryosphere: Pan-Arctic snow trends (1979–2009). *Journal of Climate*, **24**: 5691–712, doi:10.1175/jcli-d-11-00081.1.

Liston, G. E., *et al.* (2007). Simulating complex snow distributions in windy environments using SnowTran-3D. *J. Glaciol.*, **53**(181): 241–56.

Liston, G. E., *et al.* (2007). Instruments and methods, simulating complex snow distributions in windy environments using SnowTran-3D. *J. of Glaciology*, **53** (181): 241–56, doi: 10.3189/172756507782202865.

Liston, G. E. and Sturm, M. (1998). A snow-transport model for complex terrain. *J. Glaciol.*, **44** (148): 498–516.

Little, C. M., Gnanadesikan, A., and Hallberg, R. (2008). Large-scale oceanographic constraints on the distribution of melting and freezing under ice shelves. *J. Phys. Oceanog.*, **38**: 2242–55.

Little, C., Gnanadesikan, A., and Oppenheimer, M. (2009). How ice shelf morphology controls basal melting. *J. Geohys. Res.*, **114**: C12007, 14.

Liu, C.-H., *et al.* (2008). Glaciers and their distribution in China. In Shi, Y.-F. (ed.), *Collection of the studies on glaciology, climate and environmental changes in China*, Beijing: Meteorological Press. pp. 170–241.

Liu, H.-X., Wang, L., and Jezek, K. C. (2006). Spatiotemporal variations of snowmelt in Antarctica derived from satellite scanning multichannel microwave radiometer and Special Sensor Microwave Imager data (1978–2004). *J. Geophys. Res.*, **111** (F1): F01003, doi:10.1029/2005JF000318.

Liu, J.-P. and Curry, J. A. (2010). Accelerated warming of the Southern Ocean and its impacts on the hydrological cycle and sea ice. *Proc. Nat. Acad. Sci.*, **107**: 1488–93, doi: 10.1073/pnas.1003336107.

Liu, J. P., Curry, J. A., and Hu, Y. Y. (2004). Recent Arctic sea ice variability: Connections to the Arctic Oscillation and the ENSO. *Geophys. Res. Lett.*, **31**: L09211, doi:10.1029/2004GL019858.

Liu, J., Li, Z., Huang, Z., and Tian, B. (2014). Hemispheric-scale comparison of monthly passive microwave snow water equivalent products. *J. of Applied Remote Sensing*, **8** (1): 084688, https://doi.org/10.1117/1.JRS.8.084688.

Liu, L., Zhang, T.-J., and Wahr, J. (2010). InSAR measurements of surface deformation over permafrost on the North Slope of Alaska. *J. Geophys. Res.*, **115**: F03023, doi:10.1029/2009JF001547.

Liu, S.-Y., *et al.* (2008). Mass and energy balance of glaciers. In Shi, Y.-F. (ed.), *Glaciers and related environments in China*, Beijing: Science Press. pp. 131–71.

Liu, X.-L., Yang, Z.-P., and Xie, T. (2006). Development and conservation of glacier tourist resources – A case study of Bogda Glacier Park. *Chinese Geog. Soc.*, **16**: 365–70.

Livingatone, D. M. (1997). Breakup dates of Alpine lakes as proxy data for local and regional mean surface air temperature. *Clim. Change*, **37**: 407–39.

Lliboutry, L. (1954). The origin of penitentes. *J. Glaciol.*, **2** (15): 331–8.

Lliboutry, L. (1965). *Traité de glaciologie. Tome H: Glaciers, variations du climat, sols gels*. Paris: Masson et Cie.

Lliboutry, L. (1968). General theory of subglacial cavitation and sliding of temperate glaciers. *J. Glaciol.*, **7** (49): 21–58.

Lliboutry, L. (1975). La catastrophe du Yungay (Pérou). *Proceedings of Snow and Ice Symposium*, Moscow, 1971. IAHS publication, 104: 353–63.

Lliboutry, L. (1979). Local friction laws for glaciers: a critical review and new openings. *J. Glaciol.*, **23**: 67–95.

Loewe, F. (1935). Das Klima des grönlandischen Inlandeises (The climate of Greenland's inland ice). In Koeppen, W. and Geiger, R. (eds.), *Handbuch der Klimatologie, Vol. 2, Part K, Klima des kanadischen Archipels und Grönland*, Berlin: Borntraeger. pp. K67–K101.

Loewe, F. (1936). The Greenland Ice Cap as seen by a meteorologist. *Quart. J. Roy. Met. Soc.*, **62**(266): 359–78.

Lopatin, I. (1876). Some facts about icy layers in eastern Siberia. *Izvestia Akad. Nauk Supplement*, **29**: 4–31 (In Russian).

Lopez, L. S., Hewitt, B. A., and Sharma, S. (2019). Reaching a breaking point: How is climate change influencing the timing of ice breakup in lakes across the northern hemisphere? *Limnol. Oceanogr.*, **64**(6): 2621–31, ASLO, doi.org/10.1002/lno.11239.

Lopez-Moreno, J. I., *et al.* (2008). Sensitivity of the snow energy balance to climatic changes: Prediction of snowpack in the Pyrenees in the 21st century. *Climate Res.*, **36**: 203–17.

Loriaux, T. and Casassa, G. (2013). Evolution of glacial lakes from the Northern Patagonia Icefield and terrestrial water storage in a sea level rise context, Global Planet. *Change*, **102**: 33–40.

Lourens, L. J., *et al.* (2010). Linear and non-linear response of late Neogene glacial cycles to obliquity forcing and implications for the Milankovitch theory. *Quat. Sci. Rev.*, **29**: 352–65.

Louis, J. F. (1979). A parametric model of vertical eddy fluxes in the atmosphere. *Boundary Layer Meteorol.*, **66**: 281–301.

Lucchita, B. K. and Ferguson, H. M. (1986). Antarctica: Measuring glacier velocity from satellite images. *Science*, **234**(4779): 1105–8.

Lucchita B. K. and Rosanova, C. E. (1998). Retreat of northern margins of George VI and Wilkins ice shelves. *Ann. Glaciol.*, **27**: 41–6.

Lucchita, B. K., Rosanova, C. E., and Mullins, K. F. (1995). Velocities of Pine Island Glacier, West Antarctica, from ERS-1 SAR images. *Ann. Glaciol.*, **21**: 277–83.

Lüdecke, C. (1995). Die deutsche Polarforschung seit der Jahrhundertwende und der Einfluss Erich von Drygalski. *Polar Berichte*, **158**: 340 pp + Appx. 72 pp.

Lukovich, J. V. and Barber, D. G. (2007). On the spatiotemporal behavior of sea ice concentration anomalies in the Northern Hemisphere. *J. Geophys. Res.*, **112**(D13): D13117, 12, doi: 10.1029/2006JD007836.

Lunardini, V. J. (1978). Theory of n-factors and correlation of data. In Permafrost. Proceedings of the third international conference on permafrost. Ottawa: National Research Council of Canada. Vol. 1, pp. 40–6.

Lunardini, V. J. (1995). Permafrost formation time. CRREL Report 95–8. Hanover, NH: US Army Corps of Engineers, Cold Regions Research & Engineering Laboratory. 44 pp.

Lundqvist, J. (2004). Glacial history of Sweden. In Ehlers, J. and Gibbard, P. L. (eds.), *Quaternary glaciations – extent and chronology*, New York: Elsevier. pp. 402–12.

Lundqvist, J., *et al.* (2013). Lower forest density enhances snow retention in regions with warmer winters: A global framework developed from plot-scale observations and modeling. *Water Resources Res.*, **49**: 1–15, doi: 10.1002/wrcr.20504.

Lundy, C., *et al.* (2001). A statistical validation of the SNOWPACK model in a Montana climate. *Cold Reg. Sci. Technol.*, **33**: 237–46.

Lunt, D. J., *et al.* (2008). Late Pliocene Greenland glaciation controlled by a decline in atmospheric CO_2 levels. *Nature*, **454**: 1102–6.

Lunt, D. J., *et al.* (2009). The Arctic cryosphere in the Mid-Pliocene and the future. *Phil Trans. R. Soc. A*, **367**: 49–67.

Luo, D., Wu, Q., Jin, H., Marchenko, S. S., Lü, L. Z., and Gao, S. (2016). Recent changes in the active layer thickness across the northern hemisphere. *Environ. Earth Sci.*, **75**: 555.

Luojus, K., *et al.* (2010). Investigating the feasibility of the GlobSnow snow water equivalent data climate research purposes. *Geoscience and Remote Sensing Symposium (IGARSS), 2010 IEEE International (4851–4853)*, doi:10.1109/IGARSS.2010.5741987.

Luthcke, S. B. , *et al.* (2006). Recent Greenland ice mass loss by drainage system from satellite gravity observations. *Science*, **314**: 1286–9, doi.org/10.1126/science.1130776, 2006.

Luthcke, S. B., Arendt, A. A., Rowlands, D. D., McCarthy, J. J., and Larsen, C. F. (2008). Recent glacier mass changes in the Gulf of Alaska region from GRACE mascon solutions. *J. Glaciol.*, **54**: 767–77, doi:10.3189/002214308787779933.

Luthcke, S. B., *et al.* (2013). Antarctica, Greenland and Gulf of Alaska landice evolution from an iterated GRACE global mascon solution. *J. Glaciol.*, **59**: 613–31.

Lüthi, D., *et al.* (2008). High-resolution carbon dioxide concentration record 650,000–800,000 years before present. *Nature*, **453**: 379–82.

Lüthi, M. P., Bauder, A., and Funk, M. (2010). Volume change reconstruction of Swiss glaciers from length change data. *J. Geophys. Res.*, **115**(F4): F04022.

Lyon, S. W., *et al.* (2009). Estimation of permafrost thawing rates in a sub-arctic catchment using recession flow analysis. *Hydrol. Earth Syst. Sci.*, **13**: 595–604.

Lyon, W. (1961). Ocean and sea-ice research in the Arctic Ocean via submarine. *Trans. New York Acad. Sci., Series II*, **23**(8): 662–74.

Lythe, M. B., Vaughan, D. G., and the BEDMAP Consortium. (2001). BEDMAP: A new ice thickness and subglacial topographic model of Antarctica. *J. Geophys. Res.*, **106**(B6): 11,335–51.

Ma, N., Yasunari, T., and Fukushima, Y. (2002). Modeling of river ice breakup date and thickness in the Lena River. Ice in the environment, Vol. **1**, Squire, V. and Langhorne, P. (eds.), *Proc. 16th IAHR Internat. Sympos. on Ice, Int. Assoc. Hydraulic Eng. Rea.*, Dunedin, New Zealand, pp. 22–6.

MacAyeal, D. R. (1993). A low-order model of growth/purge oscillations of the Laurentide Ice Sheet. *Paleoceanog.*, **8**: 767–73.

MacAyeal, D.R. 1984. Thermohaline circulation below the Ross Ice Shelf: A consequence of tidally induced vertical mixing and basal melting. J. Geophys. Res., 89(C1): 597–606.

Macdonald, F. A., *et al.* (2010). Calibrating the Cryogenian. *Science*, **327** (5970): 1241–3.

Machatschek, F. (1914). Die Depression der eiszeitlichen Schneegrenze. *Zeit. f. Gletscherk.*, **7**: 104–28.

Macias Fauria, M. *et al.* (2009). Unprecedented low twentieth century winter sea ice extent in the western Nordic Seas since A.D. 1200. *Climate Dynam.*, **34**: 781–95, doi:10.1007/500382–009–0610–2.

Mackay, J. R. (1962). The pingos of the Pleistocene Mackenzie Delta area. *Geogr. Bull.*, **18**: 21–63.

Mackay, J. R. (1972). The world of underground ice. *Annals Assoc. Amer. Geogr.*, **62**: 1–22.

Mackay, J. R. (1973). A frost tube for the determination of freezing in the active layer above permafrost. *Canad. Geotech. J.*, **10**: 392–6.

Mackay, J. R. (1986a). Frost mounds. In French, H. M. (ed.), *Focus: Permafrost geomorphology*, Canad. Geographer **30**: 363–4.

Mackay, J. R. (1986b). Growth of Ibyuk pingo, western Arctic coast, Canada and some implications for environmental reconstruction. *Quatern. Res.*, **26**: 68–80.

Mackay, J. R. (1993). Air temperature, snow cover, creep of frozen ground, and the time of ice-wedge cracking, western Arctic coast. *Canad. J. Earth Sci.*, **30**: 1720–9.

Mackay, J. R. and Dallimore, S. R. (1992). Massive ice of the Tuktoyaktuk area, western Arctic coast, Canada. *Canad. J. Earth Sci.*, **29** (6): 1235–49.

Mackintosh, N. A. and Herdman, H. F. P. (1940). Distribution of the pack ice in the Southern Ocean. *Discovery Rep.*, **19**: 285–96, plates 69–95.

Magono, C. and Lee, C. W. (1966). Meteorological classification of natural snow crystals. *Journal of the Faculty of Science*, Hokkaido University. Series 7, Geophysics, Vol. **II** (4): 321–55

Magnuson, J. D. (2000a). Lake and river ice as a powerful indicator of past and present climates. *Veh, Int, Verein Limnol.*, **27**: 2749–56.

Magnuson, J. D., *et al.* (2000b). Historical trends in lake and river ice cover in the Northern Hemisphere. *Science*, **289**(5485): 1743–6.

Mahaffy, M. W. (1976). A three-dimensional numerical model of ice sheets: Test on the Barnes ice cap, Northwest Territories. *J. Geophys. Res.*, **81**(6): 1059–66.

Mahoney, A. (2010). Life with ice: The importance of sea ice to Arctic communities. Paper 57A207. Proceedings, Tromso Sea Ice Symposium. Int. Glaciol. Soc.

Mahoney, A. R., Barry, R. G., Smolyanitsky, V., and Fetterer, F. (2008). Observed sea ice extent in the Russian Arctic, (1933–2006). *J. Geophys. Res. (Oceans)*, **113**: C11005, 11, doi:10.1029/2008JC004830.

Mahoney, A. and Gearheard, S. (2008). *Handbook for community-based sea ice monitoring*. NSIDC Special Report 14. Boulder, CO: National Snow and Ice Data Center. 34 pp.

Mahoney, A., *et al.* (2007a). Alaska landfast sea ice: Links with bathymetry and atmospheric circulation. *J. Geophys. Res.*, **112**: C02001, doi:10.1029/2006JC003559.

Mahoney, A. R., Eicken, H., and Shapiro, L. (2007b). How fast is landfast sea ice? A study of the attachment and detachment of nearshore ice at Barrow, Alaska. *Cold Regions Sci. Technol.*, **47**: 233–55.

Mair, D. W. F., Burgess, D. O., and Sharp, M. J. (2005). Thirty-seven year mass balance of Devon Island ice cap, Nunavut, Canada, determined by shallow ice coring and melt modeling. *J. Geophys. Res.*, 110: F01011, doi:10.1029/2003JF000099.

Mair, D., Burgess, D., Sharp, M., Dowdeswell, J. A., Benham, T., Marshall, S., *et al.* (2009). Mass balance of the Prince of Wales Icefield, Ellesmere Island, Nunavut, Canada. *J. Geophys. Res.*, **114**: F02011, doi:10.1029/2008JF001082.

Male, D. H. and Granger, R. J. (1981). Snow surface energy exchange. *Water Resour. Res.*, **17** (3): 609–27.

Malkova, G. V. (2008). The last twenty-five years of changes in permafrost temperature in the European Russian Arctic. In Kane, D. L. and Hinkel, K. M. (eds.), *Ninth International Conference on Permafrost, 29 June–3 July 2008*, vol. **2**, Fairbanks, AK: University of Alaska. pp. 1119–25.

Malmgren, F. (1927). On the properties of sea ice. In Sverdrup, H. (ed.), *Scientific results of the Norwegian North Polar Expedition "Maud," 1918–1925, vol. 1 (5)*, Bergen: Geofysisk Institutt. pp. 67.

Mangerud, J., *et al.* (2008). Glaciers in the Polar Urals, Russia, were not much larger during the last global glacial maximum than today. *Quat. Sci. Rev.*, **27** (9–10): 1047–57.

Mangor, K., and Zorn, R. (1983). Iceberg conditions offshore Greenland. *Iceberg Res.* (Scot Polar Res. Inst. Cambridge), **4**: 4–20.

Mankin, J. S. and Diffenbaugh, N. S. (2015). Influence of temperature and precipitation variability on near-term snow trends. *Clim. Dynam.*, **45** (3–4): 1099–116, doi:10.1007/s00382-014-2357-4.

Mann, M. E., *et al.* (2009). Global significance and dynamical origins of the Little Ice Age and Medieval climate anomaly. *Science*, **326**: 1256–61.

Marchenko, S. S., Gorbunov, A. P., and Romanovsky, V. E. (2007). Permafrost warming in the Tein Shan mountains, Central Asia. *Global Planet. Change*, **56**: 311–27.

MARGO Project Members. (2009). Constraints on the magnitudes and patterns of cooling at the last glacial maximum. *Nat. Geosci.*, **2**: 127–32.

Markham, C. R. and Mill, H. R. (1901). In Murray, G. (Ed): *The Antarctic manual for the use of the expedition of 1901.* London, Royal Geographical Society. pp. xiv–xvi.

Marko, J. R., *et al.* (1994). Iceberg severity off eastern North America: Its relationship to sea ice variability and climate change. *J. Climate*, **7** (9): 1335–51.

Marks, D. (1988). Climate, energy exchange, and snowmelt in Emerald Lake Watershed, Sierra Nevada. PhD Thesis, University of California at Santa Barbara.

Markus, T. and Cavalieri, D. (2000). An enhancement of the NASA Team sea ice algorithm. *IEEE Trans. Geosci. Remote Sensing*, **38**: 1387–98.

Markus, T., Stroeve, J. C., and Miller, J. (2009). Recent changes in Arctic sea ice melt onset, freezeup, and melt season length. *J. Geophys. Res.*, **114** (C12): C12024.

Mars, J. C. and Houseknecht, D. W. (2007). Quantitative remote sensing study indicates doubling of coastal erosion rate in past 50 yr along a segment of the Arctic coast of Alaska. *Geology*, **35** (7): 583–6.

Marsh, P. and Prowse, T. D. (1987). Water temperature and heat flux at the base of river ice covers. *Cold Reg, Sci. Technol.*, **14**: 33–50.

Martin, M. A., *et al.* (2011). The Potsdam Parallel Ice Sheet Model (PISM-PIK)-Part 2: Dynamic equilibrium simulation of the Antarctic ice sheet. *The Cryosphere*, **5**: 727–40.

Martin, S. (1981). Frazil ice in rivers and oceans. *Annual Rev. Fluid Mechan.*, **13**: 379–97.

Martin, S. , *et al.* (2010). Kinematic and seismic analysis of giant tabular iceberg breakup at Cape Adare, Antarctica. *J. Geophys. Res.*, **115**: B06311, 17, doi:10.1029/2009JB006700.

Martin, Y. and Gerdes, R. (2007). Sea ice drift variability in Arctic Ocean model intercomparison project models and observations. *J Geophys. Res.*, **112** (C4): C04S10, 13.

Martinec, J. (1980). Limitations in hydrological interpretations of the snow coverage. *Nordic Hydrol.*, **11**: 209–20.

Martinec, J. and Rango, A. (1986). Parameter values for snowmelt runoff modelling. *J. Hydrol.*, **84**: 197–219.

Martinec, J., Rango, A., and Roberts, R. (1998). Snowmelt Runoff Model (SRM) user's manual. In Baumgartner, M. F. and Apfl, G. M. (eds.), *Geographica Bernensia Ser. P*, no. **35**. Berne: University of Berne.

Martín-Español *et al.* (2016). Spatial and temporal Antarctic Ice Sheet mass trends, glacio-isostatic adjustment, and surface processes from a joint inversion of satellite altimeter, gravity, and GPS data. *J. Geophys. Res.-Earth Surf.*, **121**: 182–200.

Martinelli, M. (1986). A test of the avalanche runout equations developed by the Norwegian Geotechnical Institute. *Cold Reg. Sci. Technol.*, **13**: 19–33.

Martinson, D. and Pitman, W. (2007). The Arctic as a trigger for glacier terminations. *Clim. Change*, **80**: 253–63.

Marty, C. (2008). Regime shift of snow days in Switzerland. *Geophys. Res. Lett.*, **35** (12): L12501, 1–5.

Marty, C., Schlögl, S., Bavay, M., and Lehning, M. (2017). How much can we save? Impact of different emission scenarios on future snow cover in the Alps. *The Cryosphere*, **11**: 517–29, https://doi.org/10.5194/tc-11-517-2017.

Marty, C., Tilg, A.-M., and Jonas, T. (2017). Recent evidence of large-scale receding snow water equivalents in the European Alps. *J. Hydrometeorol.*, **18**: 1021–31, doi:10.1175/JHM-D-16-0188.1.

Marzeion, B., Jarosch, A. H., and Hofer, M. (2012). Past and future sea-level change from the surface mass balance of glaciers. *Cryosphere*, **6**: 1295–322.

Marzeion, B., *et al.* (2014). Attribution of global glacier mass loss to anthropogenic and natural causes. *Science*, **345**: 919–20.

Marzeion, B., *et al.* (2018). Limited influence of climate change mitigation on short-term glacier mass loss. *Nat. Clim. Change*, **8**: 305–8, https://doi.org/10.1038/s41558-018-0093-1.

Masiokas, M. H., *et al.* (2009). Glacier fluctuations in extratropical South America during the past 1000 years. *Palaeogeog., Palaeoclimatol., Palaeoecol.*, **281**: 242–68.

Maslanik, J. A. and Barry, R. G. (1989). Short-term interactions between atmospheric synoptic conditions and sea ice behavior in the Arctic. *Annals Glaciol.*, **12**: 113–17.

Maslanik, J. A. and Barry, R. G. (1990). Remote sensing in Antarctica and the Southern Ocean: Applications and development. *Antarctic Sciences*, **2**: 105–21.

Maslanik, J. A., Key, J. R., and Barry, R. G. (1989). Merging AVHRR and SMMR data for remote sensing of ice and cloud in polar regions. *Internat. J. Rem. Sens.*, **10**: 1,691–6.

Maslanik, J. A., Serreze, M. C., and Barry, R. G. (1996). Recent decreases in Arctic summer ice cover and linkages to atmospheric circulation anomalies. *Geophys. Res. Lett.*, **23**(13): 1,677–80.

Maslanik, J. A., *et al.* (1995). Remotely-sensed and simulated variability of Arctic sea-ice concentrations in response to atmospheric synoptic systems. *Int. J. Remote Sensing*, **16**(17): 3,325–42.

Maslanik, J., *et al.* (2007a). On the Arctic climate paradox and the continuing role of atmospheric circulation in affecting sea ice conditions. *Geophys. Res. Lett.*, **34**: L03711, doi:10.1029/2006GL028269.

Maslanik, J. A., *et al.* (2007b). A younger, thinner Arctic ice cover – increased potential for rapid, extensive ice loss, Geophys. *Res. Lett.*, **34**: L24501, doi:10.1029/2007GL032043.

Maslin, M. A., *et al.* (2006). The progressive intensification of northern hemisphere glaciation as seen from the North Pacific. *Internat. J. Earth Sci.*, **85**: 452–65.

Mason. B. J. (1994). The shapes of snow crystals – fitness for purpose? *Quart. J. Roy. Met. Soc.*, **120**: 849–60.

Massom, R. A. (2009). Principal uses of remote sensing in sea ice research. In Eicken, H., *et al.* (eds.), *Field techniques for sea ice research*, Fairbanks, AK: University of Alaska Press. pp. 405–66.

Massom, R. A., *et al.* (2018). Antarctic ice shelf disintegration triggered by sea ice loss and ocean swell, *Nature*, Macmillan Publishers, 383, Vol. 558, https://doi.org/10.1038/s41586-018–0212-1

Masson-Delmotte, V., *et al.* (2010). EPICA Dome C record of glacial and interglacial intensities. *Quat. Sci. Rev.*, **29**: 113–28.

Matsuo, S. and Miyake, Y. (1966). Gas composition in ice samples from Antarctica. *J. Geophys. Res.*, **71** (22): 5235–41.

Matthes, F. E. (1934). Ablation of snow-fields at high altitude by radiant solar heat. *Trans. Amer. Geophys. Union*, **15**: 380–5.

Matthes, F. F. (1939). Report of the committee on glaciers. *Trans. Amer. Geophys. Union*, **20**: 518035.

Matthews, J. A. and Briffa, K. R. (2005). The "Little Ice Age": Re-evaluation of an evolving concept. *Geograf. Annal., A*, **87**: 17–36.

Matti, B., Dahlke, H. E., Dieppois, B., Lawler, D. M., and Lyon, S. W. (2017). Flood seasonality across Scandinavia-Evidence of a shifting hydrograph? *Hydrol. Process.*, **31**: 4354–70, doi:10.1002/hyp.11365.

Mätzler, C. (1994). Passive microwave signatures of landscapes in winter. *Meteorol. Atmos. Phys.*, **54**: 241–60.

Mätzler, C., Schanda, E., and Wood, W. (1982). Toward the definition of optimum sensor specifications for microwave remote sensing of snow. *IEEE Trans. Geosci. Remote Sensing*, GE-20: 57–66.

Maurer, J. (2007). Atlas of the Cryosphere. Boulder, CO, USA: National Snow and Ice Data Center, *Digital media*.**99**: 141–53.

Maurer, J. M., Schaefer, J. M., Rupper, S. , and Corley, A. (2019). Acceleration of ice loss across the Himalayas over the past 40 years. *Sci. Adv.*, **5** (6): eaav7266, doi: 10.1126/sciadv.aav7266.

Mauritsen, T. (2016). Greenhouse warming unleashed. *Nat. Geosci.*, **9**: 271–2.

Mayer, C. (2010). The early history of remote sensing of glaciers. In Pellikka, P. and Rees, W. R. (eds.), *Remote sensing of glaciers*, London: CRC Press, Taylor and Francis. pp. 67–80.

Mayewski, P. A., *et al.* (2009). State of the Antarctic and Southern Ocean climate system (SASOCS). *Rev. Geophys.*, **47**: RG1003, 38.

Maykut, G. A. (1982). Large-scale heat exchange and ice production in the central Arctic. *J. Geophys. Res.*, **87**: 7971–84.

Maykut, G. (1985). The ice environment. In Horner, R. (ed.), *Sea ice biota*, Boca Raton, FL: CRC Press. pp. 21–82.

Maykut, G. A. (1986). The surface heat and mass balance. In Untersteiner, N. (ed.), *The geophysics of sea ice*, New York: Plenum Press. pp. 395–462.

Maykut, G. A. and Untersteiner, N. (1971). Some results from a time-dependent thermodynamic model of sea ice. *J. Geophys. Res.*, **76**: 1550–75.

Mazhitova, G. G. (2008). Soil temperature regimes in the discontinuous permafrost zone in the East European Russian Arctic. *Eurasian Soil Science*, **41**: 48–62.

McCabe, G. J. and Wolock, D. M. (2010). Long-term variability in Northern Hemisphere snow cover and associations with warmer winters. *Climatic Change*, **99**: 141–53.

McCall, J. G. (1952). The internal structure of a cirque glacier. *J. Glaciol.*, **2**: 122–30.

McClung, D. M. (1981). Fracture mechanical models of dry slab avalanche release. *J. Geophys. Res.*, **86** (B11): 10783–90.

McClung, D. M. (1987). Mechanics of snow slab failure from a geotechnical perspective. Avalanche formation, movement and effects, *IAHS Publ.*, **162**: 475–508.

McClung, D. M. (2002). The elements of applied avalanche forecasting, Part II: The physical issues and the rules of applied avalanche forecasting. *Nat. Hazards*, **26**: 131–46.

McClung, D. M. (2008). Risk-based land use planning in snow avalanche terrain. In J. Locat, D. Perret, D. Turmel, D. Demers, and S. Leroueil (eds.), *Proceedings of the 4th Canadian Conference on Geohazards : From Causes to Management*, Québec: Presse de l'Université Laval, 594 p.

McClung, D. M. (2009). Dimensions of dry snow slab avalanches from field measurements. *J. Geophys. Res.*, **114**: F01006, doi:10.1029/2007JF000941.

McClung, D. M. and Lied, K. (1987). Statistical and geometric definitions of snow avalanche runout. *Cold Reg. Sci. Technol.*, **13**: 107–19.

McClung, D. M. and Mears, A. I. (1991). Extreme value prediction of snow avalanche runout. *Cold Reg. Sci. Technol.*, **19**: 163–75.

McClung, D. M. and Schaerer, P. A. (2006). *The Avalanche Handbook*, 3rd edn., Seattle, WA: The Mountaineers Books. 342 pp.

McKay, C. P., *et al.* (1985). Thickness of ice on perennially frozen lake. *Nature*, **313**: 561–2.

McKnight, D. M., *et al.* (2008). High-latitude rivers and streams. In Vincent, W. F. and Laybourn-Parry, J. (eds.), *Polar lakes and rivers: limnology of Arctic and Antarctic aquatic ecosystems*, Oxford: Oxford University Press. pp. 83–102.

McLaren, A. S. (1989). The under-ice thickness distribution of the Arctic Basin as recorded in 1958 and 1970. *J. Geophys. Res.*, **94** (C4): 4971–83.

McLaren, A. S., Barry, R. G., and Bourke, R. H. (1990). Could Arctic ice be thinning? *Nature*, **345** (6278): 762.

McLaren, A. S., Serreze, M. C., and Barry, R. G. (1987). Seasonal variations of sea ice motion in the Canada Basin and their implications. *Geophys. Res. Lett.*, **14**: 1,123–6.

McNamara, J. P., Kane, D. L., and Hinzman, L. D. (1999). An analysis of an Arctic channel network using a digital elevation model. *Geomorphol.*, **29** (3–4): 339–53.

Mears, A. I. (1976). *Guidelines and methods for detailed snow avalanche hazard investigations in Colorado.* Bulletin No. 38. Denver, CO: Colorado Geological Survey.

Meehl, G. A., *et al.* (1997). Intercomparison makes for a better climate model. *EOS*, **78**: 445–6.

Meehl, G. A., Stocker, T. F., *et al.* (2007): Global climate projections. In: Solomon, S., *et al.* (eds.). Climate Change 2007: The Physical Science Basis. Contribution of Working Group I to the 4th Assessment Report of the Intergovernmental Panel on Climate Change, Cambridge University Press., UK Chapter 10.

Meehl, G. A., Stocker, T. F., Collins, W. D., Friedlingstein, P., Gaye, A. T., Gregory, J. M., *et al.* (2007). Global Climate Projections. In: Climate Change 2007: The Physical Science Basis. Contribution of Working Group I to the Fourth Assessment Report of the Intergovernmental Panel on Climate Change [Solomon, S., D. Qin, M. Manning, Z. Chen, M. Marquis, K.B. Averyt, M. Tignor and H.L. Miller (eds.)]. Cambridge, United Kingdom and New York, NY, USA.

Meier, M. F. (1962). Proposed definitions for glacier mass budget terms. *J. Glaciol.*, **4** (33): 252–61.

Meier, M. F. (1969). Glaciers and water supply. *J. Amer Water Works Assoc.*, **61**: 8–12.

Meier, M. F. and Bahr, D. B. (1996). Counting glaciers: Use of scaling methods to estimate the number and size distribution of the glaciers of the world. In Colbeck, S. C. (ed.), *Glaciers, ice sheets and volcanoes.*

A tribute to Mark F. Meier, Hanover, NH: US Army CRREL Special Rep. 96–27. pp. 89–94.

Meier, M. F. and Post A. (1962). Recent variations in mass net budgets of glaciers in western North America. *IAHS*, **58**: 63–77.

Meier, M. F. and Post, A. (1969). What are glacier surges? *Can. J. Earth Sci.*, **6** (4): 807–17.

Meier, M. F. and Post, A. (1987). Fast tidewater glaciers. *J. Geophys. Res.*, **92** (B9): 9051–8.

Meier, M. F., *et al.* (2007). Glaciers dominate eustatic sea-level rise in the 21st century. *Science*, **317** (5841): 1064–7.

Meister, R. (2002). Avalanches: Warning, rescue and prevention. *Avalanche News*, **62**: 37–44.

Mekis, È. and Hopkinson, R. (2004). Derivation of an improved snow water equivalent adjustment factor map for application on snowfall ruler measurements in Canada. Proceedings, 14th Conference on Climatology, Seattle, WA, January 12–15, Paper 7.12, 5 pp.

Mekis, E. and Vincent, L. A. (2011). An overview of the second generation adjusted daily precipitation dataset for trend analysis in Canada. *Atmosphere-Ocean*, **49** (2): 163–77, https://doi.org/10.1080/070559 00.2011.583910.

Mekis, E. and Brown, R. (2010). Derivation of an adjustment factor map for the estimation of the water equivalent of snowfall from ruler measurements in Canada. *Atmos. Ocean*, **48** (4): 284–93.

Melling, H. (2002). Sea ice of the northern Canadian Arctic Archipelago. *J. Geophys. Res.*, **107** (C11): 3181, 21.

Melling, H. and Lewis, E. L. (1982). Shelf drainage flows in the Beaufort Sea and their effect on the Arctic Ocean pycnocline. *Deep-sea Res.*, **29**(8A): 967–85.

Melni'kov, P. A. and Street, R. B., *et al.* (1993). Terrestrial components of the

Cryosphere. In Tegart, W. J. McG. and
Sheldon, G. W. (eds.), *Climate Change
1992. The Supplementary Report to IPCC
Impacts Assessment. Australian
Government.* Publication Service,
Canberra, pp. 94–102.

Menard, P. *et al.* (2002). Simulation of ice
phenology on a large lake in the Mackenzie
River Basin (1960–2000). Proc. 59th
Eastern Snow Conference, Stowe, VT. pp.
3–12.

Ménard, P., *et al.* (2002). Sensitivity of Great
Slave Lake ice phenology to climate
change. Ice in the environment, Vol. 3,
Squire, V. and Langhorne, P. (eds.), Proc.
16th IAHR Internat. Sympos. on Ice, Int.
Assoc. Hydraulic Eng. Res., Dunedin, New
Zealand. pp 57–63.

Mercer, J. H. (1978).West Antarctic ice sheet
and CO_2 greenhouse effect: A threat of
disaster. *Nature*, **271**: 321–5.

Mermoz, S., Allain-Bailhache, S., Bernier, M.,
Pottier, E., Van Der Sanden, J., and
Chokmani, K. (2014). Retrieval of river ice
thickness From C-Band PolSAR Data.
*IEEE Trans. on Geosci. and Remote
Sensing*, **52** (6): 3052–62.

Mernild, S. H., *et al.* (2008). Jökulhlaup
observed at Greenland ice sheet. *EOS*, **89**
(35): 321–2.

Mernild, S. H., *et al.* (2008). Surface melt area
and water balance modeling on the
Greenland ice sheet 1995–2005.
J. Hydromet., **9**: 1191–211.

Mernild, S. H., *et al.* (2009). Record 2007
Greenland Ice Sheet surface melt extent and
runoff. *EOS*, **90** (2): 13–14.

Mernild, S. H., *et al.* (2010). Greenland Ice
Sheet surface mass-balance modeling in
a 131-yr perspective, 1950–2080.
J. Hydromet., **11**: 3–25.

Mesinger, F., *et al.* (2006). North American
regional reanalysis. *Bull. Amer. Met. Soc.*,
87: 343–60.

Metcalfe, R. A. and Buttle, J. M. (1995).
Controls of canopy structure on snowmelt

rates in the boreal forest. *Proc. of the 52nd
Eastern Snow Conf.*, Toronto, Ont.: 249–
57.

Meyers, S. R. and Hinnov, L. A. (2010).
Northern Hemisphere glaciation and the
evolution of Plio-Pleistocene climate noise.
Paleoceanog., **25**: PA3207, 11, doi:10.1029/
2009PA001834.

Michel, B. (1971). Winter regime of rivers and
lakes. US Army Corps of Engineers, *Cold
Regions Research and Engineering
Laboratory, Monograph.* **III**-B1a, 139pp.

Microwave. (2007). *Proceedings of
international works on earth observation
small satellites for remote sensing
applications*, EOSS 2007, 20–
23 November 2007, Kuala Lumpur,
Malaysia.

Middendorff, A. T. (1844). Bericht über den
Schergin-Schacht zu Jakutsk. *Annal. Phys.
Chem.*, **62**: 404–15.

Mikolajewicz, U., *et al.* (2005). Simulating
Arctic sea ice variability with a coupled
regional atmosphere-ocean-sea ice model.
Met. Zeit., **14**: 793–800.

Milankovitch, M. (1920). *Théorie
mathématique des phénomènes thermiques
produits par la radiation solaire.* Paris:
Gauthier-Villars.

Milburn, D. (2008). The ice cycle on Canadian
rivers. In Beltaos, S. (ed.), *River ice
breakup*, Highlands Ranch, CO: Water
Resources Publ. pp. 21–49.

Miles, M. W. and Barry, R. G. (1991). Large-
scale characteristics of fractures in
multi year Arctic pack ice. In
Axelsson, K. B. E. and Fransson, L. A.
(eds.), *10th International Conference on
Port and Ocean Engineering, under Arctic
Conditions (POAC 89) Vol. 1*, Lulea,
Sweden: University of Technology, pp.
103–12.

Miles, M. W. and Barry, R. G. (1998). A
5-year satellite climatology of winter sea ice
leads in the western Arctic. *J. Geophys.
Res.*, **103**(C10): 21,723–34.

Milillo, P., Rignot, E., Rizzoli, P., Scheuchl, B., Mouginot, J. , Bueso-Bello, J., and Prats-Iraola, P. (2019). Heterogeneous retreat and ice melt of Thwaites Glacier, West Antarctica. *Science Advances*, **5**(1): eaau3433, doi: 10.1126/sciadv.aau3433.

Millar, D. H. M. (1981). Radio-echo layering in polar ice sheets and past volcanic activity. *Nature*, **292**: 441–3.

Miller, G. H., Bradley, R. S., and Andrews, J. T. (1975). The glaciation kevel and lowest equilibrium line altitude in the High Canadian Arctic: Maps and climatic interpretation. *Arct. Alp. Res.*, **7**: 155.

Miller, G. H., *et al.* (2010). *Abrupt onset and intensification of the Little Ice Age around the northern North Atlantic: A role for volcanic forcing? Program and abstracts.* American Polar Society meeting, May 13–14, 2010. Boulder, CO: Institute of Arctic and Alpine Research. 19 pp.

Miller, J. D. and Hotzel, I. S. (1984). Iceberg flux estimation in the Labrador Sea. In Lunardini, V. J. (ed.), *Proceedings, 3rd International Offshore Mechanics and Arctic Engineering Symposium, Vol. 3*, 246–52, United States of America.

Miller, P. E., *et al.* (2009). Assessment of glacier volume change using ASTER-based surface matching of historical photography. *IEEE Trans. Geosci. Remote Sensing*, **47**(7): 1971–9.

Millerd, F. (2007). Global climate change and Great Lakes international shipping, Transportation Research Board Special Report 291. Washington, DC. 28 pp.

Millerd, F. (2011). The potential impact of climate change on Great Lakes international shipping. *Climatic Change*, **104**: 629–52.

Min, S.-K., *et al.* (2008). Human influence on Arctic sea ice detectable from early 1990s onwards. *Geophys. Res. Lett.*, **35**: L21701, 6.

Ming, J., *et al.* (2009). Black Carbon (BC) in the snow of glaciers in west China and its potential effects on albedos. *Atmos. Res.*, **92**: 114–23.

Mirrless, S. T. A. (1932). Meteorological results of the British Arctic Air Route Expedition. 1930–31. *Geophysical Memoir 7*. London: Meteorological Office.

Mitchell, J. M. Jr. (1963). On the world-wide pattern of secular temperature change, In: Changes of Climate. Proceedings of the Rome Symposium Organized by UNESCO and the World Meteorological Organization, 1961. Arid Zone Research Series No. 20, UNESCO, Paris, pp. 161–181.

Mitchell, K. A. and Tiedje, T. (2010). Growth and fluctuations of suncups on alpine snowpacks. *J. Geophys. Res.*, **115** (F4): F04039, 10.

Mitchell, T. D. and Jones, P. D. (2005). An improved method of constructing a database of monthly climate observations and associated high-resolution grids. *Int. J. Climatol.*, **25**: 693–712.

Mock, C. J. and Birkeland, K. W. (2000). Snow avalanche climatology of the western United States mountain ranges. *Bull. Amer. Met. Soc.*, **81** (10): 2367–92.

Mock, C. J., Carter, K. C., and Birkeland, K. W. (2017). Some perspectives on avalanche climatology. *Annals of the American Association of Geographers*, **107** (2): 299–308, doi:10.1080/24694452.2016.1203285.

Moeser, D., Stähli, M., and Jonas, T. (2015), Improved snow interception modeling using canopy parameters derived from airborne LiDAR data. *Water Resources Research*, https://doi.org/10.1002/2014WR016724.

Möller, M. and Schneider, C. (2010). Calibration of glacier volume-area relations from surface extent and application to future glacier change. *J. Glaciol.*, **56** (195): 33–40.

Molnia, B. F. (2007). Late nineteenth to early twenty-first century behavior of Alaskan glaciers as indicators of changing regional climate. *Global Planet. Change*, **56**: 23–56.

Mool, P., Bajracharya, S. R., and Joshi, S. P. (2001). *Inventory of glaciers, glacial lakes and glacial lake outburst floods: Monitoring and early warning systems in the Hindu Kush-Himalayan region – Nepal.* Kathmandu, Nepal: ICIMOD. 198 pp + Appendices.

Moon, T. and Joughin, I. (2008). Changes in ice front position on Greenland's outlet glaciers from 1992 to 2007. *J. Geophys. Res.*, **113**: F02022, doi:10.1029/ 2007JF000927.

Moore, R. D., *et al.* (2009). Glacier change in western North America: Influences on hydrology, geomorphic hazards and water quality. *Hydrol. Processes*, **23**: 42–61.

Morales Arnao, B. (1966). The Huascarán avalanche in the Santa Valley, Pe68: ru, In Co95: 3180lbeck, S. C. (ed.) International symposium on the scientific aspects of snow and ice avalanches, Wallingford, UK: Davos 1965. IAHS Publication 69, pp. 304–15.

Morales Maqueda, M. A., Willmott, A. J., and Biggs, N.R.T. (2004). Polynya dynamics: A review of observations and modeling, *Rev. Geophys.*, **42**: RG1004. doi:10.1029/ 2002RG000116.

Moran, K., *et al.* (2006). The cenozoic palaeoenvironment of the Arctic Ocean. *Nature*, **441** (7093): 601–5.

Morassutti, M. P. and LeDrew, E. F. (1995). Albedo and depth of melt ponds on sea-ice. *Int. J. Climatol.*, **16**: 817–38.

Morice, C. P. , Kennedy, J. J., Rayner, N. A., and Jones, P. D., (2012). Quantifying uncertainties in global and regional temperature change using an ensemble of observational estimates: the HadCRUT4 dataset. *J. Geophy. Res*, **117**: D08101, doi:10.1029/2011JD017187

Morris, E. M. (1989). Turbulent transfer over snow and ice. *J. Hydrol.*, **105**: 205–23, doi:10.1016/0022–1694(89)90105–4.

Morris, E. and Vaughan, D. (2003). Spatial and temporal variation of surface temperature on the Antarctic Peninsula. In Domack, E., *et al.* (eds.), *Antarctic Peninsula climate variability: Historical and paleoenvironmental perspectives*, Washington, DC: American Geophysical Union. pp. 61–8.

Morris, J. N., Poole, A. J., and Klein, A. G. (2006). Retreat of tropical glaciers in Colombia and Venezuela from 1984 to 2004 as measured from ASTER and Landsat images. Proc. 63rd Eastern Snow Conf., 181–91.

Mosimann, L., *et al.* (1993). Ice crystal observations and the degree of riming in winter precipitation. *Water, Air and Soil Pollution*, **68**: 29–42.

Moskalev, Yu D. (1997). Snow avalanche dynamics and snow avalanche accounts. *Proceedings, SANIGMI*, **36** (117): 232.

Mosley-Thompson, E., *et al.* (1999). Late 20th century increase in South Pole snow accumulation. *J. Geophys. Res.*, **104** (D4): 3877–86.

Moss, R. H., *et al.* (2010). The next generation of scenarios for climate change research and assessment. *Nature*, **463**: 747–56, doi:10.1038/nature08823.

Mote, P. W. and Kaser, G. (2007). The shrinking glaciers of Kilimanharo: Can global warming be blamed? *Amer. Sci.*, **95**: 318–25.

Mote, P. W., *et al.* (2005). Declining mountain snowpack in western North America. *Bull. Amer. Met. Soc.*, **86**: 39–49.

Mote, T. L. (2008). On the role of snow cover in depressing air temperature. *J. Appl. Met. Clim.*, **47**: 2008–22.

Mote, T. L. and Anderson, M. R. (1995). Variations in snowpack melt on the

Greenland ice sheet based on passive microwave measurements. *J. Glaciol.*, **17**: 51–60.

Mothes, H. (1926). Dickenmessung von Gletschereis mit seismischen Methoden. *Geol. Rundschau*, **27**: 397–400.

Mothes, H. (1929). Neue Ergebnisse der Eisseismik. *Zeit. Geophys.*, **5**: 120–44.

Motyka, R. J., Fahnestock, M., and Truffer, M. (2010). Volume change of Jakosbshavn Isbrae, West Greenland: 1985–1997–2007. *J. Glaciol.*, **56** (198): 635–46.

Mouginot, J. *et al.* (2014). Sustained increase in ice discharge from the Amundsen Sea Embayment, West Antarctica, from 1973 to 2013. *Geophys. Res. Lett.*, **41**: 1576–84.

Mudryk, L., Kushner, P., Derksen, C., and Thackeray, C. (2017). Snow cover response to temperature in observational and 20 climate model ensembles. *Geophys. Res. Lett.*, **44**: 919–26, doi:10.1002/2016GL071789.

Mountain, D. G. (1980). On predicting iceberg drift. *Cold Reg. Sci. Technol.*, **1**: 273–82.

Mueller, D. R., Vincent, W. F., and Jeffries, M. O. (2003). Breakup of the largest Arctic ice shelf and associated loss of an epishelf lake. *Geophys. Res. Lett.*, **30** (20): 2031.

Muhammad, P., Duguay, C., and Kang, K. K. (2016). Monitoring ice breakup on the Mackenzie River using MODIS data. *Cryosphere*, **10**: 569–84, https://doi.org/10.5194/tc-10-569-2016.

Mullan, D., *et al.* (2017). Climate change and the long-term viability of the World's busiest heavy haul ice road. *Theoretical and Applied Climatology*, **129** (3): 1089–108, doi:10.1007/s00704-016-1830-x.

Muller, D. E., Copland, L., and Stern, D. (2008). Examining Arctic ice shelves prior to the 2008 breakup. *EOS*, **89**(49): 502–3.

Müller, F. (1959). Eight months of glacier and soil research in the Everest region. *Mountain World*, **1958/59**: 191–208.

Müller, F. (1962). Zonation in the accumulation area of the glaciers of Axel Heiberg Island. *J. Glaciol.*, **4**: 302–13.

Müller, F., Ohmura, A., and Braithwaite, R. (1977). The North Water project (Canadian Greenland Arctic). *Geogr. Helv.*, **2**: 111–17.

Müller, J., *et al.* (2009). Variability of sea-ice conditions in the Fram Strait over the past 30,000 years. *Nat. Geosci,*, **2**: 772–6, doi:10.1038/ngeo665.

Muller, S. W. (1947). *Permafrost or permanently frozen ground and related engineering problems.* Ann Arbor, MI: J. W. Edwards. 231 pp.

Muller, S. W. (French, H. M. and Nelson, F. E., eds.) (2008). *Frozen in time: Permafrost and Engineering Problems.* Reston, VA: Amer. Soc. Civil Engineers. 280 pp.

Murphy, J. (1909). The ice question as it affects Canadian water power with special reference to frazil and Anchor ice. *Proc. Tarans. Roy. Soc, Can. 3rd Ser. Sec. III*: 143–77.

Murton, J. B. (2009). Global warming and thermokarst. In Margesin, R. (ed.), *Permafrost soils, soil biology*, vol. 16, Berlin: Springer. pp. 185–203, doi.org/10.1007/978-3-540-69371-0_13.

Murton, J. B. and French, H. M. (1994). Cryostructures in permafrost, Tuktoyatuk coastlands, western Arctic, Canada. *Canad. J. Earth Sci.*, **31**: 737–47.

Murton, J. B., *et al.* (2010). Identification of Younger Dryas outburst flood path from Lake Agassiz to the Arctic Ocean. *Nature*, **440**: 740–3.

Muskett, R. R., *et al.* (2008). Surging, accelerating surface lowering and volume reduction of the Malaspina Glacier system, Alaska, USA, and Yukon, Canada, from 1972 to 2006. *J. Glaciol.*, **54** (188): 788–800.

Muttoni, G., *et al.* (2003). Onset of major Pleistocene glaciations in the Alps. *Geology*, **31** (11): 989–92.

Myel' nikov, I. A. (1995). The Weddell ice drift station in Antarctica. *Oceanol.*, **35** (2): 286–8.

Mysak, L. A., R. G. Ingram, J. Wang, and A. van der Baaren, 1996: The anomalous sea-ice extent in Hudson bay, Baffin bay and the Labrador sea during three simultaneous NAO and ENSO episodes. Atmosphere-Ocean, 34, 313–343.

Mysak, L. A. (2008). Glacial inceptions: Past and future. *Atmos. – Ocean*, **46**: 317–41.

Naaim, M., *et al.* (2016). Impact of climate warming on avalanche activity in French Alps and increase of proportion of wet snow avalanches. *Houille Blanche*, **59** (6): 12–20, doi:10.1051/lhb/2016055.

Naaim, M., Durand, Y., Eckert, N., and Chambon, G. (2013). Dense avalanche friction coefficients: Influence of physical properties of snow. *J. Glaciol.*, **59** (216): 771–82, doi:10.3189/2013JoG12J205.

Narama, C., *et al.* (2006). Recent changes of glacier coverage in the western Terskey–Alatoo range, Kyrgyz Republic, using Corona and Landsat. *Annals Glaciol.*, **43**: 223–9.

Narama, C., *et al.* (2010). Spatial variability of recent glacier area changes in the Tien Shan Mountains, Central Asia, using Corona (~1970), Landsat (~2000), and ALOS (~2007) satellite data. *Planet. Global Change*, **71**: 42–54.

Narod, B. B., Clarke, G. K. C., and Prager, B. T. (1988). Airborne UHF radar sounding of glaciers and ice shelves, northern Ellesmere Island, Arctic Canada. *Canad. J. Earth Sci.*, **25**: 95–105.

Naruse, R., *et al.* (1995). Recent variations of calving glaciers in Patagonia, South America, revealed by ground surveys, satellite-data analyses and numerical experiments. *Annals Glaciol.*, **21**: 297–303.

NASA (National Aeronautics and Space Administration). 2016. Arctic sea ice melt. https://neptune.gsfc.nasa.gov/csb/index .php?section=54.

Nash, T., *et al.* (2007). A record of Antarctic climate and ice sheet history recovered. *EOS*, **88** (50): 557–5.

Nash, T., *et al.* (2009). Obliquity-paced Pliocene West Antarctic ice sheet oscillations. *Nature*, **458**: 322–9.

National Research Council. (1990). *Snow avalanche hazards and mitigation in the United States, Commission on Engineering and Technical Systems, Panel on Snow Avalanches.* Washington, DC: National Academy Press. 96 pp.

National Research Council. (2010). *Advancing the science of climate change.* Washington, DC: National Academy Press. 528 pp.

National Research Council (NRC). (2011). Climate Stabilization Targets: Emissions, concentrations, and impacts over decades to millennia, 190 pages, by Committee on Stabilization Targets for Atmospheric Greenhouse Gas Concentrations of NRC, National Academy Press, USA.

Nazintsev, Y. L. (1964). The heat balance of the surface of the multiyear ice cover in the central Arctic (In Russian). *Trudy Arkt. Antarkt. NauchnoIssled. Inst.*, **267**: 110–26.

Nelson, F. E. (1986). Permafrost in central Canada: Applications of a climate-based predictive model. *Annals Assoc. Amer. Geogr.*, **76** (4): 550–69.

Nelson, F. E. and Anisimov, O. A. (1993). Permafrost zonation in Russia under anthropogenic climatic change. *Permafrost Periglacial Process*, **4**: 137–48.

Nelson, F. E., Hinkel, K. M., and Paetzold, R. (1997). An active layer thermal regime at Barrow, Alaska. CMDL Summary Report 24, 1996–1997. Boulder, CO: NOAA/ESRL.

Nelson, F. E. and Outcalt, S. I. (1983). A frost-index number for spatial prediction of ground-frost zones. In *Permafrost – Fourth international conference proceedings, vol. 1.* Washington, DC: National Academy Press. pp. 907–11.

Nelson, F. E. and Outcalt, S. I. (1987). A computational method for prediction and regionalization of permafrost. *Arct. Alp. Res.*, **19**: 279–88.

Nelson, F. E., *et al.* (2008). Decadal results from the Circumpolar Active Layer Monitoring (CALM) program. In Kane, D. L. and Hinkel, K. M. (eds.), *Ninth international conference on permafrost*, Fairbanks, AK: Institute of Northern Engineering, University of Alaska Fairbanks. pp. 1273–80.

Nerem, R. S., *et al.* (2018). Climate Change Driven Accelerated Sea Level Rise Detected In The Altimeter Era. *Proc. Natl. Acad. Sci. USA*, **115**: 2022–5, https://doi.org/10.1073/pnas.1717312115, 2018.

Nesje, A. (2009). Latest Pleistocene and Holocene alpine glacier fluctuations in Scandinavia. *Quatern. Sci. Rev.*, **28** (21–22): 2119–36.

Nesje, A. and Dahl, S. O. (2000). *Glaciers and environmental change*. London: Hodder Education. 203 pp.

Newell, J. P. (1993). Exceptionally large icebergs and ice islands in Eastern Canadian Waters: A review of sightings from 1900 to present. *Arctic*, **46** (3): 205–11.

Newton, B. W., Prowse, T. D., and de Rham, L. P. (2017). Hydro-climatic drivers of mid-winter breakup of river ice in western Canada and Alaska. *Hydrol. Res.*, **48**: 945–56, https://doi.org/10.2166/nh.2016.358.

Nezhikhovskiy, R. A. (1964). Coefficients of roughness of bottom surface on slush-ice cover. *Soviet Hydrol.*, **2**: 127–50.

Nghiem, S. V. and Tsai, W.-Y. (2001). Global snow monitoring with Ku-band scatterometer. *IEEE Trans. Geosci. Remote Sens.*, **39** (10): 2118–34.

Nghiem, S., *et al.* (2001). Detection of snowmelt regions on the Greenland ice sheet using diurnal backscatter change. *J. Glaciol.*, **47**: 539–47, doi:10.3189/172756501781831738.

Nghiem, S. V., *et al.* (2007). Rapid reduction of Arctic perennial sea ice. *Geophys. Res. Lett.*, **34**: 1–6.

Nguyen, T.-N. , *et al.* (2009). Estimating the extent of near-surface permafrost using remote sensing, Mackenzie Delta, Northwest Territories. *Permafrost Periglac.Process*, **20** (2): 141–53.

Nicholls, K. W., *et al.* (2009). Ice-ocean processes over the continental shelf of the southern Weddell Sea, Antarctica: A review. *Rev. Geophys.*, **47**: RG3003, 23.

Nick, F. M., van der Veen, C. J., and Oerlemans, J. (2007). Controls on advance of tidewater glaciers: Results from numerical modeling applied to Columbia Glacier. *J. Geophys. Res.*, **112**: G03S24.

Nick, F. M., *et al.* (2009). Large-scale changes in Greenland outlet glacier dynamics triggered at the terminus. *Nat. Geosci.*, **2**: 110–14.

Nick, F. M., *et al.* (2013). Future sea-level rise from Greenland's main outlet glaciers in a warming climate. *Nature*, **497**: 235–8, doi:10.1038/nature12068.

Nicolsky, D. and Romanovsky, V. (2018). Modeling long-term permafrost degradation. *J. of Geophys. Res.: Earth Surface*, **123**: 1756–71, doi.org/10.1029/2018JF004655.

Nicolussi, K. (1990). Bilddokumente zur Geschichte des Vernagtferners im 17 Jahrhundert. *Zeit. Gletscherk. Glazialgeolog*, **26**: 97–119.

Niederer, P., *et al.* (2007). Tracing glacier wastage in the northern Tien Shan (Kyrgyzstan/ Central Asia) over the last 40 years. *Clim. Change*, **86**: 227–34.

Ning, Li, *et al.* (2009). Using remote sensing to estimate sea ice thickness in the Bohai Sea, China based on ice type. *Int. J. Rem. Sensing.*, **30** (17): 4539–52.

Nitta, T., *et al.* (2014). Representing variability in subgrid snow cover and snow depth in a global land model: Offline validation. *J. Climat.*, **27**(9): 3318–30. doi: 10.1175/JCLI-D-13-00310.1.

Niu, G.-Y., *et al.* (2007). Retrieving snow mass from GRACE terrestrial water storage change with a land surface model. *Geophys.*

Res. Lett., **34**: L15704, doi:10.1029/2007GL030413.

Niu, G. Y. and Yang, Z. I. (2003). The versatile integrator of surface atmospheric processes – Part 2: Evaluation of three topography-based runoff schemes. *Global Planet. Change*, **38**: 191–208.

Niu, G. Y. and Yang, Z. L. (2004). Effects of vegetation canopy processes on snow surface energy and mass balances. *J. Geophys. Res. Atmospheres*, **109**: D23111.

Nolan, M., *et al.* (1995). Ice-thickness measurements of Taku Glacier, Alaska, USA, and their relevance to its recent behavior. *J. Glaciol.*, **41** (139): 541–53.

Nolin, A. W., Fetterer, F. M., and Scambos, T. A. (2002). Surface roughness characterizations of sea ice and ice sheets: Case studies with MISR data. *IEEE Trans. Geosci. Remote Sens.*, **40** (7): 1605–15.

Nolin, A. W., *et al.* (2001). Cryospheric applications of MISR data. *IEEE Internat. Geosci. Remote Sensing Symposium (IGARRS) 2001. Proceedings*, **3**: 1219–21.

Notz, D. and Stroeve, J. (2016). Observed Arctic sea-ice loss directly follows anthropogenic CO_2 emission. *Science*, **354** (6313): 747–50, doi:10.1126/science.aag2345.

Notz, D. and Worster, M. G. (2009). Desalination processes of sea ice revisited. *J. Geophys. Res.*, **114**: C05006, 10.

Nötzli, J., Naegeli, B., and Vonder Mühll, D. (eds.). (2009). *PERMOS. Permafrost in Switzerland. 2004/2005 and 2006/2007. Glaciol. Rep. (Permafrost) no.6/7. Cryospheric Commission, Swiss Acad. Sci.*, Zurich: University of Zurich, Dept. of Geography. 100 pp.

Nummelin, A., Ilicak, M., Li, C., and Smedsrud, L. H. (2016). Consequences of future increased Arctic runoff on Arctic Ocean stratification, circulation, and sea ice cover. *J. Geophys. Research-Oceans*, 121: 617–37. doi: 10.1002/2015JC011156.

Nuth, C., *et al.* (2010). Svalbard glacier elevation changes and contribution to sea level rise. *J. Geophys. Res.*, **115**: F01008, doi:10.1029/2008JF001223.

Nutt, D. C. (1966). The drift of ice iceland WH-5. *Arctic*, **16**: 204–6.

NWS. (1992). *Airborne gamma radiation snow survey program and satellite hydrology program: User's guide version 4.0.* Minneapolis, MN: Office of Hydrology, National Weather Service, NOAA. 54 pp.

Nye, J. F. (1953). The flow law of ice from measurements in glacier tunnels, laboratory experiments and the Jungfraufirn borehole experiment. *Proc. Roy. Soc. London. A*, **219** (1139): 477–89.

Nye, J. F. (1958). A theory of wave formation in glaciers. International Association of Scientific Hydrology Publ. 47 (Symposium at Chamonix 1958 – Physics of the movement of the ice), pp. 139–54.

Nye, J. F. (1960). The response of glaciers and ice-sheets to seasonal and climatic changes. *Proc. Roy. Soc. London. A*, **256** (1287): 559–84.

Nye, J. F. (1961). The influence of climatic variations on glaciers. *IASH, General Assembly Helsinki, IASH Publ.*, **54**: 397–404.

Nye, J. F. (1987). On the theory of the advance and retreat of glaciers. *Geophys. J. Roy. Astron. Soc.*, **7**: 431–56.

Oberleitner, F., Thaler, K., and Spötl, C. (2009). *Glacio-meteorological investigations in an alpine ice cave* (Eisriesenwelt, Austria). Abstract.

Obu, J., Westermann, S., Bartsch, A., Berdnikov, N., Christiansen, H. H., Dashtseren, A., Delaloye, R., Elberling, B., Etzelmüller, B., and Kholodov, A. (2019). Northern Hemisphere permafrost map based on TTOP modelling for 2000–2016 at 1 km^2 scale. *Earth Sci. Rev.*, **193**: 299–316.

Obyazov, V. A. and Smakhtin, V. K. (2014). Ice regime of Transbaikalian rivers under changing climate. *Water Resour.*, **41**: 225–31, doi.org/10.1134/S0097807814030130.

O'Connor, F. M., *et al.* (2010). Possible role of wetlands, permafrost, and methane hydrates in the methane cycle under future climate change: A review. *Rev. Geophys.*, **48**: RG4005, 33.

Oedl, R. (1922). Die grosse Eishölle im Tennengebirge (Salzburg). (Eisriesenwelt). Vermessung.17/18: 63–83. *Ber. Bundeshöhlenkommission*, **3**: 5–30.

Oerlemans, J. (1989). *Glacier fluctuations and climatic change*. Dordrecht: Kluwer. 417 pp.

Oerlemans, J. (1991). A model for the surface balance of ice masses.Pt.1: Alpine glaciers. *Zeit. Gletscherk. Glazialgeol.*, **27/28**: 63–83.

Oerlemans, J. (1997). A flowline model for Nigardsbreen, Norway: Projection of future glacier length based on dynamic calibration with the historic record. *Annals Glaciol.*, **24**: 382–9.

Oerlemans, J. (2005). Extracting a climate signal from 169 glacier records. *Science*, **308**: 675–7.

Oerlemans, J. and van der Veen, C. J. (1984). *Ice sheets and climate*. Dordreche: D. Reidel Publ. Co. 217 pp.

Oerlemans, J., *et al.* (1998). Modelling the response of glaciers to climatic warming. *Clim. Dynam.*, **14**: 267–74.

Ogi, M., Yamazaki, K., and Wallace, J. M. (2010). Influence of winter and summer surface wind anomalies on summer Arctic sea ice extent. *Geophys. Res. Lett.*, **37**: L07701, doi:10.1029/2009GL042356.

Ogi, M., *et al.* (2008). Summer retreat of Arctic sea ice: Role of summer winds. *Geophys. Res. Lett.*, **35**: L24701, 5.

Ogilvie, A. E. J. (1984). The past climate and sea-ice record from Iceland, part 1: Data to AD 1780. *Clim. Change*, **6**: 131–52.

Ogilvie, A. E. and Jonsson, T. (2001). "Little Ice Age" research: A perspective from Iceland. *Clim. Change*, **48**: 9–52.

Ohata, T., Furukawa, T., and Higuchi, K. (1994). Glacioclimatological study of perennial ice in the Fuji Ice Cave, Japan.

Part 1: Seasonal variation and mechanism of maintenance. *Arct. Alp. Res.*, **26**: 227–37.

Ohmura, A. (1987). Heat budget of the climate system between the Last Glacial Maximum and the present. *Bull. Dept. Geogr., Univ. Tokyo*, **19**: 21–8.

Ohmura, A. (2001). Physical basis for the temperature-based melt-index method. *J. appl. Met.*, **40** (4): 753–61.

Ohmura, A. (2009). Completing the world glacier inventory. *Annals Glaciol.*, **50** (53): 144–8.

Ohshima, K. I. and Riser, S. C. (2010). Mapping and interannual variations of sea ice thickness in the Okhotsk Sea inferred from ocean salinity profile in spring. Paper 57A140. Proceedings, Tromso Sea Ice Symposium. Int. Glaciol. Soc. www .igsoc.org/symposia/previous.html

Ohshima, K. I., *et al.* (2006). Interannual variability of sea ice area in the Sea of Okhotsk: Importance of sea heat flux in fall. *J. Met. Soc. Japan*, **79**: 123–9.

Oller, P., Fischer, J. T., and Muntán, E. (2020). The Historic Avalanche that Destroyed the Village of Àrreu in 1803, Catalan Pyrenees. *Geosciences*, **10**(5): 169, https://doi.org/10 .3390/geosciences10050169

Olonscheck, D., Mauritsen, T., and Notz, D. (2019). Arctic sea-ice variability is primarily driven by atmospheric temperature fluctuations. *Nat. Geosci*, 12(6). doi:10.1038/s41561-019-0363-1.

Olyphant, G. A. and Isard, S. A. (1988). The role of advection in the energy balance of late-lying snowfields: Niwot Ridge, Front Range, Colorado. *Water Resour. Res.*, **24** (11): L1962–8.

Onarheim, I. H., Eldevik, T., Smedsrud, L. H., and Stroeve, J. C. (2018). Seasonal and regional manifestation of Arctic sea ice loss. *J. of Clim.*, **31**: 4917–32.

O'Neill, B. C., *et al.* (2014). A new scenario framework for climate change research: the concept of shared socioeconomic pathways. *Climatic Change*,

122: 387–400. https://doi.org/10.1007/s105 84-013-0905-2

O'Neill, B. C., *et al.* (2017). IPCC reasons for concern regarding climate change risks. *Nat. Clim. Change*, **7** (1): 28–37, doi:10.1038/nclimate3179.

Onstott, R. G. (1992). SAR and scatterometer signatures of sea ice. In Carsey, F. D. (ed.), *Microwave remote sensing of sea Ice*, Washington, DC: American Geophysical Union. pp. 73–104.

Orheim, O. (1980). Physical characteristics and life expectancy of tabular Antarctic icebergs. *Ann. Glaciol.*, **1**: 11–18.

Orheim, O. (1987). Icebergs in the Southern Ocean. *Annals Glaciol.*, **9**: 241–2.

Osmaston, H. (2005). Estimation of glacier equilibrium line altitude by the area x altitude, area x altitude balance ratio, and the area-altitude balance index methods and their validation. *Quat. Int.*, **138** (9): 22–31.

Osokin, I. M. (1973). Zonation and regime of naleds in Trans-Baikal region. Proceedings of the Second International Conference on Permafrost. USSR Contribution. Washington, DC. pp. 391–6.

Osterkamp, T. E. (1975). *Frazil ice nucleation mechanisms. Report UAGR-230.* Fairbanks: University of Alaska.

Osterkamp, T. E. (2001). Sub-sea permafrost. In Steele, J. H., Thorpe, S. A., and Turekian, K. K., (eds.), *Encyclopedia of ocean sciences*, San Diego: Academic Press. pp. 2902–12.

Osterkamp, T. E. (2008). Thermal state of permafrost in Alaska during the fourth quarter of the twentieth century. In Kane, D. L. and Hinkel, K. M. (eds.), *Ninth International Conference on Permafrost, June 29–July 3, 2008, University of Alaska Fairbanks. Proceedings, vol. 2*, Fairbanks, AK: University of Alaska. pp. 1333–7.

Osterkamp, T. E., *et al.* (2000). Observations of thermokarst and its impact on boreal forest in Alaska. *Arct. Antarct. Alp. Res.*, **32**: 303–15.

Østrem, G. (1964). Ice-cored moraines in Scandinavia. *Geograf. Annal.*, **46**: 282–337.

Østrem, G. (1966). The height of the glaciation limit in southern British Columbia and Alberta. *Geograf. Annal., A*, 48: 126–38.

Østrem, G. (1972). Height of the glaciation level in northern British Columbia and southeastern Alaska. *Geograf. Annal., A*, **54**: 76–84.

Østrem, G. and Brugman, M. (1991). Glacier mass-balance measurement. A manual for field and office work. NHRI Sci. Rep. No, 4, Saskatoon, Sas,: National Hydrology Research Institute. 224 pp.

Østreng, W. (2006). The International Northern Sea Route Programme (INSROP): Applicable lessons learned. *Polar Record*, **42**: 71–81.

Otiemo, F. and Bromwich, D. H. (2009). Contribution of atmospheric circulation to Inception of the Laurentide Ice Sheet at 116 kyr BP. *J. Climate*, **22**(1): 39–57.

Outcalt, S. I. and MacPhail, D. D. (1965). *A survey of neoglaciation in the front range of Colorado*. Study Series in Earth Sciences, No. 4. Boulder, CO: University of Colorado Press. 124 pp.

Overland, J., *et al.* (2009). International Arctic Sea Ice monitoring program continues into second summer. *EOS, Transactions, AGU*, **90** (37): 321–2.

Overland, J. E. and Wang, M. (2005). The Arctic climate paradox: The recent decrease of the Arctic Oscillation. *Geophy. Res. Lett.*, **32**: L06701, https://doi.org/10.1029 /2004GL021752

Paillard, D. (2001). Glacial cycles: Towards a new paradigm. *Rev. Geophys.* 39: 325–46.

Palacios, D. and Vázquez-Selem, L. (1996). Geomorphic effects of the retreat of Jamapa Glacier, Pico de Orizaba volcano (Mexico). *Geogr. Annal. A*, **78**: 19–34.

Pálsson, S. (Williams. R. S., Jr and Sigurðsson. O., eds.) (2004). *Icelandic ice mountains:*

draft of a physical, geographical, and historical description of icelandic ice mountains on the basis of a journey to the most prominent of them in 1792–1794. Reykjavik: Icelandic Literary Society. 183 pp.

Parajka, J., et al. (2010). A regional snow-line method for estimating snow cover from MODIS during cloud cover. J. Hydrol., 38: 203–12.

Park, H., Yabuki, H., and Ohata, T. (2012). Analysis of satellite and model datasets for variability and trends in Arctic snow extent and depth, 1948–2006. Polar Science, 6 (1): 23–37.

Parkinson, C. L. (2006). Earth's cryosphere: Current state and recent changes. Ann. Rev. Environment Resour., 31: 33–60.

Parkinson, C. L. and Cavalieri, D. J. (2008). Arctic sea ice variability and trends, 1979–2006. J. Geophys. Res., 113: C07003, 1–28.

Parkinson, C. L. and Cavalieri, D. J. (2009). Sea Ice. In Williams, R. S. and Ferrigno, J. (eds.) Satellite Image Atlas of Glaciers of the World. U.S. Geological Survey Professional Paper, 1386-A.

Parkinson, C. L., et al. (1987). Antarctic sea ice, 1973–1976: Satellite passive-microwave observations. SP 489. Washington, DC: NASA. 296 pp.

Parkinson, C. L., et al. (1999a). Arctic sea ice extents, areas, and trends, 1978–1996. J. Geophys. Res., 104(C9): 20,837–56.

Parkinson, C., Comiso, J., and Zwally, H. J. (1999). Nimbus-5 ESMR daily polar gridded brightness temperatures. Boulder, CO: National Snow and Ice Data Center. Digital media.

Parkinson, C. and Washington, W. M., Jr. (1979). A large-scale numerical model of sea ice. J Geophys. Res., 84: 311–37.

Parlee, B. and Furgal, C. (2012). Well-being and environmental change in the Arctic: A synthesis of selected research from Canada's International Polar Year program. Climatic Change, 115(1): 13–34.

Parmerter, R. R. and Coon, M. D. (1973). On the mechanics of pressure ridge formation in sea ice. Offshore Technology Conference, 1973, Houston, Texas. Paper No. 1810-MS: 10 pp.

Partington, K. C. (1998). Discrimination of glacier facies using multi-temporal SAR data. J. Glaciol., 44(146): 42–53.

Paterson, W. S. B. (1994). The physics of glaciers, 3rd ed. Oxford: Pergamon Press. 480 pp.

Paul, F. (2000). Evaluation of different methods for glacier mapping using Landsat TM, Proceedings, EARSeL-SIG Workshop, Land ice and snow, Dresden. pp. 239–45.

Paul, F., et al. (2007). Alpinewide distributed glacier mass balance modelling. In Orlove, B., et al. (eds.), Darkening peaks: Glacier retreat, science and society, Berkeley, CA: University of California Press. pp. 111–25.

Paul, F., et al. (2009). Recommendations for the compilation of glacier inventory data from digital sources. Annals Glaciol., 50 (54): 119–26.

Paul, F. and Svoboda, F. (2009). A new glacier inventory on southern Baffin Island, Canada, from ASTER data II: Data analysis, glacier change and applications. Annals Glaciol., 50 (53): 22–31.

Paulcke, W. (1938). Praktische Schnee- und Lawinenkunde. Berlin: J. Springer, Verstandliche Wissenschaft, vol. 38. 217 pp.

Pavelsky, T. M. and Smith, L. C. (2004). Spatial and temporal patterns in Arctic river ice breakup observed with MODIS and AVHRR time series. Rem. Sensing Environ., 93: 328–38.

Payne, A. J., et al. (2000). Results from the EISMINT model intercomparison: The effects of thermomechanical coupling. J. Glaciol., 46 (153): 227–38.

Pease, C. H. (1987). The size of wind-driven coastal polynyas. J. Geophys. Res., 92: 7049–59.

Pedersen, C. A., *et al*. (2010). A new sea ice albedo scheme including melt ponds for ECHAM5 general circulation model. *J. Geophys. Res.*, **114**: D08101, 15.

Pedro, J. B., *et al*. (2016). Southern Ocean deep convection as a driver of Antarctic warming events. *Geophys. Res. Lett.*, **43**: 2192–9.

Pellikka, P. and Rees, W. G. (eds.) (2010). *Remote sensing of glaciers*. London: CRC Press, Taylor and Feancis. 330 pp.

Peltier, W. R. (1994). Ice Age paleotopography. *Science*, **265**: 195–201.

Peltier, W. R. (2004). Global glacial isostasy and the surface of the Ice-Age Earth, 2004, The ICE-5G(VM2) model and GRACE. *Ann. Rev. Earth Planet. Sci.*, **32**: 111–49.

Pelto, M. S. (2016). State of the Climate in 2015. *Bull. Am. Meteor. Soc.*, **97** (8): S23–S24.

Pelto, M. S. and Hedlund, C. (2001). Terminus behavior and response time of North Cascade Glaciers, Washington, USA. *J. Glaciol.*, **47** (158): 497–506.

Pelto. M. S. and Warren, C. R. (1991). Relationship between tidewater glacier calving velocity and water depth at the calving front. *Annals Glaciol.*, **15**: 115–18.

Pelto, M. S., Beedle, M., and Miller, M. M. (2009). Mass balance measurements on the Taku glacier, Juneau Icefield, Alaska 1946–2008, www.nichols.edu/departments/gla-cier/taku.html.

Pelto, M. S., *et al*. (2008). The equilibrium flow and mass balance of the Taku Glacier, Alaska 1950–2006. *The Cryosphere*, **2** (2): 147–57.

Pepe, A. and Calo, F., (2017). A Review of Interferometric Synthetic Aperture RADAR (InSAR) Multi-Track Approaches for the Retrieval of Earth's Surface Displacements. *Appl. Sci.*, **7**: 1264, MDPI, doi:10.3390/app7121264

Perla, R. I. (1980). Avalanche release, motion, and impact. In Colbeck, S. C. (ed.), *Dynamics of snow and ice masses*, New York: Academic Press. pp. 397–462.

Perla, R. I., Cheng, T. T., and McClung, D. M. (1980). A two-parameter model of snow avalanche motion. *J. Glaciol.*, **26**: 197–207.

Perovich, D. K., *et al*. (2002). Seasonal evolution of the albedo of multiyear Arctic sea ice. *J. Geophys Res.*, **107** (C10): 8044, 13.

Perovich, D. K., Light, B., Eicken, H., Jones, K. F., Runciman, K., and Nghiem, S. V. (2007). Increasing solar heating of the Arctic Ocean and adjacent seas, 1979–2005: Attribution and role in the ice-albedo feedback. *Geophys. Res. Lett.*, **34**: L19505, doi:10.1029/2007GL031480.

Perovich, D. K., *et al*. (2008). Sunlight, water, and ice: Extreme Arctic sea ice melt during the summer of 2007. *Geophys. Res. Lett.*, **35**: L11501, 4.

Perovich, D. K., *et al*. (2009a). Transpolar observations of the morphological properties of Arctic sea ice. *J. Geophys. Res.*, **114**: C00A04, doi:10.1029/2008JC004892.

Perovich, D. K., *et al*. (2009b). Sea ice cover. Arctic Report Card 2009, www.arctic.noaa.gov/reportcard/.

Perovich, D., *et al*. (2018). Arctic Report Card: Update for 2018, effects of persistent Arctic warming continue to mount, Arctic Program, NOAA, USA.

Perovich, D. K. and Richter-Menge, J. A. (2009). Loss of sea ice in the Arctic. *Ann. Rev. Marine Sci.*, **1**: 417–41.

Perry, A. H. and Symons, L. (eds.). (1991). Highway Meteorology. *E and F N Spon*, London. 209 pp.

Perutz, M. F. (1953). The flow of glaciers. *Nature*, **172** (621): 929–31.

Perutz, M. F. and Seligman, G. (1939). A crystallographic investigation of glacier structure and the mechanism of glacier flow. *Proc. Roy. Soc. London, Ser. A*, **172**: 335–60.

Pessina, S. and Kasten-Coors, S. (2011). "In-Flight Characterization of CRYOSAT-2 Reaction Control System," Proceedings of the 22nd International Symposium on Space Flight Dynamics, February 28–March 4, 2011, Sao Jose dos Campos, SP,

Brazil, URL: www.issfd22.inpe.br/S7-Attitude.Dynamics.1-AD1/S7_P4_ISSFD22_PF_041.pdf

Peterson, B. J., *et al*. (2002). Increasing river discharge to the Arctic Ocean. *Science*, **293**: 2171–3.

Petrov, V. G. (1930). *Naledy na Amursko-Yakutskoi magistral. (Icings on the Amur-Yakustk highway)*. *Izd. Akad. Nauk, SSSR, Nauchno-Issled*. Leningrad: Avtomobil. Dorozhno. Inst. 177 pp + atlas 37 pp.

Petrovic, J. J. (2003). Mechanical properties of ice and snow. *J, Materials Sci.*, **38**: 1–6.

Petryk, S. (1995). Numerical modeling. In Beltaos, S. (ed.), *River ice jams*, Highlands Ranch, CO: Water Resources Publications. pp. 147–72.

Petty, A. A., Stroeve, J. C., Holland, P. R., Boisvert, L. N., Bliss, A. C., Kimura, N., and Meier, W. N. (2018). The Arctic sea ice cover of 2016: a year of record-low highs and higher-than-expected lows. *The Cryosphere*, **12**: 433–52.

Pfeffer, W. T. (2003). Tidewater glaciers move at their own pace. *Nature*, **426**: 602.

Pfeffer, W. T. (2007). A simple mechanism for irreversible tidewater glacier retreat. *J. Geophys. Res.*, **112**: F03S25, 12.

Pfeffer, W. T., Harper, J. T., and O 'Neel, S. (2008). Kinematic constraints on glacier contributions to 21st-century sea-level rise. *Science*, **321**: 1340–2.

Pielmeier, C. and Scchneebelli, M. (2003). Developments in the stratigraphy of snow. *Surveys Geophys.*, **24**: 389–416.

Pierce, D. W., *et al*. (2008). Attribution of declining western U.S. snowpack to human effects. *J. Climate*, **21**: 6425–44.

Pirazzinni, R. (2009). Challenges in snow and ice albedo parameterizations. *Geophysica*, **45** (1–2): 41–62, Geophysical Society of Finland, Helsinki.

Pithan, F. and Mauritsen, T. (2014). Arctic amplification dominated by temperature feedbacks in contemporary climate models.

Nat. Geosci. Lett., **7**: 181–4, doi: 10.1038/NGEO2071.

Plafker, G. and Ericksen, G. E. (1978). Nevados Huascaran avalanches, Peru. In Voight, B. (ed.), *Rockslides and avalanches*, New York: Elsevier Scientific. pp. 277–314.

Plewes, L. A. and Hubbard, B. (2001). A review of the use of radio-echo sounding in glaciology. *Progr. Phys. Geog.*, **25**: 203–36.

Plug, L. J. and West, J. J. (2009). Thaw lake expansion in a two-dimensional coupled model of heat transfer, thaw subsidence, and mass movement. *J. Geophys. Res.*, **114**: F01002, doi:10.1029é2006JF000740.

Podyakanov, S. A. (1903). Naledy vostochnoi Sibiri i prichiny ikh voznikoveniya (Icings of eastern Siberia and their origins). *Izv. Vsesoyuz.Geogr. Obshch*, **39**: 305–37.

The Polar Pathfinder Group (Maiden, M., *et al.*) (1997). The Polar Pathfinders: Data Products and Science Plans. Part II. EOS Electronic Supplement, 96149e.

Pollard, D. (2010). A retrospective look at coupled ice sheet–climate modeling. *Climatic change*, **100** (1): 173–94.

Pollard, D. and DeConto, R. M. (2009). Modelling West Antarctic ice sheet growth and collapse through the past five million years. *Nature*, **458**: 329–33.

Pollard, W. H. and Couture, N. J. (2008). Massive ground ice in the Eureka Sound Lowlands, Canadian High Arctic. In Kane, D. L. and Hinkel, K. M. (eds.), *Proceedings, Ninth International Conference on Permafrost*, Fairbanks, AK: University of Alaska, Institute of Northern Engineering. pp. 1433–8.

Pollard, W. H. and French, H. M. (1980). A first approximation of the volume of ground ice, Richards Island, Pleistocene Mackemzie delta, Northwest Territories, Canada. *Canad. Geotech. J.*, **17**: 509–16.

Polyak, L., *et al*. (2010). History of sea ice in the Arctic. *Quat. Sci. Rev*, doi:10.1016/j.quascirev.2010.02.010

Polyakov, I. and Johnson, M. A. (2000). Arctic decadal and inter-decadal variability. *Geophys. Res. Lett.*, **27**: 4097–100.

Pomeroy, J. W. (2000). Pririe and Arctic areal snow cover mass balance using a blowing snow model. *J. Geophys. Res.*, **105**(D21): 26,619–34.

Pomeroy, J. W. (2009). Centre for Hydrology, University of Saskatchewan.

Pomeroy, J. W. and Gray, D. M. (1990). Saltation of snow. *Water Resour. Res.*, **26** (7): 1583–94.

Pomeroy, J. W., Gray, D. M., and Landine, P. G. (1993). The Proirie blowing snow model: Characteristics, validation, operation. *J. Hydrol.*, **144**: 165–92.

Pomeroy, J. W. and Schmidt, R. A. (1993). The use of fractal geometry in modeling intercepted snow accumulation and sublimation. *Proc. Joint Eastern and Western Snow Conf., Quebec City, P. Q*: 1–10, 50th Eastern and 61st Western Snow Conference.

Pomeroy, J. W., *et al.* (1998). Coupled modelling of forest snow interception and sublimation. *Hydrol. Proc.*, **12**: 2317–37.

Porter, S. C. (2000). Snowline depression in the tropics during the last glaciation. *Quart. Sci. Rev.*, **20** (10): 1067–91.

Portis, D. H., *et al.* (2001). Seasonality of the North Atlantic oscillation. *J. Climate*, **14**: 2069–78.

Post, A. (1969). Distribution of surging glaciers in western North America. *J. Glaciol.*, **8** (53): 229–40.

Post, A. (1975). Preliminary hydrography and historical terminal changes of Columbia Glacier. US Geological Survey, Hydrologic Investigations Atlas HA-559.

Post, A. (2005). EPIC. Austin Post collection (Images online), https://archive .gi.alaska.edu/austin-post-collection.

Post, A. and LaChapelle, E. R. (2000). *Glacier ice* (2nd ed.). Seattle, WA: University of Washington Press.

Post, A. and Meier, M. F. (1980). A preliminary inventory of Alaskan glaciers. World Glacier Inventory. Proceedings of the Riederalp Workshop, September 1978. IAHS Publ., No.126, pp. 45–7.

Post, A. and Motyka, R. (1995). Taku and Le Conte Glaciers, Alaska: Calving speed control of late-Holocene asynchronous advances and retreats. *Phys. Geogr.*, **16**: 59–82.

Pour, H. K., *et al.* (2017). Improvement of lake ice thickness retrieval from MODIS satellite data using a thermodynamic model. *IEEE Trans. on Geosci. and Remote Sensing*, **99**: 1–10, doi:10.1109/ TGRS.2017.2718533.

Preußer, A., Heinemann, G., Willmes, S., and Paul, S., (2016). Circumpolar polynya regions and ice production in the Arctic: results from MODIS thermal infrared imagery from 2002/2003 to 2014/2015 with a regional focus on the Laptev Sea. *Cryosphere*, **10**: 3021–42, www.the-cryosphere.net/10/3021/2016/, doi:10.5194/ tc-10-3021-2016.

Price, A. G. (1988). Prediction of snowmelt rates in a deciduous forest. *J. Hydrol.*, **101**: 145–57.

Price, A. G. and Dunne, T. (1976). Energy balance computations of snowmelt in a sub-Arctic area. *Water Resour. Res.*, **12**: 686–94, doi:10.1029/WR012i004p00686.

Priscu, J. C., *et al.* (2008). Antarctic subglacial water: Origin, evolution, and ecology. In Vincent, W. F. and Laybourn-Parry, J. (eds.), *Polar lakes and rivers: limnology of Arctic and Antarctic aquatic ecosystems*, Oxford: Oxford University Press. pp. 119–35.

Pritchard, H. (2009). State of the cryosphere: Glaciers and ice sheets. E-Book. Special Publ, 60. Washington, DC, AGU.

Pritchard, H. D., *et al.* (2009). Extensive dynamic thinning on the margins of the Greenland and Antarctic ice sheets. *Nature*, **461**: 971–5, doi:10.1038/nature08471.

Pritchard, H. D., *et al.* (2012). Antarctic ice-sheet loss driven by basal melting of ice shelves, *Nature*, Macmillan Publishers, Vol **484**, 26, doi:10.1038/nature10968.

Pritchard, R. S. (ed.) (1980). *Sea ice processes and models*. Seattle, WA: University of Washington Press. 474 pp.

Pritchard, R. S., Coon, M., McPhee, M. G., and Leavitt, E. (1977). *Winter ice dynamics in the nearshore Beaufort Sea. AIDJEX Bull.37*. Seattle, WA: Applied Physics Lab, University of Washington. 37–93 pp.

Prowse, T. D. (2000). *River-ice ecology*. Saskatoon, Canada: Environment Canada. 64 pp.

Prowse, T. D. (2005). River-ice hydrology. In Anderson, M. G. (ed.), *Encyclopedia of hydrological sciences*, New York: John Wiley & Sons. Vol. 4. pp. 2657–78.

Prowse, T. D. and Beltaos, S. (2002). Climatic control of river-ice hydrology: A review. *Hydrol. Proc.*, **16**(4): 805–22.

Prowse, T. D. and Bonsal, B. R. (2004). Historical trends in river-ice breakup: A review. *Nordic Hydrol.*, 35: 281–93.

Prowse, T. D., *et al.* (2002). Trends in river-ice breakup and related controls. In Squire, V. and Langhorne, P. (eds.), *Proc. 16th IAHR International Symposium on Ice*. New Zealand: Department of Physics, University of Otago, Dunedin, 3, pp. 64–71.

Prowse, T. D., *et al.* (2007). River-ice breakup/freeze-up: a review of climatic drivers, historical trends and future predictions. *Ann. Glaciol.*, **46**: 443–51.

Prowse, T. D., Shrestha, R., Bonsal, B., and Dibike, Y. (2010). Changing spring air-temperature gradients along large northern rivers: Implications for severity of river-ice floods. *Geophys. Res. Lett.*, **37**: L19706, doi:10.1029/2010GL044878.

Pugh, H. L. D. and Price, W. I. J. (1954). Snow drifting and the use of snow fences. *Polar Rec.*, **7**: 4–23.

Pulliainen, J., Koskinen, J., and Hallikainen, M. (2001). Compensation of forest canopy effects in the estimation of snow covered area from SAR data. *IEEE Geosci. Remote Sens. Symp.*, **2**: 813–15, doi:10.1109/IGARSS.2001.976645.

Punsalmaa, B. and Nyamsuren, B. (2002). Climate change impacts on ice regime of the rivers in Mongolia. In: Ice in the environment, Vol. 1, Squire, V. and Langhorne, P. (eds.). Proc. 16th IAHR Internat. Sympos. on Ice, Int. Assoc. Hydraulic Eng. Rea., Dunedin, New Zealand. pp. 122–6.

Purdie, H. (2013). Glacier retreat and tourism: Insights from New Zealand. *Moun. Res. and Dev.*, **33**(4): 463–72, https://doi.org/10.1659/MRD-JOURNAL-D-12-00073.1.

Purves, R., *et al.* (2003). Nearest neighbours for avalanche forecasting in Scotland – development. verification and optimisation of a model. *Cold Reg. Sci. Technol.*, **37**: 343–55.

Putkonen, J. (2008). What dictates the occurrence of zero curtain effect? In Kane, D. L. and Hinkel, K. M. (eds.), *Ninth International Conference on Permafrost, 29 June–3 July 2008, University of Alaska Fairbanks. Proceedings, Vol. 2*. Fairbanks, AK: University of Alaska, pp. 1451–55.

Pyles, R. D., Weare, B. C., and Pawu, K. T. (2000). The UCD advanced canopy-atmosphere-soil algorithm: Comparisons with observations from different climate and vegetation regimes. *Quart. J. Roy. Met. Soc.*, **126** (569): 2951–80, doi:10.1002/qj.49712656917.

Qin, D.-H. (1999). *Map of glaciers resources in the Himalayas*. Beijing: Science Press.

Qin, D.-H. (2002). *Glacier inventory of China (maps)*. Xi'an, China: Xi'an Cartographic Publishing House.

Qiu, G. Q., *et al.* (2000). *The map of geocryological regionalization and classification in China (1:10,000,000)*. Xian, China: Xian Press. In Chinese and English.

Quincey, D. J. and Glasser, N. F. (2009). Morphological and ice-dynamical changes on the Tasman Glacier, New Zealand, 1990–2007. *Global Planet. Change*, **68**: 185–97.

Quincey, D. J. and Luckman, A. (2009). Progress in satellite remote sensing of ice sheets. *Progr. Phys. Geog.*, **33**: 546–67.

Rabatel, A., Dedieu, J. P., and Vincent, C. (2005). Using remote-sensing data to determine equilibrium-line altitude and mass-balance time series: validation on three French glaciers, 1994–2002. *J. Glaciol.*, **51**: 539–46.

Rabenstein, L. (2010). Sea-ice volume production in Laptev Sea polynya from January to April 2008. Paper 57A147. Proceedings, Tromso Sea Ice Symposium. Int. Glaciol. Soc.

Rabenstein, L., *et al.* (2010). Thickness and surface-properties of different sea-ice regimes within the Arctic Trans Polar Drift: Data from summers 2001, 2004 and 2007. *J. Geophys. Res.*, 115: C12059, 18.

Rachold, V., *et al.* (2007). Near-shore Arctic subsea permafrost in transition. *EOS*, **88** (13): 149–56.

Racoviteanu, A. E., *et al.* (2008a). Decadal changes in glacier parameters in the Cordillera Blanca, Peru, derived from remote sensing. *J. Glaciol.*, 54 (186): 499–510.

Racoviteanu, A., Williams, N. W., and Barry, R. G. (2008b). Optical remote sensing of glacier characteristics: A review with focus on the Himalaya. *Sensors*, 8: 3355–83.

Racoviteanu, A. E., *et al.* (2009). Challenges and recommendations in mapping of glacier parameters from space: Results of the 2008 Global Land Ice Measurements from Space (GLIMS) workshop, Boulder, Colorado. *Annals Glaciol.*, **50** (53): 17.

Radić, V. and Hock, R. (2010). Regional and global volumes of glaciers derived from statistical upscaling of glacier inventory data. *J. Geophys. Res.*, **115**: F01010, doi:10.1029/2009JF001373.

Radic, V. and Hock, R. (2011). Regionally differentiated contribution of mountain glaciers and ice caps to future sea-level rise. *Nature Geosci.*, **4**: 91–4.

Radić, V., Hock, R., and Oerlemans, J. (2008). Analysis of scaling methods in deriving future volume evolution of valley glaciers. *J. Glaciol.*, **54** (187): 601–12.

Radić, V., Bliss, A., Beedlow, A. C., Hock, R., Miles, E., and Cogley, J. G. (2014). Regional and global projections of twenty-first century glacier mass changes in response to climate scenarios from global climate models. *Clim. Dyn.*, **42**: 37–58, doi:10.1007/s00382-013-1719-7.

Radok, U. (1997). The International Commission on Snow and Ice (ICSI) and its precursors, 1894–1994. *J. Hydrol. Sci.*, **42**: 131–40.

Ragner, C. L. (2000). Northern Sea Route cargo flows and infrastructure – Present state and future potential. FNI Report 13/2000. Lysaker, Norway: Fridtjof Nansen Institute. 130 pp.

Raina, V. K. and Srivastava, D. (2008). *Glacier atlas of India*. Bangalore: Geological Society of India. 315 pp.

Räisänen, J. (2008). Warmer climate: Less or more snow? *Clim. Dyn.*, **30**: 307–19.

Ran, Y., Li, X., and Cheng, G. (2018). Climate warming over the past half century has led to thermal degradation of permafrost on the Qinghai–Tibet Plateau. *The Cryosphere*, **12**: 595–608.

Rango, A. (1993). Snow hydrology processes and remote sensing. *Hydrologic Processes*, **7**: 121–38.

Ramillien, G., *et al.* (2006). Interannual variations of the mass balance of the Antarctic and Greenland ice sheets from GRACE. *Global Planet. Change*, **53**: 198–208.

Raper, S. C. B. and Braithwaite, R. J. (2009). Glacier volume response time and its links to climate and based on a conceptual model of glacier hypsometry. *Cryosphere*, **3**: 183–94.

Rasmussen, R., *et al.* (2013). How well are we measuring snow. *Bull. Am. Meteor. Soc.*, **93** (6): 811–29, doi:10.1175/BAMS-D-11-00052.1.

Rastner, P., Bolch, T., Mölg, N., Machguth, H., and Paul, F. (2012). The first complete glacier inventory for entire Greenland. *Cryosphere*, **6**: 1483–95.

Raup, B., *et al.* (2007). The GLIMS geospatial glacier database: A new tool for studying glacier change. *Global Planet. Change*, **56** (1–2): 101–10.

Raymo, M. E., Lieseck, L. E., and Nisancioglu, K. H. (2006). Plio-Pleistocen ice volume: Antarctic climate and the global $\delta18O$ record. *Science*, **313** (3786): 492–5.

Raymo, M. E. and Huybers, P. (2008). Unlocking the mysteries of the ice ages. *Nature*, **251**: 284–5.

Raymond, A. and Metz, C. (2004). Ice and its consequences: Glaciation in the Late Ordovician, Late Devonian, Pennsylvanian-Permian, and Cenozoic compared. *J. Geol.*, **112**: 665–70.

Raymond, C. F. (1987). How do glaciers surge? A review. *J. Geophys. Res.*, **92** (B9): 9,121–34.

Rea, B. R. (2009). Defining modern day Area-Altitude Balance Ratios (AABRs) and their use in glacier-climate reconstructions. *Quat. Sci. Rev.*, **28** (3–4): 237–48.

Reeh, N. (1968). On the calving of ice from floating glaciers and ice shelves. *J. Glaciol.*, **7**: 218–32.

Reeh, N. (1994). Calving from Greenland glaciers: Observations, balance estimates of calving rates, calving laws. In Reeh, N. (ed.), *Workshop on the calving rate of West Greenland glaciers in response to climate change*, Copenhagen: Danish Polar Center, pp. 85–102.

Rees, G. H. and Collins, D. N. (2006). Regional differences in responses of flow in glacier-fed Himalayan rivers. *Hydrol. Processes*, **20**: 2157–67.

Rees, W. G. (2006). *Remote sensing of snow and ice*. London: Taylor and Francis. 312 pp.

Reid, H. F. (1896a). Glacier Bay and its glaciers. U.S. Geological Survey, 16th Annual Report, Part 1, pp. 421–61.

Reid, H. F. (1896b). The mechanics of glaciers. *J. Geol.*, **4**: 912–28.

Regensburger, K. (1963). Comparative measurements on Fedtschenko glacier. In Ward, W. (ed.), Variations of the regime of existing glaciers. Symposium of Oberurgl, Int. Assoc. Sci. Hydrol., Publ. no. 58: pp. 57–61.

Reichle, R. H., *et al.* (2017). Assessment of MERRA-2 land surface hydrology estimates. *J. Clim.*, **30** (8): 2937–60, doi:10.1175/jcli-d-16-0720.1.

Reimnitz, E., Dethleff, D., and Nürnberg, D. (1994). Contrasts in Arctic shelf sea-ice regimes and some implications: Beaufort Sea and Laptev Sea. *Mar. Geol.*, **119**: 215–25.

Reinwarth, O. and Stäblein, G. (1972). Die Kryosphre. Das Eis der Erde und seine Untersuchung. *Würzburger Geograph. Arbeit.*, **36**: 71.

Rémy, F. and Parouty, S. (2009). Antarctic ice sheet and radar altimetry: A review. *Remote Sensing*, **1**: 1212–39.

RGI Consortium, (2017). Randolph Glacier Inventory – A Dataset of Global Glacier Outlines: Version 6.0: Technical Report, Global Land Ice Measurements from Space. Digital Media, Colorado, USA, doi: https://doi.org/10.7265/N5-RGI-60.

Rhodes, J. J., Armstrong, R. L., and Warren, S. G. (1987). Mode of formation of "ablation hollows" controlled by dirt content of snow. *J. Glaciol.*, **33**: 135–9.

Richter-Menge, J. A., *et al.* (2006). Ice mass balance buoys: A tool for measuring and attributing change in the thickness of the Arctic ice cover. *Ann. Glaciol.*, **44**: 205–10.

Ricker, R., Girard-Ardhuin, F., Krumpen, T., and Lique, C. (2018). Satellite-derived sea ice export and its impact on Arctic ice mass balance. *Cryosphere*, **12**: 3017–32, https://doi.org/10.5194/tc-12-3017-2018

Rignot, E. J. (1998). Fast recession of a West Antarctic glacier. *Science*, **281**: 549–51.

Rignot, E. and Kanagaratnam, P. (2006). Changes in the velocity structure of the Greenland ice sheet. *Science*, **311**: 986–90.

Rignot, E., Koppes, M., and Velicogna, I. (2010). Rapid submarine melting of the calving faces of West Greenland glaciers. *Nature Geosci.*, **3** (3): 187–91.

Rignot, E., Rivera, A., and Casassa, G. (2003). Contribution of the Patagonia icefields of South America to area level rise. *Science*, **302** (5644): 434–7.

Rignot, E., *et al.* (1997). North and northeast Greenland ice discharge from satellite radar interferometry. *Science*, **276** (5314): 934–7.

Rignot, E., *et al.* (2004). Accelerated ice discharge from the Antarctic Peninsula following the collapse of Larsen B ice shelf. *Geophys. Res. Lett.*, 31: L18401, 4.

Rignot, E., *et al.* (2008). Mass balance of the Greenland ice sheet from 1958 to 2007. *Geophys. Res. Lett.*, **35**: L20502, 5.

Rignot, E., *et al.* (2008). Recent Antarctic ice mass loss from radar interferometry and regional climate modeling. *Nature Geosci.*, **1**: 106–10.

Rignot, E., *et al.* (2011). Ice flow of Antarctic ice sheet. *Science*, **333**(6048): 1427–30, https://doi.org/10.1126/science.1208336

Rigor, I. G. and Wallace, J. M. (2004). Variations in the age of Arctic sea ice and summer sea-ice extent. *Geophys. Res. Lett.*, 31: L09401.

Riley, J. P., Israelsen, E. K., and Eggleston, K. O. (1972). Some approaches to snowmelt prediction. *AISH Publ.*, **2** (107): 956–71.

Rinke, A., *et al.* (2003). A case study of the anomalous Arctic sea ice conditions during 1990: Insights from coupled and uncoupled regional climate model simulations. *J. Geophys. Res.*, **108**: 4275, 15.

Riseborough, D. (2007). The effect of transient conditions on an equilibrium permafrost–climate model. *Permafrost Periglac. Process.*, **18**: 21–32 (Erratum: 18 (2):215).

Riseborough, D., *et al.* (2008). Recent advances in permafrost modelling. *Permafrost Periglac. Process.*, **19** (2): 137–56.

Risebrobakken, B., *et al.* (2003). A high resolution study of Holocene paleoclimatic and paleoceanographic changes in the Nordic Seas. *Paleoceanog.*, **18**: 1017–31.

Rivera, A., *et al.* (2002). Use of remote sensing and field data to estimate the contribution of Chilean glaciers to the sea level rise. *Annals Glaciol.*, **34**: 367–72.

Robe, R. Q. (1980). Iceberg drift and deterioration. In Colbeck, S. C. (ed.), *Dynamics of snow and ice masses*, New York: Academic Press. pp. 211–59.

Roberts, M. J. (2005). Jökulhlaups: A reassessment of floodwater flow through glaciers. *Rev. Geophys.*, **43**: RG1002, 21.

Robin, G. de Q. (1975). Radio-echo sounding: Glaciological interpretations and applications. *J. Glaciol.*, **15** (73): 49–64.

Robinson, D. A. (2008). Northern Hemisphere continental snow cover extent: A 2008 update. unpublished report, Rutgers University.

Robinson, D. A. and Dewey, K. F. (1990). Recent secular variations in the extent of Northern Hemisphere snow cover. *Geophys. Res. Lett.*, **17**: 1557–60.

Robinson, D. A., Frei, A., and Serreze, M. C. (1995). Recent variations and regional relationships in Northern Hemisphere snow cover. *Ann. Glaciol.*, **21**: 71–6.

Robinson, D. A., *et al.* (1992). Large-scale patterns and variability of snow melt and parameterized surface albedo in the Arctic Basin. *J. Climate*, **5** (10): 1,109–19.

Rodrigues, J. (2008). The rapid decline of sea ice in the Russian Arctic. *Cold Regions Sci. Technol.*, **54**: 124–42.

Rodrigues, J. (2009). The increase in the length of the ice-free season in the Arctic. *Cold Regions Sci. Technol.*, **59**: 78–101.

Roe, G. H., Baker, M. B., and Herla, F. (2016). Centennial glacier retreat as categorical evidence of regional climate change. *Nature Geosci.*, **10**: 95–99, doi:10.1038/ngeo2863.

Rokaya, P., Budhathoki, S., and Lindenschmidt, K. E. (2018). Trends in the Timing and Magnitude of Ice-Jam Floods in Canada, Scientific Reports, 8:5834, doi:10.1038/s41598-018-24057-z.

Romanov, I. P. (1995). *Atlas of ice and snow of the Arctic Basin and Siberian shelf seas* (A. Tunik, translator and editor). 2nd edn. Paramus, NJ: Backbone Publishing Company. 176 pp.

Romanovskii, N. N., Afanaseo, V. E., and Koreisha, M. M. (1978). Long term dynamics of groundwater icings. Third International Conference on Permafrost. Edmonton, Alberta. Vol. 1. Part I: English translations of twenty-six of the Soviet papers. Ottawa: National Research Council of Canada. pp. 195–207.

Romanovskii, N. N., *et al.* (2004). Permafrost of the east Siberian Arctic shelf and coastal lowlands. *Quat. Sci,. Rev.*, **23** (11–13): 1359–69.

Romanovsky, V. E. and Osterkamp, T. E. (1997). Thawing of the active layer on the coastal plain of the Alaskan Arctic. *Permafrost Periglac. Processes*, **8**: 1–22.

Romanovsky, V. E., Smith, S. L., and Christiansen, H. H. (2010). Permafrost thermal state in the polar Northern Hemisphere during the International Polar Year 2007–2009: A synthesis. *Permafrost Periglac. Proc.*, **21**: 106–16.

Romanovsky, V. E., *et al.* (2007a). *Frozen ground. In Global outlook for ice and snow.* UNEP, Earthprint. 181–200 pp.

Romanovsky, V. E., *et al.* (2007b). Past and recent changes in air and permafrost temperatures in eastern Siberia. *Global Plamet. Change*, **56**: 399–413.

Romanovsky, V. E., *et al.* (2008a). Thermal state and fate of permafrost in Russia: First resuts of IPY. In Kane, D. L. and Hinkel, K. M. (eds.), *Ninth International Conference on Permafrost, 29 June– 3 July 2008, University of Alaska Fairbanks*, Proceedings, vol. **2**, Fairbanks, AK: University of Alaska. pp. 1511–18.

Romanovsky, V. E., *et al.* (2008b). Soil climate and frost heave along the permafrost/ ecological North American Arctic Transect. In Kane, D. L. and Hinkel, K. M. (eds.), *Ninth International Conference on Permafrost, 29 June–3 July 2008, University of Alaska Fairbanks*, Proceedings, vol. **2**, Fairbanks, AK: University of Alaska. pp. 1519–24.

Rooney, J. F., Jr. (1967). The urban snow hazard in the United States. *Geog. Rev.*, **57**: 538–59.

Ropelewski, C. F. (1989). Monitoring large-scale cryosphere/atmosphere interactions. *Adv. Space Res.*, **9**: 213–18.

Rosen, P. A., *et al.* (2000). Synthetic aperture radar interferometry. *Proc. IEEE*, **88** (3): 333–80.

Rosenfeld, S. and Grody, N. C. (2000). Metamorphic signature of snow revealed in SSM/I measurement. *IEEE Trans. Geosci. Remote Sensing*, **38**: 53–63.

Rosenthal, W. and Dozier, J. (1996). Automated mapping of montane snow cover at sub-pixel resolution from the Landsat Thematic Mapper. *Water Resour. Res.*, **32**: 115–30.

Roth, A. *et al.* (1993). Experiences with ERS-1 SAR compositional accuracy. IEEE Transactions Geoscience. Remote Sensing,

IGARRS Symposium, 1993. Tokyo, Japan, Proc. 3, 1450–52.

Röthlisberger, H. and Lang, H. (1987). Glacial Hydrology. Glacio-fluvial Sediment Transfer. In A. M. Gurnell and M. J. Clark (eds.). Chichester: John Wiley and Sons, pp. 207–84.

Rothrock, D. (1986). Ice thickness distribution – measurement and theory. In Untersteiner, N. (ed.), *The geophysics of sea ice*, New York: Plenum Press. pp. 551–75.

Rothrock, D. A. and Zhang, J. (2005). Arctic Ocean sea ice volume: What explains its recent depletion? *J. Geophys. Res.*, **110**: C01002, doi:10.1029/2004JC002282.

Rothrock, D. A., Yu, Y., and Maykut, G. A. (1999). Thinning of the Arctic sea-ice cover. *Geophys. Res. Lett.*, **26** (23): 3469–72.

Rothrock, D. A., Zhang, J., and Yu, Y. (2001). The arctic ice thickness anomaly of the 1990s: A consistent view from observations and models. *J. Geophys. Res.*, **108** (C3): 3083.

Rothrock, D. A., Zhang, J., and Yu, Y. (2003). The arctic ice thickness anomaly of the 1990s: A consistent view from observations and models. *J. Geophys. Res.*, **108** (C3): 3083, https://doi.org/10.1029 /2001JC001208.

Rothrock, D. A., Percival, D. B., and Wensnahan, M. (2008). The decline in arctic sea ice thickness: Separating the spatial, annual, and interannual variability in a quarter century of submarine data. *J. Geophys. Res.*, **113**: C05003. doi:10.1029/ 2007JC004252.

Rott, H. and Nagler, T. (1995). Intercomparison of snow retrieval algorithms by means of spaceborne microwave radiometry. In Choudhury, B. J. , Kerr, Y. H. , Njoki, E. G. , and Pampaloni, P. (eds.), *Passive microwave remote sensing of Land-Atmosphere Interactions*, Utrecht, The Netherlands: VSP. pp. 227–41.

Rott, H., Skvarca, P., and Nagler, T. (1996). Rapid collapse of the northern Larsen Ice Shelf. *Antarct. Sci.*, **271**: 788–92.

Rowe, C. M., Kuiven, K. C., and Jordan, R. (1995). Simulation of summer snowmelt on the Greenland ice sheet using a one-dimensional model. *J. Geophys. Res.*, **100**: 16,265–73.

Rowland, J. C., *et al.* (2010). Arctic landscapes in transition: responses to thawing permafrost. *EOS Trans.*, **31** (26): 220–30.

Roy, M., *et al.* (2004). Glacial stratigraphy and paleomagnetism of late Cenozoic deposits of the north-central United States. *Bull. Geol. Soc. Amer.*, **116**: 30–41.

Ruddiman, W. F. (2006). Orbital changes and climate. *Quat. Sci. Rev.*, **25**: 3092–112.

Ruddiman, W. F. (2010). A paleoclimatic enigma? *Science*, **328** (5980): 838–9.

Ruffieux, D., *et al.* (1995). Ice pack and lead surface energy budgets during LEADEX 1992. *J. Geophys. Res.*, **100** (C3): 4593–612.

Russell, W. E., Riggs, N. P., and Robe, R. Q. (1978). Local iceberg motion – a comparison of fluid and model studies. POAC '77. Fourth International Conference on Port and Ocean Engineering under Arctic Conditions. Newfoundland, Proceedings Vol.2: 784–98.

Rutt, I. C., *et al.* (2009). The glimmer community ice sheet model. *J. Geophys. Res.*, **114**: F02004, 22, doi:10.1029/2008JF001015.

Rutter, N., Essery, R. L. H., *et al.* (2009). Evaluation of forest snow processes models (SnowMIP2). *J. Geophys. Res.*, **114** (D6): D06111, doi:10.1029/2008JD011063.

Ryder, C. (1896). *Isforholdene I Nordhavet, 1877–1892*. Kobenhaven: Tidsskr. f. Sovaesen. 28 pp.

Sagarin, R. and Micheli, F. (2001). Climate change in nontraditional data sets. *Science*, **294**: 811.

Saito, K. and Cohen, J. (2003). The potential role of snow cover in forcing interannual variability of the major Northern Hemisphere mode. *Geophys. Res.*

*Lett.***30**(6): 1302, doi:10.1029/
2002GL016341

Salerno, F., *et al.* (2008). Glacier surface-area
changes in Sagarmatha national park,
Nepal, in the second half of the 20th
century, by comparison of historical maps.
J.Glaciol., **54**(187): 738–52.

Salm, B., Burkard, A., and Gubler, H. U.
(1990). Berechnung von Fliesslawinen; eine
Anleitung für Praktiker mit Beispielen.
Mitteilunge, Eidgenössischen Institutes für
Schnee und Lawinenforschung, No.47,
Davos.

Sangewar, C. V. and Shukla, S. P. (eds.).
(2009). *Inventory of the Himalayan
Glaciers: A contribution to the International
Hydrological Programme*. Special
Publication No. 34, Geological Survey of
India, New Delhi, India: Vedams eBooks
594 pp.

Sarnthein, M., *et al.* (2009). Mid-Pliocene
shifts in ocean overturning circulation and
the onset of Quaternary-style climates.
Clim. Past, **5**: 269–83.

Sasgen, I., *et al.* (2012). Timing and origin of
recent regional ice-mass loss in Greenland.
Earth Planet. Sci. Lett., **333**: 293–303, doi:
10.1016/j.epsl.2012.03.033

Sasgen I. *et al.* (2013). Antarctic ice-mass
balance 2003 to 2012: regional reanalysis of
GRACE satellite gravimetry measurements
with improved estimate of glacial-isostatic
adjustment based on GPS uplift rates. *The
Cryosphere*, **7**: 1499–512, https://doi.org/10
.5194/tc-7–1499-2013

Satterlund, D. R. and Haupt, H. F. (1967).
Snow catch by conifer crowns. *Water
Resour. Res.*, **3** (4): 1035–39.

Savage, S. B. (2001). Aspects of iceberg
deterioration and drift. In
Geomorphological fluid mechanics (Lecture
notes in physics, volume 582). Berlin:
Springer. pp. 279–318.

Savko, N. F. (1973). *Prediction of naleds and
ways of regulating the naled process. Second
International Conference on Permafrost.*

USSR Contribution. Washington, DC:
National Research Council, pp. 403–08.

Savoie, M. H., *et al.* (2009). Atmospheric
corrections for improved satellite passive
microwave snow cover retrievals over the
Tibet Plateau. *Remote Sensing Environ.*,
113: 2661–669.

Sawyer, C. F. and Butler, D. R. (2006).
A chronology of high-magnitude snow
avalanches reconstructed from archived
newspapers. *Disaster Prevention
Management*, **15** (2): 313–24.

Scambos, T. A. and Bindschadler, R. (1993).
Complex ice stream flow revealed by
sequential satellite imagery. *Ann. Glaciol.*,
17: 177–82.

Scambos, T. A. and Fahnestock, M. A. (1998).
Improving digital elevation models over ice
sheets using AVHRR-based
photoclinometry. *J.Glaciol.*, **44**: 97–103.

Scambos, T., Hulbe, C., and Fahnestock, M.
(2003). Climate-induced ice shelf
disintegration in the Antarctic Peninsula.
In Domack, E., *et al.* (eds.), *Antarctic
Peninsula climate variability: Historical and
paleoenvironmental perspectives.*
Washington, DC: American Geophysical
Union: pp. 79–92.

Scambos, T. A., *et al.* (1992). Application of
image cross-correlation to the
measurement of glacier velocity using
satellite image data. *Remote Sens, Environ.*,
42: 177–86.

Scambos, T. A., *et al.* (2000). The link between
climate warming and breakup of ice shelves
in the Antarctic Peninsula. *J. Glaciol.*,
46 (154): 116–30.

Scambos, T. A., *et al.* (2004). Glacier
acceleration and thinning after ice shelf
collapse in the Larsen B embayment,
Antarctica. *Geophys. Res.Lett.*, **31** (18):
L18402.

Scambos, T. A., *et al.* (2006). Impact of
megadunes and glaze areas on estimates of
East Antarctic mass balance and
accumulation rate change. EOS, Trans.

American Geophysical Union, Fall Meeting Suppl., Abstr. #C11A-1130.

Scambos, T., *et al.* (2007). MODIS-based Mosaic of Antarctica (MOA) data sets: Continent-wide surface morphology and snow grain size. *Remote Sens. Environ.*, **111**: 242–57.

Scambos, T., *et al.* (2008). Calving and ice-shelf breakup processes investigated by proxy: Antarctic iceberg evolution during northward drift. *J. Glaciol.*, **54** (187): 579–91.

Scambos, T., *et al.* (2009). Ice shelf disintegration by plate bending and hydro-fracture: Satellite observations & model results of 2008 Wilkins ice shelf break-ups. *Earth Planet. Sc. Lett.*, **280**: 51–60, doi:10.1016/j.epsl.2008.12.027

Scanlon, B. R., *et al.* (2018). *Global models underestimate large decadal declining and rising water storage trends relative to GRACE satellite data*, PNAS, www.pnas.org/cgi/doi/10.1073/pnas.1704665115

Scarchilli, C., Frezzotti, M. and Grigioni, P. (2010). Extraordinary blowing snow transport events in East Antarctica. *Clim. Dynam.*, **34** (7–8): 1195–1206.

Schaefer, J., *et al.* (2009). High-frequency Holocene glacier fluctuations in New Zealand differ from the northern signature. *Science*, **324**: 622–25.

Schaefer, K., *et al.* (2011). Amount and timing of permafrost carbon release in response to climate warming. *Tellus B*, **63**: 165–80.

Schaefer, V. J., Klein, G. J., and de Quervain, M. R. (1954). The international classification for snow (with special reference to snow on the ground), 31, The Commission of Snow and Ice of the International Association of Hydrology, Associate Committee on soil and snow mechanics. Ottawa, Ont: National Research Council of Canada.

Schaerer, P. (1988). The yield of avalanche snow at Rogers Pass, British Columbia, Canada, *J. Glaciol.*, **34** (117): 1–6.

Schannwell, C., Cornford, S., Pollard, D., and Barrand, N. E. (2018). Dynamic response of Antarctic Peninsula Ice Sheet to potential collapse of Larsen C and George VI ice shelves. *Cryosphere*, **12** (7): 2307–26.

Schanda, E., (1983). Selection of microwave bands for global detection of snow. *Adv. Space Res.*, **3** (2): 303–08.

Scherler, D., Bookhagen, B., and Strecker, M., (2011). Spatially variable response of Himalayan glaciers to climate change affected by debris cover. *Nature Geoscience*, doi:10.1038/NGEO1068

Scherer, R. P., *et al.* (1998). Pleistocene collapse of the West Antarctic ice sheet. *Science*, **281**: 82–85.

Schiefer, E., Menounos, B., and Wheate, R. (2007). Recent volume loss of British Columbian glaciers, Canada. Geophys. *Res. Lett.*, **34** (16): L16503, doi:10.1029/2007GL030780.

Schirmer, M., Lehning, M., and Schweizer, J. (2009). Statistical forecasting of regional avalanche danger using simulated snow cover data. *J. Glaciol.*, **55** (103): 761–68.

Schlüchter, C. (1988). A non-classical summary of the Quaternary stratigraphy of the northern Alpine Foreland of Switzerland, Bull. *Soc. Neuchâtel. Géogr.*, **32/33**: 143–57.

Schmidt, D. F., Grise, K. M., and Pace, M. L. (2019). High-frequency climate oscillations drive ice-off variability for Northern Hemisphere lakes and rivers. *Clim. Change*, **152**: 517–32, doi:10.1007/s10584-018-2361-5

Schmitt, C., *et al.* (2005). Atlas of Antarctic sea ice drift. http://imkhp7.physik.uni-karlsruhe.de/~eisatlas/

Schneebeli, M., Laternser, M., and Amman, W. (1997). Destructive snow avalanches and climate change in the Swiss Alps. *Eclogau Geol. Helv.*, **90**: 457–61.

Schneebeli, M., *et al.* (1998). Measurement of density and wetness in snow using time-domain-reflectometry. *Annals of Glaciology*, **26**: 69–72.

Schneeberger, C., *et al.* (2003). Modelling changes in the mass balance of glaciers of the northern hemisphere for a transient 2×CO2 scenario. *J, Hydrol.*, **282L**: 145–63.

Schneider, M., *et al.* (2007). Glacier inventory of the Gran Campo Nevado icecap in the southern Andes and glacier changes observed during recent decades. *Global Planet. Change*, **59**: 87–100.

Schneider, W. and Budeus, G. (1997). Summary of the Northeast Water Polynya formation and development (Greenland Sea). *J. Mar. Systems*, **10**: 107–22.

Schneider von Deimling, T. , *et al.* (2006). How cold was the Last Glacial Maximum? *Geophys. Res. Lett.*, **33** (L14709): 5.

Schnell, R. C., *et al.* (1989). Lidar detection of leads in Arctic sea ice. *Nature*, **339**: 530–32.

Schoof, C. (2007a). Ice sheet grounding line dynamics: Steady states, stability, and hysteresis. *J. Geophys. Res.*, **112**: F03S28, doi:10.1029/2006JF000664.

Schoof, C. (2007b). Marine ice-sheet dynamics. Part 1. The case of rapid sliding. *J. Fluid Mech.*, **573**: 27–55.

Schoof, C. (2010). Ice-sheet acceleration driven by melt supply variability. *Nature*, **468**: 803–06.

Schrama *et al.* (2014). A mascon approach to assess ice sheet and glacier mass balances and their uncertainties from GRACE data. *J. Geophys. Res.-Solid Earth*, **119**: 6048–66.

Schuenemann, K. C., Cassano, J. J. and Finnis, J. (2009). Synoptic forcing of precipitation over Greenland: Climatology for 1961–99. *J. Hydromet.*, **10**: 60–78.

Schuur, E.A.G., McGuire, A.D., Schdel, C., Grosse, G., Harden, J.W., Hayes, D.J., Hugelius, G., Koven, C.D., Kuhry, P., Lawrence, D.M., 2015. Climate Change and the Permafrost Carbon Feedback, *Nature*, 520, 171–179.

Schuur, E. A. G., *et al.* (2008). Vulnerability of permafrost carbon to climate change: implications for the global carbon cycle. *BioScience*, **58** (8): L701–14, doi:10.1641/ B580807.

Schwarzacher, W. and Hunkins, K. (1961). Dredged gravels from the central Arctic Ocean. In Raasch, G. O. (ed.), *Geology of the Arctic*. Toronto: University of Toronto Press. pp. 666–77.

Schweiger, A. J. and Barry, R. G. (1989). Evaluation of algorithms for mapping snow cover in the Federal Republic of Germany using passive microwave data. *Erdkunde*, **43**: 85–94.

Schweiger, A. J., Armstrong, R. , and Barry, R. G. (1987). Snow cover parameter retrieval from various data sources in the Federal Republic of Germany. In B. E. Goodison, R. G. Barry and J. Dozier (eds.), *Large-Scale Effects of Seasonal Snow Cover. IAHS Publ. No. 166*, Wallingford, UK: IAHS Press, 353–364.

Schweiger, A. J., *et al.* (2008). Did unusually sunny skies help drive the record sea ice minimum of 2007? *Geophys. Res. Lett.*, **35**: L10503, doi:10.1029/2008GL033463.

Schweizer, J. (1998). Laboratory experiments on shear failure of snow. *Ann. Glaciol.*, **26**: 97–102.

Schweizer, J. (2008). Snow avalanche formation and dynamics. *Cold Regions Sci, Technol.* **54**: 153–54.

Schweizer, J., *et al.* (2008). Review of spatial variability of snowpack properties and its importance for avalanche formation. *Cold Reg. Sci. Technol.*, **51**: 253–72.

Schweizer, J., Jamieson, J.B., and Schneebeli, M. (2003). Snow avalanche formation. *Rev. Geophys.*, **41** (4): 1016, 2.1–2.25, doi:10.1029/2002RG000123.

Schweizer, J., Mitterer, C., and Stoffel, L. (2009). On forecasting large and infrequent snow avalanches. *Colk Reg. Sci. Technol.*, **59**: 234–41.

Schytt, V. (1954). Glaciology in Queen Maud Land: Work of the Norwegian-British-Swedish Antarctic Expedition. *Geog. Rev.*, **44**: 70–87.

Scholander, P. F. and Nutt, D. C. (1960). Bubble pressure in icebergs, *J, Glaciol.*, **3**: 671–78.

Schubert, C. (1992). The glaciers of the Sierra Nevada de Merida (Venezuela) : A photographic comparison of recent deglaciation. *Erdkunde*, **46**: 58–64.

Schuur, E. A. G., McGuire, A. D., Schdel, C., Grosse, G., Harden, J. W., Hayes, D. J., Hugelius, G., Koven, C. D., Kuhry, P., and Lawrence, D. M. (2015). Climate change and the permafrost carbon feedback. *Nature*, **520**: 171–79.

Scoresby, W. Jr. (1820). *An account of the Arctic regions with a history and description of the northern whale-fishery*. Republished 1969. NewYork: Augustus M. Kelley. **2** vols. 551 pp. and 574 pp. (vol. 1, pp. 225–33, 238–41).

Scourse, J. D., *et al.* (2009). Growth, dynamics and deglaciation of the last British–Irish ice sheet: the deep-sea ice-rafted detritus record. *Quat, Sci. Rev.*, **28** (27–28): 3066–84.

Screen, J. A. and Simmonds, I. (2010). The central role of diminishing sea ice in recent Arctic temperature amplification. *Nature*, **464**: 1334–37.

Sedláček, J. and Mysak, L. A. (2009). Sensitivity of sea ice to wind-stress and radiative forcing since 1500: a model study of the Little Ice Age and beyond. *Clim. Dynam.*, **32**: 817–31.

Segal, R. A., Lantz, T. C., and Kokelj, S. V. (2016). Acceleration of Thaw Slump Activity in Glaciated Landscapes of the Western Canadian Arctic. *Environmental Research Letters*, **11**: 034025.

Seibert, J. and Vis, M. J. P. (2012). Teaching hydrological modeling with a user-friendly catchment runoff-model software package. *Hydrol. Earth Syst. Sc.*, **16**: 3315–25.

Seidel, K. and Martinec, J. (2004). *Remote sensing in snow hydrology. Runoff modeling, Effect of climate change*. Chichester, UK: Springer/Praxis, 150 pp.

Seligman, G. (1936). *Snow structure and ski fields: Being an account of snow and ice forms met in nature and a study on avalanches and snowcraft. (Appendix on alpine weather by C.K.M. Douglas)*. London: MacMillan and Cox. 327 pp

Semakova, E., Myakov, S., and Armstrong, R. (2009). The current state of avalanche risk analysis and hazard mapping in Uzbekistan. Proceedings of the International Snow Science Workshop. Davos, Switzerland. Davos: Swiss Federal Institute for Snow and Avalanche Research SLF, 509–513

Semtner, A. J. (1976). A model for the thermodynamic growth of sea ice in numerical investigations of climate. *J. Phys. Oceanogr.*, **6**: 27–37.

Senneset, K. (ed.). (2000). *Proceedings, international workshop on permafrost engineering, longyearbyen, svalbard*. Norway: Norwegian University of Science and Technology. 327 pp.

Sergent, C., *et al.* (1993). Experimental investigation of optical snow properties, *Ann. Glaciol.*, **17**: 281–87.

Sergienko, O. V., Macayeal, D. R., and Hulbe, C. L (2008). Flexural-gravity wave phenomena on ice shelves. Fall Meeting, Amer. Geophys. Union, C31D0536S.

Serreze, M. C. and Barry, R. G. (2005). *The arctic climate system*. Cambridge, UK: Cambridge University Press, 385 pp.

Serreze, M. C. and Barry, R. G. (2011). Processes and impacts of Arctic amplification: A research synthesis. *Global and Planetary Change*, **77**: 85–96.

Serreze, M. C. and Stroeve, J. (2015). Arctic sea ice trends, variability and implications for seasonal ice forecasting. *Philos Trans*

A Math Phys Eng Sci., **373** (2045): 20140159, doi: 10.1098/rsta.2014.0159.

Serreze, M. C., Barry, R. G., and McLaren, A. S. (1989a). Seasonal variations in sea ice motion and effects on sea ice concentrations in the Canada Basin. *M.C. J. Geophys. Res.*, **94** (8): 10,955–70.

Serreze, M. C., McLaren, A. S., and Barry, R. G. (1989b). Seasonal variations of sea ice motion in the Transpolar Drift Stream. *M.C. Geophys. Res. Letters*, **16** (8): 811–14.

Serreze, M. C., et al. (1990). Sea ice concentration in the Canada Basin during 1988: Comparisons with other years and evidence of multiple forcing mechanisms. *M.C J. Geophys. Res.*, **95** (C12): 22,253–267.

Serreze, M. C., et al. (1993). Interannual variations in snow melt over Arctic sea ice and relationships to atmospheric forcing. *Annals of Glaciol.*, **17**: 327–31.

Serreze M. C., et al. (1999). Influence of snow vertical structure on hydrothermal regime characteristics of the western United States snowpack from snowpack telemetry (SNOTEL). *Water Resour. Res.*, **35**: 2145–60.

Serreze, M. C, et al. (2003). A record minimum in Arctic sea ice extent and area in 2002. *Geophys. Res. Lett.*, **30**(3)1110: 10.1–10.4, doi: 10.1029/2002GL016407.

Serreze, M. C., et al. (2009). The emergence of surface-based Arctic amplification. *Cryosphere*, **3**: 11–19, doi:10.5194/tc-3-11-2009.

Serreze, M. C., and A. P. Barrett, 2010: Characteristics of the Beaufort Sea High. Journal of Climate, 24, 159–182.

Serson, H. (1979). Mass balance of the Ward Hunt ice rise and ice shelf: An 18-year record. Tech. Mem. 79–4 *Defense Research Establishment*, Canada: Ottawa, 14 pp.

Severinghaus, J. P. (2009). Southern see-saw seen. *Nature*, **457**: 1093–94.

Sevestre, H. (2017). Surging glaciers. *Science, AAAS 1*, **358**(6367). pp. 11–20.

Sexstone et al. (2018). Snow sublimation in mountain environments and its sensitivity to forest disturbance and climate warming. *Water Resources Research,* https://doi.org/10.1002/2017WR021172

Shahgedanova, M. et al. (2010). Glacier shrinkage and climatic change in the Russian Altai from the mid-20th century: An assessment using remote sensing and PRECIS regional climate model. *J. Geophys. Res.* **115** (D16107):1–12, doi 2009JD012976.

Shakova, N. et al. (2010). Extensive methane venting to the atmosphere from sediments of the East Siberian Arctic Shelf. *Science* **327** (597):1246–50.

Shakun, J. D. and Carlson, A. E. (2010). A global perspective on Last Glacial Maximum to Holocene climate change. *Quat. Sci. Rev.*, **29** (15–16): 1674–90.

Shangguan, D., et al. (2006). Monitoring the glacier changes in the Muztag Ata and Konggur mountains, east Pamirs, based on Chinese Glacier Inventory and recent satellite imagery. *Annals Glaciol.*, **43**: 79–85.

Sharma, S., et al. (2013). Influences of local weather, large-scale climatic drivers, and the ca. 11 year solar cycle on lake ice breakup dates; 1905–2004. *Clim. Change*, **118**: 857–870, doi.org/10.1007/s10584-012-0670-7

Sharma, S., et al., (2016). Direct observations of ice seasonality reveal changes in climate over the past 320-570 years. *Sci. Rep.*, **6**: 25061. doi:10.1038/srep25061

Sharp, M. and Wang, L.-B. (2009). A five-year record of summer melt on Eurasian Arctic ice caps. *J. Clim.*, **22**: 133–45.

Sharp, R. P. (1954). Glacier flow: A review. *Bull. Geol. Soc. Amer.*, **65**: 821–38.

Shchetinnikov, A. S. (1998). *Morfologiya i rezhim lednikov Pamiro-Alaya (Morphology and regime of the Pamir-Alai glaciers)*. Tashkent: (SANIGMI) Central Asia Hydro-Meteorological Institute. 219 pp. (in Russian).

Shea, J. M., Moore, R. D., and Stahl, K. (2009). Derivation of melt factors from glacier mass-balance records in western Canada. *J. Glaciol.*, **55** (189): 123–30.

Shen, H. T. (2010). Mathematical modeling of river ice processes. *Cold Reg. Sci. Technol.*, **62**: 3–13.

Shepherd, A. and Wingham, D. (2007). Recent sea-level contributions of the Antarctic and Greenland ice sheets. *Science*, **315** (5818): 1529–32.

Shepherd, A, *et al.* (2007). Mass balance of Devon Island ice cap, Canadian Arctic, *Annals Glaciol.* **46**: 249–54.

Shepherd, A., *et al.* (2010). Recent loss of floating ice and the consequent sea level contribution. *Geophys. Res. Lett.*, **37** (L13503): 5.

Shepherd, A., *et al.* (2012). A reconciled estimate of ice-sheet mass balance. *Science*, **338**: 1183–89.

Shi, J. and Dozier, J. (2000). Estimation of snow water equivalent using SIR-C/X-SAR, Part I: Inferring snow density and subsurface properties. *IEEE Trans. Geosci. Remote Sensing*, **38** (6): L 2465–74.

Shi, X., *et al.* (2009). SnowSTAR2002 transect reconstruction using a multilayered energy and mass balance snow model, *J. Hydromet.*, **10** (5): 1151–67.

Shi, Y.-F. (ed.-in-chief). (2008a). *Glaciers and related environments in China*. Beijing: Science Press. 539 pp.

Shi, Y.-F. (2008b). *Collection of the studies on glaciology, climate and environmental change in China*. Beijing: China Meteorological Press. 850 pp.

Shi, Y.-F. *et al.* (2008c). Impact of global warming on glaciers and related water resources in China. In Shi, Y.-F. *et al.* (eds.), *Glaciers and related environments in China*. Beijing: Science Press. pp. 507–28.

Shi, Y.-F., Zheng, B.-X., and Su, Z. (2008b). Quaternary glaciations, glacial and interglacial cycles and environmental changes. In: Shi, Y.-F. (ed.-in-chief). *Glaciers and related environments in China Vol.2*. Beijing: Science Press. pp. 436–506.

Shields, G. A. (2008). Palaeoclimate: Marinoan meltdown. *Nature Geosci.*, **1**: 351–53.

Shields, G. A. (2009). Palaeoclimate: Marinoan meltdown. *Nature Geosci.*, **1** (6): 351–53.

Shiklomanov, N. I. (2005). From exploration to systematic investigation: development of geocryology in 19th- and early–20th-century Russia. *Phys. Geog.*, **26**: 249–63.

Shiklomanov, N. I. and Nelson, F. E. (2002). Active-layer mapping at regional scales: a 13-year spatial time series for the Kuparuk region, north-central Alaska. *Permafrost Periglac. Process.*, **13** (3): 219–30.

Shiklomanov, N. I., *et al.* (2010). Decadal variations of active-layer thickness in moisture-controlled landscapes, Barrow, Alaska. *J. Geophys. Res.*, **115**: G00I04.

Shiklomanov, N. I., Streletskiy, D. A., and Nelson, F. E., (2012). Northern Hemisphere Component of the Global Circumpolar Active Layer Monitoring (CALM) Program, 377–382, *Proceedings of the Tenth International Conference on Permafrost*, Salekhard, Russia, June 25–29, 2012.

Shil'nikov, V. L. (1965). Volume and number of icebergs in the Antarctic (from 44° to 66° E). *Soviet Antarct. Exped. Info. Bull*, [translation], **3**: 21–6.

Shine, K. P., Henderson-Sellers, A., and Barry, R. G. (1984). Albedo-climate feedback: the importance of cloud and cryosphere variability. K.P. In A. Berger and C. Nicolis, (eds.), *New Perspectives in Climate Modelling*. Amsterdam: Elsevier, 135–55.

Shokr, M. and Sinha, N. (2015). *Sea ice: physics and remote sensing*, 600 pp., AGU Geophysical Monograph Series, Wiley,

ISBN-13: 978–1119027898, ISBN-10: 9781119027898

Shook, K. (1993). Fractal geometry of snowpacks during ablation. Saskatoon, Sas., Canada: University of Saskatchewan: M.Sc. thesis. 178 pp.

Shook, K. (1995). Simulation of the ablation of prairie snow covers, Ph.D. dissertation, 189 pp., Univ. of Saskatchewan, Saskatoon, Sask., Canada.

Shook, K. and Gray, D. M. (1997). Synthesizing shallow seasonal snow covers. *Water Resour. Res.*, **33** (3): 419–26.

Shrestha, K. L. (2005). Impact of climate change on Himalayan glaciers. In Muhammed, A., Mirza, M. M. Q., and Stewart, B. A. (eds.), *Climate and water resources in South Asia: Vulnerability and adaptation. (APN, START) Pakistan,* Islamabad: Asiatics Agro Dev. International. pp. 44–57.

Shul 'tz, V. L. (ed.). (1962). *Lednik Fedchecnko.* (Fedchenko glacier) (in Russian). Tashkent: Izdat, Akad, Nauk, Uzbekskoi SSR. Vol. 1. 248 pp. Vol. 2 198 pp.

Shulyakovskii, L. G. (ed.). (1966). Manual of forecasting ice-formation for rivers and inland lakes. Manual of hydrological forecasting No. 4, Central Forecasting Institute of USSR: 1963, Translated from Russian, Israel Program for Scientific Translations, Jerusalem, Israel. 245pp.

Shum, C. K., Kou, C.-Y., and Guo, J.-Y. (2008). Role of Antarctic ice mass balance in present-day sea-level change. *Polar Sci.*, **2**: 149–61.

Shumskii, P. A. (1964). *Principles of structural glaciology. The petrography of freshwater ice as a method of glaciological investigation.* (trans. D.Kraus). New York: Dover Publ. Inc. 497 pp.

Shumskiy, P. A. (1969). Glaciation. In Tolstikov, E. (ed.), *Atlas of Antarctica.* Leningard: Gidrometeoizdat. pp. 367–400.

Sibrava, V. (2010). Quaternary climatic changes in the Alpine foreland – new observation and new conclusions. *Global and Planetary Change*, **72**(4): 374–80, doi:10.1016/j.gloplacha.2010.01.013

Sicart, J. E., *et al.* (2007). Glacier mass balance of tropical Glaciar Zongo, Bolivia, comparing hydrological and glaciological methods. *Global Planet. Change*, **59** (1–4): 27–36.

Sicart, J. E., Hock, R., and Six, D. (2008). Glacier melt, air temperature, and energy balance in different climates: The Bolivian Tropics, the French Alps, and northern Sweden. *J. Geophys. Res.*, **113** (D24113): 11.

Siegert, M. J. (1999). On the origin, nature and uses of Antarctic ice-sheet radio-echo layering. *Prohr. Phys. Geog.*, **23**: 159–79.

Siegert, M. J. (2005). Reviewing the origin of subglacial Lake Vostok and its sensitivity to ice sheet changes. *Progr. Phys. Geog.*, **29**: 156–70.

Sikonia, W. G. (1982). Finite-element glacier dynamics model applied to Columbia Glacier, Alaska. U.S. Geological Survey Profess.Paper 1258-B, 74 pp.

Sillmann, J., Kharin, V. V. , Zhang, X., Zwiers, F. W. , and Bronaugh, D. (2013). Climate extremes indices in the CMIP5 multimodel ensemble: Part 1. Model evaluation in the present climate. *J. Geophy. Res. Atm.* **118**(4): 1716–33. https://doi.org/10.1002/jgrd.50203

Simojoki, H. (1940). Über die Eisverhältnisse der Binnenseen Finnlands. *Ann. Acad. Sci. Fenn.*, **A52** (6): 1–194.

Singh, P. S. and Gan, T. Y. (2000). Retrieval of snow water equivalent using passive microwave brightness temperature data. *Remote Sensing Environ.*, **74**: 275–86.

Singh, P. S. and Gan, T. Y. (2005). Modeling snowpack surface temperature in the Canadian Prairies, Hydrol. *Processes*, **19**: 3481–500.

Singh, P. S., Gan, T. Y., and Gobena, A. K. (2005). A modified temperature index approach for snowmelt modeling in the Canadian Prairies using near surface soil and air temperature. *J. Hydrol. Engineering., ASCE*, **10** (5): 405–19.

Singh, P. S., Gan, T.Y., and Gobena, A. K. (2009). Evaluating a hierarchy of snowmelt models at a watershed in the Canadian Prairies. *J. Geophys. Res.*, **114**: D04109. doi:10.1029/2008JD010597.

Sinha, N. K. (1985). Confined strength and deformation of second-year columnar-grained sea ice in Mould Bay. *Proceedings Ocean, Offshore and Arctic Engineering OMAE'85*, **2**: 209–91.

Sinha, T., Cherkauer, K. A., and Mishra, V. (2010). Impacts of historic climate variability on seasonal soil frost in the midwestern United States. *J. Hydromet.*, **11**: 229–52.

Skyllingstad, E. D., Paulson, C. A., and Perovich, D. K. (2009). Simulation of melt pond evolution on level ice. *J, Geophys. Res.*, **114**: C12019, doi:10.1029/ 2009JC005363.

Slaymaker, O. and Kelly, R. E. J. (2006). *The cryosphere and global environmental change*. Oxford, UK: Wiley-Blackwell. 272 pp.

Slater, A. G., *et al.* (2001): The representation of snow in land-surface schemes: Results from PILPS 2(d). *J. Hydrometeorol.*, **2**: 7–25.

Slobbe, D. C., Lindenbergha, R. C., and Ditmar, P. (2008). Estimation of volume change rates of Greenland's ice sheet from ICESat data using overlapping footprints. *Remote Sens, Environ.*, **112** (12): 4204–13.

Slobbe, D. C., Ditmar, P., and Lindenbergh, R. C. (2009). Estimating the rates of mass change, ice volume change and snow volume change in Greenland from ICESat and GRACE data. *Geophys. J. Int.*, **176**: 95–106.

Smedsrud, L. H., Sorteberg, A., and Kloster, K. (2008). Recent and future changes of the Arctic sea-ice cover. *Geophys. Res. Lett.*, **35** (L20503): 4.

Smedsrud, L. H., *et al.* (2010). Fram Strait sea ice area export: 1950–2010. Abstract 379363 Oslo Science Conference, IPY.

Šmejkalová, T., Edwards, M. E., and Dash, J. (2016). Arctic lakes show strong decadal trend in earlier spring ice-out. *Sci. Rep.*, **6**: 1–8. https://doi.org/10.1038/srep38449

Smith, B. E., *et al.* (2009). An inventory of active subglacial lakes in Antarctica detected by ICESat (2003–2008). *J. Glaciol.*, **54** (192): 573–95.

Smith, L. C. (2000). Time-trends in Russian Arctic river ice formation and breakup: 1917– 1994. *Phys. Geog.*, **21**: 46–56

Smith, M. W. and Riseborough, D. W. (2002). Climate and the limits of permafrost: A zonal analysis. *Permafrost Periglac. Processes*, **13**: 1–15.

Smith, S., *et al.* (2009). *Active-layer characteristics and summer climatic indices, Mackenzie Valley, Northwest Territories.* Canada: Permafrost Periglac. Proc., **10**: 201–20.

Smith, S. D. (1993). Hindcasting iceberg drift using current profiles and winds. *Cold Regions Sci. Technol.*, **22**: 34–45.

Smith, S. D., Muench, R. D., and Pease, C. H. (1990). Polynyas and leads: an overview of physical processes and environment. *J. Geophys. Res.*, **95** (C6): 9461–79.

Smith, S. L. and Riseborough, D. W. (2010). Modelling the thermal response of permafrost terrain to right-of-way disturbance and climate warming/ Cold Reg. Sci. Technol., **60**: 92–103.

Smith, S. L., *et al.* (2010). Thermal state of permafrost in North America: A contribution to the International Polar Year. *Permafrsot Periglac. Proc.*, **21**: 117–35.

Sokolov, B. L. (1973). *Regime of naleds. Second International Conference on Permafrost. USSR Contribution.* Washington, DC: National Research Council. pp. 408–11.

Sokratov, S. A. and Barry, R. G. (2002). Intraseasonal variations in the thermoinsulation effect of snow cover on soil temperatures and energy balance. *J. Geophys. Res.*, **107** (D1): 4374, doi:10.1029/2002JD001595.

Soldatova, I. I. (1993). Secular variations in river breakup dates and their relations to climate changes. *Sovuet Met. Hydrol.*, **9**: 70–76.

Sole, A., *et al.* (2008). Testing hypotheses of the cause of peripheral thinning of the Greenland Ice Sheet: is land-terminating ice thinning at anomalously high rates? *The Cryosphere*, **2**: 205–18.

Solomina, O., Barry, R., and Bodnya, M. (2005). The retreat of Tien Shan glaciers (Kyrgyzstan) since the Little Ice Age estimated from aerial photographs, lichenometric and historical data. *Geograf. Annal.*, **86A** (2): 205–15.

Solomon, *et al.* (2010). Contributions of stratospheric water vapor to decadal changes in the rate of global warming. *Science*, **327**: 1219–23.

Soloviev, P. A. (1962). Alasnyy ryelev Centralnoi Yakutii i ego proiskhozdenie. (Alas relief in central Yakutia and its origin). In *Mnogoletnemerzlyye porody i soptstvuyushchie im yavlenie na territorii YASSR.* Moscow: Izdat. Akad Nauk, SSSR. Pp. 38–53.

Solow, A. R. (1991). The nonparametric analysis of point process data: The freezing history of Lake Konstanz. *J. Climate*, **4**: 116–19.

Soruco, A., *et al.* (2009). Glacier decline between 1963 and 2006 in the Cordillera Real, Bolivia. *Geophys. Res. Lett.*, **36**: L03502, doi:10.1029/2008GL036238.

Sou, T. and Flato, G. (2009). Sea ice in the Canadian Arctic Archipeago: Modeling the past (1959–2004) and the future (2041–60). *J. Climate*, **27** (8): 2181–97.

Soulis, E. D. (1975). Modelling of drift of nearby icebergs using wind and current measurements at a fixed station. *Canad. Soc. Petrol, Geol., Memoir*, **4**: 879–89.

Speerschneider, C. I. H. (1915). Om Isforholdene i danske Farvande i aeldre of nyere Tid: Aarene 690–1860. *Medd. Danske Met. Inst.*, Nr **2** (Copenhagen): 123.

Speerschneider, C. I. H. (1927). Summary to the state of the ice in arctic seas. In Nautisk Meteorologisk Aarbog, 1916, Danske Met. Inst., (Copenhagen), xxiii–xlvii.

Speerschneider, C. I. H. (1931). The state of the ice in Davis Strait, 1820–1930. *Meddd, Danske Met. Inst.*, **8** (Copenhagen): 53.

Speloläogisches Institut. (1926). *Die Eisriesenwelt im Tennengebirge (Salzburg).* Speloläog. Monogr. **6**, 145pp. Vienna.

Spielhagen, R. F., *et al.*, (2011). Enhanced modern heat transfer to the Arctic by warm Atlantic water. *Science*, **331**: 450, doi: 10.1126/science.1197397

Spötl, C. (2007). Ein neues Forschunsproject in der Eisriesenwelt (Werfen). Alpin Untertage., Berchesgarden November 9–11, 2007. Proceedings. Dtsch. Höhlen- und Karstforscher, Munich. p. 80.

Spreen, G., Aaleschke, L., and Heygster, G. (2008). Sea ice remote sensing using AMSR-E 89 GHz channels. *J. Geophys. Res.*, **113**: C02S03. doi:10.1029/2005JC003384.

SROCC (2019). *IPCC Special Report on the Ocean and Cryosphere in a Changing Climate* [In H.-O. Pörtner, D.C. Roberts, V. Masson-Delmotte, P. Zhai, M. Tignor, E. Poloczanska, K. Mintenbeck,

M. Nicolai, A. Okem, J. Petzold, B. Rama, N. Weyer (eds.)].

St. John, K. (2008). Cenozoic ice-rafting history of the central Arctic Ocean: Terrigenous sands on the Lomonosv Ridge. *Pakeoceanog.*, **23**: PA1S05.

Stafford, H. M. (1959). History of snow surveying in the West. Proc.27th Western Snow Conf., Reno, NV. pp. 1–12.

Statham, G., *et al.* (2017). A conceptual model of avalanche hazard. *Nat Hazards*, doi:10.1007/s11069-017–3070-5

Statham, G., *et al.* (2010). The North American public avalanche danger scale. In: Proceedings of the 2010 international snow science workshop, Squaw Valley, CA, pp 117–123.

Steele, M. and Flato, G. M. (2000). Sea ice growth and modeling: A survey. In Lewis, E. L. , *et al.* (eds.), *The freshwater budget of the Arctic*. Dordrecht: Kluwer, pp. 549–87.

Stefan, J. (1890).Über die Theorie der Eisbildung, inbesondere über die Eisbildung im Polarmeere. *Sitzber. Akad. Wiss. Wien*, **7**: 98.

Steffen, K. (1985). Warm water cells in the North Water, northern Baffin Bay during winter. *J. Geophys. Res.*, **90**: 9129–36.

Steffen, K. (1986). Ice conditions of an Arctic polynya: North Water in winter. *J. Glaciol.*, **32**: 383–90.

Steffen, K., *et al.* (2008). Rapid changes in glaciers and ice sheets and their impact on sea level. In Delworth, T. L., *et al.* (eds.), *Abrupt climate change, U.S. Climate Change Science Program and Subcommittee on Global Change Research*. Washington, DC: U.S. Geological Survey

Steig, E. J., *et al.* (1998). Synchronous climate changes in Antarctica and the North Atlantic. *Science*, **282** (5386): 92–96.

Steiner, D., Zumbühl, H., and Bauder, A. (2008). Two alpine glaciers over the past two centuries. In Orlove, B., Wiegandt, E., and Luckman, B. H. (eds.), *Darkening peaks. Glacier retreat, science and society*. Berkeley, CA: University of California Press. pp. 83–99.

Stern, W. (1926). Versuch einer elektrodynamischen Dickenmessung von Gletschereis. *Gerlands Beitr. Geophysik*, **3**: 292–333.

Stewart, I. T. (2009). Changes in snowpack and snowmelt runoff for key mountain regions. *Hydrol. Proc.*, **23**: 78–94.

Stickley, C., *et al.* (2009). Evidence for middle Eocene Arctic sea ice from diatoms and ice-rafted\debris. *Nature*, **460** (7253): 376, doi:10.1038/nature08163

Stiles, W. H. and Ulaby, F. T. (1980). The active and passive microwave response to snow parameters. 1. *Wetness, J. Geophys. Res.*, **85** (C2): 1037–44.

Stolarski, S., *et al.* (2010). Representing glaciers in a regional climate model. *Clim. Dynam.*, **34**: 27–46.

Stokes, C. R., Clark, C. D., and Storrar, R. (2009). Major changes in ice stream dynamics during deglaciation of the north-western margin of the Laurentide Ice Sheet. *Quatern. Sci. Rev.*, **28**: 721–38.

Stranneo, F., *et al.* (2010). Rapid circulation of warm subtropical waters in a major glacial fjord in East Greenland. *Nature Geoci.*, **3** (3): 182–86.

Strasser, U., *et al.* (2008). Is snow sublimation important in the alpine water balance? *Cryosphere*, **2**: 53–66.

Streletskiy, D. A., Shiklomanov, N. I., and Nelson, F. E. (2008). Thirteen years of observations at Alaskan CALM Sites: Long-term active layer and ground surface temperature trends. In Kane, D. L. and Hinkel, K. M. (eds.), *Proceedings, Ninth*

International Conference on Permafrost.
Fatrbanks, AK:University of Alaska,
Institute of Northern Engineering. pp.
1727–32.

Streletskiy, D. A. *et al.* (2017). Thaw
subsidence in undisturbed tundra
landscapes, barrow, alaska,
1962–2015. *Permafrost and Periglacial
Processes*, **28** (3): 566–72, doi:10.1002/
ppp.1918.

Stroeve, J. (2010). The accelerating decline of
Arctic sea ice. Paper A57206. Proceedings,
Tromso Symposium on Sea Ice.
Cambridge, UK: Int. Glaciol. Soc.

Stroeve, J. C. and Nolin, A. W. (2002).
Comparison of snow albedo from MISR
with ground-based observations on the
Greenland ice sheet. *IEEE Trans. Geosci.
Remote Sens.*, **40**: 1616–25.

Stroeve, J. and Notz, D. (2018). Changing
state of Arctic sea ice across all
seasons. *Environ. Res. Lett.*, **13**:
103001.

Stroeve, J., *et al.* (2006). Recent changes in the
Arctic melt season. *Ann. Glaciol.*, **44**:
367–74.

Stroeve, J., *et al.* (2007). Arctic sea ice
decline: Faster than forecast. *Geophys. Res.
Lett.*, **34**(9): L09501. doi: 10.1029/
2007GL029703

Stroeve, J. C., *et al.* (2012). Trends in Arctic
sea ice extent from CMIP5, CMIP3 and
observations. *Geophysical Research
Letters*, **39**: L16502, doi:10.1029/
2012GL052676

Sturm, M. (1992). Snow distribution and heat
flow in the taiga. *Arctic Alp. Res.*, **24** (2):
145–52.

Sturm, M. (2009). Field techniques for snow
observations on sea ice. In H. Eicken *et al.*
(eds.), *Field techniques for sea ice research*.
Fairbanks, AK: University of Alaska Press.
pp. 25–47.

Sturm, M. and Benson, C. S. (2004). Scales of
spatial heterogeneity for perennial and
seasonal snow layers. *Ann. Glaciol.*, **38**:
253–60.

Sturm, M., Holmgren, J., and Liston, G. E.
(1995). A seasonal snow cover classification
system for local to global application.
J. Climate, **8** (3): 1261–83.

Sturm, M. and Massom, R. A. (2010). Snow
and sea ice. In Thomas, D. N. and
Dieckmann, G. S. (eds.), *Sea ice*. 2nd edn.
Chichester, UK: Wiley-Blackwell. pp.
153–204.

Sturm, M., *et al.* (1997). Thermal conductivity
of seasonal snow. *J. Glaciol.*, **43** (143):
26–41.

Sturm, M., *et al.* (2010). Estimating snow
water equivalent using snow depth data and
climate classes. *J. Hydromet.*, **11** (6):
1380–94.

Sturm, M., *et al.* (2017). Using an option
pricing approach to evaluate strategic
decisions in a rapidly changing climate:
Black–Scholes and climate change.
Climatic Change, **140** (3): 437–49,
doi:10.1007/s10584-016-1860-5.

Sumgin, M. I. (1927). Vechnaya merzlota
pochvy v predelach SSSR. (Perennially
frozen soils in the USSR). Izdanie
Dal'ne-Vostochnoi Geofizicheskoi
Observatorii 23. Vladivostok. Sumgin,
M. I. (1941). Naledy i naledne bugry.
(Icings and icing mounds). *Priroda*, **30**
(1): 26–33.

Sundal, A. V., *et al.* (2009). Evolution of
supra-glacial lakes across the Greenland Ice
Sheet. *Remote Sensing Environ.*, **113** (10):
2164–71.

Sundal, A. V., *et al.* (2011). Melt-induced
speed-up of Greenland ice sheet offset by
efficient subglacial drainage. *Nature*, **469**:
521–24.

Surazakov, A. B., *et al.* (2007). Glacier changes
in the Siberian Altai Mountains, Ob river
basin, (1952–2006) estimated with high
resolution imagery. *Environ. Res. Lett.*, **2**
(045017): 7.

Surdu, C. M., *et al.* (2016). Evidence of recent changes in the ice regime of lakes in the Canadian High Arctic from space borne satellite observations. *Cryosphere*, **10**(3): 941–60, doi:10.5194/tc-10-941-2016

Suyetova, I. A. (1966). The dimensions of Antarctica. *Polar Re.*, **13**(84): 344–47.

Sverdrup, H. U. (1935). Scientific results of the Norwegian-Swedish Spitsbergen Expedition in 1934. *Part IV. Geograf. Annal.*, **17**: 145–66.

Swart, S. *et al.* 2018, Return of the Maud Rise polynya: climate litmus or sea ice anomaly? [in "State of the Climate in 2017"]. *Bull. Am. Meteorol. Soc.*, **99**: S188–89.

Swithinbank, C. W. M. (1969). Giant icebergs in the Weddell Sea. *Polar Rec.*, **13** (84): 344–47.

Taber, S. (1943). Perennially frozen ground in Alaska; its origin and history. *Geol. Soc. Amer. Bull.*, **54**: 1433–548.

Tabler, R. D. (1975) Predicting profiles of snow drifts in topographic catchments, Proceedings, 43rd Annual Western Snow Conference (Coronado, CA): 87–97.

Tait, A. (1998). Estimation of snow water equivalent using passive microwave radiation data. *Remote Sens. Environ.*, **64**: 286–91.

Tajika, E. (2003). Faint young sun and the carbon cycle: Implication for the Proterozoic global glaciation. *Earth Planet. Sci. Lett.*, **214**: 443–53.

Takaia, M., *et al.* (2009). Detection of snowmelt using spaceborne microwave radiometer data in Eurasia from 1979 to 2007. *IEEE Trans. Geosci, Rem. Sensing*, **47** (9): 2996–3007.

Takala, M., *et al.* (2011). Estimating northern hemisphere snow water equivalent for climate research through assimilation of space-borne radiometer data and ground-based measurements. *Remote Sensing of Env.*, **115**: 3517–29, doi:10.1016/j.rse.2011.08.014

Tammiksaar, E. (2001). *Materiale zur Kenntnis des unvergänglichen Boden-Eises in Sibirien*. Germany, Giessen: Universitätsbibliothek, University of Giessen. 234 pp.

Tangborn, W. V. (1984). Prediction of glacier derived runoff for hydroelectric development, *Geogr. Ann.*, **66A**: 257–65.

Tangborn, W. V. (1999). A mass balance model that uses low-altitude meteorological observations and the area-altitude distribution of a glacier. *Geogr. Ann. A*, **81** (4): 753–65.

Tao, W. (ed.). (2006). *Map of the glaciers, frozen ground and deserts in China Behei*: SinoMaps Press. (ISBN: 9787503139888).

Tapley, B. D., Bettadpur, S., Ries, J. C., Thompson, P. F., and Watkins, M. M. L., (2004). GRACE measurements of mass variability in the Earth system. *Science*, **305**: 503–05, https://doi.org/10.1126/science.1099192

Tarasov, L. and Peltier, W. R. (2007). Coevolution of continental ice cover and permafrost extent over the last glacial–interglacial cycle in North America. *J. Geophys. Res.*, **112** (F2): F02S08, doi:10.1029/2006JF000661.

Tarnocai, C. (2009). Arctic permafrost soils. In Margesin, R. (ed.), *Permafrost soils*. Berlin: Springer Verlag. pp. 3–16.

Tarr, R. S. and Martin, L. (1914). *Alaskan glacier studies*. Washington, DC: National Geographic Society. 498 pp.

Taylor, K. E., Stouffer, R. J., and Meehl, G. A. (2012). An overview of CMIP5 and the experiment design, *Bull. Am. Meteorol. Soc.*, **93**: 485–98, doi:10.1175/BAMS-D-11-00094.1.

Taylor, R. G., *et al.* (2006). Recent glacial recession in the Rwenzori Mountains of East Africa due to rising air temperature. *Geophys. Res. Lett.*, **33**: L10402, doi:10.1029/2006GL025962.

Techel, F., *et al.* (2016). Avalanche fatalities in the European Alps: long-term trends and statistics. *Geogr. Helv.*, **71**: 147–59, doi:10.5194/gh-71-147-2016

Tedesco, M. (2007). A new record in 2007 for melting in Greenland. *Eos, Transactions American Geophysical Union*, **88**: 39.

Tedesco, M. and Monaghan, A. J. (2009). An updated Antarctic melt record through 2009 and its linkages to high-latitude and tropical climate variability. *Geophys. Res. Lett.*, **36**: L18502.

Tedesco, M., *et al.* (2008). Extreme snowmelt in northern Greenland during summer 2008. *Eos*, **82** (41): 391.

Tedesco, M., *et al.* (2009). Pan arctic terrestrial snowmelt trends (1979–2008) from spaceborne passive microwave data and correlation with the Arctic Oscillation. *Geophys. Res. Lett.*, **36**: L21402, doi:10.1029/2009GL039672. Pan arctic terrestrial snowmelt trends (1979–2008) from spaceborne passive microwave data and correlation with the Arctic Oscillation

Tedesco, M., *et al.* (2017). Greenland Ice Sheet, *Arctic Report Card*: Update for 2017, Arctic Program.

Teel, S. (1994). Snow and ice activities to celebrate the Alaskan cold. 10 pp. (britton .disted.camosun.bc.ca/snow/snowbook .pdf)

Teich, M. *et al.* (2012). Snow and weather conditions associated with avalanche releases in forests: Rare situations with decreasing trends during the last 41 years. *Cold Regions Science and Technology*, **83–84**: 77–88, doi:10.1016/j.coldregions.2012.06.007

Tennant, C., Menounos, B., Wheate, R., and Clague, J. J. (2012). Area change of glaciers in the Canadian Rocky Mountains, 1919 to 2006. *Cryosphere*, **6**: 1541–52.

Terzago, S., Fratianni, S., and Cremonini, R., (2013). Winter precipitation in Western Italian Alps (1926–2010). *Meteorol. Atmos. Phys.*, **119**: 125–36, https://doi.org/10.1007/s00703-012-0231-7

Thackray, G. D., Owen, L. A., and Yi, C.-L. (2008). Timing and nature of late Quaternary mountain glaciation. *J. Quatern. Sci.*, **23**: 503–08.

Thackeray, C. W., Fletcher, C. G., Mudryk, L. R., and Derksen, C. (2016). Quantifying the uncertainty in historical and future simulations of Northern Hemisphere spring snow cover. *J. Clim.*, **29**: 8647–63, doi:10.1175/JCLI-D-16-0341.1.

Thaler, K. (2008). Analyse der Temperaturverhältnisse in der Eisriesenwelt-Höhle im Tennengebirgeanhandeiner 12 jährigen Messreihe. MSc thesis, Institut für Meteorologie und Geophysik, Leopold-Franzens Universität, Innsbruck. 101pp.

Thayyen, R. J. and Gergan, J. T. (2010). Role of glaciers in watershed hydrology: a preliminary study of a "Himalayan catchment". *The Cryosphere*, **4**: 115–28.

Thiede, J., *et al.* (2001). The late Quaternary stratigraphy and environments of northern Eurasia and the adjacent Arctic seas – new contributions from QUEEN. *Global Planet. Change*, **31**: vii–x.

Thomas, D. N. and Dieckmann, G. S. (eds.). (2010). *Sea ice*, 2nd ed. Chichester, UK: Wiley-Blackwell. 621 pp.

Thomas, D. R. and Rothrock, D. A. (1993). The arctic ocean ice balance: A Kalman smoother estimate. *J. Geophys. Res.*, **98** (C6):10,053–67.

Thomas, E. R., *et al.* (2009). Anatomy of a Dansgaard-Oeschger warming transition: High-resolution analysis of the North Greenland Ice Core Project ice core. *J. Geophys. Res.*, **114**: D08102, doi:10.1029/2008JD011215.

Thomas, R. H. (1979). The dynamics of marine ice sheets. *J. Glaciol.*, **24** (90): 167–77.

Thomas, R. H. (2004). Force-perturbation analysis of recent thinning and acceleration of Jakobshavn Isbrae, Greenland. *J. Glaciol.*, **50** (168): 57–66.

Thomas, R. H., *et al.* (2006). Progressive increase in ice loss from Greenland, Geophys. *Res Lett.*, **33**: L10503, doi:10.1029/2006GL026075

Thompson, D. W. J. and Wallace, J. M. (1998). The arctic oscillation signature in the wintertime geopotential height and temperature fields. *Geophys. Res. Lett.*, **25**: 1297–300.

Thompson, L. G., *et al.* (1991). Laminated ice bodies in collapsed lava tubes at El Malpais National Monument, central New Mexico. Field Guide to Geologic Excursions in New Mexico and adjacent areas of Texas and Colorado. New Mexico Bureau of Mines and Mineral Resources, Bulletin 137, 149.

Thompson, L. G., *et al.* (1997). Tropical climate instability: The last glacial cycle from a Qinghai-Tibetan ice core. *Science*, **276** (5320): 1821–5.

Thompson, L. G., *et al.* (2009). Glacier loss on Kilimanjaro continues unabated. *Proc. Nat. Acad. Sci.*, **106** (47): 19770–5. doi: 10.1073/pnas.0906029106

Thomson, L. I., Osinski, G. R. and Ommanney, C. S. L. (2011). Glacier change on Axel Heiberg Island, Nunavut, Canada. *J. Glaciol.*, **57**: 1079–86

Thomson, S. (1966). *Icings on the Alaska Highway. Proceedings International Conference on Permafrost, (Nov. 1963 Lafayette, Indiana)*. Washington, DC: National Research Council, National Academy of Sciences. pp. 526–29.

Thorarinsson, S. (1943). Oscillations of the Icelandic glaciers in the last 250 years. *Geogr. Annal.*, **25**: 1–54.

Thorndike, A. (1992). Estimates of sea ice thickness distributions using observations and theory. *J. Geophys. Res.*, **97** (C8): 12,601–605.

Thorndike, A. S., *et al.* (1975). The thickness distribution of sea ice. *J. Geophys. Res.*, **80** (33): 4501–13.

Tietsch, S., *et al.* (2010). Rapid recovery of Arctic summer sea-ice loss. Paper 57A031. Proceedings, Tromso Sea Ice Symposium. *Int. Glaciol. Soc.*, www.igsoc.org/symposia/previous.html

Timco, G. W. and Barker, A. (2002). What is the maximum pile-up height for ice? Ice in the environment, Vol. **2**, Squire, V. and Langhorne, P. (eds.), *Proc. 16th IAHR Internat. Sympos. on Ice, Int. Assoc. Hydraulic Eng. Res.*, Dunedin, New Zealand: International Association of Hydraulic Engineering and Research. pp. 69–77.

Timco, G. W. and Frederking, R. (2009). Overview of historical Canadian Beaufort Sea information. *Tech. Rep. CHC-TR-057.* Ottawa, Canada: NRC Canadian Hydraulics Centre. 99 pp.

Timco, G. W. and Weeks, W. F. (2010). A review of the engineering properties of sea ice, Cold Regions Sci. *Technol.*, **60**: 107–29.

Timokhov, L. A. (1994). Regional characteristics of the Laptev and the East Siberian seas: climate, topography, ice phases, thermohaline regime, and circulation. In Kassens, H., Hubberten, H. W., Priamikov, S. and Stein, R. (eds.), *Russian–German Cooperation in the Siberian Shelf Seas: Geo-SysteCHC-TR-057. m Laptev Sea. Ber.* St. Petersburg, Russia: Polarforsch. **144**: 15–31.

Todd, M. C. and Mackay, A. W. (2003). Large-scale climatic controls on Lake Baikal ice cover. *J. Climate*, **16** (19): 3186–99.

Todhunter, P. E. (2007). Hydroclimatological analysis of the red rwer of the north snowmelt flood catastrophe of 1997. *J. Amer. Water Resour. Assoc.*, **37** (5): 1263–78.

Tolstikhin, O. N. (1968). The meaning and calculation of the icing processes in the balance of the underground waters in the permafrost areas. IUGG General Assembly of Bern, *Int. Assoc. Hydrol. Sci., Publ.*, **77**: 361–67.

Tomas, R. A., Deser, C., and Sun, L. T. (2016). The role of ocean heat transport in the global climate response to projected arctic sea ice loss. *Journal of Climate*, **29**: 6841–59.

Tremper B., (2008). *Staying alive in avalanche terrain.* 2nd ed. The Mountaineers, Seattle, WA.

Trenberth, K. E. (2009). An imperative for climate change planning: tracking Earth's global energy. *Current Opinion Environ, Sustain.*, **1**: 19–27.

Tramoni, F., Barry, R. G., and Key, J. (1985). Lake ice cover as a temperature index for monitoring climate perturbations. *Zeitschrift Gletscherkunde Glazialgeologie*, **21**: 43–9.

Tran, N., *et al.* (2008). Snow facies over ice sheets derived from Envisat active and passive observations. *IEEE Trans. Geoscience Remote Sensing*, **46** (11): 3694–708.

Tripati, A. K., Roberts, C. D., and Eagle, R. A. (2009). Coupling of CO2 and ice sheet stability over major climate transitions of the last 20 million years. *Science*, **326**: 1394–7.

Troll, C. (1942). Der Büßerschnee (Nieve de los Penitentes) in den Hochgebirgen der Erde. Petermanns Geographische Mitteilungen, Ergänzungsband (210): Gotha.

Trujillo, E., Ramirez, J. A., and Elder, K. J. (2007). Topographic, meteorologic and canopy controls on the scaling characteristics if the spatial distribution of snow depth fields. *Water Resour. Res.*, **43**: W07409.

Tsang, L, *et al.* (2001). Scattering of electromagnetic waves, vol. **2**, *Numerical simulations.* Hoboken, NJ: Wiley Interscience.

Tschudi, M. A., Maslanik, J. A., and Perovich, D. K. (2008). Derivation of melt pond coverage on Arctic sea ice using MODIS observations. *Rem. Sens. Env.*, **112**: 2605–14.

Tsukimoto, H. (2000). Extracting rules from trained neural networks. *IEEE Trans. Neural Network*, **11**(2): 377–89. doi: 10.1109/72.839008

Tsytovich, N. A. (1966). Permafrost problems. In Proceedings, Permafrost International Conference. Washington, DC: National Research Council, p. 7.

Tucker, W. B. (1989). An overview of the physical properties of sea ice. In *Proceedings of workshop on ice properties, Assoc. Comm. Geotech. Res., Natl. Res. Council Canada, Tech. Memorandum No. 144, NRCC 30358*, pp. 71–85.

Tucker, W. B., *et al.* (1999). Physical characteristics of summer sea ice across the Arctic Ocean. *J. Geophys. Res.*, **104**: 1489–504.

Tyndall, J. (1860). *The glaciers of the alps*, London: John Murray. 444 pp.

Tyrell, J. B. (1910). Ice on Canadian lakes. *Trans. Canad. Inst.*, **9** (20, Pt 1): 13–22.

Tzedakis, P. C., *et al.* (2009). Interglacial diversity. *Nat. Geosci.*, **2**: 751–55.

Ulaby, F. T., Stiles, W. H., and Abdelrazik, M. (1984). Snowcover influence on backscattering from terrain. *IEEE Trans. Geosci. Remote Sens.*, GE-22 (2): 126–33.

UNEP. (2007). Global outlook for ice and snow. www.unep.org/geo/geo_ice/

UNEP/WGMS. (2008). *Global glacier changes; facts and figures.* Zurich: World Glacier Monitoring Service. 45 pp.

University of Alaska. (2008). *Compendium of the Proceedings of the first nine International Conferences on Permafrost 1963–200.*

DVD. ISBN 10: 0–98001794–7.Fairbanks, AK: University of Alaska.

Untersteiner, N. (1961). On the mass and heat budget of Arctic sea ice. *Archiv Meteorol., Geophys. Bioklimatol.*, **A12**: 151–82.

Untersteiner, N. (1968). Natural desalination and equilibrium salinity profile of perennial sea ice. *J. Geophys. Res.*, **73**: 12–57.

Untersteiner, N. (ed.). (1986). *The geophysics of sea ice*. New York: Plenum Press. 1096 pp.

Untersteiner, N. and van der Hoeven, F. (2009). *International geophysical year, 1957– 1958, drifting station alpha documentary film*. Boulder, CO: National Snow and Ice Data Center. Digital media.

Untersteiner, N., *et. al.*, (2007). AIDJEX revisited: A look back at the U.S.-canadian arctic ice dynamics joint experiment 1970–78. *Arctic*, **60**: 27–36.

US Army Corps of Engineers. (1956). Snow hydrology: Summary report of the snow investigations. Portland, OR: North Pacific Div., US Army Corps of Engineers. www .navcen.uscg.gov/pdf/iip/2018_Annual_Re port_FINAL.pdf

U.S. Coast Guard, (2018). *Report of the International Ice Patrol in the North Atlantic Season of 2018*, Bulletin #104, CG 188-73. www.navcen.uscg.gov/pdf/iip/2018 _Annual_Report_FINAL.pdf

U.S. National Academy (1990). *Snow/-avalanche hazards and mitigation in the United States. Panel on Snow Avalanches, Commission on Engineering and Technical Systems.* Washington, DC. National Academy Press. 84 pp.

Van de Wal, R. S. W. and Wild, M. (2001). Modelling the response of glaciers to climate change by applying volume-area scaling in combination with a high-resolution GCM. *Clim. Dynam.*, **18**: 359–66.

van den Broeke, M. R., *et al.* (2009). Partitioning recent Greenland mass losses, *Science* **326**: 984–86.

van den Broeke *et al.* (2016). On the recent contribution of the Greenland ice sheet to sea level change, *The Cryosphere*, **10**: 1933–46, https://doi.org/10.5194/tc-10–1933-2016

van der Veen, C. J. (1996). Tidewater calving, *J. Glaciol.*, **42**: 375–85.

van der Veen, C. J. (2002). Calving glaciers. *Progr. Phys. Gog.*, **26**: 96–122.

van der Veen C. J. and Payne, A. J. (2004). Modelling land-ice dynamics. In Bamber J. A. and Payne, A .J. (eds.), *Mass balance of the cryosphere: Observations and modelling of contemporary and future change.* Cambridge: Cambridge University Press, pp. 169–225

van Everdingen, R. O. (1985). Unfrozen permafrost and other taliks. In J.Brown *et al.* (eds.), *Workshop on permafrost geophysics. CRREL Special Rep.* Hanover, NH: US Army, 101–5.

van Wessem, J. M. , *et al.,* (2018). Modelling the climate and surface mass balance of polar ice sheets using RACMO2, part 2: Antarctica (1979–2016). *The Cryosphere,* 1–35, https://doi .org/10.5194/tc-2017-202

Vare, L. L., *et al.* (2009). Sea ice variations in the central Canadian Arctic Archipelago during the Holocene. *Quatern. Sci. Rev.*, **28**: 1354–66, doi:10.1016/j. quascirev.2009.01.013.

Vasil ' chuk, Y. K. and Vasil'chuk, A. C. (1997). Radiocarbon dating and oxygen-isotope variations in Late-Pleistocene syngenetic ice wedges in northern Siberia. *Permafrost & Periglac. Proc.*, **8**: 335–45.

Vaughan, D. G. (2008). West Antarctic Ice Sheet collapse – the fall and rise of a paradigm, *Clim. Change*, **91**: 65–79.

Vaughan, D. G., *et al.* (1993). A synthesis of remote sensing data on Wilkins Ice Shelf, Antarctica. *Ann. Glaciol.*, **17**: 211–8.

Vaughan, D. G., *et al.* (2003). Acoustic impedance and basal shear stress beneath four Antarctic ice streams. *Ann. Glaciol.*, **36**: 225–32.

Vaughan, *et al.* (2013). Observations: cryosphere. In Stocker, T. F., *et al.* (eds.), *Climate change 2013: The physical science basis. Contribution of WG I to the AR5 of the intergovernmental panel on climate change, pg. 317-382.* Cambridge University Press, Cambridge, UK and New York, NY, USA.

Vavrus, S., (2004). The impact of cloud feedbacks on Arctic climate under greenhouse forcing. *Journal of Climate*, **17**: 603–15.

Vavrus, S. (2007). The role of terrestrial snow cover in the climate system. *Clim. Dyn.*, **29**: 73–88.

Vavrus, S. J., Wynne, R. H. and Foley, J. A. (1996). Measuring the sensitivity of southern Wisconsin lake ice to climate variations and lake depth using a numerical model. *Limnol. Oceanogr.*, **41** (5): 822–31.

Veatch, W., *et al.* (2009). Quantifying the effects of forest canopy cover on net snow accumulating at a continental mid-latitude site. *Ecohydr.*, **2**: 115–28.

Velicogna, I. (2009). Increasing rates of ice mass loss from the Greenland and Antarctic ice sheets revealed by GRACE, Geophys. *Res. Lett.*, **36**: L19503. doi:10.1029/2009GL040222.

Velicogna, I. and Wahr, J. (2005). Greenland mass balance from GRACE. *Geophys. Res. Lett.*, **32**: L18505. doi:10.1029/2005GL023955.

Velicogna, I. and Wahr, J. (2006a). Acceleration of Greenland ice mass loss in spring 2004. *Nature* **443**: 329–31.

Velicogna, I. and Wahr, J. (2006b). Measurements of time-variable gravity show mass loss in Antarctica. *Science*, **311**: 1754–6, https://doi.org/10.1126/science.1123785

Velicogna, I. and Wahr, J., 2013, Time-variable gravity observations of ice sheet mass balance: Precision and limitations of the GRACE satellite data, Geophysical Res. Letters, https://doi.org/10.1002/grl.50527

Velicogna, I., *et al.* (2014). Regional acceleration in ice mass loss from Greenland and Antarctica using GRACE time-variable gravity data. *Geophys. Res. Lett.*, **41**: 8130–7.

Venkatesh, S. and M. El-Tahan, M. (1988). Iceberg life expectancies in the Grand Banks and Labrador Sea. *Cold Reg. Sci. Technol.*, **15**: 1–11.

Vimeux, F., *et al.* (1999). Glacial-interglacial changes in ocean surface conditions in the Southern Hemisphere. *Nature*, **399**: 410–3.

Vilesov, E. N. and Morozova, V. I. (2005). *Degradacia oledenenia gor Yuzhnoy Dzhungarii vo vtoroj polovine 20 veka (Degradation of glaciers in Southern Djungaria mountaines in the second part of 20th century)*. Moscow: Inst. of Geography, RAS. Data Glaciol. Studies **98**: 201–6 (in Russian).

Vilesov, E. N. and Uvarov, V. N. (2001). *Evoljutsija sovremennogo oledeninja Zailijskogo Alatau v XX Veke* (Evolution of glaciers at the Zailiysky Alatau in 20th century). Almaty: Kazakh State University Press. (in Russian).

Vilesov, E. N., *et al.* (2006). Degradacia oledenenia i kryogenez na sovremennyh morenah severnogo Tian-Shania (Degradation of the glaciation and cryogenesis of modern moraines in the northern Tien Shan). *Cryosphera Zemli*, **10**: 69–73 (in Russian).

Vincent, C., *et al.* (2004). Ice ablation as evidence of climate change in the Alps over the 20th century. *J. Geophys. Res.*, **109** (D10): D10104.

Vincent, W. F., Gibson, J. A. E. and Jeffries, M. O. (2001). Ice shelf collapse, climate change, and habitat loss in the

Canadian high Arctic. *Polar Record*, **37** (201): 133–42.

Vincent, W. F., Hobbie, J. E. and Layborne-Parry, J. (2008a). Introduction to the limnology of high-latitude lake and river ecosystems. In Vincent, W. F. and Laybourn-Parry, J., (eds.), *Polar lakes and rivers: limnology of Arctic and Antarctic aquatic ecosystems*. Oxford: Oxford University Press. pp. 1–23.

Vincent, W. F., *et al.* (2008b). The physical limnology of high-latitude lakes. In Vincent, W. F. and Laybourn-Parry, J., (eds.), *Polar lakes and rivers: limnology of Arctic and Antarctic aquatic ecosystems*. Oxford: Oxford University Press. pp. 65–81

Vinje, T. (1980). Some satellite-tracked iceberg drifts in the Antarctic. *Annals Glaciol.*, **1**: 83–7.

Vinje, T. (1999). Barents Sea-ice edge variation over the past 400 years. Proceedings of the Workshop on sea-ice charts of the Arctic. Geneva: World Meteorological Organization. WMO/TD 949. pp. 4–6.

Vinje, T. (2001). Anomalies and trends of sea-ice extent and atmospheric circulation in the Nordic Seas during the period 1864–1998. *J. Clim.*, **14**: 255–67.

Vinther, B. M., *et al.* (2009). Holocene thinning of the Greenland ice sheet. *Nature*, **461**: 385–8.

Visser, P. C. (1928). Von den Gletschern am Obersten. *Indus. Zeit. Gletscherk*, **16**: 169–229.

Vizcaino, M., Rupperm S., and Chiang. J. C. H. (2010). Permanent El Niño and the onset of Northern Hemisphere glaciations: Mechanism and comparison with other hypotheses. *Paleoceanog*, **25** (PA2205): 20.

Voeikov, A. I. (1889). Permafrost in Siberia along prospective railroad route. *J. Minesterstva Putei Soobshenia*, **13**: 14–18. (In Russian).

Voellmy, A. (1955), Über die Zerstörungskraft von Lawinen. *Schweizer Bauzeitung*, **73** (12, 15, 17, 19, 37): 159–65, 212–17, 246–49, 280–85.

von Cholnoky, E. (1909). Das Eis des Baltonsees. *Geogr. Gesellschaft*, **1**(5).

Vonderthann, H. (2007). *Die Schnellberger Eishöhle 1339/26. Eine touristische Besonderheit des Berchtesgadener Landes. Berchtesgadener Alpen. Karst und Höhle 2004/2005*. Munich: Verband Deutschen Höhlen- und Karstforscher, pp. 197–211.

von Drygalski, E. (1897). *Gronland-Expedition der Gesellschaft für Erdkunde zu Berlin, 1891–1893*, vol. **1**. Berlin: W.H. Kühl. pp. 385–95

von Drygalski, E. (1983). The temperature of the iceberg. (transl. of text from German Antarctic Expedition 1901–1903, 1903). *Iceberg Res.*, No. **6** (Cambridge: Scott Polar Res. Inst.,). pp. 10–12.

von Saar, R. (1956). Eishöhlen, Ein Meteorologisch-Geophysikalisches Phänomen (Untersuchungen an der Rieseneishöhle (R. E. H.) im Dachstein, Oberösterreich). *Geogr, Annal.*, **38**: 1–63.

Vose, R. S., Oak Ridge National Laboratory. Environmental Sciences Division, U.S. Global Change Research Program, US Dept. of Energy, Office of Health and Environmental Research, Carbon Dioxide Information Analysis Center (U.S.), 1992, The Global Historical Climatology Network: Long-Term Monthly Temperature, Precipitation, Sea Level Pressure, and Station Pressure Data. Carbon Dioxide Information Analysis Center.

Vose, R. S., et al., (2012). NOAA's Merged Land–Ocean Surface Temperature Analysis. *American Meteorological Society*, **19** (11): 1677–85. doi: https://doi.org/10.1175/BAMS-D-11-00241.1

Vuglinsky, V. S. (2002a). Peculiarities of ice events in Russian Arctic rivers. *Hydrol. Proc.*, **15**: 905–13.

Vuglinsky, V. S. (2002b). Ice events on the Soberian rivers: Formation and variability. Ice in the environment, Vol. 1, Squire, V. and Langhorne, P. (eds.), Proc. 16th IAHR Internat. Sympos. on Ice, Int. Assoc. Hydraulic Eng. Res., Dunedin, New Zealand. pp. 59–66.

Vuglinsky, V. S. (2006). Ice regime in the rivers of Russia, its dynamics during last decades and possible future changes. In Saeki, H. (ed.), Proceedings of the 18th IAHR International Symposium on Ice, vol. 1, Sapporo: Nakanishi Publishing Co., pp. 93–98.

Vuglinsky, V. S., Gronskaya, T. P., and Lemeshko, N. A. (2002). Long-term characteristics of ice events and ice thickness on the largest lakes and reservoirs of Russia, Ice in the environment, Vol. 3, Squire, V. and Langhorne, P. (eds.). Proc. 16th IAHR Internat. Sympos. on Ice, Int. Assoc. Hydraulic Eng. Res., Dunedin, New Zealand. pp. 80–6.

Vuichard, D. and Zimmemann, M. (1986). The Langmoche flash-flood, Khumbu Himal Nepal. *Mountain Res. Devel.*, **6**: 90–4.

Vuille, M., *et al.* (2008). Climate change and tropical Andean glaciers: past, present & future. *Earth-Scien. Rev.*, **89**(3–4): 79–96, doi:10.1016/j.earscirev.2008.04.002

Vuyovich, C., *et al.* (2009). Monitoring river ice conditions using web-based cameras. *J. Cold Reg. Engin.*, **23**(1): 1–17.

Wadhams, P. (1998). Sea ice morphology. In *Physics of ice covered seas*, vol. **1**, M. Lepparanta (ed.), Finland: University of Helsinki, pp. 483–516.

Wadhams, P. (2000). *Ice in the oceans.* Amsterdam: Gordon and Breach. 351 pp.

Wadhams, P. (2008). How does Arctic sea ice form and decay? www.arctic.noaa.gov/ess ay_wadhams.html

Wadhams, P. and Amanatidis, G. (eds.). (2007). *Arctic sea ice thickness: past, present and future.* Brusseks: European Commission. 409 pp.

Wadhams, P. and Dobie, M. J. (2010). Sea ice thickness measurement using episodic infragravity waves from distant storms. *Cold Reg. Sci. Technol.*, **56**: 98–101.

Wadhams, P., *et al.* (1992). Relationships between sea ice freeboard and draft in the Arctic Basin and implications for ice thickness monitoring. *J. Geophys. Res.*, **97** (C12):20, 325–34.

Wagnon, P., *et al.* (1999). Annual cycle of energy balance of Zongo Glacier, Cordillera Real, Bolivia. *J. Geophys. Res.*, **104** (D4): 3907–23.

Wailer, C. (1995). A comparison of two avalanche-models with exemplary avalanches of Tyrol and Switzerland and the effects to hazard zoning. *Surveys Geophs.*, **16**(5–6): 671–9.

Waite A. H. and Schmidt, S. J. (1961). Gross errors in height indication from pulsed radar altimeters operating over thick ice or snow. Inst. Radio Engineers. *International Convention Record, Part*, **5**: 38–53.

Walker, A. E. and Davey, M. R. (1993). Observation of Great Slave Lake ice freeze-up and breakup processes using passive microwave satellite data. Proc. 16th Canadian Symposium on Remote Sensing, Sherbrooke, Quebec: 233–8.

Walker, A. E. and Goodison, B. E. (1993): Discrimination of a wet snow cover using passive microwave satellite data. *Ann. Glaciol.*, **17**: 307–11.

Walker, A. E. and Silis, A. (2002). Snow-cover variations over the Mackenzie River basin, Canada, derived from SSM/I passive-microwave satellite data. *Annals of Glaciology*, **34** (1): 8–14.

Walker, E. R. and Wadhams, P. (1979). Thick sea-ice floes. *Arctic*, **32**: 140–7.

Walker, M., *et al.* (2009). Formal definition and dating of the GSSP (Global Stratotype Section and Point) for the base of the Holocene using the Greenland NGRIP ice core, and selected auxiliary records. *J. Quatern. Sci.*, **24**: 3–17.

Wallace, A. R. (1871). The theory of glacier motion. *Nature*, **3**: 309–10.

Walland, D. J. and Simmonds, I. (1997). Modelled atmospheric response to changes in Northern Hemisphere snow cover. *Climate Dyn.*, **13**: 25–34.

Wallén, C. C. (1948). Glacial-meteorological investigations on the Kårsa Glacier in Swedish Lappland. *Geogr. Annal.*, **30**: 451–672.

Wallevik, J. E. and Sigurjónssson, H. (1998). *The Koch index: formulations, corrections and extensions. Vedurstofa Islands Report VI-G98035-UR28.* Iceland: Reyjavik. 15 pp.

Walsh, J. E., *et al.* (2008). Glaciers and ice sheets in the Arctic. The Encyclopedia of Earth. Earth Portal. www.eoearth.org/article/Glaciers_and_ice_sheets_in_the_Arctic.

Walsh, J. E., *et al.* (2016). A database for depicting Arctic sea ice variations back to 1850. *Geographical Review*, **107** (1): 89–107, doi: 10.1111/j.1931-0846.2016.12195.x.

Walsh, S., *et al.* (1998). Global patterns of lake ice phenology and climate: Model simulations and observations. *J. Geophys. Res.*, **103** (D22): 28, 825–37.

Walvoord, M. A. and Kurylyk. B. L. (2016). Hydrologic impacts of thawing permafrost – a review, *Vadose Zone Journal*, **15** (6): 1–20. doi: 10.2136/vzj2016.01.0010.

Wang, A., Xu, L., and Kong, X. (2018). Assessments of the north hemisphere snow cover response to 1.5 C and 2.0 C warming. *Earth Syst. Dyn.*, **9**: 865–77. doi:10.5194/esd-9-865-2018.

Wang, J., *et al.* (2009). Is the dipole anomaly a major driver to record lows in Arctic summer sea ice extent? Geophys. *Res. Lett.*, **36** (L05706): 5.

Wang, J., *et al.* (2010). Severe ice cover on Great Lakes during winter 2008–2009. *Eos*, **01** (5): 41–42.

Wang, L.-B., *et al.* (2005). Melt season duration on Canadian Arctic ice caps, 2000–2004. *Geophys. Res. Lett.*, **32**: L19502.

Wang, L.-B., *et al.* (2007). Melt season duration and ice layer formation on the Greenland ice sheet, 2000–2004. *J. Geophys. Res.*, **112**: F04013.

Wang, L., *et al.* (2013). Recent changes in pan-Arctic melt onset from satellite passive microwave measurements. *Geophysical Research Letters*, **40**: 522–8.

Wang, M.-Y. and Overland, J. E. (2009). A sea ice free summer Arctic within 30 years? Geophys. *Res. Lett.*, **36 L** (L07502): 5.

Wang, T., *et al.* (2020). Permafrost thawing puts the frozen carbon at risk over the Tibetan Plateau,. *Sci. Adv.*, **6**: eaaz3513.

Warren, C. and Aniya, M. (1999). The calving glaciers of southern South America. *Global Planet. Change*, **22**: 59–77.

Warren, S. G. and Town, M. S. (2009). Antarctica. In Schneider, S. H. (ed.) *Encyclopedia of climate and weather.* Oxford: Oxford University Press.

Warren, S. G., *et al.* (1998). Snow depth on Arctic sea ice. *J. Clim.*, **12**: 1814–29.

Washburn, A. L. (1973). *Periglacial processes and environments.* London: Edward Arnold, 320 pp.

Washburn, A. L. (1980). *Geocryology: a survey of periglacial processes and environments.* 2nd edn. New York: Wiley, 406 pp.

Washington, W. M. and Meehl, G. A. (1996). High-latitude climate change in a global coupled ocean-atmosphere-sea ice model with increased atmospheric CO_2. *J. Geophys. Res.*, **101** (D8): 12, 795–802.

Watanabe, T., Lamsal, D., and Ives, J. D. (2009). Evaluating the growth characteristics of a glacial lake and its degree of danger of outburst flooding: *Imja*

Glacier, Khumbu Himal, Nepal. Norsk Geogr. Tidsskr, **63**: 255–67.

Watson, C. S. *et al.* (2015). Unabated global mean sea-level rise over the satellite altimeter era. *Nat. Clim. Change*, **5**: 565–8, https://doi.org/10.1038/nclimate2635, 2015.

Weber, M., *et al.* (2011). Contributions of rain, snow and icemelt in the Upper Danube discharge today and in the future. Geaogr, Fis. Dinam. Quat, 33, 221–30, www.glaciologia.it/wp-content/uploads/Abstracts/Abstract_33_2/12_ GFDQ_33_2_Weber_Abst.pdf

Webster, M., Rigor, I., and Morison, J. (2010). Improved weather filters for analyzing sea-ice concentration. Paper 57A096. Proceedings, Tromso Sea Ice Symposium. Int. Glaciol. Soc.

Weeks, W. F. (1998). On the history of sea ice research. In M. Leppäranta (ed.), *Physics of ice-covered seas*, Vol.**1**. Helsinki: University of Helsinki Press. pp. 1–24.

Weeks, W. F. (2010). *On sea ice*. Fairbanks, AK: University of Alaska Press. 664 pp.

Weeks, W. F. and Ackley, S. F. (1986). The growth, structure, and properties of sea ice. In Untersteiner, N. (ed.), *The geophysics of sea ice*. New York: Plenum Press. pp. 9–164.

Weeks, W. F. and Lofgren, G. (1967). The effective solute distribution coefficient during the freezing of NaCl solutions. In Oura, H. (ed.), *Physics of snow and ice*. Sapporo, Japan: Institute of Low Temperature Science, Hokkaido University. pp. 579–97.

Weertman, J. (1957). On the sliding of glaciers. *J. Glaciol.*, **3**(21): 33–38.

Weertman, J. (1983). On the creep deformation of ice. *Ann. Rev. Earth Planet. Sci.*, **11**: 215–40.

Weijer, W. *et al.* (2017). Local atmospheric response to an open-ocean polynya in a high-resolution climate model. *J. Clim.*, **30**: 1629–41.

Wendisch, M., et al. (2017), Understanding causes and effects of rapid warming in the Arctic, Eos, 98, AGU, https://doi.org/10.1029/2017EO064803. Published on 17 January 2017.

Wetherald, R. T. and Manabe, S. (1988). Cloud feedback processes in a general-circulation model. *Journal of the Atmospheric Sciences*, **45**: 1397–415.

Weyhenmeyer, G. A., Livingstone, D. M., Meili, M., Jensen, O., Benson, B., and Magnuson, J. J. (2011). Large geographical differences in the sensitivity of ice-covered lakes and rivers in the Northern Hemisphere to temperature changes. *Glob. Chang. Biol.*, **17**: 268–75. https://doi.org/10.1111/j.1365–2486.2010.02249.x

Wiersma, A. P. and Jongma, J. I. (2010). A role for icebergs in the 8.2 ka climate event. *Clim. Dyn.*, **35**: 535–49.

Wiese, D. N., *et al.* (2016a). Quantifying and reducing leakage errors in the JPL RL05 M GRACE mascon solution. *Water Resour. Res.*, **52**: 7490–502, //doi.org/10.1002/2016WR019344, 2016

Wiese, D. N., *et al.* (2016b). JPL GRACE Mascon Ocean, Ice, and Hydrology Equivalent Water Height RL05 M. 1 CRI Filtered, Ver. 2, PO. DAAC, CA, USA.

Weisman, R. (1977). Snowmelt: a two-dimensional turbulent diffusion model. *Water Resour. Res.*, **13** (2): 337–42.

Weiss, J., Schulson, E. M., and Stern, H. L. (2006). Sea ice rheology from in-situ, satellite and laboratory observations: Fracture and friction. *Earth Planet. Sci. Lett.*, **255**: 1–8.

Weyhenmeyer, G. A., Meili, M., and Livingstone, D. M. (2004). Nonlinear temperature response of lake ice breakup. *Geophys Res Lett.*, **31**: L07203. doi:10.1029.2004GL019530.

Weyrick, P. P., White, K. D., Daly, S. F., Bullock, M. J., and Gagnon, J. J. (2007). CRREL's Ice Jam Database and Website. In Proc., 14th Workshop on the Hydraulics of Ice Covered Rivers, Quebec City, Quebec, Canada.

WGMS, (2017). Global glacier change bulletin no. 2 (2014–2015). In Zemp, M., Nussbaumer, S. U., Gärtner-Roer, I., Huber, J., Machguth, H., Paul, F., and Hoelzle, M. (eds.), *ICSU(WDS)/ IUGG(IACS)/UNEP/UNESCO/ WMO*, Zurich, Switzerland: World Glacier Monitoring Service, 244 pp., based on database, doi:10.5904/wgms-fog-2018-11

Wigle, T., *et al.* (1990). *Optimum operation of hydroelectric plants during the ice regime of rivers – A Canadian experience*. Ottawa, Canada: Task Force of the Subcommittee on Hydraulics of Ice-Covered Rivers, National Research Council of Canada, NRCC 31107.

Wild, G. O. (1882). Air temperature in the Russian Empire. Izdat. Russk. Geograf. Obshest. St. Petersburg (in Russian). 159 pp.

Wilhelmy. F. (1975). *Schnee und Gletscherkunde*. Berlin: Walter de Gruyter. 454 pp.

Wilken, M. and Meinert, J. (2006). Submarine glacigenic debris flows, deep-sea channels and past ice-stream behaviour of the East Greenland continental margin. *Quat. Sci. Rev.*, **25**: 784–810.

Willett, H. C. (1950). Temperature trends of the past century. In Centenary Proceedings. London: Royal Meteorological Society, pp.195–206.

Williams, G. P. (1965). Correlating freeze-up and breakup with weather conditions. *Canad. Geotech. J.*, **2**: 313–26.

Williams, R. S. Jr. and Ferrigno, J. G. (eds.). (1988). Satellite image atlas of glaciers of the world – Antarctica. (Swithinbank, C.), U.S. Geological Survey, Prof. Papers 1386-B. 290 pp.

Williams, R. S., Jr. and Ferrigno, J. G. (eds.). (1998). Satellite image atlas of glaciers of the world: Glaciers of South America. U.S. Geological Survey Professional Paper 1386-I. 206 pp.

Williams, R. S., Jr. and Ferrigno, J. G. (eds.). (2012). State of the Earth's cryosphere at the beginning of the 21st century–Glaciers, global snow cover, floating ice, and permafrost and periglacial environments: U.S. Geological Survey Professional Paper 1386–A, 546 p, (https://pubs.usgs.gov/pp/p1386a.).

Williams, G., Layman, K. L., and Stefan, H. G. (2004). Dependence of lake ice covers on climatic, geographic and bathymetric variables. *Cold Regions Sci. Technol.*, **40**: 145–64.

Williams, S. G. and Stefan, H. G. (2006). Modeling of lake ice characteristics using climate, geography, and lake bathymetry. *J. Cold Reg, Engrg.*, **87**: 140–67.

Williamson, S., *et al.* (2008). Iceberg calving rates from northern Ellesmere Island ice caps, Canadian Arctic, 1999–2003. *J. Glaciol.*, **54** (186): 391–400.

Willmott, C. J. and Robeson, S. M. (1995). Climatologically aided interpolation (CAI) of terrestrial air temperature. *Int. J. Climatol.*, **15** (2): 221–29.

Willmott, C. J. and Matsuura, K. (2009). Terrestrial precipitation: 1900–2008, Gridded Monthly Time Series. http://climate .geog.udel.edu/~climate/html_pages/Global2 Ts_2009/README.global_p_ts_2009.html.

Wilson, L., *et al.* (1999). Mapping snow water equivalent in the mountainous areas by combining a spatially distributed snow hydrology model with passive microwave remote sensing data. *IEEE Trans. Geosci. Remote Sensing*, **37**: 690–704.

Wimmer, M. (2007). Eis- und Tenperaturmessungen im Schönberg System (Totes Gebirge, Öbersterreich/Steiermark). Alpin Untertage, Berchesgarden November 9–11, 2007. Proceedings. Dtsch. Höhlen- und Karstforscher, Munich. p. 83. www.zobodat.at/pdf/BNO_0023_2_0757-07 94.pdf

Wingham, D. J., *et al.* (2006). Rapid discharge connects Antarctic subglacial lakes. *Nature*, **440**: 1033–36.

Wingham, D. J., Wallis, D. W., and Shepherd, A. (2009). The spatial and temporal evolution of Pine Island glacier thinning, 1995–2006. *Geophys. Res. Lett.*, **36** (17): L17501

Winkelmann, R., *et al.* (2011). The Potsdam parallel ice sheet model (PISM-PIK)–part 1: model description. *The Cryosphere*, **5**: 715–26.

Winsborrow, M. C. M., Clark, C. D., and Stokes, C. R. (2004). Ice streams of the Laurentide Ice Sheet. *Geogr. phys. Quatern.*, **58**: 269–80.

Winsemius H. C., *et al.* (2016). Global drivers of future river flood risk. *Nat. Clim. Change*, **6**: 381–85. doi:10.1038/nclimate2893

Winstral, A. and Marks, D. (2002). Simulating wind fields and snow redistribution using terrain-based parameters to model snow accumulation and melt over a semi-arid mountain catchment. *Hydrol. Processes*, **16**: 3585–603.

Wisshak, M., Straub, R., and Lopez Correa, M. (2005). Das Eisrohrhöhle – Bammelschacht – System (1337/118) im Kleinen Weitschartenkopf (Reiteralm). *Berchtesgadener Alpen. Karst und Höhle 2004/2005*. Munich: Verband Deutschen Höhlen- und Karstforscher, pp. 68–81.

WMO. (1986). *Intercomparison of models of snowmelt runoff. Operational Hydrology Rep. 23. WMO-No. 646.* Geneva: World Meteorological Organization. 36 pp.

Wohlleben, T. and Tivy, A. (2010). An investigation into the anomalous sea-ice conditions in Lincoln Sea and Nares Strait: 2007 and 2009. Paper 57A019. Proceedings of the Tromso Sea Ice Symposium. Int. Glaciol. Soc.

Wojtowicz, A., *et al.* (2009). 2-D modeling of ice-cover formation processes on the Athabaska River, AB. CGU HS Committee on river ice processes and the environment. 15th Workshop on river ice. St. John's, Newfoundland and Labrador, 19 pp.

Wolff, E. M., Fischer, H. and Röthlisberger, R. (2009). Glacial terminations as southern warmings without northern control. *Nature Geosci.*, **2**: 206–09.

Wolken, G. J., England, J. H., and Dyke, A. S. (2008). Changes in late-Neoglacial perennial snow/ice extent and equilibrium-line altitudes in the Queen Elizabeth Islands, Arctic Canada. *The Holocene*, **18**: 615–27. doi:10.1177/0959683608089215.

Wolken, G. J., Sharp, M., and Wang, L. (2009). Snow and ice facies variability and ice layer formation on Canadian Arctic ice caps, 1999–2005. *J. Geophys. Res.*, **114**(F3): F03011. doi:10.1029/2008JF001173.

Woo, M. and J. Valverde (1982). Ground and water temperatures of a forested mid-latitude swamp, paper presented at Canadian Hydrology Symposium '82, Can. Natl. Res. Counc., Fredericton, N. B., Canada.

Woo, M.-K., Marsh, P., and Pomeroy, J. W. (2000). Snow, frozen soils and permafrost hydrology in Canada, 1995–1998. *Hydrol. Processes.*, **14**: 1591–611.

Woo, M.-K., Mollinga, M., and Smith, S. L. (2008). Modeling maximum active layer thaw in boreal and tundra environments using limited data. In Woo, M.-K. (ed.), *Cold region atmospheric and hydrologic studies. The Mackenzie GEWEX experience, Vol. 2: Hydrologic processes.* Dordrecht: Springer-Verlag. pp.125–37.

Woo, M. K., *et al.* (2004). A two-directional freeze and thaw algorithm for hydrologic and land surface modelling. *Geophys Res Lett.*, **31**(L12501), 4.

Woodbury, A. D., *et al.* (2009). Observations of northern latitude ground- surface and surface-air temperatures. *Geophys. Res. Lett.*, **36**(L07703): 1–4. doi:10.1029/2009GL037400.

Woodgate, R., Weingartner, T., and Linsay, R. (2010). The 2007 Bering Strait oceanic heat flux and anomalous Arctic sea-ice retreat. *Geophys. Res. Lett.*, **37**: L01602. doi:10.1029/2009GL041621.

Worby, A. P. (1999). Observing Antarctic sea ice: A practical guide for conducting sea ice observations from vessels operating in the Antarctic pack ice. Antarctic Sea Ice Processes and Climate (ASPeCt) program of the Scientific Committee for Antarctic Research (SCAR) Global Change (GLOCHANT) program. Australia: Hobart, Tasmania. CD ROM.

Worby, A., *et al.* (1998). East Antarctic sea ice: a review of its structure, properties and drift. In Jeffries, M. (ed.), *Antarctic sea ice physical processes, interactions and variability. Antarctic Res. Ser. 74.* Washington, DC: American Geophysical Union, pp. 41–68.

Worby, A. P., *et al.* (2008a). Thickness distribution of Antarctic sea ice. *J. Geophys. Res.*, **113** (C05592): 14.

Worby, A., *et al.* (2008b). Evaluation of AMSR-E snow depth product over East Antarctic sea ice using in situ measurements and aerial photography. *J. Geophys. Res.*, **113** (C05S94):. 13.

Workman, W. H. (1914). Nieve penitente and allied formations in Himalaya, or surface forms of névé and ice created or modeled by melting. *Zeit, f. Gletscherk.*, **7**: 289–330.

World Meteorological Organization. (1970–2004). *Sea ice nomenclature. Volume I Terminology. Volume II Illustrated Glossary. Volume III International system of sea ice symbols. WMO No. 259.* Geneva: World Meteorological Organization.

World Meteorological Organization. (2007). *Sea ice nomenclature. WMO No. 259.* Geneva: World Meteorological Organization. 23 pp.

World Meteorological Organization. (2010). *Sea-ice information services in the world. WMO No. 574.* Geneva: World Meteorological Organization. 73 pp.

World Meteorological Organization. (2009). *The state of polar research*. Geneva: World Meteorological Organization, 12 pp.

Wouters B., *et al*. (2013). Limits in detecting acceleration of ice sheet mass loss due to climate variability. *Nat. Geosci.*, **6**: 613–16.

Wouters, B., Chambers, D., and Schrama, E. J. O. (2008). GRACE observes small-scale mass loss in Greenland. *Geophys. Res. Lett.*, **35** (L20501): 1–5, doi:10.1029/2008GL034816.

Wouters, B., Gardner, A. S., and Moholdt, G. (2019). Global glacier mass loss during the GRACE satellite mission (2002–2016). *Front. Earth Sci.*, https://doi.org/10.3389/feart.2019.00096.

Wright, A. and Siegert, M., (2012). A fourth inventory of Antarctic subglacial lakes. *Antarctic Science*, **24** (6): 659–64, https://doi.org/10.1017/S095410201200048X

Wu, Q.-B., Li, X., and Li W.-J. (2001). The response model of permafrost along the Qinghai Tibetan Highway under climate change. *J. Glaciol. Geocryol.*, **23**: 1–6.

Wu, Q.-B. and Zhang, T.-J. (2010). Changes in active layer thickness over the Qinghai-Tibetan Plateau from 1995 to 2007. *J. Geophys. Res.*, **115** (D09107), 12.

Wu, Q.-B., Zhang, T.-J., and Liu, Y.-Z. (2010). Permafrost temperatures and thickness on the Qinghai-Tibet Plateau. *Global Planet. Change*, **72**: 32–8.

Wu, X., Che, T., Li, X., Wang, N., and Yang, X. (2018). Slower snowmelt in spring along with climate warming across the Northern Hemisphere. *Geophys. Res. Lett.*, **45**: 12,331–9. doi:10.1029/2018GL079511.

Wu, Q., Hou, Y., Yun, H., and Liu, Y., (2015). Changes in active-layer thickness and near-surface permafrost between 2002 and 2012 in Alpine ecosystems, Qinghai–Xizang (Tibet) Plateau, China, *Glob. Planet. Change*, **124**: 149–55, doi:10.1016/j.gloplacha.2014.09.002

Wulder, M. A., Nelson,T. A. , Derksen, C., and Seemann. D. (2007). Snow cover variability across central Canada (1978–2002) derived from satellite passive microwave data. *Clim. Change*, **82**: 113–30.

Wunsch, C. (2004). Quantitative estimate of the Milankovitch-forced contribution to observed climate change. *Quat. Sci. Revs.*, **23** (9–10): 1001–12.

Wurbs, R. A. (1993). Reservoir-system simulation and optimization models. *J Water Resour Plan Manage*, **119** (4): 455–72.

Wynne, R. H., *et al.* (1998). Satellite monitoring of lake ice breakup on the Laurentian shield (1980–1994). *Photogram. Engin. Remote Sens.*, **64**: 607–17.

Xie, Z.-C, *et al.* (1996). Mass balance at the steady state equilibrium line altitude and its application. *Zeit. Glelscherk. Glazialgeol.*, **32**: 129

Xu, J.-C. , *et al.* (2007). The melting Himalayas. ICIMOD Technical Paper. Kathmandu, Nepal: International Centre for Integrated Mountain Development. 15 pp.

Xu, X.-K., *et al.* (2010). *Responses of two branches of Glacier No.1 to climate change from 1993–2005*, Tianshan, China. Quat. Int., doi: 10.1016/j.quaint.2010.06.013

Xue, Y., Sun, S., Kahan, D. S., and Jiao, Y., (2003), Impact of parameterizations in snow physics and interface processes on the simulation of snow cover and runoff at several cold region sites. *J. Geophy. Res.*, **108** (D22): 8859, doi:10.1029/2002JD003174.

Yachevskyi, L. A. (1889). Permafrt soils in Siberia. *Izvestiya Russ. Imperator. Geograf. Obshestva*, **25**: 341–55. (In Russian).

Yachevskyi, L. A. and Vannari, P. I. (eds.) (1912). *Instructions for studying permafrost in soils*. 2nd edn. St. Petersburg, Russia: Russian Imperial Geographical Society. (In Russian).

Yamazaki, T. and Kondo, J. (1992). The snowmelt and heat balance in snow-covered forested areas, J. *Appl. Meteor.*, **31**: 1322–7.

Yang, D., *et al.* (2000). An evaluation of the Wyoming Gauge system for snow measurement. *Water Resour. Res.*, **36** (9): 2665–77.

Yang, X., Pavelsky, T. M., and Allen, G. H., (2020). The past and future of global river ice. *Nature*, **577**: 69–73, https://doi.org/10.1038/s41586-019-1848-1

Ye, Q., Zong, J., Tian, L., Cogley, J. G., Song, C., and Guo, W. (2017). Glacier changes on the Tibetan Plateau derived from Landsat imagery: mid-1970s – 2000–13. *J. Glaciol.*, **63**: 273–87. doi:10.1017/jog.2016.137.

Yeh, W. W.-G. (1985). Reservoir management and operation models: a state-of-the-art review. *Water Resour Res.*, **21** (12): 1797–818.

Yershov, E. D. (1989). *Geokriologiya SSSR*. (Geocryology of the USSR). Moscow: Nauka. **5** volumes; in Russian).

Yershov, E. D. (1998). *General geocryology*. (English translation,Williams, P. J. (ed.)). Cambridge: Cambridge University Press. 580 pp.

Yi, D.-H., Zwally, H. J., and Robbins, J. W. (2010). Sea-ice freeboard and thickness in the Weddell Sea (2003–2009). Paper 57A160. Proceedings, Tromso Sea Ice Symposium. Int. Glaciol. Soc. www.igsoc.org/symposia/previous.html

Ying, I.-L, *et al.* (2006). Impacts of Yulong Mountain glacier on tourism in Lijiang, *J. Mountain Sci.*, **3**: 71–80.

Yoo, J.-C. and d ' Odorico, P. (2002). Trends and fluctuations in the dates of ice breakup of lakes and rivers in northern Europe: the effect of the North Atlantic Oscillation. *J. Hydrol.*, **268**: 100–12.

Yu, S.-Y., *et al.* (2010). Freshwater outburst from Lake Superior as a trigger for the cold event 9300 years ago. *Science*, **328** (5983): 1262–6

Yu, Y., Maykut, G. A., and Rothrock, D. A. (2003). Changes in the thickness distribution of Arctic sea ice between 1958–1970 and 1993–1997. *J. Geophys. Res.-Ocean*, **108** (C3): 3083, https://doi.org/10.1029/2001JC001208

Yuan, L.- l., *et al.* (2006). Impacts of Yulong Mountain glacier on tourism in Lijian. *J. Mountain Sci.*, **3**: 71–80.

Zemp, M., Hoelzle, M., and Haeberli, W. (2009a). Six decades of glacier mass-balance observations: a review of the worldwide monitoring network. *Annals Glaciol.*, **50**: 101–11.

Zemp, M., *et al.* (2009b). *ECV T6 – Glaciers and ice caps. Assessment of the status of the development of standards for the Terrestrial Essential Climate Variables*. Rome: GTOS Secretariat, 31 pp.

Zemp, M., *et al.* (2015). Historically unprecedented global glacier decline in the early 21st century. *J. Glaciol.*, **61**(228): 745–62, doi:10.3189/2015JoG15J017

Zemp, M., *et al.* (2019). Global glacier mass changes and their contributions to sea-level rise from 1961 to 2016. *Nature*, **568**: 382–86. https://doi.org/10.1038/s41586-019-1071-0

Zeng, Q.-H., *et al.* (2008). Snow and ice hazards and their control measures. In: Shi, Y.-F. (ed.-in-chief). *Glaciers and related environments in China*. Beijing: Science Press. pp. 317–85.

Zeng, X.-P., *et al.* (2009). A contribution by ice nuclei to global warming. *Quart.J. Roy. Met. Soc.*, **135** (643): 1614–29.

Zeng, X., Broxton, P., and Dawson, N. (2018). Snowpack Change from 1982 to 2016 over Conterminous United States. *Geophy. Res. Let.*, https://doi.org/10.1029/2018GL079621

Zhang, J., 2014, Modeling the Impact of Wind Intensification on Antarctic Sea Ice Volume, *J. of Climate*, 27(1):202–214, DOI: 10.1175/JCLI-D-12-00139.1

Zhang, J., *et al.* (2007). Climate downscaling for estimating glacier mass balances in northwestern North America: validation with a USGS benchmark glacier. *Geophys. Res. Lett.*, **34**: L21505. doi:10.1029/2007GL031139.

Zhang, J.-L., *et al.* (2008). What drove the dramatic retreat of arctic sea ice during summer 2007? *Geophys. Res. Lett.*, **35** (L11505): 5.

Zhang, T.-J. (2005a). Influence of the seasonal snow cover on the ground thermal regime: An overview. *Rev. Geophys.*, **43**: RG4002, doi:10.1029/2004RG000157.

Zhang, T.-J. (2005b). Historical overview of permafrost studies in China. *Phys. Geog.*, **26**: 279–98.

Zhang, T.-J. and Armstrong, R. L. (2001). Soil freeze/thaw cycles over snow-free land detected by passive microwave remote sensing. *Geophys. Res. Lett.*, **28** (5): 763–6.

Zhang, T.-J., *et al.* (1999). Statistics and characteristics of permafrost and ground ice distribution in the Northern Hemisphere. *Polar Geogr.*, **23** (2): 147–69.

Zhang, T.-J., *et al.* (2000). Further statistics on the distribution of frozen ground and permafrost. *Polar Geogr.*, **24** (2): 126–31.

Zhang, T.-J., *et al.* (2001). An amplified signal of climate change in soil temperatures during the last century at Irkutsk, Russia. *Clim. Change*, **49**: 41–76.

Zhang, T.-J., *et al.* (2003a). Ground-based and satellite-derived measurements of surface albedo on the North Slope of Alaska. *J. Hydrometeorol.*, **4** (1): 77–91.

Zhang, T.-J., *et al.* (2003b). Distribution of seasonally and perennially frozen ground in the Northern Hemisphere. In M. Phillips, S. M. Springman and L. U. Arenson (eds.). *Permafrost, Vol. 2, Proceedings of the 8th international conference on permafrost*. Lisse, Netherlands: A.A. Balkema, pp. 1289–94.

Zhang, T., *et al.* (2003c). Investigation of the near-surface soil freeze-thaw cycle in the contiguous United States: algorithm development and validation. *J. Geophys. Res*, **108**: 8860. doi:10.1029/2003JD003530.

Zhang, T., Barry, R. G., and Armstrong, R. L. (2004). Application of satellite remote sensing on frozen ground studies. *Polar Geog.*, **28** (3): 193–96.

Zhang, T.-J., *et al.* (2005). Spatial and temporal variability in active layer thickness over the Russian Arctic drainage basin. *J. Geophys. Res.*,**110** (D16): D16101, 14.

Zhang, T.-J., Baker, T. H. W., and Cheng, G. D. (2008). The Qinghai–Tibet Railroad: a milestone project and its environmental impact. *Cold Reg. Sci. Technol.*, **53** (3): 229–40.

Zhang, X., *et al.* (2001). Trends in Canadian streamflow. *Water Res. Res.*, **37**: 987–98.

Zhang, X.-D. (2010). Sensitivity of arctic summer sea ice coverage to global warming forcing: towards reducing uncertainty in arctic climate change projections. *Tellus*, **62**: 220–27.

Zhang, Y., *et al.* (2004). Sublimation from snow surface in southern mountain taiga of eastern Siberia. *J. Geophys. Res.*, **109**: D21103, doi:10.1029/2003JD003779

Zhang, Y., Chen, W., and Riseborough, D. W. (2008a). Disequilibrium response of permafrost thaw to climate warming in Canada over 1850–2100. *Geophys. Res. Lett.*, **35** (2): L02502. doi:10.1029/ 2007GL032117

Zhang, Y., Chen, W., and Riseborough, D. W. (2008b). Transient projections of permafrost distribution in Canada during the 21st century under scenarios of climate change. *Global Planet. Change*, **60** (3–4): 443–56.

Zhang, W., X. Mei, X. Geng, A. G. Turner, and F.-F. Jin, 2018: A Nonstationary ENSO–NAO Relationship Due to AMO Modulation. Journal of Climate, 32, 33–43.

Zheng G., Yang, Y., Yang, D., Dafflon, B., Lei, H. , and Yang, H. (2019). Satellite-based simulation of soil freezing/thawing processes in the northeast Tibetan Plateau. *Remote Sens. Environ.*, **231** (2019): 111269, doi.org/10.1016/j.rse.2019.111269

Zimov, S. A., Schuur, E. A. G., and Chapin III, F. S. (2006). Permafrost and the global carbon budget. *Science*, **312**: 1612–13.

Zotikov, I. A. (2006). *The antarctic subglacial lake vostok: glaciology, biology and planetology*. Chichester, U.K.: Praxis Publishing Ltd. 139 pp.

Zotikov, I. A., Zagorodnov, V. S., and Raikovsky, J. V. (1980). Core drilling through the Ross Ice Shelf (Antarctica) confirmed basal freezing. *Science*, **207** (4438):1463–65.

Zubov, N. N. (1943). *Arctic ice*. Moscow: Izdat. Glavsevmorputi. (translated 1963) San Diego, CA: US Navy Electronics Laboratory. 491 pp.

Zuerndorfer, B. and England, A. W. (1992). Radiobrightness decision criteria for freeze/ thaw boundaries. *IEEE Trans. Geosci. Remote Sens.*, **30**: 89–101.

Zwally, H. J., *et al.* (1983). *Antarctic sea ice, 1973–1976; Satellite passive-microwave observations. SP 459, NASA*, Washington, DC: National Aeronautics and Space Administration, 206 pp.

Zwally, H. J., *et al.* (2002). Surface melt-induced acceleration of Greenland ice-sheet flow. *Science*, **197**: 218–22.

Zwally, H. J., *et al.* (2005). Mass changes of the Greenland and Antarctic ice sheets and shelves and contributions to sea-level rise: 1992–2002. *J. Glaciol.*, **51** (175): 509–27.

Zwally, H. J. and Gloersen, P. (2008). Arctic sea ice surviving the summer melt: interannual variability and decreasing trend. *J. Glaciol.*, **54** (185): 279–96.

Index